HANDBOOK OF SOLID WASTE MANAGEMENT

Frank Kreith   Editor in Chief

McGraw-Hill, Inc.

New York   San Francisco   Washington, D.C.   Auckland   Bogotá
Caracas   Lisbon   London   Madrid   Mexico City   Milan
Montreal   New Delhi   San Juan   Singapore
Sydney   Tokyo   Toronto

Library of Congress Cataloging-in-Publication Data

Kreith, Frank.
    Handbook of solid waste management / Frank Kreith.
        p.   cm.
    Includes bibliographical references and index.
    ISBN 0-07-035876-1
    1. Refuse and refuse disposal.   I. Title.
TD791.K74   1994
628.4'4'068—dc20   94-6976
                                                CIP

    2 3 4 5 6 7 8 9 0   DOC/DOC   9 0 9 8 7 6 5

ISBN 0-07-035876-1

The sponsoring editor for this book was Larry S. Hager, the editing supervisor was David E. Fogarty, and the production supervisor was Suzanne W. Babeuf. This book was set in Times Roman. It was composed by McGraw-Hill's Professional Book Group composition unit.

Printed and bound by R. R. Donnelley & Sons Company.

This book is printed on acid-free paper.

HANDBOOK OF
SOLID WASTE
MANAGEMENT

Related Titles from McGraw-Hill

American Water Works Association • WATER QUALITY AND TREATMENT

Baker, Herson • BIOREMEDIATION

Brunner • HANDBOOK OF INCINERATION SYSTEMS

Brunner • HAZARDOUS WASTE INCINERATION

Corbitt • STANDARD HANDBOOK OF ENVIRONMENTAL ENGINEERING

Freeman • HAZARDOUS WASTE MINIMIZATION

Freeman • STANDARD HANDBOOK OF HAZARDOUS WASTE TREATMENT AND DISPOSAL

Jain et al. • ENVIRONMENTAL ASSESSMENT

Kolluru • ENVIRONMENTAL STRATEGIES HANDBOOK

Levin, Gealt • BIOTREATMENT OF INDUSTRIAL AND HAZARDOUS WASTES

Lund • THE MCGRAW-HILL RECYCLING HANDBOOK

McKenna & Cunneo, Technology Services Group • PESTICIDE REGULATION HANDBOOK

Maidment • THE HANDBOOK OF HYDROLOGY

Majumdar • REGULATORY REQUIREMENTS FOR HAZARDOUS MATERIALS

Waldo, Hinds • CHEMICAL HAZARD COMMUNICATION GUIDEBOOK

Willig • ENVIRONMENTAL TQM

CONTENTS

Chapter 10. Composting of Municipal Solid Wastes
L. Diaz, G. M. Savage, and C. G. Golueke   **10.1**

Chapter 11. Waste-to-Energy Conversion   **11.1**

CONTRIBUTORS

Nicholas S. Artz *Senior Environmental Engineer and Principal, Franklin Associates, Ltd., Prairie Village, Kan.* (CHAP. 14)

Jacob E. Beachey *Senior Environmental Scientist and Principal, Franklin Associates, Ltd., Prairie Village, Kan.* (CHAP. 14)

Tracy Bone *Environmental Scientist, Municipal and Industrial Solid Waste Division, Office of Solid Waste, U.S. Environmental Protection Agency, Washington, D.C.* (SEC. 9.9)

Gary R. Brenniman *Office of Solid Waste Management, Environmental and Occupational Health Sciences Division, School of Public Health, University of Illinois, Chicago, Ill.* (SECS. 9.7, 9.8)

Calvin R. Brunner *Consulting Engineer, Reston, Va.* (SEC. 11.1)

Stephen D. Cosper *Office of Solid Waste Management, Environmental and Occupational Health Sciences Division, School of Public Health, University of Illinois, Chicago, Ill.* (SEC. 9.8)

T. Randall Curlee *Energy and Global Change Analysis Section, Energy Division, Oak Ridge National Laboratory, Oak Ridge, Tenn.* (SEC. 11.4)

Richard A. Denison *Senior Scientist, Environmental Defense Fund, Washington, D.C.* (SECS. 7.1, 7.2, 7.3, 7.5)

Roger Diedrich *Sierra Club* (SEC. 7.4)

L. F. Diaz *CalRecovery, Inc., Hercules, Calif.* (CHAP. 10)

Bette K. Fishbein *Senior Fellow, Municipal Solid Waste Program, INFORM, Inc., New York, N.Y.* (SEC. 8.2)

Marjorie A. Franklin *President, Franklin Associates, Ltd., Prairie Village, Kan.* (CHAP. 3)

Ken Geiser *Univ. of Massachusetts–Lowell, Lowell, Mass.* (SEC. 8.3)

Carolyn Gelb *Researcher, INFORM, Inc., New York, N.Y.* (SEC. 8.2)

Jim Glenn *Project Manager, Gershman, Brickner and Bratton and Consulting Editor, BioCycle—Journal of Waste Management, Marysville, Pa.* (CHAP. 5)

C. G. Golueke *CalRecovery, Inc., Hercules, Calif.* (CHAP. 10)

William H. Hallenbeck *Office of Solid Waste Management, Environmental and Occupational Health Sciences Division, School of Public Health, University of Illinois, Chicago, Ill.* (SECS. 9.7, 9.8)

Floyd Hasselriis *Consulting Engineer, Forest Hills, N.Y.* (SEC. 11.2)

Frank Kreith *President, Kreith Engineering, Inc. and Professor of Engineering (Emeritus), University of Colorado, Boulder, Colo.* (EDITOR; CHAP. 1, SECS. 8.1, 9.1, 11.1)

James E. Kundell *Walter B. Hill Distinguished Public Fellow, Carl Vinson Institute of Government, University of Georgia, Athens, Ga.* (CHAP. 6)

David Laws *Doctoral Candidate, Department of Urban Studies and Planning, MIT, Cambridge, Mass.* (CHAP. 13)

James M. Lyznicki *Office of Solid Waste Management, Environmental and Occupational Health Sciences Division, School of Public Health, University of Illinois, Chicago, Ill.* (SEC. 9.7)

Philip R. O'Leary *Co-Director, Solid and Hazardous Waste Education Center, University of Wisconsin–Madison, Wisc.* (CHAP. 12)

Edward W. Repa *Environmental Industry Associations, Washington, D.C.* (CHAP. 4)

Deanna L. Ruffer *Project Director, Roy F. Weston, Inc.* (CHAP. 6)

John Ruston *Staff Economist, Environmental Defense Fund, New York, N.Y.* (SECS. 7.1, 7.2, 7.3, 7.5)

G. M. Savage *CalRecovery, Inc., Hercules, Calif.* (CHAP. 10)

Mary Sikora *President, Recycling Reseach Institute, Suffield, Conn.* (SEC. 9.6)

David B. Spencer *President, wTe Corporation, Bedford, Mass.* (SECS. 9.1–9.5, 9.7–9.12)

Lawrence Susskind *Professor of Urban and Environmental Planning, MIT and Director, MIT–Harvard Public Disputes Program, Cambridge, Mass.* (CHAP. 13)

George Tchobanoglous *Professor of Civil and Environmental Engineering, University of California at Davis* (CHAP. 12)

Aaron Teller *Senior Vice President, Air & Water Technologies Inc., Research Cottrell/Metcalf & Eddy, Sunrise, Fla.* (SEC. 11.3)

Jeffrey Tryens *Deputy Director, Center for Policy Alternatives, Washington, D.C.* (SECS. 7.1, 7.2, 7.3, 7.5)

Bruce R. Weddle *Director, Municipal and Industrial Solid Waste Division, Office of Solid Waste, U.S. Environmental Protection Agency, Washington, D.C.* (CHAP. 15)

Marcia E. Williams *President, Williams & Vanino, Inc., Los Angeles, Calif.* (CHAP. 2)

PREFACE

As the amount of municipal solid waste to be disposed of increases, the number of available landfills decreases, concerns about risks associated with waste management rise, opposition to the siting of new waste management facilities spreads, environmental regulations and federal mandates are passed on to states and municipalities and the cost of waste disposal goes up, communities and states all over the country are grappling with problems related to the management and disposal of solid waste. Many communities, states, businesses, and public interest groups have undertaken a variety of activities to dispose of solid waste in an economically responsible and environmentally viable manner. The goal of this Handbook is to provide to those involved in managing the disposal of municipal solid waste the background and technical information necessary to accomplish this task in a professional manner.

The book is organized into fifteen chapters and several appendixes. Chapter 1 gives an overview and economic perspective of municipal waste management. The next six chapters provide information on an integrated approach to municipal solid waste management, the characteristics of the solid waste stream, federal and state legislation concerning the management of solid waste, the planning of municipal solid waste management programs, and environmental perspectives. Following these chapters which provide a background from which to view and plan a municipal solid waste management program, each of the technologies in an integrated waste management structure is discussed in detail. Chapter 8 discusses ways and means of reducing the amount and the toxicity of solid waste. Chapter 9 deals with the various steps involved in recycling technologies, beginning with collection and extending to the transportation, processing, recovering, and upgrading of products and, finally, the economics of recycling. Chapter 10 treats the technology of composting, providing background on principles, technology, operating parameters, and economics. Chapter 11 covers the technologies of waste-to-energy combustion, as well as the control of gaseous emission and the disposal of ash. Chapter 12 covers all aspects of landfilling, including control of landfill gases and leachates; also, layout, design, operation, and landfill closure are discussed. Chapter 13 deals with ways and means of siting new waste management facilities. Chapter 14 covers financing of solid waste management programs and Chapter 15 deals with the role of the U.S. Environmental Protection Agency in solid waste management. A glossary of terms for solid waste management technologies and a list of state offices involved in municipal waste management are presented in appendixes.

A carefully presented index allows the reader access to any of the topics covered in the book and extensive references at the ends of chapters provide the reader with opportunity for additional information on specialized topics.

ACKNOWLEDGMENTS

This Handbook is an outgrowth of a two-day Conference on Integrated Solid Waste Management held in Breckenridge, Colorado in June of 1989 under the auspices of the U. S. Environmental Protection Agency (EPA), the American Society of Mechanical Engineers (ASME), and the National Conference of State Legislatures (NCSL). L. Dwight Connor, the Manager of the Energy, Science and Natural Resources (ESNR) Division at NCSL provided support and counsel in connection with this conference. Larry Hager, Senior Editor in McGraw-Hill's Professional Book Group, read the presentations made at the Breckenridge conference, which was attended by representatives of 43 states and two territories; he recognized the significance of the topic, as well as the need for a technical handbook on solid waste management and asked me to become its editor. He also provided counsel and support during some difficult periods in the course of editing this work. David Fogarty, the Editing Manager at McGraw-Hill assisted in the organization and technical preparation of the Handbook. Mr. Phil Shepherd, Program Manager at the National Renewable Energy Laboratory, made recommendations for potential authors and provided valuable information for Chapter 1. Some of the state legislators who participated in the Breckenridge workshop told me about the needs existing in their states and suggested topics, many of which were incorporated into the various chapters. Ms. Bev Weiler was the editorial assistant of this project and also prepared the glossary. She communicated with authors, helped in locating references, proofread the typeset copies of the chapters and shepherded the project throughout its entire development. My wife Marion helped me in keeping the organization of the Handbook under control. The final product, however, is a reflection of the dedication and hard work of the authors who have contributed to this Handbook. I want to express my thanks and appreciation to them and all of the people without whose help this project could not have come to fruition.

Frank Kreith

CHAPTER 1

INTRODUCTION

Frank Kreith, P.E.

President, Kreith Engineering, Inc. and
Professor of Engineering (Emeritus)
University of Colorado
Boulder, Colorado

Human activities generate wastes that are often discarded because they are considered useless. These wastes are normally solid and, the meaning of the word waste suggests that the material is useless and unwanted. But, many of these waste materials can be reused and thus, can become a resource for industrial production or energy generation, if properly managed.

Currently, the most prevalent method of disposing of solid wastes is to first collect them from the source and then bury them in a landfill. However, landfills have created environmental problems and the number of operational landfills has decreased from 20,000 in 1978 to 6000 in 1988 and is currently estimated to be 3300. Moreover, new federal legislation to protect the environment has increased the cost of building new landfills and siting them has become increasingly difficult because the public opposes having such a facility nearby.

Solid waste management has become a major concern in the United States. Industry, private citizens and state legislatures are searching for means to reduce the growing amount of waste that American homes and businesses discard and to dispose of it safely and economically. In recent years, state legislatures have passed more laws dealing with solid waste management than with any other topic on their legislative agenda.[1] Waste management has become one of the most significant problems of our time because the American way of life produces enormous amounts of waste and most people want to preserve their lifestyle, while also protecting the environment and public health.

Historically, waste management has been an engineering function. It is related to the evolution of a technological society, which, along with the benefits of mass production, has also created problems that require the disposal of its wastes. The schematic diagram in Fig. 1.1 shows the flows of materials in a technological society and the resulting waste generation.[2] Wastes are generated during the mining and production processes of raw materials, such as the tailings from a mine or the discarded husks from a cornfield. After the raw materials have been mined, harvested, or otherwise procured, more wastes are generated during subsequent steps of the process which generates goods for consumption by society from these raw materials.

It is apparent from the diagram in Fig. 1.1 that the most effective way to ameliorate the solid waste disposal problem is to reduce the generation and the toxicity of waste.

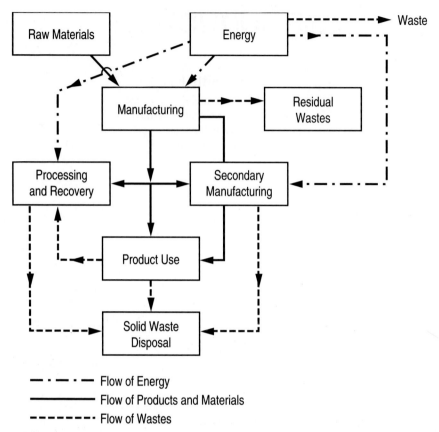

FIGURE 1.1 Flow of materials and wastes in an industrial society.

But, as people search for a better life and higher standard of living, they tend to consume more goods and generate more waste. Consequently, society is searching for improved methods of waste management and ways to reduce the amount of waste that needs to be landfilled.

According to recent public opinion polls, a majority of Americans support an integrated waste management system.[3] This consists of reducing the amount and toxicity of wastes at the source, recycling, reusing, or composting as much of the waste as is economically reasonable, burning the waste that cannot be economically recycled to generate heat in waste-to-energy facilities that reduce the need for fossil and nuclear fuels, and finally, landfilling the rest in an environmentally acceptable manner. Recycling and waste reduction will play important parts in any future waste management strategy. But, engineering analysis clearly shows that these options alone cannot solve the solid waste problem. By the year 2000, according to EPA, Americans will throw away more than 220 million tons of garbage annually. At the same time, according to best estimates, it may be possible to reach a recycling and composting rate of about 30 percent. But, to meet this goal, new recycling technologies must be developed, additional markets must be found, and industry must produce more products that are easy to recycle. All the same, even if all of these steps are successfully taken, more than 160 million tons of solid waste will still have to be treated by other means, such as waste-to-energy combustion and landfilling.

Solid waste management is a difficult process because it involves many disciplines. These include technologies associated with the control of generation, storage, collection, transfer and transportation, processing, marketing, incineration, and disposal of solid wastes. All of these processes have to be carried out within existing legal and social guidelines that protect the public health and the environment and are aesthetically and economically acceptable. They must be responsive to public attitudes and the disciplines included in the disposal process include administrative, financial, legal, architectural, planning, and engineering functions. For a successful integrated solid waste management plan, it is necessary that all these disciplines communicate and interact with each other in a positive interdisciplinary relationship. This handbook is devoted to facilitate this process.

1.1 PHILOSOPHY AND ORGANIZATION

The philosophy of the handbook is that the integrated waste management approach is not a hierarchical scheme, but is synergistic in nature, as shown in Fig. 1.2. In other words, an appropriate design of a toxicity reduction and/or recycling program, which removes heavy metals from the waste stream, in particular lead, mercury, and cadmium, should not be considered merely a reduction or recycling function because it also assists the waste-to-energy incineration function which benefits from the absence of heavy metals and batteries. Recycling is not a complete process unless the legal and institutional framework can create markets for the recycled products that can beneficially utilize the materials picked up from the curb. The technical and engineering function of waste management cannot function in a vacuum, but must be aware of the political and social ramifications of its action.

Another philosophic underpinning of the book is that there is no single prescription for an integrated waste management program that will successfully work in every instance. Each situation must be analyzed on its own merit, an appropriate integrated waste management plan must be developed from hard data, and social attitudes and the legal framework must be taken into account. The waste management disposal field

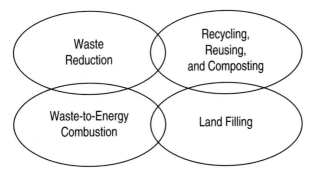

FIGURE 1.2 Synergism of integrated solid waste management.

is in a constant state of flux and appropriate solutions should be innovative, as well as technically and economically sound.

The organization of this handbook reflects the realities of the situation, as well as the philosophy of its editorship. Chapter 2, "Integrated Solid Waste Management," provides an overview of solid waste disposal in the United States and Canada. Chapter 3, "Solid Waste Stream Characteristics," gives the background on what the constituency of waste is today and a projection of what it will be at the end of the century. Chapter 4 deals with federal laws and regulations that impact the different solid waste management schemes. But, over the last few years, the federal government has passed authority and responsibility for waste management to the states and many states in turn have passed their responsibilities on to municipalities. The fifth chapter, therefore, provides an overview of the state legislation within which any waste management plan needs to be devised. Chapter 6 gives an outline of how to plan an integrated municipal waste management program. But the plans themselves have to be devised within the existing views of the public and its spokespeople. Chapter 7, therefore, provides an outline of the social context on waste management from the perspectives of spokespeople for organizations devoted to protection of the environment.

Chapters 8, 9, 10, 11, and 12 are devoted to the major technologies for an integrated waste management scheme. Starting with source reduction in Chap. 8, and continuing to recycling in Chap. 9, composting in Chap. 10, waste-to-energy combustion in Chap. 11, and landfilling in Chap. 12, the technologies are analyzed and their economic costs are discussed.

It is clear, though that, irrespective of what combination of technologies is employed in a waste management scheme, new facilities will have to be sited and financed. The old approach, when technical experts determined the best location for a waste management site, then announced their decision and defended it to the public, is not working. The confrontational results of the "decide, announce, and defend" strategy must be replaced by an interactive procedure in which the public participates in the siting process as a full partner. Therefore, Chap. 13 is devoted to a recommended procedure for the siting of new facilities. Chapter 14 deals with the financing options for waste management facilities. The last chapter presents the role of the U.S. Environmental Protection Agency (EPA) in solid waste management.

Appendices contain a glossary of terms, conversion factors, a list of organizations active in solid waste management, and a list of state offices responsible for waste disposal in each state. Lists of companies that can provide technical help and equipment in the development and operation of an integrated solid waste management scheme are integrated into the chapters on specific technologies.

1.2 STATUS OF THE MAJOR WASTE MANAGEMENT OPTIONS

Today, 69 to 73 percent of municipal solid waste (MSW) is landfilled. Landfill gas is recovered for energy in little more than 100 of the nation's larger landfills and 17 percent of it is burned, most of it with energy recovery. There is no uniformly accepted definition of what constitutes recycling, and consequently estimates of the percentage of MSW that is recycled vary significantly. The U.S. Environmental Protection Agency and the Office of Technology Assessment (OTA) have published estimates ranging from 10 to 17 percent. Composting accounts for only a small percentage of waste treatment, but the amount is expected to increase significantly. The EPA has set a national voluntary goal of reducing the quantity of MSW by 25 percent through source reduction and recycling.

Curbside Separation, Mixed Waste Separation, and Recycling

Curbside separation, mixed waste separation, and recycling can reduce the amount of waste that must be handled by other MSW options. There are five steps in recycling:

1. Separating reusable materials from other waste
2. Transporting and processing (including remanufacturing) the separated materials for use as replacements for virgin materials
3. Managing the wastes from separation and recycling
4. Returning the materials to commerce
5. Selling the recycled products

At present, most recycling efforts focus on newsprint, cardboard, glass, steel, aluminum, tin cans, and some plastics.[3]

Some of the statistics that indicate that recycling now manages 17 percent or more of the nation's MSW are based on estimates of the amounts of material diverted from the local landfill by separate collection of recyclables, bottle deposit laws, and separate collection of yard waste for composting. Information on the amount of MSW that is actually remanufactured and returned to commerce is mostly anecdotal; however, it is clearly much lower than the total quantities collected because some of the material is used as fuel, some is lost during remanufacturing, and when market conditions are poor, some is landfilled after collection for reuse.[4]

Communities that wish to include recycling in their MSW management strategies have several options for separating recyclables from other waste. They can offer convenient sites where residents can receive payment for containers (e.g., buy-back centers), provide drop-off centers that may accept a wide range of recyclable and compostable materials, implement curbside collection of recyclable materials separated by residents from other MSW, and process mixed waste to separate recyclables.

Either mixed MSW collected in a standard packer truck or recyclables collected separately at curbside can be sent to a materials recovery facility (MRF) for further separation and consolidation of the collected materials. MRFs can be divided into "low-tech" and "high-tech" facilities, depending on the amount of manual labor required. All MRFs rely heavily on manual labor to sort and separate grades of paper and glass bottles by color and plastic bottles by resin type and color. Most MRFs also use magnets for recovering ferrous metals, and many use balers for paper, crushers for glass, and flatteners for the aluminum cans. High-tech MRFs generally also use additional shredders, screens, possibly air classifiers for separating heavy materials from lighter ones, and special eddy-current separators that can separate aluminum.

Composting

Composting is a low-temperature, partial oxidation process of the easily degradable proteins, fats, simple sugars, and carbohydrates in plant cells and animal tissues.

Composting may be either aerobic or anaerobic, but only aerobic composting technology—which occurs in the presence of oxygen—has been sufficiently developed to be implemented commercially at this time. As part of an integrated waste management scheme, composting can be applied to mixed MSW or to separately collected leaves and other yard wastes. MSW composting results in a volume reduction of about 50 percent and consumes about 50 percent of the organic mass, which is released as CO_2 and water.

Aerobic composting is a technically proven process, for which basically three systems are used:

1. Static windrows piles
2. Turned windrows
3. In-vessel composting

There are about 1400 composting programs in the United States, but 500 of them compost only leaves on a seasonal basis. Since yard waste constitutes between 18 and 20 percent of the total municipal waste generated, substantial increases in the amount of composting facilities are expected in the near future. Unfortunately, products of composting facilities are sometimes difficult to sell because there are no uniformly accepted standards for compost content and, without standards, consumers often are afraid to use waste-derived products.

Combustion with Energy Recovery

Open burning has been used for centuries to dispose of waste. In the United States, combustion of MSW to recover energy for generation of electricity was first introduced in New York City at the turn of the century. In modern plants, energy can be recovered in the form of heat, electricity, or cogeneration. Until the 1970s, MSW combustors had hardly any air pollution control equipment and those pollution control units that were used generally emitted bad odors and smoke. They were primarily used to reduce the volume of waste, but since the early 1970s increasingly stringent environmental controls have been required by law, and combustors built today meet all federal and state limits on air pollution. The two most common combustion options are mass burning or preparation and combustion of refuse-derived fuel (RDF). They differ in extent of pretreatment of the MSW before firing, the type of furnace used, and the firing conditions. (See Chap. 11.)

Figure 1.3 shows a schematic diagram of a waste-to-energy facility. In a mass-burn facility, pretreatment of MSW includes inspection and simple separation to remove oversized and noncombustible items and unacceptable components such as obviously hazardous or explosive materials. The MSW is then fed into a combustor, where it is typically supported on a grate or hearth. Air is fed below and above the grate to promote combustion. Field-erected mass burn plants can be large facilities, with capacities of 3000 tons of MSW per day or more. However, they can be scaled down to handle the waste from smaller communities and modular plants with capacities as low as 25 tons/day have been built.

RDF production begins with inspection of MSW, removal of bulky or hazardous waste and ferrous materials, and shredding of the remaining MSW, as shown in Fig. 1.4. Noncombustible materials are often separated as well. The shredded RDF can be stored or directly burned above a traveling grate. RDF preparation and direct firing cannot be performed economically in small plants, and the minimum size of an RDF

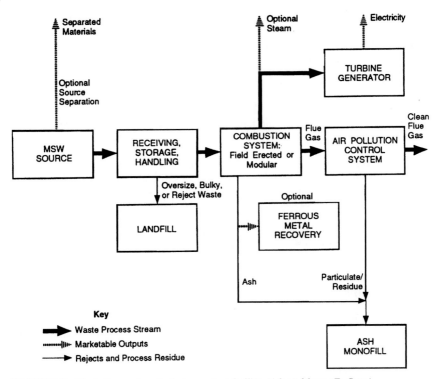

FIGURE 1.3   Block diagram for a typical mass burn facility. (*Adapted from wTe Corp.*)

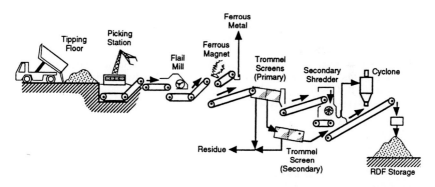

FIGURE 1.4   RDF processing system.

plant tends to be quite large. If RDF is compressed into pellets or cubes, it can be used in existing conventional furnaces with grates. A few operating facilities now produce such pellets or cubes at one location for sale or use at another.

The energy produced by both mass burning and RDF combustion is generally used to generate heat for electrical power production. MSW can thus not only reduce the volume of the MSW, but also eliminate the need to mine, burn, and dispose of the residue of some of the coal, gas, or uranium that would otherwise be needed to generate the electricity obtained from MSW.

Regulatory requirements for control of MSW combustion have grown increasingly stringent since they were first implemented in the 1970s. For both types of options, federal regulations governing all facilities with capacities greater than 250 tons per day set limits on the emission of pollutants, including acid gases, metals, and dioxins/furans (see Chap. 4). The EPA is developing comparable requirements for units with capacities of less than 250 tons/day. State and local requirements may be more stringent than federal regulations, and may apply to even smaller combustors (see Chap. 5). Current regulations for the larger MSW plants are more stringent than those governing fossil fuel plants.

The ash from MSW combustion and the residue from scrubbers used to neutralize acid gases in the gas stream must be disposed of safely. Many states require the use of separate landfills, called *ash monofills,* that contain only ash. Modern plants using good combustion practices can reduce the volume of MSW by up to 90 percent. The leachate from ash monofills is normally smaller in volume than that from ordinary landfills and the constituents of the leachate are also different. There are, so far, no federal regulations governing ash disposal.

Sanitary Landfilling

Open landfills have been used as a waste management method all over the world for centuries. In the United States, rules and regulations for construction and operation of solid waste landfills were established by the Resource Conservation and Recovery Act (RCRA) of 1976 as a way to eliminate open dumps and reduce pollution. Since then, landfill requirements have become more stringent. By carefully enclosing MSW in landfills, providing liners underneath it, covering the landfill with dirt ("daily cover") at the end of each day, installing gas collection systems, and capping the landfill when it is filled, between 30 and 85 percent of the methane, carbon dioxide, and other organic gases generated by the waste can be collected (see Chap. 12). Those gases can be burned for energy recovery if the quantity generated is large enough to justify the expense of the equipment.

Landfills will reach their capacity when they fill up or reach practical height limits. Therefore, efforts to reduce the amount of space that MSW occupies can extend the life of a landfill. Combustion and recycling programs help to reduce waste volume, and are thus synergistic options with landfilling in an integrated waste management scheme.

1.3 COMPONENTS IN AN INTEGRATED WASTE MANAGEMENT STRATEGY

As shown in Fig. 1.2, an integrated waste management scheme involves several technologies. At present, most communities use two or more of the MSW management options to dispose of their waste, but there are only a few instances where a truly integrated and optimized waste management plan has been developed. To achieve an integrated strategy for handling municipal waste, an optimization analysis combining all of the available options should be conducted. However, there is at present no proven methodology for performing such an optimization analysis, and most communities, therefore, rely on two or more of the available options to achieve a plan to dispose of their waste. Figure 1.5 shows nine of the most common combinations of technologies that are used in waste management schemes in the United States.[4] The most common in the U.S. is probably strategy 4, consisting of curbside recycling and landfilling the remaining waste. In rural communities, strategy 3, consisting of composting and landfilling is prevalent. In large cities, where tipping fees for landfilling sometimes reach and exceed $100 per ton, strategy 5, consisting of curbside recycling with the help of an MRF, followed by mass burn or combustion at an RDF facility and landfilling of the unrecyclable materials from the MRF and ash from the incinerator, is the most prevalent combination. However, as mentioned previously, each situation should be analyzed individually and the combination of technologies which fits the situation best should be selected. As a guide to the potential effect of any of the nine strategies in Fig. 1.5 on the landfill space and its lifetime, Fig. 1.6 shows the required volume of landfill per ton of MSW generated for each of the nine combinations of options displayed in Fig. 1.5.

However, apart from availability of landfill volume and space, the cost of the option combinations is of primary concern to the planning of an integrated waste management scheme. Cost estimates are difficult and cumbersome, but in the next section, a preliminary overview of the cost of various landfill options is presented. More

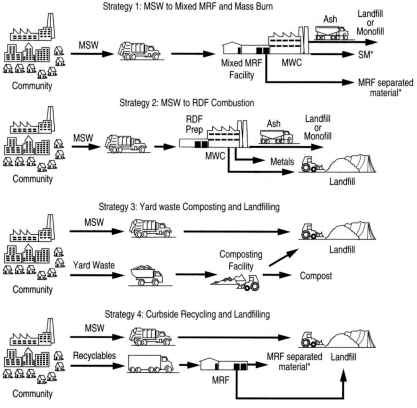

FIGURE 1.5   Nine typical waste management strategies.

Strategy 5: Curbside MRF Mass Burn and Cash Landfilling

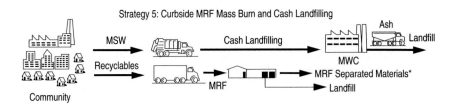

Strategy 6: Curbside Recycling RDF Combustion and Cash Landfilling

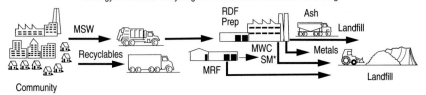

Strategy 7: Curbside Recycling, RDF Composting, and Landfilling

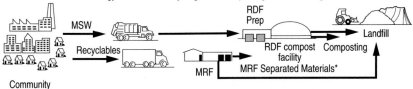

Strategy 8: Curbside Recycling, Yard Waste Composting, and Landfilling

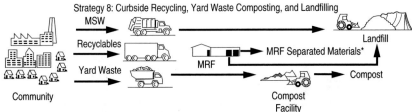

Strategy 9: Curbside Recycling, Yard Waste Composting, Mass Burning and Ash Landfilling

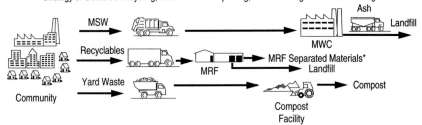

*MRF separates materials for processing by industry. These materials include paper, cardboard, glass, aluminum, steel, and plastics.

FIGURE 1.5 *(Continued)*

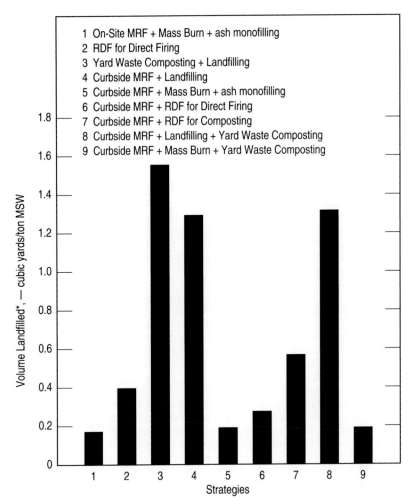

*Excludes volume required for residue from remanufacturing recyclables.

FIGURE 1.6 Landfill volume per ton of MSW generated for the nine waste management strategies in Fig. 1.5.

detailed cost analysis considering the cost of components, labor, land, and financing, is presented in some of the later chapters in the handbook dealing with specific waste management options.

1.4 PRELIMINARY COST ESTIMATES FOR MAJOR WASTE MANAGEMENT OPTIONS

At the outset, it should be noted that the only reliable way to compare the costs of waste management options is to obtain site-specific quotations from experienced contractors. However, it is often necessary to make some preliminary estimates in the early stages of designing an integrated waste management system. To assist in such preliminary costing, this section summarizes the results of a study made by SRI International for the National Renewable Energy Laboratory in 1992.[4] For this study, cost data from the literature were examined from many parts of the country and published estimates of the capital costs and operating costs for the most common municipal solid waste options (materials recycling, composting, waste-to-energy combustion, and landfilling) were correlated. All of the cost data for the individual options were converted, for this comparison, to 1991 dollars per ton of daily MSW capacity to provide a consistent basis for cost comparisons. The available data were then plotted and a straight line was faired through the data to provide a simple way to estimate the capital investment and operation and maintenance (O&M) costs as a function of capacity in tons of MSW per day.

It should be noted that cost data in the literature vary in quality, detail, and reliability. Thus, the range of capital and operating cost estimates is broad. Factors which will affect the costs reported are the year when a facility was built, the interest rate paid for the capital, the regulations in force at the time of construction, the manner in which a project was funded (privately or publicly), and the location in which the facility is located. Also, costs associated with ancillary activities, such as road improvements, pollution control, and land acquisition greatly affect the results.

In addition to the capital investment, operating costs are important in making an analysis of integrated waste management systems. Once again, it should be noted that the O&M cost data from the SRI study are only preliminary and show large variations. For a reliable estimate, a study of the conditions in the time and place of the project must be made. Operating costs are affected by local differences in labor rates, labor contracts, safety rules, and crew sizes. Also, accounting systems, especially those used by cities and private owners, and the age of landfills or incinerators can greatly affect the O&M costs. Cost data on separation, recycling, and composting are scarce and, in many cases, unreliable. It is, therefore, recommended that for a comparison of various strategies to manage MSW, costs for all systems should be built up from system components, using a consistent set of assumptions and realistic cost estimates at the time and place of operation. The most extensive and reliable data available appear to be for the combustion option. Combustion is a controlled process that is completed within a short period of time and for which there is a good deal of experience. Also, inputs and outputs can be measured effectively with techniques that have previously been used for fossil fuel combustion plants.

In addition to the externalized costs presented in this chapter, there are also social costs associated with each of the waste management options. For example, recycling will generate air pollution from the trucks used to pick up, collect, and distribute the materials to be recycled. Many steps in a recycling process, such as deinking of newspaper, create pollution whose cost must be born by society, since it is not a part of the recycling cost. Waste-to-energy combustion creates air pollution from stack emissions and water pollution from the disposal of ash, particularly if heavy metals are present. Landfilling has environmental costs due to leakage of leachates into aquifers and the generation of methane and other gases from the landfill. It has been estimated that between 60 and 110 lb of methane will be formed per ton of wet municipal waste during the first 20 years of operation of a landfill. About 9 to 16 lb of that gas will not be recovered, but will leak into the atmosphere because of limitations in the collection system and the permeability of the cover. EPA has estimated that 12 million tons of methane are released from U.S. landfills per year. New regulations, however, will reduce the environmental impact of landfilling in the future.

Collection and Separation

Costs included in this option are for curbside collection and processing (MRF). The estimates exclude the cost of efforts to encourage community participation in recycling efforts through advertising and educational programs; those costs have been estimated at $1.00 to $1.50 per household per year.[5]

Collection costs are affected more by the number of stops made than by the tonnage collected. For example, in the Hudson Valley area of New York, the cost of collecting newspaper, glass, and metals and keeping them separate was reported as $50 per ton of collected material, but other studies have reported costs as high as $60 to $80 (1989 dollars) per ton. At those cost levels, curbside collection of separated materials adds 8 to 25 percent to the total collection costs for unseparated MSW.[6]

Figures 1.7 and 1.8 show the range of capital costs for existing low-tech and high-tech MRFs that sort reusable materials, whether mixed or separately collected. The capital cost of those facilities (with an average capacity of 89 tons/day) averages about $26,000 per ton of design capacity per day. Planned facilities are larger, averaging 162 tons per day, and their average capital cost is estimated at $37,000 per ton of design capacity. More detailed data on MRF costs are provided in Ref. 4.

Operating costs for the MRFs are shown in Figs. 1.9 and 1.10. In general, low-technology MRFs have higher operating costs, averaging $65 per ton, than high-technology MRFs, which average $39 per ton, because of the greater labor intensity of the former. The figures show log-log plots that tend to suggest a narrower range of prices than the actual range. Other studies show a range of $26 to $86 per ton, with an average of $45 per ton.[7] The variations reflect inconsistencies in the sources of the estimates, rather than predictable variations based on the type of technology or the size of the facility.

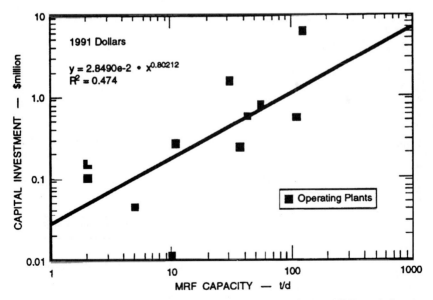

FIGURE 1.7 Effect of plant capacity on capital costs for low-technology MRFs, excluding costs associated with collection (e.g., trucks).[4]

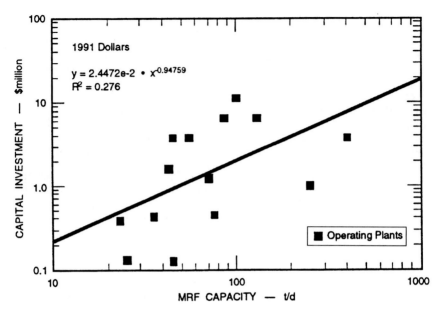

FIGURE 1.8   Effect of plant capacity on capital costs for high-technology MRFs, excluding costs associated with collection (e.g., trucks).[4]

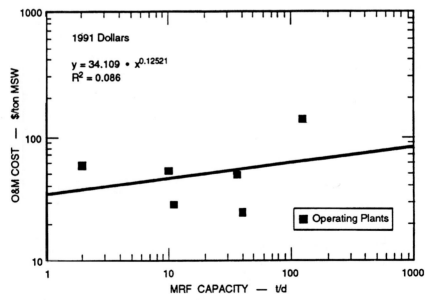

FIGURE 1.9   Effect of plant capacity on O&M costs for low-technology MRFs, excluding debt service charges and operating costs associated with collection.[4]

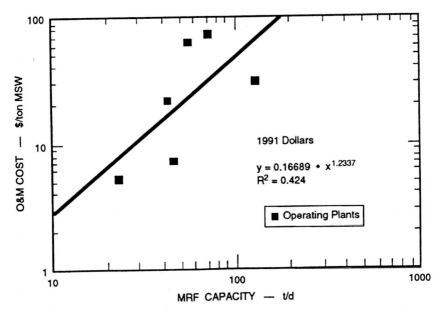

FIGURE 1.10   Effect of plant capacity on O&M costs for high-technology MRFs, excluding debt service charges and operating costs associated with collection.[4]

MSW Composting

Published data on MSW composting costs are limited. Capital costs in the range of $40,000 to $80,000 per ton of daily capacity for MSW composting and preprocessing facilities have been reported.[8]

Figure 1.11 summarizes the capital cost estimates for 12 facilities published as of late 1991 (detailed costs are presented in Ref. 4).Capital costs range from $21,600 to $73,540 per ton of MSW per day. Figure 1.12 shows the operation and maintenance cost estimates for these plants. The average O&M cost is $66 per ton of MSW processed, and the range is $30 to $70 per ton. (Tipping fees ranged from $15 to $78 per ton in 1992.)

The investment costs show no scale effects: investment is a linear function of capacity within the capacity range of 10 to 1000 tons/day. O&M costs decline linearly with capacity, but the correlation is quite poor.

Mass Burn: Field Erected

Most field-erected mass burn plants generate electricity. The average size of the 68 facilities for which useful data are available is 1200 tons/day of design capacity (with a range of 750 to 3000 tons/day). Figure 1.13 summarizes the capital cost estimates for the 68 electricity-generating plants. The average capital cost is $106,000 per ton per day of design capacity, with a range of $30,000 to $210,000.

In some studies, facilities were not differentiated by the form of energy produced; instead, all field-erected mass burn units were grouped according to the calendar years

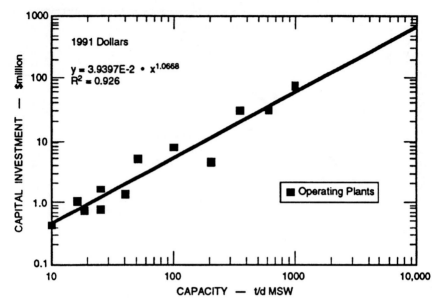

FIGURE 1.11 Effect of plant capacity on capital costs for 12 composting facilities, excluding costs associated with collection (e.g., trucks).[4]

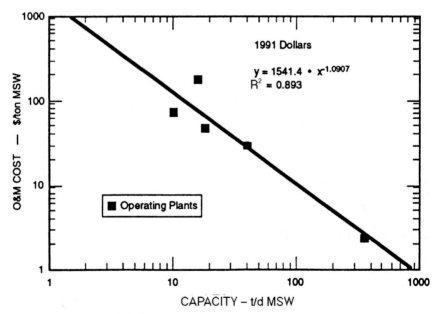

FIGURE 1.12 Effect of plant capacity on O&M costs for 12 composting facilities, excluding costs associated with collection (e.g., trucks).[4]

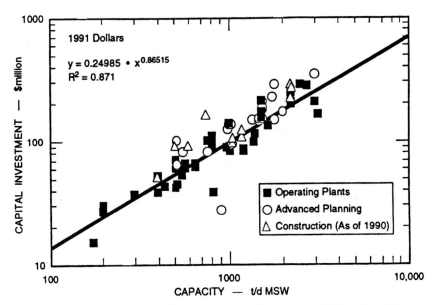

FIGURE 1.13 Effect of plant capacity on capital costs for mass burn facilities (electricity only), excluding costs associated with collection (e.g., trucks).[4]

in which construction began and ended. The capital costs reported in those studies range from $21,000 to $114,000 per daily ton of design capacity.[9]

Figure 1.14 shows the O&M cost estimates for the electricity-only mass burn plants for which data are available. The average O&M cost is $26.50 per ton of MSW processed, with a range of $9 to $48 per ton. Figures 1.15 and 1.16 provide estimates of capital and O&M costs for 20 plants producing steam and electricity. Capital costs are lower for those facilities than for electricity-producing plants, but O&M costs are about the same.

Mass Burn: Modular

Figure 1.17 summarizes the capital cost estimates for 11 modular steam and electricity generating plants that have an average capacity of 243 tons/day. The average capital cost is $95,110 per ton of design capacity per day.

Figure 1.18 shows estimated O&M costs for the modular plants for which data are available. The average O&M cost for the facilities is $32 per ton of MSW processed, with a range of $21 to $42 per ton. Reference 4 presents costs for 34 modular mass burn plants that produce steam only, and for four plants that produce only electricity. The average capital costs are lower for the steam-only plants, but the O&M costs are not. Tipping fees average $50 for the steam/electricity plants, and $25 for the steam-only plants.

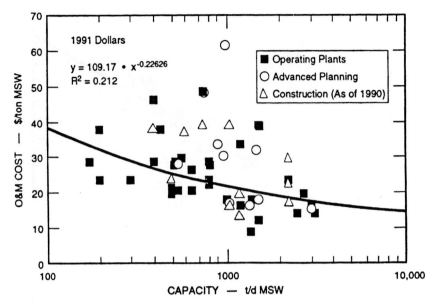

FIGURE 1.14 Effect of plant capacity on O&M costs for mass burn facilities (electricity only), excluding costs associated with collection (e.g., trucks).[4]

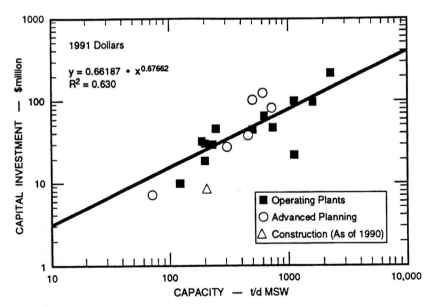

FIGURE 1.15 Effect of plant capacity on capital costs for mass burn facilities (steam and electricity), excluding costs associated with collection (e.g., trucks).[4]

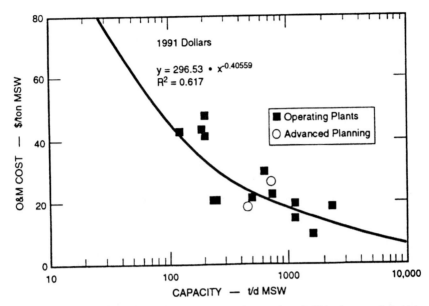

FIGURE 1.16 Effect of plant capacity on O&M costs for mass burn facilities (steam and electricity), excluding costs associated with collection (e.g., trucks).[4]

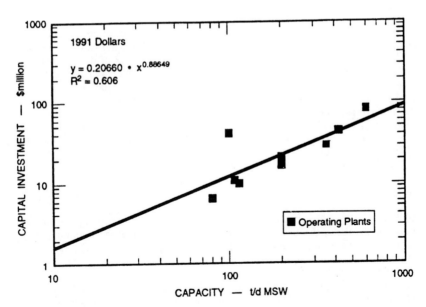

FIGURE 1.17 Effect of plant capacity on capital costs for modular facilities (steam and electricity), excluding costs associated with collection (e.g., trucks).[4]

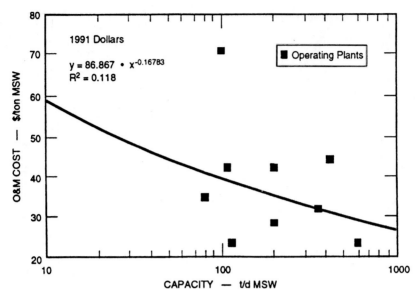

FIGURE 1.18 Effect of plant capacity on O&M costs for modular facilities (steam and electricity), excluding costs associated with collection (e.g., trucks).[4]

RDF Facilities

Figure 1.19 summarizes the capital cost estimates for 15 operating integrated RDF production/combustion facilities. The estimates are based on detailed data included in Ref. 4. The average unit capital cost is $98,000 per ton per day of design capacity. A comparison study that gave capital costs for RDF plants completed in different periods provided a range of costs from $75,000 to $102,000 per ton per day of design capacity.[9] Figure 1.20 shows the O&M cost estimates for the plants for which data are available. The average O&M cost for the facilities is $36 per ton of MSW processed, with a range of $13 to $67 per ton. Note that the averages cited above are based on wide ranges and the number of data points is small. Hence, they are only a rough estimate of future RDF facility costs.

Landfilling

Landfilling costs are difficult to come by because construction often continues throughout the life of the landfill, instead of being completed at the beginning of operations. Capital costs are, therefore, often mixed with operating costs. Capital and operating costs of landfills can be estimated by using models, but such models are valid only for a particular region. For example, a model developed for estimating costs for landfills in Michigan estimated a construction cost of $25.5 million for a 20-year, 1000-ton per day landfill. The estimate for the total project cost, $125 million, was conservative and could underestimate actual costs by as much as 100 percent, according to the model developers.[10]

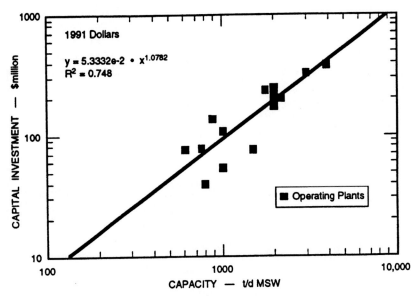

FIGURE 1.19   Effect of plant capacity on capital costs for RDF facilities, excluding costs associated with collection (e.g., trucks).[4]

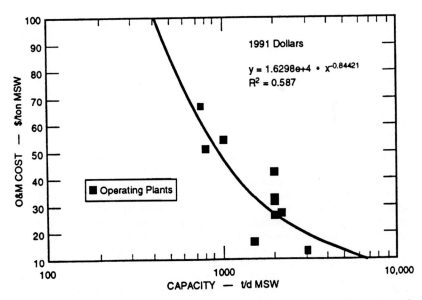

FIGURE 1.20   Effect of plant capacity on O&M costs for RDF facilities, excluding costs associated with collection (e.g., trucks).[4]

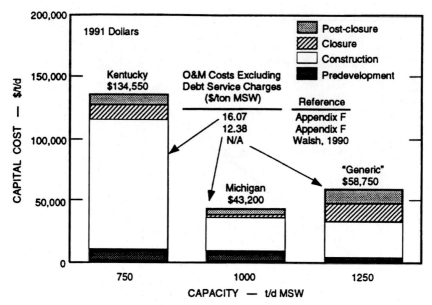

FIGURE 1.21   Landfill capital and O&M costs, excluding cost associated with collection (e.g., trucks). (*Source: SRI International.*)

Available data are few and indicate wide variability in landfill costs as a result of local conditions. These costs, separated in cost elements, are shown in Fig. 1.21. Data on ash disposal costs were found for only seven RDF plants. The costs range from $3 to $57 per ton of ash. For two new plants in the Northeast, the cost averages $26 per ton of the ash. That cost is lower than the cost for MSW disposal, and the ash amounts to 17 percent by weight of the MSW.

1.5   CONCLUDING REMARKS

The technologies to handle solid waste economically and safely are available today. This handbook describes each of them and provides a framework to coordinate them into an integrated system. However, the approach takes cognizance that the responsible disposal of our solid waste is not merely an engineering problem, but involves sociological and political factors. The public supports waste reduction, reuse, recycling, and, sometimes, composting, on the assumption that these measures are environmentally benign, economical, energy conserving, and capable of solving the waste management problem.[10] There is often, however, a lack of understanding of the limitations of these measures and the meaning of the terms. Waste reduction is not merely a matter of using less or reducing the amount of packaging. For example, doubling the life of the tires on automobiles cuts the number of tires that need to be disposed of in half. Deposit laws on bottles and batteries can be used to encourage reuse and recycling, as well as waste reduction. The public usually ignores the fact that recycling has technical and market limitations, becomes more expensive as the percentage of the waste recycled increases, and also has adverse environmental impact.

There is general agreement that waste reduction, reuse, and recycling should be supported within their respective technical and economic limits. But, even if current reduction and recycling efforts are successful, and we reduce the amount of waste that needs to be disposed by EPA's goal of 25 or even 30 percent, the amount of waste that must be disposed in the year 2000 will be as much as or more than today. Consequently, the need to site additional facilities for composting, waste-to-energy combustion, and landfilling will continue. But these technically obvious requirements for waste disposal often create opposition from the public that is politically difficult to resolve. These are factors that cannot be adequately treated in a technical handbook, but they must be kept in mind when devising an integrated waste management strategy.

It should also be pointed out that the technologies for waste disposal are in a state of flux. New and more efficient methods are being introduced. Better equipment to control air and water pollution is being developed, materials that increase the life of product and, thereby, reduce the production of waste are becoming available, and more economical ways for recycling and composting are being tried in pilot projects. For the waste management professional, it is, therefore, important to keep up with the current literature. Table 1.1 provides a list of some professional publications that describe the state-of-the-art and present new developments.

TABLE 1.1 Some Professional Journals for Solid Waste Management Issues

Solid Waste and Power HCI Publications 910 Archibald Street Kansas City, MO 64111-3046 (816) 931-1311 Fax (816) 931-2015	MSW Management 216 East Gutierrez Santa Barbara, CA 93101 (805) 899-3355 Fax (805) 899-3350
Resource Recycling P.O. Box 10540 Portland, OR 97210 (800) 227-1424 Fax (503) 227-6135	Household Hazardous Waste Management News 16 Haverville St. Andover, MA 01810 (508) 470-3044
Bio-Cycle P.O. Box 37 Marysville, PA 17053 (717) 957-4195	Journal of Air and Waste Management Assoc. P.O. Box 2861 Pittsburgh, PA 15230 (412) 232-3444 Fax (412) 232-3450
Waste Age 1730 Rhode Island Avenue, NW Washington, DC 20036 (202) 861-0708 Fax (202) 659-0925	Energy from Biomass & Waste U.S. Department of Energy, OSTI P.O. Box 62 Oak Ridge, TN 37831 (615) 576-1168
Solid Waste Management UIC, School of Public Health 2121 West Taylor Street Chicago, IL 60612-7260 (312) 996-8944	Public Works P.O. Box 688 Ridgewood, NJ 07451 (201) 445-5800 Fax (201) 445-5170
Waste Tech News 131 Madison St. Denver, CO 80206 (303) 394-2905 Fax (303) 394-3011	Resource Recovery Report 5313 38th St. NW Washington, DC 20015 (202) 362-6034

To establish responsible integrated waste management systems, we need to begin with a public education program to make everyone accept responsibility. Once the problem becomes a shared responsibility, we can work toward a solution together. Failure to site new facilities, more than anything else, could create a crisis that would lead to panic solutions that we would regret later on. Engineers must work with the public to find acceptable sites. The public must understand that waste-to-energy combustion produces electric power with less environmental impact than fossil fuel power plants, and at the same time, reduces the amount of waste that needs to be landfilled. Technically trained people should build continuous program evaluation into waste management plans and share information to improve the process in the future. Engineers, politicians, and the waste management industry will have to work to win the confidence of the public, so that technical solutions will be accepted and implemented. The technologies to handle our waste safely exist, and this handbook is an effort to provide the information to implement the safe and economic disposal of the waste generated by our industrial society.

REFERENCES

1. Kreith, F., "Solid Waste Management in the U.S. and 1989–91 State Legislation," *Energy, the International Journal,* vol. 17, pp. 427–476, 1992

2. Tchobanoglous, G., et al., *Integrated Solid Waste Management,* McGraw-Hill, New York, 1993.

3. Kreith, F., ed., *Integrated Solid Waste Management,* Genium Publishing Corp., Schenectady, N.Y., 1990

4. SRI International, *Data Summary of Municipal Solid Waste Management Alternatives,* NREL, 1992.

5. Deyle, R. E., and B. Schade, "Residential Recycling in Mid-America: The Cost Effectiveness of Curbside Programs in Oklahoma," *Resources Conservation and Recycling,* vol. 5, pp. 305–327.

6. Deyle, R. E., and N. N. Hanks, "Cost Effective Composting of Yard Waste," *Waste Age,* November 1991, p. 83.

7. Bishop, R. S., "Defining the MRF and the Role of MRFs in Residential Waste Recycling," *Resource Recovery,* October 1991, p. 37.

8. Apotheker, S., "Engineering the Nation's Largest MSW Composting Plant," *Resource Recycling,* July 1991, p. 43.

9. Kiser, J. V. L., "Comprehensive Report on Municipal Waste Combustion, Part V, MWC Project Cost Data," *Waste Age,* November 1990, p. 156.

10. Kreith, F., "Technology and Policy of Waste Management," *Proceedings of the 85th Air and Waste Management Association,* 92-20.07, Kansas City, Mo., 1992.

CHAPTER 2
INTEGRATED MUNICIPAL SOLID WASTE MANAGEMENT

Marcia E. Williams

President, Williams & Vanino, Inc.
Los Angeles, California

This chapter examines how integrated municipal solid waste management can be used to solve municipal solid waste problems. The first section of the chapter formulates the integrated waste management challenge. The second section discusses the basic tools of waste management. The third section discusses the framework for making decisions. Finally, the last section presents key factors which impact the likelihood of fully implementing a successful integrated waste management program.

2.1 THE CHALLENGE

A useful starting point is a descriptive look at what makes up the municipal solid waste (MSW) stream by weight.[1]

Paper and paperboard	37.5%
Yard waste	17.9%
Wood	6.3%
Metal	8.3%
Glass	6.7%
Food waste	6.7%
Plastics	8.3%
Other (including rubber, leather, textiles, miscellaneous inorganic waste)	8.3%
Household hazardous waste	<0.5%

According to 1990 data, as a nation we generate about 196 million tons per year of MSW.[1] This works out to be over 1500 pounds per year per person (4.3 pounds per person per day). The amount of MSW generated each year has continued to increase on both a per capita basis and a total generation rate basis. In 1960 per capita generation was about 2.65 pounds per person per day and 88 million tons per year. By 1986, per capita generation jumped to 3.58 pounds per person per day. Over the next decade, it is expected to continue to increase by about 5 percent over the 1990 level to a per capita rate of 4.5 pounds per person per day and an overall rate of 222 million tons per year.

Importantly, this rate is location specific. Rural areas tend to be lower waste generators than urban areas. Many large cities generate waste at the rate of 5 to 7 pounds per person per day while rural areas are often between 2 and 3 pounds per person per day. There is also variation in whether the waste is generated by residential or industrial/commercial sources. The national residential average is 55 to 65 percent of all MSW;[1] however, the residential component can dip below 50 percent. It is understandable that urban Americans generate three times more waste per capita than people in Manila or Lahore generate. However, it is less understandable that the United States currently generates MSW at twice the rate of our urban counterparts in Germany or Italy. The positive news, however, is that the projected rate of increase is slowing.

In addition to these large volumes of MSW, larger quantities of solid waste are not included in these national MSW totals. Certain states do classify some of these materials as MSW and regardless of what the materials are called, many of them are managed in the same facilities that manage MSW. These wastes include construction and demolition waste, agricultural waste, municipal sludges, combustion ashes including cement kiln dust and boiler ash, medical waste, contaminated soils, mining wastes, oil and gas wastes, and industrial process wastes which are not classified as hazardous waste. The national volume of these wastes is extremely high and has been estimated between 7 and 10 billion tons per year.[1] Most of the wastes are managed at the site of generation. However, if even 1 or 2 percent of these wastes are managed in MSW facilities, it can dramatically affect MSW capacity. One or two percent is probably a reasonable estimate.

Landfills have always managed a significant portion of the MSW stream, although their use fluctuates with changes in the use of alternative solid waste management methods. The use of these alternative methods tends to increase when landfill costs increase or when landfill capacity at the local level contracts. In 1990, approximately 67 percent of MSW in this country was buried, if buried is the right word.[1] The Fresh Kills landfill in New York City will be the highest point on the eastern seaboard south of Maine by the time it closes at the turn of the century. Each year, the United States landfills enough municipal waste to produce two million automobiles, enough wood to build a million homes, enough paper to produce all of our newspapers, one-half million house trailers worth of aluminum, and enough energy to drive thirty 1000-MW power plants. This percentage is down considerably from the 80 percent landfilled in 1980. However, since the overall volume of MSW has continued to increase, the absolute volumes landfilled in 1990 are slightly higher than the absolute volumes landfilled in 1980.

The 1990 rates for recycling/composting were 17 percent.[1] The recycling rates varied widely by type of waste, with overall recovery rates of aluminum at 38 percent, paper and paperboard at 29 percent, yard wastes at 12 percent, and food wastes negligible. These recovery rates are significantly higher than the 7 percent estimated in 1960 and the 13 percent estimated in 1988. The projected year 2000 recovery rates are between 25 and 35 percent.

The 1990 rates for combustion were 16 percent.[1] This is significantly higher than the 10 percent combustion in 1980. However, it is estimated that in 1960 about 30 percent of MSW was burned in combusters. The absence of energy recovery and lack of air pollution controls were largely responsible for the closing of these early facilities. However, combustion is projected to manage about 21 percent of the waste stream in the year 2000.

Again, national averages are useful but can disguise very important local differences. For example, some states and communities utilize much greater amounts of combustion than the national averages. These include Florida, Connecticut, and Minnesota. Other states, like California, combust very little of their MSW. The same variation occurs with recycling/composting rates. Some states, like New Jersey, and many cities have reached rates above 30 percent. Other states and localities are recycling at rates significantly below the national average.

The reasons for these differences are partially explained by differences in available local landfill capacity and the local cost of landfilling. Many communities are in the throes of a capacity crisis today or will be soon. Seventy-five percent of existing landfill capacity will close within 10 to 15 years. Some will close because they are full, but others will close because they are poorly located, poorly designed, or poorly operated and thus pose environmental problems. These capacity problems are strongly driven by two factors: the failure to site new capacity and the unwillingness of communities to accept waste from other communities, particularly those that are distant. Siting and permitting time frames for landfills and combustion facilities range from 4 years to 10 years if they occur at all. Even recycling and composting facilities typically take several years to site and permit. Although we have the technical knowledge to site and permit environmentally sound facilities and to transport waste to these facilities, we have failed to achieve these end points because the public and private institutions responsible for managing our waste problems do not have the public's confidence. Also, only few state and local governments have established effective dispute resolution techniques, preemption techniques, or voluntary siting/host community partnership programs.

Due to both capacity shortages and strengthened environmental regulation, the cost of municipal waste management continues to increase dramatically in most areas of the country. This has caused a crisis for many communities. The average national price for landfills has increased dramatically from $10 per ton in 1983 to around $35 per ton in 1992. Once again, averages fail to tell the whole story. In the Northeast, the prices can be over $125 per ton. On top of these prices, the new federal and state solid waste landfill controls will cost between $12 and $17 per ton on average.[3] The variation between individual landfills is large, and the incremental cost of new controls can range from as low $1 per ton at an extremely well-located landfill to over $41 per ton, according to the United States Environmental Protection Agency (EPA) in its Regulatory Impact Analysis performed to support the proposed Subtitle D rule, December 1988. These figures are strictly figures for the incremental cost of implementing the new federal solid waste landfill regulations. As such, these costs are very significant since they are similar in magnitude to the total cost of land disposal today in many parts of the country. Of course, the lack of capacity will continue to drive prices even higher because value for remaining capacity is based on the cost to create new capacity. As long as the siting process remains in a bottleneck, the value of remaining capacity continues to skyrocket.

The costs for combustion have also increased as the regulatory requirements governing this waste management technology have increased in stringency. Recycling and composting costs are highly dependent on many factors including collection methods; facility volumes; facility environmental, health, and safety controls; and the availability of markets for secondary materials.

Where capacity is limited and costs of landfilling are high, states and communities have increasingly looked to institute a proactive planning process and aggressive state legislation. These forces pushing proactive planning will continue to accelerate for the following reasons:

- Stringent new federal and state regulations on landfill design and operation will increase the cost of landfilling and will result in early closure of some existing facilities.
- Stringent new federal and state regulations on combustion design and operation will increase the cost of incineration and will result in closure of some existing units.
- New regulations covering recycling/composting permit requirements, facility design, and operation will increase the cost of these functions.
- New regulations covering non-municipal solid waste (including industrial process waste, construction and debris waste, contaminated soils, agricultural waste, municipal sludges, and combustion ash) will result in many of these materials going to higher-quality management facilities including MSW management facilities.
- Increased cleanup of existing landfills and recycling facilities will lead to additional capacity needs and create further public distrust of waste management facilities in general, further slowing down new siting activities.
- Increased attention to the air and water pollution impacts of all facilities including recycling/composting facilities will make it difficult to site any new facility.
- Pressure to manage wastes locally or with the full acceptance of the nonlocal receiving community will continue to build.
- Increased cost of transporting wastes and recyclables will establish pressure for examining a full range of waste management options.
- The cost-sensitive nature of most waste management facilities to waste volume managed will continue to create pressure for fewer but larger regional facilities.

As states and communities confront the rapidly changing MSW management landscape in their planning efforts, there are many factors which affect the development of quality management strategies. Each of these is discussed below.

Clear Definitions

To date, the lack of clear definitions has been a significant impediment to the development of sound waste management strategies. At a fundamental level, it has resulted in confusion as to what constitutes MSW and what capacity exists to manage it. Three examples are provided. First, many municipal landfills accept petroleum-contaminated soils, municipal sludges, combustion ash, and construction/demolition wastes. However, since these are often not called MSW, the MSW generation rates do not consider these volumes. Thus, available landfill capacity for MSW can be significantly overstated, resulting in MSW management plans which fail to plan for needed landfill capacity.

Another example concerns recycling rates. Some communities count construction/demolition waste and automobiles in their projected recycling rates. However, these materials are not considered in their MSW generation rates. Therefore, the use of these high recycling rates can overstate the amount of waste managed by recycling and underestimate the amount of alternative capacity needed.

A third example focuses on states which have set specific recycling standards, such as 25 percent. By not counting certain items in those recycling rates, the state can provide a disincentive to recycle certain commodities because those items "don't count."

Consistent definitions form the basis for a defensible measurement system. They allow an entity to track progress and to compare its progress with other entities. They facilitate quality dialogue with all affected and interested parties. Moreover, we manage what we measure. If we don't measure it, it is unlikely to receive careful management attention. Waste management decision makers should give significant attention to definitions at the front end of the planning process. Since all future legislation, regulations, and public dialogue will depend on these definitions, decision makers should consider an open public comment process to establish appropriate definitions early in the strategy development (planning) process.

Quality Data

It is difficult to develop sound integrated MSW management strategies without good data. It is even more difficult to engage the public in a dialogue about the choice of an optimal strategy without these data. While the federal government and some states have focused on collecting better waste generation and capacity data, these data are still weaker than they should be. Creative waste management strategies often require knowledge of *who* generates the waste, not just what volumes are generated. Often a large percentage of the waste is generated by a small percentage of the generators. This is particularly true for many specific subsets of the waste which lend themselves to cost-effective source reduction, recycling, or composting.

The environmental, health, and safety impacts and the costs of alternatives to land-filling and combustion are another data weakness. Landfilling and combustion have been studied in depth, although risks and costs are usually highly site-specific. Source reduction, recycling, and composting have received much less attention. While these activities can often result in reduced EHS impacts over landfilling, they do not always. Again, the answer is often site- and/or commodity-specific.

MSW management strategies developed without quality data on risks and costs of all available options under consideration are not likely to optimize decision making and may, in some cases, result in unsound decisions. Because data are often costly and difficult to obtain, decision makers should plan for an active data collection stage before making critical strategy choices. While this approach may appear to result in slower progress in the short term, it will result in true long-term progress characterized by cost-effective and environmentally sound strategies.

Clear Roles and Leadership in Federal, State, and Local Government

Historically, MSW has been considered a local government issue. That status has become increasingly confused over the past 5 years as environmental, health, and safety concerns have increased and more waste has moved outside the generator's locality for management. At the present time, federal, state, and local governments are developing location, design, and operating standards for waste management facilities. State and local governments are controlling facility permitting for a range of issues including air emissions, storm-water runoff, and surface/groundwater discharges in addition to solid waste management. These requirements often result in the involvement of multiple agencies and multiple permits. While product labeling and product design have traditionally been regulated at the federal level, state and local governments have increasingly looked to these areas as they attempt to increase source reduction and recycling of municipal waste.

As a result, the current regulatory situation is becoming increasingly less efficient. Unless there is increased cooperation among all levels of government, the current trends will continue. However, a more rational and cost-effective waste management framework can result if roles are clarified and leadership is embraced. In particular, federal leadership on product labeling and product requirements is important. It will become increasingly unrealistic for multinational manufacturers to develop products for each state. The impact will be particularly severe on small states and for small businesses operating nationally. Along with the federal leadership on products, state leadership will be crucial in permit streamlining. The cost of facility permitting is severely impacted by the time-consuming nature of the permitting process, although a long process does nothing for increased environmental protection. Moreover, the best waste management strategies become obsolete and unimplementable if waste management facilities and facilities using secondary materials as feedstocks cannot be built or expanded. Even source reduction initiatives often depend on major permit modifications for existing manufacturing facilities.

Even and Predictable Enforcement

The public continues to distrust both the individuals who operate waste facilities and the regulators who enforce proper operation at those facilities. One key contributor to this phenomenon is the fact that state and federal enforcement programs are perceived to be understaffed or weak. Thus, even if a strong permit is written, the public lacks confidence that it will be enforced. Concern is also expressed that governments are reluctant to enforce against other government-owned or -operated facilities. Whether or not these perceptions are true, they are crucial ones to address in order to achieve consensus on a sound waste management strategy.

There are multiple approaches which decision makers can consider. They can develop internally staffed state-of-the-art enforcement programs designed to provide a level playing field for all facilities regardless of type, size, or ownership. If they involve the public in the overall design of the enforcement program and report on inspections and results, public trust will increase. If internal resources are constraining, decision makers can examine more innovative approaches, including use of third-party inspectors, public disclosure requirements for facilities, or separate contracts on performance assurance between the host community and the facility.

Resolution of the Intercounty, Interstate, and Intercountry Waste Issue for MSW and its Components

This has been a continuous issue over the last few years, as communities without sufficient local capacity ship their wastes to other locations. While a few receiving communities have welcomed the waste because it has resulted in a significant income source, most receiving communities have felt quite differently. These communities have wanted to preserve their existing capacity, knowing they will also find it difficult to site new capacity. Moreover, they do not want to be a dumping ground for other community waste because they believe the adverse environmental impacts of the materials outweigh any short-term financial benefit.

This dilemma has resulted in the adoption of many restrictive ordinances and subsequent court challenges. While the current federal legislative framework, embodied in the interstate commerce clause, makes it difficult for any state or local official to uphold state and local ordinances which prevent the inflow of nonlocal waste, the federal legislative playing field can be changed. At this writing, it is still expected that Congress will address the issue in the near future. However, this is a difficult issue in part because:

- Most communities and states export some of their wastes (e.g., medical wastes, hazardous wastes, radioactive wastes).

- New state-of-the-art waste facilities are costly to build and operate and require larger volumes than what can typically be provided by the local community in order to cover their costs.

- Waste facilities are often similar in environmental effects to recycling facilities and manufacturing facilities. If one community will not manage wastes from another community, why should one community have to make chemicals or other products which are ultimately used by another community?

- While long-distance transport (over 200 mi) of MSW usually indicates the failure to develop a local waste management strategy, shorter interstate movements (less than 50 mi) may provide the foundation for a sound waste management strategy. Congress should be careful to avoid overrestricting options.

2.2 MUNICIPAL WASTE MANAGEMENT OPTIONS

Source reduction, recycling and composting, waste-to-energy facilities, and landfills are the four basic approaches to waste management. They are discussed in considerable detail in the following chapters of this book.

Source Reduction

Source reduction focuses on reducing the volume and/or toxicity of generated waste. Source reduction includes the switch to reusable products and packaging, the most familiar example being returnable bottles. However, bottle bills only result in source reduction if bottles are reused once they are returned. Other good examples of source reduction are grass clippings that are left on the lawn and never picked up and modified yard plantings that do not result in leaf and yard waste. The time to consider source reduction is at the product/process design phase.

Source reduction can be practiced by everybody. Consumers can participate by buying less or using products more efficiently. The public sector (government entities at all levels: local, state, and federal) and the private sector can also be more efficient consumers. They can reevaluate procedures which needlessly distribute paper (multiple copies of documents can be cut back), require the purchase of products with longer life spans, and cut down on the purchase of disposable products. The private sector can redesign its manufacturing processes to reduce the amount of waste generated in the manufacturing process. This may require closed-loop manufacturing processes, the use of different raw materials and/or different production processes. Finally, the private sector can redesign products by increasing their durability, substituting less toxic materials, or increasing product effectiveness. However, while everybody can participate in source reduction, it digs deeply into how people go about their business—something that is difficult to mandate through regulation without getting mired in the tremendous complexity of commerce.

Source reduction is best encouraged by making sure that the cost of waste management is fully internalized. *Cost internalization* means pricing the service such that all the costs are reflected. For waste management, the costs that need to be internalized include pickup and transport, site and construction, administrative and salary, and environmental controls and monitoring. This is true whether the product is ultimately managed in a landfill, combustor, recycling facility, or composting facility. Regulation can aid cost internalization by requiring product manufacturers to provide public disclosure of the costs associated with these aspects of product use and development.

Recycling and Composting

Recycling is perhaps the most positively perceived and doable of all the waste management practices. Recycling will return raw materials to market by separating reusable products from the rest of the municipal waste stream. The benefits of recycling are many. It saves precious finite resources, lessens the need for mining of virgin materials, which lowers the environmental impact for mining and processing, and reduces the amount of energy consumed. Moreover, recycling can help stretch landfill capacity. Recycling can also improve the efficiency and ash quality of incinerators and composting facilities by removing noncombustible materials, such as metals and glass.

Recycling can also cause problems if it is not done in an environmentally responsible manner. Many Superfund sites are what is left of poorly managed recycling operations. Examples include deinking operations for newsprint, waste oil recycling, solvent recycling, and metal recycling. In all of these processes, toxic contaminants that need to be properly managed are removed. Composting is another area of recycling that can cause problems without adequate location controls. For example, groundwater can be contaminated if grass clippings, leaves, or other yard wastes that contain pesticide or fertilizer residues are composted on sandy or other permeable soils. Air contamination by volatiles can also result.

Recycling will flourish where economic conditions support it, not just because it is mandated. For this to happen, the cost of landfilling or resource recovery must reflect its true cost and must be at least $40 per ton or higher. Successful recycling programs also require stable markets for recycled materials. Examples of problems in this area are not hard to come by; a glut of paper occurred in Germany in the 1984 to 1986 time frame due to a mismatch between the grades of paper collected and the grades required by the German papermills. Government had not worked with enough private industries to find out whether the mills had the capacity and equipment needed to deal with low-grade household newspaper. In the United States, we have also seen a loss of markets for paper. Prices have dropped to the point where it may even cost money to dispose of collected newspapers in some parts of the country.

Stable markets also require that stable supplies are generated. This supply-side problem has been problematic in certain areas of recycling, including metals and plastics. Government and industry must work together to address the market situation. It is critical to make sure that mandated recycling programs do not get too far ahead of the markets.

Even with a good market situation, recycling and composting will flourish only if they are made convenient. Examples include curbside pickup for residences on a frequent schedule and easy drop-off centers with convenient hours for rural communities and for more specialized products. Product mailback programs have also worked for certain appliances and electronic components.

Even with stable markets and convenient programs, public education is a critical component for increasing the amount of recycling. At this point, the United States must develop a conservation, rather than a throwaway, ethic. We saw the United States do this back in the energy crisis of the 1970s. Recycling presents the next opportunity for cultural change. It will require us to move beyond a mere willingness to collect our discards for recycling. That cultural change will require consumers to purchase recyclable products and products made with recycled content. It will require businesses to utilize secondary materials in product manufacturing and to design new products for easy disassembly and separation of component materials.

Incineration

The third piece of the integrated waste management program is incineration, or waste-to-energy, facilities. These facilities are attractive because they do one thing very well; they reduce the volume of waste dramatically, up to ninefold. Incinerators can also recover useful energy either in the form of steam or in the form of electricity. Depending on the economics of energy in the region, this can be anywhere from profitable to unjustified. Volume reduction alone can make the high capital cost of incinerators attractive when landfill space is at a premium, or when the landfill is distant from the point of generation. For many major metropolitan areas, new landfills must be located increasingly far from the center of the population. Moreover, incinerator bottom ash has a promise for reuse as a building material. Those who make products from cement or concrete may be able to utilize incinerator ash.

The major constraints of incinerators are their cost, the relatively high degree of sophistication needed to operate them safely and economically, and the fact that the public is very leery of their safety. The public is concerned about both stack emissions from incinerators and the toxicity of ash produced by incinerators. EPA addressed both of these concerns through the development of new regulations for solid waste incinerators/waste-to-energy plants and improved landfill requirements for ash. These regulations will ensure that well-designed, well-built, and well-operated facilities will be fully protective from environmental and health standpoints.

Landfills

Landfills are the one form of waste management that nobody wants but everybody needs. There are simply no combinations of waste management techniques that do not require landfilling to make them work. Of the four basic management options, landfills are the only management technique that is both necessary and sufficient. Some wastes are simply not recyclable, many recyclable wastes eventually reach a point where their intrinsic value is completely dissipated and they no longer can be recovered, and recycling itself produces residuals.

The technology and operation of a modern landfill can assure protection of human health and the environment. The challenge is to ensure that all operating landfills are designed properly and are monitored once they are closed.

It is critical to recognize that today's modern landfills do not look like the old landfills that are on the current Superfund list. Today's operating landfills do not continue to take hazardous waste. In addition, they do not receive bulk liquids. They have gas control systems, liners, leachate collection systems, extensive groundwater monitoring systems, and perhaps most importantly, they are better sited and located in the first place to take advantage of natural geological conditions.

Landfills can also turn into a resource. Methane gas recovery is occurring at many landfills today and CO_2 recovery is being considered. After closure, landfills can be used for recreation areas such as parks, golf courses, or ski areas. Some folks are looking at landfills as repositories of resources for the future; in other words, today's landfills might be able to be mined at some time in the future when economic conditions warrant. This could be particularly true of monofills, which focus on one kind of waste material like combustion ash or shredded tires.

2.3   FRAMEWORK FOR DECISION MAKING

The previous section has presented some of the basics on the four waste management tools—source reduction, recycling, waste-to-energy, and landfilling. With that as a background, we must map out a framework for making decisions. In a world without economic constraints, the tools for waste management could be ordered by their degree of apparent environmental desirability. Source reduction would clearly be at the top, since it prevents waste from having to be managed at all. Recycling, including composting, would be the next best management tool, because it can return resources to commerce after the original product no longer serves its intended purpose. Waste-to-energy follows because it is able to retrieve energy that otherwise would be buried and wasted. Finally, landfilling, while often listed last, is really not any better or worse than incineration, since it too can recover energy. Moreover, waste-to-energy facilities still require landfills to manage their ash.

In reality, however, every community and region will have to customize its integrated management system to suit its environmental situation and to consider its economic constraints. A small, remote community such as Nome, Alaska, has little choice but to rely solely on a well-designed and -operated landfill. At the other extreme, New York City can easily and effectively draw on some combination of all the elements of the waste management hierarchy. Communities that rely heavily on groundwater that is vulnerable, such as Long Island, New York, and many Florida communities, usually need to minimize landfilling and look at incineration, recycling, and residual disposal in regions where groundwater is less vulnerable. Communities that have problems with air quality usually avoid incineration to minimize more atmospheric pollutants. Sometimes these communities can take extra steps to ensure that incineration is acceptable by first removing metals and other bad actors out of the waste stream. In all communities, the viability of recycling certain components of the waste stream is linked to volumes, collection costs, available markets, and the environmental consequences of the recycling and the reuse operations.

Long-term planning at the local, state, and even regional level is the only way to come up with a good mix of management tools. It must address both environmental concerns and economic constraints. As discussed earlier, planning requires good data. This fact has long been recognized in fields such as transportation and health-care planning. However, until recently, databases for solid waste planning were not available, and, even now, they are weak.

There are a number of guidelines that planners should embrace. First, it is critical to look to the long term. Today's outrageous spot market prices are a symptom of crisis conditions where new facilities are simply not being sited. Examples already exist of locations where current prices are significantly reduced from their highs as new capacity options have emerged.

Second, planners must make sure that all costs are reflected in each option. Municipal accounting practices sometimes hide costs. For example, the transportation department may purchase vehicles while another department may pay for real estate, and so on. Accurate accounting is essential.

Third, skimping on environmental controls brings short-term cost savings with potentially greater liability down the road. It is always better to do it right the first time. This is true of recycling and composting facilities, as well as incineration facilities and landfills.

Fourth, planners should account for the volatility of markets for recyclables. The question becomes: can, in a given location for a given commodity, a recycling program survive the peaks and valleys of recycling markets without going broke in between.

Fifth, planners must consider the availability of efficient facility permitting and siting for waste facilities using recycled material inputs, and for facilities which need permit changes to implement source reduction.

Finally, planners should look beyond strictly local options. When we forget about political boundaries, different management combinations may become possible at reasonable costs. Potential savings can occur in the areas of procurement, environmental protection, financing, administration, and ease of implementation. Regional approaches include public authorities, nonprofit public corporations, special districts, and multicommunity cooperatives.

The process of formulating a good integrated solid waste strategy is time-consuming and difficult. Ultimately, the system must be holistic; each of its parts must have its own purpose and work in tandem with all the other pieces like a finely crafted, highly efficient piece of machinery. Like a piece of machinery, it is unlikely that an efficient and well-functioning output is achieved unless a single design team, understanding its objective and working with suppliers and customers, develops the design. The successful integrated waste plan drives legislation; it is not driven by legislation. More legislation does not necessarily lead to more source reduction and recycling. In fact, disparate pieces of legislation or regulation can work at cross purposes. Moreover, our free market system works best when there is some sense of stability and certainty. This encourages risk taking because it is easier to predict expected market response. The faster a holistic framework for waste management is stabilized, the more likely public decision making will obtain needed corporate investment.

The first stage of planning involves carefully defining terminology including what wastes are covered, what wastes are not covered, and what activities constitute recycling and composting. It also requires the articulation of clear policy goals for the overall waste management strategy. Is the goal to achieve the most cost-effective strategy that is environmentally protective or to maximize diversion from landfills? There are no absolute right or wrong answers. However decision makers should share the definitions, key assumptions, and goals with the public for their review and comment.

The second stage involves identification of the full range of possible options and the methodical collection of environmental risks and costs associated with each option. Data collection is best done before any strategy has been selected. The cost estimates for recycling and composting can be highly variable depending on what assumptions are made about market demand and what actions are taken to stimulate markets. These differing assumptions about markets can also impact the assumptions on environmental risks, since some types of reuse scenarios have more severe environmental impacts than others. The stringency of the regulatory permitting and enforcement programs which set and enforce standards for each type of waste management facility, including recycling facilities and facilities which use recycled material inputs in the manufacturing process, will also impact the costs and environmental risks associated with various options. Finally, the existence of product standards for recycled materials will impact the costs and risks of various recycling and composting strategies. The costs of all management strategies will be volume dependent. Once this information is collected, the public should have the opportunity for meaningful input on the accuracy of the assumptions. Acceptance by the public at this stage can foster a smoother and faster process in the long run.

The final step involves examining the tradeoffs between available options so that an option or package of options can be selected. At the core, these tradeoffs involve risk and cost comparisons. However, they also involve careful consideration of implementation issues such as financing, waste volumes, enforcement, permit time frames, siting issues, and likely future behavior changes.

Some examples of implementation issues are useful. Pay-by-the-bag disposal programs may result in less garbage because people really cut back on their waste generation when they can save money. On the other hand, there has been some indication that pay-by-the-bag systems have actually resulted in the same amount of garbage generated, but an increase in burning at home or illegal dumping. Another example is the need to assess the real effect of bottle bills. Bottle bills may be very effective if collected bottles are reused, or if markets exist so that the collected bottles can be recycled. However, in some locations bottle bills result in a double payment; once to collect the bottles and then again to landfill the bottles because a viable market strategy is not in place. A final example concerns flow control. Flow control promises a way to ensure that each of the various solid waste facilities has enough waste to run efficiently. On the other hand, if governments use flow control to send a private generator's waste to a poorly designed or operated solid waste facility, then the government may be tampering with the generator's Superfund liability, or the government may increase the amount of waste that is going to an environmentally inferior facility.

Some computerized decision models have been developed which compare the costs of various strategies.[4] However, these often require considerable tailoring before they accurately fit a local situation. It is often useful to develop a final strategy in an iterative manner by first selecting one or two likely approaches and then setting the exact parameters of the selected approach in a second iteration. Public involvement is critical throughout the selection process.

It may also be useful to develop an integrated waste management strategy by formulating a series of generator-specific strategies. Residential generators are one group of MSW generators. Another important group includes the public sector, including municipalities and counties, who generate their own waste streams. Finally, there are numerous specific industry groups such as the hotel industry, the restaurant industry, petrochemical firms, the pulp and paper industry, and the grocery industry. In each case, the character of the solid waste generated will vary. For some groups, all the waste will fall into the broad category of MSWs. For other groups, much of the waste will include industrial, agricultural, or other non-MSW waste. In some cases, the variation within the generator category will be significant. However, in other cases, the within-group waste characterization is likely to be relatively uniform. Industry-by-industry strategies, focusing on the largest waste generator categories, may result in more implementable and cost-effective strategies.

2.4 KEY FACTORS FOR SUCCESS

Arriving at successful solid waste management solutions requires more than just good planning. The best technical solution may fail if politicians and government officials do not consider a series of other important points. This section attempts to identify some of these points.

Credibility for Decision Makers

It is absolutely critical to work to protect the credibility of those individuals who must ultimately make the difficult siting and permitting decisions. Proper environmental standards for all types of facilities, including recycling, can help give decision makers necessary support. Credible enforcement that operates on a level playing field is also crucial. Operator certification programs, company-run environmental audit programs, company-run environmental excellence programs, government award programs for outstanding facilities, and financial assurance provisions can also increase the public's level of comfort with solid waste management facilities. Finally, clear-cut siting procedures and dispute resolution processes can provide decision makers with a critical support system.

Efficient Implementation Mechanisms Including Market Incentives

A number of things can be done to help facilitate program implementation. Expedited permitting approaches for new facilities and expedited permit modification approaches for existing facilities can be helpful. Approaches such as class permits or differential requirements based on the complexity of the facilities are examples. Pilot programs can be particularly helpful in determining whether a program which looks good on paper will work well in real life. However, some of the most efficient implementation mechanisms involve the consideration of market incentives.

Much of today's federal and state legislation and regulation has focused on a command and control strategy. Such a strategy relies on specified mandates that cover all parties with the same requirements. These requirements are developed independent of market concepts and other basic business incentives, and, as a result, these approaches are often slower and more expensive to implement for both the regulated and the regulators.

Market approaches can significantly cut the cost of achieving a fixed amount of environmental protection, energy conservation, or resource conservation when compared with a traditional command and control approach. The concept behind this is simple. Determine what total goal is needed. Then, let those who can achieve the goal most cost-effectively do so. They can sell extra credits to those who have a more difficult time meeting the goal. Other market approaches rely on using market pricing to strongly encourage desirable behaviors.

Another incentive-type program which has achieved major environmental benefits in a cost-effective way has been the implementation of the Federal Emergency Planning and Community Right-to-Know Act (1986) emissions reporting program.[5] The law does not mandate specific reductions in emissions to air, water, and land. However, it does require affected facilities to publicly report quantities of chemicals released. The mere fact of having to publicly report has resulted in a dramatic lowering of emissions.

The types of programs which decision makers at the state or federal level could examine include the following:[6]

- An overall program to reduce average per capita waste generation rates through the use of a marketable permit program. There would be several ways to implement this type of program. A fixed per capita figure could be established throughout a state. Whichever municipalities (or counties) could achieve it most efficiently could sell extra credits to other affected municipalities. Other alternatives would set the per capita rate by size of municipality or require all municipalities to achieve a fixed percent reduction from established baseline rates.
- A marketable permit program to implement recycling goals. Rather than require all municipalities/counties to achieve the same recycling rates, let those municipalities/counties that can achieve the recycling rates most cost-effectively sell any extra credits to other affected parties.
- A program which would develop differential business tax rates based on the amount of recycling (or source reduction) which the company achieves. The tax rates could be based on fixed rate standards (for example, source reduction of 10 percent or recycling of 25 percent) or percentage improvements over a baseline year.

- A program that would develop differential property taxes for homes that recycle or reduce their disposed waste by a given percentage. The percentage could be increased gradually each year in order to maintain the tax break.
- Product and service procurement preferences for those companies who have high overall recycling rates or who utilize a high percentage of secondary materials.
- Differential business tax rates or permit priority for companies who use recycled material inputs in production processes or who buy large quantities of recycled materials for consumption.
- Differential water rates for companies who use large volumes of compost or who reduce their green waste.
- Information disclosure requirements that require certain types/sizes of businesses to provide the public with information on their waste generation rates, their recycling rates, their procurement of secondary materials, and their waste management methods. (Good examples would be hotels and other types of consumer businesses.) The state could also compile state average values by industry group and require that these rates be posted along with the company-specific rates.

Significant Attention on Recycling Markets

Recycling will not be sustainable in the long term unless it is market-driven, so that there is a market demand for secondary materials. The market incentive discussion provided some ideas as to how market incentives can be utilized broadly to drive desirable interpreted waste strategies by influencing the behavior of affected entities. Some of these behaviors may lead to the creation of market demand for specific secondary materials. However, it is also important to examine secondary material markets on a commodity-specific basis, particularly in the subset of materials that compose a large fraction of the MSW stream.

There is a wide range of policy choices that can impact market demand. These include commodity-specific procurement standards, entity-specific procurement plans, equipment tax credits, tax credits for users of secondary materials, mandated use of secondary materials for certain government-controlled activities (such as landfill cover or mine reclamation projects), use of market development mechanisms in enforcement settlements, recycled content requirements for certain commodities, manufacturer take-back systems, virgin material fees, and labeling requirements. Whether any of these actions are needed and if so, which ones, can only be determined after a careful analysis of each commodity.

If such actions are needed, two cautions are in order. First, it is often better to discuss the need for market strengthening with affected parties before mandating a specific result. If, after a fixed time frame, the market does not improve, a regulated outcome can be automatically implemented. That hammer often provides the needed impetus for action without regulatory involvement. Second, while the first six examples of market demand approaches can be implemented at the federal or state level, the last four examples of market demand approaches are best implemented at the federal level.

Public Involvement

As mentioned previously, the best technical solution is unlikely to work unless the public is active in helping to reach the final choice of options. Public involvement must be just that; it cannot be a one-way street, but rather the public must be involved in two-way discussions. There must be a give and take on the final solution. Included in this dialogue must be serious discussion about the tradeoff between risk reductions and cost. This public involvement is best done with multiple opportunities for both formal and informal inputs.

Continuous Commitment to High-Quality Operations for All Facilities

Today's solid waste solutions require a commitment to high-quality operations. In the past, solid waste management, as with many other government services, was often awarded to the lowest bidder. This approach needs to be seriously reconsidered, given the environmental liabilities associated with poorly managed solid waste.

Evaluation of the Effectiveness of the Chosen Strategy

In developing specific legislation and regulations, it is important that the full impact of individual legislative or regulatory provisions be monitored after the program has been implemented. MSW planning is a process, not a project. That process must continually ensure that the plan mirrors reality and that implementation obstacles are addressed expeditiously.

2.5 SUMMARY

In closing, four critical points need to be stressed. First, the biggest step forward in finding solutions to waste management problems is getting everyone to accept responsibility for ownership of the solid waste problem and its solution. Second, the lack of facility siting, more than anything else, is pushing us closer to a widespread crisis that can easily breed panic solutions, guaranteed to fail in the long run. Third, a successful solid waste program requires a focus on both planning and execution. Continuous program evaluation is important if the system is to function properly. And finally, the public and the private sectors have to win the confidence of the public back by insisting on first-rate environmental protection. This can happen only through strong regulations and their deliberate enforcement.

While the focus of this chapter has been MSW, there is a need for states to get ahead of the curve and begin to look at industrial solid waste. The volume of industrial solid waste is an order of magnitude larger than the volume of MSW (7 to 10 billion metric tons versus 196 million metric tons per year). At the same time, the environmental control systems in place for these wastes are probably less developed than those for MSW. As a result, the industrial solid waste situation may be a greater problem than the MSW situation and certainly deserves serious attention.

REFERENCES

1. U.S. Environmental Protection Agency, *Characterization of Municipal Solid Waste in the United States: 1992 Update,* July 1992, EPA-SW-92.

2. Congress of the United States, Office of Technology Assessment, *Managing Industrial Solid Wastes,* February 1992, OTA-BP-0-82.

3. Private communication with Browning-Ferris Industries.

4. California Council on Science and Technology, *Science and Technology Research Priorities for Waste Management in California,* November 1992.

5. CFR, Parts 370 and 372.

6. California Council on Science and Technology, *Science and Technology Research Priorities for Waste Management in California: Supplement,* November 1992.

CHAPTER 3

SOLID WASTE STREAM CHARACTERISTICS

Marjorie A. Franklin

President, Franklin Associates, Ltd.
Prairie Village, Kansas

No matter what method of solid waste management is being considered or implemented, an understanding of the characteristics of the waste stream is a must. Good planning goes beyond developing a "snapshot" of current waste composition; the long-term trends in waste stream characteristics are also important. If future quantities and components of the waste stream are under- or overestimated, then facilities may be over- or undersized, and project revenues and costs can be affected.

Most of the data presented in this chapter are taken from a report published by the U.S. Environmental Protection Agency, *Characterization of Municipal Solid Waste in the United States: 1992 Update.*[1] This report, the latest in a series published periodically since the 1970s, includes 30 years of historical data on municipal solid waste generation and management, supplemented with projections to the year 2000.

3.1 MUNICIPAL SOLID WASTE DEFINED

The definition of municipal solid waste (MSW) used in this chapter is the same as that used in the U.S. Environmental Protection Agency (EPA) reports. This definition states that MSW includes wastes from residential, commercial, institutional, and some industrial sources.

Three other definitions are important in this chapter:

Generation refers to the amount of materials and products in MSW as they enter the waste stream before any materials recovery, composting, or combustion take place.

Recovery refers to removal of materials from the waste stream for recycling or composting. Recovery does not automatically equal recycling.

Discards refers to the MSW remaining after recovery. The discards are generally combusted or landfilled, but they could be littered, stored, or disposed onsite, particularly in rural areas.

Types of Wastes Included

The sources of municipal solid waste are

Source	Examples
Residential	Single-family homes, duplexes, town houses, apartments
Commercial	Office buildings, shopping malls, warehouses, hotels, airports, restaurants
Institutional	Schools, medical facilities, prisons
Industrial	Packaging of components, office wastes, lunchroom and rest room wastes (but not industrial process wastes)

The wastes from these sources are categorized into durable goods, nondurable goods, containers and packaging, and other wastes. These categories are defined in detail in the following sections.

Types of Wastes Excluded

As defined in this chapter, municipal solid waste does *not* include a wide variety of other nonhazardous wastes that often are landfilled along with MSW. Examples of these other wastes are municipal sludges, combustion ash, nonhazardous industrial process wastes, construction and demolition wastes, and automobile bodies.

3.2 METHODS OF CHARACTERIZING MUNICIPAL SOLID WASTE

There are two basic methods for characterizing MSW—sampling and the material flows methodology used to produce the data referenced in this chapter. Each method has merits and drawbacks, as shown below:

Material flows	Sampling
Characterizes residential, commercial, institutional, and some industrial wastes	Characterizes wastes received at the sampling facility
Characterizes MSW nationwide	Is site-specific
Characterizes MSW generation as well as discards	Usually characterizes only discards as received
Characterizes MSW on an as-generated moisture basis	Usually characterizes wastes after they have been mixed and moisture transferred
Provides data on long-term trends	Provides only one point in time (unless multiple samples are taken over a long period of time)
Characterizes MSW on an annual basis	Provides data on seasonal fluctuations (if enough samples are taken)
Does not account for regional differences	Can provide data on regional differences

Not mentioned above is the fact that on-site sampling can be very expensive, especially if done with large enough samples and with the frequency required for reasonable accuracy. To date, only the material flows method has been used to characterize the MSW stream nationwide.

The idea for the material flows methodology was developed at the U.S. Environmental Protection Agency in the early 1970s. The methodology has been further developed and refined for EPA and other organizations over the past two decades. The basic methodology is described below.

Data on domestic production of the materials and products in municipal solid waste provide the basis of the material flows methodology. Every effort is made to obtain a data series that is consistent from year to year rather than a single point in time. This allows the methodology to provide meaningful historical data that can be used for establishing time trends. Data sources include publications of the U.S. Department of Commerce and statistical reports published by various trade associations. Numerous adjustments are made to the raw data, as follows:

- Deductions are made for converting/fabrication scrap, which is classified as industrial process waste rather than MSW.

- Where imports and/or exports are a significant portion of the products being characterized, adjustments are made, usually using U.S. Department of Commerce data. For example, more than half of the newsprint consumed in the United States is imported.

- Adjustments are made for various diversions of products from disposal as MSW. Examples include toilet tissue, which goes into sewer systems rather than solid waste, and paperboard used in automobiles. Such materials are not classified as MSW when disposed.

- Adjustments are made for product lifetimes. It is assumed that all containers and packaging and most nondurables are disposed the same year they are produced. Durable products such as appliances and tires, however, are assigned product lifetimes and "lagged" before they are assumed to be discarded. Thus, a refrigerator is assumed to be discarded 20 years after production.

While the basis of the material flows methodology is adjusted production data, it is necessary to use the results of sampling studies to determine the generation of food wastes, yard trimmings, and some miscellaneous inorganic wastes. A wide variety of sampling studies from all regions of the country over a long time period have been scrutinized to determine the relative percentages of these latter wastes in MSW. Since production data are as-generated rather than as-disposed, data on food, yard, and miscellaneous inorganic wastes are adjusted to account for the moisture transfer that occurs before these wastes are sampled.

3.3 MATERIALS IN MUNICIPAL SOLID WASTE BY WEIGHT

The materials generated in municipal solid waste over a 30-year period (1960 through 1990) are shown in Table 3.1 (by weight), Table 3.2 (by percentage), and in Figs. 3.1 and 3.2. Projections are shown for the years 1995 and 2000. The projections are based on trend line analysis and on data about the industries involved in production of the materials and products.

TABLE 3.1 Materials Generated* in the Municipal Waste Stream, 1960 to 2000
(In millions of tons)

Materials	1960	1965	1970	1975	1980	1985	1990	1995†	2000†
Paper and paperboard	29.9	38.0	44.2	43.0	54.7	61.5	73.3	79.2	84.7
Glass	6.7	8.7	12.7	13.5	15.0	13.2	13.2	13.6	13.5
Metals:									
Ferrous	9.9	10.1	12.6	12.3	11.6	10.9	12.3	12.0	12.1
Aluminum	0.4	0.5	0.8	1.1	1.8	2.3	2.7	3.1	3.6
Other nonferrous	0.2	0.5	0.7	0.9	1.1	1.0	1.2	1.4	1.5
Total metals	10.5	11.1	14.1	14.3	14.5	14.2	16.2	16.5	17.1
Plastics	0.4	1.4	3.1	4.5	7.8	11.6	16.2	20.0	24.8
Rubber and leather	2.0	2.6	3.2	3.9	4.3	3.8	4.6	5.9	6.5
Textiles	1.7	1.9	2.0	2.2	2.6	2.8	5.6	5.9	6.7
Wood	3.0	3.5	4.0	4.4	6.7	8.2	12.3	13.5	16.0
Other	0.1	0.3	0.8	1.7	2.9	3.4	3.2	3.4	3.7
Total materials in products	54.3	67.5	84.1	87.5	108.5	118.7	144.6	158.0	172.9
Other wastes:									
Food wastes	12.2	12.7	12.8	13.4	13.2	13.2	13.2	13.2	13.2
Yard trimmings	20.0	21.6	23.2	25.2	27.5	30.0	35.0	33.7	32.9
Miscellaneous inorganic wastes	1.3	1.6	1.8	2.0	2.2	2.5	2.9	3.0	3.1
Total other wastes	33.5	35.9	37.8	40.6	42.9	45.7	51.1	49.9	49.2
Total MSW generated	87.8	103.4	121.9	128.1	151.4	164.4	195.7	207.9	222.1

*Generation before materials recovery or combustion. Details may not add to totals due to rounding.

†Projected data.

Source: Adapted from U.S. Environmental Protection Agency, *Characterization of Municipal Solid Waste in the United States: 1992 Update*, EPA/530-R-92-013, July 1992.

TABLE 3.2 Materials Generated* in the Municipal Waste Stream, 1960 to 2000
(*In percent of total generation*)

Materials	1960	1965	1970	1975	1980	1985	1990	1995†	2000†
Paper and paperboard	34.1	36.8	36.3	33.6	36.1	37.4	37.5	38.1	38.1
Glass	7.6	8.4	10.4	10.5	9.9	8.0	6.7	6.5	6.1
Metals:									
Ferrous	11.3	9.8	10.3	9.6	7.7	6.6	6.3	5.8	5.4
Aluminum	0.5	0.5	0.7	0.9	1.2	1.4	1.4	1.5	1.6
Other nonferrous	0.2	0.5	0.6	0.7	0.7	0.6	0.6	0.7	0.7
Total metals	12.0	10.7	11.6	11.2	9.6	8.6	8.3	7.9	7.7
Plastics	0.5	1.4	2.5	3.5	5.2	7.1	8.3	9.6	11.2
Rubber and leather	2.3	2.5	2.6	3.0	2.8	2.3	2.4	2.8	2.9
Textiles	1.9	1.8	1.6	1.7	1.7	1.7	2.9	2.9	3.0
Wood	3.4	3.4	3.3	3.4	4.4	5.0	6.3	6.5	7.2
Other	0.1	0.3	0.7	1.3	1.9	2.1	1.6	1.7	1.6
Total materials in products	61.8	65.3	69.0	68.3	71.7	72.2	73.9	76.0	77.8
Other wastes:									
Food wastes	13.9	12.3	10.5	10.5	8.7	8.0	6.7	6.3	5.9
Yard trimmings	22.8	20.9	19.0	19.7	18.2	18.2	17.9	16.2	14.8
Miscellaneous inorganic wastes	1.5	1.5	1.5	1.6	1.5	1.5	1.5	1.4	1.4
Total other wastes	38.2	34.7	31.0	31.7	28.3	27.8	26.1	24.0	22.2
Total MSW generated	100.0	100.0	100.0	100.0	100.0	100.0	100.0	100.0	100.0

*Generation before materials recovery or combustion. Details may not add to totals due to rounding.
†Projected data.
Source: Adapted from U.S. Environmental Protection Agency, *Characterization of Municipal Solid Waste in the United States: 1992 Update*, EPA/530-R-92-013, July 1992.

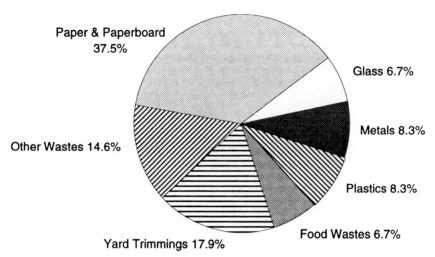

FIGURE 3.1   Materials generated in municipal solid waste, 1990. (*U.S. Environmental Protection Agency, Characterization of Municipal Solid Waste in the United States: 1992 Update, EPA/530-R-92-013, July 1992.*)

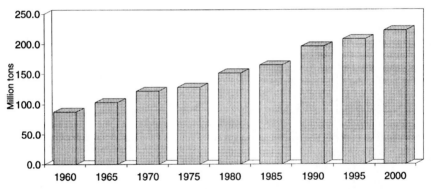

FIGURE 3.2   Materials generated in municipal solid waste, 1960 to 2000. (*U.S. Environmental Protection Agency, Characterization of Municipal Solid Waste in the United States: 1992 Update, EPA/530-R-92-013, July 1992.*)

Generation of MSW in the United States has grown steadily from about 88 million tons in 1960 to about 196 million tons in 1990.* Generation is projected to grow to 222 million tons in the year 2000. (Management of this tonnage of MSW by source reduction, recycling, composting, combustion, and landfilling is discussed elsewhere in this chapter and this handbook.)

Data on the most prominent materials in MSW are summarized in the following sections.

---

*Units of weight in this chapter are expressed in short tons. SI units can be obtained by multiplying short tons by 907.2 to obtain kilograms.

TABLE 3.3   Paper and Paperboard Products in Municipal Solid Waste, 1990

Product category	Generation, millions of tons	Percent
Nondurable goods:		
Newspapers	12.9	6.6
Books	1.0	0.5
Magazines	2.8	1.4
Office papers*	6.4	3.3
Telephone books	0.5	0.3
Third-class mail	3.8	2.0
Other commercial printing	5.5	2.8
Tissue paper and towels	3.2	1.6
Paper plates and cups	0.7	0.3
Other nonpackaging paper†	3.8	2.0
Total paper and paperboard nondurable goods	40.7	20.8
Containers and packaging:		
Corrugated boxes	23.9	12.2
Milk cartons	0.5	0.3
Folding cartons:		
Bleached	1.5	0.8
Unbleached and recycled	2.8	1.4
Subtotal folding cartons	4.3	2.2
Other paperboard packaging	0.3	0.1
Bags and sacks	2.4	1.2
Wrapping papers	0.1	0.1
Other paper packaging	1.0	0.5
Total paper and paperboard containers and packaging	32.6	16.7
Total paper and paperboard	73.3	37.5
Total MSW generation	195.7	100.0

*Includes high-grade papers and an adjustment for files removed from storage.

†Includes tissue in disposable diapers, paper in games and novelties, cards, etc.

Details may not add to totals due to rounding.

Source: Adapted from U.S. Environmental Protection Agency, *Characterization of Municipal Solid Waste in the United States: 1992 Update,* EPA/530-R-92-013, July 1992; American Paper Institute, *1991 Statistics of Paper and Paperboard* (Ref. 2); and National Office Paper Recycling Project, *Supply of and Recycling Demand for Office Waste Paper, 1990 to 1995* (Ref. 3).

Paper and Paperboard

For the entire historical period documented, paper and paperboard have been the largest component of the municipal solid waste stream, always comprising more than one-third of total generation. In 1990, paper and paperboard were 37.5 percent of total generation.

Paper and paperboard are found in a wide variety of products in two categories of MSW—nondurable goods and containers and packaging (Table 3.3). In the non-durables category, newspapers comprise the largest portion at nearly 7 percent of total MSW generation. Other important contributions in this category come from office papers, third-class mail (including catalogs), and other commercial printing, which includes advertising inserts in newspapers, reports, brochures, and the like. Some paper components of MSW that attract considerable attention, such as telephone books and disposable plates and cups, amount to less than 1 percent of total MSW generation.

In the containers and packaging category, corrugated boxes are by far the largest contributor. Indeed, corrugated boxes, at over 12 percent of total MSW generation, are the largest single product category for all materials. Other paper and paperboard contributors to MSW include folding cartons (e.g., cereal boxes), paper bags and sacks, and other kinds of packaging.

Glass

Glass in MSW is found primarily in glass containers, although a small portion is found in durable goods (Table 3.4). The glass containers are used for beer and soft drinks, wine and liquor, food products, toiletries, and a variety of other products. Glass containers were about 6 percent of total MSW generation in 1990, with the total contribution of glass at less than 7 percent.

As a percentage of MSW generation, glass containers grew throughout the 1960s and into the 1970s. In the 1970s, however, first aluminum cans and then plastic bottles encroached on the markets for glass bottles. As a result, glass as a percentage of MSW has dropped in the 1980s and 1990s, and this is projected to continue.

Ferrous Metals

Overall, ferrous metals made up about 6 percent of MSW generation in 1990 (Table 3.5). The most significant sources of ferrous metals in MSW generation are durable goods, including major appliances, furniture, tires, and miscellaneous items such as small appliances. These sources contributed enough ferrous metals to make up nearly 5 percent of MSW generation in 1990.

Steel cans contribute the remainder of ferrous metals in MSW, about 1.5 percent of total MSW in 1990. The steel cans mainly package food, with small amounts of steel found in beverage cans and other packaging such as strapping. The pattern of steel in containers and packaging is much like that of glass. Steel was commonly used in beverage cans until the 1970s, when aluminum cans became popular. Steel has also been displaced in many applications by plastics.

TABLE 3.4   Glass Products in Municipal Solid Waste, 1990

Product category	Generation, millions of tons	Percent
Durable goods*	1.3	0.7
Containers and Packaging:		
Beer and soft drink bottles	5.7	2.9
Wine and liquor bottles	2.1	1.1
Food and other bottles and jars	4.1	2.1
Total glass containers	11.9	6.1
Total glass	13.2	6.7
Total MSW generation	195.7	100.0

*Glass as a component of appliances, furniture, consumer electronics, etc.
Details may not add to totals due to rounding.

Source: Adapted from U.S. Environmental Protection Agency, *Characterization of Municipal Solid Waste in the United States: 1992 Update,* EPA/530-R-92-013, July 1992.

Aluminum

Most aluminum in MSW is found in containers and packaging, primarily in beverage cans (Table 3.5). Some aluminum is also found in durable and nondurable goods. Overall, aluminum amounted to an estimated 1.4 percent of MSW generation in 1990.

Other Nonferrous Metals

Other nonferrous metals (lead, copper, zinc) are found in MSW, primarily in durable goods. These metals have totaled less than one percent of MSW over the entire period quantified. The major source of nonferrous metals in MSW is lead in automotive batteries. In 1990, this lead amounted to an estimated 0.4 percent of all MSW generated (Table 3.5).

Plastics

Plastics are used very widely in the products found in municipal solid waste; in 1990, plastics composed over 8 percent of MSW generation (Table 3.6). Use of plastics has

TABLE 3.5   Metal Products in Municipal Solid Waste, 1990

Product category	Generation, millions of tons	Percent
Durable goods:		
Ferrous metals*	9.4	4.8
Aluminum†	0.6	0.3
Batteries (lead)	0.8	0.4
Other nonferrous metals‡	0.4	0.2
Total metals in durable goods	11.2	5.7
Nondurable goods:		
Aluminum	0.2	0.1
Containers and packaging:		
Steel:		
Beer and soft drink cans	0.1	0.1
Food and other cans	2.5	1.3
Other steel packaging	0.2	0.1
Total steel packaging	2.9	1.5
Aluminum:		
Beer and soft drink cans	1.6	0.8
Food and other cans	Neg.	Neg.
Foil and closures	0.3	0.2
Total aluminum packaging	1.9	1.0
Total metals in containers and packaging	4.8	2.4
Total metals	16.1	8.2
Total MSW generation	195.7	100.0

*Ferrous metals in appliances, furniture, tires, and miscellaneous durables.

†Aluminum in appliances, furniture, and miscellaneous durables.

‡Other nonferrous metals in appliances, lead-acid batteries, and miscellaneous durables.
Neg. = Negligible (less than 0.1 percent or 100,000 tons).
Details may not add to totals due to rounding.

Source: Adapted from U.S. Environmental Protection Agency, *Characterization of Municipal Solid Waste in the United States: 1992 Update,* EPA/530-R-92-013, July 1992.

TABLE 3.6   Plastic Products in Municipal Solid Waste, 1990

Product category	Generation, millions of tons	Percent
Durable goods*	4.94	2.5
Nondurable goods:		
Plastic plates and cups	0.32	0.2
Trash bags	0.78	0.4
Disposable diapers†	0.30	0.2
Clothing and footwear	0.17	0.1
Other misc. nondurables‡	2.67	1.4
Total plastics nondurable goods	4.24	2.2
Containers and packaging:		
Soft drink bottles:		
PET bottles	0.38	0.2
HDPE base cups	0.06	Neg.
Subtotal soft drink bottles	0.44	0.2
Milk bottles (HDPE)	0.36	0.2
Other containers		
HDPE	1.00	0.5
PET	0.17	0.1
Polystyrene	0.20	0.1
Other resins	0.46	0.2
Subtotal other containers	1.83	0.9
Bags and sacks	0.94	0.5
Wraps	1.53	0.8
Other plastic packaging	1.97	1.0
Total plastic containers and packaging	7.07	3.6
Total plastics	16.25	8.3
Total MSW generation	195.73	100.0

*Plastics as a component of appliances, furniture, lead-acid batteries and miscellaneous durables. Adjustments have been made for lifetimes of products.

†Does not include other materials in diapers.

‡Trash bags, eating utensils and straws, shower curtains, etc.
PET—polyethylene terephthalate; HDPE—high-density polyethylene.
Details may not add to totals due to rounding.

Source: Adapted from U.S. Environmental Protection Agency, *Characterization of Municipal Solid Waste in the United States: 1992 Update,* EPA/530-R-92-013, July 1992; and *Modern Plastics,* January 1991 (Ref. 4).

grown rapidly, with plastics in MSW increasing from less than 1 percent in 1960. This growth is projected to continue, with plastics making up about 11 percent of MSW in 2000.

Because plastics are relatively light, no one plastic product makes up a large portion of MSW. In 1990, plastics in durable goods (mainly appliances, carpeting, and furniture) amounted to 2.5 percent of MSW generation. Plastics in nondurable goods made up just over 2 percent of MSW generation in 1990. The plastics in nondurables are found in plastic plates and cups, trash bags, and many other products.

The largest source of plastics in MSW is containers and packaging, where plastics amounted to an estimated 3.6 percent of MSW generation in 1990. Containers for soft drinks, milk, water, food, and other products were the largest portion of plastics in containers and packaging. The remainder of plastic packaging is found in bags, sacks, wraps, closures, and other miscellaneous packaging products.

Other Materials in Products

In addition to the materials in products summarized above, other materials making up lesser percentages of MSW generation are described in this section.

Rubber and Leather.   In 1990, rubber and leather made up an estimated 2.4 percent of MSW generation (Table 3.2). Most of the rubber and some of the leather was found in the durable goods category in products such as tires, furniture and furnishings, and carpets. Both rubber and leather were found in clothing and footwear in MSW.

Textiles.   Textiles comprised almost 3 percent of MSW generation in 1990 (Table 3.2). The primary sources of textiles in MSW are clothing and household items such as sheets and towels. Textiles are also found, however, in such items as tires, furniture, and footwear.

Wood.   Wood is a surprisingly important component of MSW, amounting to over 6 percent of MSW generation in 1990 (Table 3.2). The wood is found in durable goods such as furniture and cabinets for electronic goods and in the containers and packaging category in wood shipping pallets and boxes.

Other Materials.   Since the material flows methodology is essentially a materials balance, some materials that cannot be classified into one of the basic material categories of MSW are put into an *Other* category in order to account for all the components associated with a product. This *Other* category amounted to 1.6 percent of MSW generation in 1990 (Table 3.2).

Most of the materials in this category are associated with disposable diapers, including the fluff (wood) pulp used in the diapers and the feces and urine that are disposed along with the diapers. The electrolyte in automotive batteries is also included in this category.

Food Wastes

Food wastes in MSW include uneaten food and food preparation wastes from residences, commercial establishments (e.g., restaurants), institutions (e.g., schools, hospitals), and some industrial sources (e.g., factory cafeterias or lunchrooms). In 1990 food wastes made up an estimated 6.7 percent of MSW generation (Table 3.2).

As described above, the only source of data on food wastes is sampling studies conducted around the country. These studies show that a declining percentage of MSW is composed of food wastes; the decline has been from about 14 percent of total in 1960 to the present amount of less than 7 percent of total MSW. It appears that the tonnage of food wastes in MSW is staying about constant, which means that the generation of food wastes per capita in MSW is declining. The reasons for this decline include increased use of food disposals, which put the food wastes into the wastewater system rather than MSW, and increased use of preprocessed and packaged food both at home and in commercial and institutional food service. When food is processed and packaged at, for instance, a meat packing plant or a frozen food plant, the wastes are classified as industrial process wastes rather than MSW.

Yard Trimmings

Yard trimmings include grass, leaves, and tree and brush trimmings from residential, commercial, and institutional sources. Yard trimmings made up nearly 18 percent of MSW generation in 1990 (Table 3.2). Like food wastes, yard trimmings in MSW are estimated from sampling study data. It appears that the amount of yard trimmings generated has been increasing slowly over the years. Although other components of MSW have been increasing more rapidly, yard trimmings have been declining as a percentage of total MSW generation.

Miscellaneous Inorganic Wastes

This relatively small category, which includes soil, bits of stone and concrete, and the like, was estimated to be 1.5 percent of MSW generation in 1990 (Table 3.2). The estimates are derived from sampling studies, where the items in the category would usually be classified as "fines."

Municipal Solid Waste Generation on a per Capita Basis

For planning purposes, municipal solid waste generation on a per person per day basis is often important. Some of the MSW generation data in Table 3.1 is converted to a per capita basis in Table 3.7. This table reveals some interesting trends. Overall, MSW generation per person has been increasing over the 30 years for which historical data are available. Generation went from 2.6 pounds per person per day in 1960 to 4.3 pounds per person per day in 1990, with a projected increase to 4.5 pounds per person per day in 2000. Much of the increase can be attributed to paper and plastics, while other materials in MSW have tended to stay about constant or even decrease on a per capita basis.

TABLE 3.7   Materials Generated* in the Municipal Solid Waste Stream, 1960 to 2000
(In pounds per person per day)

Materials	1960	1970	1980	1990	2000†
Paper and paperboard	0.9	1.2	1.3	1.6	1.7
Glass	0.2	0.3	0.4	0.3	0.3
Metals	0.3	0.4	0.3	0.4	0.3
Plastics	Neg.	0.1	0.2	0.4	0.5
Rubber and leather	0.1	0.1	0.1	0.1	0.1
Textiles	0.1	0.1	0.1	0.1	0.1
Wood	0.1	0.1	0.2	0.3	0.3
Other	Neg.	Neg.	0.1	0.1	0.1
Total materials in products	1.6	2.2	2.6	3.2	3.5
Food wastes	0.4	0.3	0.3	0.3	0.3
Yard trimmings	0.6	0.6	0.7	0.8	0.7
Miscellaneous inorganic wastes	Neg.	Neg.	0.1	0.1	0.1
Total MSW generated	2.6	3.2	3.7	4.3	4.5

*Generation before materials recovery or combustion. Details may not add to totals due to rounding.
†Projected data.
Neg. = negligible (less than 0.05 pounds per person per day).
   Source: Adapted from U.S. Environmental Protection Agency, *Characterization of Municipal Solid Waste in the United States: 1992 Update*, EPA/530-R-92-013, July 1992.

3.4 PRODUCTS IN MUNICIPAL SOLID WASTE BY WEIGHT

The materials in municipal solid waste are found in products that are used and discarded. These products can be classified as durable goods, nondurable goods, and containers and packaging. To these categories, food wastes, yard trimmings, and miscellaneous inorganic wastes are added to obtain total MSW. (Note that these totals are by definition the same as the totals of all materials in MSW.) Products in MSW are summarized in Table 3.8 (by weight), in Table 3.9 (by percentage), and in Figs. 3.3 and 3.4.

In 1990, containers and packaging accounted for over 64 million tons of MSW, or almost 33 percent of generation. Nondurable goods in MSW generation weighed over 52 millions tons, or nearly 27 percent of generation. Durable goods were almost 28 millions tons, or over 14 percent of generation. The remainder of MSW was food wastes, yard trimmings, and other miscellaneous wastes, as previously discussed.

Durable Goods

Durable goods are generally defined as products having lifetimes of 3 years or more. This category includes major appliances, furniture and furnishings, carpets and rugs, rubber tires, lead-acid automotive batteries, and miscellaneous durables such as small appliances and consumer electronics. Historical and projected data on durable goods in MSW are shown in Tables 3.10 and 3.11.

Most durable goods would be called "oversize and bulky" items by solid waste managers. They typically would not be counted in a sampling survey, but they must nevertheless be managed, though perhaps in a somewhat different manner from other wastes.

On a weight basis, the miscellaneous durables are the largest line item in the durable goods category, at an estimated 12.5 million tons generated in 1990. This category includes small appliances such as radios and kitchen items, electronics such as televisions and videocassette recorders, and the like.

Furniture and furnishings comprise the second largest segment of durable goods, at an estimated 7.4 million tons generated in 1990. Furniture from both residences and commercial buildings such as offices is counted as MSW. Items such as bedding are also included in this category.

Major appliances ("white goods") are the third largest item in the durables category. These appliances include refrigerators, washing machines, stoves, etc. They have long lifetimes before they are finally discarded.

Finally, rubber tires, carpets and rugs, and lead-acid batteries all contribute similar tonnages to MSW, each at less than 2 million tons in 1990.

Nondurable Goods

Nondurable goods are generally defined as those having lifetimes of less than 3 years. The majority of these products are, however, discarded the same year they are manufactured. Paper products account for a large portion of nondurables, with plastics and textiles accounting for most of the remainder (Tables 3.12 and 3.13).

Newspapers, at almost 13 million tons generated in 1990, are the largest single item in nondurables, with office papers second at over 6 million tons generated in 1990, and other commercial printing third at 5.5 million tons.

TABLE 3.8 Products Generated* in the Municipal Waste Stream, 1960 to 2000
(*In millions of tons*)

Products	1960	1965	1970	1975	1980	1985	1990	1995†	2000†
Durable goods	9.4	11.1	15.1	17.5	19.7	21.5	27.9	30.3	33.8
Nondurable goods	17.6	22.2	25.5	25.6	36.5	42.6	52.3	58.6	64.4
Containers and packaging	27.3	34.2	43.5	44.4	52.3	54.6	64.4	69.1	74.7
Total product‡ wastes	54.3	67.5	84.1	87.5	108.5	118.7	144.6	158.0	172.9
Other wastes:									
Food wastes	12.2	12.7	12.8	13.4	13.2	13.2	13.2	13.2	13.2
Yard trimmings	20.0	21.6	23.2	25.2	27.5	30.0	35.0	33.7	32.9
Miscellaneous inorganic wastes	1.3	1.6	1.8	2.0	2.2	2.5	2.9	3.0	3.1
Total other wastes	33.5	35.9	37.8	40.6	42.9	45.7	51.1	49.9	49.2
Total MSW generated	87.8	103.4	121.9	128.1	151.4	164.4	195.7	207.9	222.1

*Generation before materials recovery or combustion. Details may not add to totals due to rounding.

†Projected data.

‡Other than food products.

Source: Adapted from U.S. Environmental Protection Agency, *Characterization of Municipal Solid Waste in the United States: 1992 Update*, EPA/530-R-92-013, July 1992.

TABLE 3.9 Products Generated* in the Municipal Waste Stream, 1960 to 2000
(*In percent of total generation*)

Products	1960	1965	1970	1975	1980	1985	1990	1995†	2000†
Durable goods	10.7	10.7	12.4	13.7	13.0	13.1	14.3	14.6	15.2
Nondurable goods	20.0	21.5	20.9	20.0	24.1	25.9	26.7	28.2	29.0
Containers and packaging	31.1	33.1	35.7	34.7	34.5	33.2	32.9	33.2	33.6
Total product‡ wastes	61.8	65.3	69.0	68.3	71.6	72.2	73.9	76.0	77.8
Other wastes:									
Food wastes	13.9	12.3	10.5	10.5	8.7	8.0	6.7	6.3	5.9
Yard trimmings	22.8	20.9	19.0	19.7	18.2	18.2	17.9	16.2	14.8
Miscellaneous inorganic wastes	1.5	1.5	1.5	1.6	1.5	1.5	1.5	1.4	1.4
Total other wastes	38.2	34.7	31.0	31.7	28.3	27.8	26.1	24.0	22.2
Total MSW generated	100.0	100.0	100.0	100.0	100.0	100.0	100.0	100.0	100.0

*Generation before materials recovery or combustion. Details may not add to totals due to rounding.
†Projected data.
‡Other than food products.

Source: Adapted from U.S. Environmental Protection Agency, *Characterization of Municipal Solid Waste in the United States: 1992 Update,* EPA/530-R-92-013, July 1992.

3.15

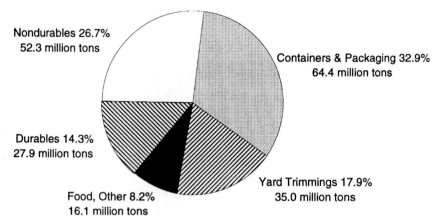

FIGURE 3.3   Products generated in municipal solid waste, 1990. (*U.S. Environmental Protection Agency, Characterization of Municipal Solid Waste in the United States: 1992 Update, EPA/530-R-92-013, July 1992.*)

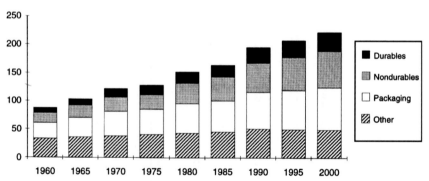

FIGURE 3.4   Products generated in municipal solid waste, 1960 to 2000. (*U.S. Environmental Protection Agency, Characterization of Municipal Solid Waste in the United States: 1992 Update, EPA/530-R-92-013, July 1992.*)

Some disposable nondurable products that are highly visible and consequently attract much attention are actually fairly minor constituents of the municipal waste stream. Thus, third-class mail (often called junk mail) is about 2 percent of MSW generation, paper and plastic plates and cups combined (including the "clamshells" used in fast food restaurants) are about 0.5 percent of MSW, and disposable diapers (including the waste products contained within them) are about 1.4 percent of MSW generation.

Containers and Packaging

The containers and packaging category includes both primary packaging (the containers that directly hold food, beverages, toiletries, and a host of other products) and sec-

TABLE 3.10 Durable Goods Generated* in the Municipal Waste Stream, 1960 to 2000
(In millions of tons)

Durable goods	1960	1965	1970	1975	1980	1985	1990	1995†	2000†
Major appliances	1.5	1.0	2.7	2.6	2.8	2.7	2.8	3.2	3.4
Furniture and furnishings	2.1	2.7	3.4	4.1	5.1	5.8	7.4	7.7	9.1
Carpets and rugs‡							1.7	2.3	2.8
Rubber tires	1.1	1.4	1.9	2.5	2.6	1.9	1.8	2.3	2.4
Batteries, lead acid	Neg.	0.7	0.8	1.2	1.5	1.5	1.7	2.0	2.2
Miscellaneous durables	4.7	5.4	6.3	7.1	7.7	9.6	12.5	12.8	13.9
Total durable goods	9.4	11.1	15.1	17.5	19.7	21.5	27.9	30.3	33.8

*Generation before materials recovery or combustion. Details may not add to totals due to rounding.

†Projected data.

‡Not estimated separately prior to 1990.

Source: Adapted from U.S. Environmental Protection Agency, *Characterization of Municipal Solid Waste in the United States: 1992 Update,* EPA/530-R-92-013, July 1992.

TABLE 3.11 Durable Goods Generated* in the Municipal Waste Stream, 1960 to 2000
(In percent of total generation)

Durable goods	1960	1965	1970	1975	1980	1985	1990	1995†	2000†
Major appliances	1.7	1.0	2.2	2.0	1.8	1.6	1.4	1.5	1.5
Furniture and furnishings	2.4	2.6	2.8	3.2	3.4	3.5	3.8	3.7	4.1
Carpets and rugs‡							0.9	1.1	1.3
Rubber tires	1.3	1.3	1.6	2.0	1.7	1.2	0.9	1.1	1.1
Batteries, lead acid	Neg.	0.6	0.7	0.9	1.0	0.9	0.9	1.0	1.0
Miscellaneous durables	5.4	5.2	5.2	5.5	5.1	5.8	6.4	6.2	6.3
Total durable goods	10.7	10.8	12.4	13.7	13.0	13.1	14.3	14.6	15.2

*Generation before materials recovery or combustion. Details may not add to totals due to rounding.

†Projected data.

‡Not estimated separately prior to 1990.

Source: Adapted from U.S. Environmental Protection Agency, *Characterization of Municipal Solid Waste in the United States: 1992 Update,* EPA/530-R-92-013, July 1992.

3.17

TABLE 3.12 Nondurable Goods Generated* in the Municipal Waste Stream, 1960 to 2000
(In millions of tons)

Nondurable goods	1960	1965	1970	1975	1980	1985	1990	1995†	2000†
Newspapers	7.1	8.3	9.5	8.8	11.0	12.5	12.9	14.1	15.1
Books and magazines	1.9	2.2	2.5	2.3	3.4	4.7			
Books‡							1.0	1.1	1.2
Magazines‡							2.8	3.3	3.8
Office papers	1.5	2.2	2.7	2.6	4.0	5.7	6.4	7.5	8.1
Telephone books‡							0.5	0.6	0.7
Third-class mail‡							3.8	4.2	4.6
Other commercial printing	1.3	1.8	2.1	2.1	3.1	3.2	5.5	5.9	6.5
Tissue paper and towels	1.1	1.5	2.1	2.1	2.3	2.7	3.2	3.5	3.8
Paper plates and cups	0.3	0.3	0.4	0.4	0.6	0.6	0.7	0.7	0.7
Plastic plates and cups§					0.2	0.3	0.3	0.5	0.6
Trash bags‡							0.8	1.1	1.3
Disposable diapers	Neg.	Neg.	0.3	1.2	2.3	2.9	2.6	2.8	2.9
Other nonpackaging paper	2.7	3.9	3.6	3.5	4.2	3.5	3.8	3.9	4.1
Clothing and footwear	1.3	1.5	1.5	1.7	2.3	2.7	3.7	3.9	4.5
Towels, sheets, and pillowcases‡							1.0	1.1	1.2
Other miscellaneous nondurables	0.4	0.5	0.8	0.9	3.1	3.8	3.2	4.4	5.5
Total nondurable goods	17.6	22.2	25.5	25.6	36.5	42.6	52.3	58.6	64.4

*Generation before materials recovery or combustion. Details may not add to totals due to rounding.

†Projected data.

‡Not estimated separately prior to 1990. Some categories used in previous years have been reallocated.

§Not estimated prior to 1980.

Neg = negligible (less than 50,000 tons).

Source: Adapted from U.S. Environmental Protection Agency, *Characterization of Municipal Solid Waste in the United States: 1992 Update*, EPA/530-R-92-013, July 1992.

TABLE 3.13 Nondurable Goods Generated* in the Municipal Waste Stream, 1960 to 2000
(*In percent of total generation*)

Nondurable goods	1960	1965	1970	1975	1980	1985	1990	1995†	2000†
Newspapers	8.1	8.0	7.8	6.9	7.3	7.6	6.6	6.8	6.8
Books and magazines	2.2	2.1	2.1	1.8	2.2	2.9			
Books‡							0.5	0.5	0.5
Magazines‡							1.4	1.6	1.7
Office papers	1.7	2.1	2.2	2.0	2.6	3.5	3.3	3.6	3.6
Telephone books‡							0.3	0.3	0.3
Third-class mail‡							2.0	2.0	2.0
Other commercial printing	1.5	1.7	1.7	1.6	2.0	1.9	2.8	2.8	2.9
Tissue paper and towels	1.3	1.5	1.7	1.6	1.5	1.6	1.6	1.7	1.7
Paper plates and cups	0.3	0.3	0.3	0.3	0.4	0.4	0.3	0.3	0.3
Plastic plates and cups§					0.1	0.2	0.2	0.2	0.3
Trash bags‡							0.4	0.5	0.6
Disposable diapers	Neg.	Neg.	0.2	0.9	1.5	1.8	1.4	1.3	1.3
Other nonpackaging paper	3.1	3.8	3.0	2.7	2.8	2.1	1.9	1.9	1.9
Clothing and footwear	1.5	1.5	1.2	1.3	1.5	1.6	1.9	1.9	2.0
Towels, sheets and pillowcases‡							0.5	0.5	0.5
Other miscellaneous nondurables	0.5	0.5	0.7	0.7	2.0	2.3	1.6	2.1	2.5
Total nondurable goods	20.0	21.5	20.9	20.0	24.1	25.9	26.7	28.2	29.0

*Generation before materials recovery or combustion. Details may not add to totals due to rounding.
†Projected data.
‡Not estimated separately prior to 1990. Some categories used in previous years have been reallocated.
§Not estimated prior to 1980.

Source: Adapted from U.S. Environmental Protection Agency. *Characterization of Municipal Solid Waste in the United States: 1992 Update*, EPA/530-R-92-013, July 1992.

ondary and tertiary packaging, which contains the packaged products for shipping and display. By definition, it is assumed that all containers and packaging are discarded the same year they are manufactured (with a few exceptions, such as reusable wood pallets). Containers and packaging generation is shown in Table 3.14 (by weight) and Table 3.15 (in percentage).

By far the dominant material in this category is paper and paperboard, which accounted for just over half of the weight of containers and packaging generated in 1990. Corrugated boxes, at nearly 24 million tons generated in 1990, are the single largest product line item in MSW.

The second largest material category in containers and packaging, by weight, is glass bottles and jars, at nearly 12 million tons generated in 1990. Wood is in third place at almost 8 million tons generated in 1990, with plastics fourth at 7 million tons. Steel and aluminum cans and other packaging occupy relatively minor positions in MSW generation.

Other Wastes

Food wastes, yard trimmings, and miscellaneous inorganic wastes are added to the products categories of durable goods, nondurable goods, and containers and packaging to obtain total MSW generation. These other wastes were discussed in Sec. 3.3.

3.5 MUNICIPAL SOLID WASTE MANAGEMENT

Once municipal solid waste is generated, it must be managed somehow. In the United States, the usual management alternatives are recovery for recycling or composting, combustion, or landfilling. (The source reduction alternative is discussed elsewhere in this handbook. Chapter 3 deals with MSW as generated, after any source reduction measures have been applied.)

Recovery for Recycling and Composting

Recovery of MSW for recycling and composting in 1990 is estimated by weight and by percent of generation in Tables 3.16 through 3.19. (Note that the estimates are for *recovery*. The EPA data source referenced in this chapter does not attempt to determine whether materials recovered are actually recycled, and in fact, some materials collected for recycling or composting are unrecyclable and become residues that must be disposed.)

Materials Recovery. Of all materials recovered from MSW (Table 3.16), paper and paperboard comprised by far the largest tonnage—nearly 21 million tons out of a total of 33 million tons recovered. Yard trimmings represented the second highest tonnage recovered, at over 4 million tons, with glass third at 2.6 million tons. In terms of percentage of generation recovered, *Other nonferrous metals* were the highest, at nearly 67 percent recovered. This is almost entirely due to the high rate of recovery of lead in automotive batteries. Aluminum had the second highest recovery percentage in 1990 at 38 percent of generation, due to the relatively high recovery rates of aluminum beverage cans. Recovery of paper and paperboard ranked third, at over 28 percent of generation recovered.

TABLE 3.14 Containers and Packaging Generated* in the Municipal Waste Stream, 1960 to 2000
(*In millions of tons*)

Containers and packaging	1960	1965	1970	1975	1980	1985	1990	1995†	2000†
Glass packaging:									
Beer and soft drink bottles	1.4	2.6	5.6	6.3	6.7	5.7	5.7	5.7	5.6
Wine and liquor bottles	1.1	1.4	1.9	2.0	2.5	2.2	2.1	2.2	2.2
Food and other bottles & jars	3.7	4.1	4.4	4.4	4.8	4.2	4.1	4.2	4.1
Total glass packaging	6.2	8.1	11.9	12.7	14.0	12.1	11.9	12.1	11.9
Steel packaging:									
Beer and soft drink cans	0.6	0.9	1.6	1.3	0.5	0.1	0.1	0.1	0.1
Food and other cans	3.8	3.6	3.5	3.4	2.9	2.6	2.5	2.4	2.3
Other steel packaging	0.2	0.3	0.3	0.2	0.2	0.2	0.2	0.2	0.2
Total steel packaging	4.6	4.8	5.4	4.9	3.6	2.9	2.9	2.7	2.6
Aluminum packaging:									
Beer and soft drink cans	0.1	0.1	0.3	0.5	0.9	1.3	1.6	1.7	2.0
Other cans	Neg.	Neg.	0.1	Neg.	Neg.	Neg.	Neg.	0.1	0.1
Foil and closures	0.1	0.2	0.2	0.3	0.3	0.3	0.3	0.4	0.4
Total aluminum packaging	0.2	0.3	0.6	0.8	1.2	1.6	1.9	2.2	2.5
Paper and paperboard packaging:									
Corrugated boxes	7.3	10.0	12.7	13.5	17.0	19.0	23.9	25.3	27.0
Milk cartons‡					0.6	0.5	0.5	0.5	0.5
Folding cartons‡					3.7	4.0	4.3	4.5	4.7
Other paperboard packaging	3.8	4.5	4.8	4.4	0.3	0.4	0.3	0.3	0.3
Bags and sacks‡					3.4	3.1	2.4	2.5	2.5
Wrapping papers‡					0.2	0.1	0.1	0.1	0.1
Other paper packaging	2.9	3.3	3.8	3.3	0.8	1.3	1.0	1.0	1.0
Total paper and board packaging	14.0	17.8	21.3	21.2	26.0	28.4	32.6	34.3	36.2

TABLE 3.14 Containers and Packaging Generated* in the Municipal Waste Stream, 1960 to 2000 (Continued)
(In millions of tons)

Containers and packaging	1960	1965	1970	1975	1980	1985	1990	1995†	2000†
Plastics packaging:									
Soft drink bottles‡						0.4	0.4	0.6	0.7
Milk bottles‡						0.3	0.4	0.5	0.5
Other containers	0.1	0.3	0.9	1.3	0.9	1.2	1.8	2.8	3.5
Bags and sacks‡					0.4	0.6	0.9	1.2	1.4
Wraps‡					0.8	1.0	1.5	1.7	2.0
Other plastics packaging	0.1	0.7	1.2	1.4	0.8	1.0	1.9	2.1	2.6
Total plastics packaging	0.2	1.0	2.1	2.7	3.4	4.5	7.0	8.7	10.7
Wood packaging	2.0	2.1	2.1	2.0	3.9	4.9	7.9	8.9	10.6
Other misc. packaging	0.1	0.1	0.1	0.1	0.2	0.2	0.2	0.2	0.2
Total containers and packaging	27.3	34.2	43.5	44.4	52.3	54.6	64.4	69.1	74.7

*Generation before materials recovery or combustion. Details may not add to totals due to rounding.

†Projected data.

‡Not estimated prior to 1980.

Neg. = negligible (less than 50,000 tons).

Source: Adapted from U.S. Environmental Protection Agency, *Characterization of Municipal Solid Waste in the United States: 1992 Update,* EPA/530-R-92-013, July 1992.

TABLE 3.15 Containers and Packaging Generated* in the Municipal Waste Stream, 1960 to 2000
(In percent of total generation)

Containers and packaging	1960	1965	1970	1975	1980	1985	1990	1995†	2000†
Glass packaging:									
Beer and soft drink bottles	1.6	2.5	4.6	4.9	4.4	3.5	2.9	2.7	2.5
Wine and liquor bottles	1.3	1.4	1.6	1.6	1.7	1.3	1.1	1.1	1.0
Food and other bottles and jars	4.2	4.0	3.6	3.4	3.2	2.6	2.1	2.0	1.9
Total glass packaging	7.1	7.8	9.8	9.9	9.2	7.4	6.1	5.8	5.4
Steel packaging:									
Beer and soft drink cans	0.7	0.9	1.3	1.0	0.3	0.1	0.1	0.1	0.1
Food and other cans	4.3	3.5	2.9	2.7	1.9	1.6	1.3	1.1	1.0
Other steel packaging	0.2	0.3	0.2	0.2	0.1	0.1	0.1	0.1	0.1
Total steel packaging	5.2	4.6	4.4	3.8	2.4	1.8	1.5	1.3	1.2
Aluminum packaging									
Beer and soft drink cans	0.1	0.1	0.2	0.4	0.6	0.8	0.8	0.8	0.9
Other cans	Neg.	Neg.	0.1	Neg.	Neg.	Neg.	Neg.	Neg.	Neg.
Foil and closures	0.1	0.2	0.2	0.2	0.2	0.2	0.2	0.2	0.2
Total aluminum packaging	0.2	0.3	0.5	0.6	0.8	1.0	1.0	1.1	1.1
Paper and paperboard packaging:									
Corrugated boxes	8.3	9.7	10.4	10.5	11.2	11.6	12.2	12.2	12.2
Milk cartons‡					0.4	0.3	0.3	0.2	0.2
Folding cartons‡					2.4	2.4	2.2	2.2	2.1
Other paperboard packaging	4.3	4.4	3.9	3.4	0.2	0.2	0.1	0.1	0.1
Bags and sacks‡					2.2	1.9	1.2	1.2	1.1
Wrapping papers‡					0.1	0.1	0.1	0.1	0.1
Other paper packaging	3.3	3.2	3.1	2.6	0.5	0.8	0.5	0.5	0.5
Total paper and board packaging	15.9	17.2	17.5	16.5	17.2	17.3	16.7	16.5	16.3

TABLE 3.15 Containers and Packaging Generated* in the Municipal Waste Stream, 1960 to 2000 (*Continued*)
(*In percent of total generation*)

Containers and packaging	1960	1965	1970	1975	1980	1985	1990	1995†	2000†
Plastics packaging									
Soft drink bottles‡					0.2	0.2	0.2	0.3	0.3
Milk bottles‡					0.2	0.2	0.2	0.2	0.2
Other containers	0.1	0.3	0.7	1.0	0.6	0.7	0.9	1.3	1.6
Bags and sacks‡					0.3	0.4	0.5	0.6	0.6
Wraps‡					0.6	0.6	0.8	0.8	0.9
Other plastics packaging	0.1	0.7	1.0	1.1	0.5	0.6	1.0	1.0	1.2
Total plastics packaging	0.2	1.0	1.7	2.1	2.2	2.7	3.6	4.2	4.8
Wood packaging	2.3	2.0	1.7	1.6	2.6	3.0	4.0	4.3	4.8
Other misc. packaging	0.1	0.1	0.1	0.1	0.1	0.1	0.1	0.1	0.1
Total containers and packaging	31.1	33.1	35.7	34.7	34.5	33.2	32.9	33.2	33.6

*Generation before materials recovery or combustion.

†Projected data.

‡Not estimated prior to 1980.

Neg. = negligible (less than 0.05 percent).

Source: Adapted from U.S. Environmental Protection Agency, *Characterization of Municipal Solid Waste in the United States: 1992 Update*, EPA/530-R-92-013, July 1992.

TABLE 3.16 Materials Generated, Recovered, and Discarded* in the Municipal Waste Stream, 1990

Materials	Generation, million tons	Recovery, million tons	Recovery, % of generation	Discards,* million tons	Discards, % of total
Paper and paperboard	73.3	20.9	28.6	52.4	32.3
Glass	13.2	2.6	19.9	10.6	6.5
Metals:					
Ferrous	12.3	1.9	15.4	10.4	6.4
Aluminum	2.7	1.0	38.1	1.6	1.0
Other nonferrous	1.2	0.8	66.7	0.4	0.2
Total metals	16.2	3.7	23.0	12.5	7.7
Plastics	16.2	0.4	2.2	15.9	9.8
Rubber and leather	4.6	0.2	4.4	4.4	2.7
Textiles	5.6	0.2	4.3	5.3	3.3
Wood	12.3	0.4	3.2	11.9	7.3
Other	3.2	0.8†	23.8	2.4	1.5
Total materials in products	144.6	29.2	20.2	115.4	71.1
Other wastes:					
Food wastes	13.2	Neg.	Neg.	13.2	8.1
Yard trimmings	35.0	4.2	12.0	30.8	19.0
Miscellaneous inorganic wastes	2.9	Neg.	Neg.	2.9	1.8
Total other wastes	51.1	4.2	8.2	46.9	28.9
Total MSW generated	195.7	33.4	17.1	162.3	100.0

*Discards after recovery for recycling or composting. Details may not add to totals due to rounding.

†Recovery of electrolytes in batteries. May not be recycled.

Source: Adapted from U.S. Environmental Protection Agency, *Characterization of Municipal Solid Waste in the United States: 1992 Update,* EPA/530-R-92-013, July 1992.

Durable Goods Recovery.   Since it is actually products in MSW that are recovered, these estimates are more enlightening. Recovery of durable goods is shown in Table 3.17. Recovery of lead from lead-acid automotive batteries was estimated to be at nearly a 97 percent level in 1990, the highest recovery rate of all products in MSW. The other significant recovery in this category is ferrous metals from major appliances. The estimated ferrous recovery was estimated to be 32 percent of the total weight of the appliances. Some rubber—about 200,000 tons—was recovered from tires in 1990.

Nondurable Goods Recovery.   Recovery of nondurable goods in 1990 is shown in Table 3.18. Newspapers have a long history of recovery, and they were recovered at an estimated rate of over 42 percent of generation in 1990 (5.5 million tons recovered). Office papers were recovered at an estimated rate of 26.5 percent, with other paper products recovered at lower rates. The only other significant recovery identified was some recovery of textiles for export. (Reuse of textile products, e.g., clothing, in the United States was not considered to be recycling.)

Containers and Packaging Recovery.   Recovery of containers and packaging (almost 17 million tons in 1990) comprised over half of all MSW recovery in that year. This is largely due to recovery of corrugated containers at an estimated 48 percent, or 11.5 million tons. The significance of this level of recovery can be illustrated by the point that corrugated containers were over 12 percent of MSW *generation* in 1990 (Table 3.15), but their discards after recovery were less than 8 percent of total *discards* (Table 3.19).

Aluminum beverage cans were recovered at a high rate (over 63 percent in 1990), but this recovery represents only 1 million tons of MSW. Glass beverage containers were recovered at an estimated rate of 33 percent, or nearly 2 million tons. A number of other package types were also recovered in 1990, but none totaling more than 1 million tons.

Combustion

In 1990 about 32 million tons of MSW were combusted, with well over 90 percent of that amount sent to energy recovery facilities. Thus combustion was the management of choice for over 16 percent of MSW generated, or about 19 percent of MSW discarded after recovery.[1]

Landfilling

Over 130 million tons of MSW were landfilled in 1990, according to the EPA data source.[1] This was close to 67 percent of MSW generated, or about 81 percent of MSW discarded after recovery. (Note that some of the amount of MSW assumed to be landfilled may in fact be littered, self-disposed, or otherwise disposed. These amounts are not estimated, but are thought to be relatively small.)

Trends in MSW Management

When 30 years of data from the EPA database are combined with projections from the same report, some interesting trends are demonstrated (Tables 3.20 and 3.21 and Fig. 3.5). Generation of MSW has increased steadily over the entire period except for some

TABLE 3.17 Durable Goods Generated, Recovered, and Discarded* in the Municipal Waste Stream, 1990

Durable goods	Generation, million tons	Recovery, million tons	Recovery, % of generation	Discards,* million tons	Discards, % of MSW total
Major appliances	2.8	0.9	32.4	1.9	1.2
Furniture and furnishings	7.4	Neg.	Neg.	7.4	4.6
Carpets and rugs	1.7	Neg.	0.2	1.7	1.0
Rubber tires	1.8	0.2	11.6	1.6	1.0
Batteries, lead acid	1.7	1.6	96.6	0.1	Neg.
Miscellaneous durables	12.5	0.4	3.0	12.1	7.5
Total durable goods	27.9	3.1	11.2	24.8	15.3

*Discards after recovery for recycling or composting. Details may not add to totals due to rounding. Neg. = negligible.

Source: Adapted from U.S. Environmental Protection Agency, *Characterization of Municipal Solid Waste in the United States: 1992 Update,* EPA/530-R-92-013, July 1992.

TABLE 3.18 Nondurable Goods Generated, Recovered, and Discarded* in the Municipal Waste Stream, 1990

Nondurable goods	Generation, million tons	Recovery, million tons	Recovery, % of generation	Discards,* million tons	Discards, % of MSW total
Newspapers	12.9	5.5	42.5	7.4	4.6
Books	1.0	0.1	10.3	0.9	0.5
Magazines	2.8	0.3	10.7	2.5	1.5
Office papers	6.4	1.7	26.5	4.7	2.9
Telephone books	0.5	0.1	9.3	0.5	0.3
Third-class mail	3.8	0.2	5.2	3.6	2.2
Other commercial printing	5.5	1.1	19.4	4.5	2.7
Tissue paper and towels	3.2	Neg.	Neg.	3.2	2.0
Paper plates and cups	0.7	Neg.	Neg.	0.7	0.4
Plastic plates and cups	0.3	Neg.	Neg.	0.3	0.2
Trash bags	0.8	Neg.	Neg.	0.8	0.5
Disposable diapers	2.6	Neg.	Neg.	2.6	1.6
Other nonpackaging paper	3.8	Neg.	Neg.	3.8	2.3
Clothing and footwear	3.7	0.2	5.0	3.6	2.2
Towels, sheets, and pillowcases	1.0	Neg.	Neg.	1.0	0.6
Other miscellaneous nondurables	3.2	Neg.	Neg.	3.2	2.0
Total nondurable goods	52.3	9.1	17.4	43.2	26.6

*Discards after recovery for recycling or composting. Details may not add to totals due to rounding. Neg. = negligible (less than 50,000 tons).

Source: Adapted from U.S. Environmental Protection Agency, *Characterization of Municipal Solid Waste in the United States: 1992 Update,* EPA/530-R-92-013, July 1992.

TABLE 3.19 Containers and Packaging Generated, Recovered, and Discarded* in the Municipal Waste Stream, 1990

Containers and packaging	Generation, million tons	Recovery, million tons	Recovery, % of generation	Discards,* million tons	Discards, % of MSW total
Glass packaging:					
Beer and soft drink bottles	5.7	1.9	33.2	3.8	2.3
Wine and liquor bottles	2.1	0.2	10.0	1.9	1.2
Food and other bottles and jars	4.1	0.5	12.7	3.6	2.2
Total glass packaging	11.9	2.6	22.0	9.3	5.7
Steel packaging:					
Beer and soft drink cans	0.1	Neg.	24.7	0.1	0.1
Food and other cans	2.5	0.6	23.4	1.9	1.2
Other steel packaging	0.2	Neg.	5.0	0.2	0.1
Total steel packaging	2.9	0.6	22.1	2.3	1.4
Aluminum packaging:					
Beer and soft drink cans	1.6	1.0	63.2	0.6	40.0
Other cans	Neg.	Neg.	4.0	Neg.	Neg.
Foil and closures	0.3	Neg.	7.1	0.3	0.2
Total aluminum packaging	1.9	1.0	53.3	0.9	0.5
Paper and paperboard packaging:					
Corrugated boxes	23.9	11.5	48.0	12.5	7.7
Milk cartons	0.5	Neg.	Neg.	0.5	0.3
Folding cartons	4.3	0.3	7.9	4.0	2.4
Other paperboard packaging	0.3	Neg.	Neg.	0.3	0.2
Bags and sacks	2.4	0.2	8.2	2.2	1.4
Wrapping papers	0.1	Neg.	Neg.	0.1	0.1
Other paper packaging	1.0	Neg.	Neg.	1.0	0.6
Total paper and board packaging	32.6	12.0	36.9	20.6	12.7
Plastics packaging:					
Soft drink bottles	0.4	0.1	31.5	0.3	0.2
Milk bottles	0.4	Neg.	6.9	0.4	0.2
Other containers	1.8	Neg.	1.2	1.8	1.1
Bags and sacks	0.9	Neg.	3.1	0.9	0.6
Wraps	1.5	Neg.	2.0	1.5	0.9
Other plastics packaging	1.9	Neg.	0.9	1.9	1.2
Total plastics packaging	7.0	0.3	3.7	6.7	4.1
Wood packaging	7.9	0.4	5.0	7.5	4.6
Other misc. packaging	0.2	Neg.	Neg.	0.2	0.1
Total containers and packaging	64.4	16.9	26.2	47.4	29.2

*Discards after recovery for recycling or composting. Details may not add to totals due to rounding.
Neg. = negligible (less than 50,000 tons or 0.05 percent).

Source: Adapted from U.S. Environmental Protection Agency, *Characterization of Municipal Solid Waste in the United States: 1992 Update,* EPA/530-R-92-013, July 1992.

recession years. Recovery for recycling and composting was quite modest until the late 1980s, when the level of activity increased markedly, reaching 17 percent of generation in 1990.

Combustion of MSW in the United States exhibits a different pattern. In 1960 an estimated 30 percent of MSW generated in the U.S. was combusted, mostly in old-fashioned incinerators without energy recovery. When pollution controls began to be

TABLE 3.20 Management of Municipal Solid Waste, 1960 to 2000
(In millions of tons)

Materials	1960	1965	1970	1975	1980	1985	1990	1995‡	2000§
Generation	87.8	103.4	121.9	128.1	151.5	164.4	195.7	207.9	222.1
Recovery for recycling	5.9	6.8	8.6	9.9	14.5	16.4	29.2	40.8	50.9
Recovery for composting	0.0	0.0	0.0	0.0	0.0	0.0	4.2	11.1	15.8
Total recovery	5.9	6.8	8.6	9.9	14.5	16.4	33.4	51.9	66.7
Discards after recovery*	81.9	96.6	113.3	118.2	137.0	148.1	162.3	156.0	155.4
Combustion with energy recovery	0.0	0.2	0.4	0.7	2.7	7.6	29.7	35.4	46.2
Combustion without energy recovery	27.0	26.8	24.7	17.8	11.0	4.1	2.2	0.0	0.0
Total combustion	27.0	27.0	25.1	18.5	13.7	11.7	31.9	35.4	46.2
Discards to landfill, other disposal†	54.9	69.6	88.2	99.7	123.3	136.4	130.4	120.6	109.2

*Does not include residues from recycling/composting processes.

†Does not include residues from recycling, composting, or combustion processes.

‡Total recovery of 25 percent assumed.

§Total recovery of 30 percent assumed.
Details may not add to totals due to rounding.

    Source: Adapted from U.S. Environmental Protection Agency, *Characterization of Municipal Solid Waste in the United States: 1992 Update,* EPA/530-R-92-013, July 1992.

TABLE 3.21 Management of Municipal Solid Waste, 1960 to 2000
(In percent of total generation)

Materials	1960	1965	1970	1975	1980	1985	1990	1995‡	2000§
Generation	100.0	100.0	100.0	100.0	100.0	100.0	100.0	100.0	100.0
Recovery for recycling	6.7	6.6	7.1	7.7	9.6	9.9	14.9	19.6	22.9
Recovery for composting	0.0	0.0	0.0	0.0	0.0	0.0	2.1	5.3	7.1
Total recovery	6.7	6.6	7.1	7.7	9.6	9.9	17.1	25.0	30.0
Discards after recovery*	93.3	93.4	92.9	92.3	90.4	90.1	82.9	75.0	70.0
Combustion with energy recovery	0.0	0.2	0.3	0.5	1.8	4.6	15.2	17.0	20.8
Combustion without energy recovery	30.8	25.9	20.3	13.9	7.3	2.5	1.1	0.0	0.0
Total combustion	30.8	26.1	20.6	14.4	9.0	7.1	16.3	17.0	20.8
Discards to landfill, other disposal†	62.5	67.3	72.4	77.8	81.4	82.9	66.6	58.0	49.2

*Does not include residues from recycling/composting processes.

†Does not include residues from recycling, composting, or combustion processes.

‡Total recovery of 25 percent assumed.

§Total recovery of 30 percent assumed.
Details may not add to totals due to rounding.

    Source: Adapted from U.S. Environmental Protection Agency, *Characterization of Municipal Solid Waste in the United States: 1992 Update,* EPA/530-R-92-013, July 1992.

required, the old incinerators that did not meet the new standards were phased out. By 1985, it was estimated that only about 7 percent of MSW was combusted. Since then, there has been a steady increase in the amount of MSW going to combustion units—an estimated 16 percent in 1990.

    With combustion declining in the 1960s and recycling still at relatively low levels, discards of MSW to landfills grew rapidly in the 1970s and 1980s. These discards appeared to peak in about 1985 at over 136 million tons landfilled, or about 83 percent of generation that year. Since then, increased recovery and combustion of MSW have

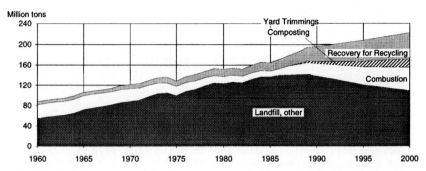

FIGURE 3.5   Generation and management of municipal solid waste, 1960 to 2000. (*U.S. Environmental Protection Agency, Characterization of Municipal Solid Waste in the United States: 1992 Update, EPA/530-R-92-013, July 1992.*)

caused a decline in the estimated tonnage landfilled. In 1990, landfilled tonnage was just over 130 million tons, or about 67 percent of generation.

The EPA characterization report projected continued growth of MSW generation. Recovery of MSW was projected in scenarios, with 25 percent recovery projected as likely in 1995, and 30 percent in 2000. Combustion of MSW was also projected to grow to about 21 percent of MSW generation in 2000. The effect of these projections was to lower projected landfilling of MSW to about 109 million tons in 2000, or less than 50 percent of generation.

3.6   DISCARDS OF MUNICIPAL SOLID WASTE BY VOLUME

Since weighing on a set of scales is quick and convenient, MSW is most often quantified in tons. Also, the material flows methodology for characterizing weight nationwide relies on data that are most often expressed in tons. Measurement of MSW by volume (cubic yards in the United States) is also relevant, however, since space occupied by the waste is an important consideration.

Unfortunately, density factors for MSW have been difficult to obtain. A 1990 study[5] did provide a uniform set of experimental data for landfill densities of many of the products in MSW (Table 3.22), and these factors have been used to calculate the relative amounts of MSW discards as landfilled for the 1990 and 1992 EPA characterization reports (Table 3.23). Note that the data reported by EPA are for MSW *discards* after recovery (not generation), as these are the quantities corresponding most closely to MSW landfilled.

The information in Table 3.23 is primarily useful as an indicator of the *relative* density of materials in a landfill. In the real world, the materials in a landfill are mixed together before they are compacted, with the effect that air spaces are filled by small objects. Thus, the overall density in a particular landfill is probably greater than the sum of the individual components measured separately, and the total volume is probably less than the sum of the individual components.

A relative comparison of the weights and volumes of materials can be obtained by taking the ratio of volume percentage to weight percentage (the right-hand column in

TABLE 3.22 Summary of Density Factors for Landfilled
Materials
(In pounds per cubic yard)

Products	Density
Durable goods*	520
Nondurable goods:	
Nondurable paper	800
Nondurable plastic	315
Disposable diapers:	
Diaper materials	795
Urine and feces	1350
Rubber	345
Textiles	435
Misc. nondurables (mostly plastics)	390
Packaging:	
Glass containers:	
Beer and soft drink bottles	2800
Other containers	2800
Steel containers:	
Beer and soft drink cans	560
Food cans	560
Other packaging	560
Aluminum:	
Beer and soft drink cans	250
Other packaging	550
Paper and paperboard:	
Corrugated	750
Other paperboard	820
Paper packaging	740
Plastics:	
Film	670
Rigid containers	355
Other packaging	185
Wood packaging	800
Other miscellaneous packaging	1015
Food wastes	2000
Yard trimmings	1500

*No measurements were taken for durable goods or plastic coatings.

Source: U.S. Environmental Protection Agency, *Characterization
of Municipal Solid Waste in the United States: 1992 Update,* EPA/530-R-
92-013, July 1992.

Table 3.23). A ratio of 1.0 means that the material occupies the same proportion of
volume as weight in the landfill. Paper and paperboard exhibit this characteristic. A
ratio greater than 1.0 shows that the material occupies a larger proportion by volume
than by weight. Four materials have ratios of approximately 2.0 or higher: plastics,
rubber and leather, textiles, and aluminum. Materials that are relatively dense and
occupy proportionately less volume compared to their weight include glass, food
wastes, and yard trimmings.

TABLE 3.23   Volume of Materials Discarded in Municipal Solid Waste, 1990

Materials	1990 discards,* million tons	Weight,* % of MSW total	Landfill density,† lb/yd³	Landfill volume, million yd³	Volume, % of MSW total	Ratio, vol %/ wt %
Paper and paperboard	52.4	32.3	784	133.6	31.9	1.0
Plastics	15.9	9.8	359	88.5	21.1	2.2
Yard trimmings	30.8	19.0	1500	41.1	9.8	0.5
Ferrous metals	10.4	6.4	560	37.2	8.9	1.4
Rubber and leather	4.4	2.7	346	25.6	6.1	2.2
Textiles	5.3	3.3	400	26.7	6.4	1.9
Wood	11.9	7.3	840	28.4	6.8	0.9
Food wastes	13.2	8.1	2000	13.2	3.2	0.4
Other†	5.7	3.5	2000	5.7	1.4	0.4
Aluminum	1.6	1.0	366	9.0	2.2	2.1
Glass	10.6	6.5	2268	9.3	2.2	0.3
Totals	162.3	100.0	776	418.3§	100.0§	1.0

*From Table 3.16. Discards after materials recovery and landfilling, before combustion and landfilling.

†Composite factors derived by Franklin Associates, Ltd. for the source report.

‡This assumes that all waste is landfilled, but some is combusted and otherwise disposed.

§This density factor and volume are derived by adding the individual factors. Actual landfill density may be considerably higher (see discussion in text).

Source: U.S. Environmental Protection Agency, *Characterization of Municipal Solid Waste in the United States: 1992 Update,* EPA/530-R-92-013, July 1992.

3.7   THE VARIABILITY OF MUNICIPAL SOLID WASTE GENERATION

The data presented in this chapter relate to MSW generation and management in the United States as a whole, and these data can provide useful guidelines for local planning. Local planners should, however, take care when adapting general data for local use. Some of the factors affecting variability in the waste stream are discussed in this section.

Commercial versus Residential Waste

Average people are most conscious of the wastes coming from their own homes, whether single-family residences, apartment buildings, or other residential options. Large amounts of waste are, however, also generated where people work, shop, travel, attend classes, or engage in other activities. These latter wastes are generally classified as commercial. To add to the confusion, waste haulers often classify wastes collected from apartment buildings as "commercial," although the nature of the wastes may be very similar to that from single-family residences.

The EPA report used as a source for much of this chapter includes a classification of MSW into residential and commercial fractions. The range of residential wastes is estimated to be between 55 and 65 percent of MSW generation, with commercial wastes estimated to range between 35 and 45 percent of generation. (MSW from multifamily residences was classified as residential, not commercial.)

Local/Regional Variability

Municipal solid waste managers generally agree that there are variations in the amount and characteristics of MSW around the country, although it is not easy to generalize with any degree of reliability. Some observations based on experience can be made, however.

First, there is some agreement that residential wastes vary less from location to location than do commercial wastes.[6] People across the country tend to buy much the same kinds of goods, whether they live in rural or urban areas or in different climates. Exceptions to this generalization include:

- *Yard trimmings.* Yard trimmings tend to be much more plentiful in warmer, moister parts of the country. Also, there are marked differences in how yard trimmings are managed. In rural areas and small towns, yard trimmings often are not hauled to landfills or compost facilities, while in suburban and urban areas, they usually are handled off-site.

- *Food wastes.* Discards of food wastes in MSW will vary according to the prevalence of food disposers, which put the food wastes into the wastewater treatment system. Use of food disposers may not be allowed (e.g., in New York City). Also, food disposers typically cannot be used in rural areas that are not on a municipal sewer system.

- *Newspapers.* Newspapers, which are mostly discarded from residences, vary greatly in size, and thus contribute to regional and urban/rural variations in MSW generation. As an example, annual per capita generation of daily newspapers varies from about 120 pounds per person in states like California, Massachusetts, and Florida to 30 or 40 pounds per person in less densely populated states like Wyoming, Kansas, and South Dakota.

Generation of MSW in a particular locality will be strongly influenced by commercial activity in the area. A concentration of office buildings will produce office papers and other wastes. Shopping malls, warehouses, and factories generate large amounts of corrugated containers and other wastes as well. Schools, hospitals, airports, train and bus stations, hotels and motels, and sports facilities all contribute to the commercial waste stream. Thus, small towns and rural areas without concentrations of commercial activities will typically generate less MSW per person than urban areas.

Seasonal Variations

Another well-known phenomenon in municipal waste management is seasonal variations in waste generation. Yard trimmings are generally the important variable for most communities, with seasonal cleanup of yards and garages often contributing to peak generation weeks. Late spring and autumn are peak generation periods in many communities, while generation of yard trimmings may approach zero in winter months in cold climates. As a rule of thumb, MSW generation may vary around 30 percent above or below the average in many communities.

Changes over Time

Municipal solid waste generation has increased steadily in the United States, both in tonnage and in per capita generation. This does not mean, however, that generation of each material and product in MSW has grown at the same rate. In fact, generation of some materials and products has grown rapidly, while others have had slow growth or actually declined. An understanding of this phenomenon is especially important in making projections of MSW generation and in planning waste management facilities.

Some factors tending to increase MSW generation are

- *Increasing population.*   Obviously, more people use and throw away more things. One preliminary analysis indicates that about one-half of the growth of MSW generation over a 15-year period can be attributed to population growth.[7]

- *Increasing levels of affluence.*   There is a rather strong correlation between generation of MSW and gross domestic product. Generation of paper and paperboard products is especially sensitive to economic activity. As an example, a plot of paper and paperboard generation (Fig. 3.6) shows declines in recession years such as 1975, 1982, and 1985. The reasons are obvious: when orders for goods go down, fewer boxes and other packages are ordered for shipping. Also, advertising in newspapers and magazines declines during a recession.

- *Changes in lifestyles.*   Changes in lifestyles are somewhat related to affluence. The United States has increasing numbers of individuals living alone, families with two wage-earners, and single-parent families. People in these situations tend to buy more prepackaged food and to eat out more, often at fast-food establishments using disposable packaging. They may also do more shopping through catalogs, which increases the amounts of third-class mail received and discarded at home. Also, each new household, however small, must have some appliances and furnishings.

- *Changes in work patterns.*   Over a 15-year period, the number of office workers increased 72 percent, while manufacturing jobs declined.[7] At the same time, offices added personal computers, high-speed copiers, and facsimile machines, resulting in an increase in office papers generated.

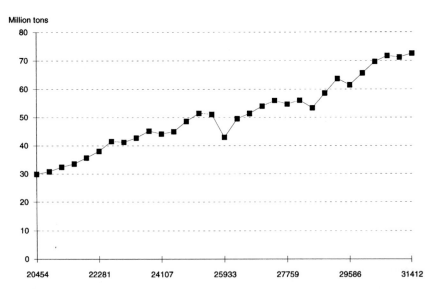

FIGURE 3.6   Generation of paper and paperboard in municipal solid waste, 1960 to 1990. (*Adapted from U.S. Environmental Protection Agency, Characterization of Municipal Solid Waste in the United States: 1992 Update, EPA/530-R-92-013, July 1992.*)

- *New products.*   New products may increase the amounts of MSW generated. Disposable diapers are an example of this phenomenon.

  While the overall pattern has been an increase, some factors tend to decrease MSW generation. Some of these factors include:

- *Redesign of products.*   Some products in MSW have actually grown lighter over the years. Appliances such as refrigerators are one example, due largely to changes in insulation and use of more lightweight plastics. Another example is rubber tires, which have not only been made smaller but last longer. The result is that discards of tires (by weight) have decreased in the last decade. Also, many kinds of packaging have been made lighter over the years, often to save on transportation costs.

- *Materials substitution.*   Especially in packaging, there has been a tendency to substitute lighter materials in many applications. Thus, aluminum cans have largely replaced steel cans in beverage packaging, and plastic bottles have been substituted for glass. This is reflected in declining or flat generation of steel and glass packaging, while aluminum and plastics have shown rapid growth. Plastics have also substituted for paper in many applications. For example, even though generation of paper packaging has grown overall, generation of paper bags and sacks has declined (Figure 3.7). The decline is primarily due to increased use of plastic bags, which are much lighter.

- *Changes in food processing.*   A review of numerous sampling studies indicates that food wastes have been declining as a percentage of MSW generation, and in fact are probably not increasing in tonnage.[1] The reasons for this were discussed in Sec. 3.3.

FIGURE 3.7   Generation of paper bags and sacks in municipal solid waste, 1960 to 1990. (*Adapted from U.S. Environmental Protection Agency, Characterization of Municipal Solid Waste in the United States: 1992 Update, EPA/530-R-92-013, July 1992.*)

Trends in MSW generation are thus quite complex and difficult to quantify. Planners need to look at what is happening in their communities and nationwide when making projections affecting waste management facilities.

REFERENCES

1. Franklin Associates, Ltd., *Characterization of Municipal Solid Waste in the United States: 1992 Update,* U.S. Environmental Protection Agency, Washington, EPA/530-R-92-013, July 1992.

2. American Paper Institute, *1991 Statistics of Paper Paperboard & Wood Pulp,* New York, 1992.

3. Franklin Associates, Ltd., *Supply of and Recycling Demand for Office Waste Paper, 1990 to 1995,* National Office Paper Recycling Project, Washington, July 1991.

4. "Resins 1991: Trouble in the Pipeline," *Modern Plastics,* January 1991.

5. Hunt, R. G., et al., *Estimates of the Volume of MSW and Selected Components in Trash Cans and Landfills,* Franklin Associates, Ltd. with The Garbage Project for The Council for Solid Waste Solutions, February 1990.

6. SENES Consultants Limited and First Consulting Group, *Waste Stream Quantification and Characterization Methodology Study,* Solid Waste Management Division, Environment Canada, January 1992.

7. Franklin Associates, Ltd., *Analysis of Trends in Municipal Solid Waste Generation, 1972 to 1987,* The Procter & Gamble Company, Browning-Ferris Industries, General Mills, and Sears, January 1992.

CHAPTER 4
FEDERAL REGULATION

Edward W. Repa, Ph.D.

Environmental Industry Associations
Washington, D.C.

4.1 INTRODUCTION

There is a plethora of regulations at the federal level that impact the management and disposal of municipal solid waste. These regulations have been authorized by numerous pieces of legislation including the Resource Conservation and Recovery Act (RCRA) and amendments to it under the Hazardous and Solid Waste Amendments of 1984 (HSWA), Clean Air Act (CAA), Clean Water Act (CWA), Occupational Safety and Health Act (OSHA), Superfund, and Toxic Substances Control Act (TSCA). This chapter covers only a few of the more important regulations that have been developed over the past few years that affect solid waste management. Readers are encouraged to refer to the environmental statutes for a complete list of laws and regulations affecting the industry or contact any of the organizations listed in Appendix D.

4.2 RESOURCE CONSERVATION AND RECOVERY ACT

On October 21, 1976, Congress passed the Resource Conservation and Recovery Act, which has been amended numerous times over the years since passage. RCRA for the first time divided the management of waste into two main categories: Subtitle C—Hazardous Waste and Subtitle D—Non-Hazardous Waste. RCRA directed the U.S. Environmental Protection Agency (EPA) to promulgate criteria within a year for determining which facilities should be classified as sanitary landfills and which should be classified as open dumps. The regulations were promulgated on September 13, 1979 by EPA and are contained in 40 CFR part 257, "Criteria for Classification of Solid Waste Disposal Facilities and Practices."

The act was later amended by the Hazardous and Solid Waste Amendments of 1984 (HSWA). Under HSWA, EPA was directed to develop minimum criteria for solid waste management facilities which may receive household hazardous waste or small quantity hazardous waste exempted from the Subtitle C requirements. EPA promulgated these criteria on October 9, 1991 under 40 CFR part 258, "Criteria for Municipal Solid Waste Landfills."

The part 257 criteria that were developed in 1979 remain in effect for all the disposal facilities that accept nonhazardous waste except for municipal solid waste landfills (MSWLFs) subject to the revised criteria contained in part 258. A MSWLF is defined under RCRA as:

> A discrete area of land or an excavation that receives household waste, and that is not a land application unit, surface impoundment, injection well, or waste pile, as those terms are defined in this section. A MSWLF unit also may receive other types of RCRA Subtitle D wastes, such as commercial solid waste, nonhazardous sludge, and industrial solid waste. Such a landfill may be publicly or privately owned. A MSWLF may be a new MSWLF unit or a lateral expansion.

Furthermore, RCRA defines household waste as follows:

> Any solid waste (including garbage, trash, and sanitary waste in septic tanks) derived from households (including single and multiple residences, hotels and motels, bunkhouses, ranger stations, crew quarters, campgrounds, picnic grounds, and day-use recreation areas).

These definitions are important because they form the basis of the Subtitle D program. The origin of the waste, whether or not it was derived from a household, will determine the applicable regulations that must be complied with when the waste is landfilled.

Part 257—"Criteria for Classification of Solid Waste Disposal Facilities and Practices"

The disposal of most nonhazardous solid waste occurs in landfills that meet the part 257 criteria. These criteria apply to all nonhazardous waste streams except the following:

- Household wastes (as defined above)
- Sewage sludge
- Municipal solid waste incinerator ash
- Agricultural wastes
- Overburden resulting from mining operations
- Nuclear wastes

The federal criteria contained in this part are minimal and encompass seven general areas. The requirements of each of these areas is described below.

Floodplains. Facilities or practices in floodplains must not restrict the flow of the base flood, reduce the temporary water storage capacity of the floodplain, or result in the washout of solid waste that could pose a hazard to human life, wildlife, or land or water resources.

Endangered Species. Facilities or practices must not cause or contribute to the taking of any endangered or threatened species or result in the destruction or adverse modification of the critical habitat of endangered or threatened species.

Surface Water. Facilities or practices must not cause a point-source or non-point-source discharge of pollutants into waters of the United States in violation of the Clean Water Act. Also, facilities are not allowed to cause a discharge of dredged materials or fill materials into waters in violation of the CWA.

Groundwater. A facility or practice is prohibited from contaminating an underground drinking water source beyond the solid waste boundary or an alternative boundary established in court. An alternative boundary can be established only if the change would not result in contamination of groundwater that may be needed or used for human consumption.

Disease Vectors. The facility must control on-site populations of disease vectors through the periodic application of cover material or other techniques as appropriate to protect public health.

Air. Facilities are prohibited from engaging in the open burning of solid waste. The requirement does not apply to the infrequent burning of agricultural wastes, silvicultural wastes from forest management practices, land clearing debris, diseased trees, debris from emergency cleanup operations, and ordnance. Also, a facility is not permitted to violate any applicable requirements of a State Implementation Plan (SIP) developed under the Clean Air Act.

Safety. The areas covered by the safety requirements include the control of explosive gases, prevention of fires, control of bird hazards, and control of public access. Facilities are required to control the concentration of explosive gases generated by the operation so that they do not exceed 25 percent of the lower explosive limit (LEL) for the gas in facility structures and LEL at the property boundary. Fires are to be controlled so that they do not pose a hazard to the safety of persons or property.

Facilities disposing of putrescible wastes that may attract birds and are located within 10,000 ft of an airport used by turbojet aircraft or 5000 ft of an airport used by piston-type aircraft must be operated in a manner so as not to pose a bird hazard to aircraft. Finally, the facility must not allow uncontrolled public access, which would expose the public to potential health and safety hazards at the disposal site.

Facilities or practices that fail to meet the requirements established in this part are considered to be *open dumps* under RCRA. States are responsible for developing and enforcing programs within their state to ensure all applicable facilities are complying with the criteria. Operations not in compliance can be sued under the citizen suit provisions of RCRA.

Part 258—"Criteria for Municipal Solid Waste Landfills"

On October 9, 1991, EPA promulgated its long-awaited regulations for municipal solid waste landfills (MSWLFs). Both existing and new MSWLFs were affected by the rules which became effective on October 9, 1993. The Subtitle D criteria set forth minimum criteria in the following six areas:

- Location restrictions
- Operations
- Design
- Groundwater monitoring and corrective action

- Closure and postclosure care
- Financial assurance

Each of these criteria is summarized in detail in the following sections.

Structure of Rule.   The rule's structure allows for the requirements to be self-implementing or for states to receive approval from EPA to implement and enforce its provisions. In those states that do not receive or seek approval, the owner/operator is responsible for ensuring the facility is in compliance with all provisions of the rule. Owners/operators must document compliance and make this documentation available to the state on request. Enforcement of the Subtitle D criteria in unapproved states can occur through the citizen suit provisions of RCRA or, if EPA finds the state program wholly inadequate, the agency itself may provide enforcement.

To encourage states to adopt this rule, EPA added greater flexibility for approved states to specify alternative requirements and schedules. These flexibilities are discussed in further detail in later sections as they occur in context of the rule.

Applicability.   The revised Subtitle D criteria apply to any landfill that accepts household waste (as defined above), including those that receive sewage sludge or municipal waste combustion ash. The criteria in part 258 do not affect landfills that accept only industrial and special waste and construction and demolition waste.

The requirements vary, depending on the landfill's closure date, according to the following breakdown:

- The rule does not apply to any MSWLF that ceased receipt of waste prior to October 9, 1991.
- Only the final cover requirements apply to any MSWLF that ceases receipt of waste after October 9, 1991 but before October 9, 1993.
- The entire Subtitle D requirements apply to any MSWLF that receives waste on or after October 9, 1993.

Additionally, a small landfill exemption is available in two cases for owners/operators of landfills disposing of less than 20 tons/day and where there is no evidence of existing groundwater contamination. These exemptions are as follows:

Alaska provision.   Landfills located in areas where there is an annual interruption of at least three consecutive months of surface transportation that prevents access to a regional facility.

Arid provision.   Landfills located in areas that annually receive less than 25 in of precipitation and where no practicable waste management alternative exists.

MSWLFs meeting the above requirements can be exempted from the design and corrective action requirements contained in these criteria. However, if an owner/operator has been exempted and receives knowledge of groundwater contamination, the exemption no longer applies and the landfill must comply with the design and corrective action requirements. (*Note:* On May 7, 1993, the United States Court of Appeals vacated the small landfill exemption relating to groundwater monitoring as promulgated in the Subtitle D criteria. EPA is in the process of developing a groundwater monitoring program specifically tailored to these landfills.)

Location Restrictions.   The criteria establish restrictions or bans on locating and operating new and existing MSWLF in six unsuitable areas. Restrictions regarding airports and floodplains existed in the part 257 criteria, while the remainder are new to part 258. The restricted areas in the final rule include:

- *Airports.* New, existing, and lateral expansions to existing MSWLFs that are located within 10,000 ft of an airport runway end used by turbojet aircraft or within 5000 ft of a runway end used by piston-type aircraft must demonstrate that they are designed and operated so as not to pose a bird hazard to aircraft. Also, new MSWLFs and lateral expansions at existing MSWLFs within a 5-mi radius of any airport runway end must notify the affected airport and the Federal Aviation Administration (FAA).

- *Floodplains.* New, existing, and lateral expansions at MSWLFs located in the 100-year floodplain must demonstrate that they will not restrict the flow of the 100-year flood, reduce the temporary water storage capacity of the floodplain, or result in washout of solid waste so as to pose a hazard to human health and the environment.

- *Wetlands.* New and lateral expansions of MSWLFs into wetlands are prohibited in unapproved states. However, an approved state may allow siting, provided the landfill can demonstrate that a practicable alternative is not available; construction and operation will not violate any other local, state, or federal law or cause or contribute to significant degradation of the wetland; and steps have been taken to achieve no net loss of wetlands.

- *Fault areas.* New and lateral expansions to MSWLFs are prohibited within 200 ft of a fault that has had displacement in Holocene time (i.e., last 9000 years). Approved states may allow an alternative setback of less than 200 ft if the facility can demonstrate that the structural integrity of the MSWLF will not be damaged and it will be protective of human health and the environment.

- *Seismic impact zones.* New and lateral expansions of MSWLFs are prohibited in seismic impact zones, unless it can be demonstrated to an approved state that all containment structures, including liners, leachate collection systems, and surface water control systems, are designed to resist the maximum horizontal acceleration in lithified earth material for the site.

- *Unstable areas.* New, existing, and lateral expansions of MSWLFs located in unstable areas are required to demonstrate engineering measures that have been incorporated into the MSWLF design that ensure that the integrity of the structural components will not be disrupted. Unstable areas can include poor foundation conditions, areas susceptible to mass movements, and karst terrains.

Existing MSWLFs that cannot make the demonstration pertaining to airports, floodplains, or unstable areas are required to close (according to the closure provisions) within 3 years of the effective date (i.e., October 9, 1996). This deadline can be extended up to 2 years in approved states if the facility can demonstrate that there is no alternative disposal capacity and no immediate threat to human health and the environment.

Operating Criteria. The revised Subtitle D criteria impose 10 operating requirements on MSWLFs. Of the 10, five were carryovers from the part 257 criteria. The operating criteria are

1. *Procedures for excluding the receipt of hazardous waste.* Landfill owners/operators are required to implement a program for detecting and preventing the disposal of regulated hazardous wastes and PCBs at their facilities. The program must include random inspections of incoming loads, maintaining records of inspections, training facility personnel to recognize regulated hazardous wastes and PCB wastes, and notification procedures to the regulatory authority if such wastes are discovered.

2. *Cover material requirements.* MSWLFs must be covered with 6 in of earthen material at the end of each operating day, or more frequently, to control disease vectors, fires, odor, blowing litter, and scavenging. States with approved programs may allow alternative materials that meet or exceed the performance standard and may grant temporary waivers from the requirements when extreme seasonal climatic conditions make the requirements impractical.

3. *Disease vector control.* Landfills are required to prevent or control on-site populations of disease vectors using techniques appropriate for the protection of human health and the environment.

4. *Explosive gas control.* Landfill owners/operators must ensure through a routine methane monitoring program that the concentration of methane gas generated by the facility does not exceed 25 percent of the lower explosive limit in facility structures and does not exceed the lower explosive limit at the property boundary. If the landfill exceeds these limits, a remediation program must be immediately implemented.

5. *Air criteria.* Landfills are required to meet applicable requirements developed under a State Implementation Plan pursuant to Sec. 110 of the Clean Air Act. Also, open burning of solid waste is prohibited except for the infrequent burning of agricultural wastes, silvicultural wastes, land-clearing debris, diseased trees, and debris from emergency cleanup operations.

6. *Access requirements.* Owners/operators of MSWLFs must control public access and prevent unauthorized traffic and illegal dumping of wastes through the use of artificial barriers, natural barriers, or both.

7. *Runoff/run-on controls.* MSWLFs must design, construct and maintain a run-on and runoff control system on the active portion of the landfill capable of handling the peak discharge from a 24-h, 25-year storm event.

8. *Surface water requirements.* MSWLFs must not cause a discharge of pollutants into waters violating any requirements of the Clean Water Act, including the National Pollution Discharge Elimination System (NPDES), or cause a nonpoint source of pollution to waters that violates any requirement of an area-wide or statewide water quality management plan approved under Sec. 208 or 319 of the Clean Water Act.

9. *Liquid restrictions.* MSWLFs are prohibited from accepting bulk and non-containerized liquid wastes unless the waste is a household waste. Leachate and gas condensate derived from a MSWLF unit can be recirculated back into that unit as long as the unit is designed with a composite liner.

10. *Record keeping requirements.* Owners/operators are required to record and retain any location restriction demonstrations; inspection records, training procedures, and notification procedures; gas monitoring results; design documentation for gas condensate and leachate recirculation; monitoring, testing, and analytical data; and closure and postclosure care plans.

Design Criteria. The design criteria establish a specific engineering design for those states that are not approved and an alternative design standard based on performance that approved states can allow. The part 258 design criteria are as follows:

- *Unapproved states.* A composite liner with a leachate collection system that is capable of maintaining less than 30-cm (12-in) depth of leachate over the liner must be installed. The composite liner system must consist of two components: an upper component composed of a flexible membrane liner (FML) at least 30 mils thick and a lower component composed of at least 2 ft of compacted soil with a hydraulic conductivity of no more than 1×10^{-7} cm/s. If the FML component consists of high-density polyethylene (HDPE), its thickness must be at least 60 mils. Alternative designs meeting the performance standard below are allowed in unapproved states but require that a demonstration be made to the unapproved state and that the state review and petition EPA for concurrence. This option is only available until EPA promulgates the State and Tribal Implementation Rule (discussed later).

- *Approved states.* A design that ensures the concentration of 24 organic and inorganic constituents in the uppermost aquifer at the point of compliance do not exceed maximum contaminant levels (MCLs). The point of compliance can range from the waste management unit boundary to 150 m from the boundary, depending on local hydrogeologic conditions.

These new design standards apply to new MSWLFs and to new units and lateral expansions to existing MSWLFs. Existing units are not required to be retrofitted with liners and leachate collection systems.

Groundwater Monitoring and Corrective Action. The groundwater monitoring program requirements apply to all MSWLF units unless the owner/operator can demonstrate that no potential for migration of hazardous constituents from the unit to the uppermost aquifer exists during the active life of the unit and the postclosure care period.

As with the design criteria, the compliance schedule for the groundwater monitoring provisions of the rule are or can be different, depending on whether or not the state has an approved program. The following compliance schedule applies to existing units and lateral expansions in unapproved states:

- Units less than 1 mi from a drinking water intake (surface or subsurface)—October 9, 1994

- Units greater than 1 mi but less than 2 mi from a drinking water intake—October 9, 1995

- Units greater than 2 mi from a drinking water intake—October 9, 1996

New MSWLF units are required to be in compliance before waste can be placed in the unit.

States with approved programs can adopt an alternative schedule. This schedule must ensure that 50 percent of all existing MSWLF units and lateral expansions are in compliance by October 9, 1994 and all units are in compliance by October 9, 1996.

The groundwater monitoring program for an MSWLF consists of four steps:

Groundwater Monitoring System. The first step is the establishment of a groundwater monitoring system. The regulations require that a system be installed that has a sufficient number of wells, installed at appropriate locations and depths, to yield groundwater samples from the uppermost aquifer. The system must include wells that represent the quality of background groundwater that has not been affected by leakage from the unit, as well as groundwater that has passed under and is downgradient of the unit.

Detection Monitoring Program. The second step is the establishment of a detection monitoring program. This step consists of semiannual monitoring of wells for both water quality and head levels during the active life of the facility through the postclosure care period. The minimum detection monitoring program includes the monitoring of 15 heavy metals and 47 volatile organics (Table 4.1). A minimum of four indepen-

TABLE 4.1 Constituents for Detection Monitoring*

Common name†	CAS RN‡	Common name†	CAS RN‡
Inorganic constituents:		(35) 1,1-Dichloroethylene;	
(1) Antimony	Total	1,1-dichloroethene; vinylidene	
(2) Arsenic	Total	chloride	75-35-4
(3) Barium	Total	(36) cis-1,2-Dichloroethylene;	
(4) Beryllium	Total	cis-1,2-dichloroethene	156-59-2
(5) Cadmium	Total	(37) trans-1,2-Dichloroethylene;	
(6) Chromium	Total	trans-1,2-dichloroethene	156-60-5
(7) Cobalt	Total	(38) 1,2-Dichloropropane;	
(8) Copper	Total	propylene dichloride	78-87-5
(9) Lead	Total	(39) cis-1,3-Dichloropropene	10061-01-5
(10) Nickel	Total	(40) trans-1,3-Dichloropropene	10061-02-6
(11) Selenium	Total	(41) Ethylbenzene	100-41-4
(12) Silver	Total	(42) 2-Hexanone; methyl	
(13) Thallium	Total	butyl ketone	591-78-6
(14) Vanadium	Total	(43) Methyl bromide;	
(15) Zinc	Total	bromomethane	74-83-9
Organic constituents:		(44) Methyl chloride;	
(16) Acetone	67-64-1	chloromethane	74-87-3
(17) Acrylonitrile	107-13-1	(45) Methylene bromide;	
(18) Benzene	71-43-2	dibromomethane	74-95-3
(19) Bromochloromethane	74-97-5	(46) Methylene chloride;	
(20) Bromodichloromethane	75-27-4	dichloromethane	75-09-2
(21) Bromoform; tribromomethane	75-25-2	(47) Methyl ethyl ketone;	
(22) Carbon disulfide	75-15-0	MEK; 2-butanone	78-93-3
(23) Carbon tetrachloride	56-23-5	(48) Methyl iodide; iodomethane	74-88-4
(24) Chlorobenzene	108-90-7	(49) 4-Methyl-2-pentanone;	
(25) Chloroethane;		methyl isobutyl ketone	108-10-1
ethyl chloride	75-00-3	(50) Styrene	100-42-5
(26) Chloroform;		(51) 1,1,1,2-Tetrachloroethane	630-20-6
trichloromethane	67-66-3	(52) 1,1,2,2-Tetrachloroethane	79-34-5
(27) Dibromochloromethane;		(53) Tetrachloroethylene; tetra-	
chlorodibromomethane	124-48-1	chloroethene; perchloro-	
(28) 1,2-Dibromo-3-chloropro-		ethylene	127-18-4
pane; DBCP	96-12-8	(54) Toluene	108-88-3
(29) 1,2-Dibromoethane;		(55) 1,1,1-Trichloroethane;	
ethylene dibromide; EDB	106-93-4	methylchloroform	71-55-6
(30) o-Dichlorobenzene; 1,2-		(56) 1,1,2-Trichloroethane	79-00-5
dichlorobenzene	95-50-1	(57) Trichloroethylene; trichloro-	
(31) p-Dichlorobenzene;		ethene	79-01-6
1,4-dichlorobenzene	106-46-7	(58) Trichlorofluoromethane;	
(32) trans-1,4-Dichloro-2-butene	110-57-6	CFC-11	75-69-4
(33) 1,1-Dichloroethane;		(59) 1,2,3-Trichloropropane	96-18-4
ethylidene chloride	75-34-3	(60) Vinyl acetate	108-05-4
(34) 1,2-Dichloroethane;		(61) Vinyl chloride	75-01-4
ethylene dichloride	107-06-2	(62) Xylenes	1330-20-7

*This list contains 47 volatile organics for which possible analytical procedures are provided in EPA Report SW-846 "Test Methods for Evaluating Solid Waste," November 1986, as revised December 1987, includes Method 8260; and 15 metals for which SW-846 provides either Method 6010 or a method from the 7000 series of methods.

†Common names are those widely used in government regulations, scientific publications, and commerce; synonyms exist for many chemicals.

‡Chemical Abstracts Service registry number. Where "Total" is entered, all species in the ground water that contain this element are included.

dent samples from each well must be collected and analyzed during the first semiannual sampling event. During subsequent sampling events, at least one sample from each well must be collected and analyzed for the above parameters.

Approved states are permitted to establish an alternative list of inorganic indicator parameters (i.e., in lieu of some or all of the heavy metals) as long as it provides a reliable indication of inorganic releases from the landfill to the groundwater. Also, an approved state can specify an appropriate alternative frequency for repeated sampling and analysis during the active life and postclosure care period. However, this frequency cannot be less than annual during the active life and closure period.

Results of sampling events must be statistically analyzed to determine if a statistically significant increase over background has occurred for one or more of the monitored constituents. A number of statistical methods are permitted to analyze the data collected. If a statistically significant increase occurs over background, the owner/operator must establish an assessment monitoring program (i.e., the next step in the sequence) and demonstrate that a source other than the landfill is the cause, or show that the increase is the result of sampling, analysis, statistical error or natural variation.

Assessment Monitoring Program. An assessment monitoring program is required whenever a statistically significant increase over background has been detected for one or more of the constituents in the detection monitoring program. Within 90 days of triggering an assessment monitoring program, and annually thereafter, an owner/operator is required to sample and analyze groundwater for some 213 organic and inorganic constituents (Table 4.2). A minimum of one sample from each downgradient well must be analyzed during this sampling event. For any constituents detected, a minimum of four independent samples from each well are required to be collected and analyzed to establish background for these constituents.

After the initial sampling, owners/operators must sample all wells twice a year for detection monitoring parameters and those assessment monitoring parameters that were detected in the first assessment sampling. Also, all 213 assessment monitoring parameters must be sampled and analyzed annually.

The owner/operator is required to establish a groundwater protection standard for any constituents found in the assessment monitoring program. This standard must be based on an MCL, background concentration established during the assessment monitoring program for those constituents for which an MCL has not been promulgated, or background concentration for constituents where the background level is higher than the MCL.

States with approved programs are permitted to establish a number of alternative standards in the assessment monitoring program. Also, approved states are allowed to specify an appropriate subset of wells to be sampled and analyzed, delete any monitoring parameters that are not reasonably expected to be derived form the waste contained in the unit, specify an appropriate alternative frequency for repeated sampling and analysis, and establish an alternative groundwater protection standard based on an appropriate health-based level.

Results of the sampling must be statistically analyzed by using an appropriate procedure. On the basis of the statistical analysis, the owner/operator:

- May return to detection monitoring if a demonstration can be made that a source other than the MSWLF caused the contamination, or that the increase resulted from an error in sampling, analysis, statistical evaluation, or natural variation in groundwater quality

- May return to detection monitoring if the concentration of all assessment monitoring constituents are shown to be at or below background values for two consecutive sampling events

TABLE 4.2 List of Hazardous Inorganic and Organic Constituents[1]

Common name[2]	CAS RN[3]	Chemical abstracts service index name[4]	Suggested methods[5]	PQL[6] (μg/L)
Acenaphthene	83-32-9	Acenaphthylene, 1,2-dihydro-	8100 / 8270	200 / 10
Acenaphthylene	208-96-8	Acenaphthylene	8100 / 8270	200 / 10
Acetone	67-64-1	2-Propanone	8260	100
Acetonitrile; methyl cyanide	75-05-8	Acetonitrile	8015	100
Acetophenone	98-86-2	Ethanone, 1-phenyl-	8270	10
2-Acetylaminofluorene; 2-AAF	53-96-3	Acetamide, N-9H-fluoren-2-yl-	8270	20
Acrolein	107-02-8	2-Propenal	8030	5
Acrylonitrile	107-13-1	2-Propenenitrile	8260 / 8030	100 / 5
Aldrin	309-00-2	1,4:5,8-Dimethanonaphthalene, 1,2,3,4,10,10-hexachloro-1,4,4a,5,8,8a-hexahydro-(1α,4α,4aβ,5α,8α, 8aβ)-	8260 / 8080	200 / 0.05
Allyl chloride	107-05-1	1-Propene, 3-chloro-	8270 / 8010	10 / 5
4-Aminobiphenyl	92-67-1	[1,1'-Biphenyl]-4-amine	8260 / 8270	10 / 20
Anthracene	120-12-7	Anthracene	8100 / 8270	200 / 10
Antimony	Total	Antimony	6010 / 7040 / 7041	300 / 2000 / 30
Arsenic	Total	Arsenic	6010 / 7060 / 7061	500 / 10 / 20
Barium	Total	Barium	6010 / 7080	20 / 1000
Benzene	71-43-2	Benzene	8020 / 8021 / 8260	2 / 0.1 / 5

Name	Systematic name	CAS number	Method	Value
Benzo[a]anthracene; benzanthracene	Benz[a]anthracene	56-55-3	8100	200
			8270	10
Benzo[b]fluoranthene	Benz[e]acephenanthrylene	205-99-2	8100	200
			8270	10
Benzo[k]fluoranthene	Benzo[k]fluoranthene	207-08-9	8100	200
			8270	10
Benzo[ghi]perylene	Benzo[ghi]perylene	191-24-2	8100	200
			8270	10
Benzo[a]pyrene	Benzo[a]pyrene	50-32-8	8100	200
			8270	10
Benzyl alcohol	Benzenemethanol	100-51-6	8270	20
Beryllium	Beryllium	Total	6010	3
			7090	50
			7091	2
alpha-BHC	Cyclohexane, 1,2,3,4,5,6-hexachloro-, (1α,2α,3β,4α,5β,6β)-	319-84-6	8080	0.05
			8270	10
beta-BHC	Cyclohexane, 1,2,3,4,5,6-hexachloro-, (1α,2β,3α,4β,5α,6β)-	319-85-7	8080	0.05
			8270	20
delta-BHC	Cyclohexane, 1,2,3,4,5,6-hexachloro-, (1α,2α,3α,4β,5α,6β)-	319-86-8	8080	0.1
			8270	20
gamma-BHC; lindane	Cyclohexane, 1,2,3,4,5,6-hexachloro-, (1α,2α,3β,4α,5α,6β)-	58-89-9	8080	0.05
			8270	20
Bis[2-chloroethoxy)methane	Ethane, 1,1'-[methylenebis(oxy)bis(2 chloro-	111-91-1	8110	5
			8270	10
Bis(2-chloroethyl) ether; dichloroethyl ether	Ethane, 1,1'-oxybis[2-chloro-	111-44-4	8110	3
			8270	10
Bis-(2 chloro-1-methylethyl) ether; 2,2'-dichlorodisopropyl ether; DCIP[7]	Propane, 2,2'-oxybis[1-chloro-	108-60-1	8110	10
			8270	10
Bis(2-ethylhexyl) phthalate	1,2-Benzenedicarboxylic acid, bis(2-ethylhexyl) ester	117-81-7	8060	10
			8270	20
Bromochloromethane; chlorobromomethane	Methane, bromochloro-	74-97-5	8021	0.1
			8260	5
Bromodichloromethane; dibromochloromethane	Methane, bromodichloro-	75-27-4	8010	1
			8021	0.2
			8260	5

TABLE 4.2 List of Hazardous Inorganic and Organic Constituents[1] (*Continued*)

Common name[2]	CAS RN[3]	Chemical abstracts service index name[4]	Suggested methods[5]	PQL[6] (μg/L)
Bromoform; tribromomethane	75-25-2	Methane; tribromo-	8010	2
			8021	15
			8260	5
4-Bromophenyl phenyl ether	101-55-3	Benzene, 1-bromo-4-phenoxy-	8110	25
			8270	10
Butyl benzyl phthlate; benzyl butyl phthlate	85-68-7	1,2-Benzenedicarboxylic acid, butyl phenylmethl ester	8060	5
			8270	10
Cadmium	Total	Cadmium	6010	40
			7130	50
			7131	1
Carbon disulfide	75-15-0	Carbon disulfide	8260	100
Carbon tetrachloride	56-23-5	Methane, tetrachloro-	8010	1
			8021	0.1
			8260	10
Chlordane	See Note 8	4,7-Methano-1H-indene, 1,2,4,5,6,7,8,8-octochloro-2,3,3a,4,7,7a-hexahydro-	8080	0.1
			8270	50
p-Chloroaniline	106-47-8	Benzenamine, 4-chloro-	8270	20
Chlorobenzene	108-90-7	Benzene, chloro-	8010	2
			8020	2
			8021	0.1
			8260	5
Chlorobenzilate	510-15-6	Benzeneacetic acid, 4-chloro-α-(4-chlorophenyl-α-hydroxy-, ethyl ester	8270	10
p-Chloro-m-cresol; 4-chloro-3-methylphenol	59-50-7	Phenol, 4-chloro-3-methyl-	8040	5
			8270	20
Chloroethane; ethyl chloride	75-00-3	Ethane, chloro-	8010	5
			8021	1
			8260	10
Chloroform; trichloromethane	67-66-3	Methane, trichloro-	8010	0.5

2-Chloronaphthalene	91-58-7	8021	0.2
Naphthalene, 2-chloro		8260	5
2-Chlorophenol	95-57-8	8120	10
Phenol, 2-chloro-		8270	10
		8040	5
		8270	10
4-Chlorophenyl phenyl ether	7005-72-3	8110	40
Benzene, 1-chloro-4-phenoxy-		8270	10
Chloroprene	126-99-8	8010	50
1,3-Butadiene, 2-chloro-		8260	20
Chromium	Total	6010	70
		7190	500
		7191	10
Chrysene	218-01-9	8100	200
		8270	10
Cobalt	Total	6010	70
		7200	500
		7201	10
Copper	Total	6010	60
		7210	200
		7211	10
m-Cresol, 3-methylphenol	108-39-4	8270	10
o-Cresol, 2-methylphenol	95-48-7	8270	10
p-Cresol, 4-methylphenol	106-44-5	8270	10
Cyanide	57-12-5	9010	200
2,4-D; 2,4-dichlorophenoxyacetic acid	94-75-7	8150	10
Acetic acid, (2,4-dichlorophenoxy)-			
4,4'-DDD	72-54-8	8080	0.1
Benzene 1,1'-(2,2-dichloroethylidene)bis-[4-chloro-		8270	10
4,4'-DDE	72-55-9	8080	0.05
Benzene, 1,1'-(dichloroethyenylidene)bis[4-chloro-		8270	10
4,4'-DDT	50-29-3	8080	0.1
Benzene, 1,1'-(2,2,2-trichloroethylidene)bis[4-chloro-		8270	10
Diallate	2303-16-4	8270	10
Carbamothioic acid, bis(1-methylethyl)-,S-(2,3-dichloro-2-propenyl) ester			

TABLE 4.2 List of Hazardous Inorganic and Organic Constituents[1] (*Continued*)

Common name[2]	CAS RN[3]	Chemical abstracts service index name[4]	Suggested methods[5]	PQL[6] (μg/L)
Dibenz[a,h]anthracene	53-70-3	Dibenz[a,h]anthracene	8100	200
			8270	10
Dibenzofuran	132-64-9	Dibenzofuran	8270	10
Dibromochloromethane; chlorodibromomethane	124-48-1	Methane, dibromochloro-	8010	1
			8021	0.3
			8260	5
1,2-Dibromo-3-chloropropane; DBCP	96-12-8	Propane, 1,2-dibromo-3-chloro-	8011	0.1
			8021	30
			8260	25
1,2-Dibromoethane; ethylene dribromide; EDB	106-93-4	Ethane, 1,2-dibromo-	8011	0.1
			8021	10
			8260	5
Di-n-butyl phthlate	84-74-2	1,2-Benzenedicarboxylic acid, dibutyl ester	8060	5
			8270	10
o-Dichlorobenzene; 1,2-dichlorobenzene	95-50-1	Benzene, 1,2-dichloro-	8010	2
			8020	5
			8021	0.5
			8120	10
			8260	5
			8270	10
m-Dichlorobenzene; 1,3-dichlorobenzene	541-73-1	Benzene, 1,3-dichloro-	8010	5
			8020	5
			8021	0.2
			8120	10
			8260	5
			8270	10
p-Dichlorobenzene; 1,4-dichlorobenzene	106-46-7	Benzene, 1,4-dichloro-	8010	2
			8020	5
			8021	0.1

Compound	CAS No.	Chemical name	Method	PQL
3,3'-Dichlorobenzidine	91-94-1	[1,1'-Biphenyl]-4,4'-diamine, 3,3'-dichloro-	8120	15
trans-1,4-Dichloro-2-butene	110-57-6	2-Butene, 1,4-dichloro, (E)-	8260	5
Dichlorodifluoromethane; CFC 12	75-71-8	Methane, dichlorodifluoro-	8270	10
			8260	20
			8021	100
			8260	0.5
1,1-Dichloroethane; ethyldidene chloride	75-34-3	Ethane, 1,1-dichloro-	8010	5
			8021	1
			8260	0.5
1,2-Dichloroethane; ethylene dichloride	107-06-2	Ethane, 1,1-dichloro-	8010	5
			8021	0.5
			8260	0.3
1,1-Dichloroethylene; 1,1-dichloroethene; vinylidene chloride	75-35-4	Ethene, 1,1-dichloro-	8010	5
			8021	1
			8260	0.5
cis-1,2-Dichloroethylene; cis-1,2-dichloroethene	156-59-2	Ethene, 1,2-dichloro-, (Z)-	8021	5
			8260	0.2
trans-1,2-Dichloroethylene trans-1,2-dichloroethene	156-60-5	Ethene, 1,2-dichloro-, (E)-	8010	5
			8021	1
			8260	0.5
2,4-Dichlorophenol	120-83-2	Phenol, 2,4-dichloro-	8040	5
			8270	5
2,6-Dichlorophenol	87-65-0	Phenol, 2,6-dichloro-	8270	10
1,2-Dichloropropane; propylene dichloride	78-87-5	Propane, 1,2-dichloro-	8010	10
			8021	0.5
			8260	0.05
1,3-Dichloropropane; trimethylene dichloride	142-28-9	Propane, 1,3-dichloro-	8021	5
			8260	0.3
2,2-Dichloropropane; isopropylidene chloride	594-20-7	Propane, 2,2-dichloro-	8021	5
			8260	0.5
1,1-Dichloropropene	563-58-6	1-Propene, 1,1-dichloro-	8021	15
			8260	0.2
cis-1,3-Dichloropropene	10061-01-5	1-Propene, 1,3-dichloro-, (Z)	8010	5
			8260	20

TABLE 4.2 List of Hazardous Inorganic and Organic Constituents[1] (*Continued*)

Common name[2]	CAS RN[3]	Chemical abstracts service index name[4]	Suggested methods[5]	PQL [6] (μg/L)
trans-1,3-Dichloropropene	10061-02-6	1-Propene, 1,3-dichloro-, (E)-	8260	10
			8010	5
			8260	10
Dieldrin	60-57-1	2,7,3,6-Dimethanonaphth[2,3-b]oxirene, 3,4,5,6,9,9-hexa-chloro-1a,2,2a,3,6,6a,7,7a-octahydro-, (1aα,2β,2aα,3β, 6β,6aα,7β,7aα)-	8080	0.05
			8270	10
Diethyl phthlate	84-66-2	1,2-Benzenedicarboxylic acid, diethyl ester	8060	5
			8270	10
0,0-Diethyl 0-2-pyrazinyl phosphoro-thioate; thionazin	297-97-2	Phosphorothioic acid, 0,0-diethyl 0-pyrazinyl ester	8141	5
			8270	20
Dimethoate	60-51-5	Phosphorodithioic acid, 0,0-dimethyl S-[2-(methylamino)-2-oxyethyl] ester	8141	3
			8270	20
p-(Dimethylaminc)azobenzene	60-11-7	Benzenamine, N,N-dimethyl-4-(phenylazo)-	8270	10
7,12-Dimethylbenz[a]anthracene	57-97-6	Benz[a]anthracene, 7,12-dimethyl	8270	10
3,3¹-Dimethylbenzidine	119-93-7	[1,1¹-Biphenyl]-4,4¹-diamine, 3,3,¹-dimethyl-	8270	10
2,4-Dimethylphenol; m-xylenol	105-67-9	Phenol, 2,4-dimethyl	8040	5
			8270	10
Dimethyl phthalate	131-11-3	1,2-Benzenedicarboxylic acid, dimethyl ester	8060	5
			8270	10
m-Dinitrobenzene	99-65-0	Benzene, 1,3-dinitro-	8270	20
4,6-Dinitro-o-cresol 4,6-dinitro-2-methylphenol	534-52-1	Phenol, 2-methyl-4,6-dinitro	8040	150
			8270	50
2,4-Dinitrophenol	51-28-5	Phenol, 2,4-dinitro-	8040	150
			8270	50
2,4-Dinitrotoluene	121-14-2	Benzene, 1-methyl-2,4-dinitro-	8090	0.2
			8270	10
2,6-Dinitrotoluene	606-20-2	Benzene, 2-methyl-1,3-dinitro-	8090	0.1
			8270	10
Dinoseb; DNBP; 2-sec-butyl-4,6-dinitrophenol	88-85-7	Phenol, 2-(1-methylpropyl)-4,6-dinitro	8150	1
			8270	20

4.16

Compound	CAS	Chemical name	Method	Value
Di-n-octyl phthalate	117-84-0	1,2-Benzenedicarboxylic acid, dioctyl ester	8060	30
			8270	10
Diphenylamine	122-39-4	Benzenamine, N-phenyl-	8270	10
Disulfoton	298-04-4	Phosphorodithioic acid, O,O-diethyl S-[2-(ethylthio)ethyl] ester	8140	2
			8141	0.5
			8270	10
Endosulfan I	959-98-8	6,9-Methano-2,4,3-benzodioxathiepin, 6,7,8,9,10,10-hexachloro-1,5,5a,6,9,9a-hexahydro-, 3-oxide	8080	0.1
			8270	20
Endosulfan II	33213-65-9	6,9-Methano-2,4,3-benzodioxathiepin, 6,7,8,9,10,10-hexachloro-1,5,5a,6,9,9a-hexahydro-, 3 oxide, (3α,5aα,6β,9β,9aα)-	8080	0.05
			8270	20
Endosulfan sulfate	1031-07-8	6,9-Methano-2,4,3-benzodioxathiepin, 6,7,8,9,10,10-hexachloro-1,5,5a,6,9,9a-hexahydro-3,3-dioxide	8080	0.5
			8270	10
Endrin	72-20-8	2,7:3,6-Dimethanonaphth[2,3-b]oxirene, 3,4,5,6,9,9-hexachloro-1a,2,2a,3,6,6a,7,7a-octahydro-, (1aα, 2β,2aβ,3α,6α,6aβ,7β,7aα)-	8080	0.1
			8270	20
Endrin aldehyde	7421-93-4	1,2,4-Methenocyclopenta[cd]pentalene-5-carboxaldehyde, 2,2a,3,3,4,7-hexachlorodecahydro-, (1α,2β,2aβ,4β, 4aβ,5β,6aβ,6bβ, 7R*)-	8080	0.2
			8270	10
Ethylbenzene	100-41-4	Benzene, ethyl-	8020	2
			8221	0.05
			8260	5
Ethyl methacrylate	97-63-2	2-Propenoic acid, 2-methyl-, ethyl ester	8015	5
			8260	10
			8270	10
Ethyl methanesulfonate	62-50-0	Methanesulfonic acid, ethyl ester	8270	20
Famphur	52-85-7	Phosphorothioic acid, O-[4-[(dimethylamino)sulfonyl]phenyl] O,O-dimethyl ester	8270	20
Fluoranthene	206-44-0	Fluoranthene	8100	200
			8270	10
Fluorene	86-73-7	9H-Fluorene	8100	200
			8270	10
Heptachlor	76-44-8	4,7-Methano-1H-indene, 1,4,5,6,7,8,8-heptachloro-3a,4,7,7a-tetrahydro-	8080	0.05
			8270	10

TABLE 4.2 List of Hazardous Inorganic and Organic Constituents[1] (*Continued*)

Common name[2]	CAS RN[3]	Chemical abstracts service index name[4]	Suggested methods[5]	PQL[6] (μg/L)
Heptachlor epoxide	1024-57-3	2,5-Methano-2H-indeno[1,2-b]oxirene, 2,3,4,5,6,7,7-hepta-chloro-1a,1b,5,5a,6,6a-hexahydro-, (1aα,1bβ, 2α,α,5aβ, 6β, 6aα)	8080 8270	1 10
Hexachlorobenzene	118-74-1	Benzene, hexachloro-	8120 8270	0.5 10
Hexachlorobutadiene	87-68-3	1,3-Butadiene, 1,1,2,3,4,4-hexachloro-	8021 8120 8260	0.5 5 10
Hexachlorocyclopentadiene	77-47-4	1,3-Cyclopentadiene, 1,2,3,4,5,5-hexachloro-	8270	10
Hexachloroethane	67-72-1	Ethane, hexachloro-	8120 8120 8260	5 10 0.5 10
Hexachloropropene	1888-71-7	1-Propene, 1,1,2,3,3,3-hexachloro-	8270	10
2-Hexanone; methyl butyl ketone	591-78-6	2-Hexanone	8270	10
Indeno(1,2,3-cd)pyrene	193-39-5	Indeno(1,2,3-cd)pyrene	8260 8100	50 200
Isobutyl alcohol	78-83-1	1-Propanol, 2-methyl-	8270 8015	10 50
Isodrin	465-73-6	1,4,5,8-Dimethanonaphthalene, 1,2,3,4,10,10-, hexachloro-1,4,4a,5,8,8a hexahydro-(1α,4α,4aβ,5β,8β,8aβ)-	8240 8270	100 20
Isophorone	78-59-1	2-Cyclohexen-1-one, 3,5,5-trimethyl-	8260 8090	10 60
Isosafrole	120-58-1	1,3-Benzodioxole,5-(1-propenyl)-	8270	10
Kepone	143-50-0	1,3,4-Metheno-2H-cyclobuta[cd]pentalen-2-one 1,1a,3,3a,4,5,5a,5b,6-decachlorooctahydro-.	8270	20
Lead	Total	Lead	6010 7420 7421	400 1000 10

Compound	Systematic name	Total CAS	Method	Value
Mercury	Mercury		7470	2
Methacryloinitrile	2-Propenenitrile, 2-methyl-	126-98-7	8015	5
			8260	100
			8270	100
Methapyrilene	1,2-Ethanediamine, N.N-dimethyl-N¹-2-pyridinyl-N1/2-thienyl-methyl)-	91-80-5		
Methoxychlor	Benzene,1,1'-(2,2,2,trichloroethylidene)bis[4-methoxy-	72-43-5	8080	2
			8270	10
Methyl bromide; bromomethane	Methane, bromo-	74-83-9	8010	20
			8021	10
Methyl chloride; chloromethane	Methane, chloro-	74-87-3	8010	1
			8021	0.3
3-Methylcholanthrene	Benz[j]aceanthrylene, 1,2-dihydro-3-methyl-	56-49-5	8270	10
Methyl ethyl ketone; MEK; 2-butanone	2-Butanone	78-93-3	8015	10
			8260	100
Methyl iodide; iodomethane	Methane, iodo-	74-88-4	8010	40
			8260	10
Methyl methacrylate	2-Propenoic acid, 2-methyl-, methyl ester	80-62-6	8015	2
			8260	30
Methyl methanesulfonate	Methanesulfonic acid, methyl ester	66-27-3	8270	10
2-Methylnaphthalene	Naphthalene, 2-methyl-	91-57-6	8270	10
Methyl parathion; parathion methyl	Phosphorothioic acid, 0,0-dimethyl 0-(4-nitrophenyl) ester	298-00-0	8140	0.5
			8141	1
			8270	10
4-Methyl-2-pentanone; methyl isobutyl ketone	2-Pentanone, 4-methyl-	108-10-1	8015	5
			8260	100
Methylene bromide; dibromomethane	Methane, dibromo-	74-95-3	8010	15
			8021	20
			8260	10
Methylene chloride; dichloromethane	Methane, dichloro-	75-09-2	8010	5
			8021	0.2
			8260	10
Naphthalene	Naphthalene	91-20-3	8021	0.5
			8100	200
			8260	5
			8270	10

TABLE 4.2 List of Hazardous Inorganic and Organic Constituents[1] (*Continued*)

Common name[2]	CAS RN[3]	Chemical abstracts service index name[4]	Suggested methods[5]	PQL[6] (μg/L)
1,4-Naphthoquinone	130-15-4	1,4-Naphthalenedione	8270	10
1-Naphthylamine	134-32-7	1-Naphthalenamine	8270	10
2-Naphthylamine	91-59-8	2-Naphthalenamine	8270	10
Nickel	Total	Nickel	6010	150
			7520	400
o-Nitroaniline; 2-nitroaniline	88-74-4	Benzenamine, 2-nitro-	8270	50
m-Nitroaniline; 3-nitroaniline	99-09-2	Benzenamine, 3-nitro	8270	50
p-Nitroaniline; 4-nitroaniline	100-01-6	Benzenamine, 4-nitro	8270	20
Nitrobenzene	98-95-3	Benzene, nitro-	8090	40
			8270	10
o-Nitrophenol; 2-nitrophenol	88-75-5	Phenol, 4-nitro-	8040	5
			8270	10
p-Nitrophenol; 4-nitrophenol	100-02-7	Phenol, 4-nitro-	8040	10
			8270	50
N-Nitrosodi-n-butylamine	924-16-3	1-Butanamine, N-butyl-N-nitroso-	8270	10
N-Nitrosodiethylamine	55-18-5	Ethanamine, N-ethyl-N-nitroso-	8270	20
N-Nitrosodimethylamine	62-75-9	Methanamine, N-methyl-N-nitroso-	8070	2
N-Nitrosodiphenylamine	86-30-6	Benzenamine, N-nitroso-N-phenyl-	8070	5
N-Nitrosodipropylamine; N-nitroso-N-dipro-pylamine; Di-n-propylnitrosamine	621-64-7	1-Propanamine, N-nitroso-N-propyl-	8070	10
N-Nitrosomethylethalamine	10595-95-6	Ethanamine, N-methyl-N-nitroso-	8270	10
N-Nitrosopiperidine	100-75-4	Piperidine, 1-nitroso	8270	20
N-Nitrosopyrrolidine	930-55-2	Pyrrolidine, 1-nitroso-	8270	40
5-Nitro-o-toluidine	99-55-8	Benzenamine, 2-methyl-5-nitro-	8270	10
Parathion	56-38-2	Phosphorothioic acid, 0,0-diethyl 0-(4-nitrophenyl) ester	8141	0.5
			8270	10
Pentachlorobenzene	608-93-5	Benzene, pentachloro-	8270	10
Pentachloronitrobenzene	82-68-8	Benzene, pentachloronitro-	8270	20
Pentachlorophenol	87-86-5	Phenol, pentachloro-	8040	5
			8270	50

Name	CAS	Chemical name	Method	Value
Phenacetin	62-44-2	Acetamide, N-(4-ethoxyphenl)	8270	20
Phenanthrene	85-01-8	Phenanthrene	8100	200
			8270	10
Phenol	108-95-2	Phenol	8040	1
p-Phenylenediamine	106-50-3	1,4-Benzenediamine	8270	10
Phorate	298-02-2	Phosphorodithioic acid, 0,0-diethyl S-[(ethylthio)methyl] ester	8140	2
			8141	0.5
			8270	10
Polychlorinated biphenyls; PCBs; aroclors	See Note 9	1,1'-Biphenyl, chloro derivatives	8080	50
			8270	200
Pronamide	23950-58-5	Benzamide, 3,5-dichloro-N-(1,1-dimethyl-2-propynyl)-	8270	10
Propionitrile Ethyl cyanide	107-12-0	Propanenitrile	8015	60
			8280	150
Pyrene	129-00-0	Pyrene	8100	200
			8270	10
Safrole	94-59-1	1,3-Benzodioxole, 5-(2-propenyl)-	8270	10
Selenium	Total	Selenium	8010	750
			7740	20
			7741	20
Silver	Total	Silver	6010	70
			7760	100
			7761	10
Silvex; 2,4,5-TP	93-72-1	Propanoic acid,2-(2,4,5-trichlorophenoxy)-	8150	2
Styrene	100-42-5	Benzene, ethenyl-	8020	1
			8021	0.1
			8260	10
Sulfide	18496-25-8	Sulfide	9030	4000
2,4,5-T; 2,4,5-trichlorophenoxyacetic acid	93-76-5	Acetic acid (2,4,5-trichlorophenoxy)-	8150	2
1,2,4,5-Tetrachlorobenzene	95-94-3	Benzene, 1,2,4,5-tetrachloro-	8270	10
1,1,1,2-Tetrachloroethane	630-20-6	Ethane, 1,1,1,2-tetrachloro-	8010	5
			8021	0.05
			8260	5
1,1,2,2-Tetrachloroethane	79-34-5	Ethane, 1,1,2,2-tetrachloro-	8010	0.5
			8021	0.1
			8260	5

TABLE 4.2 List of Hazardous Inorganic and Organic Constituents[1] (*Continued*)

Common name[2]	CAS RN[3]	Chemical abstracts service index name[4]	Suggested methods[5]	PQL[6] (μg/L)
Tetrachloroethylene; tetrachloroethene; perchloroethylene	127-18-4	Ethene, tetrachloro-	8010 / 8021 / 8260	0.5 / 0.5 / 5
2,3,4,6-Tetrachlorophenol	58-90-2	Phenol, 2,3,4,6-tetrachloro-	8270	10
Thallium	Total	Thallium	6010 / 7840 / 7841	400 / 1000 / 10
Tin	Total	Tin	6010	40
Toluene	108-88-3	Benzene, methyl-	8020 / 8021 / 8260	2 / 0.1 / 5
o-Toluidine	95-53-4	Benzenamine, 2-methyl-	8270	10
Toxaphene	See Note 10	Toxaphene	8080	2
1,2,4-Trichlorobenzene	120-82-1	Benzene, 1,2,4-trichloro-	8021 / 8120 / 8260 / 8270	0.3 / 0.5 / 10 / 10
1,1,1-Trichloroethane; methylchloroform	71-55-6	Ethane, 1,1,1-trichloro-	8010 / 8021 / 8260	0.3 / 0.3 / 5
1,1,2-Trichloroethane	79-00-5	Ethane, 1,1,2-trichloro-	8010 / 8260	0.2 / 5
Trichloroethylene; trichloroethene	79-01-6	Ethene, trichloro-	8010 / 8021 / 8260	1 / 0.2 / 5
Trichlorofluoromethane; CFC-11	75-69-4	Methane, trichlorofluoro-	8010 / 8021 / 8260	10 / 0.3 / 5
2,4,5-Trichlorophenol	95-95-4	Phenol, 2,4,5-trichloro-	8260 / 8270	5 / 10

Substance	CAS No.	Common name	Method	PQL
2,4,6-Trichlorophenol	88-06-2	Phenol, 2,4,6-trichloro-	8040	5
			8270	10
1,2,3-Trichloropropane	96-18-4	Propane, 1,2,3-trichloro-	8010	10
			8021	5
			8260	15
0,0,0-Triethyl phosphorothioate	126-68-1	Phosphorothioic acid, 0,0,0-triethylester	8270	10
sym-Trinitrobenzene	99-35-4	Benzene, 1,3,5-trinitro-	8270	10
Vanadium	Total	Vanadium	6010	80
			7910	2000
			7911	40
Vinyl acetate	106-05-4	Acetic acid, ethenyl ester	8260	50
Vinyl chloride, chloroethene	75-01-4	Ethene, chloro-	8010	2
			8021	0.4
			8260	10
Xylene (total)	See Note 11	Benzene, dimethyl-	8020	5
			8021	0.2
			8260	5
Zinc	Total	Zinc	6010	20
			7950	50
			7951	0.5

[1]The regulatory requirements pertain only to the list of substances; the right-hand columns (Methods and PQL) are given for informational purposes only. See also footnotes 5 and 6.

[2]Common names are those widely used in government regulations, scientific publications, and commerce; synonyms exist for many chemicals.

[3]Chemical Abstracts Service registry number. Where "Total" is entered, all species in the groundwater that contain this element are included.

[4]CAS index names are those used in the 9th Collective Index.

[5]Suggested methods refer to analytical procedure numbers used in EPA Report SW-846 "Test Methods for Evaluating Solid Waste," 3d ed., November 1986, as revised, December 1987. Analytical details can be found in SW-846 and in documentation on file at the agency. CAUTION: The methods listed are representative SW-846 procedures and may not always be the most suitable method(s) for monitoring an analyte under the regulations.

[6]Practical Quantitation Limits (PQLs) are the lowest concentrations of analytes in groundwaters that can be reliably determined within specified limits of precision and accuracy by the indicated methods under routine laboratory operating conditions. THE PQLs listed are generally stated to one significant figure. PQLs are based on 5 mL samples for volatile organics and 1 L samples for semivolatile organics. CAUTION: The PQL values in many cases are based only on a general estimate for the method and not on a determination for individual compounds; PQLs are not a part of the regulation.

TABLE 4.2 List of Hazardous Inorganic and Organic Constituents[1] (*Continued*)

[7]This substance is often called bis(2-chloroisopropyl) ether, the name Chemical Abstracts Service applies to its noncommercial isomer, propane, 2,2″-oxybis[2-chloro-(CAS RN 39638-32-9).

[8]Chlordane: This entry includes alpha-chlordane (CAS RN 5103-71-9), beta-chlordane (CAS RN 5103-74-2), gamma-chlordane (CAS RN 5566-34-7), and constituents of chlordane (CAS RN 57-74-9 and CAS RN 12789-03-6). PQL shown is for technical chlordane. PQLs of specific isomers are about 20 μg/L by method 8270.

[9]Polychlorinated biphenyls (CAS RN 1336-36-3); this category contains congener chemicals, including constituents of Aroclor 1016 (CAS RN 12674-11-2), Aroclor 1221 (CAS RN 11104-28-2), Aroclor 1232 (CAS RN 11141-16-5), Aroclor 1242 (CAS RN 53469-21-9), Aroclor 1248 (CAS RN 12672-29-6), Aroclor 1254 (CAS RN 11097-69-1), and Aroclor 1260 (CAS RN 11096-82-5). The PQL shown is an average value for PCB congeners.

[10]Toxaphene: This entry includes congener chemicals contained in technical toxaphene (CAS RN 8001-35-2), i.e., chlorinated camphene.

[11]Xylene (total): This entry includes o-xylene (CAS RN 96-47-6), m-xylene (CAS RN 108-38-3), p-xylene (CAS RN 106-42-3), and unspecified xylenes (dimethylbenzenes) (CAS RN 1330-20-7). PQLs for method 8021 are 0.2 for o-xylene and 0.1 for m- or p-xylene. The PQL for m-xylene is 2.0 μg/L by method 8020 or 8260.

- Must continue assessment monitoring if the constituent concentrations are above background values, but are below the groundwater protection standard.
- Must initiate a corrective action program if one or more of constituents are detected at statistically significant levels above the groundwater protection standard.

Corrective Action Program. The corrective action program requires that owners/operators characterize the nature and extent of any release, assess the corrective action measures, select an appropriate corrective action, and implement a remedy. The major components of each of these steps is summarized below:

1. Characterize the nature and extent of a release by installing additional monitoring wells as necessary to characterize the release fully and notify all persons who own the land or reside on the land that directly overlies any part of the plume if contaminants have migrated off-site.

2. Assess appropriate corrective measures within 90 days of finding a statistically significant increase exceeding the groundwater protection standard.

3. Select a remedy that is protective of human health and the environment, attains the groundwater protection standard, controls the source of releases, and complies with RCRA standards for waste management.

4. Implement the selected corrective action remedy and take any interim measures necessary to ensure the protection of human health and the environment.

An approved state can determine that remediation of a release from a MSWLF is not necessary if the owner/operator can demonstrate that:

- Groundwater is additionally contaminated by substances originating from another source and cleanup of the MSWLF releases would not provide a significant reduction in risk to actual or potential receptors caused by such substances.
- The contaminants are present in groundwater that is not currently or reasonably expected to be a source of water, and not hydraulically connected with waters where contaminant migration would exceed the groundwater protection standard.
- Remediation of the release is technically impracticable.
- Remediation will result in unacceptable cross-media impacts.

Closure and Postclosure Care. The final closure and postclosure care requirements impose significant new requirements on landfill owners/operators.

Closure Criteria. The closure criteria require owners/operators to install a final cover system that is designed to minimize infiltration and erosion, prepare a written closure plan and place it in the operating records, notify the state when closure is to occur, and make a notation on the deed to the landfill that landfilling has occurred on the property. The final cover system must comprise an erosion layer underlaid by an infiltration layer meeting the following specifications:

- An erosion layer of a minimum of 6 in of earthen material that is capable of sustaining native plant growth.
- An infiltration layer of a minimum of 18 in of earthen material that has a permeability less than or equal to the permeability of any bottom liner system, natural soils present, or a permeability not greater than 1×10^{-5} cm/s, whichever is less.

An owner/operator is required to begin closure activities of each MSWLF unit not later than 30 days after the date the unit receives its last known final receipt of waste.

However, if the unit has remaining capacity and there is a reasonable likelihood that the unit will receive additional waste, the unit may close no later than 1 year after the most recent receipt of wastes.

In approved states, alternative final cover designs may be permitted if the design provides equivalent infiltration reduction and erosion protection. Also, an extension beyond the 1-year deadline for beginning closure may be granted in approved states if the owner/operator demonstrates that the unit has additional capacity and has taken all necessary steps to prevent threats to human health and the environment.

Postclosure Care Requirements. Following closure of the unit, the owner/operator must conduct postclosure care for 30 years. This must be performed in accordance with a prepared postclosure care plan. The postclosure care requirements include:

- Maintaining the integrity and effectiveness of any final cover systems
- Maintaining and operating the leachate collection system
- Monitoring groundwater and maintaining the groundwater monitoring system
- Maintaining and operating the gas monitoring system

In approved states, an owner/operator may be allowed to stop managing leachate if a demonstration can be made that leachate no longer poses a threat to human health and the environment. Also, an approved state can decrease or increase the postclosure care period as appropriate to protect human health and the environment.

Financial Assurance Criteria. Owners/operators of MSWLFs are required to show financial assurance for closure, postclosure care, and known corrective actions. The requirement applies to *all* owners/operators except state and federal government entities whose debts and liabilities are the debts and liabilities of a state or the United States. The requirements for financial assurance are effective 30 months after promulgation of the rule.

The rule requires that the owner/operator have a detailed written estimate, in current dollars, of the cost of hiring a third party to perform closure, postclosure care, and any known corrective action. The cost estimates must be based on a worst-case analysis (i.e., most costly) and be adjusted annually. The owner/operator must increase or may decrease the amount of financial assurance on the basis of these estimates.

The allowable mechanisms for demonstrating financial assurance for closure, postclosure care, and known corrective actions are

Trust fund

Surety bond guaranteeing payment or performance

Letter of credit

Insurance

Corporate financial test

Local government financial test

Corporate guarantee

Local government guarantee

An owner/operator can satisfy the requirements by establishing one or more of the financial mechanisms listed above. In an approved state, an owner/operator may be able to satisfy the requirements by obtaining other mechanisms that ensure funding and are approved by the state.

Status of State Adequacy Determinations under Part 258. Section 4005(c)(1) of the Resource Conservation and Recovery Act requires states to develop and implement permit programs to ensure that municipal solid waste landfills are in compliance with the revised federal criteria contained in part 258. These permit programs were required to be in place not later than 18 months after the promulgation date of the criteria (i.e., by April 9, 1993). Also, under Sec. 4005(c)(1), EPA is required to determine whether states have an adequate permit program.

Although not mandated under RCRA, EPA has drafted, and will propose and promulgate, a State/Tribal Implementation Rule that contains the procedures by which the agency approves or partially approves a state/tribal MSWLF permit program. The proposal of the STIR has been delayed by the Office of Management and Budget (OMB) because of a dispute over aquifer protection and classification. EPA expected that the STIR would not be promulgated prior to the effective date of Subtitle D on October 9, 1993. However, the agency used the draft STIR to determine adequacy and provide approval to state permit programs prior to the effective date so that states could adopt rules/regulations incorporating the flexibilities allowable in the Subtitle D criteria.

As of September 1993, only six states had received final program determinations of adequacy for their municipal solid waste landfill permit programs and another 18 had received tentative adequacy determinations (Table 4.3). For Virginia and Wisconsin, the determinations provided only partial approval of their programs. The partial approval process allows states to receive approval on those portions of their

TABLE 4.3 Status of State Adequacy Determinations

State of approval	Tentative	Final approval date	Type	State of approval	Tentative	Final approval date	Type
VA	8-18-92	2-3-93	Part	MS	8-12-93	10-7-93	Full
WI	9-25-92	12-29-92	Part	AR	8-25-93	11-9-93	Full
KY	5-4-93	7-1-93	Full	KS	8-25-93	10-7-93	Part
ID	5-10-93	9-21-93	Full	TX	8-25-93	12-17-93	Part
OR	5-18-93	10-7-93	Part	OK	9-16-93	12-28-93	Full
MN	5-28-93	8-16-93	Full	MT	9-23-93	12-21-93	Part
CA	6-29-93	10-7-93	Part	NE	10-5-93	12-17-93	Part
SD	7-20-93	10-8-93	Full	IL	10-20-93	1-3-94	Full
IN	7-23-93	11-8-93	Part	PA	11-4-93		Full
CO	7-26-93	10-1-93	Full	DE	11-15-93	3-4-94	Full
ND	7-29-93	10-5-93	Part	MO	11-19-93		Full
SC	8-3-93	9-16-93	Full	AL	12-17-93	3-2-94	Part
NC	8-5-93	10-7-93	Full	MI	12-22-93	3-10-94	Part
GA	8-5-93	9-21-93	Full	NV	12-30-93	3-7-94	Full
TN	8-5-93	9-16-93	Full	WA	1-13-94		Part
CT	8-6-93	12-15-93	Full	OH	1-14-94		Full
LA	8-9-93	11-4-93	Full	FL	3-2-94		Full
UT	8-12-93	10-8-93	Part	HI	3-7-94		Full
WY	8-12-93	10-8-93	Part	PR	3-23-94		Full

(31) Final approvals—VA(p), WI(p), KY(f), MN(f), SC(f), TN(f), ID(f), OR(p), CA(p), SD(f), CO(f), NC(f), GA(f), UT(p), WY(p), MS(f), KS(p), ND(p), LA(f), IN(p), AR(f), TX(p), NE(p), IL(f), CT(f), OK(f), MT(p), AL(p), DE(f), NV(f), MI(p).

(7) Tentative approvals—PA(f), MO(f), WA(p), OH(f), FL(p), HI(p), PR(f).

regulations that meet or exceed the part 258 criteria and receive approval at a later date on those sections not complying after rule adoption or legislative changes are completed. Through the partial approval process, states can take advantage of the regulatory flexibilities in approved areas prior to receiving full approval.

Currently, EPA plans to propose in the STIR that all partial approvals will expire in October 1995. The expiration of a partial approval would mean that the less flexible, minimum part 258 criteria apply to all landfills in the state, not the state standards that may have taken advantage of the flexibilities.

EPA estimates the average time period for the approval process, i.e., from the date of state application to EPA to date of publication in the *Federal Register* of final determination of adequacy, to be 6 months. In the case of Virginia, the process took slightly more than 8 months and for Wisconsin the process took about 5 months. Table 4.4 provides the details of these states' application process.

According to a recent EPA survey of state regulatory agencies, all but one state is expected to have made an application for full or partial approval before October 9, 1993. If a state does not receive approval prior to October 9, 1993, landfills must be in compliance with the minimum Subtitle D criteria on this date until such date that the state's program becomes approved.

While the states are preparing applications for approval, many landfills are deciding whether or not they should stop receiving wastes prior to Subtitle D's effective date and close their facility. The number of landfills expected to close is not known precisely. However, the number of closures is expected to be high in some states. For example, in a recent *Solid Waste Report* (February 11, 1993), Utah authorities stated that they thought more than 60 percent of the state's landfills would close as a result of Subtitle D implementation. This represents a loss of more than 100 of the state's 164 existing landfills.

On a national level, more than 1600 (31 percent) municipal solid waste landfills are expected to close prior to the effective date of Subtitle D according to a survey performed by Cambridge Environmental Group, Inc. (*Landfill Price Digest*; STATS, May 1992. Table 4.5 provides landfill closures by region of the United States.

From the table, the number of landfills closing ranges from a high of 42 percent in the Atlantic region to a low of 15 percent in the Pacific region. These data probably provide a lower limit to the number of landfills that will close because in some areas of the United States the impact of the part 258 criteria are not well-understood. As the part 258 criteria become effective and states amend their solid waste regulations and apply for Subtitle D approval, landfills will be making decisions on whether to close or upgrade to meet the new requirements. The result of all this activity will almost assuredly be fewer regional landfills that are designed and operated to a higher standard than the average landfill today.

TABLE 4.4   Schedule of Application Process for Virginia and Wisconsin

	Date of action	
Action	Virginia	Wisconsin
State submits application to EPA region	May 29, 1992	July 27, 1992
EPA publishes a tentative determination of adequacy	August 18, 1992	September 25, 1992
EPA publishes a final determination of adequacy	February 3, 1993	December 29, 1992

TABLE 4.5   Number of Landfills Closed as a Result of Subtitle D

Region	No. of landfills	% of total	No. closing
Atlantic	703	42	295
South	1174	27	317
Midwest	1357	33	448
Pacific	553	15	83
Western	1489	34	506
Total	5276	31	1649

Source: Adapted from *Landfill Price Digest*, 1992.

RCRA Reauthorization

The 102d Congress (1991–1992 session) began work on the reauthorization of RCRA but did not complete its work before the session ended. However, both houses' authorizing committees undertook activities that would have resulted in significant change for the waste management industry.

Bills were introduced and discussed that addressed:

- Restrictions on the interstate transportation and disposal of municipal solid waste
- State solid waste management planning
- Municipal solid waste recycling requirements
- Management of batteries and scrap tires
- Industrial nonhazardous waste reporting requirements
- Marketing claims pertaining to the environment

Only Senate bill 2877 cleared the floor, and it contained a narrow set of provisions for restricting the interstate movement of municipal solid waste. The House never called the bill for a floor vote because House members wanted to pass a comprehensive RCRA bill, not pieces of bills.

With the start of the 103d Congress (1993–1994 session), indications are that a comprehensive RCRA reauthorization bill will not pass in 1993 because other environmental issues have priority (e.g., Superfund reauthorization, Clean Water Act reauthorization). However, there appears to be enough support in Congress to break out the interstate provisions of RCRA and pass a stand-alone bill on this issue. Bills authorizing restrictions on the movement of interstate waste were introduced in early 1993 in both the Senate and the House.

4.3   CLEAN AIR ACT

The regulation of gas emissions from solid waste management facilities for all practical purposes was ignored until the late 1970s. However, with growing public concern in the United States over solid waste disposal and widespread nonattainment of national ambient air quality standards, the United States Congress and the United States Environmental Protection Agency have initiated a number of programs that will control gas emissions, both inorganic and organic, at municipal solid waste landfills and combustors. This section summarizes the future regulatory or legislative actions that are pending at the federal level.

Guidelines for Control of Existing Sources and Standards of Performance for New Stationary Sources

The EPA proposed new source performance standards (NSPS) for new MSWLFs under Sec. 111(b) of the CAA and emission guidelines for existing MSWLFs under Sec. 111(d) on May 30, 1991. This action was in response to EPA's findings that MSWLFs can be a major source of air pollution which contributes to ambient ozone problems, airborne toxic gas concerns, global warming, and potential explosion hazards. The regulations are scheduled for promulgation in late 1993.

As part of its regulatory analysis, EPA developed a baseline emission estimate for nonmethane organic compounds (NMOC) and methane using the Scholl Canyon model. In order to make the predictions, EPA needed to estimate the values for the methane generation rate constant k, potential methane generation capacity of the refuse L_0, and NMOC. These values were estimated from information in three publicly available sources.

The concentrations of NMOC reported in the data vary widely from 237 ppm to 14,294 ppm. EPA was not able to develop any apparent correlation between landfill gas composition and site-specific factors. However, they concluded that the highest NMOC concentrations were measured at sites with a known history of codisposing hazardous waste (i.e., prior to EPA restrictions on such activity).

The compounds occurring most frequently in landfill gas included trichlorofluoromethane, trichloroethylene, benzene, vinyl chloride, toluene, and perchloroethylene. Four of these compounds are known or suspected carcinogens. However, the compounds with highest average concentration included toluene, ethylbenzene, propane, methylene chloride, and total xylenes. Only one of these, methylene chloride, is a known or suspected carcinogen, and it is also a possible laboratory contaminant.

EPA's data on uncontrolled air emissions at MSWLFs were limited to seven landfills using active gas collection systems. The NMOC mass emission rates (based on inlet flow measurements) ranged from 43 to 1853 mg/year with no apparent correlation between landfill design and operation parameters.

From its database and other assumptions, EPA estimated that the baseline (1987) emissions from the 7124 existing landfills in the United States was 300,000 Mg/year of NMOC and 15,000,000 Mg/year of methane. These predictions do not include emissions from some 32,000 landfills closed prior to 1987.

Regulatory Approach.   Because of the air emissions outlined above, EPA proposed regulations to control gas emissions from MSWLFs under Sec. 111 of the CAA. The development of the regulations according to EPA will respond to several health and welfare concerns including:

* Contribution of NMOC emissions in the formation of ozone
* Contribution of methane to possible global warming
* Cancer and other potential health effects of individual compounds emitted
* Odor nuisance associated with emissions
* Fire and explosion hazard concerns

Because of all these concerns, EPA is considering the regulation of municipal landfill gas emissions in total, rather than regulation of the individual pollutants or class of pollutants emitted. This approach has several advantages, according to EPA, including control of all air emissions with the same control technology, less expense for the regulated community, and easier enforcement and implementation for the regulatory agencies.

The regulatory approach proposed by EPA is twofold: a maximum landfill design capacity value coupled with a landfill-specific emission rate greater than a designated level. This format requires emission controls to be installed when NMOC emissions exceed a designated mass emission rate. The mass emission rate approach has a number of advantages, according to EPA, including:

- Control of landfills with the greatest emissions
- A high level of cost efficiency in terms of national cost per ton of NMOC emission reduction
- Relative ease to understand, implement, and enforce

EPA will be setting the stringency level for controlling emissions at MSWLFs when the regulations are finalized. However, the maximum design capacity level being considered is 111,000 tons and an emission rate set at 167 tons of nonmethane organics per year. At this level, EPA expects that 621 existing landfills will be affected. The methane and NMOC emission reductions expected are 266 million Mg and 10.6 million Mg respectively.

Under the proposed regulatory format, EPA is considering a tiered approach to determine the site-specific landfill emissions. The tiered approach, according to EPA, would reduce or eliminate unnecessary source testing because costly monitoring would be replaced with a conservative emission estimation technique. Figure 4.1 illustrates the tiered approach that EPA is considering.

In the first tier, an MSWLF owner/operator would compute a landfill emission rate using the Scholl Canyon model with readily available site-specific information and conservative values of k, NMOC concentrations in the landfill gas, and L_0. Because conservative values are used for k, L_0, and NMOC concentrations, the emission rate calculated is likely to be overestimated. Therefore, owners/operators with calculated emission levels greater than the stringency level will be given the option to determine emissions more precisely under Tier 2 or install emission controls.

In the second tier, the same emission estimation techniques as used in Tier 1 are used; however, the owner/operator would determine the site-specific NMOC emission rate. EPA is considering this approach because NMOC emission rates can vary widely from landfill to landfill, and are easier and less expensive to measure than k. As in Tier 1, if the calculated emissions are greater than the emission threshold, the owner/operator will have the option to install emission controls or move on to Tier 3 and collect more site-specific data.

In the final tier, the landfill owner/operator will have to determine the landfill-specific methane generation rate constant k, as well as the NMOC emission rate, and calculate the landfill emission rate. As before, if the calculated emission rates are greater than the stringency levels, the owner/operator will be required to install and operate emission controls.

The emission controls installed at a landfill that exceeds the stringency levels will be required to achieve a 98 percent reduction in collected NMOC emissions. These control devices must be operated until all of the following conditions are met:

- The landfill is no longer accepting waste and is closed permanently.
- The collection and control system has been in continuous operation for a minimum of 15 years.
- The calculated NMOC emission rate has been less than 167 tons/year on three successive test dates that must be no closer together than 3 months but no longer than 6 months apart.

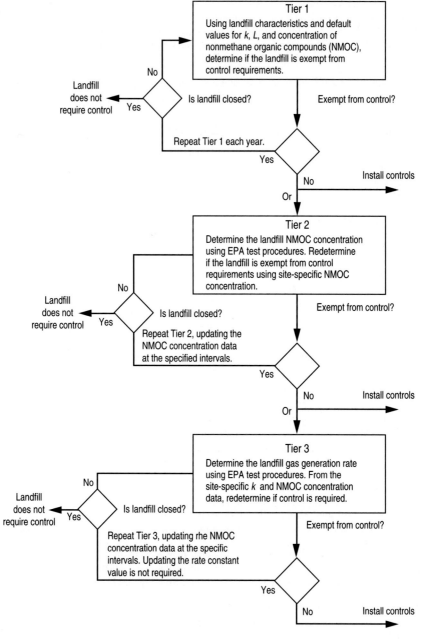

FIGURE 4.1   Overall three-tiered approach for determination of control requirements.

Clean Air Act Amendments of 1990

The Clean Air Act Amendments of 1990 require EPA to promulgate rules regulating air emissions from a variety of sources, including many associated with solid waste management (e.g., air emissions from landfills, transportation of waste). Title I, the General Provisions for Non-attainment Areas, expand the requirements for State Implementation Plans to include pollutants primarily responsible for urban air pollution problems. The primary causes of this air pollution include ozone, carbon monoxide, and particulate matter.

Presently, some 100 urban areas do not meet the minimum ozone health standard and must be classified into one of five categories: marginal, moderate, serious, severe, and extreme. The requirements for revised SIPs could include:

Mandatory transportation controls, e.g., programs for inspection and maintenance of vehicle emission control systems, programs to limit or restrict vehicle use in downtown areas, employer requirements to reduce employee work-trip-related vehicle emissions

Specified reductions in ozone levels, e.g., controlling volatile organic compounds (VOC) emissions from stationary sources

Offsets and/or reductions of existing sources prior to allowing new or modified sources in an area

Regulation of small sources as "major" sources in certain categories (i.e., for the extreme category a major source could be defined as one emitting 10 tons of VOCs per year)

Improved emission inventories

Also, the CAA Amendments of 1990 require the implementation of reasonably achievable control technology (RACT) at all major sources.

These new requirements could affect the ability of an industry (e.g., landfill) to operate or expand in certain areas and may require extensive control systems to be installed for the control of gas emissions (e.g., VOCs). Additionally, the requirements could restrict the access of collection vehicles in nonattainment areas to certain times of the day or certain roadways. The development of the CAA requirements under Title I could dramatically affect solid waste management operations as the states revise SIPs and reduce and control emissions in nonattainment areas.

Title III (Air Toxics) of the Clean Air Act Amendments of 1990 also affects the management of solid waste. Under this title, requirements are set forth for hazardous air pollutants, air emissions from municipal waste combustion, and ash management and disposal.

Under Title III, the CAA amendments set forth a new regulatory program for hazardous air pollutants from major stationary and area sources and establish a list of 189 regulated pollutants (Table 4.6). Major stationary sources are those operations that emit:

- 10 tons/year of any single hazardous air pollutant
- 25 tons/year of any combination of hazardous air pollutants

Area sources are considered smaller sources that had not been previously controlled under the act.

On June 21, 1991, EPA proposed a preliminary draft list of categories of major and area sources of hazardous air pollutants by industry (those sources considered small sources that had not been previously controlled under the act). Waste treatment and disposal was listed as an industry group, and the categories listed under the group include:

TABLE 4.6 List of Hazardous Air Pollutants

CAS number	Chemical name
75070	Acetaldehyde
60355	Acetamide
75058	Acetonitrile
98862	Acetophenone
53963	2-Acetylaminofluorene
107028	Acrolein
79061	Acrylamide
79107	Acrylic acid
107131	Acrylonitrile
107051	Allyl chloride
92671	4-Aminobiphenyl
62533	Aniline
90040	o-Anisidine
1332214	Asbestos
71432	Benzene (including benzene from gasoline)
92875	Benzidine
98077	Benzotrichloride
100447	Benzyl Chloride
92524	Biphenyl
117817	Bis(2-ethylhexyl)phthalate (DEHP)
542881	Bix(chloromethyl)ether
75252	Bromoform
106990	1,3-Butadiene
156627	Calcium cyanamide
105602	Caprolactam
133062	Captan
63252	Carbaryl
75150	Carbon disulfide
56235	Carbon tetrachloride
463581	Carbonyl sulfide
120809	Catechol
133904	Chloramben
57749	Chlordane
7782505	Chlorine
79118	Chloroacetic acid
532274	2-Chloroacetophenone
108907	Chlorobenzene
510156	Chlorobenzilate
67663	Chloroform
107302	Chloromethyl methyl ether
126998	Chloroprene
1319773	Cresols/cresylic acid (isomers and mixture)
95487	o-Cresol
108394	m-Cresol
106445	p-Cresol
98828	Cumene
94757	2,4-D, salts and esters
3547044	DDE
334883	Diazomethane
132649	Dibenzofurans
96128	1,2-Dibromo-3-chloropane

TABLE 4.6 List of Hazardous Air Pollutants (*Continued*)

CAS number	Chemical name
84742	Dibutylphthalate
106467	1,4-Dichlorobenzene(p)
91941	3,3-Dichlorobenzidene
111444	Dichloroethyl ether (bis(2-chloroethyl)ether)
542756	1,3-Dichloropropene
62737	Dichlorvos
111422	Diethanolamine
121697	N,N-Diethyl aniline (N,N-dimethylaniline)
64675	Diethyl sulfate
119904	3,3-Dimethoxybenzidine
60117	Dimethyl aminoazobenzene
119937	3,3'-Dimethyl benzidine
79447	Dimethyl carbamoyl chloride
68122	Dimethyl formamide
57147	1,1-Dimethyl hydrazine
131113	Dimethyl phthalate
77781	Dimethyl sulfate
534521	4,6-Dinitro-o-cresol, and salts
51285	2,4-Dinitrophenol
121142	2,4-Dinitrotoluene
123911	1,4-Dioxane (1,4-Diethyleneoxide)
122667	1,2-Diphenylhydrazine
106898	Epichlorohydrin (1-chloro-2,3-epoxypropane)
106887	1,2-Epoxybutane
140885	Ethyl acrylate
100414	Ethyl benzene
51796	Ethyl carbamate (urethane)
75003	Ethyl chloride (chloroethane)
106934	Ethylene dibromide (dibromoethane)
107062	Ethylene dichloride (1,2-dichloroethane)
107211	Ethylene glycol
151564	Ethylene imine (aziridine)
75218	Ethylene oxide
96457	Ethylene thiourea
75343	Ethylidene dichloride (1,1-dichloroethane)
50000	Formaldehyde
76448	Heptachlor
118741	Hexachlorobenzene
87683	Hexachlorobutadiene
77474	Hexachlorocyclopentadiene
67721	Hexachloroethane
822060	Hexamethylene-1,6-diisocyanate
680319	Hexamethylphosphoramide
110543	Hexane
302012	Hydrazine
7647010	Hydrochloric acid
7664393	Hydrogen fluoride (hydrofluoric acid)
123319	Hydroquinone
78591	Isophorone
58899	Lindane (all isomers)
108316	Maleic anhydride

TABLE 4.6   List of Hazardous Air Pollutants (*Continued*)

CAS number	Chemical name
67561	Methanol
72435	Methoxychlor
74839	Methyl bromide (bromomethane)
74873	Methyl chloride (chloromethane)
71556	Methyl chloroform (1,1,1-trichloroethane)
78933	Methyl ethyl ketone (2-butanone)
60344	Methyl hydrazine
74884	Methyl iodide (iodomethane)
108101	Methyl isobutyl ketone (hexone)
624839	Methyl isocyanate
80626	Methyl methacrylate
1634044	Methyl tert butyl ether
101144	4,4'-Methylene bis(2-chloroaniline)
75092	Methylene chloride (dichloromethane)
101688	Methylene diphenyl diisocyanate (MDI)
101779	4,4-Methylenedianiline
91203	Naphathalene
98953	Nitrobenzene
92933	4-Nitrobiphenyl
100027	4-Nitrophenol
79469	2-Nitropropane
684935	N-Nitroso-N-methylurea
62759	N-Nitrosodimethylamine
59892	N-Nitrosomorpholine
56382	Parathion
82688	Pentachloronitrobenzene (quintobenzene)
87865	Pentachlorophenol
108952	Phenol
106503	p-Phenylenediamine
75445	Phosgene
7803512	Phosphine
7723140	Phosphorus
85449	Phthalic anhydride
1336363	Polychlorinated biphenyls (aroclors)
1120714	1,3-Propane sultone
57578	beta-Propiolactone
123386	Propionaldehyde
114261	Propoxur (Baygon)
78875	Propylene dichloride (1,2-dichloropropane)
75569	Propylene oxide
75558	1,2-Propylenimine (2-methyl aziridine)
91225	Quinoline
106514	Quinone
100425	Styrene
96093	Styrene oxide
1746016	2,3,7,8-Tetrachlorodibenzo-p-dioxin
79345	1,1,2,2-Tetrachloroethane
127184	Tetrachloroethylene (perchloroethylene)
7550450	Titanium tetrachloride
108883	Toluene
95807	2,4-Toluene diamine

TABLE 4.6 List of Hazardous Air Pollutants (*Continued*)

CAS number	Chemical name
584849	2,4-Toluene diisocyanate
95534	o-Toluidine
8001352	Toxaphene (chlorinated camphene)
120821	1,2,4-Trichlorobenzene
79005	1,1,2-Trichloroethane
79016	Trichloroethylene
95954	2,4,5-Trichlorophenol
88062	2,4,6-Trichlorophenol
121448	Triethylamine
1582098	Trifluralin
540841	2,2,4-Trimethylpentane
108054	Vinyl acetate
593602	Vinyl bromide
75014	Vinyl chloride
75354	Vinylidene chloride (1,1-dichloroethylene)
1330207	Xylenes (isomers and mixture)
97576	p-Xylenes
108383	m-Xylenes
106423	p-Xylenes
0	Antimony compounds
0	Arsenic compounds (inorganic including arsine)
0	Beryllium compounds
0	Cadmium compounds
0	Chromium compounds
0	Cobalt compounds
0	Coke oven emissions
0	Cyanide compounds[1]
0	Glycol ethers[2]
0	Lead compounds
0	Manganese compounds
0	Mercury compounds
0	Fine mineral fibers[3]
0	Nickel compounds
0	Polycyclic organic matter[4]
0	Radionuclides (including radon)[5]
0	Selenium compounds

Note: For all listings above which contain the word *compounds* and for glycol ethers, the following applies: unless otherwise specified, these listings are defined as including any unique chemical substance that contains the named chemical (i.e., antimony, arsenic, etc.) as part of that chemical's infrastructure.

[1]'X'CN where X = H' or any other group where a formal dissociation may occur. For example KCN or Cal(CN)$_2$.

[2]Includes mono- and diethers of ethylene glycol, diethylene glycol, and triethylene glycol R-(OCH2CH2)$_n$-OR' where

    n = 1, 2, or 3

    R = alkyl or aryl groups

    R' = R, H, or groups which, when removed, yield glycol ethers with the structure: R-(OCH2CH)$_n$-OH. Polymers are excluded from the glycol category.

[3]Includes mineral fiber emissions from facilities manufacturing or processing glass, rock, or slag fibers (or other mineral-derived fibers) of average diameter 1 μm or less.

[4]Includes organic compounds with more than one benzene ring, and which have a boiling point greater than or equal to 100°C.

[5]A type of atom which spontaneously undergoes radioactive decay.

Solid waste disposal—open burning

Sewage sludge incineration

Municipal landfills

Groundwater cleaning

Hazardous waste incineration

Tire burning

Tire pyrolysis

Cooling water chlorination for steam electric generators

Wastewater treatment systems

Water treatment purification

Water treatment for boilers

For these listed sources, the CAA Amendments require EPA to establish a maximum achievable control technology (MACT) for each category. MACT standards can differ, depending on whether the source is an existing source or a new source:

- *New sources.* MACT must be established as the degree of emission reduction that is achievable by the best controlled similar source and may be more stringent where feasible
- *Existing sources.* MACT can be less stringent than standards for new sources; however, it must be at least as stringent as:

    The average emission limitations achieved by the best-performing 12 percent of the existing sources where there are 30 or more sources

    The average emission limitation achieved by the best-performing three sources where there are fewer than 30 sources

On September 24, 1992, EPA proposed a schedule for the promulgation of emission standards for categories and subcategories of hazardous air pollutants:

Hazardous waste incineration	November 15, 2000
Municipal landfills	November 15, 1997
POTW emissions	November 15, 1995
Sewage sludge incineration	November 15, 1997
Site remediation	November 15, 2000
TSDFs	November 15, 1994

The new standards, once promulgated, should result in emission reductions of approximately 80 to 90 percent below current levels.

Section 129 of the act required EPA to revise the new source performance standards and emission guidelines (EG) for new and existing municipal waste combustion (MWC) facilities. The revised NSPS will eventually apply to all solid waste combustion facilities according to category:

- Solid waste incineration units with capacity greater than 250 tons/day
- Solid waste incineration units with capacity equal to or less than 250 tons/day or combusting hospital waste, medical waste, or infectious waste
- Solid waste incineration units combusting commercial or industrial waste
- All other categories of solid waste incineration

The revised standards require the greatest degree of emission reduction achievable through the application of best available control technologies and procedures that have been achieved in practice, or are contained in a state or local regulation or permit by a solid waste incineration unit in the same category, whichever is more stringent. The performance standards established are required to specify opacity and numerical emission limitations for the following substances or mixtures:

Particulate matter (total and fine)

Sulfur dioxide

Hydrogen chloride

Oxides of nitrogen

Carbon monoxide

Lead

Cadmium

Mercury

Dioxins and dibenzofurans

The EPA is allowed to promulgate numerical emissions limitations, or provide for the monitoring of postcombustion concentrations of surrogate substances, parameters, or periods of residence times.

For existing facilities, EPA is required to promulgate guidelines that include:

- Emission limitations (as defined above)
- Monitoring of emissions and incineration and pollution control technology performance
- Source separation, recycling, and ash management requirements
- Operator training requirements

EPA sent a draft of the regulations to the Office of Management and Budget for review in July 1993. The regulations will probably be finalized in 1994.

Despite the requirements above, Title V does not restrict states from incorporating their own standards or emission limitations prior to the time a permit is issued. States may also use permits to impose new standards and emission limitations. None of the provisions of the Clean Air Act may have as great an impact on the waste management industry as the general operating permit provisions of Title V.

Title V establishes a program for issuing operating permits to all major sources (and certain other sources) of air pollutants in the United States. These permits will collect in one place all applicable requirements, limitations, and conditions governing regulated air emissions. Whereas in the past, air regulations governed specific air emission sources, beginning in November 1993, the law required states and localities to regulate emissions from all major stationary sources that directly emit or have the potential to emit 100 tons or more of any pollutant, 10 tons or more of a single hazardous air pollutant, or 25 tons or more of two or more hazardous air pollutants.

The applicability of these provisions is to major sources which are variously defined in Secs. 112 and 302 and Part D of Title I of the act. The generally accepted definition is one having "the potential to emit." Such sources will be defined in the same way EPA has defined major sources under the Prevention of Significant Deterioration of Air Quality (PSD) and nonattainment New Source Review (NSR) permit programs. The term *potential to emit* means:

...the maximum capacity of a stationary source to emit any air pollutant under its physical and operational design. Any physical or operational limitation on the capacity of a source to emit an air pollutant, including air pollution control equipment and restrictions on hours of operation or on the type or amount of material combusted, stored, or processed, shall be treated as part of its design if the limitation is enforceable by the Administrator.

The Title V permit program must satisfy certain federal standards (40 CFR) Part 70, 57 *Fed. Reg.* 32249, July 21, 1992), and will be administered by state and local air pollution control authorities. Under the terms of Title V, state and local authorities are required to submit their own operating permit programs to EPA for review and approval by November 15, 1993. If such authorities fail to submit and implement an approvable permit program by this date, Title V directs EPA to impose severe sanctions on states, including the withdrawal of federal highway assistance funds (80 percent of state highway budgets comes from federal highway assistance funds), and the imposition of a minimum 2-to-1 offset ratio for emissions from new or modified sources in certain nonattainment areas. In addition, Title V directs EPA to establish and administer a federal permit program where state and local programs are deemed to be inadequate. States and localities will have 1 year after the submittal of their programs to EPA to issue permits.

The immediate impact of Title V is that any source fitting the above description must apply for an operating permit. Facilities that are the least bit uncertain about their status should conduct a site-wide air emission inventory for those substances listed in §111 and §112 of the act. For landfills, this means those substances identified in §111(b) and (d) and identified in EPA's New Source Performance Standards. Municipal solid waste (MSW) combustors, recycling centers, material recovery facilities (MRFs), transfer stations, hazardous waste depots, and treatment and disposal facilities which emit substances listed under §112 are also likely to be subject to this title and to various sections under the act whether or not they are currently subject to specific regulations.

For those facilities that are required to obtain an operating permit, Sec. 70.3(c)(1) states that a permit for a major source must contain all applicable requirements for each of the source's regulated emission units. Therefore, if a source is listed as a major source for a single pollutant—say methane—all other emissions from the site are subject to regulation under the permit. Section 504(a) requires the permit to include all applicable implementation plans (e.g., state, tribal, or federal implementation plans), and, where applicable, monitoring, compliance plans and reports, and information that is necessary to allow states to calculate permit fees.

Included in the implementation plans are National Ambient Air Quality Standards (NAAQS) which deal with non-emission-related control strategies, such as collection access limitations, roadway access limitations, odd-even day operation requirements, etc. Thus, as states move forward to implement the CAA requirements under Title I (nonattainment), solid waste management operations could be severely impacted.

Additionally, all permits are judicially reviewable in state court and are to be made available for public review. Each state's program must include civil and criminal enforcement provisions, including fines for unauthorized emissions. Fines are to begin at a rate of $10,000 per day per violation. A failure to submit a "timely and complete" permit application is subject to civil penalties. A complete application is one that includes information "sufficient to evaluate the subject source and its application, and to determine all applicable requirements." Because EPA did not adopt a standard application form, the requirements of each state in which a facility is located will have to be fulfilled in order to satisfy this provision. Section 70.5(c) of the act lists the minimum requirements that are to be included in permit applications, including:

- Company information (e.g., name, address, phone numbers including those for emergencies, nature of business)
- A plant description (e.g., size, throughput, special characteristics)
- Emission sources and emission rates, as well as emission control equipment
- A description of applicable Clean Air Act requirements and test methods for determining compliance
- Information necessary to allow states to calculate permit fees, which are anticipated to be $25 per ton of actual emissions of regulated pollutants

Permits must include requirements for emission limitations and standards, monitoring, record keeping, reporting, and inspection and entry to assure compliance with applicable emission limitations. Section 70.6(a)(3) states that, where periodic emissions monitoring is not required by applicable emission standards, the permit itself must provide for "periodic monitoring sufficient to yield reliable data for the relevant time period that are representative of the source's compliance with the permit." However, EPA's rules indicate that in some cases record keeping may be sufficient to satisfy this requirement. Additionally, monitoring reports are to be submitted at least every 6 months and must be maintained for at least 5 years. Sources deemed to be out of compliance with any applicable provision of the act must also submit semiannual progress reports. Finally, the permit must contain a certificate of compliance which must be signed by a responsible official.

Last, Sec. 608 of the CAA required EPA to develop a regulatory program to reduce chlorofluorocarbon (CFC) and hydrochlorofluorocarbon (HCFC) emissions from all refrigeration and air-conditioning sources to the "lowest achievable level." Also, this section prohibits individuals from knowingly releasing ozone-depleting compounds into the atmosphere. Penalties for violating this prohibition on venting CFCs and HCFCs can be assessed up to $25,000 per day per violation by EPA. These fines can be levied against any person in the waste management process if a refrigeration unit's charge is not intact and the possessor cannot verify that the refrigerants were removed in accordance with these regulations. Haulers, recyclers, and landfill owners can face significant penalties for noncompliance with these rules.

On May 14, 1993, EPA promulgated final regulations that established:

- Restrictions on the sale of refrigerants to only certified technicians
- Service practices for the maintenance and repair of refrigerant-containing equipment
- Certification requirements for service technicians and equipment and reclaimers
- Disposal requirements to ensure that refrigerants were removed from equipment prior to disposal

The final regulations establish safe disposal requirements to ensure the recovery of the refrigerants from equipment that is disposed with an intact charge; however, they do not require that the recovery take place at any specific point along the disposal route. The recovery of the refrigerant can be done at the place of use prior to disposal (e.g., a consumer's home), at an intermediate processing facility, or at the final disposal site. In order to ensure that the refrigerant is properly removed, the regulation requires the "final processor" (e.g., landfills) to (1) verify that the refrigerant has been removed or (2) remove the refrigerant themselves prior to disposal. EPA final rules do not establish any type of specific markings to be placed on equipment that has had its refrigerant recovered or removed, although they are recommended. However, the regulations require some form of verification. This verification may include:

A signed statement with the name and address of the person delivering the equipment and the date the refrigerant was removed

The establishment of contracts for removal with suppliers such as those presently used for PCB removal

Regardless of who supplies the service, the service provider must recover at least 90 percent of the refrigerant in the unit and must register the removal equipment with the appropriate EPA regional office.

Global Warming

Global warming, its cause, effects, and prevention, is one of the major environmental concerns of the decade and will likely be an issue readdressed by the 103d Congress. One of the sources commonly listed and recommended for control is methane emissions from MSWLFs.

There is growing consensus in the scientific community that changes and increases in the atmospheric concentrations of the "greenhouse" gases (carbon dioxide, methane, chlorofluorocarbons, nitrous oxide, and others) will alter the global climate by increasing world temperatures. The atmospheric greenhouse gases naturally absorb heat radiated from the earth's surface and emit part of the energy as heat back toward the earth, warming the climate. Increased concentrations of these gases on a global basis can intensify the greenhouse effect.

The specific rate and magnitude of future changes to the global climate caused by human activities is hard to predict. However, EPA predicts that if nothing is done, global temperatures may increase as much as 10°C by the year 2100. Reportedly, global warming of just a few degrees would present an enormous change in climate. For example, the difference in mean annual temperature between Boston and Washington, D.C. is only 3.3°C and the difference between Chicago and Atlanta is 6.7°C. The total global warming since the peak of the last ice age (18,000 years ago) was only about 5°C, a change that shifted the Atlantic Ocean inland by about 100 mi, created the Great Lakes, and changed the composition of forests throughout the continent.

Many human activities contribute to the greenhouse gases currently accumulating in the atmosphere. The most important gas is carbon dioxide (CO_2), followed by methane (CH_4), chlorofluorocarbons, and nitrous oxide (N_2O). The breakdown of greenhouse gases as of the 1980s is as follows:

CO_2	49%
CH_4	18%
N_2O	6%
CFCs	14%
Other	13%

Carbon dioxide is a primary by-product of burning fossil fuels such as coal, oil, and gas, and is also released as a result of deforestation. The largest source of methane is decaying organic matter in the absence of oxygen. CFCs are predominantly produced by the chemical industry. Nitrous oxide sources are not well-characterized but are assumed to be related to soil processes such as nitrogenous fertilizer use.

The sources of methane emissions to the atmosphere can be broken into six broad categories: natural resources, rice production, domestic animals, fossil fuel production, biomass burning, and landfills. The annual estimated methane emissions (in million tons) by source are as follows:

Natural sources	130–382	(32%)
Rice production	67–191	(20%)
Domestic animals	73–112	(15%)
Fossil fuel production	56–107	(15%)
Biomass burning	56–112	(10%)
Landfills	34–79	(8%)

The largest source is naturally occurring and is derived from the decomposition of organics in environments such as swamps and bogs. Rice production contributions result from the anaerobic decomposition of organics that occur when rice fields are flooded. The top three rice-producing countries are India, China, and Bangladesh.

Domestic animals, such as beef and dairy cattle, produce methane as a by-product of enteric fermentation, a digestive process by which grasses are broken down by microorganisms in the animal's stomach. The top three countries in domestic animal utilization are India, the former Soviet Union, and Brazil. Methane releases through fossil fuel production are primarily related to the mining of coal. The United States, the former Soviet Union, and China are the largest coal-producing countries.

Biomass burning results in the production of methane through the burning related to deforestation and shifting cultivation, burning agricultural waste, and fuelwood use.

The smallest source, landfills, generates methane through the decomposition of organic refuse. EPA predicts that landfilling is not expected to increase very much in countries such as the United States in the future, but can be expected to increase dramatically in developing countries.

The total contribution to the global warming problem that is directly attributed to MSWLFs is less than 2 percent (i.e., 18 percent attributable to methane $\times$ 8 percent of the methane attributable to MSWLFs = 1.4 percent of the total greenhouse gases attributable to MSWLFs). The actual amount of methane attributable to landfills in the United States is unknown. However, one researcher has estimated it to be between 14 to 32 million tons per year. Based on these estimates, the contribution to global warming by U.S. landfills is about 0.6 percent.

The contribution of landfill-generated methane to the overall greenhouse gases is relatively small, but landfills are one of the few sources that potentially can be controlled. The question remains as to whether control of such a small source of emissions is economically justifiable. The answer to this question will likely be the discussion and debate of many hearings in Congress as they address the issue of global warming and how it should be controlled.

4.4 CLEAN WATER ACT

Until recently, the Clean Water Act only required waste management facilities to obtain permits for their point-source discharges under the National Pollution Discharge Elimination System. These regulations have been extensively described in older texts, and readers should refer to them for greater detail.

On November 16, 1990, the EPA promulgated regulations requiring an NPDES permit for storm water discharges associated with industrial activities. The EPA defined storm water discharge as a discharge from any conveyance which is used for collecting and conveying storm water. The word *conveyance* has a very broad meaning and includes almost any natural or man-made depression that carries storm water runoff, snow melt runoff, and surface runoff and drainage (i.e., not process wastewater). These conveyances are required to be permitted and must achieve CWA 301 best available technology/best control technology (BAT/BCT) and water-quality-based limitations.

The types of waste management activities covered by the regulations include:

- Transportation facilities [Standard Industrial Classifications (SIC) 40, 41, 42 (except 4221–25), 43, 44, 45, and 5171)] which have vehicle maintenance shops, equipment cleaning operations, and refueling and lubrication operations. These classifications cover most haulers/transporters of solid and hazardous waste.

- Material recycling facilities classified under SIC 5015 and 5093.

- Landfills, land application sites, and open dumps that receive any industrial wastes, where industrial wastes are defined very broadly.

- Steam electric power-generating facilities such as waste-to-energy facilities.

- Hazardous waste treatment, storage, and disposal facilities.

The only exempted facilities are those that hold an NPDES permit that incorporates storm water runoff, have no storm water runoff that is carried through a conveyance (i.e., all sheet flow), or discharge all runoff to a sewage treatment facility.

The regulations allowed existing regulated industrial activities to apply for permits through one of three methods:

1. An individual permit that would be specific to that facility and the permit conditions would be based on water-quality test results from that facility. Individual permits had to be received by EPA by October 9, 1993.

2. A general permit that EPA would develop based on existing information for a general class of facilities. These permit conditions were promulgated in the *Federal Register* (vol. 57, no. 175, p. 41236) on September 9, 1992 by EPA. Applicants seeking coverage under the general permits had to submit a notice of intent within 180 days from rule promulgation.

3. Group permits that would include a grouping of a number of similar facilities. Groups were required to sample storm water runoff from a fraction of the facilities to obtain unbiased results (e.g., 100 facilities may only have to sample 10 sites). EPA would base the permit conditions for the whole group on the limited sampling. To date EPA has not finished the group applications.

New facilities will presumably be able to take advantage of any of the three application methodologies since they can elect to apply as an individual, under one of the general permits developed by EPA for a general class of facilities, or under one of the group permits developed by EPA for a specific group of facilities.

Because the NPDES program is a federal permit program that states can seek delegation under, many states have adopted different programs for application or modified the federal program. Table 4.7 lists the states that presently have delegated authority for all or part of the NPDES permit program. Facilities seeking storm water discharge permits should check with a delegated state to ascertain the availability of permits.

4.5 FEDERAL AVIATION ADMINISTRATION GUIDELINES

Federal Aviation Administration (FAA) directive 5200.5A establishes guidance concerning the siting, construction, and operation of municipal solid waste facilities (i.e.,

TABLE 4.7 NPDES General Permitting Authorities

| | NPDES states | | non-NPDES states* |
State	With general permitting authority	Without general permitting authority	
Alabama	X		
Alaska			Region 10
Arizona			Region 9
Arkansas	X		
California	X		
Colorado	X		
Connecticut	X		
Delaware	X		
District of Columbia			Region 3
Florida			Region 4
Georgia	X		
Hawaii	X		
Idaho			Region 10
Illinois	X		
Indiana	X		
Iowa	X		
Kansas		X	
Kentucky	X		
Louisiana			Region 6
Maine			Region 1
Maryland	X		
Massachusetts			Region 1
Michigan		X	
Minnesota	X		
Mississippi	X		
Missouri	X		
Montana	X		
Nebraska	X		
Nevada	X		
New Hampshire			Region 1
New Jersey	X		
New Mexico			Region 6
New York	X		
North Carolina	X		
North Dakota	X		
Ohio	X		
Oklahoma			Region 6
Oregon	X		
Pennsylvania	X		
Rhode Island	X		
South Carolina	X		
South Dakota			Region 8
Tennessee	X		
Texas			Region 6
Utah	X		
Vermont	X		
Virginia	X		
Washington	X		

TABLE 4.7   NPDES General Permitting Authorities (*Continued*)

State	NPDES states With general permitting authority	Without general permitting authority	Non-NPDES states*
West Virginia	X		
Wisconsin	X		
Wyoming	X		
American Samoa			Region 9
Guam			Region 9
Northern Marihara Islands			Region 9
Puerto Rico			Region 2
Virgin Islands		X	

*Permitting in non-NPDES states is done by the EPA regional office indicated.

landfills, recycling facilities, transfer stations) on or in the vicinity of FAA-regulated airports. The directive states that these types of facilities are considered incompatible land-use activities near airports and provides for a zone of notification to be added so that FAA has the opportunity to comment on such land uses.

The FAA directive basically bans the operation of municipal solid waste management facilities, particularly landfills:

- Within 10,000 ft of any runway used by turbine powered aircraft (e.g., jets)
- Within 5000 ft of any runway used by piston-powered aircraft
- Within 5 mi of a runway that attracts or sustains hazardous bird movements (i.e., movements that bring birds in contact with aircraft)

Because the FAA's directive has not proceeded through the federal rule-making process, it is not a regulation. Rather it is guidance and not enforceable except at facilities that receive FAA funds (airports). However, many states have codified the requirements into their regulatory programs and, as mentioned previously, the Subtitle D criteria require a notification to FAA if a landfill is expanded or built within 5 mi of a public-use airport.

The FAA is presently in the process of updating the directive so that the provisions are not as onerous on municipal solid waste management facilities that do not attract birds (e.g., construction and demolition fills). Individual states should be contacted when a facility is being proposed to determine any siting restrictions that may have resulted from the existing or modified directive.

CHAPTER 5
SOLID WASTE STATE LEGISLATION

Jim Glenn

*Project Manager, Gershman, Brickner and Bratton
and Consulting Editor,
BioCycle—Journal of Waste Management*

Since the late 1970s, how municipalities deal with solid waste has changed dramatically. U.S. EPA estimated that in 1980 about 81 percent of all municipal solid waste (MSW) was landfilled. Ten years later that figure was 67 percent. Both recycling and incineration are becoming more important in MSW management. In 1990, about 17 percent of the nation's 195.7 million tons of MSW was recycled and 16 percent was incinerated. While the root cause of this shift may be found in people's concern about landfilling, it is state legislation that has been directly responsible. By the end of 1992, 36 states had waste reduction goals. Additionally, 27 states required municipalities to develop recycling programs and 40 had banned some waste materials from landfills and/or incinerators.

5.1 INTRODUCTION

Since the 1960s, the level of sophistication in solid waste management laws has grown. Requirements on where disposal and processing facilities can be located, how they must be constructed, and how they are to operate are becoming increasingly stringent. Beyond the restrictions that were placed on the actual facilities, state solid waste management laws also began to require that municipalities and counties start planning for the proper disposal of their solid waste.

Following the lead set by the 1976 Federal Resource Conservation and Recovery Act (RCRA), states began to see municipal solid waste in a broader context. Rather than just viewing what is being thrown away as waste to be disposed of, it is increasingly recognized that it contains a considerable amount of resources that can be utilized if they are put back into the economy in some fashion. In the late 1970s and early 1980s, the principal direction this took was waste-to-energy projects. However, the passage of Oregon's 1983 "Opportunity to Recycle" legislation ushered in the era of waste reduction legislation that focuses on recycling, composting, and source reduction.

5.2   TRENDS IN MUNICIPAL WASTE GENERATION AND MANAGEMENT

According to the U.S. Environmental Protection Agency (U.S. EPA), 195.7 million tons of municipal solid waste (MSW) was generated in this country in 1990. That figure represents a 122 percent increase over the amount of MSW the U.S. EPA estimates was generated in 1960. By the year 2000, U.S. EPA estimates that the amount generated will increase to 222 million tons.[1]

In terms of per capita generation, in 1960, the rate was 2.7 lb per person per day of MSW. By 1990, the rate had increased to 4.3 lb per person per day. By 2000, the U.S. EPA expects the rate to be 4.5 lb per person per day, better than double the 1960 rate.[1]

Another study of the amount of waste generated throughout the United States based on information provided by solid waste officials in all the states and the District of Columbia typically exhibits MSW figures larger than the U.S. EPA estimates.[2,4–8,10] There are several reasons for the difference in MSW generation rates in these two studies. The U.S. EPA definition for MSW "includes wastes such as durable goods, nondurable goods, containers and packaging, food scraps, yard trimmings, and miscellaneous inorganic wastes from residential, commercial, institutional, and industrial sources." It does not include wastes from other sources, such as construction and demolition wastes, municipal sludges, combustion ash, and industrial process wastes that might also be disposed of in municipal waste landfills or incinerators. However, the definition used by some states to characterize MSW varies from that definition and includes such things as construction and demolition waste and municipal sewage sludge. Another reason is that several states base the rate on disposal facility records, which receive non-MSW wastes. According to the figures, the amount of MSW generated in the United States was approximately 250,000,000 tons in 1988.[2] By 1992, the amount had risen to 291,742,000 tons (see Table 5.1).

By either measure, there is a significant amount of MSW which must be managed. Landfilling is still the way most MSW is managed. According to U.S. EPA figures, in 1960, approximately 62 percent of all MSW was landfilled. That percentage increased until 1980, when it reached roughly 81 percent, and has declined to 67 percent in 1990.[1]

The figures generated by the *BioCycle* surveys also show a steady decline in the reliance on landfills. The 1989 survey showed that approximately 85 percent of the MSW waste landfilled in 1988.[2] By the end of 1992, that number had declined to 72 percent.[10]

Stimulation provided by state waste reduction legislation in recent years has been causing the landfilling rate to decline and increased the use of alternatives such as recycling, composting, and incineration. In 1989, the sum of recycling and yard waste composting rate was approximately 7 percent and the incineration rate was 8 percent.[2] By the end of 1992, the recycling and composting rate was estimated to total 17 percent, while the incineration rate had climbed to 11 percent.[10]

5.3   THE WASTE REDUCTION LEGISLATION MOVEMENT

One of the factors in the push toward waste reduction legislation was the perception that landfill space is dwindling. While such an assertion is debatable, it is clear that the number of landfills in this country continues to decline. At the end of 1988, there

TABLE 5.1 Waste Generation and Disposal and Methods of Disposal (by State)

State	Solid waste, tons/year	Recycled, %	Incinerated, %	Landfilled, %
Alabama	5,200,000	12	8	80
Alaska	500,000	6	15	79
Arizona*	4,147,000	7	0	93
Arkansas*	2,154,000	10	5	85
California†	44,535,000	11	2	87
Colorado	3,500,000	26	1	73
Connecticut	2,900,000	19	57	24
Delaware*	790,000	16	19	65
District of Columbia	919,000	30	59	11
Florida‡	19,400,000	27	23	50
Georgia	6,000,000	12	3	85
Hawaii*	1,300,000	4	42	54
Idaho	850,000	10	0	90
Illinois†	14,140,000	11	2	87
Indiana§	8,400,000	8	17	75
Iowa	2,088,000	23	2	75
Kansas	2,400,000	5	0	95
Kentucky	4,650,000	15	0	85
Louisiana	3,484,000	10	0	90
Maine	1,246,000	30	37	23
Maryland	5,000,000	15	17	68
Massachusetts	6,600,000	30	47	23
Michigan§	13,000,000	26	17	57
Minnesota	4,270,000	38	35	27
Mississippi	1,400,000	8	3	89
Missouri	7,500,000	13	0	87
Montana	744,000	5	2	93
Nebraska	1,400,000	10	0	90
Nevada	2,300,000	10	0	90
New Hampshire§	1,138,000	10	26	64
New Jersey	7,513,000	34	21	45
New Mexico	1,487,000	6	0	94
New York‡	22,800,000	21	17	62
North Carolina†	7,788,000	4	1	95
North Dakota	466,000	17	0	83
Ohio†	16,400,000	19	6	75
Oklahoma	3,000,000	10	8	82
Oregon*	3,350,000	23	6	71
Pennsylvania	8,984,000	11	30	59
Rhode Island	1,200,000	15	0	85
South Carolina*	5,000,000	10	5	85
South Dakota*,¶	800,000	10	0	90
Tennessee†	5,800,000	10	8	82
Texas	14,469,000	11	1	88
Utah	1,500,000	13	7	80
Vermont	550,000	25	3	72
Virginia	7,600,000	24	18	58
Washington	5,708,000	33	2	65
West Virginia	1,700,000	10	0	90

TABLE 5.1   Waste Generation and Disposal and Methods of Disposal by State) *(Continued)*

State	Solid waste, tons/year	Recycled, %	Incinerated, %	Landfilled, %
Wisconsin	3,352,000	24	4	72
Wyoming*	320,000	4	0	96
Total	291,742,000	17	11	72

*Includes some industrial waste.
†Includes significant industrial waste.
‡Includes C & D waste.
§Includes C & D waste and sewage sludge.
¶Data from 1992 "The State of Garbage in America" survey.
Source: Adapted from Robert Steuteville and Nora Goldstein, "The State of Garbage in America," *BioCycle*, May 1993.

were at least 7924 landfills operating in this country.[2] By the end of 1992, that figure had dropped to 5386 (see Table 5.2).

Although the absolute number of landfills is decreasing steadily, it does not necessarily follow that there is a comparable decline in disposal capacity. In Pennsylvania, for example, at the end of 1988, the state had 75 landfills which could accept municipal solid waste, while by the end of 1992, there were only 46. But while the number of landfills declined, the disposal capacity in the state rose from something less than 5 years to approximately 15 years.[10]

At the end of 1991, approximately 12.5 years of landfill capacity remained in the country. Regionally, the southern states appear to be the worst off, with an average of only 5 years of remaining capacity, followed by the Great Lakes states and the New England states, which have an average of less than 10 years of capacity. The mid-Atlantic states have more than 12 years of remaining capacity. The western half of the country appears to be in the best shape, with more than 20 years of capacity remaining.[8]

Another factor stimulating waste reduction legislation has been the escalation of disposal prices. Although there was some easing of prices in 1991 and 1992, which may be caused by the general economic recession, with new requirements imposed by U.S. EPA, costs are expected to rise generally in the next several years. The average tipping fee was $26.50 per ton in 1991. The highest tipping fees in the country were in the New England states, where they averaged $46.83 per ton, and the mid-Atlantic states, where they were on average $45.86 per ton. The lowest average fees are in the Rocky Mountain states ($8.71 per ton) and the middle west ($12.76 per ton).[8]

The highest state average in the country is in New Jersey, where landfill costs average $74 per ton. Connecticut, Massachusetts, Minnesota, New Hampshire, and New York all have average tipping fees at or above $50 per ton. While it does not always hold true, the states most likely to pursue waste reduction have been those with high tipping fees.[8]

With the exception of Oregon's 1983 legislation, the states that initially developed waste reduction laws, such as Florida, New Jersey, New York, Pennsylvania, and Rhode Island, were ones which had relatively high disposal fees. In most cases, these same states also had limited amounts of disposal capacity. For instance, Pennsylvania and Rhode Island both had an average of less than 5 years of remaining capacity, while New York's was less than 10 years.

Because of both disposal cost and limited capacity, the first wave of waste reduction legislation tended to be concentrated in the mid-Atlantic and New England states. But since then the pattern has not been so clear-cut. As illustrated above, most of the

TABLE 5.2   Disposal Capacity, Number of Facilities, and Tipping Fees (by State)

	Landfills			Incinerators		
State	Number	Average tipping fee, $	Remaining capacity years	Number	Average tipping fee, $	Daily capacity, tons/day
Alabama	92	20	9	2	30	1,050
Alaska	740	40	n/a	3	n/a	200
Arizona	92	22	n/a	0		
Arkansas	120	16	6	5	n/a	380
California	322	n/a	16	3	n/a	2,600
Colorado	125	10	n/a	1	n/a	n/a
Connecticut	40	65	n/a	7	74	5,033
Delaware	3	n/a	17	1	56	600
District of Columbia*						
Florida	140	n/a	n/a	14	n/a	17,000
Georgia	181	30	5	1	n/a	533
Hawaii	13	35	12	1	n/a	2,000
Idaho	79	n/a	n/a	0		
Illinois	106	n/a	n/a	1	n/a	618
Indiana	65	21	5	2	18	2,650
Iowa	82	n/a	24	1	n/a	125
Kansas	115	12	n/a	0		
Kentucky	30	25	4	0		
Louisiana	29	15	n/a	0		
Maine	185†	25	n/a	4	45	1,967
Maryland	30	43	10	4	49	4,000
Massachusetts	122	65	3	9	65	8,767
Michigan	67	n/a	n/a	7	n/a	n/a
Minnesota	45	50	10	8	84	1,600
Mississippi	75	15	n/a	1	n/a	150
Missouri	77	20	9	0		
Montana	87	15	20+	1	12	70
Nebraska	36	n/a	n/a	0		
Nevada	90	10	n/a	0		
New Hampshire	46	50	n/a	15	n/a	n/a
New Jersey	12	74	n/a	4	93	4,323
New Mexico	113	4	51	0		
New York	101	62	5	17	75	14,000
North Carolina	145	27	n/a	2	47	600
North Dakota	43	18	n/a	0		
Ohio	96	26	8	11	n/a	2,560
Oklahoma	110	12	12–15	3	n/a	1,150
Oregon	86	40	24+	2	63	547
Pennsylvania	46	48	15	6	n/a	7,300
Rhode Island	5	24	20	0		
South Carolina	59	17	11	2	25	820
South Dakota‡	100	8	10	0		
Tennessee	105	20	n/a	5	45	1,300
Texas	390	13	20	4	n/a	n/a
Utah	162	25	10+	1	50	300
Vermont	25	n/a	n/a	2	60	n/a

TABLE 5.2   Disposal Capacity, Number of Facilities, and Tipping Fees (by State) *(Continued)*

	Landfills			Incinerators		
State	Number	Average tipping fee, $	Remaining capacity years	Number	Average tipping fee, $	Daily capacity, tons/day
Virginia	254	25	n/a	10	35	7,000
Washington	45	47	n/a	4	n/a	n/a
West Virginia	40	4	n/a	0		
Wisconsin	138	30	10	5	n/a	1,000
Wyoming	77	10	n/a	0		
Total	5,386			169		

n/a—figure not available.

*Municipal solid waste from the District of Columbia is handled by a landfill and incinerator outside its boundaries.

†Source: Waste Dynamics of New England.

‡Data taken from 1992 SOG survey.

Source: Robert Steuteville and Nora Goldstein, "The State of Garbage in America," *BioCycle,* May 1993.

southern states don't have high tipping fees at present but are facing a lack of disposal capacity. In the middle west and the Rocky Mountain states, the tipping fees are not high, nor is there a lack of capacity. However, while most of the middle western states have passed some form of legislation, most of the Rocky Mountain states have yet to pass comprehensive laws.

5.4   THE EFFECT OF LEGISLATION

The passage of legislation is just the first step in developing a statewide solid waste management strategy. The evidence that progress is being made lies in the amount of material that goes to alternative uses. At the end of 1992, 20 states landfilled less than 75 percent of their waste (see Table 5.1). Five of those, Connecticut, Maine, Massachusetts, Minnesota, and New Jersey, along with the District of Columbia, land-filled less than 50 percent. That is a considerable change since 1988 when only six states (Connecticut, Delaware, Florida, Maine, Massachusetts, and Minnesota) staked that claim.[2]

There were 11 states, in 1992, which estimated their waste reduction rate to be greater that 25 percent. The District of Columbia, Maine, Massachusetts, Minnesota, New Jersey, and Washington all estimated that their rates are 30 percent or higher. Another eight states exceeded the average nationwide recycling rate of 17 percent (see Table 5.1). In 1988, no state had a waste reduction rate of 25 percent or higher; the highest was Washington, with 22 percent.[2] Eight states had incineration rates of 25 percent or greater in 1992. In 1988, only half of those had rates above 25 percent.[10]

Another measure of the effect of legislation is the number of projects that have been developed over the last several years. While it is difficult to gauge how many waste reduction projects have become part of the country's solid waste management system, some indicators are relatively easy to track.

One way to chart the trend in recycling is to keep track of the most visible of the various collection techniques, curbside recycling. In 1981, there were fewer than 300 known curbside recycling projects in all the United States.[2] By the end of 1988, there were an estimated 1042 curbside programs collecting recyclables in the United States. By the end of 1992, that figure had increased almost 420 percent to 5404.[6] Fourteen states had 100 or more programs in operation (see Table 5.3). In 1988, only three states (New Jersey, Pennsylvania, and Oregon) had 100 or more functioning programs.[2] Back then New Jersey, which had just initiated its mandatory recycling legislation, had 439 curbside recycling programs—better than three times the next closest state.[2] By the end of 1992, Pennsylvania, with 709 programs, topped the list. Rounding out the top five are Minnesota (571), New Jersey (515), Wisconsin (450), and California (446).[10]

By the end of 1992, curbside recycling programs served more than 77 million people in the United States.[10] Fifteen states have programs that serve in excess of 1 million people (see Table 5.3). In 1988, only four states (California, New Jersey, Oregon, and Pennsylvania) had programs serving at least that many people.[2] California estimated that in 1992, 15,200,000 of its residents have access to curbside service. That top figure is followed by New York (14,100,000), Pennsylvania (7,500,000), New Jersey (7,200,000), and Florida (6,500,000) (see Table 5.3). The common thread among the programs is that all these states are in the process of implementing waste reduction legislation.

By the end of 1992, 2981 facilities were composting some part of the yard waste stream in the United States. That figure is almost 360 percent higher than the 651 sites known to be operating in 1988.[10] Eight states have at least 100 sites composting yard waste. Minnesota has 397 sites, followed by Pennsylvania (300), New Jersey (270), Massachusetts (265), Wisconsin (213), Michigan (200), New York (200), and Indiana (128) (see Table 5.4). All these states, with the exception of New York, have passed legislation or regulations banning yard waste from disposal facilities.

Beyond sheer numbers, the most dramatic change in yard waste composting over the past several years has been in the number of facilities that are composting brush and grass as well as leaves. In 1988, there weren't more than a handful of sites which handled any significant amount of grass and brush. By the end of 1991, at least 750 sites were composting a full range of yard waste.[10] That's 35 percent of all sites operating that year. This growth has largely been in states such as Illinois and Minnesota, where all yard waste is banned from other disposal options.

5.5 STATE MUNICIPAL SOLID WASTE LEGISLATION

The approach to municipal solid waste management legislation varies significantly from state to state. For some states, such as Minnesota and Illinois, passage of MSW-related legislation is an annual affair. In fact, in some cases, several laws are passed each year. For instance, in 1991, 94 solid waste related bills were introduced in the Illinois legislature, at least 15 of those bills, including ones that dealt with procurement of recycled products, household hazardous waste collection, and establishment of a tire recycling fund, became law.[9]

In other cases, states work on solid waste legislation periodically, revisiting the issue every 5 to 10 years. As an example, Pennsylvania passed omnibus solid waste legislation in 1968. Twelve years later, that legislation was updated, with particular emphasis

TABLE 5.3 Curbside Recycling Programs (by State)

State	Number	Population served	Mandatory	Voluntary
Alabama	25	587,600	1	24
Alaska	0			
Arizona	15	n/a	0	15
Arkansas	12	209,000	1	11
California	446	15,200,000	n/a	n/a
Colorado	65	1,000,000	0	65
Connecticut	169	3,056,000	169	0
Delaware	*			
District of Columbia	1	130,000	1	0
Florida	215	6,500,000	n/a	n/a
Georgia	80	n/a	5	75
Hawaii	0			
Idaho	n/a			
Illinois	260	2,750,000	0	260
Indiana	54	875,600	n/a	n/a
Iowa	240	n/a	n/a	n/a
Kansas	10	n/a	0	10
Kentucky	40	n/a	n/a	n/a
Louisiana	21	283,605	n/a	n/a
Maine	18	235,000	6	12
Maryland†	40	2,405,000	0	40
Massachusetts	109	n/a	n/a	n/a
Michigan‡	200	n/a	n/a	n/a
Minnesota	571	3,051,246	n/a	n/a
Mississippi	15	n/a	n/a	n/a
Missouri	40	n/a	n/a	n/a
Montana	3	n/a	0	3
Nebraska§	4	400,000	0	4
Nevada	7	250,000	0	7
New Hampshire	25	244,445	7	18
New Jersey	515	7,200,000	515	0
New Mexico	7	250,000	0	7
New York	250	14,167,210	n/a	n/a
North Carolina	95	n/a	n/a	n/a
North Dakota	10	80,000	3	7
Ohio	144	1,534,243	n/a	n/a
Oklahoma	14	200,000	0	14
Oregon	118	2,150,000	0	118
Pennsylvania	709	7,466.923	488	221
Rhode Island	20	734,000	n/a	n/a
South Carolina	27	750,000	1	6
South Dakota	1	n/a	0	1
Tennessee	29	n/a	n/a	n/a
Texas	90	1,600,000	0	90
Utah	13	750,000	0	13
Vermont	16	n/a	0	16
Virginia	62	433,035	4	58
Washington¶	98	2,000,000	0	98
West Virginia	50	360,000	15	35

TABLE 5.3 Curbside Recycling Programs (by State) *(Continued)*

State	Number	Population served	Mandatory	Voluntary
Wisconsin	450	1,000,000	n/a	n/a
Wyoming	1	480	0	1
Total	5,404	77,603,387	1216	1249

n/a—figure not available.
*Delaware is served by a statewide network of drop-off sites.
†Source: Maryland Municipal League.
‡Source: "County Collection Information," Michigan Departments of Commerce and Natural Resources.
§Data from the 1992 SOG survey.
¶Source: 1992 Solid Waste Survey, Association of Washington Cities.
Source: Robert Steuteville and Nora Goldstein, "The State of Garbage in America," *BioCycle,* May 1993.

TABLE 5.4 Yard Waste Composting Programs (by State)

State	Number	State	Number
Alabama	12	Missouri	50
Alaska	0	Montana	9
Arizona	2	Nebraska	15
Arkansas	17	Nevada	1
California	26	New Hampshire	78
Colorado	5	New Jersey	270
Connecticut	84	New Mexico	1
Delaware	2	New York	200
District of		North Carolina	75
Columbia	1	North Dakota	5
Florida	20	Ohio	78
Georgia	88	Oklahoma	2
Hawaii	5	Oregon	20
Idaho*	6	Pennsylvania	300
Illinois	96	Rhode Island	16
Indiana	128	South Carolina	25
Iowa	30	South Dakota*	3
Kansas	30	Tennessee	4
Kentucky	26	Texas	75
Louisiana	13	Utah	1
Maine	22	Vermont	12
Maryland	8	Virginia	19
Massachusetts	265	Washington	15
Michigan	200	West Virginia	n/a
Minnesota	397	Wisconsin*	213
Mississippi	8	Wyoming	3
		Total	2981

n/a—figure not available.
*From the 1992 SOG survey.
Source: Robert Steuteville and Nora Goldstein, "The State of Garbage in America," *BioCycle,* May 1993.

on hazardous waste management, with Act 97 of 1980. In 1988, the state legislature passed Act 101, the Municipal Waste Planning and Waste Reduction Act, which, among other things, mandated some municipalities to establish recycling programs.

5.6 STATE PLANNING PROVISIONS

MSW management planning provisions in legislation generally direct that planning be conducted on two levels. Numerous laws, including those in Alabama, Minnesota, Montana, and Washington, direct that a state agency develop a state solid waste management plan. Beyond this statewide planning, which often serves as a guide for local governments, laws also dictate that local governments or counties develop solid waste management plans on a periodic basis.

A state planning requirement that is representative of most of those passed in the late 1980s and early 1990s is contained in New Mexico's Solid Waste Act of 1990.[9] That law required that a comprehensive and integrated solid waste management plan for the state be developed by the Dec. 31, 1991. The plan had to rank management techniques, placing source reduction and recycling first, environmentally safe transformation second, and landfilling third. As laid out in the law, the basis for developing the plan was information provided by each county and municipality.

The plan was required to establish a goal of diverting 50 percent of all solid waste from disposal facilities by July 1, 2000. Other elements of the plan include waste characterization, source reduction, recycling and composting, facility capacity, education and public information, funding, special waste, and siting.

The content of a state solid waste management plan often dictates what type of solid waste management planning activities are undertaken by local and county governments. For instance, in South Carolina, each local solid waste management plan has to be designed to achieve the recycling and waste reduction goals established in the state plan. This type of "top-down" planning is not always the case. In North Dakota, local plans are being used to formulate the comprehensive statewide solid waste management plan.

Like state plans, recent local planning requirements established by legislation follow the same general pattern. Plan contents typically include a description of the current solid waste management situation, both physical and institutional, and how adequate processing and disposal capacity will be made available over a 10- to 20-year period. Most plans now are required to have a waste reduction element that is aimed at achieving the state's reduction goal.

Most states place the primary responsibility for planning on counties, although there are other approaches. For instance, Connecticut requires all its municipalities to submit 20-year plans. Nevada also requires municipalities to plan. In Alabama, counties are given the planning responsibility unless the local municipality chooses to retain it. In several states, including North Dakota, Ohio, and Vermont, laws require that separate solid waste management districts be formed to plan and implement solid waste programs.

5.7 PERMITTING AND REGULATION REQUIREMENTS

The permitting and regulations provisions of state laws vary significantly from state to state. At the most fundamental, laws simply direct that a state agency develop a means of permitting and regulating municipal waste management activities. In other cases, the law establishes a regulatory framework. At the most extreme, lawmakers will actually write into law the requirements that facilities must meet. One case of the latter approach occurred in Illinois, where, when the legislature amended the Solid Waste Management Act in 1988, it put into law requirements on how yard waste composting facilities were to be sited and operated. This strategy necessitates that any adjustments to the requirements be formulated, debated, and passed by the legislature. For instance, in 1991, it was the legislature that had to increase yard waste composting facility setback provisions from 200 to 660 ft from the nearest residence, it couldn't be done by the Illinois Environmental Protection Agency, the state regulatory agency.

The breadth of legislative direction in permitting and regulation can be illustrated by looking at laws passed, respectively, in North Dakota and South Carolina. H.B. 1060, passed in North Dakota during 1991, directs that the Department of Health and Consolidated Laboratories, "adopt and enforce rules governing solid waste management." Additionally, it is to adopt rules to establish standards and requirements for various categories of solid waste management facilities, establish financial assurance requirements, and conduct an environmental compliance background review of any permit applicant.

South Carolina's Solid Waste Policy and Management Act of 1991 goes into much more detail on how solid waste facilities are to be permitted and regulated. It spells out how the permitting process is to take place and what minimum requirements must be met by different types of facilities, including landfills, incinerators, processing facilities, and land application facilities.

While it's beyond the scope of this chapter to detail the solid waste regulations in individual states, information on those rules can be obtained from the state solid waste programs. The appendix at the end of this chapter provides a listing of the appropriate state agencies.

5.8 WASTE REDUCTION LEGISLATION

While most changes in solid waste management laws over the last 20 to 30 years can be described as evolutionary, waste reduction provisions are better characterized as revolutionary. In 1980, the most far-reaching waste reduction initiatives being implemented were mandatory deposit legislation on certain beverage containers. Ten years later, probably every legislature in the country that was at least giving serious consideration to bills targeting 25 to 50 percent of the municipal waste stream for reduction.

The approaches states have taken are not uniform. Some states, like Pennsylvania and New Jersey, have opted for laws that ultimately require waste generators to participate. Others, like Arizona and Washington, require only that recycling programs be made available to citizens, while other methods, like those used in Florida and Iowa, require local governments to reach a certain goal.[9]

The strategies do not stop with the various ways of developing waste reduction programs. Most states have banned outright the disposal of some materials such as yard waste, oil, and white goods. They are also taxing hard to dispose of items such as tires, and increasingly are requiring manufacturers and retailers to take responsibility for their products.

5.9 ESTABLISHING WASTE REDUCTION GOALS

Perhaps the most fundamental provision in any solid waste legislation as it relates to waste reduction is the establishment of a statewide goal. By the end of 1992, 36 states

TABLE 5.5 Statewide Solid Waste Management Goals

State	Source Reduction, %	Recycling,[a] %	Composting, %	Other	Mandated, %	Deadline
Alabama		.........25.........			Yes	1991
Arkansas		.........40.........			Yes	2000
California		.........50[b].........			Yes	2000
Connecticut	[c]	25			Yes	1991
Delaware		21		50[e]	No	2000
District of Columbia		45			Yes	1994
Florida		....30.........			Yes	1995
Georgia		.........25.........			Yes	1996
Hawaii		.........50.........			Yes	2000
Illinois		25			Yes	2000
Indiana		.........50.........			Yes	2000
Iowa		.........50.........			Yes	2000
Kentucky		.........25.........			Yes	1997
Louisiana		.........25.........			Yes	1992
Maine		.........50.........			No	1994
Maryland		.........20[d].........			Yes	1994
Massachusetts	10	31	15	48[e]	Yes	2000
Michigan	8–12	20–30	8–12	35–45[e] 4–6[f]	No	2005
Minnesota		[g]			Yes	1996
Mississippi		.........25.........			Yes	1996
Missouri		.........40.........			Yes	1998
Montana		.........25.........			No	1996
Nebraska		.........25.........			Yes	n/a
Nevada		.........25.........			Yes	1994
New Hampshire		.........40.........			Yes	2000
New Jersey		.........60.........			Yes	1995
New Mexico		.........50.........			Yes	2000
New York		.........60.........			No[i]	2000
North Carolina		.........25.........			Yes	1993
North Dakota		.........40.........			Yes	2000
Ohio		.........25.........			Yes	1994
Oregon		.........50.........			Yes	2000
Pennsylvania	[j]	25			Yes	1997
Rhode Island[k]		.........70.........			Yes	
South Carolina		.........30.........			Yes	1997
South Dakota		.........50.........			No	2005
Tennessee		.........25.........			Yes	1996
Texas		.........40.........			Yes	1994
Vermont		.........40.........			No[i]	2000
Virginia		25			Yes	1995
Washington		.........50.........			Yes	1995
West Virginia		.........50.........			Yes	2010

[a]Includes yard waste composting.

[b]May include 10 percent waste transformation.

[c]Goal is no change in waste generation rate.

[d]15 percent goal for counties under 100,000; 20 percent goal for counties over 100,000.

[e]Incineration.

[f]Reuse.

[g]45 percent goal in the seven-county, Twin Cities area; 30 percent in greater Minnesota.

[h]Does not include leaf composting as part of the goal. In 1990, a solid waste management task force recommended a 60 percent recycling goal, although this goal is currently not mandated by law.

TABLE 5.5   Statewide Solid Waste Management Goals *(Continued)*

[i]Goal was developed pursuant to the state Solid Waste Management Plan.

[j]Goal is to reduce the amount of waste generated.

[k]Rhode Island's ultimate goal is to recycle as much as possible.

Source: Adapted from Robert Steuteville, Nora Goldstein, and Kurt Grotz, "The State of Garbage in America, Part II," *BioCycle,* June 1993.

had put some type of waste reduction goals on the books (see Table 5.5). Four years earlier, only eight states (Connecticut, Florida, Maine, Maryland, New Jersey, New York, Pennsylvania, and Rhode Island) had such goals.[3]

When states first started to establish goals, they were primarily recycling goals. New Jersey, one of the first states to put a goal in legislation back in 1987, didn't even allow the composting of leaves to count for part of the 25 percent. While most goals aren't quite so restrictive, some such as those in Florida and North Carolina, allow no more than half of the goal to be met by yard waste composting. No more than 40 percent of South Carolina's 25 percent recycling goal can be met by yard waste, land clearing debris, and construction and demolition debris.[9]

Over the last several years, the focus of goals has changed from strictly recycling and yard waste composting to overall waste reduction, which may also include other forms of composting and source reduction. While South Carolina has a 25 percent recycling goal, its overall goal is to reduce by 30 percent the amount of solid waste received at MSW landfills and incinerators. In West Virginia, the year 2010 goal is to reduce the disposal of MSW by 50 percent.[9]

As the focus of the goals has expanded, so has the amount of waste expected to be diverted. Recycling goals, as in Connecticut, Illinois, Maryland, and Rhode Island, typically ranged from 15 to 25 percent. The more general waste reduction laws have rates from 25 to 70 percent. States at the lower end of the scale include Alabama, Louisiana, and Ohio. The high end includes New Jersey and Rhode Island. It should be noted that most states that have lower goals have placed deadlines for achieving them sooner than those states with higher goals. The deadlines for states with 25 percent goals range from 1991 to 1996. Deadlines for 50 and 70 percent goals usually stretch from the year 2000 and beyond. New Jersey appears to have the most ambitious goal—70 percent by 1995 (see Table 5.5).

Although a majority of states have established some form of waste reduction goal, for some it is just that, a goal. For instance, in West Virginia there is no tie between the goal established in the law and the development of waste reduction programs at the local level. Any county in South Carolina which meets the state goal is to be rewarded for that effort by sharing in a special bonus grant program. For others, the goal becomes part of requirements which local governments must reach. In Tennessee, which has a 25 percent waste reduction goal, each municipal solid waste region must meet that goal or ultimately face fines of up to $5000 per day.

5.10   LEGISLATING LOCAL GOVERNMENT RESPONSIBILITY

For the better part of the last decade, waste reduction legislation has centered on requiring municipalities to develop recycling programs. In general, three approaches have been utilized. One type of law mandates that municipalities require generators to separate recyclables, and in some cases compostables, from waste and have them collected and further processed. The most prevalent form of legislation mandating municipal involvement is that which requires local governments to reach specified goals. The third model is legislation which requires local governments to provide some form of waste reduction system to citizens.

Mandatory Recycling Laws

As of the end of 1992, so-called mandatory recycling laws are employed by six states (see Table 5.6). In four of those states (Connecticut, New Jersey, New York, and Rhode Island) every municipality must pass an ordinance requiring municipal waste generators to recycle certain materials. For instance, in New Jersey, municipalities must collect a minimum of three recyclables. Connecticut and Rhode Island both require collection of a more extensive list of recyclables including, among other things, newspaper, glass containers, metal cans, and some plastic bottles. Connecticut also includes leaves in its list of recyclables which must be collected.

The two remaining states, Pennsylvania and West Virginia, limit the number of municipalities that are required to pass ordinances. In Pennsylvania, initially, only those municipalities with a population of 10,000 or greater needed to comply. As of September 1991, those municipalities with a population between 5000 and 10,000 and a population density of 300 or more people per square mile were also required to comply. West Virginia limits its mandate to those municipalities with a population of 10,000 or greater. In both cases, waste generators have to recycle at least three recyclables.

In addition to requiring municipalities to pass ordinances, these laws compel them to establish recycling programs which meet certain criteria. For instance, West Virginia municipalities must establish curbside programs which collect at least on a monthly basis. The program must also include a comprehensive public information and education element.

"Opportunity to Recycle" Laws

The prototype of recycling legislation in this country was Oregon's "Opportunity to Recycle" Act. In this type of legislative scheme, municipalities are required to provide

TABLE 5.6   States with Legislation Requiring Municipalities to Pass Mandatory Ordinances

State	Municipalities involved	Deadline	Act ID
Connecticut	All	1/91	PA 90-220
District of Columbia	n/a	4/90*	7-226—1989
New Jersey	All	8/88	P.L. 1987, C.107
New York	All	9/92	Chap. 70—1988
Pennsylvania	Population of 5000 or greater†	9/26/91	101—1988
Rhode Island	All	‡	23-18—1986
West Virginia	Population of 10,000 or greater	10/93	S.B. 18—1991

*The 4/90 deadline was for residents only. The deadline for recycling at commercial establishments was 10/90.

†All municipalities with 10,000 and above must pass mandatory ordinances. For municipalities with populations between 5000 and 10,000, only those that have a population density of 300 people per square mile must pass ordinances.

‡Deadline based on the implementation schedule for each municipality.

Source: Jim Glenn, "The State of Garbage in America, Part II," *BioCycle,* May 1992.

some form of recycling program, but the municipalities are not required to pass ordinances requiring participation by waste generators.

In Oregon's first recycling law, curbside recycling programs (which collected at least monthly) had to be put in place in every municipality with a population of 4000 or more. Additionally, every MSW disposal facility had to provide a drop-off program. In 1991, the Oregon legislature saw fit to modify this earlier legislation. Some of the improvements in this update include requirements for weekly collection of recyclables, distribution of home storage containers for recycling, and an expanded education and promotion program.

Besides Oregon, 12 other states have passed legislation requiring some forms of local government to develop recycling programs (see Table 5.7). In six of those states (Alabama, Arkansas, Maryland, Minnesota, Nevada, and South Carolina), counties are charged with the responsibility. Nevada's law applies only to counties with populations of more than 25,000. Three states (Arizona, California, and Washington) targeted both cities and counties. In addition to Oregon, only Virginia's legislation puts the requirement at the municipal level. In Vermont, solid waste management districts, which are generally groups of municipalities, are responsible. And in North Carolina, it's "designated local governments," of which 90 are counties and 15 are municipalities.

Oregon's law to the contrary, most "Opportunity to Recycle" legislation was constructed so that local governments could establish programs that were right for them. In Nevada, counties (with populations greater than 40,000) are required to make available a program for "the separation at the source of recyclable material from other solid waste originating from the residential premises where services for collection of solid

TABLE 5.7   State Legislation Requiring Local Government Units to Develop Recycling Programs

State	Local government units involved	Deadline	Act ID
Alabama	Counties[a]	5/92	824—1989
Arizona	Cities and counties	Not set	H.B. 2574—1990
Arkansas	Counties	7/92	H.B. 1447—1991
California	Cities and counties	[b]	A.B. 939—1989
Maryland	Counties	1/94	H.B. 714—1988
Minnesota	Counties	10/90	115A—1989
Nevada	Counties over 40,000	Not set	A.B. 320—1991
North Carolina	[c]	7/91	S.B. 111—1989
Oregon[d]	Municipality	7/92	S.B. 66—1991
South Carolina	Counties	Not set	H.B. 388—1991
Vermont	SW management districts	Not set	78—1987
Virginia	Municipalities	[e]	1743—1989
Washington	Cities and counties	[f]	E.S.H.B. 1671—1989

[a]Local municipalities can develop programs on their own if they choose.

[b]No deadline for establishing programs, but each city and county must reach a 25 percent waste diversion goal by 1995.

[c]Designated local government, of which 90 are counties and 15 are municipalities.

[d]Oregon's original legislation (S.B. 405) was effective July 1, 1986.

[e]No deadline for establishing programs, but each municipality must reach an interim recycling goal of 10 percent for the 1991 calendar year.

[f]7/91 for Spokane, Snohomish, King, Pierce, and Kipsap counties, 7/92 for all other counties west of the Cascade Mountains, 7/94 for all counties east of the Cascade Mountains.

Source: Jim Glenn, "The State of Garbage in America, Part II," *BioCycle,* May 1992.

waste are provided." Additionally, those counties are required to establish a recycling center, if none are available. Counties with between 25,000 and 40,000 people only have to make sure a recycling center is available. In South Carolina, the legislation states only that counties may include curbside collection, drop-offs, or multifamily systems in their recycling programs.

Arkansas' statute defines the opportunity to recycle as the "availability of curb side pick-up or collections centers for recyclable materials at sites that are convenient for persons to use." It's up to the county or regional solid waste board to determine the number and type of facilities needed and what type of recyclables are to be collected. But it does require the board to develop a public education program and establish a yard waste composting program.

Required Goals

Beyond the fact that most legislatures generally feel that it isn't prudent to dictate what type of recycling program will work best in a particular locality, one reason most states did not require a certain type of program be established is that they also specified in the law a goal which the program must reach. In fact, 10 of the 13 states that have "Opportunity to Recycle" laws also require local governments to reach certain waste reduction goals. These include Alabama, California, Maryland, Minnesota, Nevada, North Carolina, Oregon, Vermont and Virginia. Additionally, Connecticut, New Jersey, Rhode Island, and South Carolina, which have mandatory recycling laws, also have goal requirements. Of the twenty-one states that put goal requirements on local governments, only eight (Florida, Georgia, Hawaii, Illinois, Iowa, Louisiana, Ohio, and Tennessee) do not combine them with some other form (see Table 5.8).

In most cases, the goal local government is to reach is identical to the state goal established in the law. The major exception to that is in Oregon, where different groups of counties have different goals to meet, ranging from 15 to 45 percent, by 1995. The goal each county is to meet by the year 2000 is to be determined after 1995. Other legislation has allowances for local governments that cannot meet a goal, either providing it additional time to comply, as is the case in Tennessee, or allowing the state agency that oversees the program to reduce or modify the goal if circumstances warrant.

What happens if a local government does not meet the goal varies widely. In some cases, it is not clear if anything will occur. In Oregon, if a recovery rate is not achieved, the municipality must take steps to upgrade its program. In Tennessee, fines can ultimately be levied for not complying with the law.

5.11 DISPOSAL BANS

There is probably no more direct approach to waste reduction than banning specific types of waste from disposal facilities. For MSW, bans were first ushered in back in 1984 when Minnesota passed legislation banning the disposal of tires in landfills.[3] Since then 42 additional states have passed bans on one or more waste materials (see Table 5.9).

Over the years, lawmakers have focused particularly on materials coming from vehicles, such as batteries, tires, and oil. Oregon goes so far as to ban the disposal of discarded vehicles. The most popular product to ban is vehicle batteries. By the end of

1992, 34 states had passed such restrictions. Twenty-five states have banned the disposal of, at least, whole tires and 15 ban motor oil. Of those states that ban the disposal of tires at least three (Minnesota, Vermont, and Wisconsin) ban any form of tire from being landfilled. Another, Ohio, has banned the disposal of tires in MSW landfills, but will permit them to be buried in tire "monofills."

The ban which can have the greatest effect on reducing the amount of waste being disposed of is a ban on yard waste. In all, 23 states have put yard waste bans on the books. For 21 of those states, the ban is on all types of yard waste. In the remaining two, it applies only to a portion of the yard waste stream. In New Jersey (which in 1987, was the first state to pass a yard waste ban) and Pennsylvania the prohibition against disposal does not include grass. There are instances where the bans are not

TABLE 5.8 State Legislation Requiring Local Government Units to Reach Specified Waste Reduction Goals

State	Goal, %	Deadline	Type of government unit	Act ID
Alabama	25	5/92	County/city	824—1989
California	50[a]	1/2000	County/city	A.B. 939—1989
Connecticut	25	10/91	Municipality	P.A. 91-92
Florida	30	12/31/94	County	S.B. 1192—1988
Georgia	25	1996	County/city	S.B. 533—1990
Hawaii	50	1/2000	County	H.B. 954—1991
Illinois	25	1/2000[b]	County	P.A. 85—1198
Iowa	50	1/2000	County/city	H.F. 753—1989
Louisiana	25	12/31/92	Parish[c]	185—1989
Maryland	20[d]	1994	County	H.B. 714—1988
Minnesota	35[e]	12/31/93	County	H.F. 1;S.S.89
Nevada	25	Not set	County[f]	A.B. 320—1991
New Jersey	25	1/90	Municipality	P.L. 1987, C.107
North Carolina	25	1/93	[g]	S.B. 111—1989
Ohio	25	6/24/94	SW Planning District	H.B. 592—1988
Oregon	[h]	1/96	County/city	S.B. 66—1991
Rhode Island	15	[i]	Municipality	23-18, 23-19, and 37-15—1986
South Carolina	30	5/97	County	H.B. 388—1991
Tennessee	25	1/96	County	H.B. 1252—1991
Vermont	40	1/2000	SW Management District	78—1987
Virginia	25	12/31/95	Municipality	H.B. 1743—1989

[a]May include 10 percent waste transformation.

[b]Counties with populations of more than 100,000 must reach goal by 3/97; all others must reach the goal by 1/2000.

[c]Also targets cities with more than 50,000 population.

[d]20 percent recycling rate for counties with populations of more than 100,000, 15 percent recycling rate for counties with populations of less than 100,000.

[e]35 percent recycling rate for the seven counties in the Twin Cities Metro area, 25 percent recycling rate for the remainder of Minnesota's counties.

[f]Only applies to counties with populations greater than 40,000.

[g]Designated local government, of which there are 90 counties and 15 municipalities.

[h]Varies from 45 percent in Portland area counties to 7 percent in the most rural counties.

[i]Within 3 years of a program's implementation.

Source: Jim Glenn, "The State of Garbage in America, Part II", *BioCycle,* May 1992.

TABLE 5.9 Disposal Bans for Selected Waste Materials

State	Vehicle batteries	Tires	Yard Waste	Motor Oil	White goods	Others
Arkansas	x	x	x			
Arizona	x	x				
Connecticut						a
Florida	x	x	x	x	x	b
Georgia	x	x	x			
Hawaii	x					
Idaho	x	x				
Illinois	x	x	x		x[c]	
Indiana			x			
Iowa	x	x	x	x		d
Kansas		x				
Louisiana	x	x			x	
Maine			x			e
Maryland		x	x			
Massachusetts	x	x	x		x	f
Michigan	x		x			
Minnesota	x	x	x	x	x	g
Mississippi	x					
Missouri	x	x	x	x	x	
Nebraska	x	x	x	x	x	
New Hampshire	x		x			
New Jersey	x		x[h]			
New York	x					
North Carolina	x	x	x	x	x	
North Dakota	x			x	x	
Ohio	x	x	x			
Oklahoma		x				
Oregon	x	x		x	x	i
Pennsylvania	x		x[j]			
Rhode Island	x					
South Carolina	x	x	x	x	x	
South Dakota	x	x	x	x	x	k
Tennessee	x	x	x	x		
Texas	x	x		x		
Utah	x					
Vermont	x	x		x	x	e
Virginia	x					
Washington	x			x		
West Virginia	x	x	x			
Wisconsin	x	x	x	x	x	l

[a] Mercury oxide batteries.
[b] Demolition debris.
[c] White goods containing CFC gases, mercury switches, and PCBs.
[d] Nondegradable grocery bags and carbonated beverage containers.
[e] Household batteries.
[f] Glass and metal containers, recyclable paper, and single polymer plastics.
[g] Nickel-cadmium rechargeable batteries.
[h] Leaves.
[i] Discarded vehicles.

TABLE 5.9   Disposal Bans for Selected Waste Materials *(Continued)*

---

*j*Leaves and brush.

*k*Office and computer paper, newsprint, corrugated and paperboard, glass, plastic, steel, and aluminum containers.

*l*Metal, glass, and plastic containers, and recyclable paper.

Source: Adapted from Robert Steuteville, Nora Goldstein, and Kurt Grotz, "The State of Garbage in America, Part II," *BioCycle,* June 1993.

absolute. In Pennsylvania and South Carolina, the bans pertain only to loads that are primarily yard waste. In both Florida and North Carolina, yard waste is banned from incinerators and certain classes of landfills.

Another ban that can have significant effect on the amount of disposal is white goods. By the end of 1992, white goods had been banned in 14 states.[11] In 13 of the 14 states, the ban applies to all the appliances that are discarded. However, in Illinois, it applies only to those appliances which have not had CFC gases, mercury switches, and/or PCBs removed.

Since 1991, another item that has become the target of bans is dry cell batteries. In H.B. 7216, Connecticut put a disposal ban on mercury oxide batteries. Minnesota, through S.B. 793, prohibits the disposal of rechargeable nickel-cadmium batteries. The Vermont legislature passed a ban on mercuric oxide, silver oxide, nickel-cadmium, and sealed lead acid batteries used in commercial applications. Additionally, the law, H.B. 124, bans the disposal of retail nickel-cadmium batteries and bans alkaline batteries from incinerators. By the end of 1992, five states had passed bans on at least some household batteries.[11]

While many states have legislatively banned multiple materials from disposal, none has developed as extensive a list as Wisconsin. In its 1989 recycling act, the legislature banned from disposal: appliances, waste oil, automotive batteries, yard waste, cardboard boxes, glass containers, newspapers, plastic bottles, office paper, magazines, steel cans, tires, aluminum cans, and foam polystyrene packaging. (Waste oil, yard waste, and tires are allowed to be incinerated if there is energy recovery.) Wisconsin estimates that those items account for about 60 percent of the discarded waste in the state. Other states with extensive lists of banned items include Massachusetts, Oregon, and South Dakota (see Table 5.9).

5.12   MAKING PRODUCERS AND RETAILERS RESPONSIBLE FOR WASTE

Over the past several years it has become increasingly popular to attempt shifting some of the burden of waste disposal and recovery of materials back to the manufacturers of products. One approach in doing this is to require them to take back products or packages after their useful life has expired.

Beverage Container Deposits

The first attempts at making manufacturers responsible for products came in the 1970s with the passage of mandatory deposits on selected beverage containers. The first mandatory deposit law was passed in Oregon in 1971. Since then, another eight states have passed mandatory deposit legislation, the last being New York in 1983 (see Table 5.10).

The principal focus of these laws is the packages that are used for soft drink and beer containers, and to a lesser extent, mineral water and liquor. Maine has the most extensive list of containers. It includes the four mentioned previously as well as juice, water, and tea. In nine states the legislation includes glass, steel, aluminum, and plastic containers. Delaware has exempted aluminum cans from the deposits.

Another approach similar to mandatory deposits for beverage containers was passed in California in 1987. California's law is different from the other nine in that no actual deposit is paid by the consumers; instead distributors pay either 2 or 4 cents per container (based on size) into a state-administered fund. Consumers returning the containers to state-approved redemption centers receive 2.5 cents for each container under 24 oz and 5 cents for those over 24 oz.

Other Mandatory Deposits

Recently, deposit requirements have been applied increasingly to other products, most notably vehicle batteries. The first vehicle battery deposit law was passed by the Rhode Island legislature in 1987. Since then six other states (Arkansas, Idaho, Michigan, Minnesota, South Carolina, and Washington) have put deposits on auto batteries. All have pegged the deposit at $5 except Arkansas, which has a $10 deposit (see Table 5.10).

Beyond putting an added value on the batteries in question, these also require that retailers take back at least as many old batteries as a person buys new ones. And they require that wholesalers accept old batteries from the retailers.

Take-Back Provisions

What has become even more prominent than deposits on batteries is requiring retailers and then wholesalers in turn to take back products. As of the end of 1992, a total of 13 states had laws requiring retailers to accept auto batteries from consumers (see Table 5.10). These actions have all come about since 1988 when Pennsylvania was the first to pass a "take-back" provision without including a deposit.

But the first state to require retailers to accept what they sell was Minnesota. In 1985, it applied that concept to tires. Arizona, Indiana, Kansas, and Nevada have since passed similar requirements. Additionally, Connecticut passed legislation in 1991 mandating retailers take back mercuric oxide batteries.

Mandating Manufacturer Responsibility

In 1991, legislatures developed another approach to requiring manufacturers to become responsible for their products. In Minnesota, S.B. 793 requires that the manufacturers must come up with a system to collect nickel-cadmium batteries by 1994. Vermont's H.B. 124 also requires that manufacturers of dry cell batteries containing mercuric oxide electrode, silver oxide electrode, nickel-cadmium, or sealed lead, "ensure that a system for the proper collection, transportation and processing" be put in place. As part of that system, a manufacturer has to develop a link between the con-

TABLE 5.10 State Mandatory Deposit and "Take-Back" Laws

State	Type of product	Deposit or take-back	Act ID	Effective Year	Date
Arizona	Vehicle batteries	Take-back	H.B. 2012	1990	9/90
Arkansas	Vehicle batteries	Deposit	H.B. 1170	1991	7/92
Connecticut	Beverage containers	Deposit	Sec. 22A-243-246	1978	1980
	Mercury oxide batteries	Take-back	H.B. 7216	1991	1/92
Delaware*	Beverage containers	Deposit	Title 7, Chap.60	1979	1982
Idaho	Vehicle batteries	Deposit	H.B. 122	1991	7/91
Illinois	Vehicle batteries	Deposit	PA86-723	1989	9/90
Iowa	Beverage containers	Deposit	Chap. 445C	1978	1979
Kansas	Tires	Take-back	H.B. 2407	1991	5/91
Louisiana	Vehicle batteries	Take-back	185	1989	8/89
Maine	Beverage containers	Deposit	P.L. 1975, C. 739 (as amended)	1975	1978
Massachusetts	Beverage containers	Deposit	301 CMR 4.00	1981	1983
Michigan	Beverage containers	Deposit	M.C.L. 445.571-.576 (as amended)	1976	1978
	Vehicle batteries	Deposit	P.A. 20	1990	1/93
Minnesota	Vehicle batteries	Deposit	115A.9561	1989	10/89
Mississippi	Vehicle batteries	Take-back	S.B. 2985	1991	7/91
Missouri	Vehicle batteries	Take-back	S.B. 530	1990	1/91
Nevada	Tires	Take-back	A.B. 320	1991	1/92
New Jersey	Vehicle batteries	Take-back	S.B. 2700	1991	10/91
New York	Beverage containers	Deposit	Title 10, C. 200	1982	1983
	Vehicle batteries	Take-back	Chap. 152	1990	1/91
North Carolina	Vehicle batteries	Take-back	H.B. 620	1991	10/91
North Dakota	Vehicle batteries	Take-back	H.B. 1060	1991	1/92
Oregon	Beverage containers	Deposit	O.R.S. 459.810-.890	1971	1972
	Vehicle batteries	Take-back	H.B. 3305	1989	1/90
Pennsylvania	Vehicle batteries	Take-back	101	1988	9/88
Rhode Island	Vehicle batteries	Deposit	23-60-1	1987	7/89
South Carolina	Vehicle batteries	Deposit	S.B. 366	1991	5/92
Texas	Vehicle batteries	Take-back	S.B. 1340	1991	9/91
Utah	Vehicle batteries	Take-back	H.B. 146	1991	1/92
Vermont	Beverage containers	Deposit	Title 10, C. 53	1972	1973
Washington	Vehicle batteries	Deposit	E.S.H.B. 1671	1989	8/89
Wisconsin	Vehicle batteries	†	335	1990	1/91
Wyoming	Vehicle batteries	Take-back	W.S. 35-11-509-513	1989	6/89

*Any container that holds a carbonated beverage, except aluminum cans.

†Retailers are required to accept old lead acid batteries when a person purchases a new one and may place up to a $5 deposit on a battery which is sold.

Source: Adapted from Robert Steuteville, Nora Goldstein, and Kurt Grotz, "The State of Garbage in America, Part II," *BioCycle,* June 1993.

sumer and itself, and accept the batteries it produces at its manufacturing facility.[9] In 1992, two other states applied the same type of requirements to other products. Maryland passed S.B. 37, which required manufacturers of mercury oxide batteries to set up a collection program and Minnesota passed a law requiring telephone directory manufacturers to set up recycling programs.[11]

5.13   ADVANCED DISPOSAL FEES

Advanced disposal fees (ADF), where the cost of disposal is at least partially paid for upfront, have been considered for a wide variety of products and packages, but states have put fees on only a narrow range of those products. Of the 28 states with some form of ADF legislation, all have put them on tires (see Table 5.11). Motor oil is taxed in another three states (Rhode Island, South Carolina, and Texas), and fees are placed on white goods and vehicle batteries in two (Maine and South Carolina).

Only five states have put advanced disposal fees on multiple products. In Florida, there is a $1 fee on tires, a 10 cent per ton fee on newsprint, and a 1 cent fee on each glass, metal, and plastic container. Starting in 1992, the fee for newsprint not made with at least 50 percent recycled fiber went to 50 cents per ton. Additionally, the fee on the containers doubled in 1992 for material not at a 50 percent recycling level.

Other states with fees on multiple products include Maine (tires, white goods, brown goods, and vehicle batteries), Rhode Island (tires, motor oil, antifreeze, and organic solvents), South Carolina (tires, motor oil, white goods, and vehicle batteries) and Texas (tires and motor oil).

Twenty-five states require that the fees on tires be collected at the retail level. All but two of those place flat fees on each new tire sold. Those fees on tires range from 25 cents per tire in California to $2 per tire in Texas. Wisconsin's fee is $2 per tire on each new vehicle sold. Arizona and North Carolina charge a percentage on each tire sold. Arizona's is 2 percent per tire and North Carolina's 1 percent. In Rhode Island, in addition to the fees placed on individual items, a $3 fee is put on each new car sold to cover the cost of materials that are hard to dispose of.

The two states that do not collect the fee at the retail level, Michigan and Minnesota, place it on the title transfer when a vehicle is sold. Michigan charges 50 cents and Minnesota charges $4 per transfer. In South Carolina, the state has opted to collect all the fees it charges at the wholesale level. While this hides the cost of the fee from the consumer, it makes collection easier.

Perhaps one reason for ADF's popularity is that they provide a substantial funding source. In most cases, tire fees are utilized to fund tire cleanup and recovery programs. However, in others as in South Carolina and Texas, the funds will be used to help finance all state waste reduction efforts.

5.14   SPECIAL WASTES LEGISLATION

Beyond waste reduction, another trend in state legislation is an increasing awareness that certain waste products need special consideration when it comes to developing a legislative package. This is particularly true for such products as scrap tires, used motor oil, and household hazardous waste. States are going beyond simply taxing a material and/or banning it from disposal sites and are developing comprehensive management programs. For instance, 1991 saw Arkansas pass laws which developed a permit program for waste tire facilities, required solid waste management districts to establish collection sites, and provided grants for the cleanup and processing of waste tires. South Carolina and Texas have laws which require the registration of used oil haulers and processors and give out grants for a variety of public education and collection activities. Pennsylvania has established a grants program for municipalities to establish hazardous waste collection programs.

TABLE 5.11   States That Have Enacted Packaging and Product Taxes or Fees

State	Type of product or package	Type of fee or tax	Act ID	Effective Year	Date
Arizona	Tires	2%/new tire*	H.B. 2687	1990	9/90
Arkansas	Tires	$1.50/new tire	H.B. 1170	1991	7/91
California	Tires	$0.25/new tire	A.B. 1843	1989	7/70
Florida	Tires	$1/new tire	S.B. 1192	1988	1/90
	Newsprint	$.10/ton†			1/89
	Glass, Metal, and plastic containers	$0.01/container‡			10/92
Georgia	Tires	$1/new tire	H.B. 1385	1992	1/93
Idaho	Tires	$1/new tire	H.B. 352	1991	7/91
Illinois	Tires	$1/new tire	H.B. 989	1991	7/92
Kansas	Tires	$0.50/new tire	S.B. 310	1990	7/90
Kentucky	Tires	$1.00/new tire	H.B. 32	1990	1/91
Louisiana	Tires	Not yet set	Act 185	1989	n/a
Maine	Tires	$1/new tire	Chap. 585 (as amended)	1989	7/90
	White goods	$5 each			7/90
	Brown goods	$5 each			7/90
	Vehicle batteries	$1 each			7/90
Maryland	Tires	$1/new tire	H.B. 1202	1991	2/92
Michigan	Tires	$0.50/title transfer	Act 133	1990	1/91
Minnesota	Tires	$4/title transfer	Chap. 654	1984	9/84
Mississippi	Tires	$1/new tire	S.B. 2985	1991	1/92
Missouri	Tires	$0.50/new tire	S.B. 530	1990	1/91
Nebraska	Tires	$1/new tire	L.B. 163	1990	10/90
Nevada	Tires	$1/new tire	A 320	1991	1/92
North Carolina	Tires	1% sales tax	S.B. 111	1989	1/90
Oklahoma	Tires	$1/new tire	H.B. 1532	1989	7/89
Oregon	Tires	$1/new tire	H.B. 2022	1987	1/88
Rhode Island	Tires	$0.50/new tire	H 5504§	1989	1/90
	Motor oil	$0.05/qt.			1/90
	Antifreeze	$0.10/gal.			1/90
	Organic solvents	$0.0025/gal.			1/90
South Carolina	Tires	$1/new tire	S.B. 388	1991	11/91
	Motor oil	$0.08/gal.			11/91
	White goods	$2 each			11/91
	Vehicle batteries	$2 each			11/91
Texas	Tires	$2/new tire	S.B. 1340	1991	9/91
	Motor oil	$0.02/qt.			9/91
Utah	Tires	$1/new tire	H.B. 34	1990	1/91
Virginia	Tires	$0.50/new tire	H.B. 1745	1989	1/90
Washington	Tires	$1/new tire	E.S.H.B. 1671	1989	10/89
Wisconsin	Tires	$2/tire on new vehicles	W.A. 110	1987	5/88

*Tax cannot exceed $2 per tire.

†Increases to 50 cents per ton in 1992 if newsprint is not made with at least 50 percent recycled fiber.

‡If a 50 percent recycling rate is not achieved for these containers by 1992; the fee will increase to 2 cents per container if a 50 percent rate is not reached by 1995.

§$3 for each new vehicle purchased to cover all materials that are hard to dispose of.

Source: Adapted from Robert Steuteville, Nora Goldstein, and Kurt Grotz,"The State of Garbage in America, Part II," *BioCycle,* June 1993.

Tires

The disposal of used tires has been a vexing problem for years. Whole tires cause operational problems at landfills (they tend to work their way to the surface after being buried). And with the exception of burning the tires as fuel source, there has been only limited success at utilizing scrap tires. The result is thousands of tire piles, some numbering in the millions, throughout the United States.

The tire management laws in most states do not ban disposal outright. In all, 25 states have bans (see Table 5.9). All but three (Minnesota, Vermont, and Wisconsin) allow disposal if the tires have been shredded, chipped, or halved. To keep track of who is collecting and transporting tires some states also have put permitting or registration requirements into legislation. For instance, Florida and Iowa register haulers. Iowa requires haulers to be bonded. In Washington, legislation requires that all used tire haulers be licensed and pay a $250 per year fee. Perhaps most importantly, haulers have to document where the tires were delivered. Georgia has established a manifest and tracking system for scrap tires.

Disposal and processing facilities are also becoming the target of legislative oversight. In Kentucky, H.B. 32, which passed in 1990, required that piles with more than 100 tires had to be registered with the Department of Environmental Protection. Missouri dictates that anyone storing more than 500 tires for longer than 30 days must be permitted by the state.

Other state statutes are taking scrap tire management one step further. To ensure that collection and disposal facilities are available, some are making it the responsibility of local government. In Idaho, each county has to establish a waste tire program. Counties in North Carolina have had to provide collection sites, as do regional solid waste management authorities in Arkansas. A.B. 1843 in California includes provisions for a system of designated landfills that will accept and store shredded tires.

Another element to state tire management programs is the provision of grants to perform a variety of tasks. Arizona's tire disposal law provides that grant monies be used for counties and private companies to establish tire processing facilities, and for counties to establish collection centers and contract for hauling and processing services. A number of states, including Kentucky, Michigan, Minnesota, and Oklahoma, use grant monies to help with the cleanup of existing disposal sites. To help develop utilization programs, some states, such as Georgia and Illinois, use monies to fund innovative technology development.

Used Oil

A second special waste to be tackled by state legislators is used motor oil. Unlike tires, there are no significant accumulations of used oil around the country. But that is not to say that poor used oil disposal practices don't cause environmental harm. Less than two-thirds of the oil used in this country is accounted for. The speculation is that much of the remainder gets dumped in the drain or onto the ground.

Much of state used oil legislation is directed at providing "do-it-yourselfers" with sufficient collection alternatives and ensuring that once collected, the oil is properly handled and processed. In states such as Texas, the program consists of the voluntary establishment of used oil collection sites by private industry and local government. South Carolina has a similar provision, although its Department of Highways and Public Transportation is ultimately responsible for ensuring that at least one collection facility exists in each county. Additionally, both statutes require, as many other states do, that retailers post signs about a citizen's responsibilities and where to get additional information. In other states, like Arkansas, New York, and Wisconsin, retailers of motor oil must establish collection sites for used oil. Tennessee law requires that by 1995, each county has to provide a collection site for not only used oil but other automotive fluids such as antifreeze, as well as scrap tires and batteries.

But today states are going far beyond simply encouraging the development of collection programs. As with scrap tires, a growing number of states are seeking to control who collects used oil and what is done with it. South Carolina requires that used oil transporters register with the Department of Health and Environmental Control. They must submit annual reports to the Department and have liability insurance. Used oil recycling facilities in the state must be permitted. In Arizona, used oil transporters have to register as a hazardous waste transporter. The transporter must also manifest any used oil shipments.

Household Hazardous Waste (HHW)

The first state effort to manage HHW was probably in Florida. In the mid-1980s, it provided a series of temporary collection opportunities throughout the state. Since then other states have begun to develop comprehensive HHW management programs. For instance, in 1988, Pennsylvania's Act 101 established the "Right-Way-to-Throw-Away" program. In addition to initiating a grant program of local governments wanting to develop HHW collection programs, it also required the Department of Environmental Resources to register the programs, establish operational guidelines, and inspect sites.

Biomedical Waste

One area that is receiving increasing attention is waste coming from hospitals and other medical facilities. Typically, the preferred method of dealing with such waste is incineration. In many states, to date, little attention is being paid to medical waste incinerators and the only requirements put on them are what covers all incinerators. However, concern about the dangers that medical waste incineration may pose is prompting a number of states to develop specific rules.

As of 1991, 12 states had some form of medical waste incineration requirements beyond those that apply to all incinerators. These include Connecticut, District of Columbia, Illinois, Kansas, Maine, Michigan, Missouri, New Hampshire, Pennsylvania, South Carolina, Texas, and Wisconsin.[12] Most of these requirements mandate specific temperature and residence times, usually 1800° and 1 to 2 s, respectively, as well as particulate and opacity thresholds. Illinois and Kansas also have special ash handling requirements.[12]

Table 5.12 describes the regulatory schemes developed by some states to control the disposal of hospital wastes.

5.15 MARKET DEVELOPMENT INITIATIVES

States have come to recognize that developing recycling programs has to encompass more than the supply side of the equation. Markets hold the key to sustain growth in recycling. Therefore, over the last several years states have been including market development provisions in waste reduction legislation. The earliest market develop-

TABLE 5.12   State Requirements for Incineration of Hospital Wastes

State	Summary of requirements
Connecticut	There are guidelines for hospital incinerators. All incinerators must be permitted. Incinerators are required to meet BACT (best available control technology) standards, which include the following for hospital incinerators: particulate emissions may not exceed 0.15 gr/dry st ft^3 (corrected to 12 percent CO_2); there must be a 90 percent reduction of hydrogen chloride or less than 4 lb/h hydrogen chloride emitted, whichever is less; carbon dioxide emissions may not exceed 100 ppmdv (parts/million dry volume) (corrected to 7 percent O_2); and the combustion efficiency must be at least 99.8 percent. In addition, the primary-chamber temperature must be at least 1800°F, and the secondary-chamber temperature must be at least 2000°F with a retention time of at least 2 s. The visual emissions may not exceed 10 percent capacity. There are other requirements including monitoring and recording procedures. For more information call the Department of Environmental Protection, Solid Waste Management Unit, in Hartford at (203) 566-5847.
District of Columbia	There are specific air pollution control requirements for pathological waste incinerators. Incinerators must be of multiple-chamber design. Particulate emissions may not exceed 0.03 gr/dry st ft^3 (corrected to 12 percent CO_2). Pathological waste incinerators may only be operated between the hours of 10:00 A.M. and 4:00 P.M. Two interlocking devices are required: one that prevents operation of the primary chamber when the secondary-chamber outlet is less than 1800°F and another that prevents operation of the primary chamber when the primary-chamber charging door is not closed. There are several other requirements. For more information call the Department of Consumer and Regulatory Affairs, Environmental Control Division, at (202) 767-7370.
Illinois	There is a rule specifically for hospital incinerators. Requirements include obtaining an air pollution control permit and special ash disposal procedures. The particulate emissions standard for incinerators with capacities less than 2000 lb/h is 0.1 gr/dry st ft^3 (corrected to 12 percent CO_2). Carbon monoxide emissions may not exceed 500 ppmdv (parts/million dry volume). Other requirements, including secondary-chamber temperature and retention time, are determined on a case-by-case basis and specified in the permits. For more information call the Environmental Protection Agency, Division of Air Pollution Control, in Springfield at (217) 782-2113.
Kansas	Two permits are required for hospital incinerators: a solid waste processing permit and an air quality permit. There are many solid waste processing requirements, including waste handling and storage, and ash disposal requirements. Air quality regulations include particulate emissions standards which range from 0.10 to 0.30 gr/dry st ft^3 (corrected to 12 percent CO_2) depending on the capacity of the incinerator. The visual emissions standard is 20 percent opacity. For more information call the Department of Health and Environment, Solid Waste Management Section in Topeka at (913) 296-1590 and the Air Quality and Radiation Section at (916) 296-1572.
Maine	There are no regulations specifically for hospital incinerators; however, there are BPT (best practical technology) guidelines for hospital incinerators. The secondary-chamber temperature must be at least 1800°F with a retention time of 1 s. If antineoplastics are treated, the secondary-chamber temperature must be at least 2000°F with a retention time of 2 s. The particulate emissions rate may not exceed 0.1 gr/dry st ft^3 (corrected to 12 percent CO_2). Opacities in excess of 10 percent may trigger a stack test requirement. There are other

TABLE 5.12   State Requirements for Incineration of Hospital Wastes *(Continued)*

State	Summary of requirements
	requirements. For more information call the Department of Environmental Protection, Bureau of Air Quality Control, in Augusta at (207) 289-2437.
Michigan	All incinerators must be permitted. The term *pathological* refers to carcasses and body parts whereas *infectious* refers to other contaminated wastes including needles and plastic containers. Infectious waste incinerators must have a minimum secondary chamber temperature of 1800°F with a retention time of 1 s. There is no temperature requirement for pathological incinerators. All incinerators must meet particulate standards which depend on the incinerator capacity and air toxics in the emissions. There are many additional requirements. Off-site facilities must be part of a county waste management plan. For more information call the Department of Natural Resources, Air Quality Division, in Lansing at (517) 373-7023.
Missouri	All incinerators must be permitted. Infectious waste incinerators must have a minimum secondary-chamber temperature of 1800°F with a retention time of 0.5 s. This temperature requirement does not apply to pathological incinerators. The particulate emissions standard is 0.3 gr/dry st ft^3 (corrected to 12 percent CO_2) for incinerators smaller than 200 lb/h and 0.2 gr/dry st ft^3 (corrected to 12 percent CO_2) for larger incinerators. Visual emissions may not exceed 20 percent opacity. There are special requirements for off-site facilities. The regulations will be revised within 1 year. For more information call the Department of Natural Resources, Air Pollution Control Program, in Jefferson City at (314) 751-4817.
New Hampshire	There are no regulations specifically for hospital incinerators. Hospital incinerators with capacities greater than 200 lb/h must be permitted. An air quality analysis is required during the permitting process. Incinerators with capacities less than or equal to 200 lb/h may not emit particulate matter at a rate greater than 0.3 gr/dry st ft^3 (corrected to 12 percent CO_2), and visual emissions may not exceed 20 percent opacity. There are numerous requirements for larger incinerators including secondary-chamber temperature and retention time and air toxics emissions limits. These requirements are determined on a case-by-case basis and specified in the permits. For more information call the Department of Environmental Services, Division of Air Resources, in Concord at (603) 271-1390.
Pennsylvania	Both air quality and waste management permits are required for hospital incinerators. There are BAT requirements for hospital incinerators. Facilities with capacity less than or equal to 500 lb/h must meet a particulate emissions standard of 0.08 gr/dry st ft^3 (corrected to 7 percent O_2). Facilities rated between 500 and 2000 lb/h must meet a standard of 0.03 gr/dry st ft^3 (corrected to 7 percent O_2), and larger facilities must meet a standard of 0.015 gr/dry st ft^3 (corrected to 7 percent O_2). Hydrogen chloride emissions from incinerators with capacities less than or equal to 500 lb/h must not exceed 4 lb/h or shall be reduced by 90 percent. Hydrogen chloride emissions from larger incinerators may not exceed 30 ppmdv (corrected to 7 percent O_2) or shall be reduced by 90 percent. Carbon monoxide emissions may not exceed 100 ppmdv (corrected to 7 percent O_2). Visual emissions may not exceed 10 percent opacity for a period or periods aggregating more than 3 min in any hour and may never exceed 30 percent opacity. There are many other requirements including secondary-chamber temperature, ash residue testing, and monitoring and recording procedures. For more information call the Department of Environmental Resources, Bureau of Air Quality Control at (717) 787-9256 and the Bureau of Waste Management in Harrisburg at (717) 787-1749.

TABLE 5.12   State Requirements for Incineration of Hospital Wastes *(Continued)*

State	Summary of requirements
South Carolina	There are no regulations specifically for hospital incinerators; however, there is a hospital incinerator policy which all new or modified units must comply with. Incinerators must be of dual-chamber design. The secondary-chamber temperature must be at least 1800°F with a retention time of 2 s. The visual emissions standard is 10 percent opacity. New regulations for hospital incinerators are expected to go into effect in 1 year. For more information call the Department of Health and Environmental Control, Bureau of Air Quality Control, in Columbia at (803) 734-4750.
Texas	All incinerators burning general hospital waste must be permitted. Incinerators with capacity less than 200 lb/h burning only carcasses and body parts, blood, and nonchlorinated containers do not need permits. Incinerators must be of dual-chamber design with a secondary-chamber temperature of at least 1800°F and a retention time of 1 s. The visual emissions standard is 20 percent opacity. If hydrogen chloride emissions exceed 4 lb/h, the incinerator must be equipped with a scrubber. Other requirements, including air toxics emissions limits, are determined on a case-by-case basis and specified in the permits. For more information, call the Air Pollution Control Board, Combustion Section, in Austin at (515) 451-5711.
Wisconsin	All infectious waste incinerators must be permitted. The particulate standards for incinerators with capacities less than 200 lb/h is 0.00 gr/dry st ft^3 (corrected to 7 percent O_2). This standard also applies to infectious waste incinerators with capacities no greater than 400 lb/h which are operated no more than 6 h/day. The particulate standard for incinerators rated between 200 and 1000 lb/h is 0.03 gr/dry st ft^3 (corrected to 7 percent O_2), and the standard for larger incinerators is 0.015 gr/dry st ft^3 (corrected to 7 percent O_2). Incinerators larger than 200 lb/h may not emit more than 50 ppmdv (parts/million dry volume) hydrogen chloride or 4 lb/h, whichever is less restrictive. All incinerators must limit carbon monoxide emissions to 75 ppmdv. The visual emissions standard is 5 percent opacity. Stack testing is required for all facilities. For more information call the Department of Natural Resources, Air Management Bureau, in Madison at (608) 266-7718.

Source: C. R. Brunner, *Handbook of Incineration System Design,* McGraw-Hill, New York, 1991, chap. 14.

ment efforts were directed at providing financial incentives to companies willing to convert to the use of recycled feedstock. In the mid-1970s, Oregon instituted a series of tax credits that companies could use if they made recycling-related capital improvements.

New Jersey was the first state to incorporate a full range of financial incentives in legislation. Its 1987 omnibus recycling law included a market development package that contained tax credits, loans, grants, and sales tax exemption (see Table 5.13). Since then numerous states have followed suit. Seventeen states have tax credit programs. In 1992 alone, six states, Arizona, Iowa, Kansas, New York, Pennsylvania, and Virginia, passed such legislation.[11] The amount of the credit given on income tax typically ranges from Colorado's 20 percent to Louisiana's 50 percent.

Montana put an innovative piece of tax legislation on the books in 1991. In addition to providing for a 25 percent income tax credit, the state now gives a tax deduction to encourage businesses to purchase recycled goods. The law, S.B. 111, allows for a deduction of 5 percent "of the taxpayer's expenditures for the purchase of recy-

TABLE 5.13 State Financial Incentives to Produce Goods Made with Recycled Materials

State	Tax credits	Loans	Grants	Other
Arkansas	Yes			
Arizona	Yes			
California	Yes	Yes	Yes	
Colorado	Yes			
Florida				Sales tax exemption
Illinois		Yes	Yes	*
Iowa				Sales tax exemption
Kansas	Yes			
Louisiana	Yes			
Maine		Yes		
Maryland	Yes	Yes	Yes	
Massachusetts		Yes		
Michigan		Yes	Yes	
Minnesota		Yes	Yes	
Missouri				†
Montana	Yes			
New Jersey	Yes	Yes	Yes	Sales tax exemption
New Mexico	Yes			
New York	Yes	Yes	Yes	
North Carolina	Yes			
Oklahoma	Yes			
Oregon	Yes‡		Yes	
Pennsylvania	Yes	Yes	Yes	
Vermont		Yes	Yes	
Virginia	Yes			Personal property tax exemption
Wisconsin		Yes	Yes	Sales tax exemption

*Allows local governments to grant property tax abatement.

†Missouri has committed $1 million per year over the next five years from its disposal fees to fund market development but has not determined the exact nature of the programs.

‡Oregon has three separate tax credits that pertain to market development.

Source: Adapted from Robert Steuteville, Nora Goldstein, and Kurt Grotz, "The State of Garbage in America, Part II," *BioCycle*, June 1993.

cled material that was otherwise deductible by the taxpayer as a business-related expense."

Because the effectiveness of tax credits diminishes when they are applied to firms just starting out, some states also give out low-interest loans and grants. In all, 12 states give out grants and 11 have loan programs.[11]

In addition to financial and market incentives, a number of states recognize that advances in market development require the coordinated action of many players in both the public and private sector. To help in that coordination, numerous states have put together market development councils. For instance, South Carolina's Solid Waste Policy and Management Act established a council that was to analyze existing and potential markets and make recommendations on how to increase the demand for recovered materials. In Tennessee, not only was a markets council established, but its Department of Economic and Community Development had to set up an office of cooperative marketing. In 1991, Kentucky passed legislation which created an authority that helps recyclers market materials.

Minimum Content Standards

The latest approach states have taken to assist in the development of markets is to directly intervene in the market. In 1989, California and Connecticut passed statutes that required newspaper publishers to utilize newsprint made with recycled paper or face fines (see Table 5.14). Since then another 9 states have passed newsprint "minimum content" legislation, and newspaper publishers in another 11 states have voluntarily agreed to increase purchases of newsprint containing recycled fiber.[11] As a result of these actions numerous paper mills in both the United States and Canada are converting to the use of deinked fiber.

The content standards varies significantly in the statutes, ranging from Oregon's 7.5 percent of postconsumer fiber to West Virginia's requirement that 80 percent of the newsprint used by newspaper publishers contain the highest postconsumer recycled paper content practicable. For the most part the standards are to be phased in through the year 2000.

Now that content standards for newsprint have started to take hold around the country, lawmakers are beginning to utilize them to tackle other products as well. In 1991, Maryland and Oregon passed laws which require phone directories to have recycled content of 40 percent and 25 percent, respectively. Oregon also is pushing recycled content standards for plastic and glass containers. In the case of glass containers, the Oregon law (S.B. 66) requires that each glass container manufacturer use a minimum of 50 percent recycled glass by Jan. 1, 2000. Rigid plastic containers have to either contain at least 25 percent recycled content by Jan. 1, 1995, be recycled at a 25 percent rate by the same date, or be a reusable package.

TABLE 5.14 Recycled Content Standards

State	Material	Deadline	Act ID
Arizona	Newsprint	2000	
California	Newsprint	2000	
	Glass containers	2005	
Connecticut	Newsprint	2000	
Illinois	Newsprint	2000	
Maryland	Newsprint	1998	H.B. 1148
	Phone directories	2000	H.B. 1148
Missouri	Newsprint	2000	
North Carolina	Newsprint	1997	H.B. 1224
Oregon	Newsprint	1995	S.B. 66
	Phone directories	1995	S.B. 66
	Glass containers	2000	S.B. 66
	Plastic containers	1995	S.B. 66
Rhode Island	Newsprint	2001	H.B. 5638
Texas	Newsprint	2001	S.B. 1340
West Virginia	Newsprint	1997	S.B. 18

Source: Robert Steuteville, Nora Goldstein, and Kurt Grotz, "The State of Garbage in America, Part II," *BioCycle,* June 1993.

Procurement Provisions

Using the purchasing power of a state was another early market development tool. Procurement initiatives go back well before most states began to seriously consider developing comprehensive recycling programs. Initially, procurement provisions were directed at paper products, but more recently they have begun to be used in conjunction with a wide variety of products from plastics to compost.

Over the years, virtually every state in the country has passed some form of legislation encouraging the governmental purchase of products made from recycled materials. The legislation tends to focus on two things: eliminating any biases against recycled products and price preferences, particularly for paper and paper products. Additionally, some states have begun to direct their agencies to make specific purchases of recycled products.

One such law (H.B. 2020), passed in Illinois in 1991, requires that by July 1, 2000, 50 percent of the "total dollar value of paper and paper products" must be recycled. In Arkansas, H.B. 1170 establishes a progressive goal which aims to reach 60 percent of paper purchases by calendar year 2000. Oregon requires that by Jan. 1, 1993, no less than 25 percent of paper product purchases be made from recycled paper, increasing to 35 percent in 1995. West Virginia now has a goal which directs the state "to achieve a recycled product mix on future purchases" of 20 percent by the end of 1993 and 40 percent 2 years later.

Procurement requirements are going far beyond paper these days. In addition to merely telling procurement agencies they have to give a preference to recycled products, states are now targeting what materials have to be procured. For instance, Oregon's S.B. 66 requires the purchase of re-refined oil by both state and other public agencies. Illinois mandates that recycled cellulose insulation be used in weatherization projects done with state funds. Texas now can grant a 15 percent life-cycle price preference for rubberized asphalt. Maine passed a bill which requires compost to be used on all public land maintenance and landfill closures that use state funds.

5.16 STATE FUNDING

Where states come up with the money to administer their solid waste management programs has changed significantly over the years. Traditionally, the vast majority of the funds have come from an appropriation from the legislature. However, in recent years, as programs have become more diverse, so too have the funding sources. In addition to advanced disposal fees discussed previously, one of the most popular new sources of funds is the landfill tipping fee surcharge. By the end of 1992, 21 states had passed a disposal surcharge. The lowest rate in the country is Arizona's 25 cents per ton. Vermont's surcharge of $6 per ton is the highest. Most are in the $1 to $2 per ton range.[9,11]

Another funding source is a surcharge on collection bills used in Minnesota, North Dakota, and Washington. Minnesota's fee is 6 percent of the garbage collection fee. Washington's rate is 1 percent of the residential fee. North Dakota's collection tax ranges from 20 cents per month for accounts charged $10 or less to 1 percent of the gross receipts for those over $500 per month.[9,11]

APPENDIX: STATE SOLID WASTE REGULATORY AGENCIES

Alabama
Engineering Services Branch
Department of Environmental Management
1751 Dickinson Dr
Montgomery, AL 36130
205-275-7735

Alaska
Hazardous and Solid Waste Management
Department of Environmental Conservation
410 Willoughby
Juneau, AK 99801
907-465-5133

Arizona
Office of Waste Programs
Department of Environmental Conservation
2005 N Central Ave.
Phoenix, AZ 85001
602-257-6816

Arkansas
Solid Waste Division
Department of Pollution Control & Ecology
P.O. Box 8913
Little Rock, AR 72219
501-562-7444

California
Integrated Waste Management Board
1020 9th St, Suite 300
Sacramento, CA 95814
916-327-9373

Colorado
Waste Management Division
Department of Health
4210 E 11th Ave.
Denver, CO 80220
303-331-4830

Connecticut
Waste Management Bureau
Department of Environmental Protection
165 Capitol Ave.
Hartford, CT 06106
203-566-8478

Delaware
Solid Waste Authority
P.O. Box 455
Dover, DE 19903
302-739-5361

District of Columbia
Department of Public Works
65 K St NE, Lower Level
Washington, D.C. 20002
202-727-5856

Florida
Bureau of Solid Waste Management
Department of Environmental Regulation
2600 Blairstone Rd
Tallahassee, FL 32399-2400
904-922-0300

Georgia
Environmental Protection Division
Department of Natural Resources
205 Butler St, SW
Atlanta, GA 30334
404-656-2836

Hawaii
Solid & Hazardous Waste Management
  Branch
Department of Health
P.O. Box 3378
Honolulu, HI 96801
808-543-8244

Idaho
Division of Environmental Quality
Department of Health & Welfare
State House
Boise, ID 83720
202-334-5879

Illinois
Land Pollution Control
Environmental Protection Agency
P.O. Box 19278
Springfield, IL 62704
217-782-6760

Indiana
Solid & Hazardous Waste Management
  Office
Department of Environmental
  Management
P.O. Box 6015
Indianapolis, IN 46206-6015
317-232-3210

Iowa
Surface & Groundwater Protection
  Bureau
Department of Natural Resources
900 E Grand Ave.
Des Moines, IA 50319
515-281-8176

Kansas
Solid Waste Management Section
Department of Health & Environment
Forbes Field
Topeka, KS 66620
913-296-1595

Kentucky
Division of Waste Management
Department of Environmental Protection
Frankfort, KY 40601
502-564-6716

Louisiana
Solid Waste Management Bureau
Department of Environmental Quality
P.O. Box 44307
Baton Rouge, LA 70804
504-342-1216

Maine
Waste Management Agency
State House, Station 154
Augusta, ME 04333
207-289-5300

Maryland
Hazardous & Solid Waste Management
Department of the Environment
2500 Broening HWY
Baltimore, MD 21224
301-631-3304

Massachusetts
Recycling Director
Division of Solid Waste Management
Department of Environmental Protection
1 Winter St, 4th Floor
Boston, MA 02108
617-292-5915

Michigan
Waste Management Division
Department of Natural Resources
P.O. Box 30241
Lansing, MI 48909
517-373-4743

Minnesota
Groundwater and Solid Waste Division
Pollution Control Agency
620 Lafayette Rd N.
St. Paul, MN 55155
612-296-7777

Mississippi
Solid Waste Management
Department of Environmental Quality
P.O. Box 10385
Jackson, MS 39289
601-961-5171

Missouri
Waste Management Program
Department of Natural Resources
P.O. Box 176
Jefferson City, MO 65102
314-751-3176

Montana
Solid & Hazardous Waste Bureau
Department of Health & Environmental
 Sciences
836 Front St
Helena, MT 59620
406-444-2821

Nebraska
Land Quality Division
Department of Environmental Control
P.O. Box 98922
Lincoln, NB 68509-8922
402-471-2186

Nevada
Environmental Protection Division
Department of Conservation and Natural
 Resources
Capitol Complex
Carson City NV 89710
702-687-5872

New Hampshire
Waste Management Division
Department of Environmental Services
6 Hazen Dr
Concord, NH 03301
603-271-3712

New Jersey
Solid Waste Management Division
Department of Environmental Protection
 & Energy
CN 414
Trenton, NJ 08625
609-530-4001

New York
Division of Solid Waste
Department of Environmental
 Conservation
50 Wolf Rd
Albany, NY 12233
518-457-7337

North Carolina
Division of Solid Waste Management
Department of Environment, Health &
 Natural Resources
P.O. Box 27687
Raleigh, NC 27611-7687
919-733-0692

North Dakota
Division of Solid Waste Management
Department of Health
P.O. Box 5520
Bismarck, ND 58502-5520
701-224-2366

Ohio
Solid & Hazardous Waste Management
Environmental Protection Agency
P.O. Box 1049
Columbus, OH 43224
614-644-2917

Oklahoma
Waste Management Services
Department of Health
P.O. Box 53551
Oklahoma City, OK 73152
405-271-7213

New Mexico
Solid Waste Bureau
Environmental Improvement Division
1190 St Francis Dr
Santa Fe, NM 87504
505-827-2892

Oregon
Hazardous & Solid Waste Division
Department of Environmental Quality
811 SW 6th Ave.
Portland, OR 97204
503-229-5356

Pennsylvania
Bureau of Waste Management
Department of Environmental Resources
P.O. Box 2063
Harrisburg, PA 17053
717-787-7381

Rhode Island
Air & Hazardous Material Division
Department of Environmental Management
210 Promenade St
Providence, RI 02908
401-277-2808

South Carolina
Solid Waste Permitting
Department of Health & Environmental
 Control
2600 Bul St
Columbia, SC 29201
803-734-5200

South Dakota
Division of Environmental Regulation
Office of Solid Waste Management
523 Capitol
Pierre, SD 57501
605-773-3153

Tennessee
Division of Solid Waste Management
Department of Health & Environment
4th Floor, Customs House
Nashville, TN 37219
615-741-3424

Texas
Bureau of Solid Waste Management
Department of Health
1100 W 49th St
Austin, TX 78756-3199
512-458-7271

Utah
Division of Environmental Health
Department of Health
288 N 1460 W
Salt Lake City, UT 84116
801-538-6170

Vermont
Solid Waste Management Division
Agency of Natural Resources
103 S Main St
Waterbury, VT 05676
802-244-7831

Virginia
Division of Technical Services
Department of Waste Management
11th Floor, Monroe Bldg.
101 N 14th St
Richmond, VA 23219
804-225-2667

Washington
Solid & Hazardous Waste
Department of Ecology
PV-11
Olympia, WA 98504-8711
206-459-6316

West Virginia
Waste Management Division
Department of Natural Resources
1900 Washington St E
Charleston, WV 25305
304-348-3370

Wisconsin
Bureau of Solid Waste Management
Department of Natural Resources
P.O. Box 7921
Madison, WI 53707
608-268-0803

Wyoming
Solid Waste Management Program
Department of Environmental Quality
122 W 25th St
Cheyenne, WY 82002
307-777-7752

REFERENCES

1. *Characterization of Municipal Solid Waste in the United States: 1992 Update,* U.S. EPA Office of Solid Waste and Emergency Response (OS-305). EPA 530-R-92-019. Washington, D.C., July 1992.
2. Glenn, Jim, and David Riggle, "Where Does the Waste Go?" *BioCycle,* vol. 30, no. 4, pp. 34–39, April 1989.
3. Glenn, Jim, and David Riggle, "How States Make Recycling Work," *BioCycle,* vol. 30, no. 5, pp. 47–49, April 1989.
4. Glenn, Jim, "The State of Garbage in America, Part I," *BioCycle,* vol. 31, no. 3, pp. 48–53, March 1990.
5. Glenn, Jim, "The State of Garbage in America, Part II," *BioCycle,* vol. 31, no. 4, pp. 34–41, April 1990.
6. Glenn, Jim, and David Riggle, "The State of Garbage in America, Part I," *BioCycle,* vol. 32, no. 4, pp. 34–38, April 1991.
7. Glenn, Jim, and David Riggle, "The State of Garbage in America, Part II," *BioCycle,* vol. 32, no. 5, pp. 30–35, May 1991.
8. Glenn, Jim, "The State of Garbage in America, Part I," *BioCycle,* vol. 33, no. 4, pp. 46–55, April 1992.
9. Glenn, Jim, "The State of Garbage in America, Part II," *BioCycle,* vol. 33, no. 5, pp. 30–37, May 1992.
10. Steuteville, Robert, and Nora Goldstein, "The State of Garbage in America, Part I," *BioCycle,* vol. 34, no. 5, pp. 42–50, May 1993.
11. Steuteville, Robert, Kurt Grotz, and Nora Goldstein, "The State of Garbage in America, Part II," *BioCycle,* vol. 34, no. 6, pp. 42–50, June 1993.
12. Brunner, C. R., *Handbook of Incineration System Design,* McGraw-Hill, New York, 1991.

CHAPTER 6
PLANNING MSW MANAGEMENT PROGRAMS

James E. Kundell

Walter B. Hill Distinguished Public Fellow
Carl Vinson Institute of Government
University of Georgia

Deanna L. Ruffer

Project Director
Roy F. Weston, Inc.

Planning for the management of municipal solid waste becomes increasingly important as the complexity of management needs expands and the tools and procedures for addressing these needs require greater sophistication. Additionally, as the roles and responsibilities of states and their subdivisions in the management of solid waste have evolved, both state and local or regional solid waste planning is required.

6.1 STATE SOLID WASTE MANAGEMENT PLANNING

Some states may have planned for solid waste management needs in the past, but there is little evidence that such planning occurred prior to the passage of the federal Solid Waste Disposal Act of 1965. The federal focus for solid waste management planning has been at the state level, and the form and substance of state solid waste planning have been responsive to federal directives. Although some local governments were provided planning grants for demonstration projects, the plans called for under federal legislation were designed to show EPA that states had the authority and capability to oversee the management of solid waste within their borders. In recent years, however, state and local governments have found solid waste management planning necessary absent any federal directives or incentives to carry out such planning.

Historic Perspective: State Planning

The federal Solid Waste Disposal Act of 1965, like other environmental laws passed in the 1960s, did not establish a federal permit requirement for solid waste management facilities but focused initially on the provision of "financial and technical assistance and leadership in the development, demonstration, and application of new and improved methods and processes to reduce the amount of waste and unsalvageable materials and to provide for proper and economical solid waste disposal practices."[1] In addition to supporting research and demonstration projects and efforts toward regional solid waste management solutions, the federal law identified planning for solid waste disposal as an important component. The Secretary of Health, Education and Welfare was directed to encourage regional solid waste management planning[2] and to provide 50 percent matching grants to states to make surveys of solid waste disposal practices and problems within their jurisdictions and to develop solid waste disposal plans.[3] In order to consider all aspects essential to statewide planning for the proper and effective disposal of solid waste, the law identified factors to be considered in planning such as "population growth, urban and metropolitan development, land use planning, water pollution control, air pollution control, and the feasibility of regional disposal programs."[4]

The emphasis of the 1965 law was, in part, to generate a database on existing solid waste disposal efforts and problems. It must be remembered that this law predated the creation of the U.S. Environmental Protection Agency (EPA) and state environmental agencies. Consequently, most states had not assigned to a state agency the responsibility to oversee solid waste disposal practices. Since permits and reports were not routinely required by states, states were in the position of not knowing what they were dealing with. Surveys to build a database and a better understanding of practices and problems thus became an important first step for state solid waste management planning. Beyond generating information, the plans were not very well done by today's standards.[5]

The Solid Waste Disposal Act was amended in 1970 with the passage of the Resource Recovery Act.[6] This law provided funds for planning and development of resource recovery facilities and other solid waste disposal programs. States were eligible for 75 percent federal, 25 percent state grants for conducting surveys of solid waste disposal practices and problems and for "developing and revising solid waste disposal plans as part of regional environmental protection systems for such areas, providing for recycling or recovery of materials from wastes whenever possible and including planning for the reuse of solid waste disposal areas and studies of the effect and relationship of solid waste disposal practices on areas adjacent to waste disposal sites."[7] Additionally, funds were allowed to be used to plan for the removal and processing of abandoned motor vehicle hulks.

Grants were also allowed for planning and demonstration of resource recovery systems or for construction of new or improved solid waste disposal facilities. Interestingly, mass burn steam and electric power generating systems were not considered by federal officials at the time to be resource recovery systems. Instead, resource recovery meant refuse derived fuel (RDF) systems involving front-end automated materials recovery followed by combustion.[8] Most of this grant money was used to fund resource recovery facility demonstration projects.

This early solid waste legislation was replaced by the passage of the Resource Conservation and Recovery Act (RCRA) in 1976.[9] The thrust of RCRA was to remove most of the hazardous waste, principally industrial chemical waste, from the solid waste stream and to establish a separate management program for the hazardous waste. For the first time, Subtitle D of RCRA provided legislative guidance on the preparation of state solid waste management plans. Factors to be considered in state planning included:

1. Geologic, hydrologic, and climatic circumstances and the protection of ground and surface waters
2. Collection, storage, processing, and disposal methods
3. Methods for closing dumps
4. Transportation
5. Profile of industries
6. Waste composition and quantity
7. Political, economic, organizational, financial, and management issues
8. Regulatory powers
9. Types of systems
10. Markets for recovered materials and energy

In addition to this guidance, requirements for plan approval were also established.[10] To be approved by EPA, each state plan had to comply with the following requirements:

1. Identify state, local, and regional authorities responsible for plan implementation.
2. Prohibit the establishment of new dumps.
3. Provide for the closing or upgrading of existing dumps.
4. Provide for the establishment of state regulatory powers.
5. Allow for long-term contracts to be entered into for the supply of solid waste to resource recovery facilities.
6. Provide for resource conservation or recovery and for disposal of solid waste in environmentally sound facilities such as sanitary landfills.

When RCRA was amended in 1984,[11] these provisions were not altered, and as a result, this is the latest guidance provided by Congress for state solid waste management planning.

Current State Solid Waste Management Planning

State solid waste management planning reemerged in the late 1980s and early 1990s as a result of state initiatives rather than federal directives. With the 1984 amendments to RCRA, Congress directed EPA to develop environmentally protective landfill standards. These Subtitle D standards were released in draft form in 1988 and in final form in 1991. The message sent by the draft regulations was that, although greater assurance would be provided that landfills would not result in environmental degradation, the cost of landfilling would increase dramatically. It was this perceived increased cost plus the difficulty in siting new disposal facilities that caused local government officials to turn to their state legislatures for help. Between 1988 and 1991 states across the country enacted legislation to help resolve the solid waste problems facing local governments. Of note is that although these laws were in part the result of federal action (Subtitle D regulations), there was no new federal legislative guidance for states to address their overall solid waste management concerns. As a result, the legislation enacted by states varied, but owing to commonalities in problems and alternative solutions and interplay among the states, similarities emerged as state after state enacted comprehensive solid waste management legislation.

One common theme identified was the need for state and local or regional solid waste management planning. One analysis found this to be less evident in those states that were among the first to enact their legislation, but common among those states that were able to build on the experience of other states.[12] Also, the nature and extent of the planning and requirements varied considerably from state to state. State policy makers understood the need for planning, but the type of planning needed was more nebulous. As a result, states either adopted planning provisions built on historic solid waste management planning guidance as identified in 1976 by RCRA or developed planning requirements tailored to meet perceived needs.

There are four reasons why states have undertaken solid waste management planning:

1. To meet federal solid waste management planning requirements

2. To inventory and assess the solid waste management facilities and procedures in the state to determine capacity needs

3. To provide guidance to local governments and the private sector on solid waste management matters

4. To set forth the state's policies and strategy for managing solid waste

All these are valid reasons for states to plan, but the emphasis has varied with time and from state to state. All current state solid waste management plans contain three components (i.e., facility and program inventory and assessments, provision of guidance, and formation of state policy and strategy), but considerable variation exists in the emphasis placed on each component. For this reason, it is possible to categorize state solid waste management plans based on their emphasis.

Inventory and assessment documents tend to follow the historic model for state solid waste management plans. They attempt to quantify the status of programs and facilities in the state and identify problems that must be addressed. The plans prepared in Alabama and Rhode Island are of this type. Plans taking the form of technical assistance documents are generally designed to identify problem areas with solid waste management and to provide guidance to local officials (and others) on how to address the problems. The plans developed in Indiana and Tennessee emphasize this approach. The third type of plan appearing recently takes the form of policy documents which set forth the state strategy for reducing and managing solid waste and presents the strategy for doing so. The plans developed for Georgia and New York are policy and strategy documents.

Inventory and assessment has been a big part of state solid waste management planning since the 1960s. Consequently it is not surprising that some states have continued this approach into the current round of planning. It is interesting, however, that state plans have appeared that vary from the historic model because they are designed to meet identified state needs rather than federal directives. Plans which provide technical assistance to local governments are addressing perceived needs. Since local governments are the ones that have been faced with financing increasingly expensive solid waste facilities, attempting to site facilities that no one wants near them, and responding to the concerns of irate citizens, it is not surprising that states would use the state solid waste management plan as a mechanism to provide local governments with guidance. Consequently, the use of state solid waste management plans as vehicles for providing technical assistance to local governments has increased in recent years.

The approach which differs most from the traditional model for state solid waste management plans, however, is using the plan as a policy and strategy document. The historic role of the state in solid waste management has been to provide guidance and technical assistance and to regulate disposal activities. These responsibilities were generally assigned to one agency and had little direct impact on other units of state government. With the increased complexity of integrated solid waste management, however, multiple state agency involvement is now the norm. Although one agency still retains regulatory authority, other agencies may be involved in planning, recycling programs, market development efforts, procurement of products made from recovered materials, education, enforcement, and so forth. With this complexity comes the need to clearly articulate the policies and goals of the state and to set forth a strategy that assigns responsibilities to agencies and identifies the actions necessary for goal attainment.

The appearance of these new solid waste management planning efforts underscores the recognition of the complexity and interrelatedness of efforts to reduce and effectively manage solid waste. Integrated solid waste management is multifaceted, and decisions made to address one matter will likely affect other components of the system. Thus a systems management orientation is emerging which requires a continuous loop of planning and feedback. The result is a stronger commitment to solid waste management planning by state and local officials.

A New Model for State Solid Waste Management Planning

The *North Carolina Recycling and Solid Waste Management Plan* exemplifies the type of planning that is designed to address state needs. It is composed of three volumes: Volume I is an assessment of local and regional infrastructure and resources, Volume II is the state strategy for reducing and managing solid waste, and Volume III provides guidance and technical assistance to local governments.[13] Volume II of the North Carolina plan is a policy document that identifies state goals and the actions necessary to achieve those goals. Some of the elements of the strategy are derived directly from existing legislative mandates and some were developed as a result of the research conducted as part of the planning process. Each section of the strategy discusses a certain aspect of solid waste management, including:

1. Solid waste reduction, reuse, recycling, and composting
2. Waste processing and disposal
3. Illegal disposal of solid waste
4. Education and technical assistance
5. Planning and reporting
6. Resources

Since the intent of the strategy is to forge a clear path to meet state and local solid waste management needs, it clearly states goals and the actions necessary to meet those goals. A total of 29 goals were identified. For example, nine goals were presented for solid waste reduction, five for waste processing and disposal, and so forth. Each goal statement focuses on a major effort required to effectively implement a principal component of the state solid waste strategy. Generally the goals are based on policies established in legislation or that require actions by the General Assembly to implement or alter them. Goals are ordered to be consistent with the hierarchy of decision making in an integrated solid waste management program, not necessarily in the order of priority for implementation.

A total of 185 specific actions were identified to implement the 29 goals. Implementation actions identify the specific state agencies responsible for taking the action. Since time had elapsed between the enactment of the comprehensive state legislation and the development of the state plan, considerable effort had been exerted to implement portions of the act. Consequently, progress toward implementing each goal is also presented. Progress made reflects the resources available and the priorities of the implementation agencies.

Once a goal has been established, actions necessary to reach that goal identified, and attempts made to implement the identified actions, problems and issues associated with the goal or the actions may become apparent. To provide policy makers and agency personnel with insights into how these potential problems might be addressed, for each goal there is a section on future issues and guidance.

The plan recognizes that it was not possible for the state to achieve all 29 goals or implement all 185 actions at one time. Consequently it was necessary to prioritize the goals and actions so that the most important ones would be achieved first and less important ones would be implemented when resources became available. Priorities were set for both goals and actions based on four criteria:

1. Protection of public health and environment
2. Waste reduction
3. Promotion of integrated solid waste management
4. Formalization of organizational arrangements and responsibilities

The forcing mechanism for effectively reducing and managing municipal solid waste is the need to protect public health and the environment. Therefore, in setting priorities, those goals and actions designed to protect public health and the environment were given greater preference. Most of the major requirements to ensure that disposal facilities are environmentally benign, however, had already been adopted through rules for the design, construction, operation, closure, and postclosure care of disposal facilities. As a result, the environmental protection goals and actions included in the plan, although important, were in some cases of lower priority than other goals and actions.

Waste reduction is one way of reducing environmental risk from disposal. If the waste is never generated, it cannot pose an environmental threat when disposed. Second priority was given to those goals and actions designed to reduce the amount of waste being disposed through source reduction, reuse, recycling, and composting.

Third priority was given to those goals and actions that support an integrated approach to solid waste management. It is through an integrated approach that local governments will be better able to avoid actions that have unforeseen consequences and balance priorities of goals and actions.

Fourth priority was given to those goals and actions designed to formalize organizational arrangements and responsibilities. The early solid waste management legislation generally assigned all related responsibilities to one agency. As solid waste reduction and management have become more complex and integrated, this is no longer possible. Many agencies have roles to play in the solid waste arena relating to in-house recycling and waste reduction, public education, market development, curriculum development, and so forth. It is important that agency roles and responsibilities be formalized through the use of memoranda of agreements (MOAs) and other mechanisms so that each agency understands its specific functions and its working relationship with other agencies, local governments, associations, industry, and the public.

The drafters of the plan also found that it was necessary to separate goals and actions in order to prioritize them. The goals are more general policy statements that may differ from the specific actions when viewed in light of the criteria for prioritizing. For example, it may be that the greatest return on investment can be achieved by taking a specific action but it may not relate to the highest priority goal. Thus both goals and actions were separately prioritized. In prioritizing implementation actions, it was found that they were often interconnected. Sometimes it is difficult to proceed on one action without another one's being done (e.g., even though formalizing organizational arrangements is the fourth priority, MOAs may be needed before an agency is assured of its role and/or ability to take other actions).

From this discussion of the North Carolina solid waste management plan it is apparent that the form and substance of such planning is quite different from what was proposed in the Resource Conservation and Recovery Act of 1976. Recent planning efforts were undertaken without federal directives or financial assistance. It is this type of experimentation that leads to planning being relevant and of value to the states.

Reauthorization of RCRA

Currently (1994) under consideration by Congress is the reauthorization of RCRA. Draft RCRA amendments alter the minimum requirements for state plans. The proposed amendments to RCRA expand the solid waste planning requirements for states, particularly in the area of source reduction and recycling, including market development and procurement of products made from recovered materials. If these planning requirements will be enacted remains to be seen. If they are, they will require states to go through another round of solid waste management planning that will include updating both the previous plans submitted to EPA and the plans adopted pursuant to state comprehensive solid waste management legislation.

6.2 LOCAL AND REGIONAL SOLID WASTE MANAGEMENT PLANNING

At the local and regional level, integrated solid waste management planning involves the definition for later development of a wide variety of programs, facilities, strategies, procedures, and practices (elements) of a complete system of management. Beginning with the federal Solid Waste Disposal Act of 1965, planning requirements envisioned that very detailed and comprehensive plans would be prepared, with many of the federal planning requirements placed on states finding their way into the guidelines for the preparation of local plans. Yet until recently few local and regional plans actually met these expectations.

It is interesting to note that current planning efforts are now attempting to do what was called for in the 1965 Solid Waste Disposal Act (i.e., reduce the amount of waste being discarded and effectively dispose of the remainder). In the early 1970s as efforts were instituted to move from open dumps to sanitary landfills, emphasis was placed on increasing recycling and waste reduction. It was determined, however, that landfilling was still the least expensive alternative, and recycling efforts declined. With the new federal standards for landfills, again the interest has focused on waste reduction and recycling. The major difference now is that there is a greater understanding of what it will take to reduce the waste stream and a widespread commitment to do so. Central to this effort is sound solid waste management planning. The planning efforts that have occurred to date have provided a greater understanding of the focus and value of such planning.

Historic Perspective: Local and Regional Planning

In the 1960s and early 1970s, local solid waste management planning was primarily a community exercise in learning to understand and view solid waste collection and disposal practices and facilities as a total system. However, all too often local government leaders gave little priority to these planning efforts and actual decision making was seldom incorporated into, or followed, the planning process. Thus these planning efforts were little more than academic exercises or project plans used to define and justify the development of a specific program or facility.

Over the past 5 years, due to the increasing complexity and interrelatedness of integrated solid waste management, local and regional planning has taken on renewed importance in the management of solid waste. No longer is one type of program or facility adequate, or acceptable to all parties, to manage the entire waste stream. This is due in large part to the fact that solid waste is not simply solid waste anymore. In today's management system solid waste must be considered and managed as multiple waste or material streams. While one waste stream is suitable for recycling, another is more suited to composting or energy recovery, and still others will be landfilled.

As a result, planning for the management of solid waste must take into consideration the commonalities, differences, and interrelationships between the various programs, facilities, and procedures to be used. For example, specialized handling, processing, or segregation of materials may be required with some management approaches. Other aspects of the management system may require the establishment of ordinances and fee structures or the development of educational programs. Thus planning for the management of solid waste no longer involves a simple comparison of technical options and costs but includes consideration of how multiple waste streams are to be handled and the interrelationship between management practices as well as consideration of business risks and requirements, public policy, and social decision making. Furthermore, since an integrated solid waste management system contains multiple facilities, processes, programs, and procedures, it is unlikely that all aspects of the management system will be developed at one time. More likely, it will be many years before all elements of the system are implemented and that, as implementation proceeds, the plan will require some degree of modification.

State and Federal Planning Guidance

It has become increasingly common for states to require local governments to plan for solid waste management. In many instances state planning requirements have focused on defining how local governments will accomplish specific state objectives such as regionalization, the provision of adequate disposal capacity, or waste reduction and recycling.

Both the federal government and states have provided local governments with guidance on the objectives and priorities to be used in solid waste management planning.[14] EPA helped all solid waste planners when it released its report, *The Solid Waste Dilemma: An Agenda for Action.*[15] In this document, EPA stated that the elements of integrated solid waste management should be prioritized as follows:

1. Reduce the generation of solid waste.
2. Recycle (including composting) for productive reuse as much as is practicable.
3. Combust and recover energy for productive use.
4. Landfill the remainder.

EPA later revised its position and stated that the third and fourth priorities were in fact equal in priority. This statement of priorities, as simple a concept as it is, enabled states, local governments, planners, and citizens alike to focus their efforts and thus develop plans and strategies that are based on a rationale that can be clearly stated and defended.

In the North Carolina plan discussed above, explicit goals were set for state implementation activities and for local government planning and implementation activities. In addition, in recognition that it would not be possible for all goals and actions to be implemented at one time, criteria were established for setting state and local government priorities. The North Carolina plan, which requires that local integrated solid waste management plans be prepared, also includes guidance and direction for those local plans including specific goals, mandatory requirements, and recommended management programs.

In Ohio, local governments must form solid waste districts consisting of a population of at least 120,000 people. Each solid waste management district must develop and adopt a solid waste management plan that describes its existing facilities and its ability to accommodate the area's solid waste.[16] In comparison, Pennsylvania's Municipal Waste Planning, Recycling and Waste Reduction Act requires each county to develop a plan for municipal waste generated within its boundaries with emphasis on integrating recycling into existing disposal activities. The approach taken by the state of Georgia was to develop very specific planning standards and procedures to be used by local governments in demonstrating how they intend to meet the two overall objectives of assuring 10 years of disposal capacity and reducing the amount of waste (on a per capita basis) requiring disposal by 25 percent.

Planning Responsibility

The local government entity responsible for planning has varied from state to state depending on the state's planning objectives and the regulatory assignment of responsibility for solid waste management within the state. At least one state requires that regional solid waste districts be formed if a minimum population is not present within one local government.[17] Other states place the responsibility for planning with those units of government that own or operate solid waste management facilities.

Whoever is responsible for preparing the plan influences what topics are addressed, how the planning and implementation responsibilities are allocated, and the relationships that are established with other organizations and individuals involved in solid waste management in the planning area. At a minimum, solid waste management should be of significance to the entity responsible for preparing the plan, staff should have the capability to conduct the necessary analysis, and there should be a clear understanding of the authority of the entity during both plan preparation and implementation.

Individual local governments must assess and define how solid waste planning and management should be accomplished based on applicable state requirements and their own unique local circumstances. However, in determining who will be responsible for planning and how solid waste will be managed, local governments and regional entities must consider the public expectations for a safe, reliable, and cost-effective solid waste management system. At a minimum, local government must exercise overall responsibility for planning for municipal solid waste management and for the provision of municipal solid waste management services.[18]

Regardless of who is given the specific responsibility for planning, the following concepts should be embraced by those developing the plan.[19]

1. *Understanding needs.* The idea is to learn more about current problems and needs and future prospects before deciding on a course of action to accomplish objec-

tives. By this means, decisions become more rational, more objective, and based on more reliable information.

2. *Commitment to solid waste management.*   Some local governments make a decision to plan for solid waste management because they are committed to addressing the issue in a logical and comprehensive manner. Others simply develop plans because it is required. A plan can easily be written down on paper, but for a plan to work, the local government must be as committed to decision making and implementation as they are to planning. This includes paying enough attention to the planning process to assure that the plan can be implemented.

3. *Leadership.*   More often than not there is a single jurisdiction, agency, or individual that is deeply committed to seeing the planning process through to fruition and in many cases carrying on to lead implementation. When interest begins to taper off, need is considered to be less critical, or tough decisions need to be made, these leaders push on.

4. *Public involvement.*   A successful planning process not only defines programs but also opens up lines of communication, often among parties that rarely spoke to one another before (see Chap. 13). This communication results in consensus building. As a result it helps define what management practices are really needed and which are most likely to succeed. Effective public involvement in integrated solid waste management planning and program development provides the mechanism for addressing public concerns and values at each stage of the planning and decision-making process.

The Planning Process

The development of the local or regional integrated solid waste management plan should follow a clearly defined, rational process as shown in Fig. 6.1. This process should evolve through a sequence of analysis from the definition of goals and objectives to decision making on how the goals and objectives will be achieved. The steps in this process need to allow for continuous information flow, feedback, and adjustments to the planning process. The following six-step planning process accomplishes these objectives.

1. *Goals and objectives.*   The first step in the planning process should be to identify and prioritize goals and objectives for solid waste management in the planning service area. A goal statement should specify the direction and desired outcome of the solid waste management system as defined by the philosophies, values, ideals, and constraints of the community. Goal setting gives an overall, explicit purpose to the system and specific programs, facilities, and management practices in terms of the desired end result. Objectives, on the other hand, are ways in which solid waste management goals are measured. Objectives provide incremental information, or milestones, for gauging how well the system and specific programs are attaining the stated goals.

Goals and objectives serve as the foundation of the plan and management system. When possible, goals and objectives should be developed in a public process. Goals and objectives should be realistic and achievable, but also challenging. It may be necessary to reassess the goals and objectives at various points during the development of the plan. In addition, regular evaluation of the goals and objectives after plan adoption should be a routine part of the evaluation of the management system.

2. *Inventory and assessment.*   The foundation of the plan is the inventory of what resources are currently available and the assessment of the sufficiency of these

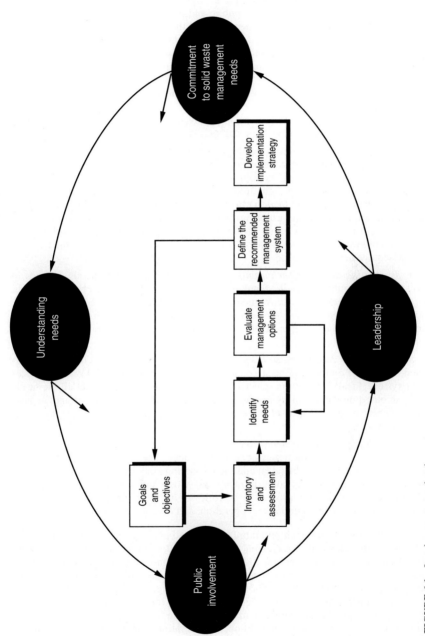

FIGURE 6.1 Local government planning process.

resources to meet federal, state, and local goals for a projected period of time. The inventory and assessment should consider all aspects of the existing solid waste management infrastructure as well as both public and private resources. It may be helpful to organize the inventory and assessments around the principal functional components of an integrated solid waste management system, which include waste characteristics, collection (recyclables and compostables as well as solid waste), reduction, disposal, administration, education, and financing. At minimum a cursory assessment should also be made of the infrastructure and resources that may be available beyond the planning area borders. The inventory and assessment should evaluate factors that affect waste stream generation and existing management practices in the planning area. These will vary from area to area but may include such factors as population, economic conditions, major industries, and tourism.

3. *Identifying needs.*   Based on the inventory and assessment, a determination should be made of what is needed to meet federal, state, and local management goals and objectives. For example, if existing waste reduction efforts have reduced disposal needs by 8 percent but a state goal requires a 25 percent reduction in disposal, one need would be to increase reduction efforts to achieve an additional 17 percent reduction in disposal needs. Each local government or regional entity should develop a list of solid waste management needs before beginning to define its desired solid waste management system. Often such an explicit process will bring out needs which would otherwise have been overlooked.

4. *Evaluating management options.*   For each of the defined needs, options to meet the need should be identified and evaluated. The feasibility of each option should be evaluated on technical, environmental, and economic grounds. Each option may have a number of components or combination of components. Each option will also have impacts on other aspects of the management system which must be taken into consideration in the evaluation.

5. *Defining the recommended management system.*   Once options have been evaluated, a series of options can be selected to form the basis of the solid waste management system. Ideally the option selected can be integrated and will meet all the needs set forth based on the inventory and assessment. If this does not occur, it may be necessary to revisit some aspects of the evaluation of options before finalizing the management system.

6. *Developing an implementation strategy.*   Once the management system is selected, an implementation strategy can be developed. The implementation strategy is the road map of actions to be taken and the measuring stick by which progress can be evaluated. It defines who is going to take what action and when. This strategy must take into consideration the process to be followed for procurement, development of facilities, funding, administration and operation, and decision making.

Consideration of how each program (existing and planned) affects other aspects of solid waste management is the essence of integrated solid waste management planning. Changing a management practice in one area almost always affects some other aspect of solid waste management. In its most basic sense this is exemplified by the planning priorities set by EPA. The focus of reducing, reusing, and recycling is to reduce the amount of solid waste requiring disposal. Integrated solid waste management requires that an examination be made of how the flow of waste will be altered by implementing certain options. In addition, consideration must be given to the impacts of waste management strategies on finances, personnel, public participation, and other policies.

Regionalization

In many instances, intergovernmental cooperation during the planning stage can be instrumental for the development of regional management systems. Through such cooperation local government decision makers can evaluate the potential for implementing solid waste management programs on a regional level. This can include:

1. Examining geographic patterns of waste generation, waste management activities, material flow, and markets for recovered materials
2. Identifying areas providing or planning for duplicative or competitive services
3. Identifying available resources
4. Evaluating alternative strategies for allocating responsibilities through existing governmental units and/or new entities

The criteria to be used in evaluating the potential for regionalization will vary from situation to situation. It is likely that economics will be a major criterion utilized in evaluating most regional options. Other factors that may be considered include:

1. Geographic pattern of need compared with the location and capacity of proposed services
2. Level and consistency of service provided
3. Availability and condition of transportation routes
4. Presence of physical and natural barriers
5. Presence of a population center
6. Need for new facilities vs. the ability to utilize existing facilities
7. Institutional, legislative, and regulatory requirements and time frame
8. Consistency with short- and long-term management objectives individually and collectively

It is logical that these factors be considered during the planning process to determine if and how regional solid waste management options should be implemented.

Plan Implementation

An annual program of monitoring and evaluation should be established to ensure that the strategy laid out in the plan is being accomplished. As part of this annual process it may be necessary to reassess program goals and objectives to adapt to the evaluation of what has been accomplished and what is still required to achieve the goals and objectives. It is extremely challenging to tailor activities to existing conditions as more is learned about what works well and what does not.

Once implementation of the plan begins, the problems and complexities seem to grow. Since an integrated solid waste management system includes multiple facilities, processes, programs, and procedures, it is unlikely that the entire system can be developed and put into place at the same time. Elements of the system will probably be developed and implemented over a period of many years. This means that the complete, integrated system as envisioned in the plan may not be in place for several years after plan adoption, if indeed it ever is. Since the waste stream is changing over time, markets for recovered materials fluctuate, and processing technologies continue to evolve, it is highly likely that by the time some elements of the plan are implemented some changes will be required to the programs, facilities, and procedures defined in that or other elements of the plan. This natural evolution of the management system will increase the importance of annual updates to the plan to assure its continued usefulness as a management tool.

6.3 CONCLUSIONS

As solid waste reduction and management become more complex and management options become more sophisticated, planning becomes increasingly important. This planning is evolving over time to truly meet the solid waste reduction and management needs of both state and local governments.

Integrated solid waste management planning is not simple. It involves what seems to be an infinite number of combinations and interactions of programs, all of which are changing constantly. Because of this, it may be more appropriate to refer to the product of the process as an integrated solid waste management strategy rather than an integrated solid waste management plan. Indeed, it is important to recognize the importance of the planning process itself in the development of a worthwhile, broadly acceptable and implementable plan. It must also be recognized that there must be a clear, explicit, and logical rationale for the approaches, actions, and strategies that the plan proposes. At the same time, the decision-making environment in which most solid waste management planning is done makes it a complex political process. Thus success is often not simply a function of the clarity, completeness, and quality of the technical analysis. Rather, the planning process and the plan itself must produce relevant products that policy makers can use within a context which may be much broader than the solid waste management system itself.

REFERENCES

1. P.L. 89-272, Sec. 202(6).

2. P.L. 89-272, Sec. 205.

3. P.L. 89-272, Sec. 206.

4. P.L. 89-272, Sec. 206(2).

5. Lewis, Stephen G., "Integrated Waste Management Planning (Chinese Checkers Planning)," Roy F. Weston, Inc., Mar. 24, 1992.

6. P.L. 91-512.

7. P.L. 91-512, Sec. 207.

8. Lewis, Stephen G., "Integrated Waste Management Planning (Chinese Checkers Planning)," Roy F. Weston, Inc., Mar. 24, 1992.

9. P.L. 94-580.

10. P.L. 94-580, Sec. 2.

11. P.L. 98-616.

12. Kundell, James E., "Ten Commandments for Developing State Solid Waste Policies," Conference of Southern County Associations, 1991.

13. *North Carolina Recycling and Solid Waste Management Plan,* Department of Environment, Health, and Natural Resources, Raleigh, N.C., February 1992.

14. Lewis, Stephen G., "Integrated Waste Management Planning (Chinese Checkers Planning)," Roy F. Weston, Inc., Mar. 24, 1992.

15. *The Solid Waste Dilemma: An Agenda for Action,* U.S. Environmental Protection Agency, EPA/530-SW-80-019, February 1989.

16. Mishkin, Andrew E., "Recent Federal and State Mandates and Regulatory Initiatives and Their Effects on Solid Waste Management Planning for Local Governments," in *1989 National Solid Waste Forum on Integrated Municipal Waste Management,* Association of State and Territorial Solid Waste Management Officials, July 1989, pp. 259–270.

17. Mishkin, Andrew E., "Recent Federal and State Mandates and Regulatory Initiatives and Their Effects on Solid Waste Management Planning for Local Governments," in *1989 National Solid Waste Forum on Integrated Municipal Waste Management,* Association of State and Territorial Solid Waste Management Officials, July 1989, pp. 259–270.

18. "The Role of the Public Sector in the Management of Municipal Solid Waste," The Solid Waste Association of North America, Technical Policy Position, August 1990.

19. McDowell, Bruce D., "Approaches to Planning," in Frank S. So, Irving Hand, Bruce D. McDowell (eds.), *The Practice of State and Regional Planning,* American Planning Association, 1986, pp 3–22.

CHAPTER 7
ENVIRONMENTAL PERSPECTIVES

Richard A. Denison, Ph.D.
Senior Scientist, Environmental Defense Fund
Washington, D.C.

John Ruston
Staff Economist, Environmental Defense Fund*
New York, N.Y.

Jeffrey Tryens
Deputy Director
Center for Policy Alternatives
Washington, D.C.

Roger Diedrich†
Sierra Club

7.1 INTRODUCTION

When the federal Resource Conservation and Recovery Act was enacted by Congress in 1976, concerns were beginning to be voiced about stack emissions from municipal incinerators and groundwater contamination from landfills, but many said they were nothing that couldn't be solved with good science and engineering. Unfortunately, science and engineering alone cannot always fix what is wrong with the solid waste sys-

---

*EDF staff scientist Jackie Prince contributed to the composting section of this chapter. EDF staff attorney Barbara Olshansky contributed to the green marketing section.
†Roger Diedrich contributed Sec. 7.4 on landfilling.

tem. Today, the United States finds itself in a difficult situation regarding solid waste because new landfills and incinerators are the focus of intense opposition almost everywhere and states are struggling to meet solid waste recycling and reduction goals which were adopted in the late 1980s.

This chapter offers some general environmental perspectives on waste management, focusing on several topics of current interest. It examines major issues associated first with source reduction, recycling, and composting and then with incineration. Finally, landfilling issues are treated.

The oft-cited "causes" of the solid waste crisis—dwindling landfill capacity, groundwater contamination, rapidly escalating costs, and intense public opposition to siting landfills or incinerators—are only its manifestations. The roots of the crisis penetrate far deeper into our social, economic, and institutional attitudes toward managing the plethora of materials we use and discard in our daily lives. Maximizing the exploitation of cheap and abundant raw materials, while disregarding the "downstream" costs, has been one of the most important forces driving our economy for the last 50 years.

Use of raw materials depends on an array of factors such as availability, performance characteristics, and ease of use. Economic theory does not account for the intrinsic value of such things as scenic vistas, clean air, and groundwater or unique habitats. In theory a raw material is used until the price goes so high that a substitute comes on the market at a lower price. It is up to governments to preserve nonquantifiable environmental "benefits."

From an environmental perspective, extracting raw materials from the earth for use in manufacturing should be the path of last resort. Only after maximum use has been extracted from materials through recovery and reentry into the manufacturing loop should new raw materials be introduced into the system. A forest not cut, a mine not opened, or an oil well untapped are opportunities to avoid environmental pollution and, in many cases, preserve nonrenewable resources for future generations.

For example, a study by the U.S. Forest Service found that achieving a 39 percent waste paper recycling rate nationwide by 2040 would result in 175 million fewer trees being cut in 2010 and would cause a drop in demand for old growth timber.[1] Mining for raw materials, instead of reusing and recycling materials, generates significant amounts of waste in both volume and environmental impact. Runoff from mining operations, which generate over a billion tons (907.2 billion kg) of mineral processing waste every year,[2] adversely affects over 180,000 acres (730 km^2) of lakes and reservoirs and 12,000 miles (19,300 km) of rivers and streams in the United States.[3]

For some metals the waste-to-product ratio can be extremely high. For example, the extraction, beneficiation, and processing of lead ores leaves behind a legacy of 33 tons of solid wastes for every ton of lead extracted[4]. Reduction or recycling of lead thereby not only reduces its direct impacts but also avoids the "upstream" generation of waste.

Traditionally, the waste cycle is considered to begin at the point when a product or material has fulfilled its intended function and is discarded. The subsequent steps of collecting, transporting, processing, and disposing of these materials institute our present view of municipal solid waste (MSW) management. Yet the safety, efficiency, and economics of waste management are dramatically affected by decisions made by manufacturers concerning the chemical and physical composition, packaging, marketing, and distribution of their products. Components or characteristics of a product that are useful or incidental to its function may increase the cost or hazards, or constrain the available means, of managing the product once it becomes waste.

As one example, the use of cadmium-based pigments as colorants in many consumer plastics is a major contributor of cadmium to MSW.[5] A manufacturer's decision to use cadmium-based pigments, which is based largely on increasing a product's attractiveness to a potential consumer, is made entirely without regard to its postconsumer impacts. Yet the cadmium present in such plastics is liberated during incineration of the MSW, increasing the toxicity of—and costs of managing—incinerator air emissions and/or ash residues. Fortunately, some manufacturers of plastics have begun to respond to public concerns on these issues by phasing out the use of cadmium in certain consumer plastics.[6]

Rational waste management should avoid a monolithic approach and instead focus on designing management strategies that take different materials' properties into account. This strategy begins with a thorough understanding of waste composition (see Chap. 3.) Source-based measures to reduce the amount and toxicity of materials entering the waste stream represent the most efficient means of waste management (see Chaps. 1, 2, and 8). As just discussed, restricting the use of heavy metal colorants in manufacturing consumer plastics is one such measure. Next, every opportunity to maximize segregation of the waste stream into its various components should be utilized. Developing methods include household-based materials separation programs supported by curbside collection and mixed-waste processing technologies. Waste segregation provides the means to direct individual materials to the management methods most appropriate for them.

7.2 ISSUES IN SOURCE REDUCTION, RECYCLING, AND COMPOSTING

Recycling and recovery of materials that can yield substantial reductions in energy and raw materials requirements and in wastes associated with manufacturing should be the options of first resort. Adoption of measures to stabilize and expand markets for such recovered materials is critical to their success. With respect to waste management, large-scale separation and recycling can provide the dual benefit of reducing the amount of waste that must be managed and increasing the safety of landfilling or incineration by removing materials that should not be buried or burned.

Unfortunately, in most cases, U.S. waste management practices still rely heavily on single management methods that collect materials in a commingled fashion and are disposal-based. The traditional method has been landfilling, supplanted in some communities in recent years by mass burn incinerators that accept an undifferentiated waste stream and recently by mass composting and mixed MSW composting (see Chap. 10).

Source Reduction and Reuse

While source reduction and reuse are identified as the highest-priority option in the majority of state solid waste management plans, reality is very different. Only 7 of the 38 states with waste reduction goals specifically establish source reduction goals as part of their plans.[7]

One of the first states to take action to get its own house in order was Connecticut, which passed a law in 1989 encouraging state agencies to use reusable materials and to practice source reduction in a wide range of state procurement activities. Illinois, Rhode Island, Wisconsin, and New Jersey have followed suit.[8] States have also begun to encourage institutions under their control to initiate source reduction measures. Other voluntary measures which states are instituting include awards for innovative businesses, public education, direct assistance for businesses, and financial assistance, such as grants and tax rebates.

One of the most popular source reduction tools developed by states so far is the ban on the landfilling of specific materials, ranging from yard waste to white goods. A 1990 survey by the Center for Policy Alternatives found 74 instances of a state banning a specific substance from landfills in 31 states.

Most data indicate that packaging waste is somewhere between one-third and one-half of the total municipal solid waste stream by volume, yet it has been one of the most difficult issues to tackle in source reduction and reuse. One of the earliest policies aimed at promoting this type of waste reduction was the bottle bill. The nine states with beverage container deposit system laws have return rates of 85 to 90 percent and have cut litter from beverage containers by 75 to 85 percent, according to the U.S. Public Interest Research Group. While these numbers are dramatic, most of the bottles returned are eventually crushed for recycling, not reused.

Recently legislative initiatives have focused, unsuccessfully, on requiring manufacturers to make their packaging more environmentally sound. Led by the public interest research groups in each state, voters in Oregon and Massachusetts were asked to require manufacturers to make certain types of packaging either reusable or made of a certain percentage of recycled materials, or recycled at a certain rate. Violators would be fined and otherwise penalized.

While both of these initiatives failed, interest in making manufacturers ultimately responsible for the packaging waste they create is growing. A so-called manufacturers' responsibility bill has recently been introduced in California with the support of Californians Against Waste.

Reducing Heavy Metals in Packaging and Products

Experience in addressing the environmental impacts of waste management options has made clear that the most efficient manner of reducing such impacts is at the earliest stages of the life cycle of consumer products and packaging: in their design and formulation. Given the prominence of heavy metals as contaminants of concern in waste management (especially incineration and composting), efforts at toxics use reduction have targeted this class of substances.

During 1989–1990, model legislation addressing the presence of heavy metals in packaging was developed by the Source Reduction Council of the Coalition of Northeastern Governors (CONEG), a unique consensus-building body with membership drawn from industry, state government, and the nonprofit environmental community.[9] This body was established by the governors of the nine northeast states for the purpose of identifying and implementing voluntary and legislative solid waste source reduction measures, with an initial focus on packaging.

The CONEG model legislation addresses four heavy metals—lead, cadmium, mercury, and hexavalent chromium—known to be toxic to humans and other organisms; present in municipal solid waste (MSW) in significant quantities; present in products of MSW management (e.g., landfill leachate, incinerator ash, mixed MSW compost); and contributed in part by packaging. The legislation bans the intentional introduction of these metals into any packaging, effective 2 years after enactment.

This legislation has now been enacted in almost 20 states, including numerous states outside of the northeast region. Given that industries that distribute products nationally will in all likelihood institute new practices for all of their packaging, rather than develop and maintain two different infrastructures, the widespread adoption of the legislation should lead de facto to national implementation of the ban by such companies.

Building on the momentum created by adoption by a number of states of the toxics-in-packaging legislation, similar legislation has been introduced in both houses of Congress. Passage at the national level would ensure compliance by all packaging nationwide, aid in resolving any interstate variations, and bring more enforcement resources to bear on the problem.

Following upon this success, several additional efforts are being pursued. First, available studies clearly demonstrate that, beyond packaging, other products and uses of these heavy metals contribute substantially to their total amounts in MSW. These include products such as batteries, consumer electronics, ceramic glazes, paint, dyes, pigments, fungicides, wood preservatives, and plastic stabilizers.

While bans on the use of the metals will not always be an appropriate strategy, for sources already known to be significant, several actions are being pursued. Where appropriate, especially for nondurable products, extension of the toxics-in-packaging legislative restrictions may be warranted. Use of toxic metals such as lead and cadmium as coloring agents is just as common in such products as in packaging, and studies indicate that in many cases alternatives are available or the heavy metal serves merely a nonessential (e.g., cosmetic) function. In addition to products, applications of heavy metals that may be amenable to similar restrictions include their use in dyes, pigments, paints, wood preservatives, and fungicides, which are employed in highly dispersive uses and in conjunction with many different materials or products that are not amenable to separate collection and management as hazardous waste. Minnesota recently enacted, as an extension of the toxics-in-packaging legislation, a restriction on the use of four heavy metals in dyes, paints, and fungicides.[10]

Second, data need to be gathered or generated on the availability and environmental and health characteristics of likely alternatives to uses for which restrictions are proposed. This will ensure that new problems are not created by replacing one toxic substance with another. The CONEG model toxics legislation contains a requirement that implementing agencies collect and report on replacement substances, a provision that should be included or even expanded in future efforts.

A third approach entails an expansion of the "right-to-know" concept to include toxic substances contained in consumer products. Currently, the Congressionally mandated Toxics Release Inventory (TRI) requires manufacturers to report annually their releases of a specified list of chemicals in the form of air emissions, water discharges, and solid waste. In many cases, however, the largest "release" is the product itself, which ultimately becomes a waste at the end of its useful life. The public has a legitimate right to know about these uses; moreover, such a reporting requirement would have the added benefit of strengthening incentives for industries to reduce their use of toxic chemicals, just as the current TRI provides incentives to reduce air, water, and solid waste discharges.

Finally, other toxic chemicals or pollutant precursors present in MSW and the by-products of its management warrant attention. For example, materials or waste components which contain pollutant precursors include chlorinated plastics (PVC, PVDC), which when incinerated create hydrochloric acid and may contribute to the formation of chlorinated dioxins and furans; yard waste, which increases incinerator nitrogen oxide emissions; and possibly rubber products, some of which contain sulfur and may increase incinerator sulfur dioxide emissions. In addition, some plastics utilize toxic organic plasticizing agents that have some leaching potential under landfill conditions (see Chap. 8 for more information).

Recycling

According to the EPA, a solid waste hierarchy (see Chap. 15)—reduce, reuse, recycle and compost, burn, landfill—reducing waste at its source is considered the best way to handle solid waste, while landfilling and incineration are the last options to consider. To implement this system of preference, funding should follow preference. But public spending on MSW reduction and management follows the exact opposite path. A 1989 study by the Northeast-Midwest Institute found that the 18 states in that region planned to spend 10 times more money to burn trash than to recycle it during a 5-year period.[11]

Some economic studies have shown that recycling is less expensive than either incineration with ash disposal or long-distance landfilling.[12] However, current local funding practices for solid waste management do not follow the EPA solid waste management hierarchy.

At the local level the disposal options of landfilling and incineration, like favored children, are provided with large allowances through local tipping fees which can provide a guaranteed income stream to operators of these facilities. Incinerators also have an added advantage in many areas with "put or pay" clauses in contracts which lock in municipalities to pay for specific amounts of garbage for combustion whether the garbage is generated or not.

Financial Issues.   Recycling programs, on the other hand, are expected to generate revenues to the municipality to cover collection and processing costs by selling the collected materials on commodities markets. While this anachronistic view is now less common, the current fiscal policy of most states and cities is still oriented toward spending hundreds of millions of dollars on incinerator and landfill projects, often funded by independent bonding authorities, while at the same time requiring that recycling eke out a much smaller budgetary allotment in competition with other government services.

Municipal accounting practices also distort the picture in many instances where landfills are the current mode of solid waste disposal. To determine the true cost of depleting the current landfill, the municipality should determine what the anticipated cost of replacing that space would be when the existing facility is closed. Such a calculation may show that extending the life of an existing facility is worth far more than the cost of setting up significant solid waste diversion programs through recycling, source reduction, and composting.

Financial advantages for disposal over reduction methods are numerous at the state level. Flow control laws, such as those in Florida, Maine, and Missouri, are the underpinning of local "put or pay" agreement allowing operators to require a guaranteed flow of garbage. California's Integrated Waste Management Act has been interpreted to deny recyclers an exclusive franchise to pick up a city's recyclables, although it allows exclusive franchise contracts for refuse collection.

In every state, bonding authorities, which issue tax-exempt bonds to finance new construction for local governments, make "cheap money" available for capital-intensive projects like landfills and incinerators. Because of their scale and their labor-intensive nature, recycling and compost are much less likely to be helped by this subsidy.

Some states also provide indirect subsidies for incinerators. For instance, Illinois allows electric utilities to receive tax credits for purchasing incinerator generated power which have undepreciated capital costs, providing in essence an interest-free loan to new MSW incinerators. Sometimes costs of landfills or incinerators are subsidized by the provision of free or underpriced land.

Some federal policies favor disposal options through subsidies to raw material extractors and haulers. For example, the Forest Service builds roads into national forests which lumber companies can use to clear-cut timber priced by the Forest Service at below market rate. Eventually some of this subsidized virgin wood fiber finds its way into head-to-head competition with recycled fiber in the paper manufacturer's marketplace. Not surprisingly, the subsidized virgin fiber has an advantage.

Also mining subsidies can distort the marketplace. A combination of "depletion allowance" income tax breaks and virtual giveaways of federal land under the General Mining Act of 1872 has cost U.S. taxpayers billions over recent decades. The Worldwatch Institute estimates that lost taxes alone added up to a $5 billion subsidy for the U.S. mining industry over the past decade.[13]

The financial mainstays of incineration are public expenditures, guarantees, and financial risk-minimization measures designed to make massive private investments attractive and profitable. Tipping fees paid under contract to incinerators and landfills are nondiscretionary, and outside of the annual budgetary allocation process. In contrast, the nuts and bolts of most local recycling programs—sanitation workers, outreach materials, educational staff, truck maintenance, and the like—are discretionary items that must be funded on an annual budgetary cycle, along with education, police, fire, housing, welfare, and other municipal services. If recycling programs were afforded the same fiscal and contractual support as incinerators, recycling would be both more cost-effective and more widely available.

New York City, for example, issued plans in 1984 for an ambitious program to build five incinerators (these plans were replaced in 1992 with proposals for much greater recycling). The key to the New York City Department of Sanitation's plans for incineration has always been a 3000 tons per day (2,721,600 kg per day) plan slated for the Brooklyn Navy Yard, which, as of mid-1993, had been issued state permits for construction.

In 1988, the city estimated that a private bond issue of $562 million would be necessary to construct the proposed Brooklyn Navy Yard incinerator, which would now surely be much larger. To ensure that these bonds will be paid off, the firm that constructs and operates the plant is counting on a number of public subsidies and risk-reduction measures. To begin with, the plant will receive a guaranteed flow of waste that is delivered free of certain specified materials, such as construction and demolition debris and hazardous waste. The incinerator vendor will also receive a guaranteed market for energy produced by the facility, ash disposal services provided by the Department of Sanitation, contract clauses that pass certain unexpected costs to the city, and most importantly, a substantial "monthly service charge," or tipping fee, for disposing of waste. The incinerator was also slated for a direct $47 million state grant from the 1972 New York State Environmental Quality Bond Act, although $19 million of this sum was redirected by the state legislature to the city's capital budget for recycling in 1992. Nothing approaching the extent and value of these risk-reducing measures is available to private or nonprofit recyclers in New York City and most other municipalities.

The current imbalances in funding for solid waste management projects must be reversed if recycling is to grow to its full potential. That is, recycling, landfilling, and incineration must be funded on a "level playing field." Only with equitable access to public funds will the economic advantages of recycling take effect in the marketplace, attracting private-sector financing of major recycling program elements.

The most direct means of allowing market forces a greater role in the selection of solid waste alternatives would be for municipalities (and states, where appropriate) to revise the rules that guide the selection of solid waste management facilities—specifically, to allow direct competition among landfilling, recycling, and incineration. Under this approach, for example, instead of issuing an RFP (request for proposals) for a 3000 tons per day (2,721,600 kg per day) incinerator and accepting bids only from incinerator vendors, a municipality would ask for proposals to provide 3000 tons per day (2,721,600 kg per day) of waste management capacity. Recyclers would be allowed to bid on material streams, and the contract would go to the lowest responsible bidder(s). If delivery of segregated materials by the municipality was necessary, the costs of doing so would have to be factored into the bid-selection equation.

By allowing consortia of recyclers to compete with incinerator vendors for municipal waste materials and tipping fees, private financing arrangements with investment banks would become feasible, just as they have been devised for incinerator proposals. Under such arrangements, very large material recovery complexes could be built using private revenue bonds. Repayment to the bondholders would be guaranteed by a municipal tipping fee paid for every ton of material processed, which would also cover the processing plant's operating expenses.

Recyclable materials would be delivered to processing centers by networks of drop-off and buy-back centers, municipal curbside collection programs, private waste haulers, or a combination of these. Revenue from sales of recovered materials would be split between the municipality and the plant operator on a formula basis. This approach, which is becoming common for material recovery facilities, partially decouples the successful operation of the intermediate processing center from swings in commodity market prices but also preserves an incentive to engage in market development and seek the highest prices for materials.

If unable or unwilling to allow recycling to compete head-to-head with incinerators and landfills for tipping fees, a state could instead require that no funds be spent on incineration until the marginal cost of recycling (i.e., the incremental cost of recycling the next ton of materials) was equal to or greater than the average cost per ton of operating an incinerator of a given size. This principle is similar to a commonly used feature of electrical utility regulation. In many states, before a utility is allowed a rate increase to finance the construction of a new power plant, it must show, before a public commission, that it has exhausted the possibilities of less expensive means of "generating" electricity, such as conservation, alternative sources, and demand management.

A recycling law with similar features (though not a comparable regulatory structure) is the 1988 New York State Solid Waste Management Act. The act requires that by 1992, municipalities in the state must have programs to collect for recycling all materials for which there are "economic markets." In the law, "economic markets" are defined to exist when the net cost of collecting, processing, and selling recovered materials is less than the cost of otherwise disposing of them.

Refuse Collection Cost Savings.   Since recycling collection is usually by far the most expensive part of a municipal recycling program, another critical need is to implement cost-effective collection programs. Curbside collection of residential recyclables can be quite expensive if it means adding a duplicate collection system to existing collection systems for regular refuse. However, one often overlooked aspect of recycling in general, and curbside recycling in particular, is its potential ability to substantially reduce the costs of conventional garbage collection. By definition, the more material that is collected for recycling, the less that will have to be picked up by regular garbage "packer" trucks. Where a well-established recycling program is diverting a significant quantity of material, regular trash collection trucks should be able to travel longer routes, make more stops, and require less maintenance. If sanitation engineers adjust their routing systems to account for the effect of recycling (route optimization is, after all, a major part of their responsibility), fewer conventional garbage collection trucks are required.

At first, especially if small fractions of material are diverted for recycling, conventional cost savings may not be especially great (routes may not be reconfigured, and costs like garages and administrative staff remain fixed over the short term). As more and more materials are collected, and as time goes on, savings in conventional collection costs should grow. In the first year of the San Jose, Calif., citywide program city officials estimated that for every 2 tons (1814 kg per day) of recyclables collected, the cost of picking up 1 ton (907 kg per day) of regular garbage is saved.[14] Other cities have experienced higher offsets after longer periods of time.[15] In densely populated cities, garbage collection takes place not once but two to four times each week. Under these conditions, substantial recycling might eliminate the need for an entire collection run, achieving close to 1:1 cost savings.[16]

Another factor that has a significant effect on the costs of curbside collection is the design and use of collection vehicles. While some of the most widely publicized curbside recycling programs are notable for their fleets of shiny new recycling trucks, buying specialized vehicles for separate collection of recyclables is not always necessary. Smaller cities may be able to use existing vehicles not fully employed by the local public works department to tow trailers with bins for separated recyclables. Other cities have tried a variety of approaches, including using regular packer trucks to pull recycling trailers, mounting bins underneath packer truck bodies to collect newspaper, or adapting beer trucks or other vehicles to recycling. Small towns can also pool resources to buy a fleet of recycling vehicles that they share.

Even when using new, specialized recycling trucks, the design of the truck itself and the way it is used can have major cost ramifications. For example, when the New York City Department of Sanitation (DOS) began its pilot curbside collection program for newsprint in 1986, it purchased a fleet of 28 trucks with a fairly small capacity for carrying collected materials, about 12 yd^3 (9.2 m^3). Because the city lacked recent waste composition data, it underestimated the amount of available newsprint by twofold. In moderate-density neighborhoods like Greenwich Village, the new trucks filled up with newspaper in about 3 h; in less dense areas such as Staten Island, the trucks filled up in 5 h. By the time the trucks were full, street traffic was heavy; without nearby intermediate processing centers or recycling transfer stations, operating costs mounted as the trucks sat in DOS garages, waiting for another shift of drivers to take them to paper dealers in Brooklyn.

This situation could be avoided with the construction of recycling mini-transfer stations or intermediate processing centers that would allow trucks to make two collection runs per shift, or by buying trucks with greater capacity. DOS is now experimenting with a 40 yd^3 (30.6 m^3) recycling truck, and for the last 2 years has been using a comparably large front-loading "EZ Pack" compactor truck to collect newspaper from high-rise apartment buildings. Many more innovations in truck design and routing can be expected as the recycling industry continues to grow.

Many other factors, such as labor costs, union rules, topography, housing density, and real estate values, influence the costs of residential recycling collection. However, these factors tend to influence the costs of regular garbage collection in the same way, making savings in conventional collection costs all the more important. In addition to a growing number of consulting firms, several publications and computer models are now available to help recycling planners evaluate the costs of different types of collection in their communities.[17]

Developing Markets for Recovered Materials.   The changing economics of solid waste management are having a major impact on markets for recovered materials, changes that have been exacerbated by the recent recession. Commodity markets that once reflected the interests of paperstock dealers who make their living buying, processing, and selling recovered paper have been entered by municipalities and private waste haulers. The primary concern of these new players is not the price of the recovered material but rather the avoidance of tipping fees. For most recovered commodities, in the late 1980s and early 1990s market prices have fallen while market volume—the amount of material bought, sold, and ultimately used in manufacturing—has increased significantly.

Recyclers' experience with the market for old newspapers—which felt a major price crash in late 1989—reinforces the recognition that public agencies must invest in market development along with recycling collection programs. The glut of old newspapers appears to be well on the way to being resolved. Newsprint manufacturers, despite the recession and overcapacity in overall newsprint production, have made significant investments in deinking facilities in the last several years. These investments have been spurred by the availability of inexpensive recovered fiber, demand from newspaper publishers, the more widespread use of flotation deinking technology, voluntary agreements with state governments, and in some cases, state legislation. In North America, about 20 newsprint mills are undergoing conversion to allow the greater use of recycled fiber between 1989 and 1994. As a result, according to the American Forest and Paper Association, the capacity of U.S. newsprint producers to consume recovered fiber is expected to increase by 958,000 annual tons (869,097,600 kg) from 1991 to 1995 (recovered fiber supplied 31 percent of the fiber needs of the U.S. newsprint industry in 1991, projected to grow to 43 percent in 1995). Old newspapers are also exported and used to make recycled folding cartons, construction paper, tissue, cellulose insulation, animal bedding, and molded pulp products such as egg cartons and packaging materials.

In this context, several states have recognized that stabilizing markets for recovered materials is an economic development problem. For example, New York State has more than 10 staff within the Department of Economic Development working on promoting growth within New York industries that use recovered materials in their manufacturing processes. The department has allocated state grants for private research and development and is adapting the state's low-interest loan and loan guarantee programs to help support industrial expansion. Similar policies are also at work in California, Massachusetts, New Jersey, Pennsylvania, and Washington.

State governments can play a significant role in long-term recycling market development as well. Preferential procurement of products containing recycled content—incorporating a 10 percent price preference—can be added to state procurement policies. In addition, states should develop mechanisms through which state procurement officers can learn of the important role they can play in resolving the solid waste crisis. Such education can help promote more enthusiastic compliance with the new elements of the procurement mandate.

"Green" Marketing and Advertising

Americans' environmental concerns encompass topics as diverse as ozone depletion, toxic waste, and tropical rainforests. When it comes time to take action, however, we first look to activities in our daily lives, like shopping or taking out the trash. For this reason, numerous surveys and opinion polls indicate that significant numbers of consumers now consider environmental factors in their purchasing decisions.

As a result, manufacturers and marketers are rapidly increasing advertising claims about the environmental benefits of their products and packaging. For example, 15 percent of all new products introduced in 1991 were accompanied by environmental claims.

The rise of "green marketing" demonstrates that consumers collectively hold enormous potential power to affect manufacturers' behavior but also that we need accurate and reliable information in order to translate our environmental concerns into purchases that truly benefit the environment. Faced with the confusing array of environmental information displayed on product labels and in other advertising, and armed with very limited power to verify the truth and significance of these claims, consumers are compelled to rely heavily on industry representations when making their purchasing decisions.

Without clear standards and enforcement actions to discourage fraudulent or misleading claims, green marketing is becoming a quagmire of confusing and even misleading information. Each manufacturer is free to ascribe its own distinct and potentially misleading definitions of "recycled," "recyclable," "biodegradable," or "compostable" to its products. As a result, consumers are becoming discouraged, and companies that actually are making environmental improvements are finding it hard to distinguish themselves. Unless these issues are quickly and definitively addressed, manufacturers will have little incentive to make genuine environmental improvements in the products that they sell. Given the importance of this issue, a number of initiatives have occurred or are underway to promote responsible use of environmental marketing and curb abuses.

The Federal Trade Commission (FTC) holds the primary authority to prevent deceptive claims in all types of advertising. In July 1992, the FTC issued its long-awaited industry guides for environmental marketing claims. Although the guides articulate a number of key principles, they are not always consistently applied, nor do they ensure that claims—even where factually accurate—represent significant environmental benefits of the products about which they are made. As a result, the guides will still allow some types of deceptive claims to be made.

One major limitation of the guides is that, in defining how terms like "recycled," "recyclable," "biodegradable," or "compostable" may be used, the guides adhere very closely to the FTC's traditional, legally circumscribed role of limiting deceptive advertising and do not venture into areas that could be viewed as setting environmental policy.

For example, while the FTC guides delineate conditions under which a product or package may be labeled as "compostable," they do not address technical issues, such as whether this attribute of a material in fact is relevant and environmentally beneficial in the specific settings in which it will be disposed. Nor do they address critical environmental policy issues, such as whether promotion of such a product as compostable in a mixed waste composting facility (i.e., a plant that shreds and composts mixed garbage) provides a genuine environmental benefit, relative to composting facilities designed to process only clean, organic material.

For these reasons, many observers are advocating federal legislation to provide the U.S. Environmental Protection Agency (EPA) with the statutory mandate to promulgate comprehensive regulations governing environmental claims—regulations that would address both technical and environmental policy concerns raised by such claims. In addition, several states have adopted, and several others are considering, their own legislation to address environmental advertising. In the absence of federal action, states have had no choice but to take action on their own.

Over the past several years, both the FTC and a task force of state attorneys general have launched a number of investigations into specific cases of misleading environmental claims. Many of these have resulted in enforcement actions and consent orders leading to cessation of the original claims and delineating conditions under which future claims may be made.

The first state-level legislation addressing green marketing was adopted by California in late 1990. This statute sets standards and definitions for claims of biodegradability and photodegradability, requires a threshold of access to recycling facilities for claims of recyclability, and requires that products or packages labeled "recycled" must contain at least 10 percent postconsumer content.

In an effort to nullify the statute and chill legislative activity by other states and Congress, a coalition of 10 industry trade associations filed a lawsuit in Federal District Court against the state of California seeking to prevent enforcement of the act and to have it declared an unconstitutional infringement of free speech.[18] In July 1992, several environmental organizations formally intervened in the lawsuit in order to aid the California attorney general in defending the statute.

In December 1992, the court ruled that the California law does not infringe upon advertisers' free speech rights.[19] The court's decision recognizes the crucial right of states to set standards regarding the use of environmental marketing claims when a substantial public interest such as the protection of the environment is at stake. Plaintiffs have appealed the decision to the Ninth Circuit Court of Appeals.

Strengthening the Scientific Basis for Sound Environmental Policy

Many environmental claims compare the attributes of competing products, such as polystyrene vs. paper coffee cups or disposable vs. reusable cloth diapers. While competition is a natural feature of a market economy, in many cases (though not all), environmental comparisons between different products can be complex and difficult to assess. This is especially true if different types of impacts are being compared (toxic water pollution vs. energy use, for example).

One emerging method for the technical evaluation of the environmental impacts of a given product is known as "life cycle assessment" (LCA). LCA is a formal methodology being developed to examine the full energy and environmental impacts of products and processes, from the acquisition of raw materials through production, distribution, and use, to ultimate disposal. LCA has assumed a high profile over the past few years owing to its enormous potential both for product improvement and for abuse as a marketing tool by manufacturers and trade associations seeking to gain an advantage for their products. Data uncertainties and the lack of an accepted methodology for conducting life cycle assessments raise serious questions about both the use of LCA as a basis for product claims and our ability to base policy decisions on LCA.

These concerns over the deficiencies and abuses of LCA have led to calls for oversight through mechanisms such as peer review and establishment of a professional code of ethical conduct, use of validated, publicly available data, and requirements for clear and complete presentation of data, assumptions, and limitations. Multisector initiatives organized by EPA and the Society for Environmental Toxicology and Chemistry (SETAC) have begun to address these issues through research and the development of guidance documents.[20]

Composting

For appropriate waste stream components, composting can be the preferred waste management approach (see Chap. 10). It not only diverts materials from the solid waste stream but also converts it into a high-quality product that can be used as a soil amendment or mulch. At present, the United States is composting less than 10 percent of yard and food wastes. Given that yard trimmings constitute roughly a fifth of MSW and food wastes contribute another 7 percent, composting of these organics has the potential to play a significant role in MSW management. However, *composting of mixed solid waste* may undermine environmentally sound materials use and waste management practices, for two major reasons:

1. *Contaminants.* Noncompostable materials in MSW contaminate the final compost product, limiting its quality and utility. Examples of such contaminants include both hazardous and inert materials: toxic heavy metals used in printing or in metal products, household hazardous waste, plastic or rubber remnants, and broken glass. In addition to rising health and environmental concerns when such compost is used, the presence of both toxic and inert contaminants limits the availability of sustainable end markets. The best means of assuring a consistent, quality compost product is to exercise careful control over the inputs to the composting process (see Chap. 10).

Heavy metals are of particular concern because composting of a heterogeneous waste stream can concentrate heavy metals, which are not destroyed in the decomposition process. Existing data clearly demonstrate much higher levels of a host of toxic heavy metals in MSW compost relative to levels found in source-separated, organic compost. Some of the available data are summarized in Tables 7.1 and 7.2.

2. *Loss of Valuable Resources.* Resources that could be recovered and recycled are squandered in a "mass-composting" approach that attempts to circumvent the need for source separation and preprocessing of recyclable materials. For materials other than yard and food waste, and possibly certain fractions of nonrecyclable paper, recycling is a preferred option over composting because recycling not only produces a useful product but also reduces the need to replace virgin materials, thereby conserving the energy and material resources used in, and reducing the environmental impacts arising from, production of the virgin materials. Mixed MSW composting interferes with recycling by discouraging the development or expansion of recycling infrastructures, relieving pressure on companies to redesign products to enhance their recyclability, and diverting resources away from improving recycling markets.

Based on very little study and information, communities across the country are contemplating funding mixed MSW composting projects, attracted by the "turnkey" approach of composting an unseparated waste stream. As of March 1993, 22 mixed

TABLE 7.1 Heavy Metal Content in Composts

	Metals in compost from	
	Mixed MSW ppm	Source separated organics ppm
Pb	290–2850	62
Hg	1.2–8	0.5
Cd	1.8–14	0.5
Cr	11–220	55
Cu	80–240	47
Zn	565–1255	198
Ni	20–73	14

Source: Bernd Franke, "Composting Source Separated Organics," *Biocycle,* vol. 33, no. 8, p. 40, July 1987.

TABLE 7.2   Impact of Separation on Heavy Metal Contents

Metal	Central separated MSW, ppm	Source separated MSW, ppm	Agricultural separated and prunings, ppm
Pb	513	133	27
Hg	2.4	<1	ND
Cd	5.5	1	<1
Cr	71	36	16
Cu	274	33	22
Zn	1570	408	80
Ni	45	29	21

Source: Clarence Golueke and Luis Diaz, "Source Separation and MSW Compost Quality," *Biocycle,* vol. 32, no. 5, p. 70, May 1991.

MSW composting facilities in 12 states were operating, with several dozen more in various phases of development. Two facilities have recently closed down because of operational problems. Agripost, Inc., closed an 800 tons per day (725,760 kg per day) mixed MSW facility in Dade County, Florida, in early 1991. Riedel Oregon Compost Co., a 600 tons per day (544,320 kg per day) mixed MSW facility in Portland, Ore., received orders to shut its doors in January 1992, less than 1 year after it began operations, owing to chronically poor compost quality and odor problems.

In Europe, where composting facilities are more common, localities are moving away from mixed composting because the quality of the final compost exceeds existing standards (e.g., in Germany) or the European Community's proposed 1992 standards. Instead, towns are experimenting with separation systems, such as a wet-dry separation in which the wet (or green) bag contains compostable materials and the dry (or black) bag contains recyclables along with other, noncompostable materials.

In sum, it is critical to recognize that, as with any waste management approach, only certain well-defined fractions of MSW that are kept separate, or are separated from, mixed MSW are appropriately managed through composting. Composting of materials derived from the mixed MSW stream produces a poor-quality product and undermines recycling. The United States needs to develop a municipal and/or private composting infrastructure that is designed—first and foremost—to provide a consistent, high-quality compost product by exerting a high degree of selectivity with respect to materials composted.

7.3   INCINERATION ISSUES

As of the end of 1992, about 160 municipal solid waste incinerators were operating in the United States, with the capacity to burn approximately 110,000 tons of waste per day (99,792,000 kg per day.) Four facilities were under construction, and another 42 in various stages of planning.[21]

Although incineration is often referred to as a waste disposal method, it is more accurately described as a waste processing technology. While it provides the important benefit of reducing the amount (particularly the volume) of waste requiring disposal, it creates a range of air pollution concerns and leaves behind its own substantial burden of toxic ash residues that must be managed and disposed of properly (see Chap. 11).

Unfortunately, indiscriminate use of incineration may severely limit other options for waste management. Without careful advanced planning, present or future opportunities to reduce, recycle, or recover components of MSW—management options that are rapidly gaining in both political and economic acceptance—may conflict with the contractual arrangements and operating requirements of incinerators that were sized or designed without regard to such opportunities.

Many existing incinerator contracts require that a municipality or region guarantee a minimum tonnage of waste for delivery to the incinerator, often for the several decades over which it will operate. Clearly, incentives to reduce the amount of waste generated through changes in consumption patterns or to implement more aggressive recycling programs may be compromised or entirely eliminated by such long-term contractual obligations.

For all these reasons, as well as the risk considerations presented below, incineration has at best a limited role to play in a sound municipal solid waste management system. Unfortunately, this approach is *not* consistent with current or planned practices in many municipalities. In general, some solid waste managers have yet to consider recycling a serious tool of waste management. Rather, the emphasis has most often been on attempting to implement mass burn incineration as a wholesale alternative to landfilling.

Comprehensive Risk Assessment

Despite the clear perception that incineration poses health and environmental risks, the debate is only just beginning to encompass the full range of such risks. Public and regulatory concern has focused on dioxins almost to the exclusion of other toxic constituents of incinerator by-products. There has been a similar fixation on cancer to the virtual exclusion of other adverse health effects, despite the fact that several of the major pollutants released by incinerators (e.g., lead and mercury) are of primary concern because of their noncancer health effects.

Incinerators continue to be viewed primarily as stationary sources of toxic air pollutants that affect ambient air quality. Only recently have risk analyses for proposed incinerators begun to assess pathways of exposure to air emissions in addition to direct inhalation, or the even larger number of pathways of exposure to incinerator ash.

Ignorance of the true scope of these risks is not bliss. Risks that are not accurately identified will not be addressed through use of appropriate control strategies, whether before, during, after, or in lieu of incineration.

Most risk assessments have also failed to consider ongoing exposures to environmental pollutants from other local sources. Such omissions are particularly critical in evaluating incineration, because at least two of the major toxins released by incinerators—lead and dioxins—are persistent toxins with universal exposure. Virtually all U.S. residents now carry measurable levels of these pollutants in their bodies; in some populations, existing levels of exposure and body burdens are already in a range associated with detectable adverse impacts.[22] Thus the evaluation of incremental inputs of these substances—even if they appear small when viewed in isolation—must be of concern. Consideration of the cumulative nature of both exposure to and the health effects induced by many pollutants from incinerators may be a key factor in the siting of such facilities.

Health and Environmental Risks

Incineration yields products that take three forms: energy, gases, and solid residues. The latter two present a variety of environmental concerns. In modern incinerators, gases exit almost exclusively via the stack, with or without prior conditioning before release into the atmosphere. The solids are of two basic types: bottom ash, which is the unburned residue remaining on the grates of the burn chamber, and fly ash, which is initially carried out of the burn chamber by hot gases. Fly ash may fall out in the heat-transfer areas of the incinerator, or it may be deliberately collected by particle-trapping devices within the stack.

Incineration differs from other methods of waste management, such as compaction and direct landfilling, in that the waste stream is physically and chemically transformed during the combustion process. The by-products of this process—both solids and gases—differ markedly from the original waste in their environmental and biological behavior, as discussed below.

Potential releases of these by-products can occur in numerous ways. At the facility site itself, opportunities for release of the solid products arise from the moment of their generation and continue during onsite management, handling, storage, and transport of ash.[23] Further releases may occur both during disposal (e.g., dust generation during landfilling, runoff or wind dispersal from uncovered ash) and after disposal (e.g., accidental or deliberate discharge of leachate or disposal of leachate treatment residues). Many or most of these pathways of exposure to waste products are unique to or far more significant for incineration than for other forms of waste management.

The preceding discussion also illustrates another fundamental point: The impacts of incineration on air quality and the nature and amounts of solid residue generated are inextricably linked. In order to condition stack emissions to meet specific health and regulatory goals, modern incinerators are increasingly required to be equipped with sophisticated air pollution control devices. Such devices, however, increase the amount and change the nature of the solids retained. In particular, both the concentration and leachability of several toxic metals in the fly ash increase.[24]

Disregard for this essential linkage between air quality control and ash toxicity results in "risk transfer," which merely moves risks around without reducing them. Such exercises do little to promote overall environmental quality.

Behavior of Metals in Incineration.   Quantitatively, the compounds of greatest environmental and health concern are the toxic heavy metals. Of primary concern are lead and cadmium, along with arsenic, mercury, selenium, and beryllium, among others.[25]

Metals, being basic chemical elements, are neither created nor destroyed by incineration. Their amounts in the waste stream before incineration are identical to those found in air emissions and ash after incineration; said another way, incineration of metals is a zero-sum game. Unfortunately, however, metals are particularly difficult to manage once they have been transformed into incinerator by-products. Although high-efficiency incineration can greatly reduce and detoxify the organic fraction of garbage, there are no improvements of the incineration process itself that can alter the total output of metals.

Further, incineration is uniquely unsuited for managing metals. Incineration essentially destroys the bulky matrix of MSW—paper, plastics, or other materials—which contains metals and which acts to retard their dispersion into the environment. In this respect, incinerators can be compared with metal smelters: the burning of combustible materials liberates metals, which are subsequently released in air emissions or ash residues in a form more readily taken up by living organisms (i.e., highly "bioavailable"). For example:

- The small particles of ash are light and more likely to be transported by air or by water. This tendency has been demonstrated for MSW incinerators and is well known for metal-containing particles released by other stationary and mobile sources.[26]
- Metals may also be released as fumes. Several metals—in particular mercury—exist largely in the vapor form and escape out the stack even when gas scrubbing equipment is present and properly operating.[27]
- The physical and chemical states of metals present on fly ash particles enhance their ability to leach into water. Metals are present at or near the surface of such particles[28] and are therefore more available for leaching. In addition, the high chlorine content of MSW results in the transformation of many toxic metals from their original chemical forms into metal chloride complexes,[29] which are generally much more soluble in water.

Health Effects of Metals Mobilized by Incineration. Many of the heavy metals present in MSW incinerator by-products have well-defined health effects that have been demonstrated in numerous studies of exposed human populations.[30] Several are carcinogenic (cancer-causing), but these and others can also exert a broad spectrum of severe neurological and other adverse effects in humans, animals, and plants. Arsenic, cadmium, beryllium, and lead are carcinogenic metals; arsenic, lead, vanadium, cadmium, and mercury are neurotoxic; zinc, copper, and mercury are acutely toxic to aquatic life.[31]

Because of their elemental (and therefore, permanent) nature, heavy metals accumulate both in environmental compartments—such as soils, surface dusts, and biota—and within the human body. Thus long-term releases even at low levels have the potential to increase substantially both environmental and human metal burdens. For example, the strong correlation in the United States between automobile lead emissions and body lead burdens[32] vividly demonstrates how a large number of small but widely dispersed releases can significantly affect the general population's exposure level. Indeed, as a result of extensive past uses of lead in paint, gasoline, and other items, the background lead level in a large portion of the general U.S. population is already high enough to cause adverse effects.[33] Accordingly, *even* marginal increases in lead exposure are of major public health concern.

Relevant Routes of Exposure. It is widely recognized that uncontrolled incinerators can release to the air a wide range of pollutants at levels that may pose significant risks through direct inhalation of the emissions. However, comprehensive exposure assessment must go beyond calculations of direct inhalation exposure. Particularly for metals and environmentally persistent organic pollutants, it is clear that the majority of opportunities for human contact occur *after* metal-bearing particulates are deposited on soils, surface water, food, and dust, or accumulate in the buildings where people live or work.

Likewise, only rarely have risk assessments for proposed MSW incinerators discussed the equally important and diverse pathways of exposure to ash residues. For these wastes, exposure assessments must consider not only ultimate disposal, but all earlier steps: onsite handling from the time of generation, storage, transport, and handling and depositing at the landfill until time of final cover. In addition, postdisposal exposure can occur as a result of direct ground or surface water contamination by leachate (for example, as a result of the deliberate discharge of leachate, or through a breach in containment and collection systems).

Needed Controls and Regulations

The risks associated with incineration require comprehensive analysis in order to develop appropriate guidelines for the use of this technology and to restrict its application to those portions of the waste stream where its benefits (e.g., volume reduction) are not overwhelmed by its risks. Techniques for comprehensive exposure assessment are still relatively new and unvalidated, and many risk assessments provided by incinerator vendors or municipalities do not adequately address all of the noninhalation or postdeposition routes of exposure for air emissions or ash residues.

The need to manage MSW more effectively without simply transferring risks, thereby continuing to incur long-term environmental and public health problems, requires that these risks be clearly recognized and addressed. Both short- and long-term solutions need consideration. In the short term, appropriate technological and operating controls should be imposed on both incinerators and ash disposal sites.

With respect to air emissions, regulations are needed which mandate the use of "best available control technology" (BACT) for both combustion and air pollution control devices (see Chap. 12). Maximally feasible on-line (continuous) monitoring should be required to provide a complete record of combustion conditions and gaseous emissions. For emissions of metals and organic compounds, where technology allowing continuous monitoring does not exist, frequent periodic stack testing should be conducted. Such monitoring should be coupled to demonstration of compliance with specific emissions standards for individual pollutants. Given the many uncertainties associated with predicting and monitoring incinerator emissions, these standards should be set at the more stringent of limits derived from health- and technology-based evaluations.

For ash residues, a "cradle-to-grave" regulatory framework is needed, analogous if not identical to that specified for hazardous wastes under the federal Resource Conservation and Recovery Act (RCRA).[34] Regulations governing all phases of ash management—handling, storage, transportation, treatment, and disposal—must be developed to ensure minimal exposure to personnel, the general population, and the environment.

Another concern in ash management is the growing interest in ash utilization.[35] Because ash utilization (e.g., in roadbuilding or construction activities) allows the placing of ash or ash-derived products into the general environment, this activity involves potential exposures that extend well beyond those from ash disposal, in both magnitude and duration. Moreover, opportunities for long-term control over ash or for remedial action are lost in most utilization applications. In the absence of sufficient demonstration of safety, ash utilization may merely transfer, disperse or postpone, rather than eliminate, exposure.

Reducing the Prevalence of Toxic Metals in the Waste Stream.   The preceding discussion of means for reducing the hazards of ash is clearly within the scope of what is normally considered to be ash management. All of them, however, involve management at the back end, that is, *after* hazardous ash has been generated. Another approach may at first glance appear to be beyond the scope of standard approaches to waste management. If, however, steps are taken to remove metal sources from MSW prior to incineration, management of the resulting cleaner ash can be accomplished in a manner that is more protective than disposal of toxic ash even in a state-of-the-art landfill. Targeted removal of items which are major sources of toxic metals can decrease ash toxicity. At least for lead and cadmium, a fairly clear picture of their major sources in MSW is emerging.[36]

Unfortunately, more than 100,000 tons (90,720,000 kg) of lead still finds its way into the solid waste stream each year in the form of lead-acid automobile batteries. In the late 1980s, such batteries accounted for about 75 percent of the lead used in the United States, and approximately 65 percent of the lead in MSW. Consumer electronics were identified as the next largest source of lead in the waste stream, followed by glass, ceramics, and plastics where metals are used as pigments or stabilizers. Nickel-cadmium batteries accounted for about half of the cadmium in solid waste, followed by plastics (30 percent), and consumer appliances (10 percent).[37]

One way to reduce the presence of these metals in the waste stream would be to ban them where substitutes are available and to require collection for recycling where substitutes are not available, such as lead-acid automotive batteries. Several states, including Florida, New Jersey, and Minnesota, and local entities including Suffolk County, New York, have begun to regulate the disposal of some types of batteries. (Also see the earlier discussion of legislation banning the use of four heavy metals in packaging, now adopted in about 20 states.)

In several European countries, government initiatives have been directed toward reducing the use of certain metals (e.g., cadmium) in consumer products, particularly those (such as disposable plastic items) that are used and discarded in large amounts after limited use.[38] Given the use of toxic metals in the manufacture of a very broad range of consumer materials (e.g., printing inks, plastic stabilizing agents) as well as in more easily identified materials such as lead-acid batteries, only comprehensive source-based strategies are likely to prove successful in achieving significant reductions in the metal content and toxicity of MSW and incineration by-products.

An alternative or supplementary approach would be to repeal the percentage depletion allowance in the tax code for lead mining, and implement a market-based approach such as marketable depletion rights or a tax on lead imports and virgin lead mined in the United States. Higher prices for virgin lead ore would improve the competitive position of lead battery recycling operations and would prompt more extensive and better-controlled recycling. This approach would decrease lead contamination of the environment from careless disposal and compel substitution for lead in the many products where substitutes are available, while at the same time reducing the total volume of new lead entering the environment.

7.4 ENVIRONMENTAL CONCERNS OF LANDFILLING*

In the past, landfills involved the placement of waste in an unlined cell or possibly one with a simple clay liner. There would be a periodic light cover and eventually a final earthen cover placed on top. With a small landfill, the amount of the inevitable leachate did not create a serious problem. As the size of landfills and variety of disposed materials increased, regulatory standards became necessary.

From the passage of the Solid Waste Disposal Act in 1965 to approval of new landfill standards under Subtitle D of the Resource Conservation and Recovery Act in 1991, there has been a steady progression of federal regulatory activity. Details of this history are provided in Chaps. 4 and 15.

As a response partially to regulatory changes but also to economics and technology, there have been significant shifts in various aspects of landfill operations. Since the early 1980s there has been a trend toward including some type of liner in the design even before they were required by regulations. Operators had begun to respond

---

*Contributed by Roger Diedrich.

to public pressure and mounting evidence of environmental problems. In the short term, this is an environmental plus, however, as discussed later, this development could have negative long-term environmental implications.

About that same time, and partially because of the addition of liners, there has been a dramatic increase in the average size of newly permitted landfills. *Waste Age* magazine reported that "New landfills generally have greater capacities and may be providing a net increase in capacity."[39] At least two factors are likely to be responsible: a scarcity of available sites due to increasing urbanization, and simple economies of scale. Having fewer but larger landfills should result in easier enforcement by typically overextended state regulators. Of course, if there is a problem, such as groundwater contamination, it will be a much larger, more serious and more expensive problem. Large landfills also suggest the possibility of oversizing. Sizing of the landfill should occur after a sound long-range solid waste management plan is completed and a recycling program is instituted. Such steps will allow an informed estimate to be made of the disposal needs for future years. The size should be limited to the required capacity to serve the target area for perhaps 30 years. Beyond that time frame, the disposal needs and technologies may change considerably. Any oversizing can be a problem similar to (but less acute than) that for an oversized MSW combuster with a put-or-pay contract. That is, the need to pay the high capital cost of the landfill will work against investment in advanced waste reduction or recycling programs. While there has been a trend toward long-hauling waste to remote landfill sites, this practice separates the waste generators from the consequences.

Another change that appears to be occurring is the replacement of the publicly operated landfill with privately owned and operated facilities. The consequences of such a structural change are not well established, but of primary interest is the relationship of each type of owner-operator with state regulators.

Under this umbrella of regulatory and structural change, several environmental concerns remain. Among these concerns are issues of siting, facility design and operation, and long-term liability.

Siting

This concern was substantially addressed under the new Subtitle D regulations; however, only the most obvious limitations have been imposed on site selection. Individual siting proposals will continue to require oversight to evaluate relative merits of specific alternatives, presuming agreement exists on the need for new capacity. All the possible pathways to exposure and cumulative effects must be evaluated, as discussed in the section on incinerators in this chapter. Consideration must be given to the geology, topography, access, and surrounding land use. Land use considerations should include the issue of environmental justice for the poor and minorities. This reflects a concern that some may be asked to accept a disproportionate share of undesired land uses. This concept was recognized by the EPA in a report by the Environmental Equity Work Group (1992). While solutions have not yet emerged, continued interest in this issue can be expected.[40]

Design and Operation

One attribute of the new Subtitle D regulations is the requirement for a composite liner for sanitary landfills. In constructing the liner, the ground surface is layered with sand, clay, and one or more synthetic liners. The flexible membrane liners (FMLs) are made of synthetic polymeric materials, the most common being high-density polyethylene (HDPE). The membranes are vulnerable to errors in installation, punctures, aging, and chemical degradation. HDPE is known to be prone to stress cracking. It is also susceptible to deterioration when exposed to various household substances.[41,42]

After the landfill is filled with waste and intermediate cover, it is capped with layers of dirt and another synthetic membrane. The intended purpose of the cap is to minimize rainwater intrusion and hopefully minimize the quantities of leachate and methane that would be generated. The cap is vulnerable to breaching by plant roots, small animals, settling, and human blundering. While the timing may be highly uncertain, it can be assumed that eventually the cap and liner systems will fail.[43] The other part of a composite liner, the underlying layer of clay, will also permit the passage of leachate. This will occur even in a tightly packed liner through the process of diffusion, which works similar to the process of osmosis.[44]

Because the waste is contained in a dry state after capping, the process of biological breakdown or decay is dramatically slowed or stopped. This is presumed to be a solution, since during the time the water is kept out, there is less generation of harmful leachate or landfill gas. The reason it is a problem, not a solution, is the above-noted failure—all landfills eventually leak. Subtitle D requirements can only cause a delay in, but not prevent, the generation and migration of leachate and methane.

The leachate collection system consists of a layer of sand and a network of perforated PVC pipe that collects leachate, whereupon it is sent to a wastewater treatment plant. Unfortunately, such collection systems can also fail. The systems can clog from silt in the landfill, the growth of microorganisms, or the precipitation of minerals in the pipes, geonets, or sand filters. They can be weakened by chemicals in the landfill and crushed by the weight of garbage above. Such clogging will cause leachate to back up, increasing pressure on any leak that may have developed.

Contributing weaknesses of the new Subtitle D regulations are the failure to mandate leak-detection systems, lax requirements on groundwater monitoring in terms of distance from the landfill and well spacing, failure to address air emissions, and allowing too much flexibility to states to determine their own programs. A leak-detection system would immediately detect problems, unlike groundwater monitoring wells which may not indicate leakage until groundwater is contaminated, in which case it may be irreversible. Unfortunately, point source leaks from lined landfills may be more difficult to detect than diffuse leakage from unlined landfills. This is because these leaks will likely emanate in fingerlike plumes that can pass between the few wells required by EPA regulations, thus further delaying detection. At this point, remediation or otherwise correcting the problem will be difficult or impossible.

Liability

The above noted delay of the release of leachate is of particular concern because of other features of Subtitle D, limiting of groundwater monitoring and financial assurance to 30 years postclosure. That is, the delay in the arrival of the inevitable environmental impacts is likely to exceed this postclosure period. Responsible parties may cease monitoring if no groundwater contamination has been detected within 30 years. They are then released of future liability for groundwater remediation. No funds are required at any point to be held in reserve for corrective action; such funds are required only after contamination has been discovered (if it is within 30 years). Hence the landfill will have been transformed into a time bomb with a 30+ year fuse, poised to release toxic pollutants at a time when the community will be least prepared. It will then be the community's problem, the operator having left with the profits.

Potential Solutions

The inevitability of the described situation, brought about by EPA regulations, will be difficult to correct as long as we continue to landfill—a proposition that most agree is necessary to some extent. At last count, we were still landfilling 65 percent of a growing waste stream. The potential solutions might be placed in three general categories: (1) tighten the EPA requirements, (2) adopt a wet-cell design and operation, and (3) move to a strict toxic reduction regime.

Tighten EPA Subtitle D Requirements.   This approach could include a variety of elements, some of which are listed below.

- Extend the postclosure monitoring period. Unfortunately, there is not yet a clear definition of a sufficient postclosure period. California has required monitoring for as long as wastes present a threat to groundwater.
- Improve groundwater monitoring requirements, such as more wells, located closer to the landfill. This will aid in earlier detection. but it still will only identify widespread contamination, not the start of a problem.
- Add a requirement for a leak-detection system.
- Significantly increase financial assurance requirements for private operators to cover long-term groundwater monitoring and require a corrective action sinking fund. (Public operators have the power to tax and should not be required to post financial assurance.)

Taken together, these types of fixes may sufficiently mitigate the problem, at least in the short term. Indeed, it may be all that is possible in the short term, but these measures will not prevent the discharge and contamination that appear so likely to occur.

Adopt a Wet-Cell Design and Operation.   This approach to accelerated decay of the organic fraction of MSW is targeted at avoiding the problem of releases postponed past the postclosure period. This is because the material will be stabilized in a finite period (estimates range from 10 to 50 years).[45,46] It also increases the methane yield in earlier years, making recovery more feasible. Finally, this approach presents the potential for reuse of the physical property of the landfill. The latter would require excavation, reclamation of recyclable materials, and washing to reduce contamination. It could be expected that the decayed organic material will contain impurities since it was essentially a mixed waste operation. Inert material, if separable, could be disposed of in an unlined cell. Elements of this approach are not fully developed, including knowledge of how to establish and maintain the proper level and distribution of moisture and reclamation procedures. It holds promise for at least a good interim solution.

Move to a Strict Toxic Reduction Regime.   This follows the law identified by Barry Commoner, paraphrased as "If you don't put a pollutant into the environment, you won't find it there." It would involve identification of every toxic element in the waste stream and taking vigorous action to keep them out of the landfill. This would also require diversion of the organic fraction, since their decay is the source of acidic leachate and methane. Because the leachate is capable of mobilizing other toxics such as metals that might otherwise remain in place, the organics are indirectly a source of toxics. Actions to be taken under this approach are the following.

- Removal of toxic materials at the landfill by banning substances or establishing a preprocessing or sorting operation
- Removal of toxics from products through regulation
- Setting standards for diversion of biodegradable materials to composting operations

Using strict toxics and biomass elimination programs, only inert material would remain for landfilling. If these things were done properly, consideration could be given to relaxing many of the new Subtitle D requirements, including the postclosure period.

7.5 CONCLUSION

The ancient Chinese malediction "may you live in interesting times" seems to have been wished upon legislators and public officials involved with solid waste management. Until recently, garbage was viewed as an unpleasant but basically noncontroversial—even boring—topic. Today, however, communities throughout the nation are faced with difficult and controversial choices that involve a variety of complex technical, economic, and environmental factors.

Because recycling is often the least-cost option—in both environmental and economic terms—for large portions of the waste stream, that option has earned a place near the top of the waste management hierarchy, along with source reduction. Recycling is not a one-stop-shopping solution, however. Rather, it is a three-step process: collection of recyclables, remanufacturing of used materials into new goods, and purchase by consumers of those goods. While the first step is within the traditional domain of waste management, the latter two are not. Nonetheless, they are critical if recycling is to fulfill its promise as a waste management strategy.

In addition, governments must keep in mind that recycling, while highly beneficial as a waste management strategy, is not always environmentally benign. Appropriate environmental controls must be established as part and parcel of recycling programs for materials that are not inherently benign.

As a practical matter, significant fractions of the waste stream will remain to be dealt with by means other than recycling in the short term. The reasons are not so much technological—a large fraction of the waste stream is recyclable with current technologies—but rather social and institutional. In particular, business-as-usual attitudes in the waste management and product design communities, along with the current lack of aggressive efforts to develop markets for recycled materials, are impediments that will have to be overcome if recycling is to realize its full potential.

For the time being, though, major debates will continue over what to do with materials that are not recycled. The proper role of incineration (accompanied by energy recovery) as a means of reducing waste volume will continue to be hotly debated; so will the proper technological standards for incinerator emissions, for ash disposal facilities, and for municipal solid waste landfills.

The recycling facilities, landfills, and incinerators built today will be with us for decades to come. They will profoundly affect both the quality of the environment and the fiscal health of communities and states that host them. Both the potential hazards and the potential riches in trash are becoming steadily more apparent, resulting in fundamental and permanent changes in waste management strategies.

REFERENCES

1. Haynes, Richard W., *An Analysis of the Timber Situation in the United States: 1989–2040*, General Technical Report, U.S. Forest Service, 1989.

2. Environmental Protection Agency, *Wastes from the Extraction and Beneficiation of Metallic Ores, Phosphate Rock, Asbestos, Overburden from Uranium Mining and Oil Shale*, December 1985.

3. Kleinmann, Robert L. P., *Acid Mine Drainage in the United States*, U.S. Bureau of Mines, Pittsburgh, Pa., 1990.

4. Denison, unpublished calculations based on data from the following sources: Woodbury, W., *Minerals Yearbook: Lead*, U.S. Department of the Interior, Bureau of Mines, 1985, 1989. Franklin Associates, *Characterization of Products Containing Lead and Cadmium in Municipal Solid Waste in the United States, 1970 to 2000*, EPA Office of Solid Waste, Washington, D.C., January 1989. EPA, *Report to Congress on Wastes from the Extraction and Beneficiation of Metallic Ores, Phosphate Rock, Asbestos, Overburden from Uranium Mining, and Oil Shale*, Office of Solid Waste and Emergency Response, EPA/530-SW-033, 1985, p. ES-6. EPA, *Draft Report to Congress on Solid Waste from Selected Metal Ore Processing Operations*, Office of Solid Waste, Draft for Red Border Review, July 15, 1988, Exhibit 3-13, pp. 3–43. EPA, 54 *Federal Register*, Apr. 17, 1989, p. 15344. EPA, *Report to Congress on Special Wastes from Mineral Processing*, Office of Solid Waste and Emergency Response, EPA/530-SW-90-070C, July 1990, p. 10-4.

5. Franklin Associates, *Characterization of Products Containing Lead and Cadmium in Municipal Solid Waste in the United States, 1970 to 2000*. EPA Office of Solid Waste: Washington, D.C., January 1989.

6. "Mobay to Eliminate Cadmium," *Plastics News*, Apr. 17, 1989, p. 22.

7. Fishbein, Bette K., and Caroline Gelb, *Making Less Garbage: A Planning Guide for Communities*, INFORM, New York, 1992.

8. Fishbein, Bette K., and Caroline Gelb, *Making Less Garbage: A Planning Guide for Communities*, INFORM, New York, 1992.

9. Coalition of Northeastern Governors, "Source Reduction Council of CONEG Progress Report," Mar. 29, 1990, pp. 1–2, Appendix D.

10. State of Minnesota, "Prohibitions on Selected Toxics in Packaging" (Section 50) and "Toxics in Packaging and Products" (Section 51), H.F. No. 303, signed into law on June 4, 1991.

11. *Northeast-Midwest Economic Review*, Sept. 5, 1989. Northeast-Midwest Institute, Washington, D.C.

12. Camp, Dresser and McKee, Inc., *North Central Florida Comprehensive Solid Waste Management Master Plan*. CDM, Tampa, Fla., December 1987. Ruston, John, and Daniel Kirshner, *A Comparative Economic Analysis of a 3,400 Ton per Day Recycling Program for New York City and the Proposed Brooklyn Navy Yard Resource Recovery Facility*. New York, EDF, May 1988. Seattle Solid Waste Utility, 1988. *Final Environmental Impact Statement; Waste Reduction and Disposal Alternatives*. Seattle, Seattle Solid Waste Utility, 1988.

13. Young, John, "Mining the Earth." Worldwatch Paper 109, Worldwatch Institute, Washington, D.C., July 1992.

14. Gertman, Richard, 1988, personal communication, (former) director of recycling, Office of Environmental Management, city of San Jose. Currently with R. W. Beck and Associates.

The city's estimate of cost savings was not necessarily shared by its private waste collection contractor.

15. Environmental Defense Fund, *To Burn or Not to Burn: The Economic Advantages of Recycling over Garbage Incineration for New York City,* New York, EDF, Appendix C, 1985.

16. New York City Department of Sanitation, *White Paper—New York City Recycling Strategy,* New York, DOS, January 1988. Appendix B.

17. For example, the Tellus Institute (formerly the Energy Systems Research Group) now offers a model called "WastePlan." 89 Broad Street, Boston, MA 02110. Peter Andersen, of New Paths, Inc., offers "RecycleWare," a program that reportedly optimizes recycling efficiency, 2701 Packers Ave., Madison, WI 53704. Jeffery Morris, an economist at Sound Resources Management Group, also offers maximum recycling and optimization services, 7220 Ledriot Court, Seattle, WA 98136. For older models and commentary, see EDF, *Coming Full Circle,* Appendix D.

18. Association of National Advertisers, Inc., et al. v. Lundren et al., complaint filed in No. C-92-0660 (MHP), United States District Court for the Northern District of California, Feb. 5, 1992.

19. Association of National Advertisers v. Lungre, 809 F. Supp. 747 (N.D. Cal. 1992).

20. For more information see Fava, J. A., F. Consoli, R. A. Denison, K. Dickson, T. Mohin, and B. Vigon (eds.), *A Conceptual Framework for Lifecycle Impact Analysis,* Society for Environmental Toxicology and Chemistry and SETAC Foundation for Environmental Education, Pensacola, Fla., 1993. U.S. Environmental Protection Agency, *Life-Cycle Assessment: Inventory Guidelines and Principles,* Risk Reduction Engineering Laboratory, Office of Research and Development, Cincinnati, Ohio, EPA Report EPA/600/R-92/036, November 1992. Denison, Richard A., "Toward a Code of Ethical Conduct for Lifecycle Assessments," Environmental Defense Fund, Washington, D.C., Sept. 2, 1992. Fava, J. A., R. A. Denison, B. Jones, M. A. Curran, B. Vigon, S. Selke, and J. Barnum (eds.), *A Technical Framework for Life-Cycle Assessment,* Society for Environmental Toxicology and Chemistry and SETAC Foundation for Environmental Education, Washington, D.C., 1991. Fava, J. A., F. Consoli, and R. A. Denison, "Analyses of Product Life-Cycle Assessment Application," presented at the SETAC-Europe Workshop on Life-Cycle Assessment, Leiden, The Netherlands, Dec. 2–3, 1991

21. "WTE in North America: Managing 110,000 Tons per Day." *Solid Waste & Power,* June 1993.

22. Agency for Toxic Substances and Disease Registry, *The Nature and Extent of Lead Poisoning in Children in the United States,* A Report to Congress, U.S. Public Health Service, Atlanta, Ga., 1988. Schecter, A., and T. Gasiewicz, "Health Hazard Assessment of Chlorinated Dioxins and Dibenzofurans Contained in Human Milk." *Chemosphere 16,* pp. 2147–2154, 1987. Schecter, A., and T. Gasiewicz, "Human Breast Milk Levels of Dioxins and Dibenzofurans and Their Significance with Respect to Current Risk Assessments," in *Solving Hazardous Waste Problems: Learning from Dioxins,* Exner, J. H. (ed.), American Chemical Society Symposium Series, No. 191. Washington, D.C., American Chemical Society Symposium, 1987.

23. Hutton, M., A. Wadge, and P. J. Milligan, "Environmental Levels of Cadmium and Lead in the Vicinity of a Major Refuse Incinerator," *Atmospheric Environment,* vol. 22, p. 411, 1988.

24. Environment Canada, *The National Incinerator Testing and Evaluation Program: Air Pollution Control Technology, Summary Report* (EPS 3/UP/2), Ottawa, Ontario, 1986. Sawell, S., T. Bridle, and T. Constable, "Leachability of Organic and Inorganic Contaminants in Ashes from Lime-Based Air Pollution Control Devices on a Municipal Waste Incinerator," paper presented at annual meeting of the Air Pollution Control Association, June 1981, at New York.

25. Environmental Protection Agency, *Municipal Waste Combustion Study: Emission Data Base for Municipal Waste Combustors,* EPA/530-SW-87-021b, Government Printing Office, Washington, D.C., 1987. Knudson, J. C. *Study of Municipal Incineration Residue and Its*

Designation as a Dangerous Waste, Solid Waste Section, Department of Ecology, state of Washington, Olympia, Wash., 1986. Brunner, P. H., and H. Monch, "The Flux of Metals through Municipal Solid Waste Incinerators," *Waste Management Research,* vol. 105, p. 4, 1986. Vogg, H., H. Braun, M. Metzger, and J. Schneider, "The Specific Role of Cadmium and Mercury in Municipal Solid Waste Incineration," *Waste Management Research,* vol. 4, p. 65, 1986. NUS Corporation, *Characterization of Municipal Waste Combustor Ashes and Leachates from Municipal Solid Waste Landfills, Monofills, and Codisposal Sites,* 7 volumes. EPA Contract 68-01-7310, U.S. Environmental Protection Agency, Government Printing Office, Washington, D.C., October 1987.

26. Berlincioni, M., and A. di Domencio, "Polychlorodibenzo-p-dioxins and Poly-chlorodibenzofurans in the Soil near the Municipal Incinerator Florence, Italy," *Environmental Science & Technology,* vol. 21, p. 1063, 1987. Environmental Protection Agency, *Air Quality Criteria for Lead,* EPA-600/8-83028aF-dF. Research Triangle Park, N.C., 1986. Harper, M., K. R. Sullivan, and M. J. Quinn, "Wind Dispersal of Metals from Smelter Waste Tips and Their Contribution to Environmental Contamination," *Environmental Science & Technology,* vol. 21, p. 481, 1987. Nriagu, J. O. (ed.), *Changing Metal Cycles and Human Health,* Springer-Verlag, Berlin, 1984. Roberts, T. M., T. C. Hutchinson, J. Paciga, A. Chattopadhyay, R. E. Jervis, J. VanLoon, and D. K. Parkindon, "Lead Contamination around Secondary Smelters: Estimation of Dispersal and Accumulation by Humans," *Science,* vol. 1986, p. 1120, 1974. Hutton et al., Ref. 23.

27. Environment Canada, *The National Incinerator Testing and Evaluation Program: Air Pollution Control Technology. Summary Report* EPS 3/UP/2, Ottawa, Ontario, 1986. Brunner, P. H., and H. Monch, "The Flux of Metals through Municipal Solid Waste Incinerators," *Waste Management Research,* vol. 4, p. 105, 1986. Vogg, H., H. Braun, M. Metzger, and J. Schneider, "The Specific Role of Cadmium and Mercury in Municipal Solid Waste Incineration," *Waste Management Research,* vol. 4 , p. 65, 1986.

28. Wadge, A., and M. Hutton, "The Leachability and Chemical Speciation of Selected Trace Elements in Fly Ash from Coal Combustion and Refuse Incineration," *Environmental Pollution,* vol. 48, p. 65, 1987. Norton, G. A., E. L. DeKalb, and K. L. Malaby, "Elemental Composition of Suspended Particulate Matter from the Combustion of Coal and Coal/Refuse Mixtures," *Environmental Science & Technology,* vol. 20, p. 604, 1986.

29. Bridle, T. R., P. L. Cote, T. W. Constable, and J. L. Fraser, "Evaluation of Heavy Metal Leachability from Solid Wasters," *Water Science & Technology,* vol. 19, p. 1029, 1987. Brunner, P. H., and P. Baccini, "The Generation of Hazardous Waste by MSW-Incineration Calls for New Concepts in Thermal Waste Treatment," presented at Second Conference on New Frontiers for Hazardous Waste Management, Pittsburgh, Pa., Sept. 27–30, 1987. Brunner, P. H. and H. Monch, "The Flux of Metals through Municipal Solid Waste Incinerators," *Waste Management Research,* vol. 4, p. 105, 1986.

30. Friberg, L., G. F. Nordberg, and V. B. Vouk, *Handbook of the Toxicology of Metals,* vol. 1, Elsevier, Amsterdam, 1986.

31. Friberg, L., G. F. Nordberg, and V. B. Vouk, *Handbook of the Toxicology of Metals,* vol. 2, Elsevier, Amsterdam, 1986. Callahan, M., M. Slimak, N. Gabel, I. May, C. Fowler, J. Freed, P. Jennings, R. Durfie, F. Whitemore, B. Maestri, W. Maybey, B. Holt, and C. Gould, "Water Related Environmental Fate of 129 Priority Pollutants." Report 440/4-79-0029, U.S. Environmental Protection Agency, Washington, D.C., 1979.

32. Annest, J. L., J. L. Pirkle, D. Makuc, J. W. Neese, D. D. Bayse, and M. G. Kovar, "Chronological Trend in Blood Lead Levels between 1976–1980," *New England Journal of Medicine,* vol. 308, p. 1373, 1983.

33. Bellinger, D., A. Leviton, C. Waterneaux, H. Needleman, and M. Rabinowitz, "Longitudinal Analysis of Prenatal and Postnatal Lead Exposure and Early Cognitive Development," *New England Journal of Medicine,* vol. 316, p. 1037, 1987. Dietrich, K., K. Krafft, M. Bier, P. A. Succop, O. Berger, and R. L. Bornschein, "Early Effects of Fetal Lead Exposure: Neurobehavioral Findings at 6 Months," *International Journal of Biosocial Research,* vol. 8, p. 151, 1986. Dietrich, K. N., K. M. Krafft, R. Shukla, R. L. Bornschein, and P. A. Succop, "Neurobehavioral Effects of Prenatal and Early Postnatal Lead Exposure," in Schroeder, S.

R. (ed.), *Toxic Substances and Mental Retardation: Neurobehavioral Toxicology and Teratology,* AAMD Monograph Series, No. 8, Washington, D.C., 1987, p. 71. Rutter, M., and R. R. Jones (eds.), *Lead versus Health: Sources and Effects of Low Level Lead Exposure,* Wiley, London, 1983. Agency for Toxic Substances and Disease Registry, *The Nature and Extent of Lead Poisoning in Children in the United States: A Report to Congress,* U.S. Public Health Service, Atlanta, Ga., 1988.

34. 42 U.S.C. section 6901 et seq.

35. Denison, R. A., "Ash Utilization: An Idea before Its Time?" *Proceedings of a Conference on Ash Utilization,* Philadelphia, Pa., Oct. 13–14, 1988.

36. Franklin, Ref. 4.

37. Franklin, Ref. 4.

38. Bothen, M., U.-B. Fallenius, *Cadmium: Occurrence, Uses and Stipulations,* National Swedish Environmental Protection Board, Report SNV PM 1615, Solna, Sweden. Ministry of Housing, Physical Planning, and Environment, "Chemical Substances Act: Cadmium Decree," *Government Gazette* 60, Leidschendam, The Netherlands, 1987. Commission of the European Communities, "Environmental Pollution by Cadmium: Proposed Action Programme," *Report COM(87)* 165, Brussels, Belgium, April 1987.

39. Repa, E. W., and S. K. Sheets, "Landfill Capacity in North America," *Waste Age,* May 1992.

40. Committee on Environmental Law, American Bar Association, series of four articles, *Environmental Law,* vol. 12, no. 1, pp. 1–6, fall/winter 1992–1993.

41. Lee, G. F., and R. A. Jones, *Municipal Solid Waste Management in Lined, "Dry Tomb" Landfills: A Technologically Flawed Approach for Protection of Groundwater Quality,* El Marcero, Calif., March 1992.

42. Rachel's Hazardous Waste News 217, "Why Plastic Landfill Liners *Always* Fail," *Environmental Research Foundation,* Washington, D.C., Jan. 23, 1991.

43. Johnson, R. L., J. A. Cherry, and J. F. Pankanow, "Diffusive Contaminant Transport in Natural Clay: A Field Example and Implications for Clay-lined Waste Disposal Sites," *Environmental Science and Technology,* vol. 23, pp. 340–349, March 1989.

44. Montague, Peter, "The Limitations of Landfilling," in Piasecki, Bruce (ed.), *Beyond Dumping, New Strategies for Controlling Toxic Contamination,* Quorum Books, Westport, Conn., 1984.

45. "Let It Rot," *Garbage Magazine,* September-October 1991, pp. 28–30.

46. Lee, G. F., and R. A. Jones, *Garbage Magazine,* September-October 1991.

CHAPTER 8

SOURCE REDUCTION

Bette K. Fishbein

Senior Fellow, Municipal Solid Waste Program, INFORM, Inc.
New York, N.Y.

Ken Geiser

University of Massachusetts-Lowell
Lowell, Massachusetts

Caroline Gelb*

Researcher, INFORM, Inc.
New York, N.Y.

8.1 INTRODUCTION†

As discussed in Chap. 1, the most straightforward strategy to cope with the problems related to municipal solid waste management is to prevent the generation of waste in the first place. OTA defines MSW prevention as "activities that reduce the toxicity and/or quantity of discarded products *before* these products are purchased, used or discarded." The EPA uses the term "source reduction."

The two basic routes to MSW prevention are for the manufacturers to change the design of products and the ways in which they are packaged, and for consumers to change their purchasing decisions and the way in which they use and discard products. Source reduction can be achieved by reducing the toxicity of manufactured products and their quantity. Reducing toxicity means finding benign substitutes for substances that pose risks when they are ultimately discarded. An example of reducing quantity is changing the design of the product so that less MSW is generated when the product or its residuals are discarded. For example, when the life of automobile tires is increased from 20,000 miles on the average to 40,000 miles on the average, only half the number of tires will be discarded per unit time. The substitution of biodegradable inks for newspaper print using heavy metals would facilitate the recycling or incineration of newspapers.

---

*Currently at New York City Department of Sanitation.
†Contributed by Frank Kreith.

Over the past years, some reduction in MSW toxicity and quantity has occurred, but many more possibilities exist. New products can be designed to make MSW management easier and increase their useful life. Just as toxicity and quantity should be considered when new products are designed, also their recyclability and degradability should be part of good engineering design in the future. However, it is difficult to target products for quantity reduction efforts because people have become accustomed to the convenience of packaging and single-use products, such as disposable diapers. On the other hand, packaging performs critical functions, such as decreasing food spoilage and preventing pilferage or tampering with medications. Even the increased use of plastics, which is criticized by many people, can contribute to improved product use and increased lifetime. These are considerations that will enter into the decision of how far it is useful to carry source reduction. In this chapter, many opportunities for source reduction in both quantity and toxicity reduction are discussed. Since the approaches taken to quantity reduction and toxicity reduction are quite different, the chapter is divided into two parts dealing with each of these source reduction approaches separately.

8.2   REDUCING THE AMOUNT OF GARBAGE*

Understanding Source Reduction

The increasing amounts and toxicity of municipal solid waste, increasing disposal costs, and increasing opposition to siting waste management facilities all point to the importance of source reduction in alleviating this country's solid waste problem. From 1960 to 1990, United States garbage more than doubled, rising from 88 to 196 million tons per year. Per capita, waste generation also increased: the average U.S. resident is producing 59 percent more waste now than 30 years ago, or 4.3 lb per person per day. Even as we are producing ever-increasing amounts of garbage, our ability and willingness to bear the environmental and economic costs of managing and disposing of this garbage are decreasing.

Source reduction—by reducing the amount and/or toxicity of waste actually generated—offers promising environmental and economic opportunities. It means that communities need to collect, process, and dispose of less waste, thereby reducing both their waste management costs and the environmental impacts of waste management facilities and activities. The case for source reduction is the case for prevention rather than remediation: the old saying that an ounce of prevention is worth a pound of cure is as true for garbage as it is for health.

In fact, source reduction tops the list of solid waste policy options adopted by the U.S. Environmental Protection Agency and endorsed by most states and localities.[1] Often referred to as "the hierarchy," these options are as follows:

1. Source reduction and reuse
2. Recycling of materials, including composting of yard and some food waste
3. Waste combustion (incineration)
4. Landfilling

Although most states and localities have endorsed the hierarchy, until recently source reduction has been hailed in principle but neglected in practice, as evidenced by the

*Contributed by Bette K. Fishbein and Caroline Gelb.

ever-growing per capita waste generation rates. Why? Because solid waste officials, while recognizing the benefits of source reduction, have not had a clear picture of how to make it happen.

This chapter is designed to help incorporate source reduction in solid waste plans. This information was developed after a 2-year study that culminated in a book published by INFORM in January 1993: *Making Less Garbage: A Planning Guide for Communities.*[2]

What Source Reduction Is.   Much confusion exists as to what source reduction is and how it differs from "waste reduction" and "recycling." INFORM uses the term "source reduction" to mean "a reduction in the amount and/or toxicity of waste entering the waste stream—or waste prevention" (this is consistent with the definition used by the EPA).

Waste is generated at "the source," the point where it enters a waste collection system. After waste is generated, it must be collected and managed, through composting, recycling, incineration, or landfilling. Only initiatives that reduce the amount of waste actually generated are source reduction; anything that happens to waste after it is put out for collection is not source reduction because the waste must be managed.

There are, however, some gray areas in defining source reduction, which generally relate to reuse. Activities that require some collection outside the public collection system and that lead to reuse without remanufacturing can be considered source reduction. For example, reuse of clothing and equipment through donations to charitable organizations or swaps is source reduction since no remanufacturing is required. Similarly, returning beverage containers for refilling is source reduction, but returning them for recycling is not because they must be remanufactured into a new product.

In defining source reduction, it is important to understand its dual meaning: it refers not only to the amount of waste generated but also to its toxicity. Sometimes reductions in amount may be accompanied by increases in toxicity. In light weighted packages (packages that use lighter-weight materials), for example, the lighter-weight plastics may contain more toxic constituents than the glass they replace. Such potential trade-offs between reductions in amount and toxicity are difficult to quantify, but preferable source reduction strategies would not reduce either amount or toxicity at the expense of the other.

Source reduction of municipal solid waste can be accomplished in many ways: through manufacturer redesign of products and packages; through consumer purchases of less wasteful products and reuse of products; and through institutional changes in practices (such as using paper on both sides) and purchases of more durable and less toxic products. Extending product life accomplishes source reduction because, for example, a refrigerator that lasts 20 years creates half the waste of a refrigerator of the same size that lasts 10 years. Product life can be extended through design and manufacturing processes as well as through better maintenance and repair.

What Source Reduction Is Not.   To implement an effective policy, determining what source reduction is not is as important as determining what it is. It is not recycling or buying products with recycled content. It is not municipal composting. It is not household hazardous waste collection. It is not beverage container deposit and return systems, as accomplished by currently operating "bottle bills" in the United States when they lead to recycling, not refilling.

Recycling is a strategy for managing materials that would otherwise be treated as waste, but it does not reduce the amount of waste generated. It reduces the amount of waste requiring disposal (incineration and landfilling) and is a vital strategy for conserving energy and natural resources. But recyclables must be collected, processed, and remanufactured into new materials or products. These processes incur costs and may themselves cause pollution. Growing concerns about the high costs of recycling and the problems in developing markets for recyclable materials point to the necessity of maximizing source reduction.

Packaging provides an example of how the implications of source reduction policy differ from those of recycling. Manufacturers often claim that a package is "environmentally friendly" because it is made of recycled content or is recyclable. However, recycling cannot justify overpackaging (such as the use of extra, unneeded layers of packaging), if the hierarchy that puts source reduction before recycling is taken seriously. Following the hierarchy would require that packaging be reduced as much as possible before moving to a recycling strategy. For example, consider a product such as shampoo or a bath gel that is sold in a bottle that is wrapped in corrugated paper and then inserted in a box that in turn is shrink-wrapped. A source reduction strategy would aim to eliminate the excess layers of packaging: the box, corrugated paper, and shrink wrap—leaving the product in a single container. A recycling strategy alone would require that all the layers be recyclable or made of recycled materials, but it would not eliminate them.

Composting also illustrates the distinctions between source reduction and recycling. Municipal composting is a form of recycling. It requires the collection and processing of grass clippings and leaves, and sometimes marketing as well. Backyard composting, however, is source reduction. When individuals keep grass clippings and leaves in their own backyards and then reuse the decomposed material, nothing has entered the recycling system or waste stream. The municipality has not collected, processed, or disposed of the material, so source reduction has been achieved.

Separate collection of household hazardous wastes is often cited as a source reduction initiative. While this activity is commendable, it is not source reduction. The hazardous waste must still be collected and disposed of—often in special hazardous waste facilities. True hazardous waste source reduction would substitute nontoxic materials for toxics in the manufacturing process or reuse materials, such as old paint, so that they do not enter the waste stream at all. Replacing a battery-operated pencil sharpener by a manually operated one is source reduction because it eliminates some batteries from the waste stream.

Source Reduction Planning and Infrastructure

Planning is central to developing effective source reduction programs. Before source reduction planners start developing specific source reduction initiatives for their communities, it is extremely important that they know what they are trying to reduce, how much reduction they want to achieve, and how they will measure their results. Municipal solid waste plans need to include an explicitly stated source reduction policy, clearly defined goals, and meaningful measurement strategies. Without these, planners will have difficulty evaluating the effectiveness of their programs.

Implementing a source reduction program also involves developing an infrastructure to support it. Specifically, an effective program requires independent leadership, authority, appropriate staffing, and an adequate budget.

Source Reduction Policy.   The first step in planning for source reduction is a clear statement of policy, including a definition of terms that clarifies what source reduction means so that it can be differentiated from other waste management options, such as recycling. In other words, instead of a policy of "diversion from landfills," which leaves ambiguity as to whether the strategy should be source reduction, recycling, or (in some cases) incineration, a clear policy would state explicitly that its aim is source reduction, include a definition of that term, and then specify goals and measurement methodology.

Setting Source Reduction Goals and Establishing Measurement Methodologies. The next steps in source reduction planning are setting goals and establishing measurement methodologies. Goals and measurement systems are important for effective source reduction programs because they help communities establish program priorities, track and evaluate progress, and recognize accomplishments and target areas for further efforts.

As of mid-1992, only seven states had specific, separate source reduction goals: Connecticut, Maine, Massachusetts, Michigan, New Jersey, New York, and Pennsylvania. Rhode Island and Wisconsin are also actively pursuing source reduction, although they deliberately have not set separate goals. In comparison, 23 states and the District of Columbia had recycling goals, and 17 states had waste reduction goals (these do not differentiate between the amount designated for recycling and that designated for source reduction).

To most effectively set goals and establish measurement methodologies, communities need to take four steps.

1. Establish an overall source reduction goal that is separate from the recycling goal with specification of:

   - The baseline year
   - Target year
   - Type of reduction to be measured (from the current total waste generation levels, from current per capita generation levels, or from the projected increase)

2. Determine separate goals desired for:

   - Generating sectors (residential, commercial, and institutional)
   - Materials (paper, glass, plastics, organics, etc.)
   - Products (Styrofoam cups, glass bottles, tires, cardboard boxes, newspapers, etc.)

3. Select unit of measurement:

   - Weight
   - Volume
   - Weight and volume (preferable, if possible)

4. Select measurement methodology:

   - Waste audits
   - Sampling (including weighing-in places such as transfer stations)
   - Surveys
   - Purchases (tracking sales)

Information Needs for Measuring Source Reduction. Good data collection is vital for measuring source reduction, since communities need to know which sources are generating which types of waste materials, and how much they are generating. Thus, at a minimum, communities need to collect data on:

- Amount of residential waste
- Amount of commercial waste
- Residential population

- Total employment
- Projections of population change
- An index of economic activity

The Importance of Waste Composition.   Knowledge of the composition of the waste stream is very helpful in setting realistic goals because it allows communities to set source reduction priorities. Materials can be targeted for source reduction if they constitute a major proportion of the waste stream, are easy to reduce, or are major contributors to pollution during disposal. Since the waste stream varies from community to community, in-depth information about it requires a waste audit—an actual sampling of waste generated to determine its composition by material, product, and generating sector.

Yard waste, for example, is a good target for source reduction. When burned, it creates emissions of nitrogen oxides (NO_x); it represents a large component of the waste stream (approximately 18 percent nationally);[3] and it can quite readily be reduced through backyard composting. A suburban community with a large proportion of yard waste can set a higher overall source reduction goal than a densely populated city with a small proportion of yard waste.

Administration and Budget.

Administration and Budget.   Departments charged with managing solid waste have traditionally been staffed by officials knowledgeable primarily about waste disposal and, more recently, about recycling. Their responsibilities have been the collection, transport, and disposal of waste, and the processing and marketing of recyclable materials. Their key concerns have been diminishing disposal capacity, siting new facilities, and controlling costs.

Implementing source reduction programs involves vastly different staff skills and concerns. It requires staff with a broader, long-term view of the use of materials in society and an understanding of how behavior can be changed to optimize the use of resources and minimize the waste generated. Staff members need diverse skills so they can work on planning, program development, technical assistance, education, outreach, legislation, data collection, program evaluation, waste audits, and enforcement. Their concerns must encompass broad issues (such as impacts on economic development) that go well beyond questions of how to manage garbage.

Administration.   Efforts to provide independence and authority for source reduction are essential if it is to become a viable policy option. For the most effective administrative structure, source reduction would be separate from and independent of waste management functions. The head of the source reduction effort would have authority at least equal to that of the individuals in charge of recycling and disposal, and would have a commitment to minimizing the amount of materials actually entering the waste stream.

Source reduction is much broader in scope than recycling or disposal and is, in fact, resource management rather than waste or material management. That is, it involves decisions about what products and packages are made, how they are made, and how they are used. An effective source reduction program deals with producers, distributors, and consumers. It can thus be argued that source reduction does not belong in sanitation or solid waste departments at all, and should not be a function of waste managers. Theoretically, it might make more sense to place source reduction activities in a department of economic development. On a more practical level, however, the motivation to promote source reduction is generally the need to reduce waste, so it is likely to remain in the purview of solid waste departments.

If source reduction functions are placed in a solid waste department, they need some independence from the recycling functions because the immediate, everyday demands of recycling can tend to overwhelm the longer-term, more complex source reduction activities. Larger budgets and more personnel are required for recycling because it includes collection, processing, and marketing; the scale and urgency of these management tasks may result in eclipsing the attention given to source reduction.

Despite the virtually universal endorsement of source reduction as the top priority, in no instance did INFORM find an administrative structure that reflects this, in terms of either independence or authority. In fact, INFORM has identified only three localities and two states that have government employees who spend most of their time on source reduction: New York City, Seattle, and Olmstead County (Minn.), at the local level, and Minnesota and New Jersey at the state level. Often described as waste prevention specialists, most of these individuals are in departments that deal with recycling, spend some of their time on recycling activities, and have very limited resources.

Budget. Source reduction does not require the costly collection and processing operations involved in waste management options, but it is not free and it cannot be accomplished without an adequate budget. The costs of source reduction are in the form of an up-front investment in data collection, waste audits, legislative development, education, technical assistance, equipment, and planning.

The payoff for investment in source reduction is not immediate, but it can be very large. In the northeast, for example, the cost of collection and disposal can exceed $200 per ton, and New York City estimates its recycling costs at about $300 per ton. In its 20-year solid waste management plan, New York City estimates the cost of some source reduction activities at $20 per ton and avoided costs from source reduction in the $140 to $150 per ton range. Therefore, the plan states, the cost of prevention programs could increase by as much as $120 to $130 per ton before costs would exceed benefits.[4]

New York City's plan further estimates that in the year 2000, if the city reduces 670,000 tons of waste (approximately 8 percent of its waste stream), total avoided waste management costs due to source reduction would be approximately $90 million: $30 million in avoided collection costs and $60 million in avoided processing costs.[4] The New York City plan states that "prevention programs become increasingly cost-effective as prevented percentages increase. The reason for this is that larger prevented tonnages allow relatively greater reductions in truck shifts and facility capacity; conversely, when reductions are smaller, fewer savings are captured through reduced collection and facility costs."[4]

A barrier to funding source reduction is that results may not happen immediately, so there may be no return on the investment in the budget year in which the expense is incurred. In order to assure continual and adequate funding, source reduction could be funded from a designated income stream. This might be a portion of the funds raised from charging residents for the amount of waste disposed or other waste collection fees, environmental taxes or fees, or possibly unreturned beverage container deposits. Source reduction could also be funded as a specific percentage of a recycling budget. For instance, if 5 percent were chosen, a recycling budget of $10 million would mean source reduction would be allocated $500,000.

Specific Source Reduction Initiatives

The potential for source reduction is enormous. Virtually every individual and every organization can play a role and become part of the solution to the nation's solid waste problem. The strategies discussed here all relate to reducing the amount of waste. Reducing the toxicity of waste is addressed in Sec. 8.3.

INFORM has identified dozens of examples of source reduction initiatives that are already successfully reducing waste. They come from every sector of our society: state and local government; businesses of all sizes; public institutions such as schools, hospitals, and parks; citizens' groups; individual consumers; and nonprofit organizations. These specific successful programs are described in detail in *Making Less Garbage*,[2] and some are summarized here.

For ease of reference, these initiatives can be organized into six categories:

1. Government source reduction programs (procurement and operations)

2. Institutional source reduction programs

3. Government assistance programs (technical assistance, backyard composting and leave-on-lawn assistance, grants, pilot programs, clearinghouses, awards and contests, and reuse programs)

4. Education (in households and in schools)

5. Economic incentives and disincentives (variable waste disposal fees, taxes, deposit and refund systems, tax credits, and financial bonuses)

6. Regulatory measures (required source reduction plans, labeling, bans, and packaging)

Government Source Reduction Programs. Government—federal, state, or local—employs one out of every six workers in the United States, a total of 19 million people. Successful efforts to reduce the waste generated by this work force not only could have a great impact on the municipal solid waste stream but could also provide a model for businesses, institutions, and consumers.

Exact figures on how much waste government workers produce are not available. However, given that the nonresidential sector generated about 78 million tons of waste in 1990, it can be estimated that government generated approximately 13 million tons of waste annually.* As part of this total, government agencies generate almost 1.7 million tons of office paper waste each year, over 20 percent of all paper waste from offices throughout the country.[5]

Strategies for implementing source reduction in government agencies (as well as businesses) fall into two main categories: (1) changing procurement policies and (2) modifying operations.

Procurement. Government purchases of goods and services at the federal, state, and local levels account for approximately 20 percent of the gross national product,[6] or about $1 trillion a year. Hence government as a whole has great power as a customer. Changes in government procurement policies to favor source reduction could have impact both within and beyond government (1) by reducing the amount of waste generated by government; (2) by setting an example for the private sector; and (3) by encouraging manufacturers to develop less wasteful products and packages which would then be available to all purchasers.

While the federal government and many states have adopted procurement policies favoring recycled goods, these agencies have rarely used procurement policy to achieve source reduction objectives. Yet procurement guidelines could require the purchase of reusable, refillable, repairable, more durable, and less toxic items. They could also require minimal and reusable packaging. Such policies would not only

---

*Communication with Nicholas Artz of Franklin Associates, Ltd., in March 1992 indicates that Franklin estimates that the commercial sector generates 40 percent of the national municipal solid waste stream by weight.

reduce waste but would in many cases save money by reducing purchasing, mailing, and disposal costs.

One state that is working on government source reduction procurement is Minnesota.[7] A 1980 state statute required the Minnesota Department of Administration to develop and implement policies to recycle and reduce waste by changing what it purchases. This program now focuses on source reduction.

The department is producing a guidebook and training program for state officials to teach them about source reduction. It has also completed an assessment of energy usage of certain equipment, and plans to start an analysis of cost vs. waste issues for state purchases. Within the Department of Administration, the Materials Management Division is preparing a procurement model and routinely reviews its purchasing criteria to encourage the purchase of durable, reusable, and repairable items. The division is also buying used or surplus equipment.

The state of Wisconsin is developing a system of life-cycle costing for procurement. Life-cycle costing is helpful in comparing the costs of durable and reusable products with the costs of disposables, because it assesses the annual cost of products over their useful life. Durable products generally cost more initially, but the annual cost over their lifetime may be lower than that of disposables.

The state procurement recycling coordinator is developing cost assessment formulas for the products agencies buy. For example, the formula for lawnmowers will include fuel use and durability. The switch to equipment that will generate less waste will be facilitated by the formulas.

There is an abundance of opportunities for reducing waste through procurement policies. Some additional options are:

- Setting a price preference for reusables, refillables, durables, and equipment that reduces waste, such as double-sided copy machines

- Requiring companies that ship goods to government agencies to package them in reusable shipping containers and/or to take back the packaging; for example, furniture that can be delivered in reusable shipping blankets

- Negotiating for longer and more comprehensive warranties and service contracts when purchasing durable goods

- Leasing equipment instead of buying it to provide manufacturers with an incentive to keep it in good repair

- Purchasing items that can reduce paper use, such as double-sided photocopy machines and laser printers and equipment and computer software that permits faxing from a computer to reduce printouts

Government Operations. Government operations could also be changed to promote source reduction. For example, government offices with lawns and campuses can compost yard waste on site and leave grass clippings on lawns. Employees can be educated to reduce paper use and reliance on disposables, and to reuse materials that they might otherwise discard, such as paint. Government agencies can begin source reduction programs on their own, or they may be mandated by a mayor or governor.

Office paper is an excellent candidate for source reduction in government agencies as well as all businesses. It is an important segment of the waste stream, and organizations have a relatively high degree of control over its use and disposal. Office paper is one of the fastest-growing segments of municipal solid waste in the United States. In 1960, 1.5 million tons were generated, or 1.7 percent of municipal solid waste. This rose to 7.3 million tons in 1988 (4.1 percent of municipal solid waste) and is projected at 16 million tons by 2010 (6.4 percent of municipal solid waste).[8]

Government's use of office paper comprises over 20 percent of all office paper used in the country, or 1.5 million tons in 1988.[5] Thus office paper reductions made in government agencies could greatly reduce the waste stream. If offices across the country (both government and private) increased the rate of two-sided photocopying from the current rate of 26 percent to 60 percent, and reduced the amount of copies made by one-third (from 446 billion to 299 billion),[9] close to $1 billion a year could be saved in avoided disposal costs and paper purchases, according to INFORM's calculations.[10] This would save 890,000 tons of paper annually and over 15 million trees.*

A good example of reducing paper waste comes from AT&T which has a company goal of decreasing paper use 15 percent by 1994 from 1990 levels. AT&T estimates annual paper savings, if double-sided copying is increased to 50 percent, of 77 million sheets of paper. This would reduce annual purchasing costs by $385,000.

Some paper reductions can be achieved solely by increasing double-sided (duplex) photocopying. Even greater reductions can be made by also reducing the number of copies made and increasing the intensity of use. *A document that is double-spaced and single-sided uses four times as much paper as a document that is single-spaced and double-sided* (duplexed). Some additional strategies for reducing paper and waste include:

- Eliminating fax cover sheets
- Editing and careful proofreading on the computer before printing
- Storing files on computer disks
- Loading laser printer paper trays with paper used on one side, for drafts
- Reducing direct mail by targeting audiences as narrowly as possible
- Using small pieces of paper for short memos

Institutions.   Institutions—organizations such as correctional, health care, educational, and cultural facilities—can also play significant roles in reducing the municipal solid waste stream. Local governments could advance institutional source reduction by implementing programs in the facilities they operate and by encouraging privately run institutions to replicate these efforts. A variety of source reduction programs are currently being implemented in correctional facilities, schools, and hospitals around the country.

New York City Hospital Corporation.   The New York City Health and Hospitals Corporation (HHC) is the largest health care institution in the nation, with more than 10,600 patient beds and 50,000 employees. As of early 1992, it had implemented or considered a variety of specific source reduction strategies.
Completed

- Switched from disposable corrugated cardboard boxes and disposable bag liners to a reusable container to hold and ship regulated medical waste for disposal. As a result of this measure, HHC expects to see a 3.4 percent reduction in its total waste stream, reducing over 1250 tons per year. The new contracts require carters to supply and clean the reusable containers.

---

*National Wildlife Federation's "Citizens Action Guide" estimates savings of 17 trees for every ton of paper reused.

In process:

- Replacing paper towel dispensers with hot air dryers
- Reducing use of disposable linens and disposable food service items in patient rooms

These measures are expected to reduce HHC's waste by 6.7 percent, nearly 2500 tons annually.

Planned:

- Establish an HHC corporate product packaging evaluation committee. The goal of the committee will be to change purchasing practices and warehouse product handling procedures to reduce waste. The committee will also be responsible for developing reduced packaging criteria.
- Convert cafeterias to reusable tableware.

The HHC study estimated the cost and waste savings from selected source reduction strategies for a 1000-bed hospital, as shown in Table 8.1.

The HHC study also estimated that assigning costs to departments based on the amount of waste they generate could reduce waste generation up to 20 percent, with savings of $500,000 annually. In addition, the study cited the need for improved control of the discarding of unused products. Improper ordering methods and hoarding often result in perishable products' becoming outdated and thrown out before being used. A 10 percent reduction in the number of unused products discarded would save about $100,000 annually for a 1000-bed hospital.

Government Programs to Stimulate Source Reduction. Government can play an important role in motivating the private sector (businesses and residents) and helping

TABLE 8.1 Estimated Annual Source Reduction Savings for a 1000-Bed Hospital*

	Savings	
Strategy	Tons per year	$ per year
Replacing paper towels with air dryers	100	$45,000
Replacing disposable food service items with reusables	200	$500,000
Eliminating use of plastic trash can liners in administrative areas	7	$20,000
Replacing disposable linens with reusables	150	$200,000
Replacing disposable admissions kits (water pitchers, glasses, and bedpans) with reusables in patient rooms	20	$150,000
Switching from disposable to reusable containers for sharp medical instruments	17	$175,000

*Savings include washing, laundering, and other service costs but do not include capital costs.

Source: New York City Department of Sanitation, *A Comprehensive Solid Waste Management Plan for New York City and Draft Generic Environmental Impact Statement,* Mar. 30, 1992, pp. 7–13, 14.

to develop programs to reduce waste. Strategies to achieve this include technical assistance and assistance in conducting waste audits; assistance to businesses and residents for backyard composting, grants, pilot programs, clearinghouses, awards and contests; and sponsorship of reuse programs.

   Technical Assistance, Waste Audits, and Materials Assessments.   Technical assistance programs are designed first to help businesses recognize opportunities for waste reduction measures, and then to implement them. In addition, government publicity about successful money-saving source reduction programs can encourage innovation by other businesses.

   The first step in any business source reduction program is conducting a waste audit and a materials assessment (or procurement audit). The waste audit identifies the materials that end up in the trash can or recycling bin. The materials assessment or procurement audit identifies the supplies, food, and other materials purchased by the company and its employees that are brought into its facilities. This allows an analysis to be done of what materials can be eliminated, reduced, replaced, and reused as well as recycled. It also allows companies to identify which materials may end up in the trash of other companies or consumers.

   In addition to helping conduct waste audits and materials assessments, technical assistance programs can provide businesses with information and other assistance through the multistep process of setting up a source reduction program. The steps for setting up a program include:

- Getting upper management support and distributing a policy statement to all employees

- Creating a source reduction task force with representatives from different departments (i.e., administration, janitorial, purchasing, and professional staff)

- Gathering basic waste generation and material usage information from each department and reporting current purchases (amounts and costs) to develop baseline data

- Setting goals (companywide, departmental, or by material)

- Setting an agenda, by putting out suggestion boxes for employees, gathering ideas from task force members, prioritizing strategies based on those that are easiest to implement and those that will reduce the greatest amount of waste and save the most money, and discussing obstacles and ways to overcome them.

- Implementing programs through task force members

- Expanding and evaluating the program by discussing problems and developing solutions; continuing to test new strategies; and measuring, evaluating, and documenting cost and disposal savings

- Encouraging participating businesses to publicize their source reduction activities in trade magazines to stimulate other businesses to adopt source reduction practices on their own

   Backyard Composting and Leave-on-Lawn Programs.   Government assistance can be used to reduce the amount of yard waste in the waste stream by promoting backyard composting and leave-on-lawn programs. Yard waste, which consists of leaves, branches, and grass clippings, is the second largest material category in the U.S. waste stream—17.9 percent by weight in 1990 (after paper, at 37.5 percent). Food waste accounts for an additional 6.7 percent of the waste stream.[11] These materials can be reduced at the source by backyard composting, leaving grass clippings on the lawn, and mulching grass and leaves. To reduce the amount of these organics in the waste stream, governments can educate residents and businesses, develop demonstration sites, give composting workshops, and distribute free composting bins.

Reuse Programs. Governments can sponsor, promote, and publicize reuse programs that distribute used goods. These programs keep materials that would otherwise be discarded out of the waste stream and make items available at lower costs. Examples of reuse programs are yard sales; food distributions to drug treatment centers, day-care centers, and homeless shelters; donations of furniture and other materials to schools, hospitals, etc.; and thrift shops that sell used home furnishings, supplies, and clothing.

Two programs in New York City illustrate the benefits of reuse programs. City Harvest collects food that would otherwise go to waste and donates it to charitable organizations such as homeless shelters and day-care centers. Food is donated by restaurants, retail stores, corporate cafeterias, and individuals. City Harvest delivers more than 10,000 lb of food per day. Total 1991 deliveries of 4.6 million lb helped the poor and reduced waste. Another program, Material for the Arts (MFA), matches donations from businesses and individuals with nonprofit arts organizations and public art projects. Materials collected from approximately 1000 businesses and individuals include furniture, apparel, beads, buttons, and art and film supplies. MFA collects an average of 32 to 35 tons per month.

Government Pilot Projects, Clearinghouses, Awards, Contests, and Grants. Government-sponsored programs can provide an opportunity to test and demonstrate source reduction strategies, disseminate information, and reward innovative programs. Through pilot programs, they can encourage businesses, institutions, and manufacturers to adopt source reduction practices. With clearinghouses they can increase consumer awareness about source reduction opportunities. And through awards and grants, governments can give recognition to businesses, individuals, or organizations that achieve significant reduction in the amount of waste they generate, develop product designs that accomplish source reduction, or change their packaging in a way that reduces waste. Governments can also hold contests, such as slogan contests, to generate citizen involvement in source reduction campaigns.

The Minnesota Office of Waste Management, for example, is conducting a pilot program in Itasca County that is targeting a variety of sectors and will become a comprehensive countywide program. Demonstration programs have reduced waste and saved money in a hospital, at a newspaper company, in a county courthouse, and at city garages.

In the grant area, the Seattle Solid Waste Utility allocated 10 percent of its solid waste grant program money to source reduction projects in 1991. Some of the projects included funding for a seven-store food co-op to purchase reusable shipping containers for transporting produce, for a community group to set up and give tours of model waste reduction homes, and for a school district to purchase worm bins so classrooms can do food composting.

Education. In 1990, the residential sector generated approximately 117 million tons of municipal solid waste—60 percent of the U.S. total.* Education is the key to encouraging source reduction in this sector.

Educating Consumers. Local governments can publish and distribute a variety of materials that give individual consumers specific recommendations on how to reduce waste. These educational materials can appear in brochures, pamphlets, booklets, or posters. The more specific the recommendations, the more likely consumers are to follow them.

Some specific suggestions governments can include in educational materials are bringing reusable shopping bags to stores, returning hangers to dry cleaners, donating

---

*Franklin Associates, Ltd., estimate. Personal communication: Nicholas Artz, Franklin Associates, to Caroline Gelb, INFORM, March 1992.

items no longer needed to charitable and social service organizations, reusing containers and other materials, maintaining items for longest life (e.g., keeping proper tire pressure, cleaning air conditioners), and buying perishable items only as needed and checking expiration dates to avoid the need to dispose of spoiled or unused items.

In 1990, San Francisco won an award from the California Department of Conservation for its outstanding education program. The three major source reduction components of this campaign are public education, environmental shopping, and backyard composting. In the area of public education, the city worked with Safeway, Inc. (a national supermarket chain) to develop a promotional campaign. Safeway paid for billboards, bus signs, and transit shelter posters that promote the reduction of waste. The campaign also designed an environmental shopping guide that encourages shoppers to buy goods with little or no packaging, reusable or refillable products, and repairable and durable items. Since May 1990, more than 70,000 environmental shopping guides have been distributed through local grocery stores and through the recycling hotline. The third effort is a brochure on how to home compost. More than 15,000 brochures have been distributed since 1990.

Educating Students.   Schools are an effective place to begin source reduction programs because reaching children at an early age is a good way to encourage good environmental habits. Further, since school waste represents 2.6 to 4 percent of U.S. waste,* reducing it can have a significant impact on the overall waste stream. School activities can include:

- Developing waste education curricula that focus on reduction and reuse, including such activities as conducting mini-composting projects and collecting daily trash to visualize the amount generated and to develop opportunities for elimination, reduction, and reuse

- Establishing an extracurricular club to focus on identifying and implementing source reduction opportunities in classrooms, cafeterias, and school grounds.

Economic Incentives and Disincentives.   Local governments can provide economic incentives or disincentives to encourage businesses and consumers to produce less waste. Such programs include variable waste disposal fees, advance disposal fees, taxes, tax credits, deposit and refund systems, and financial bonuses.

Variable Waste Disposal Fees.   Variable waste disposal fees that charge residents based on the amount of waste set out for disposal create an incentive to reduce and recycle since the more garbage a household generates, the more it must pay. As currently used, a fee is generally charged for garbage, but recyclables are picked up free. While these variable rate systems are now being used to encourage reductions in waste, some cities have been using them for years as their standard way of paying for garbage disposal. San Francisco has been using such charges for garbage service since 1932, and Olympia, Wash., since 1954.

Variable waste disposal fee systems are generally based on volume. Currently, the two main systems are a prepaid tag or sticker system or a per can volume-based system. With the tag or sticker programs, residents purchase bags, tags, or stickers at local stores or other specified locations and place them on bags that are to be collected. Garbage is picked up only if the official bag or tag is used. Under the per can sys-

---

*Tellus Institute (a Massachusetts environmental consulting firm) estimates the waste generation factor per student to be 0.08 tons per year; the Washington Department of Ecology, 0.12; and R. W. Beck (a Colorado environmental consulting group), 0.09. The U.S. Census Bureau states approximately 60 million students were enrolled in kindergarten through college in 1990.

tem, residents are charged for the number of cans (charges may differ depending on the can size) set out for collection.

Some communities (including Seattle, Durham, N.C., and Farmington, Minn.) are beginning to explore weight-based systems because volume-based systems have led to compacting and are based on the size of the container rather than the actual garbage put out. For example, the charge for a 30-gal container is the same if it is half full or completely full.

The weight-based systems would equip trucks with scales and bar coding equipment that could weigh garbage at curbside and record the amount discarded by each household. While the cost of this equipment is now $5000 to $10,000 per truck, this is expected to decrease if such systems become more widespread. Two advantages of such weight-based systems are: (1) they provide a greater incentive to reduce waste since the customer is charged for the actual weight of waste discarded (not the size of the bag or can, which may or may not be filled) and (2) they can provide useful data for measuring source reduction.

Taxes. Taxes on products and packages can be designed to encourage consumers to purchase products that generate less waste, help pay for the disposal of products, encourage manufacturers to adopt source reduction measures, and raise funds for other solid waste activities. Governments can place taxes on waste-producing products, such as nonrefillable containers. They can also place taxes on products that use unnecessary packaging (whether it is virgin or recycled material). For example, packages that exceed a specified product-to-package ratio or that have multiple layers of packaging. As of mid-1992, Florida and Maine had imposed taxes on lead-acid batteries, North Carolina taxed disposable diapers, and 18 states had taxes on tires.

Deposit and Refund Systems. Deposit and refund laws place fees on products when they are purchased and give the fee back to the consumer upon return of the container or product. Governments currently aim deposit and refund systems at recycling, but few use them to promote source reduction. They could do so by encouraging local manufacturers to develop deposit and refund systems on a larger scale, and educating consumers to participate in these systems.

Deposit systems for beverage containers (bottles and cans), commonly called bottle bills, are the most common form of deposit and refund system in place in the United States. As of mid-1992, nine states had beverage container deposit legislation: Connecticut, Delaware (not including cans), Iowa, Maine, Massachusetts, Michigan, New York, Oregon, and Vermont. California had a 5-cent redemption law, but deposits are not required.

Bottle bills have return rates of between 72 and 98 percent. Most of the bottles collected in this way are recycled; few are refilled. Michigan and Oregon are the only states that attempt to promote refillable over one-way containers, thus promoting source reduction over recycling. In Michigan, bottles that can be refilled carry a 5-cent deposit, while one-way bottles and cans have a 10-cent deposit. Oregon requires only a 2-cent deposit on refillables and a 5-cent deposit on one-way beverage containers.

Tax Credits. Governments can give tax credits to businesses and institutions that take steps to reduce waste at the source, although INFORM's research has not identified any governments that have done so. For example, credits can be given for investing in the equipment needed to switch from disposables to reusables such as dishwashers and washing machines. Credits can also be given to businesses that buy waste-reducing equipment such as double-sided copy machines, reusable tableware in cafeterias, and plain paper fax machines.

Financial Bonuses. Governments can offer financial incentives to communities that reduce the amount of waste they bring to incinerators or landfills. Tonnage grant programs that award bonuses based on the tonnage a community recycles do not encourage source reduction, but grants based on a reduction in the tonnage sent to disposal encompass both source reduction and recycling. While such incentives often encourage recycling and source reduction equally, they can be designed to give priority to source reduction.

Regulatory Measures. Regulations can be adopted on the local, state, or federal levels that mandate or encourage reduced waste. They can take a variety of forms; requirements that businesses conduct waste audits and develop source reduction plans; labeling schemes that give consumers information to encourage less wasteful purchasing decisions; bans on the sale or disposal of specific materials or products; and legislation requiring manufacturers to reduce packaging.

Local governments can require businesses and institutions to conduct waste audits and materials assessments and develop source reduction plans. Conducting an audit and writing a plan often reveal to businesses opportunities to save money as well as to reduce waste. Rhode Island requires every business with more than 50 employees to submit a recycling and waste reduction plan.

Bans. As of mid-1992, 33 states and the District of Columbia had adopted bans on the disposal of specific materials, including yard waste, lead-acid batteries, used oil, tires, and major appliances. Such bans can encourage manufacturers to develop alternatives that create less waste because consumers may stop buying items they cannot dispose of in their trash. Some states and several cities have also instituted retail bans, banning the sale or use of specific products and packaging. Bans can be reinforced with consumer education to encourage source reduction.

Governments can ban materials and items from disposal that either cause problems when landfilled or incinerated because they are major contributors to the size and toxicity of the waste stream, or can easily be eliminated, reduced, reused, or repaired. Examples of these materials are batteries, tires, major appliances, drain cleaners, solvents, paints, and grocery bags. Bans need to be publicized and consumers educated about why items are banned and what they can do to keep these items out of the waste stream and out of recycling systems. For instance, governments can distribute pamphlets on backyard composting and donating unused paints. Consumers can also be educated about alternatives that generate less waste such as rechargeable batteries, nontoxic or home-made cleaners, and cloth diapers.

Packaging. On average, each resident of the United States generated about 515 lb of packaging waste in 1990 (Ref. 3, p. 2-33); packaging comprises one-third of the national waste stream by weight and is a prime target for source reduction.

While some packaging is essential for containing, protecting, transporting, and marketing products, significant reductions in waste generation and costs can be achieved by eliminating unnecessary packaging, reusing and refilling, and designing more efficient packages.

Legislation aimed at reducing packaging waste has not fared well in the United States. No action has been taken at the federal level, and the model Massachusetts bill was defeated in a referendum in November 1992 after industry waged a massive campaign against it.

However, much can be learned from other countries, where dramatic steps are being taken to reduce packaging waste. Strategies in Europe are based on the principle that "the polluter pays," which makes producers of packaging responsible for the management of packaging waste. This, in effect, "internalizes" the cost of waste management and provides strong incentives for source reduction which can reduce the costs industry must bear for collecting and recycling packaging waste.

Germany's Packaging Ordinance, passed in May 1991, virtually removes all packaging materials from the public waste stream. It also requires that retailers provide bins so consumers can leave secondary packaging (defined as packaging that promotes marketing or reduces theft) at the store, thereby providing a strong incentive for source reduction. Retailers have put pressure on manufacturers to deliver products without the extra layers of packaging that the retailer will otherwise have to manage.

While the German legislation focuses on recycling and does not set any source reduction goals, it does contain powerful incentives for source reduction that have already begun to reduce packaging waste. For example, many manufacturers have dropped the boxes for toothpaste shipped into the German market and now market their product in a tube only. Companies are designing reusable shipping containers, and transport packaging is declining. INFORM's book, *Germany, Garbage, and the Green Dot,*[12] provides a detailed description of Germany's new legislation and its implications for source reduction.

In the Netherlands, a covenant signed by government and industry in June 1991 set goals of sending no packaging waste to landfills by 2000 and reducing generation of packaging waste to below the 1986 level by 2000.

The covenant also contains specific source reduction initiatives for quantity reduction:

- No free bags at supermarkets as of July 1991
- No liquor to be sold in gift boxes or wrapping by the end of 1991
- Within 1 year, two-thirds of detergents to be concentrates
- End of small toothpaste tubes by September 1992
- Avoidance of multiple packaging and "excessive" packaging
- One guilder ($0.50) deposit on plastic bottles for water and soft drinks by October 1991
- A ban on advertising of beverages in containers that are not refillable (not in the covenant, but already announced)
- Stimulation of refill system for liquid detergent
- Wine to be sold in deposit bottles in 2000 shops, with some providing tap wine from a case

Europe is also ahead of the United States in refilling beverage bottles. Denmark, in 1977, mandated that all beer and soda be sold in refillable bottles, and the new German law requires maintenance of the current refill rate of 17 percent for milk and 72 percent for all other beverages. The refill rate for beer and soft drinks in the United States, in comparison, has dropped from 84 percent in 1964 to below 10 percent. In Denmark and Germany, beer and soft drink bottles are refilled on average about 35 times.

The Canadian Packaging Protocol adopted in 1989 set milestone targets to reduce packaging waste. By the end of 1992, packaging sent to disposal was to be no greater than 80 percent of the amount sent in 1988. By the end of 1996, this is to drop to 65 percent, and by the end of 2000 to 50 percent of the 1988 level. Half of these diversion goals are to be achieved by source reduction and the other half by recycling.

While local packaging legislation may not be an efficient policy approach, it can be effective. Packaging legislation in one large locality or state can reduce packaging waste throughout the country or even worldwide. For example, in response to a law in Iowa banning Styrofoam made with chlorofluorocarbons, American Telephone and Telegraph Company (AT&T) changed the packaging of its telephones worldwide. The company had been considering packaging changes, but the Iowa law was the catalyst for action. Since AT&T has a worldwide distribution system for telephones, it decided that producing a different package for the Iowa market would be very impractical and costly. Instead of using a large package with Styrofoam and corrugated cardboard, AT&T now packages telephones in a smaller folding cardboard box that is made of recycled material and that is itself recyclable. The company says it is pleased with the new package, which reduces both waste and costs, but it did not quantify these savings to INFORM. This example illustrates the substantial leverage a local or state packaging law can have. A smaller locality would probably have to coordinate legislation with other localities in order to have such leverage.

State and local governments can also identify positive corporate packaging initiatives in their areas and aid in their replication through technical assistance, grants, and awards programs. Many private companies have adopted strategies to reduce packaging waste. The automobile industry, for example, has put pressure on its suppliers to ship all products in reusable and returnable containers. The Herman Miller furniture company's switch to reusable blankets for shipping furniture was in part a response to customers who did not want to have to dispose of a lot of plastic and corrugated packaging. A forthcoming INFORM study on less wasteful products and packaging will identify private sector initiatives that reduce waste, including case studies on companies that are refilling bottles and shifting to reusable shipping containers.

Conclusion.   In the United States, which generates far more per capita waste than other countries with comparable standards of living, there is great potential for source reduction. It can help preserve our country's natural resources, reduce pollution generated by waste management activities, and in many cases save money.

The many successful source reduction initiatives documented by INFORM in *Making Less Garbage: A Planning Guide for Communities* demonstrate that source reduction can work and is in fact being implemented in both the public and private sectors. Replicating programs such as these and developing new ones can make a major contribution to solving the garbage crisis.

8.3   DETOXIFICATION: OPTIONS FOR REDUCING THE HAZARDS OF MSW*

The current crisis posed by municipal solid waste disposal involves two factors: volume and toxicity. Since 1960, the volume of municipal solid waste has grown from 87.5 million lb per year to a projected 167 million lb in 1990. By the close of the century the federal Environmental Protection Agency projects that the country may be generating over 216 million lb of trash per year.[13]

The toxicity of solid waste is more difficult to measure. Toxic materials have always appeared in household wastes. But, since the middle of the century, as synthetic materials began to replace many traditional materials, the proportion of synthetically derived toxic materials in waste has increased appreciably.

The toxic constituents in solid waste include heavy metals, particularly lead, cadmium, nickel, and mercury, chlorinated hydrocarbons, such as perchloroethylene, trichloroethylene, and methylene chloride, aromatic compounds such as naphthalene and toluene, pesticides and other biocides, and used motor oil.

Some of these toxic materials enter municipal solid waste streams because they are waste products from domestic or commercial processes. Waste oil from automobile service stations is such an example. Some toxic materials are toxic products discarded

---

*Contributed by Ken Geiser.

once a portion of the product has been used. Waste paints are a good example. But most of the toxic materials appear in solid waste as constituents of commercial products whose useful life is over. For example, over 2 billion dry cell batteries are sold each year in the United States. These include batteries containing mercury or mercuric oxide, magnesium, zinc, silver oxide, nickel and cadmium, and lithium. Many of these metals are not dangerous in the battery itself. But a dry cell has a useful life of somewhere between a few hours and several months, after which it is discarded. When a battery is disposed of in a landfill it eventually deteriorates. During this deterioration the metals can be released to the ground and groundwater. When a battery is incinerated, some of the metals fall out in the bottom ash and some are released in incinerator gases. Incinerator filters will collect some of the metals as fly ash. Both bottom ash and fly ash must be disposed of in landfills where again the ground and groundwater may become contaminated. During the 1980s the Swedish Environmental Protection Board estimated that up to one-third of the background levels of mercury in the environment came from the incineration of batteries.[14]

Toxic Trash

The Toxicity of Trash. Batteries are only one of the conventional constituents of municipal trash that lead to its toxicity. Table 8.2 lists a number of common toxic materials found in municipal solid waste, their sources in products, and their known health effects.

How Toxic Is Trash? The toxicity of trash can be estimated using two basic approaches: sampling or modeling. Sampling involves sampling various waste streams, sorting

TABLE 8.2 Common Toxic Materials in Municipal Solid Waste

Substance	Sources	Health effects
Cadmium	Batteries, inks, paints	Carcinogen, ecotoxin, reproductive effects
Lead	Batteries, varnishes, sealants, hair dyes	Neurotoxin, reproductive effects
Mercury	Batteries, paints, fluorescent lamps	Ecotoxin, neurotoxin, reproductive effects
Methylene chloride	Paint, paint strippers, adhesives, pesticides	Carcinogen
Methyl ethyl ketone	Paint thinner, adhesives, cleaners, waxes	Neurotoxin, reproductive effects
Perchloroethylene	Rug cleaners, spot removers, fabrics	Carcinogen, ecotoxin, reproductive effects
Phenol	Art supplies, adhesives	Ecotoxin, developmental effects
Toluene	Paint, nail polish, art supplies, adhesives	Ecotoxin, mutagen, reproductive effects
Vinyl chloride	Plastics, apparel	Carcinogen, mutagen, reproductive effects

Source: Ref. 15

each waste stream into its specific components, and weighing the components. One study conducted by the federal Environmental Protection Agency found that hazardous waste from households made up less than 1 percent of the solid waste by weight.[16] The most systematic effort to gather empirical data on municipal solid waste has involved the sampling of landfills in several communities across the country. This study found that household maintenance products made up the largest percentage by weight of the household hazardous substances. This was followed by batteries, cosmetics, cleaners, automobile, and yard maintenance products.[6] While these studies provide some data on the hazardous constituents of household trash, they do not provide evidence on the toxic materials generated by nonhousehold sources.

Municipal solid waste, or trash, is made up of the discards from domestic, commercial, and industrial settings. Businesses that generate less than 100 kg of hazardous waste per month are allowed to deposit the waste into the municipal solid waste stream. It is estimated that there are some 450,000 of these so-called very small generators. They include conventional retailers, bakers, beauty shops, dentists, dry cleaners, photography labs, printers, restaurants, schools, and vehicle maintenance shops. Some of the hazardous waste generated by very small generators is flushed into the municipal sewer system and some of it is discarded as solid waste. All together, it is estimated that very small generators produce roughly 197,000 tons of hazardous waste each year, an unknown portion of which is released as municipal solid waste.[18]

The modeling approach requires estimating the material flows through each waste stream. These material flows are calculated from materials production data and adjusted for imports and exports, materials recovery, energy conversions, and losses during production or use. The remaining volumes are then assumed to enter the solid waste stream.

Simple materials flow analyses have been done on some toxic metals (lead, cadmium, and mercury). The federal Bureau of Mines found that most cadmium was used in coatings and plating and in batteries, most lead was used in storage batteries, and most mercury was used in electrical equipment.[19] Franklin Associates found 65 percent of lead was used in auto batteries and 54 percent of cadmium was used in household (dry cell) batteries.[20]

Each approach requires identifying the toxic constituents. This is limited by the available research. Toxic chemicals are those that scientific studies have shown to cause serious health effects. Today, some 70,000 chemicals are in common use. Many are toxic. Yet the National Research Council estimates that less than 10 percent have been tested for toxicity. Of the 17,000 chemicals used in food, cosmetics, pesticides, and drugs less than 30 percent have been tested.[21]

The federal Environmental Protection Agency has found over 100 substances classified as hazardous under the federal Resource Conservation and Recovery Act (RCRA) in common household products.[22] Most of these are disposed of in municipal solid waste.

Is Trash Toxicity a Problem?   When products containing toxic chemicals are disposed of, the toxic constituents enter municipal landfills and incinerators. From these disposal facilities the toxic chemicals are dispersed into the environment as underground leachate, waste water effluents, air emissions, or hazardous waste. Once released to the environment toxic chemicals may threaten ecological systems, wildlife, or public health.

The toxic constituents of trash in landfills have a history of concern. Volatile organics can be released to the air. The leaking and leaching of halogenated hydrocarbons and heavy metals can threaten underground water supplies.

Toxic materials in solid waste streams destined for municipal incinerators also pose serious concerns. Incineration breaks down the paper, plastics, fibers, and containers of the municipal waste stream and liberates and concentrates the heavy metals contained in consumer products. Toxic metals such as lead, cadmium, arsenic, mercury, selenium, and beryllium, among others, remain in the postincineration ash. Organic pollutants of concern in incinerator air emissions include hydrogen chloride, dioxins, and furans. Incineration tends to volatilize some metals that then condense onto small fly ash particulates. Other metals such as mercury are easily converted to gaseous states. Still other metals may react with the organics to form complex compounds such as metal chlorides. Once released from incinerators on particulates or as gases the metals are easily mobilized and readily available for ecological uptake. Because the metals are released from conventional matrices and easily dispersed by air or water currents, there is an increased potential for direct (inhalation) or indirect (food chain contamination) human exposure.[23]

Most of the metals released from solid waste treatment facilities are neurotoxins; many, such as lead, cadmium, arsenic, and beryllium, are human carcinogens; some, such as lead and mercury, are recognized human reproductive toxins; and some others, such as mercury, copper, and zinc, are acutely toxic to aquatic life.[24] One chemical of particular concern is chlorine because it can be involved in the formation of hydrogen chloride, dioxins, and other chlorinated organics during incineration. Dioxins are among the most toxic compounds known to science. Chlorine occurs in many products including solvents, biocides, bleaches, disinfectants, paper, and plastics. Waste paper and plastics appear to be the major source of chlorine in municipal solid waste.[25]

Where municipalities have introduced recycling programs, the toxic nature of the trash has continued to remain a problem. Workers in recycling centers may be exposed to the toxic constituents of the materials they separate for recycling. Mismanaged recycling centers can pollute soil and groundwater with the toxic materials in the stored products. Finally, municipalities that recycle toxic products may incur liabilities for the future handling of those materials.

Reducing the Toxicity of Trash. Reducing the toxicity of trash requires policies that are well targeted, efficient, and cost-effective. Because the toxicity of trash is directly linked to the toxicity of consumer products some of the most effective policies for reducing risks involve the redesign of products and the processes that produce them.

There are three broad policy approaches to reducing the toxic constituents of solid waste. The first involves improving waste management practices to reduce the amount of toxic waste that is ultimately disposed, primarily through recycling. This is the most common approach. Its immediate effects are countered by its long-term limits. The second focuses on changing the material constituents of the products that are used in domestic and commercial activities. While this approach has more long-term impacts, the immediate prospects are less promising. The third seeks to change the processes of industrial production in order to reduce toxic inputs. Again, this is a long-term but potentially highly effective approach. Table 8.3 presents these approaches.

TABLE 8.3 Policy Strategies for Reducing the Toxicity of Trash

Strategy	Feasibility	Effectiveness
Waste management	Immediate	Modest
Product management	Intermediate	Significant
Production management	Longer-term	Significant

Waste Management Policy

The most immediate approach for reducing the toxicity of solid waste involves improving the municipal waste management programs. Once products containing toxic materials have been mixed into conventional municipal refuse, the costs and problems of safely managing the waste dramatically increase. Therefore, most programs that seek to manage the toxic constituents of solid waste begin by separating the materials or by keeping materials separate from the first point of disposal.

Toxic Waste Collection. Products containing toxic materials can be diverted from the municipal waste stream by separate waste collection programs. There are two types of programs: those directed at specific products such as batteries or tires, and those directed at specific waste generators such as households or commercial offices.

The product specific approach is well illustrated by battery collection programs. Basically there are two types of batteries—household (dry cell) batteries and automobile (lead-acid) batteries—and each provides a different program approach. Lead-acid batteries are handled by a private market dependent on the price of reprocessed lead. Although lead-acid batteries were classified under RCRA as a hazardous waste in 1985, there are only a few state or local programs to encourage the collection of lead-acid batteries for resmelting. Over 40 percent of the lead used in the United States comes from reprocessing discarded batteries at secondary smelters, but the depressed price of lead still leaves many batteries as discards. An estimated 98 million batteries (representing 900,000 tons of lead) went unrecovered between 1980 and 1986.[26]

Governments have been more assertive in encouraging household battery collection. There are household battery collection programs in operation in the United States, Japan, and at least 11 European countries.[27] In the United States most local programs such as those in operation in Bellingham, Wash.; Andover, Mass.; and Hennepin County, Minn., encourage battery retailers to collect used batteries returned by customers. Local programs in Japan and Europe rely more on voluntary government collection programs at special recycling centers. In 1983, Sweden began a voluntary collection program in 27 cities. In 1987, when the 75 percent collection goal had yet to be reached, the Swedish government made the program a national effort with heavy emphasis on education and convenient recycling centers.[28]

The generator-specific approach can be illustrated by "household hazardous waste collection days." So-called household hazardous waste is exempt from some federal hazardous waste regulations,[29] but still must be handled by a licensed hazardous waste treatment operator. Typical programs are set up by municipalities as periodic 1-day drop-off programs where residents bring in products ranging from solvents and paints to pesticides and explosives. A survey of local programs conducted by the federal Environmental Protection Agency in 1986 found that participation in such programs was low, often less than 1 percent, with unit costs for collection running as high as $18,000 per ton.[30]

Some local governments are establishing permanent facilities for household hazardous waste collection. San Bernardino County runs two permanent collection centers open on a daily basis where wastes are collected, identified, stored in specially designed storage units, and periodically shipped off for disposal at licensed hazardous waste treatment facilities.[31]

Toxic Waste Recycling. Collecting toxic materials before they enter the municipal waste stream is an important prerequisite to toxicity reduction, but if those materials are not recycled back into products the overall toxicity of the solid waste stream is not reduced. For instance, most household batteries collected in the United States in 1989 were sent to at least two commercial processing facilities. While mercuric and silver oxide batteries were then processed to recover mercury and silver for remarketing, lithium was simply treated to make it less reactive and then sent to a landfill. Carbon zinc and alkaline batteries were sent directly to hazardous waste landfills.[32]

Fluorescent lamp bulbs contain small quantities of mercury. While over 50 million fluorescent lamps are collected each year in Europe, programs to collect and process fluorescent bulbs in the United States remain limited. The largest program is in California where three firms reclaim 600,000 lamps each month. While the metal and glass can usually be sold for reuse, the mercury is often sent to a landfill. Because the current price of mercury is low, efforts to capture and recycle the mercury require a high processing cost.[33]

Used oil is another candidate for recycling. Approximately 1.2 billion gal of used vehicle or lubricant oils are generated in the United States each year. About 360 million gal are generated by home oil changes, most of which is disposed of in the trash. Yet only 100 million gal per year are re-refined into reusable oils.[34] This low level of reprocessing is primarily due to the low cost of virgin oil and the environmental problems of reprocessing. Contaminants that appear in used oil, particularly the additives that have been added to virgin oil since the 1970s, produce a hazardous sludge that is expensive to treat. The wastewater from the distilling operations is contaminated with hydrocarbons, as well. Thus roughly two-thirds of all used oil is recycled by burning it as a fuel.

Much effort has been put into programs for recycling plastics. Packaging is the largest single source of plastic waste, accounting for 40 percent of total plastic waste per year.[35] Typical plastic packaging contains a host of toxic materials. There are toxic chemicals in the resin. Polyvinyl chloride (PVC) has been linked to the formation of dioxins in incinerators. The additives are often toxic. Antioxidants include phenolics, phosphites, and thioesters. Colorants include titanium dioxide, lead chromate, chromium oxide, and cadmium, selenium, and mercury compounds. Alumina trihydrate and halogenated compounds are used as flame retardants. Heat and light stabilizers such as organotin mercaptides, methyl and butyl tins, and cadmium-zinc and barium-cadmium are frequently added. Finally, there are toxic compounds in the surface printing and treatment. Metallic inks and dyes are often used for decoration and labels.[36]

Most plastic recycling programs do not account for these various toxic constituents. Little research has been done on chemical exposure from recycled plastics. The federal Food and Drug Administration has refused to allow recycled plastic material to be used in food containers owing to uncertainties about contaminants and the difficulties of sterilizing plastics. The high cost of transporting postconsumer plastics and the absence of a reprocessing infrastructure have limited the recycling of plastics. Today, just over 1 percent of plastics in the waste stream are recycled.

Because the material structure of plastics degrades during reprocessing, recycled plastic typically goes into low-grade products like carpet fibers, fiberfill for pillows and jackets, industrial paints, and nonstructural "lumber" products. While these "second uses" for the recycled plastics prevent the plastics from disposal at that moment, many of these "second use" items are disposed of eventually. Thus recycling generally delays but does not eliminate the possibility of environmental release of the toxic constituents in plastics.

Reducing the toxicity of municipal waste by collecting, diverting, and recycling toxic materials can be readily implemented today and it is an increasingly common practice. It provides a large opportunity to educate consumers about toxic materials and it can be a significant community building activity. Yet the recycling of toxic materials in the waste stream is fraught with technical and economic limitations. It is further limited by the low level of solid waste recycling nationally—only about 10 percent of municipal solid waste is currently recycled.[37] Thus, even if recycling were to become a significant approach to reducing the toxicity of trash, it still could not account for a substantial reduction without a significant change in waste management practices.

Product Management Policy

A second general approach to detoxifying trash focuses on the toxic materials contained in products. Instead of focusing on better management of the toxic materials in waste, this approach seeks to reduce the toxicity of the waste stream by reducing the toxic constituents of the products "thrown out" as waste. While the focus on improved waste management may have more immediate effects on the toxic materials entering disposal facilities, the focus on products offers more long-term efficiencies, because less toxic products mean a reduced need for highly selective waste management techniques. By focusing policy attention on products, the emphasis is shifted to an earlier point in the life cycle of a toxic material.

Life Cycle Analysis.   Trash is a product of the linear process that supplies domestic and commercial establishments with consumable goods. In considering the environmental effects of trash, it is important to consider the entire life cycle of a product from synthesis or manufacture through distribution and use to waste and disposal. Environmental and human health effects are associated with each stage of the life cycle. Effective environmental protection requires that improvements in the environmental performance at one stage not worsen the effects at another stage.

This broader perspective on the role of a product in the environment has been incorporated in a new technique called "life-cycle analysis."[38] Ideally, a life cycle analysis is composed of an inventory of resource inputs and waste outputs for each stage and an assessment of risks associated with each of these inputs and outputs. Such life cycle analyses have been used in solid waste management for comparing plastic vs. paper packaging and disposable vs. reusable diapers. While the methodology is limited by the product focus, the large amounts of data requirements, and the necessity to set boundaries, the concept opens up a broad awareness of the environmental impacts of products before they become waste.

Product Bans.   Governments may use their authority to ban products or activities as a means of reducing toxicity in trash. The effectiveness of a ban depends in part on where in the life cycle the ban is placed. Bans may be placed on production, trade, use, or disposal.

Many states have experimented with bans at the point of disposal. Thirty-seven states have adopted bans on the disposal of some products including lead-acid batteries, used oil, tires, and major appliances. Table 8.4 identifies the states that have adopted product bans and indicates the products covered. By diverting toxic materials from disposal facilities, these disposal bans may reduce the amount of toxic materials ending up in landfills or incinerators, but they do little to discourage the amount of toxic material in the solid waste stream.

Some states have tried to target toxic materials directly by focusing on bans at the point of use or trade. California, Oregon, Minnesota, New York, New Jersey, Vermont, and Connecticut have passed legislation banning the use of mercury in dry cell batteries by 1996. The New Jersey law passed in 1992 sets a standard for mercury in dry cell batteries and prohibits the sale of batteries unable to meet the standard after 1996.

TABLE 8.4 Product Disposal Bans by State

State	Vehicle batteries	Tires	Motor oil	White goods	Others
Arkansas	X	X			
Arizona	X	X			
Connecticut					X[a]
Florida	X	X	X	X	X[b]
Georgia	X				
Hawaii	X				
Idaho	X	X			
Illinois	X	X		X[c]	
Iowa	X	X	X		X[d]
Kansas		X			
Louisiana	X	X		X	
Maine	X				
Maryland		X			
Massachusetts	X	X		X	X[e]
Michigan	X				
Minnesota	X	X	X	X	X[f]
Mississippi	X				
Missouri	X	X	X	X	
New Hampshire	X				
New Jersey	X				
New York	X				
North Carolina	X	X	X	X	
North Dakota	X		X	X	
Ohio	X	X			
Oregon	X	X	X	X	X[g]
Pennsylvania	X				
Rhode Island	X				
South Carolina	X	X	X	X	
South Dakota	X	X			
Tennessee	X	X			
Texas	X	X	X		
Utah	X				
Vermont	X	X	X	X	X[h]
Virginia	X				
Washington	X		X		
West Virginia	X	X			
Wisconsin	X	X	X	X	X[i]

[a]Mercury oxide batteries, [b]Demolition debris, [c]White goods containing CFC gases, mercury switches, and PCBs, [d]Nondegradable grocery bags and carbonated beverage containers, [e]Glass and metal containers, [f]Nickel-cadmium rechargeable batteries, [g]Discarded vehicles, [h]Various dry cell batteries, [i]Metal, glass, and plastic containers, and recyclable paper.

Source: Adapted from Jim Glenn, "The State of Garbage in America," *Biocycle,* May 1992, p. 33.[39]

Packaging Bans.   It is estimated that packaging materials account for more than one-third of municipal solid waste, by volume, in the United States.[40] The packaging industry is the largest user of plastic, accounting for over one-third of plastic resin use annually. Concern over the volume of packaging in landfills and the environmental hazards of incinerating plastics has led some local governments to try to ban the use of plastic packaging.

In 1988, Suffolk County, New York, passed a highly controversial packaging ban. The law, which was scheduled to take effect in the summer of 1989, banned polystyrene foam in food packaging including produce and meat trays, grocery bags, and fast food "clamshells." The law has been delayed by a legal suit filed by several plastics trade groups before the state Supreme Court.[41]

Minneapolis and St. Paul have passed ordinances that can ban the use of nonrecyclable plastic food packaging.[42] Maine, Minnesota, and Rhode Island have passed legislation banning the use of polystyrene foam food packaging made with ozone-depleting chlorofluorocarbons.[43]

The most prominent packaging conversion involved the use of public pressure to force the substitution of the plastic food packaging used by the McDonald's chain of fast food restaurants. Concern over the toxic emissions from the incineration of plastic packaging led the Citizens' Clearinghouse on Hazardous Waste to launch a 3-year national campaign to pressure McDonald's to convert to paper packaging. That campaign set the conditions for the Environmental Defense Fund and McDonald's in 1990 to negotiate a well-publicized phase-out of plastic packaging and the substitution of paper.[44]

Product Labeling. Like product and packaging use bans, product labeling seeks to reduce the use of toxic materials by changing consumer patterns. Product labels that reveal the toxic constituents of products are likely to affect the purchasing decisions of those consumers who read labels and do comparative shopping where alternative products are available. More significantly, product labeling may affect the material selection decisions of those manufacturers who fear that product labeling will affect consumer decisions.

In 1986, California citizens passed a ballot initiative (Proposition 65) that required that warnings on product labels must be posted on products containing chemicals that can cause cancer or adverse reproductive effects. In order to avoid such labeling, several firms reformulated their products to remove the toxic chemicals of concern. The Gillette Corporation removed trichloroethylene from its Liquid Paper typewriter correction fluid, Dow Chemical reformulated NKCV Spot-lifter to eliminate perchloroethylene, and Pet, Inc., accelerated the elimination of lead from its food cans.

Since the early 1980s several new product labeling programs have been established to provide information on environmental compatibility. There are national programs in Germany, Canada, Japan, and the Nordic countries. Two certifying and labeling programs are run as private operations in the United States. Each of these programs provides the right to use a special "eco-label" when a product is found to meet a set of environmental criteria which may include recyclability, stratospheric ozone impact, toxicity, energy input, and pollution. While these labels may have some notable impacts on toxicity, the high degree of generality involved in each label leaves these eco-labels as a limited instrument.

Product Substitutes. During this past decade there has been a growing awareness among consumers about the environmental effects of the products they use and dispose. Retailers across the country have found that consumers will respond to literature, educational materials, and warnings about products. Educational campaigns in schools and manuals and guides for environmentally conscious shopping have raised further the selective capacity of consumers to choose environmentally sound products.

Bans, labeling, and educational programs have all been used to target products containing toxic chemicals. The measurable results have been limited. In part this is due to the relatively small number of initiatives. In part this is because many of the actual product changes have resulted from the more indirect sensitivities of the market in which there are no simple causal connections.

A focus on changes in products yields a more fundamental approach to reducing the toxicity of solid waste than a focus on better waste management alone. A product management approach provides significant opportunities to educate consumers and raise awareness about toxic chemical exposure. Eliminating toxic product use in the community will clearly reduce the volume of toxic materials disposed of. But there is only so much that a local community can do to change the product mix. Ultimately it will take changes in the production systems that produce products to fully relieve products of their toxic constituents.

Production Management Policy

Detoxifying industrial production systems provides a third approach to reducing the toxicity of solid waste. Products are produced by industrial or agricultural production systems. Where those systems are dependent on toxic chemicals, the products are likely to be toxic.

Clean Production. In Europe this approach is called "clean or cleaner production." In 1989, the United Nations Industry and Environment Office set up a Clean(er) Production Programme to promote environmentally sound production. The Programme defined the concept of clean production for processes to mean "conserving raw materials and energy, eliminating toxic raw materials, and reducing the quantity and toxicity of all emissions and wastes before they leave the process."[45] Both the Netherlands and Denmark have set up special government funded clean technologies programs.

Clean production implies more than better waste management or pollution control. The essence of clean production is to fundamentally change industrial production processes in order to manufacture in a more environmentally sound manner.

There have been important initiatives in the United States as well, particularly at the state level and among leading firms. Much of this is referred to as "pollution prevention" and is guided by evidence that preventing pollution can both reduce industrial operating costs and improve environmental performance. In 1990, Congress passed the Pollution Prevention Act to promote pollution prevention.[46]

Design for the Environment. One of the most effective points in a product's life cycle for considering the future use of toxic substances is during the initial design period. At the time that new products are undergoing concept development and materials specification, attention to alternative nontoxic materials can reduce the costs of alternative development further into a product life cycle.

The idea of incorporating environmental criteria into the initial product design phase has been called "design for the environment." This term, first coined in the American Electronics Association and heavily incorporated by AT&T in its product development research centers, has been recently promoted through conferences and publications.[47] At the design step, products can be developed that can be recycled more easily, last longer, can be repaired more easily, contain no toxic material, or require no toxic material during manufacture. For instance, plastic products or containers can be limited to one type of plastic to improve recyclability. Durable goods can be designed for take-back, disassembly, and reuse of components. Electronic equipment can be designed as a set of components that can easily be repaired by removing and replacing malfunctioning elements.

The reduction in heavy metals in printing inks provides a good illustration. Traditional printing inks contain several metals, including lead. Consumer pressure, concern about occupational exposures, and efforts to reduce hazardous waste have led newspapers to switch to low-lead inks. In the mid-1970s, the American Newspaper Publishers Association prohibited the use of lead in inks approved by the association and established a logo to identify acceptable inks. Procter and Gamble have eliminated the use of all metal-based inks for printing on packaging.[48]

Unfortunately, little attention is given to the toxicity of products as waste when firms design new products. Although manufacturers have incentives to reduce the costs and liabilities of the hazardous wastes they generate, they have little incentive to consider the disposal costs of the products they make. Some of this may be changing because of European initiatives such as the German take-back laws. Germany is currently considering extending its take-back laws to other products including automobiles. In anticipation, several auto manufacturers, including BMW and Volkswagen, have begun to explore automobile designs that enhance disassembly. Both firms have built pilot plants based on the take-back and reuse principle. BMW hopes to have an automobile that is 100 percent recyclable by the year 2000.[49]

Toxics Use Reduction. Since 1989, 14 states have passed laws that promote toxics use reduction. These states are identified in Table 8.5. Toxics use reduction is a form of pollution prevention that focuses on reducing or eliminating toxic chemicals in industrial production as a means to reducing the toxicity of industrial waste streams. Most of these laws require or encourage firms to prepare plans demonstrating how they would reduce the use of toxic chemicals or the generation of toxic wastes. Typically, these toxics use reduction programs encourage firms to adopt one or several of a set of techniques including substitution of the chemical inputs, changes in production equipment or processes, redesigning products to reduce toxic chemical use, improvements in production operations and maintenance, and installing closed-loop recycling systems.[50]

While these laws were enacted to reduce the generation of hazardous waste and the chemical risks of industrial production, several of the techniques can reduce the toxicity of the products as well.

For instance, a firm may redesign a product to reduce the requirement for a known toxic constituent, or a firm may change the chemicals used to manufacture the product thus reducing the residual toxic chemicals that may remain in or on a product, or a firm may change the production of a product to reduce the generation of waste toxic scrap, small amounts of which may have been disposed of as municipal solid waste.

TABLE 8.5 States with Toxics Use Reduction Laws

Arizona	Minnesota
Colorado	New Jersey
Georgia	Oregon
Illinois	Tennessee
Indiana	Texas
Maine	Vermont
Massachusetts	Washington

Source: Ref. 51.

The Polaroid Corporation has eliminated the use of mercury and reduced the use of cadmium in the batteries used in its film cassettes by developing a carbon-zinc cell with a zinc anode designed by the Rayovac Corporation. By so doing, the camera company was able to cut the mercury in the film pack by 50 percent in 1987 and eliminate it totally in 1988. This project was initiated in anticipation of new regulations such as those in Switzerland that now require labeling and set limits on allowable concentrations of metals in batteries.[52]

Integrated Pesticide Management. Yard, home, and agricultural activities that rely on toxic biocides may contribute to the toxic constituents of solid waste. Each year thousands of pounds of chemical pesticides, herbicides, rodenticides, termiticides, fungicides, and fertilizers are sold to domestic customers. While much of this is used on lawns, gardens, basements, garages, and backyard orchards, some of it is also sent off as solid waste. Out-of-date product, unused portions of opened containers, and residuals in the bottom of "empty" containers may be set out as trash. In addition, some portion of used pesticides may be discarded as solid waste on grass clippings and other yard wastes.

A more significant contributor of toxic agricultural products in municipal solid waste may be small farms, nurseries, and agricultural product transporting firms that meet the RCRA definition of "very small generators" of hazardous waste. There is little research on the contribution of toxic agricultural products to solid waste.

Today, there are a host of new pest management practices that can reduce or eliminate the use of toxic chemical products. While there are specific safer products that can simply replace the more toxic products, in general, the preferred approach is to change the processes of yard or farm management. Knowing when and how to intervene in order to control pests is as important as the range of substances used. This new approach is often called "integrated pest management," because like clean production, it requires a rethinking of the production system itself.[53]

In the yard integrated pest management relies on natural controls (pathogens, parasites, predators, and repellents), improved yard management (increased sanitation, cultivation, aeration, and manual grooming), and selection of pest-resistant plantings. In buildings integrated pest management means natural controls (pathogens; predators, e.g., cats; and repellents), improved household management (increased sanitation), and architectural remedies (barriers and dampness prevention).

Sweden, Denmark, and the Netherlands have each adopted comprehensive national pesticide reduction programs over the past 10 years. Each of these programs establishes national goals for the reduction in use of various categories of active pesticide ingredients (up to 50 percent reduction over 5 years) and then employs a combination of regulation, education, financial incentives, and research to assist pesticide users to change to more integrated forms of pest management.[54]

While integrated pesticide management is not without some reliance on toxic chemical use, the general thrust is to minimize that use. Reducing the use of toxic pest controls will further reduce the toxic materials disposed of as waste from homes and farms.

A Sustainable Economy

Reducing the toxicity of materials in the municipal waste stream is the most fundamental way to reduce the health and environmental risks associated with trash management. Wherever communities seek a more sustainable future, there will need to be a focus on managing solid waste. That waste stream may prove to be a rich material resource for recycling and reuse, but the toxic constituents will surely inhibit the best of efforts.

Although simple collection and recycling programs may have the effect of reducing some of the materials destined for landfills and incinerators in the short run, a focus on products and production processes in the longer term is likely to prove to be the most effective.

For now, there are a wide array of government policy options that could promote reduction in the toxicity of trash. For example, local, state, or federal governments could:

- Develop a comprehensive database on toxic materials in the municipal solid waste stream.
- Support more research on environmentally sound materials, processes, and products.
- Promote better waste management by:

  Mandating manufacturer take-back and recycling of products containing highly toxic materials.

  Expanding and developing programs that separate and collect toxic and hazardous products from household wastes.

  Mandating recycled content in frequently disposed products that contain toxic substances.

  Taxing toxic emissions from waste management facilities.

- Promote better product management by:

  Instituting public media campaigns on the toxicity of consumer products and safer substitutes.

  Developing and promoting product labeling programs that identify products containing toxic substances.

  Taxing products containing highly toxic substances.

  Using government procurement authorities to promote environmentally sound products.

- Promote better production management by:

  Offering awards to firms for progress in eliminating or reducing toxicity.

  Encouraging firms or trade associations to set up guidelines for reducing the toxic constituents of products or production processes.

  Providing more technical assistance to firms who are attempting to reduce toxic materials in their production processes.

None of these options alone will suffice. There is only so much that a government can do. Reducing toxicity in trash will take significant effort from businesses and consumers as well. Several of these options may prove controversial. Whatever directions are pursued, it will be necessary to proceed even where information is limited and science has not reached consensus. Progress in reducing toxic materials in solid waste will need to move forward when the weight of evidence is compelling and the technical and economic costs can reasonably be overcome.

REFERENCES

1. U.S. Environmental Protection Agency, *The Solid Waste Dilemma: An Agenda for Action,* Washington, D.C., February 1989, pp. 16–19.

2. Fishbein, Bette, *Making Less Garbage: A Planning Guide for Communities,* INFORM, New York, January 1993.

3. U.S. Environmental Protection Agency, *Characterization of Municipal Solid Waste in the United States: 1990 Update,* Washington, D.C., June 1990, p. ES-4.

4. New York City Department of Sanitation, *A Comprehensive Solid Waste Management Plan for New York City and Final Generic Environmental Impact Statement,* August 1992, p. 17. 2–3.

5. Franklin Associates, Ltd., *National Office Paper Recycling Project: Supply of and Recycling Demand for Office Waste Paper, 1990 to 1995,* July 1991, p. B-19.

6. *Statistical Abstract 1990,* p. 425.

7. Minnesota Department of Administration, Materials Management Division, *Resource Recovery Biannual Report: FY 1989 and 1990,* submitted to the Legislative Commission on Waste Management, January 1991.

8. U.S. Environmental Protection Agency, *Characterization of Municipal Solid Waste in the United States: 1990 Update,* Washington, D.C., June 1990.

9. Banking Information Systems (BisCap), Norwell, Mass. (a consultant to the office machine industry).

10. Graff, Robert, and Bette Fishbein, *Reducing Office Paper Waste,* INFORM, New York, November 1991, p. 15.

11. U.S. EPA, p. ES-4.

12. Fishbein, Bette, *Germany, Garbage, and the Green Dot,* INFORM, New York, 1994.

13. U.S. Environmental Protection Agency, Office of Solid Waste and Emergency Response, *Characterization of Municipal Solid Waste in the United States, 1990 Update,* Washington, D.C., June 1990, p. 79.

14. U.S. Congress, Office of Technology Assessment (U.S. OTA), *Facing America's Trash: What Next for Municipal Solid Waste?,* OTA-0-24, Washington, D.C., October 1989, p. 158.

15. Adapted from Nancy Lilienthal, Michele Ascione, and Adam Flint, *Tackling Toxics in Everyday Products,* INFORM, New York, 1991.

16. U.S. Environmental Protection Agency, *A Survey of Household Hazardous Wastes and Related Collection Programs,* EPA/530-SW-86-038, Washington, D.C., 1986.

17. Wilson, Douglas, and William Rathje, "Quantities and Composition of Household Hazardous Wastes: Report on a Multi-Community, Multi-Disciplinary Project," paper presented at the Third National Conference on Household Hazardous Waste Management, Boston, Mass., Nov. 2–4, 1988.

18. Abt Associates, *National Small Quantity Hazardous Waste Generator Survey: Final Report,* report prepared for the U.S. Environmental Protection Agency, Office of Solid Waste, Cambridge, Mass., 1985.

19. U.S. Bureau of Mines, *Mineral Facts and Problems, 1985 Edition,* Bulletin 675, Washington, D.C., 1985.

20. Franklin Associates, *Characterization of Products Containing Lead and Cadmium in Municipal Solid Waste in the United States, 1970–2000,* report prepared for the U.S. Environmental Protection Agency, Municipal Solid Waste Program, Prairie Village, Kans., 1989.

21. National Research Council, *Toxicity Testing: Strategies to Determine Needs and Priorities,* Washington, D.C., 1984.

22. U.S. Environmental Protection Agency, *Sources of Toxic Compounds in Household Wastewater,* EPA 600/2-80-128, Cincinnati, Ohio, 1980.

23. Florini, Karen, Richard Dennison, and John Ruston, "An Environmental Perspective on Solid Waste Management," in Frank Kreith (ed.), *Integrated Solid Waste Management: Options for Legislative Action,* Genium, Schenectady, N.Y., 1990, pp. 179–181.

24. Ref. 23, p. 181.

25. Churney, K. L., A. Ledrod, Jr., S. Bruce, and E. Domalski, *The Chlorine Content of Municipal Solid Waste from Baltimore County, MD and Brooklyn, NY,* NBSIR 85-3213, Gaithersburg, Md., National Bureau of Standards, 1985.

26. Palmer, J., *A Cleaner Environment: Removing the Barriers to Lead-Acid Battery Recycling,* St. Paul, Minn.: GNB Inc., October 1988

27. Ref. 14, pp. 158–159.

28. Ref. 14, p. 158.

29. U.S. Resource Conservation and Recovery Act, 40 CFR Ch. 1, Section 3001.

30. U.S. Environmental Protection Agency, *A Survey of Household Hazardous Wastes and Related Collection Programs,* EPA/530-SW-86-038, Washington, D.C., 1986.

31. Gage, Kathryn, "Permanent Site," *Summary of the Second National Conference on Household Hazardous Waste Management,* Center for Environmental Management, Tufts University, Medford, Mass., 1987, pp. 73–75.

32. Ref. 14, p. 157.

33. Watson, Tom, "Fluorescent Lamps—A Bright New Recyclable," *Resource Recycling,* March 1992.

34. Ref. 14, p. 167.

35. "Resins '88," *Modern Plastics,* January 1988, pp. 63–105.

36. Wolf, Nancy, and Ellen Feldman, *Plastics: America's Packaging Dilemma,* Island Press, Covelo, Calif., 1991, pp. 109–111.

37. Franklin Associates, Ltd., *Characterization of Municipal Solid Waste in the United States,* 1960–2000, Final Report, prepared for the U.S. Environmental Protection Agency, Prairie Village, Kans., March 1988.

38. Society for Environmental Toxicology and Chemistry, *A Technical Framework for Life-Cycle Analysis,* Washington, D.C., January 1991.

39. Glenn, Jim, "The State of Garbage in America," *Biocycle,* May 1992, pp. 30–37.

40. U.S. Environmental Protection Agency, Office of Solid Waste and Emergency Response, *Characteristics of Municipal Solid Waste in the United States, 1960–2000,* Washington, D.C., 1988, pp. 15–16.

41. Ref. 36, p. 97.

42. See "An Ordinance Amending Title 10 of Chapter 204 of the Minneapolis Code of Ordinances," March 1989; and "An Ordinance to Prohibit the Use of Certain Packaging Materials for Food and Beverages Sold at Retail in St. Paul," Ch. 236, Sec. 1 of the St. Paul Legislative Code, March 1989.

43. See Maine Statutes, "Sale of Consumer Products Affecting the Environment," Title 38, Sec. 1603, 1989; Minnesota Statutes, "CFC-Processing Packaging," Sec. 116.72, 1988; and Rhode Island Public Law, "An Act Relating to CFC-Processed Products," Ch. 81, Bill 9487, 1990.

44. Dennison, Richard et al., "Good Things Come in Smaller Packages: The Technical and Economic Arguments in Support of McDonald's Decision to Phase Out Polystyrene Foam Packaging," Washington, D.C., Environmental Defense Fund, Dec. 6, 1990.

45. United Nations Environment Programme, Industry and Environment Programme, *Executive Summary of the Workshop on Country-Specific Activities to Promote Cleaner Production (Paris, France, Sept. 17–19, 1992),* Paris: United Nations Environment Programme, 1992.

46. "Pollution Prevention Act of 1990," 42 U.S.C.A., Ch. 133, Secs. 13101–13109.

47. Allenby, Braden, "Design for Environment: A Tool Whose Time Has Come," *SSA Journal,* September 1991; Werner Glautshnig, "Design for Environment: A Systematic Approach to Green Design in a Concurrent Engineering Environment," paper presented at the First International Congress on Environmentally-Conscious Design and Manufacturing, Boston, Mass., April 1992.

48. Ref. 14, pp. 104–105.

49. U.S. Congress, Office of Technology Assessment, *Green Products by Design: Choices for a Cleaner Environment,* OTA-E-541, Washington, D.C., 1992, p. 59.

50. Rossi, Mark, Michael Ellenbecker, and Kenneth Geiser, "Techniques in Toxics Use Reduction," *New Solutions,* fall 1991.

51. For references to state laws, see William Ryan and Richard Schader with Mike Derezin and Richard Regan, *An Ounce of Toxic Pollution Prevention: State Toxic Use Reduction Laws, Second Edition,* Washington, D.C., National Environmental Law Center and Center for Policy Alternatives, January 1993.

52. Ref. 14, p. 105.

53. Gipps, Terry, *Breaking the Pesticide Habit: Alternatives to 12 Hazardous Pesticides,* Minneapolis, Minn., International Alliance for Sustainable Agriculture, 1987.

54. A good review of these programs can be found in Peter Hurst, *Pesticide Reduction Programmes in Denmark, the Netherlands and Sweden,* Gland, Switzerland, World Wildlife Fund International, November 1992.

CHAPTER 9
RECYCLING

David B. Spencer

PART A

9.1 Overview
9.2 Collection of Recyclables
9.3 Processing Equipment for Recycling Facilities
9.4 Processing Recyclables
9.5 Products and Markets

Mary B. Sikora

PART B

9.6 Options for Managing and Marketing Scrap Tires

Gary R. Brenniman, Stephen D. Cosper, William H. Hallenbeck, James M. Lyznicki

PART C

9.7 Automotive and Household Batteries
9.8 Used Oil

Tracy Bone

PART D

9.9 Household Hazardous Waste Management

David B. Spencer

PART E

9.10 Environmental Impacts
9.11 Energy Production Requirements and Energy Savings
9.12 Integration with Other Technologies
9.13 MRF Economics
9.14 Overall Comparative Recycling Economics
9.15 Summary and Conclusions

INTRODUCTION*

There is no official definition of recycling, but a widely accepted view is that recycling constitutes "the beneficial reuse" of products that would otherwise be disposed of. The main concept of recycling municipal solid waste is to somehow pick up or collect the waste generated by people in their daily lives and then sort the waste so that it can be used in the manufacture of a new product. This chapter deals primarily with this approach to recycling. But, beyond this simple definition, recycling can also include

---

*Introduction by Frank Kreith.

recovery of batteries, refining used oil, recovering energy from used tires to fuel cement kilns, and refining mineral spirits for industry reuse, among other processes.

The bulk of this chapter has been contributed by David B. Spencer, the President and CEO of wTe Corporation, a waste processing and recycling services company, who described and analyzed the various technologies available for curbside pickup, sorting and managing MRFs, as well as the design of recycling programs and their economics. Important contributions were also made to this chapter by Tracy Bone, an environmental scientist at EPA, whose section deals with hazardous solid waste disposal; by Gary R. Brenniman, Stephan D. Cosper, William H. Hallenbeck and James M. Lyznicki, a team of scientists from the Office of Solid Waste Management, School of Public Health, University of Illinois at Chicago, who contributed two sections on automotive wastes (oil and batteries); and by Mary B. Sikora, the President of the Recycling Research Institute, who wrote the section on tire disposal. The chapter is designed to provide the information and background needed for developing a technically and economically sound recycling program and for analyzing its performance and economics.

PART A

David B. Spencer

President, wTe Corporation
Bedford, Massachusetts

9.1 OVERVIEW

During the 1980s, recycling took on much greater significance than just providing an alternative method for treatment of our solid waste. Recycling became an American philosophy, a public mandate. Source reduction and recycling became the only popularly accepted methods for dealing with America's solid waste crisis.

However, to place a large part of the responsibility of solid waste management for a community on recycling alone puts an undue burden on recycling and could damage a strong, sound recycling initiative if it results in excessive cost or excessive contamination of high-value products. Just as the sanitary landfill came to be viewed as a disposal panacea at midcentury—only to be later discredited—the euphoria over recycling will need to be tempered with a strong record of tangible results.[1]

Generation of Recyclables

Based on projections for recycling by respective industries manufacturing the major commodity components of municipal solid waste (MSW), the goal of 25 percent recycling by the mid-1990s may actually be achievable.[2] One significant trend is the emergence of a greater number of mandatory or voluntary programs for the source separation of recyclable materials. These so-called curbside programs require the participation of residents to separate recyclable materials into one or more fractions for collection. In 1989, 1042 curbside programs existed in 35 states.[1] There has been considerable growth since that time, with the implementation of ambitious programs in New York, Florida, California, Ohio, and other states. By 1991, the number of curbside programs had grown to 4000. The distribution of these programs throughout the United States and the number which are mandatory as compared with voluntary is detailed in Chap. 5.

Quantities and Composition of Recyclables

Between 1920 and 1965, U.S. refuse production rose from 2.8 to 4.5 lb per day per capita. For the period from 1968 to 1988, solid waste generation increased almost 30 percent, from 140 to 180 million tons per year. This estimate includes residential, commercial, and institutional solid waste, plus some similar types of waste from industrial sources.[3] While per capita daily generation rates have now leveled off, population increases will continue to increase overall volumes of waste produced into the future. MSW is estimated to be growing at rates of 0.75 to 1.5 percent per year (i.e., between the same rate and double the rate of population growth). However, it should be noted that some estimates indicate that waste generation actually dropped some 12 million tons in 1991.

Growth in the quantities of waste generated is not the only problem contributing to the present solid waste crisis. The composition and complexity of materials in the current waste stream may contribute more to the crisis than the volume or weight produced. Recycling must deal with not only the vast quantity of bottles, cans, and containers present in the affluent U.S. society but also the considerable complexity of these highly engineered products.

In 1985, 47 billion lb of plastics were produced in the United States. Of the 39 billion lb consumed domestically, 33 percent was used for packaging. These plastics, although they represented only 8 percent or less by weight of the waste stream, represented 20 percent of the solid waste stream by volume. These packages are a complex composite of many materials, making recycling in the 1990s a highly complex discipline, especially when the purity of the finished product is a critical limitation to market demand for recyclables.

Ways to Recycle

There are many ways to implement a recycling program. The program can be either voluntary or mandatory. The materials to be recycled can include paper (newspaper, cardboard, mixed paper, etc.), glass (amber, green, and/or flint), cans (aluminum, ferrous, bimetal), and plastics (PET, HDPE, PS, PVC, PP, LDPE, etc.), as well as other items.

The many recycling alternatives include:

- Return of bottle bill containers or use of reverse vending machines
- Drop boxes, drop-off centers, or buy-back centers for recyclables
- Curbside collection of homeowner-separated materials
- Curbside separation of homeowner-commingled recyclables
- Materials recycling facilities for the separation of commingled recyclables (collected commingled at curbside, commingled in drop boxes, or collected in special "blue bags") using low-tech or high-tech processing systems
- Mechanically assisted hand separation of recyclables from raw waste (front end processing or mixed waste processing)
- Fully automated separation of recyclables from raw waste

The collection process itself can occur in many different ways. Materials can be collected at curbside in a multicompartment recycling truck either with or without compaction of the various segregated materials, or they can be collected commingled in either a dump truck or a packer truck.

Figure 9.1 illustrates the extent of recycling typical of early programs in which many different materials were set out at curbside for collection. Figure 9.2 shows collection of paper and commingled beverage containers of many types. When items are commingled, they must be separated before they can be delivered to end markets. This can be done by the homeowner, by the collector at curbside, or by workers at a central processing plant. The more materials in the program, the greater the collection and processing problems encountered. The space required to deal with these materials seems to grow geometrically with the number of items recycled.

When one accounts for all the different types of materials which can be included in a recycling program, the various methods for segregation, and the various means and methods of collection, as well as the types of processing and separation systems which

FIGURE 9.1 Curbside collection program—separate set-out of many different recyclable materials.

can be used, the combinations and permutations seem endless and confusing. Specific expertise is required to evaluate the optimum method for a given community based upon its population, geographic location, and proximity to markets. At this time, the information does not exist to make the selection and program definition easy. There are too many factors and too little information on which to make an informed judgment.

This chapter describes the various methods of recycling, providing information on methods of collection, process unit operations and processing systems, products and markets, and economics. It is hoped that this perspective will aid in the implementation of sound recycling systems in which expectations and results are responsibly managed. Responsible management will then form the foundation for broad implementation of a local and national recycling strategy.

9.2 COLLECTION OF RECYCLABLES

Methods of Collection

There are three main separation and collection approaches to recycling:

1. Source separation by either the generator or the collector with consolidation for transport to markets
2. Commingled recyclables collection with processing at centralized materials recycling facilities (MRFs)
3. Mixed municipal solid waste collection with processing for recovery of the recyclable materials from the waste stream at mixed waste processing or "front-end processing" facilities

FIGURE 9.2 Curbside collection program—separate set-out of commingled containers and paper (trash collected on the same day).

Source Separation.   The first method involves the separation of recyclable materials into individual components either by the generator or at curbside by the collector. The set-out of source-separated recyclables at curbside is depicted in Fig. 9.1. The separate components can be collected individually in single-compartment trucks, or more commonly, they are collected at the same time in a specially designed multicompartment recycling vehicle (Fig. 9.3). The segregated components are then transported to a consolidation site where each component is stored until a sufficient amount can be accumulated for either further processing or shipment to markets (Fig. 9.4).

Usually, in the case of small communities, there is no further processing at the consolidation site. Processes such as can flattening, glass bottle pulverizing, and paper baling are performed by local scrap and paper dealers or recyclers who are the markets

FIGURE 9.3 Typical multicompartment collection vehicles for source-separated recyclables set out at curbside.

for the products collected and who prepare the materials as necessary for final markets (Fig. 9.5). In larger communities, each component may be further processed at the consolidation site and/or directly marketed to an end user when the materials meet buyers' specifications. Drop-off centers, buy-back centers, and "bottle bill" return stations are variations of the source separation approach.

Commingled Collection. Figure 9.6 depicts recyclable materials set out at curbside for commingled collection. Here the generator only needs to separate recyclable mate-

FIGURE 9.4 Collection vehicles deposit source-separated recyclables into separate containers. Several loads are consolidated to provide whole truckload or roll-off container quantities for delivery to intermediate markets.

rials from nonrecyclables. Newspapers are often kept separate from the rest of the commingled recyclables to prevent contamination and to improve collection vehicle efficiency. A typical collection vehicle for commingled recyclables is shown in Fig. 9.7.

The recyclables are transported to a centrally located MRF (Fig. 9.8) where they are segregated into each recyclable component—glass, metal cans, plastic bottles, etc.

FIGURE 9.5 Source-separated recyclables (cans, plastic bottles, and corrugated) are processed to densify them in preparation for shipping. (*Casella Waste Management, Rutland, Vt.*)

FIGURE 9.6 Set-out of commingled recyclable containers at curbside.

MRF process operations vary. A "low-tech" MRF, primarily dependent on the manual hand sorting of materials, is shown in Fig. 9.9. A "high-tech," highly automated MRF is depicted in Fig. 9.10.

A variation of the commingled collection approach to recycling is the use of "blue bags" in a mixed waste collection program. The color blue was chosen because it is distinctly different from the typical black or green trash bag and studies have shown that the blue bag can be easily identified in a mixture of trash bags. Commingled recyclables are placed in the blue bags by the generators. The blue bags are taken along with trash bags to a central processing plant where the blue bags are hand separated from the trash and sent into a commingled recyclables processing facility for materials recovery. The bags can be filled with paper, commingled metals, plastic, and/or glass depending on the design of the program. The objective of this type of program is to take advantage of the reduced collection costs of mixed waste collection while still implementing a materials recycling facility which processes only the mixed recyclables, not the entire solid waste stream.

Mixed Municipal Solid Waste Collection.   In the third approach to recycling, there is no segregation of recyclables from other waste materials. Mixed trash and recyclables are set out at curbside (Fig. 9.11), as would be done for landfilling or incineration. There is only one collection vehicle required to pick up the mixed waste—normally the familiar packer truck. The mixed waste is then transported to a central processing facility which employs complex separation equipment such as shredders, trommels, magnets, and air classifiers to recover the recyclables. This mixed waste processing method of recyclables recovery is also known as front-end processing or refuse-derived fuel (RDF) processing of MSW. A sketch of a 1000 ton per day mixed waste processing facility located in Wilmington, Del. is shown in Fig. 9.12.

Comparison of Recycling Methods.   The first of the three approaches to recycling, source separation, requires a high degree of homeowner involvement and has high col-

FIGURE 9.7   Typical collection vehicle for commingled recyclables.

FIGURE 9.8   Commingled recyclables delivered to a MRF for processing.

FIGURE 9.9   "Low-tech" MRF depends primarily on hand sorting and densification.

FIGURE 9.10 "High-tech" MRF utilizes automated separation systems and densification equipment.

lection costs but low processing costs. The second approach, commingled collection, requires an intermediate amount of generator effort, an intermediate amount of added collection cost, and processing costs somewhere between those for source separation and the third approach, mixed waste collection and processing. The third approach requires no extra effort by the generator and results in no incremental collection costs, but it is accompanied by high processing costs plus some risk regarding technology, operating costs, and market economics due to uncertain capital and operations costs and potentially low recovery efficiency and material purity.

The quantity and quality of recyclable materials separated, collected, processed, and recycled can depend to a large extent on which of the above approaches a community selects. Each method can affect attitudes regarding the mental and physical work required by a recycling program, and thus the extent of generator participation. Also, each method has a different capital and operating cost requiring varying levels of community financial commitment. Finally, each produces materials of differing composition or quality and thus can impact the amount of residue generated and the markets for products produced. Resident participation can, of course, also be affected by legislative actions, such as mandatory recycling. For example, motivation to recycle can be impacted by the requirement to participate in recycling in order to receive trash collection services, or by the levying of fines and penalties.

Defining Recyclables

Determining the quantity of recyclables generated, by whatever method of separation, first requires a determination of what is to be considered a recyclable and how one measures recycling performance. Unfortunately, there are no standard rules. Recycling is an elusive concept about which everyone thinks they have a clear understanding until they begin to practice it.

FIGURE 9.11   Mixed trash and recyclables set out at curbside.

The issues are complex. Several considerations are worth noting as they underline the problems of measuring recycling performance. Typically, curbside nonrecyclables and residuals are counted as recyclables even though they are not actually recycled. These nonrecycled or "lost" materials picked up at curbside do not save landfill space or reduce the need for incineration capacity. Paper or plastic set out and collected when the market is temporarily saturated may often be landfilled or alternatively burned to recover their energy value; yet, because they are collected with other recyclables, they are counted as recyclable material.

Industrial and plant scrap can often be included as recyclables in community generation and recycling statistics. Other items confusing the definition of recyclables include metals recovered from junk yards and automobile shredding operations; tires that are collected and sold to a recapper by the tire dealer, and used tires that are burned; demolition waste; yard waste processed as compost and perhaps used as cover

FIGURE 9.12   A complex mixed waste or RDF-type processing plant.

material on a landfill; and old clothing donated or sold to thrift shops or service organizations.

There is no common agreement either in industry or in government that can be
used as an accepted standard in measuring recycling performance. Recycling programs can be compared if data are standardized, but they are not. Consistent, standard,
and meaningful measurement terminology is needed if communities are to effectively
plan and assess their recycling programs.[4] It would be wise to state at the outset of a
program what will be counted as recyclables—e.g., all of the materials collected at
curbside or only those actually sold to market; all recyclables collected and processed
at a MRF or only those sold to market after separation and processing with the
residues generated at the MRF subtracted from the total.

Measures of Recycling Performance

Although difficulties remain in quantitatively measuring the performance of recycling
programs on a consistent, standard basis, four useful performance measures have been
defined: capture rate, participation rate, recycling rate, and diversion rate.

The term "capture rate" denotes the weight percent of an eligible material in the
total refuse stream actually separated out for recycling. Capture rate applies to a single
material, not recyclables in general.[5] This performance measure is of greatest importance in measuring the success of a separation and collection program. Thus, for
example, a capture rate for aluminum would describe how much aluminum is captured
by the community's curbside program versus how much is captured through the bottle
bill program.

The term "participation rate" denotes the percent of households (or businesses)
which *regularly* set out recyclables. For example, in a particular community, on a
monthly basis, 75 percent of the citizens *participate* in the curbside program. This sta-

tistic may be different on a weekly basis than on a monthly basis since fewer residents may participate weekly. The term "set-out rate" denotes the percent of households that set out on a particular collection day. Neither participation rate nor set-out rate indicates quantities of materials recycled or what materials were recycled. These terms may actually provide misleading information regarding the success or failure of a recycling program, but they do provide some useful measure of the extent of household involvement in the community's recycling program.

The term "recycling rate" is sometimes used to denote the quantity of recyclables collected per household per unit of time (e.g., 35 lb per household per month). The recycling rate normally addresses what was collected without regard to whether the material was actually sold or what amount of contamination was present in the recyclables. This term is sometimes quoted as a percentage of the total quantity of waste generated in the community.

Another performance factor in gauging the success of a recycling program is "diversion rate," which represents the weight of total refuse that is *not* landfilled (or *not* incinerated). Thus, if the objective of the program is to minimize the weight of refuse (including processing residues and incinerator ash) sent to landfill through a combination of strategies (such as source reduction and recycling), the ultimate performance measure is the net diversion rate. Again, this is often reported as a percent rather than in lb or cubic yards. It may be more useful to determine the net volumetric diversion rate since it is a better measure than weight to estimate the savings in landfill life achieved by the integrated program. "Landfills fill up long before they get too heavy."[5]

Factors Affecting Participation in Recycling Programs

It is widely believed that the easier it is for the public to participate in a recycling program, the higher will be the diversion rate or recycling rate—all other things being equal. A program that accomplishes recycling without any changes to the residents' disposal patterns, such as a mixed waste processing system (which automatically separates recyclables from mixed trash), will achieve 100 percent participation, by definition. The diversion rate will be a function of the design of the program and the efficiency of processing equipment.

The degree of source separation required can be expected to have a direct impact on participation rate and capture rate. A very complex recycling program in which many items are recycled and in which the resident separates each material, removes every label, and washes every container before setting it out to the curb fully segregated (Fig. 9.1) is more difficult for the resident than one in which materials are set out commingled, unwashed, and with labels, in a single container (Fig. 9.6).

As a general rule, a well-designed program for collection of recyclables from homes will:

1. Provide weekly collection
2. Distribute a household storage container
3. Pick up recyclables on the same day as waste collection
4. Promote the program vigorously

However, the commonly held, commonsense belief that making the separation and recycling effort easier on the resident will increase participation is not always borne out in actual practice. A study was conducted by the New Jersey Office of Recycling on 12 New Jersey recycling programs in communities with populations ranging from 5000 to 300,000. Researchers looked at the relative costs and diversion rates of "complete separation" programs versus "commingled" separation programs. Of the 12 communities studied, five of which utilized a commingled approach, the separation programs, not commingled programs, *resulted in greater waste diversion.*[6] For the household separation programs, the average recovery rate was 171 lb per capita per year, which was 15 percent greater than the recovery rate for commingled programs. Moreover, resident separation programs were reported to cost less than commingled programs, although collection costs were reported to be $10 to $15 per ton higher. Processing costs for commingled programs added $63 per ton to the collection costs, ranging from $46 to $86 per ton for the communities surveyed.

The extent of resident participation is dependent on many factors beyond the complexity level of effort required in a source separation program. Public education is a substantial factor, as are the demographic characteristics of the community including income, education, and location (suburban vs. urban). A study of participation in Seattle's curbside recycling collection program[7] showed that people who participated in the program tended to have a college degree, a household income above $30,000 (1992 dollars), and adequate storage space in their homes. Moreover, those with greater knowledge about the program and with a positive attitude about recycling recovered greater quantities of recyclables.

When taking part in a separation program, households have to learn some new, and unlearn some old, behavior patterns. Learning new behavior involves the expenditure of time, mental and physical effort, and sometimes money.[8] These patterns are reinforced when education, psychic satisfaction, and ease of program implementation are encouraged. However, generally speaking, as a program proceeds and recycling becomes more of a habit, the *perceived effort* to accomplish recycling *diminishes.* In the course of time, less mental effort is needed to separate the domestic waste and households become more positive regarding the costs and benefits of recycling.

Same day collection of both recyclables and municipal solid waste (trash) was found in test studies to significantly improve the level of household participation. Moreover, the collection of yard debris, preferably on the same day as the other collections, was required to achieve high recovery rates exceeding 30 percent.[9]

To summarize the issue, on the one hand, with a high-cost system such as mixed waste processing, participation is high because everyone participates. Incremental collection cost is negligible and recovery efficiency is also high. However, the potential for product contamination is higher than for source separated recycling. Mixed waste processing is expensive, often requiring tens of millions of dollars in one-time capital cost and millions per year in operating costs. Further, automated front-end processing programs do not focus the efforts of the public on a continuing basis in dealing with the solid wastes they generate.

On the other hand, the "good feeling" of recycling is lost when there is little or no household involvement. In a source separation program, although collection at curbside is more expensive and the percentage of participation may be lower, the ability to recover materials which will meet product quality specifications can be much higher. The resident will receive satisfaction from involvement in the program, and the diversion rate for the individual recyclables can still be high.

A study by R. W. Beck and Associates for the city of Orlando, Fla., found the total program costs for curbside recycling collection involving a curbside sort of separated recyclables was more economically attractive than the available commingled option. Although collection costs alone were 20 percent less for the commingled option as compared with complete source separation by the resident, the savings in collection cost were more than offset by the lower processing expense and higher material revenues due to less breakage and contamination. On a comparative cost basis, source separation cost Orlando $1.95 per month per household while commingled collection cost each household $2.43 per month. However, the study noted that one reason for the higher cost was that the central materials recycling facility (MRF) was not fully utilized. The difference of $0.50 per household could be reversed as the program becomes enlarged and more successful such that the processing plant is fully utilized, spreading its fixed costs over its full design capacity.

Recycling Case Study

The complexity of establishing a recycling program and then measuring the success of that program on a standardized basis is highlighted by a case study[4] comparing two different communities: Somerset County, a rural New Jersey community; and Islip, N.Y., a suburban community on Long Island.

Both Islip and Somerset have energetic residential recycling programs. Islip commingles glass, paper, metal, and plastic during collection and has a separate yard waste program. Somerset County commingles glass and metal but collects paper and plastic separately; yard waste is handled by the county's 21 towns on an individual basis. Further, Somerset adheres to New York State's bottle bill legislation.

Both communities process recyclables in a centralized MRF. The processing systems, however, are somewhat different. Islip generates 7.3 lb of waste per capita daily, while Somerset generates 6.4, both of which are more than 50 percent higher than the approximate 4 lb per day national average. However, both communities calculate their waste generation differently and include different categories in their MSW definitions.

In 1989, Islip reported a recycling rate of 35.6 percent while Somerset County reported a 14.5 percent rate. "Yet these reported recycling rates neither provide information about the effectiveness or scope of the individual programs nor permit direct comparison of them, either to each other or to programs in other communities around the country." These reported rates, however, are actually materials collection rates and do not reflect either the variety of materials collected or the percentage of material actually recycled. *Since the process of recycling is not complete until postconsumer materials become new commodities, a true recycling rate would refer only to collected materials actually sent to market and manufactured into new products. However, all the materials that communities collect in their recycling programs do not become new products; instead, some become "lost materials."*[1] (Emphasis added.)

Table 9.1 compares data on the two programs. As indicated, Islip "lost" 20 percent of the materials collected for recycling (which later increased to over 50 percent). The losses were mainly composed of mixed paper, which Islip continues to collect but for which it has no market. In the same 1989 period, Somerset lost only 3 percent of the recyclable materials it collected. *Somerset collected less material for recycling than Islip but actually sent more to market.*

Losses can occur both before recyclables are collected from households and after the recyclables are "marketed." For example, households could include garbage with the recyclable material they set out, thus contaminating the materials (e.g., coffee grounds and paper). Also, recyclables could be damaged during set-out, rendering them of lesser value or without any value (e.g., broken glass and wet paper). Moreover, mixed paper recyclables in one collection program may be contaminants in another program which has only markets for old newspaper. Solid waste managers may send collected recyclable materials to disposal when disposal is less expensive than marketing (e.g., Islip burns its low-value mixed paper in its incinerator); or they may send recyclable materials to incinerators because they want to meet tonnage guarantees and cannot provide waste from other sources, or when recyclables are worth more as incinerator feedstock than in the marketplace.

TABLE 9.1   Lost Materials in Islip and Somerset County, 1989

	Islip	Somerset
Population	308,549	233,172
Solid waste generated, tons	392,106	270,360
Recyclables collected, tons		
(glass, metal, plastic, and paper for MRF;		
excludes yard waste and bulky materials)	28,770	25,792
Materials marketed, tons	23,016	24,987
Lost materials, tons	5,754	805
Lost materials, %	20	3
Reported "recycling rate," percent	35.6	14.5

Source: deKadt, M., "Evaluating Recycling Programs: Do You Have the Data?" *Resource Recycling,* June 1992, p. 28.

Losses after marketing of the products by the community can occur when brokers dispose of materials sent to them if these materials are not of the appropriate quality, or if disposal is less expensive than additional marketing. Losses can also occur during the manufacturing process because residue is generated during processing.

Table 9.1 highlights the difficulty in evaluating and reporting on recycling success. This can be especially confusing when a program is mandatory and/or must comply with certain applicable governmental regulations. The case study also points out the problems in measuring the success of a recycling program on both a relative and an absolute basis. Islip reports higher recycling rates but Somerset recycles higher tonnage per household.

Moreover, in terms of total waste diversion, recycling may not achieve the goals and objectives that are reported. Diversion rates are not likely to match recycling rates. Accordingly, the alternative methods of waste disposal (e.g., landfill and incineration) may see greater utilization than projected based on actual diversion rates rather than reported recycling figures.

Mandatory versus Voluntary

A discussion of the impact of state recycling legislation is provided in Chap. 5. Clearly, if a community is unable to achieve its recycling goals through a voluntary recycling program, the alternative remains to make recycling mandatory. While mandatory programs should in theory increase the participation level of households, there is no evidence to indicate that a well-implemented, well-communicated voluntary program cannot achieve the same levels of participation and the same waste diversion as a mandated program.

Bottle Bill Legislation and Recycling

Before the first deposit laws were enacted in 1970, there was virtually no recycling of aluminum cans or plastic bottles; and glass bottles were recycled at just 1 percent. Bottle bill programs were implemented, not so much for their impact on materials recycling but rather because of the very positive impact they have on litter control. However, such legislation has proved to be a part of many successful recycling programs.

As shown in Chap. 5, 10 states currently have mandatory bottle deposit legislation: California, Connecticut, Delaware, Iowa, Maine, Massachusetts, Michigan, New York, Oregon, and Vermont. In each case, the consumer pays a deposit on each beverage container purchased and receives that amount as a refund when the container is returned for recycling or refilling. These bottle deposit bills primarily affect soft drink beverage containers, but in some cases, as in the state of Maine, other items such as wine and juice containers are included in the deposit system.

The 10 beverage container deposit states, with just 30 percent of the U.S. population, account for over 90 percent of all plastic soft drink containers recycled, over 70 percent of glass recycled, and nearly half of the aluminum cans recycled. In all these states, according to a study by the U.S. General Accounting Office, public approval of the bottle law is high, with over 63 percent strongly approving and only 6 percent strongly disapproving.[10] Further, it should be noted that, while there have been many efforts to prevent the passage of new bottle bills, not one of the bottle bills originally passed has ever been canceled.

The arguments against bottle bills put forth by the beverage industry over the last 20 years have included loss of jobs, higher cost to the consumer, and the concern that a bottle bill would compete with other recycling alternatives. Clearly, the stores that have to comply with the storage and handling requirements associated with a bottle bill are concerned because valuable store space is consumed (Fig. 9.13). There is also the potential for insects and rodents which can be attracted from storage of bottles which have not been cleaned or washed out. Moreover, maintaining the bottle return system requires significant clerical effort.

When a community is operating under a state bottle bill, there is always some concern regarding its impact on the amount of material which will be recovered under a curbside recycling program as compared with other communities or states which do not have a bottle bill. There is substantial evidence that demonstrates the value of deposit laws working in tandem with curbside recycling collection programs. Seattle

FIGURE 9.13   Supermarket bottle bill materials return area employing Envipco reverse vending machines.

recently completed a study on this subject and concluded that "the presence of a bottle bill would increase recycling levels of beverage containers and reduce the City's overall solid waste management costs."[10]

Drop-Off Programs

Drop-off centers are centralized locations where a specified class of waste generators, typically residential generators, may voluntarily bring certain recyclable materials (Fig. 9.14). One of the largest advantages of drop-off centers is that they are inexpensive to implement. A drop-off center can be as simple as several small-capacity containers that temporarily store the materials for regular pickup and transportation to market or a central consolidation facility or it can consist of the central consolidation facility itself.

Because programs of this nature are voluntary, participation can often be poor. However, there can be notable exceptions, especially where curbside waste collection is not performed and citizens must take their trash to a disposal facility where they may also drop off their recyclables. Moreover, participation is enhanced by public education and by ordinances that increase the difficulty to otherwise dispose of recyclable materials. Economic incentives could also be a factor. For example, a variation of the drop-off center is the buy-back center where the generators are financially compensated for materials. Both buy-back centers and drop-off centers seldom capture more than 10 percent of the waste stream.[11]

The physical layout of a drop-off center varies by the volume and number of recyclable materials processed, site characteristics, and level of supervision. A conventional drop-off center would be centrally located within a service area and provide bins or compartmentalized containers for waste generators to deposit recyclable materials (see

FIGURE 9.14 Recycling drop-off center utilized by Wellesley, Mass., residents.

FIGURE 9.15   Signs assist residents at unattended drop-off center, Wellesley, Mass.

Fig. 9.15). To ensure material quality and public safety as well as to prevent scavenging, many drop-off centers have controlled access and limited hours of operation, and are monitored by attendants. Once a sufficient quantity of a material has been collected, it can be shipped to end users or intermediaries in the container in which it was collected or alternatively transferred to a larger container or truck. Correct sizing and type of containers are key design features to address, along with traffic access and security.

The smallest drop-off center might be a neighborhood "kiosk-like" or igloo container, unattended and conveniently located to maximize its use. The state of Delaware has implemented a drop-off program which utilizes this type of igloo system very successfully. This method of recycling is particularly applicable to heavily rural areas. However convenient these unattended containers are, they must be inspected frequently to determine if they are full, present an unsightly litter problem, or have contaminated contents. Drop-off centers are also vulnerable to odors, vectors, and vandalism, aside from incurring transportation and handling costs.

The state of Delaware's drop-off program is funded by a $2 per ton surcharge collected at all Delaware Solid Waste Authority solid waste management centers which control all disposers in the state. Each site is designed with an area 80 ft by 10 ft in size. The igloos are 5 ft in diameter and approximately 5 ft high. Each igloo has a capacity of 3, 4, or 5 yd^3. The materials collected initially included old newspaper; aluminum and steel beverage and food cans; PET and clear HDPE; green, brown, and clear glass bottles and jars; household batteries; and at selected locations, used motor oil. The program has evolved to include phone books and all resins of plastic bottles. Delaware's drop-off program is one of the largest programs in the country. The collection is performed by a private hauler, as is the marketing of output products. Table 9.2 shows the evolution of the number of drop-off sites and the tons of material collected during the first 11 months of the program.[12]

Drop-off containers should be located in areas that normally have a high volume of traffic and are well known to the local populace such as schools, shopping centers, and fire stations. This makes it easier for citizens to recycle when they are shopping, picking up their children, or running errands. Participation can be improved through advertising, special events, and local mailings to citizens informing them of the location of drop-off boxes and advising them what to do and what to recycle.

Drop-off centers have low capital and operating costs, little or no technical risk, limited changes in waste generator behavior (provided they are conveniently located), and are flexible to changes in waste composition or participation rates as well as the targeted recyclable materials. On the other hand, they suffer from lower participation rates because they require the citizen to store materials and physically drive them to a remote location, and the products can often be low-quality if there is no supervision. In fact, some products may be unmarketable owing to the high degree of contamination that can frequently occur.

A limited survey conducted by *BioCycle* in 1988[13] is reproduced as Tables 9.3, 9.4, and 9.5, illustrating the scope and performance of selected drop-off programs nationwide. Convenient siting, more efficient equipment, public education, and economic incentive programs are cited as key elements in successful programs.

Wellesley, Mass., is a community which provides a case study of a long-standing, successful drop-off center operation. A community of 27,000 people, located outside of Boston, Wellesley has established a drop-off center adjacent to the town's transfer station to which residents take their trash. No municipal collection is provided. At the drop-off center, residents recycle old newspaper, old corrugated cardboard, mixed paper, three colors of glass, aluminum cans, ferrous bimetal cans, HDPE, and PET and HDPE; as well as waste motor oil, tires, batteries (automotive and household), scrap metals, wood, yard wastes, books, clothing, and bulky wastes. The center is comprised of assorted bins and roll-off containers (Fig. 9.15). In 1989, Wellesley recycled approximately 19 percent of their wastes from the adjacent transfer station through their drop-off program. By any measure, this is a very successful program. However, citizen hauling of trash to a transfer station is an unusual method of waste collection, and thus the success achieved in Wellesley may not be replicable in other locations.

TABLE 9.2 Recycle Delaware, Actual and Projected Collections*

Month	Number of drop-off sites	Newspaper	Plastic bottles	Glass containers	Metal cans	Household batteries	Total tons collected	Used motor oil†
December 1990	19	51	2	2	0	0.3	55	150
January 1991	23	163	9	117	9	0.4	298	1,250
February 1991	36	231	13	144	23	0.4	411	1,650
March 1991	48	341	14	118	4	0.5	478	1,300
April 1991	64	407	29	116	25	0.2	577	2,527
May 1991	78	613	37	191	39	1.1	881	4,170
June 1991	79	607	45	229	43	0.6	925	5,764
July 1991	89	608	58	240	43	1.0	950	7,739
August 1991	96	739	37	272	27	1.8	1,077	3,870
September 1991	100	696	69	225	39	1.2	1,030	7,230
October 1991	100	741	56	231	46	1.3	1,075	5,955
Total (11 months)		5,197	369	1,885	298	8.8	7,757	41,605
Projected (11/91 through 6/93)		14,800	1,200	4,600	900	25	21,525	120,000

*All figures are reported in tons, except for used motor oil.
†Figures for used motor oil are in gallons.

Sources: For actual collections, Delaware Solid Waste Authority, 1991; for projections, Browning-Ferris Industries, 1991. Sanderlin, G. H., "Delaware Does Drop Off," *Resource Recycling*, January 1992, p. 46.

9.23

TABLE 9.3 Drop-off Programs—General Program Characteristics

Location	Population served (Estimated)	No. of sites	Population served per site	Materials collected	Participation rate, %
Champaign Co., Ill.	171,000	15	3000–20,000	N, G, A, T, HDPE, OCC, MO	18
Columbia Co., Pa.	50,000	17	3000 (av)	N, G, A, T, OCC	25–30
Cook and Lake Co., Ill.	270,000	18	N/A	N, G, A, T	N/A
Delaware Co., Pa.	500,000	50	10,000 (av)	Glass only	25
Durham Co., N.C.	120,000	10	10,000–15,000	N, G, A	8
Fairfax Co., Va.	75,000	8	N/A	N, G, A, BI-M	10
Kent and Ottawa Co., Mich.	650,000	30	N/A	N, G, A, T, HDPE, OCC	4
Santa Monica, Calif.	70,000	66	Up to 2000	N, G, A, T	28
Snohomish Co., Wash.	N/A	15	1000–2000	N, G, A	N/A
Wayne Co., N.Y.	30,000	4	N/A	N, T, OCC	N/A

Key: N = newsprint, A = aluminum, BI-M = bimetal cans, G = glass, T = tin and bimetal cans, OCC = corrugated cardboard, MO = motor oil.

Source: "New Age Drop-off Programs," *The BioCycle Guide to Collecting, Processing and Marketing Recyclables,* The JG Press, Inc., Emmaus, Pa., 1990.

Combined Collection and "Blue Bag" Programs

While the vast majority of curbside collection programs put into operation over the past several years have been based on separate collection of mixed waste and recyclables, some communities have experimented with single collection of both mixed waste and recyclables. In some cases, a single collection vehicle has been developed, with two separate compartments, to pick up and compact trash as well as recyclables, sometimes also compacting the recyclables. Others have used a single compactor collection vehicle to pick up trash, and mixed recyclables which have been placed in specially colored bags ("blue bags") by the citizens. The bags containing the recyclables would be removed later at a central transfer station and sent to a materials recycling facility for processing in much the same way that commingled recyclables would be processed.

Although some communities such as Akron, Ohio, Kittery, Me., and most notably Toronto, Canada, have worked to develop a single-packer-type collection vehicle, this method has not really received any significant success or widespread implementation. Whether this is because the equipment and hardware was never really optimized, or whether it is because of inherent inefficiencies in matching the size of the recycling compartment with the size of the trash compartment, is not reported.

Success has been noted with the "blue bag" program. Pittsburgh was the first major urban area to embrace this type of program. Residents purchase bags at local retail outlets and set out material at curbside once a week. Container glass, plastic bottles, jugs and jars, and metal cans are commingled in one blue bag, while newspapers go in another. In Pittsburgh, the entire city is involved. The program services 120,000 single-family homes, 30,000 multiunit buildings, and 1522 major institutions. Participation is reported to be 82 percent. Citizens have shown a positive attitude to the program owing to its simplicity.[14] New York City has also included blue bags in its recycling program. Citizens can choose among city bins, blue bags, or other containers. The program served 1.7 million people in 1992 and is expected to expand to all 7.5 million residents. The advantage of this program is that the initial cost of bins is eliminated.

TABLE 9.4 Drop-off Programs—Amounts of Material Recycled

Location	All materials, tons	News		Glass		Aluminum		Tin		Others	
		Tons	%	Tons	%	Tons	%	Tons	%	Tons	%
Champaign Co., Ill.	1000	750	75	160	16	5	0.5	15	1.5	70*	7
Columbia Co., Pa.	469	271	58	88	19	6	1	19	4	85	18
Cook & Lake Co., Ill.	7140	5800	81	1200	17	75	1	65	1		
Delaware Co., Pa.	1800			1800	100						
Durham Co., N.C.	1200	900	75	300	25						
Fairfax Co., Va.	1000	721	72	271	27						
Kent and Ottawa Co., Mich.	3200	2225	70	669	20	1		158	5	157	5
Santa Monica, Calif.	1398	1032	74	360	25.5						
Snohomish Co., Wash.	233	67	29	159	68	7	3				

*Corrugated cardboard.

Source: "New Age Drop-off Programs," *The BioCycle Guide to Collecting, Processing and Marketing Recyclables,* The JG Press, Inc., Emmaus, Pa., 1990.

9.25

TABLE 9.5 Drop-off Programs—Site and Collection Characteristics

Location	Type of container	Storage capacity at site, yd^3	Collection equipment	Crew size	Collection frequency
Champaign Co.	Compartment container and lugger and barrel	15–40	Multilift and lugger, truck and van	1	Once a week to once a month
Columbia Co.	Shelters	7	Van	2	2–3 times a week to once a week
Cook and Lake Co.	Compartment container	N/A	Multilift	N/A	N/A
Delaware Co.	Dome	6.6	Tractor and trailer	2	1–2 times a week
Durham Co.	Shelters	Up to 21	Flatbed and forklift	2	Once a week
Fairfax Co.	Roll-off	120	Tractor and trailer	1	Once a week
Kent and Ottawa Co.	Roll-off, bins and barrels	N/A	Straight truck and van	2	1–3 times a week
Santa Monica	Bins	6 (at least)	Truck and trailer	2	Twice a week
Snohomish Co.	Dome	16	Truck and trailer	1	Every 10–14 days
Wayne Co.	Bins	12–24	Packer	1	Once in 2 weeks

Source: "New Age Drop-off Programs," *The BioCycle Guide to Collecting, Processing and Marketing Recyclables,* The JG Press, Inc., Emmaus, Pa., 1990.

Omaha, Neb., uses a blue bag program with cocollection of mixed recyclables and solid waste. Once packer trucks dump their loads on a tipping floor, bags are put onto a feed conveyor with a front-end loader and transported to an elevated sorting line. Hand pickers pull the blue bags containing recyclables as well as corrugated cardboard from the conveyor and drop them down chutes onto conveyors which feed another elevated sorting line. The blue bags are opened by hand and then move past 20 sorting stations. Workers manually sort out corrugated cardboard, glass, plastic, and aluminum. A cross-conveyor magnet is used to remove ferrous metal. During 1992, the recovery rate was 4.4 percent of the solid waste delivered to the site. The nonrecyclables, which represent the other 95 percent of the waste, are landfilled. Participation by citizens is reported to be 52 percent.[14]

The blue bag approach is aimed at simplifying the processing of mixed waste systems to those customarily employed by commingled collection and processing programs. Participation is as easy for the resident as curbside collection of commingled recyclables except that special trash bags must be purchased. Collection costs should be the same as for waste collection alone, but the method does require substantial labor for hand sorting or an intermediate capital investment for processing of commingled recyclables.

9.3 PROCESSING EQUIPMENT FOR RECYCLING FACILITIES

This section describes the technology and systems utilized to process recyclables once they are collected. When materials are collected and fully separated at curbside (after initial separation by the resident), they still may have too low a density to be sold directly to an end user. Bottles may need to be crushed, metals flattened or baled, plastics granulated or baled, and waste paper baled.

For commingled recyclables, not fully separated before collection, more complex separation and processing systems are employed to first separate the commingled recyclables into their component materials, remove impurities from them, and then densify them or otherwise prepare them for shipment to the end user. Commingled processing facilities, called materials recycling facilities or MRFs, are typically of larger volume or tonnage than the systems required to process source separated materials. MRF equipment is also normally of larger capacity and more rugged. Very complex, heavy-duty, capital-intensive processing systems are required for mixed waste processing facilities which must recover materials from the entire solid waste stream.

Unit Operations and Equipment

The basic process system unit operations are similar regardless of whether the materials to be processed or separated are obtained from source separation, commingled collection, or mixed waste collection programs. The unit operations employed in processing recyclable materials include baling, magnetic separation, screening, size reduction, air classification, eddy current separation, and can flattening and densification. The intent of this section is to familiarize the reader with the various unit operations, give an overview of the type of equipment available, and provide some guidelines regarding the selection and operation of these unit operations and associated equipment as they particularly pertain to the business of recycling. Selected equipment manufacturers and suppliers are listed in Table 9.6.

Baling.   Most recycling facilities employ at least one baler. In addition to the traditional function of baling paper and corrugated cardboard, the baler can also serve to densify ferrous metals, aluminum, and plastics. Furthermore, it can be an efficient means of reducing the volume and thus the disposal cost of residues or rejects from various recycling operations. Balers can be categorized into two main types: (1) vertical balers and (2) horizontal balers.

All balers have the following features:

1. Feed hopper or area into which the recyclables are fed
2. One or more hydraulic or mechanically driven rams which compress the material fed
3. Compression area where the materials are densified
4. Discharge area opening from which the completed bales are ejected

Wire ties are normally utilized in either a manual or automatic configuration to wrap wire around the bale and tie it off so that on ejection from the compression chamber, the bale does not expand or break apart.

Vertical Baler.   Vertical balers are an integral part of many commercial and industrial scrap operations and are used for baling waste paper, corrugated cardboard, ferrous and nonferrous metals, steel scrap of light gauge, foam scraps, and plastic containers and bottles. (It should be noted that when baling plastic bottles, caps should be removed to release entrapped air.)

A diagram of a vertical baler is provided in Fig. 9.16. The operator feeds material into the feed hopper and closes the hopper door, locking the material in the compression chamber. The hydraulic ram compresses the material with a downward stroke. The procedure is repeated until a sufficiently large and dense bale has been formed. The entire bale is then bound with wire strapping; plastic strapping may also be used. Tying is accomplished manually while the ram is in the down position. The process of loading, compressing, wrapping and tying the wire, and ejecting the bale takes about 20 to 30 min per bale. Thus two or three bales per hour can be produced. Bale sizes can range in length from 18 to 72 in. The standard mill size, 30 by 48 by 60 in, is the most prevalent.

The cost of a vertical baler ranges from $4000 to $30,000. The price is a function of unit size, motor horsepower, and hydraulic pressure. Motor sizes range from 1 to 30 hp but are typically 10 hp. The unit is self-contained and easily transportable; but vertical balers do require about 13 to 15 ft of vertical clearance for installation and operation. These units are inexpensive, require low power consumption, and are versatile. However, they are slow, must be hand-fed, and cannot be automated for wire tying. They have the lowest overall production output of all balers and are thus appropriate for only small-sized operations.

Similar to the vertical baler is the upstroke baler, except that the compression chamber is below grade, or underground. Installation requires foundations and structure about 12 to 18 ft deep. Motor sizes range from 10 to 25 hp; the 25-hp unit is considered a high-density baler. These units utilize mechanical, chain-driven densification rams rather than the hydraulic rams of the vertical baler. Although upstroke balers continue to be produced in limited quantities, they are, for the most part, available on the used market. They are no longer in widespread use, primarily because of the high construction cost inherent in their permanent installation.

TABLE 9.6 Equipment Manufacturers and Suppliers

Company	Address	Telephone
Balers:		
American Baler Company	200 Hickory Street Bellevue, OH 44811	(419) 483-5790
Enterprise Baler Company	P.O. Box 15546 Santa Ana, CA 92705	(714) 835-0541
Harris Waste Management Group Inc.	200 Clover Reach Peachtree City, GA 30269	(404) 631-7290
International Baler Corporation	5400 Rio Grande Avenue Jacksonville, FL 32254	(904) 358-3812
Logemann Brothers Company	3150 W. Burleigh Street Milwaukee, WI 53210	(414) 445-3005
Marathon Equipment Co.	P.O. Box 1798 Vernon, AL 35592	(205) 695-9105
Maren Engineering Corporation	P.O. Box 278 South Holland, IL 60473	(708) 333-6250
Magnetic separators:		
Dings Co. Magnetic Group	4740 W. Electric Avenue Milwaukee, WI 53219	(414) 672-7830
Eriez Magnetics	P.O. Box 10608 Erie, PA 16514	(800) 345-4946
Newell Industries Inc.	530 Steves San Antonio, TX 78204	(210) 227-9090
Stearns Magnetics, Inc.	6001 South General Avenue Cudahy, WI 5311	(414) 769-8000
Screens:		
Carrier Vibrating Equipment, Inc.	P.O. Box 37070 Louisville, KY 40233-7070	(502) 969-3171
Central Manufacturing, Inc.	P.O. Box 1900 Peoria, IL 61656	(309) 387-6591
Rader Companies	P.O. Box 181048 Memphis, TN 38181	(901) 365-8855
Simplicity Engineering	212 South Oak Street Durand, MI 48429	(517) 288-3121
SWECO, Inc.	P.O. Box 1509 Florence, KY 41022	(606) 727-5180
Triple/S Dynamics, Inc.	P.O. Box 151027 Dallas, TX 75315-1027	(800) 527-2116
Shredders, pulverizers:		
American Pulverizer Co.	5540 West Park Avenue St. Louis, MO 63110	(314) 781-6100
Heil Engineered Systems	205 Bishops Way Brookfield, WI 53005	(414) 789-5533
MAC Corporation/Saturn Shredders	201 East Shady Grove Road Grand Prairie, TX 85050	(214) 790-7800
Newell Industries Inc.	530 Steves San Antonio, TX 78204	(210) 227-9090

TABLE 9.6   Equipment Manufacturers and Suppliers (*Continued*)

Company	Address	Telephone
Pallman Pulverizers Co., Inc.	820 Bloomfield Avenue Clifton, NJ 07012	(201) 471-1450
Williams Patent Crusher Co.	2701 North Broadway St. Louis, MO 63102	(314) 621-3348
Granulators:		
Cumberland Engineering Division	P.O. Box 6065 Providence, RI 02940	(401) 728-1600
Herbold Granulators USA, Inc.	12C John Rd. Sutton, MA 01590	(508) 865-7355
M.A. Industries	303 Dividend Drive Peachtree City, GA 30269	(404) 487-7761
Nelmor Co., Inc.	Rivulet Street N. Uxbridge, MA 01538	(508) 278-5584
Rapid Granulator, Inc/Condux	P.O. Box 5887 Rockford, IL 61125	(815) 399-4605
Air classifiers:		
Forsbergs, Inc.	P.O. Box 510 Thief River Falls, MN 56701	(218) 681-1927
Kice Industries, Inc.	P.O. Box 11388 Wichita, KS 67202-0388	(316) 267-4281
Kongskilde Limited	321 Thames Rd. East Exeter, ON N0M 1S3	(519) 235-0840
Rader Companies	P.O. Box 181048 Memphis, TN 38181	(901) 365-8855
Sterling Systems/Sterling Blower Co.	P.O. Box 219 Richmond, VA	(804) 525-4030
Eddy-current separators:		
Cedarapids Inc.	916 16th Street NE Cedar Rapids, IA 52402	(319) 363-3511
Eriez Magnetics	P.O. Box 10608 Erie, PA 16514	(800) 345-4946
Lindemann Recycling Equipment Inc.	42 West 38th Street New York, NY 10018	(212) 382-0630
Steinert Elektromagnetbau GMBH	136 William Street New York, NY 10038	(212) 962-4255
Bottle-sort systems:		
Automation Industrial Control	7128 Ambassador Road Baltimore, MD	(410) 944-8400
Magnetic Separation Systems, Inc.	624 Grassmere Park Drive Nashville, TN 37211	(615) 781-2669
National Recovery Technologies, Inc.		
	566 Mainstream Drive Nashville, TN 37228-1223	(615) 734-6400

Sources: Resource Recycling Equipment Directory, August 1993, *Solid Waste & Power Sourcebook,* vol. VI, no. 7, 1992.

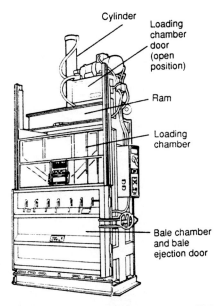

Cylinder

Loading chamber door (open position)

Ram

Loading chamber

Bale chamber and bale ejection door

FIGURE 9.16   Vertical baler. (*Source: Fig. 28.1, McGraw-Hill Recycling Handbook, 1993.*)

Horizontal Baler.   Horizontal balers are fed from the top through a feed chute. The hydraulic ram is arranged in a horizontal configuration, resulting in the ability to operate with lower roof heights than the vertical baler. Motor sizes range from 5 to 150 hp. The cost of a horizontal unit can range from $7000 to $500,000. The most expensive units are high-volume units which incorporate continuous feed and automatic tying mechanisms. "Fluffers" are frequently used on horizontal balers to loosen incoming newsprint prior to baling, thus improving the stability and integrity of the bales.

CLOSED-DOOR, MANUAL-TIE HORIZONTAL BALER.   Figure 9.17 depicts the most basic type of horizontal baler. It operates similarly to the vertical baler; however, it can be hand-fed or conveyor-fed into a charging hopper rather than into the baling chamber itself. It can also be fed continuously; the ram can be in a compression cycle while material is being fed into the charging hopper on top of the ram itself during the compression stroke. Bale size is from 42 to 72 in in length. Cross sections are normally square and range in size from 24 by 24 in to 48 by 48 in. Cost is in the $12,000 to $60,000 range. Motor size is typically 5 to 50 hp. The advantages of this baler are that it is relatively low in cost, can be fed by conveyor, and requires low headroom, little electrical power, and relatively low maintenance. The disadvantage is that while it is faster than the vertical baler, it is still relatively slow because of the manual-tie feature.

OPEN-END, AUTOMATIC-TIE HORIZONTAL BALER.   The open-end automatic-tie horizontal baler (Fig. 9.18) is similar to the horizontal unit above but is constructed on a much larger scale and operates at much higher throughput rates. Material is continuously extruded from the baler rather than manual ejection of one bale at a time from a closed door. The open-end baling chamber incorporates tension cylinders which apply varying degrees of pressure against the baling chamber walls. These tension cylinders

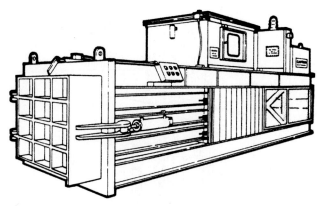

FIGURE 9.17   Closed-door horizontal baler. (*Source: Fig. 28.3, McGraw-Hill Recycling Handbook, 1993.*)

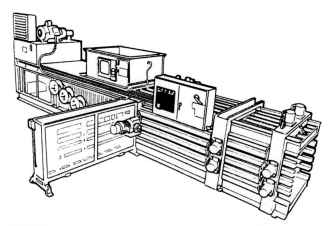

FIGURE 9.18   Open-end horizontal baler. (*Source: Fig. 28.4, McGraw-Hill Recycling Handbook, 1993.*)

are extended during the bale compression until the correct bale density is determined by the unit's pressure setting. At this point, the tension cylinders ease their pressure, allowing the compressed formation to move forward. An automatic tying mechanism wraps and ties wire around the emerging bale. Motors range in horsepower from 20 to 150 hp. Maintenance is higher owing to the degree of sophistication in these units. Feed openings from the hopper can be as large as 48 by 72 in. Thus the unit can handle large cardboard boxes. A conditioning unit can also be installed ahead of the feed hopper to puncture bottles so air can be easily released on compression, thus eliminating the need to remove caps prior to baling.

These balers are particularly suited for large-scale MRF operations. They are capable of producing bales of variable lengths such that "short bales" of various materials can be produced and the operation changed over from one material to another. Thus a good-quality bale can be produced even though there may not be enough material available to produce a "whole" bale.

Two-Ram Horizontal, Automatic-Tie Baler.   Two-ram horizontal balers (Fig. 9.19) are high-capacity, high-density balers. One ram is used to compress the material which is continuously fed into the feed hopper, and the second is used to eject the bale after it has been tied off. The bale ejection ram is situated at right angles to the compression stroke ram. These balers range in cost from $120,000 to $500,000 and are expensive to run and maintain. They are excellent units for high-volume paper and corrugated operations. Motor sizes range from 50 to 300 hp. Feed openings from the hopper are quite large, sometimes as much as 5 ft wide and 10 ft long. The major disadvantage to using the two-ram baler for high-volume operations is that it produces whole bales only; the second ram cannot eject a half-size bale.

Baler Selection Considerations.   Vertical balers are often the choice of very small recycling operations because of their low purchase price. They are much slower and operations costs are higher than for horizontal balers. Larger recycling operations and MRFs normally employ horizontal balers. The size and baling densities vary depending upon the size of the unit selected. The price increases as the capability for higher bale density is increased.

High-density, automatic-tie horizontal balers are preferred when it is necessary to generate export bales. The added density is important, especially in baling plastics (which are very difficult to bale) because they require high density to hold the bale together. High density is also desirable for improving the value of corrugated and paper bales. The labor savings and improved shipping densities for materials offset the capital costs incurred.

An MRF which has a processing line that processes only paper and corrugated cardboard, or a high-volume, paper-only facility, may best utilize a two-ram baler. However, this baler can only make whole bales. When changing over from one material to another, a mixed bale is produced. Typically, an open-end, horizontal baler is

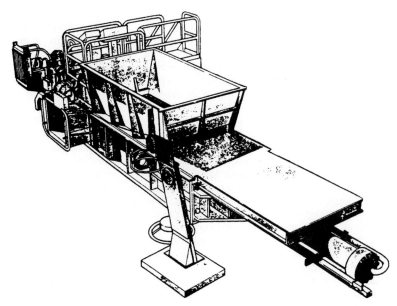

FIGURE 9.19   Two-ram horizontal baler. (*Source: Fig. 28.5, McGraw-Hill Recycling Handbook, 1993.*)

preferable to the two-ram baler when there are frequent changeovers from one type of material or one grade of paper to another. These in-line balers can make short bales and full bales of high density. They are often more appropriate for recycling operations, while the two-ram units are best for high-volume, paper-only processing facilities or lines.

Magnetic Separation. The most common method for removing ferrous metals from commingled recyclables involves the use of magnetic separation systems. Magnets can be classified as either (1) electromagnets which use electricity to magnetize or polarize an iron core or (2) permanent magnets which utilize permanently magnetized materials to create a magnetic field.

Various types of magnet configurations have been used in recycling applications. The most common, shown in Fig. 9.20, are the suspended belt magnet, the magnetic head pulley, and the suspended magnetic drum. In addition, specialized solid waste magnets have been designed which involve multiple stages of magnetic separation to shake contamination loose from tin cans while they are being separated (see Fig. 9.21).

Suspended Belt Magnets. Suspended belt magnets can be of the cross-belt or in-line configuration. Cross-belt magnets are most often used in MRF operations, especially where the magnetic materials separated from the commingled recyclables are to be lifted off the belt, much as a handpicker would do, and conveyed at a 90° angle from the primary feed conveyor into a bin or downstream item of processing equipment such as a can flattener. The in-line suspended belt magnet is discussed below together with in-line suspended magnetic drums.

The "strength" of a magnet at any location is a product of the magnetic flux density and the magnetic flux gradient. Because the magnetic core is normally smaller than the width of the feed conveyor belt, and because the flux density and flux gradient will vary at various locations across the feed conveyor, the magnetic strength will also vary. The ability to lift cans off the belt will vary accordingly. This is the reason why a lightweight, inexpensive magnet will fail to pick up ferrous metals and cans at the edges of the belt. Another reason can be that the side skirting is made of ferrous metal which may become magnetized, presenting a counteractive magnetic force. Thus, in utilizing such magnets, caution should be applied to evaluate the magnetic flux and flux gradient at various locations across the feed belt to be sure that the magnetic field strength does not fall off at the edges of the belt. Further, materials of construction in the vicinity of the magnetic field must be nonmagnetic if the magnet is to work properly. It is important to recognize that magnets not only pick up the ferrous metals, but they also magnetize the ferrous metal support structure, conveyor support structure, and chutes and hoppers which are located within the field unless these items are made of nonmagnetic materials such as nonmagnetic stainless steel, wood, plastic, etc. If magnetic materials are used and they become magnetized, they will inhibit the separation.

Cross-belt magnets lift the material to be separated. The magnet must be positioned above the feed conveyor *at least* twice the distance away from the belt as the dimension of the largest materials carried on the belt. Otherwise, materials on the belt and materials on the magnet will see mechanical or physical interference which will hinder the efficiency of separation. Moreover, magnetic metals are often covered with a burden of nonmagnetic material. This overburden can also be lifted up by the strength of force exerted between the magnetic material and the magnet itself. The entrained burden of nonmagnetics can thus be lifted and held to the magnetic belt, resulting in contamination of the magnetic product. Because of the lifting action carrying material through the air, the potential for entrapment of paper and plastic is less than for the head pulley magnet described below.

Head Pulley Magnets. The head pulley magnet is installed as an integral part of a belt conveyor. However, owing to the need for the magnet to be installed into the head

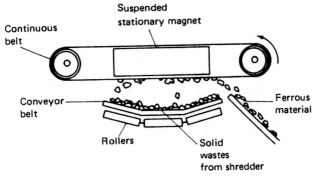

(a) **Suspended magnet**

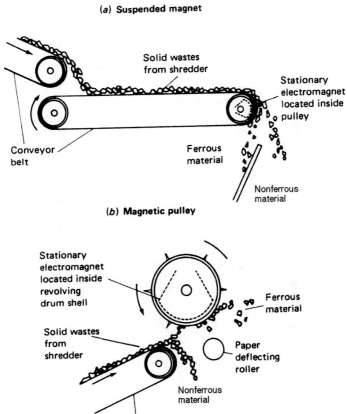

(b) **Magnetic pulley**

(c) **Suspended magnetic drum**

FIGURE 9.20 Typical magnet separators. (*Eriez Magnetics; Source: Fig. 8-18, Tchobanoglous, Solid Wastes: Engineering Principles and Management Issues, McGraw-Hill, 1977.*)

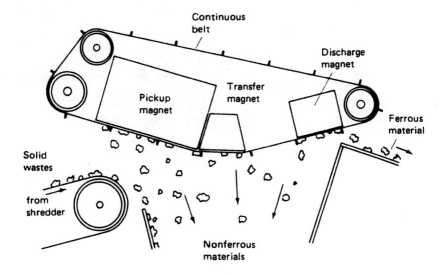

(a) Belt type magnetic separator

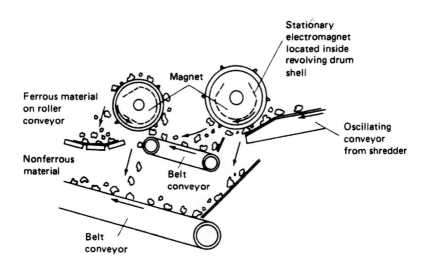

(b) Two-drum magnetic separator

FIGURE 9.21 Typical magnet separation systems used with shredded solid wastes. (*a*) (*Dings Company.*) (*b*) (*Eriez Magnetics.*) (*Source: Fig. 8-19, Tchobanoglous, Solid Wastes: Engineering Principles and Management Issues, McGraw-Hill, 1977.*)

pulley, the diameter of the head pulley is often much larger than the diameter of the normal conveyor head pulley used in recycling operations. Magnetic head pulleys are typically between 24 and 36 in in diameter but can be as small as 18 in and as large as 60 in. As material falls off the end of the conveyor, the head pulley magnetic forces hold magnetics to the belt, attracting the magnetic materials and changing their trajectory as they fall off the end of the belt (see Fig. 9.20). The advantage of head pulley magnets is that they are inexpensive and take up little space. The disadvantage is that they do not "lift" or tumble, thus liberating the magnetic product from contaminants. Underburden material, such as paper or plastic, which is entrapped under the magnetic metal will be held onto the belt by the magnetic material above it and carried over into the magnetic product, causing contamination. This is more of a problem in solid waste applications than in MRF applications.

In-Line Suspended Magnetic Drums. When the magnetic product is to be conveyed parallel to the direction of the feed conveyor or lifted off the end of a conveyor (while the nonmagetic material drops to a conveyor or bin below), an "in-line" magnetic drum separator is often used. Alternatively, an in-line suspended magnetic belt can be used. This type of magnet can produce a higher-quality finished product than the cross-belt or head pulley magnet because the separation occurs while the feed stream is suspended in air and acted upon simultaneously by momentum, gravitational forces, and magnetic forces. Since separation takes place while fully suspended in the air, the potential for entrapment of nonmagnetics is reduced. The magnet must be fed at a belt speed of about 400 ft/min. Thus this type of magnet should not be fed directly by the main conveyor belt because it moves at too slow a speed. The feed stream is actually "thrown" at the magnet. The "gap" between the magnetic material and the magnet itself is low compared with the cross-belt suspended magnet described above where the feed material just lies on the belt below the magnet. This effectively increases the strength of the magnet because the field gradient and field strength are higher near the surface of the magnet than on a belt some distance away (say 24 in) from the magnet surface. Efficiency of separation is increased accordingly. A schematic diagram of this type of magnet installation is shown in Fig. 9.20a.

Solid Waste Magnets. In applications where cans and paper are combined on the feed conveyor, and where there is a burden of contamination carried along with the magnetic product, it is desirable to "shake loose" this material from the magnetics. A special solid waste magnet (Fig. 9.21) has been designed to meet this need. The magnet is actually a combination of either two or three magnets. As the magnetic product is lifted to the first magnet, it is carried along the belt. As the magnetic force falls off near the end of the core, the magnetics are dropped from the first magnet. However, their forward momentum carries them into an area of space where they are then affected by the magnetic field of the second magnet and again attracted to the magnetic field. The same principle occurs a third time where there is a third magnet. Moreover, since the core of the magnet is one polarity and the box which holds the core serves as the return path for the magnetic flux field and thus is of opposite polarity, the material tends to flip-flop while it is being transported across the three magnets. As the material moves along this series of magnets, the metal is both dropped and flip-flopped, which tends to shake off any entrapped material, making the finished product cleaner.

Magnet Selection Considerations. The following factors should be taken into consideration in the selection of either permanent or electromagnets and in choosing the type and configuration of the magnetic installation.

1. The physical relationship between the feed conveyor and the discharge conveyor and whether the magnetic product is to be conveyed in-line or at 90° to the infeed material

2. The width of the feed conveyor and the size, weight, and cost of magnet required to effectively cover the entire conveyor width with an adequate magnetic field strength

3. The largest size of materials on the belt and/or the tendency of the feed conveyor to encounter piling and surges of flow which could affect the physical mounting arrangement and the distance of the magnet from the feed conveyor

4. The amount of contamination and the shape of contaminants which are intermixed with the magnetic product

5. Operating requirements such as electrical consumption, space requirements, structural support requirements, conveyor speeds, conveyor widths, type of magnetic cooling systems required, magnetic strength, materials of construction, maintenance, and physical access

A head pulley magnet is typically used where low-cost separation is required to remove small amounts of "tramp" magnetic particles from materials being processed. Where large quantities of highly magnetic materials are involved, permanent magnetic separators are usually employed. These can be either drum magnets or overhead suspended belts, depending on the space requirements and personal preference. For more weakly magnetic materials, electromagnetic separators can be used. These also can be selected as drum magnets or overhead suspended belts, again depending on the space requirements and personal preference. In-line magnetic separation tends to provide a higher recovery efficiency for separation of magnetic materials; however, space constraints frequently dictate the need for cross-belt magnets. In considering whether to use a belt magnet versus a drum magnet, care should be taken to identify the potential for belt damage that can result from nails, wire, and other sharp objects. Drum magnets employ a metal surface which is more resistant to damage from projectiles.

Magnetic separators typically can accomplish a recovery efficiency of 95 to 99 percent for magnetic materials, depending on the application and burden depth of materials being processed. The contamination level of the recovered magnetic fraction varies depending on the particle size and characteristics of the feed stream. Purities of 95 to 98 percent for the recovered magnetic fraction are considered typical. In mixed waste processing systems, recovery efficiencies of only 80 percent are typical, and the grade or purity of the magnetic product can be as low as 60 to 80 percent ferrous metal. Mixed waste magnetic material often requires reprocessing before it can be sold to a steel maker.

Screening.   The efficiency of a screen can be evaluated in terms of the percentage of recovery of the material desired to be separated from the feed stream by the following expression:

$$\text{Recovery } \% = \frac{U w_u}{F w_f \times 100}$$

where U = weight of material passing through screen (underflow)
   F = weight of material fed to screen
   w_u = weight fraction of material of desired size in underflow
   w_f = weight fraction of material of desired size in feed

Screens operate at the highest efficiency when the materials to be separated are either much larger or much smaller than the screen opening. That is to say, when the materials to be separated are of divergent sizes, the screening efficiency is very high. Specifically, when a spherical particle of diameter d impinges on a square hole (or round hole) of size a where $a > d$, the probability that it will pass through the hole is expressed as follows:

$$p = (1 - d/a)2 \times Q$$

The quantity Q is the ratio of the hole area to the total screen surface area. As the ratio of d/a approaches 1, the probability of material falling through the screen approaches zero; i.e., as the material to be screened approaches the size of the screen opening, the screening efficiency becomes very low and approaches zero. Meanwhile, very large materials flow on top of the screen as if it were a solid surface, and very fine materials fall through the openings of the screen with a very high efficiency or probability of passage.

The most common types of screens in recycling applications are the vibrating screen, the rotary drum screen or trommel, and the disc screen.

Vibratory Deck Screens. Vibrating screens (Fig. 9.22) typically have flat decks and are mounted on an incline to assist in material movement. The screens may be designed with one deck to make a bimodal product or may have multiple decks but normally not more than two in which three different-sized products can be produced.

The screen "cloth," which is often a wire mesh but can also be a solid metal plate with holes punched into it (a "punch plate"), is powered by an electric motor and drive

FIGURE 9.22 Typical vibrating screen. (*Source: Fig. 8-20(a), Tchobanoglous, Solid Wastes: Engineering Principles and Management Issues, McGraw-Hill, 1977.*)

mechanism which vibrates the material and "throws" it up and down on the screen so that the material impinges many times on the deck, providing numerous opportunities to pass through an opening. The gyratory motion can often be adjusted to change the throw and alter the extent of upward travel versus horizontal travel down the length of the screen. The screen is often supported on springs and a motor turns an eccentric weight which imparts motion to the material which sits on the screen deck. The throw or length of stroke, the inclination of the screen, its overall length, and its vibration frequency are selected to handle the throughput and screening efficiency.

Although vibrating screens are very efficient and cost-effective for screening fine particles such as glass, they are not very efficient for separation of large materials such as paper which tend to blind over the screen openings. The disc screen and trommel offer much better performance for these types of materials.

Rotary Screen or Trommel.   A rotary screen, also called a trommel, is shown in Fig. 9.23. The screen is normally set at a downward slope on the order of 5° so that material will flow down the screen as it is dropped and tumbled. Lifters are sometimes placed within the screen to increase the degree of lifting and dropping of material, enhancing tumbling action and thus liberation. Trommel screens can use various types of screen "cloth" and materials of construction. Both wire-type screens and punch plate screens are in use. Rubber cloth with punched holes has also been used. Wire-type screens can be troublesome in screening material which contains cloth, wire ,and stringy materials because they hang across the wire screen and plug the openings. For these stringy materials, punch plate screens are the better choice. The punch plates can be removed in sections and replaced as needed to change the size of the opening.

It is important in a rotary screen operation to avoid operating the screen at too high a rotation velocity which can centrifuge the material within the screen, drastically reducing screen efficiency. Screening should be performed at about 50 percent below this "critical velocity." Maintenance requirements for trommels normally include lubrication of bearings and removal of wire, ribbon, and cloth which wrap around the trommel periodically. In mixed waste processing, this may be required on a daily basis. Generally, however, trommels require less cleaning than any other type of screen. Screen plates normally only need to be replaced every 5 years or so depending on materials processed and severity of application.

Disc Screens.   Disc screens look very much like horizontal screens except that in place of screen cloth, there are several horizontal bars or shafts which run across

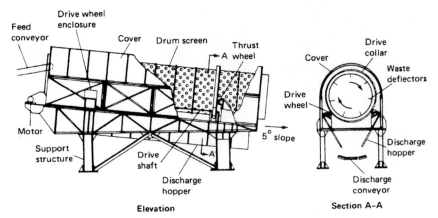

FIGURE 9.23   Rotary drum screen or trommel. (*Source: Fig. 8-20(b), Tchobanoglous, Solid Wastes: Engineering Principles and Management Issues, McGraw-Hill, 1977.*)

the screen's width arranged perpendicular to the material flow. On each shaft are several serrated or star-shaped discs spaced evenly across the width of the screen (see Fig. 9.24). As the shaft turns, it carries material across the discs and bounces it into the air. The action is not as aggressive as the trommel where the material is actually lifted and dropped, but it is much more aggressive than the horizontal vibratory screen. Long stringy objects tend to flow across the bars, while smaller objects such as glass, grit, bottles, and cans tend to fall between the discs, normally onto a take-away conveyor below, depending on the size of the spacings. In spite of the fact that these screens were designed to deal with stringy items such as wood bark, which would normally blind a vibrating screen, these items can sometimes wrap around the discs. This wrapping can eventually blind the screen and eventually entirely block the openings, especially when the opening size is set to a coarse opening of 3 to 4 in and above. This is not a very large problem when the spacing is set to remove items of a 2- to 3-in size range.

Maintenance for disc screens includes periodic unwrapping of cloth and stringy materials from around the discs. This may be required on a single-shift basis in the case of mixed waste processing. Another maintenance requirement is the periodic replacement of the discs that wear down. It is important to ensure that such replacement has been designed to be performed easily, as many of these screens use a welded construction which makes disc replacement very difficult and costly in both time and materials. Disc screens are employed in applications similar to those of trommels. The principal reasons for their acceptance are that they occupy less space and can be purchased for less total cost than a similar sized trommel. The tradeoff is that operations and maintenance costs are higher and efficiency is lower.

Screen Selection Considerations.   Factors to be considered in selection of screening equipment include:

1. Particle size, particle size distribution, bulk density, moisture content, particle shape, and potential for the material to stick together or entangle

FIGURE 9.24   Disc screen.

2. Screen design characteristics including materials of construction, size of screen openings, shape of screen openings, total surface screening area, rotational speed for rotary drum screens and oscillation rate for vibrating screens, length and width for vibrating screens, or length and diameter for rotary screens

3. Separation efficiency and overall effectiveness

4. Operational characteristics such as energy requirements, routine maintenance, simplicity of operation, reliability, noise and vibration, potential for plugging

Thus the selection of screens for a given application requires considerable attention to the characteristics of the materials being processed. Vibrating screens offer inexpensive sizing for free-flowing granular materials such as glass that do not tend to blind the screen or become entrapped with other materials. Trommel screens are better suited for large-particle applications where blinding is anticipated or where materials become entrapped and require tumbling to free them for removal by the screen. The performance of disc screens lies between that of vibratory screens and trommels, as does the cost and size of the equipment. Typically, disc screens are utilized in applications where rigid materials such as wood chips are being screened to remove the grit and dirt. Disc screens are prone to wrapping (and thus higher maintenance) when long flexible items such as wire, rope, and textiles are present in the feed stream.

Size Reduction. Several types of size-reduction equipment are utilized to rip, cut, tear, and pulverize commingled recyclables, liberating materials that are bound together so they can be separated from each other in downstream unit operations. These equipments are also utilized to densify materials prior to shipping to reduce storage, handling, and transportation costs.

Types of size-reduction equipment include:

1. Horizontal shaft hammermill

2. Vertical shaft hammermill

3. Vertical shaft ring grinder

4. Flail mill

5. Glass crusher and pulverizer

6. Granulator and knife shredder

7. Rotary shear shredder

Horizontal Shaft Hammermill.   The downrunning horizontal shaft hammermill (Fig. 9.25) is the most common type of shredder utilized in recycling operations. It is also the unit most typically utilized in automobile shredding operations, composting operations, and mixed waste processing plants. Material is fed through a feed hopper, falling into a "hammer circle." The hammers, which are attached to a rotor or shaft, impact the infeed material, crushing it, pulverizing it, and tearing it into smaller pieces. Below the hammer circle is a series of cast grates which are similar to a screen with very wide openings which may range in size from 3 by 6 to 14 by 20 in. The material remains inside the hammermill and is crushed and torn between the hammers and the grates, until its size is sufficiently reduced to pass through the grates, where it is discharged onto a belt or vibrating pan conveyor below.

Downrunning shredders require hammer changes frequently, or alternatively the hammers must be welded up by adding welding material onto the hammer itself to replace the metal which is worn off during operations. For example, a shredder having 4 rows of hammers of 9 hammers each would require welding on 36 hammers. The

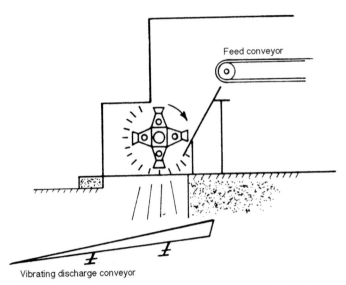

FIGURE 9.25   Downrunning horizontal shaft hammermill. (*Source: Fig. 12.37, Robinson, The Solid Waste Handbook, John Wiley & Sons, 1986.*)

frequency of hammer replacement depends on the maintenance procedures employed, the tonnage of material processed per week, and the type of material processed.

A center feed reversible horizontal shaft hammermill (Fig. 9.26) is similar to a downrunning hammermill, except the feed material is fed directly down on top of the hammer circle. The feed hopper tends to be higher in height to reduce the potential for material being "baseballed" out of the feed opening of the mill. One of the advantages of a center feed mill is that the direction of shaft rotation can be reversed. The hammermill can be kept running without having to weld up the hammer faces or reverse the hammers themselves to keep the hard or sharp face of the hammers working on the feed material. Thus the operating time between scheduled maintenance can be twice as long. A disadvantage of these mills is that they tend to have a lower throughput for similar size machines than the downrunning mills, sometimes as much as 25 percent lower, especially when operated in the reverse rotation direction. They also tend to exhibit a greater tendency for "blowback," dusting, and ejection of feed material out of the feed hopper. They are also more expensive.

A center feed mill characteristically has an impact breaker, which is sometimes adjustable either mechanically or hydraulically to allow large items to pass. It also serves as an impact plate on which materials are pulverized. This type of shredder is similar in design to a pulverizer, except that it has grates installed in the bottom to control the particle size of tough-to-shred items.

Vertical Shaft Hammermill.   A vertical shaft hammermill differs from the horizontal mill in that there is no grate. Instead, the discharge material passes through an annulus which controls the exit particle size. This shredder tends to offer less control over maximum particle size than the horizontal shaft hammermills discussed above. The infeed material is fed into the top of a chute which feeds into a breaker plate and hammer area. As the material is beat and hammered, it works its way down the cone-shaped machine (Fig. 9.27). The distance between the hammers and the breaker plate constantly decreases, thus continuously working to reduce particle size.

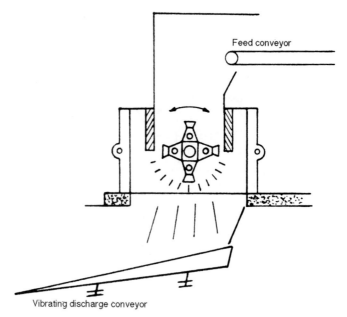

FIGURE 9.26   Reversible, center-feed horizontal shaft hammermill. (*Source: Fig. 12.38, Robinson, The Solid Waste Handbook, John Wiley & Sons, 1986.*)

Maintenance tends to be lower for the vertical mill than for the horizontal mills. However, both types are very reliable and can handle a large range of massive materials. In a vertical shaft mill, all the hammers are not changed at once, but rather new hammers are replaced on the shorter arms located at the bottom of the mill and the worn hammers are then moved up to a higher location where sharpness is not so critical. Finally, when they are moved to the top position, they must be either replaced or rewelded. Most installations utilizing vertical shaft hammermills replace the hammers rather than retip or weld the hammers when they become dull.

Vertical Shaft Ring Grinder.   A vertical shaft ring grinder externally appears similar to a vertical shaft hammermill. However, internally, rather than having hammers pinned to the end of each rotor disc on the shaft, a gear-type device is positioned in place of a hammer (Fig. 9.28). These mills provide more of a mashing and grinding action rather than the tearing, pulverizing, and ripping action of the horizontal and vertical shaft hammermills. This grinding action is particularly good in densifying materials such as metal cans and tends to produce a nuggetized metal product of quite high density. Although vertical ring grinders saw significant use in early solid waste operations, to date they have not been applied very much in recycling operations.

Flail Mill.   A flail mill is somewhat like a hammermill, but without grates. There are several types of flail mills including single and double shaft, and horizontal fed. Material is fed into the top of the single and double shaft mills through a feed chute. Horizontal fed mills are fed from a conveyor into the front of the hammer circle (Fig. 9.29). The flails which are attached to a rotating shaft function as knives. These knives can cut open bags, liberating the contents. Paper is torn and ripped. Cans pass through the mill relatively unaffected while glass is pulverized into very fine sizes. Because the flail mill does not have grates, it is not a good device for controlling particle size—especially the particle size of rags and other similar material which is very hard to shred.

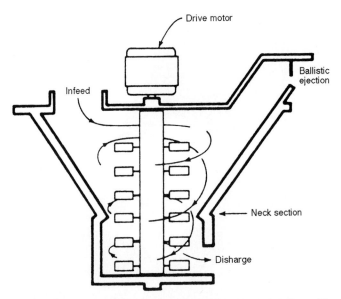

FIGURE 9.27   Vertical shaft hammermill. (*Source: Fig. 12.40, Robinson, The Solid Waste Handbook, John Wiley & Sons, 1986.*)

Pulverizer and Glass Crusher.   A pulverizer is much like a flail mill but utilizes a breaker plate and hammers rather than knives. These machines have impact bars and impact plates which assist in the pulverization of glass and other friable materials. As these materials fall into the mill, glass is struck by the hammers and thrown against the impact blocks where it is again smashed into smaller pieces. Pulverizers tend to be much smaller in size than hammermills, and normally they do not have grates. There are many different types of glass crushing and pulverizing equipment. A typical glass crusher is shown in Fig. 9.30. Noise control and dust control are important environmental considerations which must be addressed during installation of a glass processing system. The design of feed conveyors and the interface to the feed hopper and dribble chutes require experienced design knowledge if they are to be done properly.

Granulator and Knife Shredder.   A granulator or knife shredder employs very sharp, long knives for cutting materials such as rags and plastic bottles into small pieces for later separation. The knives are attached to a rotor and are positioned horizontally across the entire width of the shredder. As the shredder rotor rotates, the knives pass by an impact or cutting block at high speeds. Material caught between the impact block and the knife is cut. The cutting action is such that the material which is being size reduced is cut into particles on the order of $\frac{1}{4}$ to $\frac{3}{4}$ in in size. It is important to note that hard materials such as glass and metal should not be fed into this type of equipment, as they would damage the knives. Granulators are typically used in plastics processing operations to reduce the particle size of bottles and increase density (Fig. 9.31). They can also be used on paper products or rags.

Rotary Shear Shredder.   The rotary shear shredder is essentially a continuous rotary shear or scissor (Fig. 9.32). Two counterrotating shafts rotate in opposite directions with very close spacings between the cutters which are placed on each shaft. This type of shredder tends to cut feed material into strips which are the same dimension as the cutter width or spacing. Since the cutters tend to be not less than 1 in in size, this would be the smallest cutting dimension. Cutters have also been configured

FIGURE 9.27 (*Continued*)

up to 6 or even 8 in in size either by making larger cutters (normally not larger than 4 in) or by stacking cutters next to each other such that three 2-in cutters would produce a 6-in total cutter spacing.

The cutters are not circular, but rather oblong. Material passes down through openings which form between the top of opposing cutters from opposite shafts. The rotary shear shredder offers one significant advantage over the higher-speed mills discussed above and that is a much lower potential for explosions and dust generation from processing waste materials. Hooks are positioned on each cutter to grab material which enters the mill and pull it into the shear where it is cut. These mills are capable of cutting whole truck tires. They are also used in some solid waste shredding operations to open bags and liberate their contents.

Unlike the hammermills described above, the shear shredder operates at very slow speeds and does not pulverize glass or significantly reduce the size of cans, many of

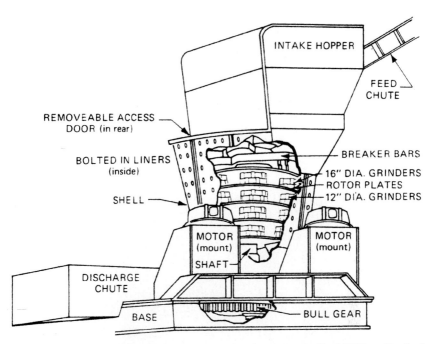

FIGURE 9.28 Vertical shaft ring grinder. (*Source: Fig. 12.41, Robinson, The Solid Waste Handbook, John Wiley & Sons, 1986.*)

which, depending upon the size of the cutters, can actually pass through the machine unaffected for a machine having 4-in cutters.

Size Reduction Equipment Selection Considerations.   The selection of size reduction equipment is dependent primarily on the characteristics of the feed stream and the process needs of the size-reduced materials. Pulverizers and crushers are preferred where friable materials are being processed and size reduction can be accomplished by impact alone. Flail mills are used when materials require only coarse shredding and reduction to a specific particle size is not a factor. Hammermills are normally employed where coarse size reduction via cutting is required and a wide-ranging particle size can be accepted, with a controlled maximum particle size passing through the grate. Vertical shaft hammermills are typically applied in place of a horizontal hammermill when the maximum particle size passing through the grate is not critical.

Pulverizers, flail mills, and hammermills tend to be noisy and are likely to generate dust during operations. They are also susceptible to explosions due to the presence of flammable materials and pressurized containers (e.g., aerosol cans, gas canisters, propane cylinders). Shear shredders are selected where the potential for explosions is high and the need to minimize dust generation is important. Because of its lower operating rotation speed, abrasion on the cutters of the shear shredder is less and the horsepower requirement is typically lower than for a hammermill. Knife mills are employed where fine particle-sized discharge is required and tight control over size distribution is important. Care must be taken to ensure that difficult-to-shred items (metals, rocks, etc.) are removed prior to entering a knife mill to prevent damage to the knives.

Air Classification.   Air classification has many applications in the processing of recyclable materials. Among many other applications, air classifiers are used to sepa-

FIGURE 9.29   Flail mill installed at Anoka County, Minnesota, waste processing facility. View through opened doors shows hammers and grate.

FIGURE 9.30   Glass crusher.

rate (1) labels from granulated plastic bottles, (2) lighter plastic bottles and cans from heavier glass bottles, (3) paper and plastic films from bottles and cans once these materials are liberated, and (4) fine glass and dirt from coarse glass.

Air classifiers are categorized into the following general types:

1. Horizontal air classifier and air knife
2. Vertical column
3. Zigzag
4. Rotary drum
5. Multiple-stage aspirators
6. Vibroelutriators

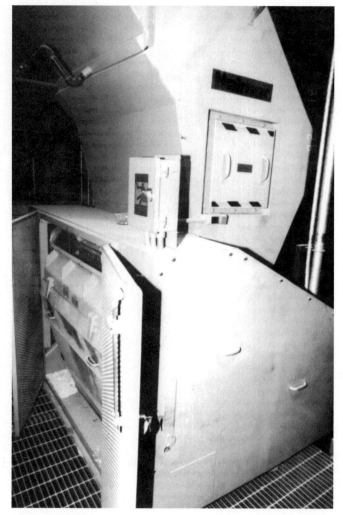

FIGURE 9.31 Granulator.

Horizontal Air Classifier and Air Knife. A horizontal air classifier is shown in Fig. 9.33. Material is fed by conveyor and dropped into the airstream over an air blower. Heavy objects which are more affected by gravitational force than by the pneumatic forces of the air current drop quickly through the air current while lighter more air-buoyant objects are carried by the airstream farther distances or are carried away with the airstream into a cyclone which acts to separate the light entrained matter from the airstream itself. (Material which is not dropped out of the airstream by the cyclone mechanical separator is later separated from the airstream using either a bag filter or a wet scrubber.)

An air knife is similar in concept to a horizontal air classifier. A diagram of an air knife at a solid waste processing facility is provided in Fig. 9.34. Here the mixed waste

(a)

(b)

FIGURE 9.32 Rotary shear shredder. (*a*) Cedarapids rotary shear shredder. (*b*) End view. (*c*) Close-up of cutters and cleaning fingers in rotary shear shredder. (*d*) Shear shredder cutting action.

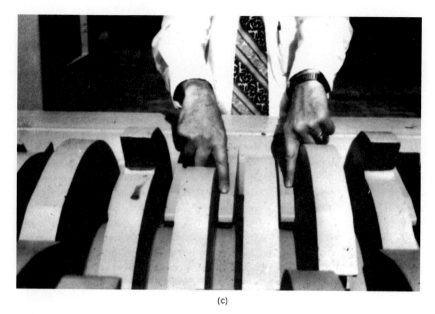

(c)

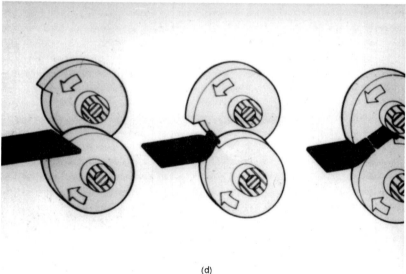

(d)

FIGURE 9.32   (*Continued.*)

sees an air separation step as it is being fed onto a disc screen. The air knife removes some of the very light paper and plastic material along with very fine glass particles which will also fly just before it is being dropped onto a screen for size separation.

Vertical Column.   The vertical column air classifier acts much like the horizontal air classifier or air knife except that the material which is fed is dropped straight down

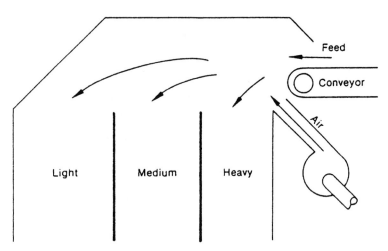

FIGURE 9.33 Horizontal air classifier. (*Source: Fig. 12.86, Robinson, The Solid Waste Handbook, John Wiley & Sons, 1986.*)

through the air column which is a vertical chamber (Fig. 9.35). The separation is bimodal in that material is either heavy or light (goes up or down), whereas in some horizontal air classifier systems, many different fractions can be recovered depending upon the trajectory of the products across the horizontal distance of the air classifier. Vertical column air classifiers have been utilized at high capacities in the solid waste and composting industries.

Zigzag. The zigzag air classifier was developed to allow multiple stages of classification to occur in order to improve the quality of separation. This type of air classifier is similar to the vertical air classifier except that material tumbles as it drops through the air current falling from one shelf onto the next (Fig. 9.36). While in principle this multiple action appears to offer great benefit, in practice much of the separation takes place in the first stage or on the first zig. Material which is liberated or separated in later stages of the air classifier must pass through the burden above before it can be separated and reclaimed as a "light." This is a very tortuous path from the lower stages to the upper stages of the air classifier and thus, unless the unit is lightly loaded, the burden falling from above will tend to take light materials freed in lower stages out with the "heavy" fraction.

Rotary Drum. A rotary drum air classifier resembles a slightly inclined large barrel on its side which rotates slowly while a gentle airstream passes through it (Fig. 9.37). The rotary drum air classifier characteristically performs its separation at much lower air velocities than other air classifiers. In essence, it acts much like the horizontal air classifier except the separation is performed many times and at lower velocities. The material to be separated is introduced near the upper end of the drum. The drum normally has lifters inside of it which lift the waste and drop it repeatedly as air is being drawn up the inclined drum and while the drum turns. The heavy material, which is more affected by gravitation forces than by the pneumatic forces, "walks" down the incline and out the bottom of the drum while the lighter material either is entrained in the air current and flies up the drum or alternatively "walks" up the drum after repeated drops in the airstream.

Experience has shown the rotary drum air classifier to be the most efficient type of air classifier available in terms of making a high-quality separation. It also has the

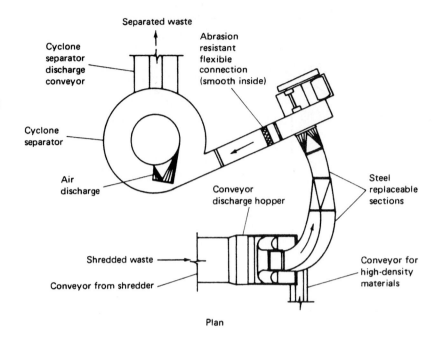

Plan

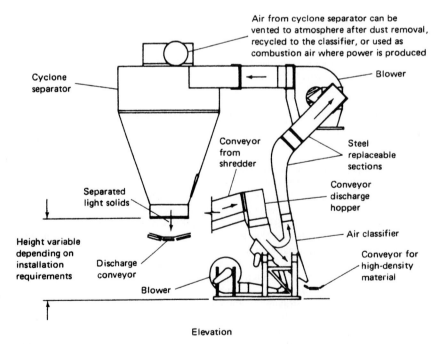

Elevation

FIGURE 9.34 Air knife. (*Source: Fig. 8-17, Tchobanoglous, Solid Wastes: Engineering Principles and Management Issues, McGraw-Hill, New York, 1977.*)

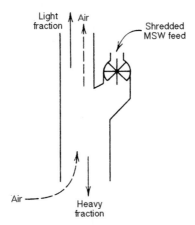

FIGURE 9.35 Vertical column air classifier. (*Source: Fig. 12.85, Robinson, The Solid
Waste Handbook, John Wiley & Sons, 1986.*)

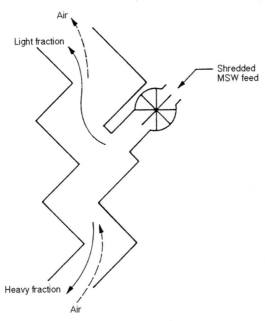

FIGURE 9.36 Vertical zigzag air classifier. (*Source: Fig.
12.90, Robinson, The Solid Waste Handbook, John Wiley &
Sons, 1986.*)

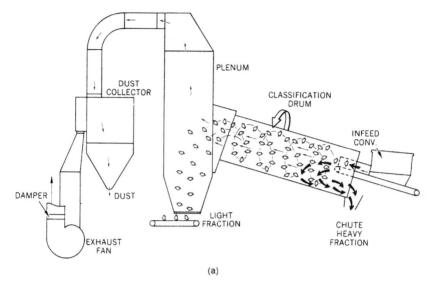

(a)

FIGURE 9.37 (*a*) Rotary drum air classifier. This equipment can be sized for use in a variety of applications. Shown are rotary drum air classifiers installed at (*b*) a large-scale mixed waste processing facility, and (*c*) a small-scale plastics recycling facility.

advantage of being able to separate materials of large particle size. The major disadvantage is its large physical size and mechanical complexity which make it more expensive to purchase and maintain than other air classifiers of similar capacity. Thus, if critical separation efficiency is important, the rotary drum is an ideal candidate for selection, but if a very rough separation is all that is required, then other types of air classifiers offer a lower-cost alternative.

Multiple-Stage Aspirator. This air classifier employs an upward flow of air to lift lighter materials away from heavier materials. The multiple-stage aspirator (Fig. 9.38) has openings along its vertical length which allow outside air to "sweep" through the cascading infeed materials at various points to lift materials away into the airstream. Heavy materials fall through the air classifier and exit via gravity at the bottom.

Vibroelutriator. The vibroelutriator (Fig. 9.39) has the characteristics of both an air classifier and a vibratory screen. Material to be separated is introduced at the upper end of a device somewhat resembling a horizontal inclined vibrating screen which has a device resembling a fume hood above it. As the material is vibrated down the screen deck, air is sucked up the hood both from the bottom discharge area and from under the screen cloth. Light material such as labels and other fine particles is lifted by the air currents and drawn up the feed hood where it is removed from the airstream by cyclones and bag filters. The heavier materials travel down the screen cloth and are discharged at the lower end. This separator has the advantage of long residence times in the separation zone and vibrating the material to provide tumbling action. The device is better for small particle sizes rather than large particle sizes because the tumbling action would not be sufficiently aggressive for large particle sizes on the order of 2 in and above.

Air Classifier Selection Considerations. The selection of an air classifier for a particular application is dependent primarily on the separation efficiency required and the money and facility space that can be committed. The least expensive equipment is

(b)

(c)

FIGURE 9.37   (*Continued.*)

a vertical column which requires a minimum of space but is the least efficient type of air classifier. The zigzag air classifier is slightly taller than a vertical column unit and has internal veins which add to the purchase price. The zigzag offers a higher separation efficiency than the vertical column but is sensitive to the infeed particle size.

The horizontal air classifier can process materials having a larger particle size than a zigzag unit; however, it requires a considerable amount of floor space. The vibroelutriator provides a higher efficiency than the horizontal air classifier because the materials are spread uniformly and the vibration tends to stratify the lighter materials on top for easy retrieval. The vibroelutriator is considerably higher in cost than the horizontal air classifier. The rotary drum air classifier is the most efficient of all air classifiers and can accommodate a wide range of particle sizes. Materials are repeatedly

FIGURE 9.38 Multiple stage aspirator.

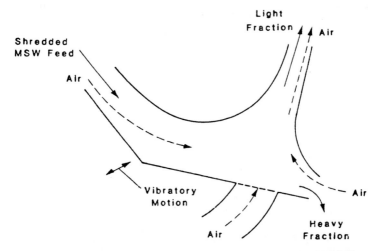

FIGURE 9.39   Vibroelutriator. (*Source: Fig. 12.89, Robinson, The Solid Waste Handbook, John Wiley & Sons, 1986.*)

tumbled and have multiple opportunities for separation. However, the rotary drum is the highest in cost and requires the greatest amount of floor space.

Eddy-Current Separation.   Eddy-current separation is utilized to separate conductors from nonconductors. The principle of separation relies upon Faraday's law of electromagnetic induction. In essence, when a magnetic field passes through a conductor (e.g., when a conductor experiences a change in an applied magnetic field), it induces in that conductor, or generates in the conductor, an electric current. That electric current also has associated with it a secondary magnetic field which always opposes the primary magnetic field.

If the conductor is in the shape of a ring, current flows in one direction and can easily be measured. If the conductor is a solid piece of metal, current is more difficult to observe and measure, owing to the complex current path, but the current is there just the same. Such currents which are closed within more or less solid pieces of metal are known as eddy currents, because they resemble eddies observed in liquids. Eddy currents give rise to a physical force that is the basis of a separation process. In general, all conductors resist changes in magnetic field strength. It has been established that the repulsive force set up by eddy currents is a function of particle size, particle geometry, and the ratio of conductivity to mass density—in addition to such factors as magnetic field strength and frequency. The opposing forces between the primary and secondary magnetic fields are utilized in eddy-current separators to make a separation between materials. The principal criteria used to make the separation have to do with conductivity and mass. Aluminum has a lower conductivity than copper but also has lower mass. Aluminum has the lower mass-conductivity ratio and thus is more affected by an eddy-current separator; that is, aluminum sees a greater trajectory than copper when subjected to an eddy-current separator. The force exerted on a particle increases with the fourth power of the radius, and thus bigger particles are much more affected than small particles. Shape and thickness can also be factors in such separations.

TABLE 9.7 Ratio of Conductivity to Density

Material	Conductivity/density
Aluminum	13.1
Copper	6.6
Zinc	2.4
Brass	1.7
Lead	0.4
Plastics, glass	0.0

Source: Dalmijn, W. L., et al., "Recovery of Aluminum Alloys from Shredded Automobiles," Department of Mineral Technology, Delft University, The Netherlands.

Eddy-current separators create forces on conductors but not on nonconductors. Thus nonmagnetic metals may be separated from plastics, wood, rubber, etc., using this principle. Table 9.7 provides the ratio of conductivity to density for various metals.

There are four principal types of eddy-current separators:

1. "Linear motor" separator
2. "Popper" or "Pulsort" separator
3. Stationary permanent magnet or sliding ramp separator
4. Active permanent magnet or rotating drum separator

In the linear motor separator, a traveling magnetic field is generated by an electromagnet energized by an electromagnetic field. In essence, this separator works like an induction motor in which the rotor has been removed and the stator has been opened up and is lying flat. The forces exerted by the stator which would normally rotate the armature or rotor are instead utilized to move conductors which pass over the flat, or linear, induction motor. The metal particles are moved laterally across the belt and roll off of one side or the other.

In the Popper separator, particles are passed over a rapidly changing magnetic induction coil capable of generating very high currents through the use of large capacitor banks which are discharged intermittently. The conductors which are over the coil when it discharges see very high forces and are "popped" off of the belt. When the popper is set for small particles and sees a large particle, very high forces (which can be dangerous) can be generated because the separation force increases to the fourth order of the particle size. Large particles can be ejected from these separators at bullet speeds.

In the sliding ramp separator, permanent magnets are used to make the separation, and thus this is called a passive separator while the others which use a change in an electromagnetic field are termed "active." In this type of separator (Fig. 9.40), magnets of alternating polarity are arranged in stripes on a flat, inclined board, or ramp. The materials to be separated are conveyed to the top of the ramp and slid down using gravitational force. As the conductors slide over the top of the magnets, they see an oscillating magnetic field similar to the active field which is generated in the stator of an electric motor. Much like the linear motor separator, the conductors see a force that moves them over to one side while the nonconductors slide down the ramp unaffected by the oscillating magnetic field. In this case, the work is done by gravity and there

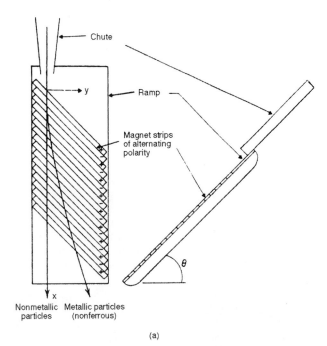

FIGURE 9.40 Eddy-current separator. (*a*) Nonferrous metal separator using fixed magnets. (*Source: Schloeman, E., "Eddy Current Separation Methods," Raytheon Corp., Waltham, Mass.*) (*b*) Eddy-current separation system in operation. (*Delaware Reclamation Project.*)

are no capacitors or any moving parts. Furthermore, the forces are never very high because large particles cannot experience forces which exceed their own gravitational force. The drum-type eddy-current separator is much like the ramp-type separator except that the magnets are attached to a drum which is in essence a continuous ramp. The permanent magnets are rotated under an outer shell or under a belt. The conductors are then moved by the eddy-current forces.

Can Flattening and Densification.   Magnetic cans recovered from a typical MRF operation are shown in Fig. 9.41a. As can be seen, the cans have not been densified. Some food remains in the cans. Labels have been removed in some instances and

(a)

(b)

FIGURE 9.41 Methods of reclaiming magnetic cans. (*a*) Cans as recovered from MRF operation. (*b*) Shredded can product from mixed waste processing facility. (*c*) High-density bales.

(c)

FIGURE 9.41 *(Continued.)*

remain on the cans in others. In order to effectively transport this material to the marketplace, these cans must be shredded, flattened, or baled. To improve the value of the finished product, the cans should also be cleaned of labels.

A photograph of a shredded can product from mixed waste processing operations is provided in Fig. 9.41*b*. The raw mixed waste processing ferrous metal product can be seen in the bucket of the front end loader and contains significant quantities of non-metallic contamination such as paper, plastic, and rags. During the course of shredding and air classification, the contamination is separated from the metal and the labels are removed. The metal is densified during the shredding operation. Alternatively, the loose metallic can product is baled (Fig. 9.41*c*) in a high-density baler. For this type of metal bale, density is so high that baling wire is not required. Still another method of increasing the density of loose metallic cans is to utilize a small can flattener. These flatteners are useful only for tin cans and should not be fed such metallic items as castings and heavy metal objects which could severely damage the equipment.

Automated Plastic Sorting Systems.   The significant differences in the material properties of the plastic resins in use today usually require that they be separated from one another before they can be reclaimed and reused. Up until recently, separation has been accomplished either through source separation practices or by manual sorting at an MRF. Manual sorting is not ideal, however, owing to its high labor and training costs and its susceptibility to error. Automated sorting systems have thus been developed in an attempt to replace manual labor with an automated technology. Costs for automated systems range from $35,000 for a small PVC detection system to over $700,000 for a multiresin and color identification system.

The major plastic resins currently used for the manufacture of containers include clear polyethylene terephthalate (PET), green PET, natural high-density polyurethane (HDPE), opaque HDPE, polyvinyl chloride (PVC), polypropylene (PP), and polystyrene (PS). A pie chart showing the relative quantities of each resin in postconsumer

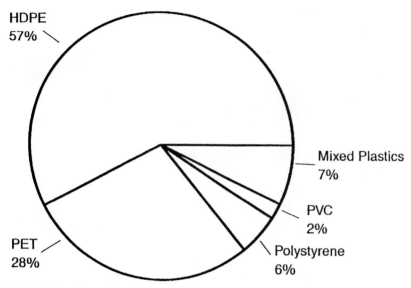

FIGURE 9.42   Composition of postconsumer plastics collected through curbside program.

packaging waste is provided in Fig. 9.42. The amount of cross contamination that can be tolerated in MRF products depends on the particular resins involved, the end use of the resin, and the capabilities of the end user's process equipment. Consequently industrywide product purity standards have not been uniformly established for the plastics products generated by primary collectors and processors such as MRFs.

From a chemical compatibility viewpoint, the major contamination concern is with PVC and PET contamination. PVC and PET are virtually incompatible with one another in the reformulation of either PVC or PET pellets. The nearly identical specific gravity of PVC and PET renders conventional gravity separation processes useless, and the visual similarity of the two resins makes manual separation difficult and impractical. The maximum allowable concentration of PVC in PET is generally in the range of 10 to 200 parts per million (ppm), depending on the end use of the PET product. At 10 ppm, this is roughly 1 bottle in 100,000 bottles, and at 200 ppm this is roughly 1 bottle in 5000 bottles. There are currently three existing commercial systems for the automated identification and separation of individual resins in a mixed plastics stream.[15–18] Each uses a combination of sensors and microprocessors to detect and identify the various resins, and air jets to eject the identified containers into product containers. These are described below.

Magnetic Separation Systems, Inc.—BottleSort System.   The basic magnetic separation systems (MSS) system uses different intensities of light to distinguish clear bottles consisting of PET and PVC, translucent bottles consisting of HDPE milk and juice jugs, polypropylene syrup and ketchup bottles, and opaque bottles consisting of detergent and shampoo bottles. Air jets are used to remove the detected bottles from the conveyor belt. Further resin separation can be accomplished using additional modules consisting of an Asoma VS-2 PVC detector, a computer-controlled HDPE color detector, an optical PET color sensor, and an optical PP sensor. The Asoma unit uses an x-ray source to detect chlorine atoms present in the PVC. The color detectors employ computer software to identify seven bottle colors, ignoring the labels and residue contamination. Each bottle is scanned by an array of 16 sensors which make 5300 readings per second to determine the bottle type. Four parallel sorting lines are claimed to be able to identify and separate 8 to 12 bottles per second (120 to 180 bottles per minute per line), providing a total capacity of 5000 lb/h. MSS claims that the basic three-way separation system can achieve a 96 to 99 percent resin separation and that over 99 percent of the PVC bottles can be captured.

National Recovery Technologies, Inc.—VinylCycle System. The National Recovery Technologies (NRT) system uses an electromagnetic sensor that separates PVC bottles from a mixed stream of whole or crushed (not granulated) HDPE, PET, and PVC bottles. The mixed stream is passed down a slide, and air jets mounted below the slide are used to blow the detected PVC bottles over a partition. Undetected bottles fall on the opposite side of the partition. This system produces two product streams; one consists of mixed bottles and one is a PVC-rich stream. It is claimed that the bottles need not be oriented or singulated before being presented to the detector.

NRT currently has three models: the highest-production model is reportedly capable of processing 10 bottles per second or 4500 lb/h and yields a product with less than 50 ppm PVC contamination. NRT guarantees a PVC contamination limit of 50 ppm if the PVC content of the infeed mixed stream does not exceed 1 percent. The PVC recovery is reportedly greater than 99.4 percent, with 100 percent PVC removal possible with two passes. According to The Vinyl Institute, 18 NRT systems have been installed worldwide.

Automation Industrial Control—PolySort System. The Automation Industrial Control (AIC) system uses infrared spectographic analysis of light transmitted through singulated bottles to distinguish containers by resin type and color. Air jets segregate the detected bottles into polypropylene, PVC, natural HDPE, colored HDPE, green PET, and clear PET product streams. Two sensors read the bottle, and the information is processed in a 486-chip microprocessor. A time of 19 ms is required to determine the bottle type and color. AIC claims the system reads 85 to 90 percent of the bottles at a process rate of 3 bottles per second or 1500 lb/h and is 100 percent accurate. The system reportedly can process either compacted or uncrushed bottles. Unread bottles remain on the belt and can be recycled through the system or discarded.

General Comments. These automated plastic sorting systems are relatively new and have seen only limited application in full-scale facilities. They demonstrate the feasibility of applying complex optical sortation techniques to the separation of recyclables, but cost and accuracy remain impediments to wide-scale implementation at present. As computing speeds increase and optical sensing becomes more rugged and reliable, the application of this technology will become more cost-effective and see more widespread use.

9.4 PROCESSING RECYCLABLES

Section 9.3 provided information on the various equipment and processes commonly in use to prepare recyclable materials for market. This section provides information on the overall processing systems which employ these various unit operations and equipments to process glass, metal, plastics, and paper.

Many different processing schemes can be devised to segregate and process recyclable materials. The selection of a particular system may depend on the size of the plant, the various materials which are collected and the form in which they are received, and the community or vendor that will operate the facility. Processing preferences and technology depend upon the design engineer or the system vendor selected to design, build, and/or operate the given recycling facility. Further, the ideal system for a given community will depend greatly on locality particular factors such as site, markets, collection practices, alternative methods of handling recyclables, traffic, transportation, applicable building regulations, codes, etc.

This section gives an overview of the types of systems available for processing source-separated recyclables, commingled recyclables, and mixed waste. General guidelines are provided for the design of these systems. The standardized design factors and considerations suggested will then be utilized as a basis for discussing market factors and developing standardized cost estimates in following sections of this chapter.

Processing of Source-Separated Recyclables

Processing systems for household separated recyclables are much the same whether the recyclables are collected by means of

- Return of bottles and containers to the store as part of a bottle bill or other type of deposit legislation
- Consumer recycling through use of drop boxes, drop-off centers, or buy-back centers
- Curbside collection of household separated materials

In all three of these cases, a glass bottle, for example, is collected with other glass bottles and not mixed with paper, cans, and plastic containers. However, the glass bottles are not just glass. They may have aluminum or plastic caps still on them. They may have paper or plastic labels. Further, the glass may or not be separated by color into clear (flint), brown (amber), or green (emerald).

Normally, the separated recyclables are taken to a central collection or consolidation center. In some cases, materials are simply consolidated into larger containers and sold "as is." Little or no processing is performed. Efforts may be made to separate the glass by color if not collected already color separated. Some quality control may be employed when separation at the source has not been properly carried out. However, little effort is made to remove labels, take off caps, crush bottles, or flatten cans. Paper is not baled but is loose. In larger facilities, source-separated containers such as those recovered through bottle bill returns are processed through crushing, flattening, baling, and label and/or cap separation operations. In either case, storage is needed for each incoming material. The storage needs must be calculated based on the overall design requirements and markets. The tipping floor, bunkers, and/or storage containers must be large enough to store *at least* 1 day's incoming material and keep it segregated from the other materials received. A weigh scale is desirable to weigh both incoming and outgoing products. Product weights can also be determined off site or by the end market purchaser.

Table 9.8 provides typical incoming densities for various materials. Processing a material may not be necessary to increase density provided a lower price can be accepted and transportation distances are not great, but some sort of consolidation is almost always required (Fig. 9.43). When processing is required as well as consolidation, processing system space requirements as well as product storage must be considered. Output product storage may have to accommodate more than 1 day's supply, since it may take time to accumulate truckload quantities of some products such as aluminum or plastic. The equipment and systems utilized for the consolidation and processing of source-separated materials—glass, metal, plastic, and paper—are described in the following sections.

TABLE 9.8   Postconsumer Material Densities

Material	Typical densities, lb/yd³
Paper:	
Newspaper	475
Corrugated	350
High grades	300–400
Glass:	
Whole bottles	500–550
Crushed	1000–1800
Aluminum:	
Whole	50
Flattened	175
Plastics:	
PET, whole	34
PET, flattened	75
HDPE (natural), whole	30
HDPE (natural), flattened	65
HDPE (colored), whole	45
HDPE (colored), flattened	90
Tin-plated steel cans:	
Whole	150
Flattened	850

Source: Romeo, E. J., Jr., "Material Recovery Design of Ocean County, New Jersey," *Proceedings of the 1992 National Waste Processing Conference,* The American Society of Mechanical Engineers, May 1992.

Glass Processing.   Glass bottles are received either packed in cardboard cases from bottle bill return locations or are received in loose form directly from collection vehicles (Fig. 9.44). They may be stored in intermediate holding containers or in gaylords. When enough glass has been received to begin processing, the material is removed from the storage containers and placed on the tipping floor. If not already accomplished, the glass may be color separated manually on the tipping floor. Care should be taken to avoid excessive breakage of the glass prior to color separation. Breakage after color separation is not a problem. While on the floor, the glass is also inspected for contaminants such as plate glass, ceramics, rocks, and/or stone which would cause impurities in the finished glass product and therefore must be removed prior to processing. In large facilities, a front-end loader is usually used for moving material. Care must be taken to avoid picking up contaminants from the floor with the loader bucket.

The glass is crushed using a pulverizer or shredder. Following crushing, small vibrating screens or trommels are used to screen away bottle caps as oversize material from the smaller pulverized glass. Labels are also largely removed by screening; or, alternatively, air classifiers can be employed. This process is repeated for each color of glass. Normally the same equipment can be utilized on a rotating basis for all three colors of glass. In higher-capacity systems, multiple parallel processing systems are installed for each color of glass. Noise and dust control are important environmental considerations which must be addressed during installation of a glass crushing and processing system. Also, the proper design of feed conveyors and the interface to the feed hopper and dribble chutes require experienced design knowledge.

FIGURE 9.43   Cans received in boxes are dumped onto conveyor to be densified.

FIGURE 9.44   Cases of glass bottles are readied for processing.

While the glass is stored prior to shipment to market, a manual inspection should be conducted as a final quality control measure to remove any obvious unwanted contamination (Fig. 9.45). A small amount of color contamination does not present a significant reduction in market value or market demand. Glass is expensive to ship relative to its market value, which is on the order of $15 to $60 per ton depending on color, cleanliness, and market conditions. Selection of the final shipping container or vehicle must be coordinated with the glass buyer. If the glass is to be used locally as an aggregate or as part of a paving mix, shipping in open dump trucks may be adequate. However, rail shipment is an important consideration if the glass is to be used as cullet in the manufacture of glass bottles and the distance to the nearest glass bottle producer exceeds 100 miles. Glass can be shipped in enclosed rail hopper cars or in roll-off containers (Fig. 9.46).

Can and Metal Processing. Generally speaking, the metal cans received from a source-separation program are of three different types: all ferrous (including tin-coated); bimetal (ferrous metal cans with aluminum "pop top" lids); and all aluminum. Collection of aluminum plate, castings, and foils may also be included within various collection programs. Moreover, large steel and castings including steel plates, metal bars and shapes, electric motors, etc., may also be thrown away by consumers into recyclable containers whether the program accepts these materials or not. If the collector does not remove them at curbside for alternative recycling and/or disposal, they must be removed on the tipping floor or from infeed conveyors. These oversize objects can destroy can flattening equipment and must be removed before processing. Normally, magnetic separators are not utilized in source-separated materials processing centers. It is relatively easy to visually identify aluminum cans and hand sort them from cans containing ferrous metal before baling or flattening.

Aluminum companies, such as Alcoa and Reynolds, often provide the recycler with a can flattener and trailer (Fig. 9.47). This equipment is provided either under a lease arrangement or as a price reduction in the finished product in exchange for a long-term purchase agreement. The aluminum buyer will normally stage the trailer on the site and replace a full trailer with an empty one as needed. The markets for steel can scrap are often more difficult than the markets for aluminum simply because the value of the finished product is so much lower. Flattening or low-density baling makes sense only if the market is a local scrap dealer rather than a steel mill or end user. The finished product most often is baled or briquetted in order to find a strong market and achieve reasonable transportation economics (Fig. 9.48). Rail shipment may be a market requirement.

It is not necessary to provide strapping for high-density bales. Also, the market is sufficiently strong for this material that it is not necessary to remove labels from the steel cans. When can scrap is shredded, the shredder automatically removes the labels. The shredded material can be stored and shipped loose because it typically has a density of 40 to 50 lb/ft^3. In some shredders, densities of 60 to 80 lb/ft^3 can be achieved by balling the product. Balled material is not suitable for detinning because the detinning reagent cannot easily penetrate the metal ball. However, balling is desirable for steelmaking applications owing to the improved density and cleanliness of the material.

Aluminum contamination of steel cans is not a problem in steelmaking applications or markets because aluminum is a valuable deoxidant for steelmaking. However, aluminum contamination can be a problem in detinning because it consumes detinning reagent. Thus, for detinning, separation of bimetal cans from tin cans may be important. It is not important to a minimill or steelmaking customer. The aluminum lost to the tin can scrap, however, will only be valued at steel prices.

FIGURE 9.45   Glass is raked to remove contaminants

FIGURE 9.46   Bulk-handling trailers used to ship glass product.

FIGURE 9.47   Aluminum trailer.

FIGURE 9.48   Flattened, baled can scrap.

Plastics Processing.

PET (Soda) Bottles. PET plastic bottles collected from bottle bill or deposit and return programs should be kept separate from PET collected from curbside programs or from drop-off centers in order to avoid the inadvertent potential for contamination by PVC. (The market value of bottle bill PET is significantly higher than that of curbside PET.)

PET is separated by color into clear and green, or alternatively, it can be sold as mixed color (Fig. 9.49). It should not be granulated unless directed by the end market buyer, the problem being that once the product is granulated, it is not possible to inspect for contamination. Technology does not yet exist to remove small particles of PVC from PET practically and efficiently at commercial throughput rates, although some promising development programs are underway (see Sec. 9.3). Some end market buyers furnish granulators in exchange for long-term supply contracts once the supplier has been certified for quality control.

PET can sometimes be sold as whole bottles without baling or granulation provided the transportation distance to the end user is quite short. Material is collected in large, reusable bags (Fig. 9.50). These bags are emptied, and the contents segregated by color and then granulated by the end user. The bags are returned to the supplier when a new shipment is received. PET bottles can be sold "as is." Automated processing systems are in place to remove caps, bottom cups, and labels, producing a high-quality finished product that can be recycled into new PET bottles (Fig. 9.51).

Homopolymer (Natural) HDPE. Homopolymer or natural HDPE consists primarily of milk and water bottles (Fig. 9.52). There is a strong market demand for this product, although price may be low at times. Natural HDPE may be processed simply by baling or granulating. The impurity issues are not as stringent for HDPE as for PET and thus granulation is often acceptable, but it should be verified by the end user. Also, granulating may be less expensive than baling and may contribute to the market

FIGURE 9.49   PET readied for market.

FIGURE 9.50 Bagged PET bottles received at consolidation center are first perforated.

FIGURE 9.51 Overview of plastics recycling facility.

(a)

(b)

FIGURE 9.52 Natural HDPE. (*a*) As-received milk and water jugs. (*b*) Tires of a front-end loader have embedded dirt and grease into the HDPE, thus lowering the value or destroying the marketability of this product.

value, since granulation must eventually be done by the end purchaser at a cost of about $0.05 per lb.

The market for natural HDPE is higher and the demand greater when caps are removed because the finished product can meet more stringent color specifications. However, cap removal requires added labor cost. Care should be taken not to contaminate the product by other plastics and to keep it clean during this processing step. Care must be taken to avoid processing materials that look like HDPE milk bottles but are not. Many forms of polypropylene look just like homopolymer HDPE. A small amount of PP can be tolerated by HDPE processors, normally less than 3 percent. If large quantities of PP are contained in the HDPE material, the finished product may not be marketable.

Mixed Color HDPE. Mixed color HDPE comprises laundry detergent bottles, bleach bottles, dishwashing detergent bottles, and the like (Fig. 9.53). Even though homopolymer HDPE milk bottles are composed of the same polymer as mixed color HDPE, the two are incompatible. The reasons for this are quite complex but in essence can be explained by the term "melt index." Homopolymer HDPE has a low melt index (sometimes described as having a "fractional melt index") because it is less than 1, while mixed HDPE has a melt index which is greater than 1 typically in the high 20s or low 30s. The higher the melt index, the more the polymer flows when melted. Mixing high melt index HDPE with fractional melt index HDPE is like mixing tar with water. It doesn't mix. For that reason, even though both polymers are HDPE, the markets for each are different. Thus they must be processed separately and stored separately as if they were different polymers. Mixed HDPE is also hard to granulate and requires more frequent granulator maintenance than does homopolymer HDPE granulation. A heavy-duty, wide-mouth granulator is desirable. HDPE should not be stored outside in sunlight for prolonged periods of time because the material is photodegradable. The market value of the finished product may be destroyed.

FIGURE 9.53   Typical mixed color HDPE containers.

FIGURE 9.54   High-density baling of old newspaper.

Paper.   Paper is received in many different forms including old newspaper (ONP), old corrugated containers (OCC), mixed paper, and high grades such as computer printout (CPO). ONP may be sold loose or baled. If the market for the product is overseas, a high-density baler is required. Low-density bales, whose ultimate market is overseas, are frequently "broken" and rebaled in high-density balers (Fig. 9.54). Thus the money and effort invested in baling the finished product may not contribute value to the recycling facility where it does not contribute value to the end user. OCC must be baled. This material is of too low a bulk density to be shipped loose more than a few miles. High-grade paper such as CPO is usually baled but may be marketed loose to local paper recyclers. Old books may also be sold loose or may be directly recycled through local book exchange programs (Fig. 9.55).

One must ensure that a market exists for mixed paper *before* it is collected or "produced" through the improper segregation of incoming materials. Hand or automated separation of mixed paper into various paper grades is not economically viable. Paper is often separated in brown paper bags or bundles tied with string. Eventually, the bundles must be untied and the bags removed. This can be done by community recyclers or alternatively at a paper recycler's operation. Although there are automated methods for opening and removing bags, currently available "bag breakers" can jam frequently. Most large-volume processors still use hand labor.

Enriched corrugated loads (such as cardboard boxes and waste collected from clothing stores, variety stores, department stores, or supermarkets) can be separated from other paper, plastics, and contaminants on a tipping floor. However, this is backbreaking work and requires significant storage and working area (Fig. 9.56). A separate sorting line with a large (minimum 5-ft-wide) conveyor is desirable. The width of this conveyor makes the sorting of other containers such as glass, metals, and plastic difficult because the operator needs to reach out such long distances across the conveyor belt.

FIGURE 9.55 Book recycling program, Wellesley, Mass., drop-off center.

FIGURE 9.56 Sufficient floor space is required for sorting and storing enriched corrugated loads. (*Courtesy of Mayfran International and Browning Ferris Industries.*)

A number of layout and design considerations are to be taken into account if corrugated is separated from other paper or wastes on a conveyor belt. First, these materials tend to be presented in surges or lumps. A variable-speed conveyor should be used and the lead (front) picker should have easy access to the speed control to slow down the conveyor when needed and speed it up when necessary. Second, it is difficult to throw material forward off the picking conveyor. It is easier to pull it toward one, swing it around, and throw it off the back of the conveyor. (This is just the opposite of bottles and cans, which are easier to throw forward.) The conveyor should therefore be high off the ground, or a takeaway conveyor should be utilized, so material separated does not interfere with the hand picking operation. Piles can be built from the floor without requiring constant front-end loader attention to remove the high volume of material being picked out for later baling. Obstructions should be kept to a minimum. Hand pickers should be spaced far enough apart so that the momentum when pulling cardboard from the pile can be used to throw the material off the conveyor. A typical corrugated operation is shown in Fig. 9.57. The baled finished product is shown in Fig. 9.58.

Central Processing of Mixed Recyclables

The term "materials recycling facility," or MRF, is one of the more ill-defined terms used in recycling. For some, it includes drop-off centers, processing plants that accept only source-separated materials, paper stock operations, and scrap yards. However, for others, to be an MRF, a facility must process commingled residential recyclables. The industry definition of an MRF is a "central operation where commingled recyclables, at least a portion of which come from the residential sector, are sorted and processed for market."[19] This is the definition of an MRF used in this chapter. The MRF, as defined, is also often referred to as an "intermediate processing center" (IPC).

In its simplest form, a "low-tech" MRF consists of not much more than a linear picking conveyor on which a number of workers pick out recyclable materials from a commingled stream and throw them into bins for shipment to customers. In its more complicated form, the "high-tech" MRF, handpicking is assisted through machine separation operations utilizing magnets, eddy-current separators, air classifiers, screens, and other devices. No fully automated MRF existing today is capable of receiving a stream of commingled recyclables and converting those materials into salable end products without some handpicking and manual quality control steps. In some cases, unique separation equipment has been developed by various system vendors to aid in the necessary separations. An example is the "chain curtain" separator (Fig. 9.59) which separates light plastic and aluminum from glass bottles. This equipment was developed by Bezner and is operated by CRInc in Rhode Island and elsewhere.

The design of an MRF depends upon many factors such as site constraints, vendor or licensed technology, methods of collection, delivery schedules, raw materials, and markets.[20] However, all MRFs normally depend upon the delivery of two separate feed streams: (1) commingled recyclables and (2) paper products. The commingled recyclables stream could include three colors of glass, tin cans and aluminum cans, foils, and plastics such as HDPE and PET. The paper products stream could be one grade of paper or several.

MRF buildings must be designed to various code requirements. Particular attention must be given to sprinklers which must be designed to meet "high-hazard" code requirements at least where bales of paper and plastic are stored. High hazard also is often the basis of design for tipping and processing areas. Floor drains and oil-water separation systems are frequently required. Automatic roll-up doors at least 27 ft high should be assumed along with a roof height in the tipping area of 30 ft to allow for tip-

FIGURE 9.57 Processing of old corrugated containers.

ping of 50 yd³ roll-off containers. Loading docks and loading dock doors should be assumed for transferring bales from intermediate storage into transfer vehicles or rail cars. At least 1 day's storage of infeed material should be provided assuming same-day processing. At least 1 day's storage should be provided for output products.

Table 9.9 provides a listing of existing and planned MRFs. The facilities are broken down into low-tech and high-tech MRFs which are operational, shut down, under construction, or in planning. Each facility is described by its location, design capacity, quantity of residue, and capital cost. Table 9.10 provides a list of owners, operators, and designers for each of the MRFs described in Table 9.9.

Small-scale (up to 100 tons per day) and large-scale (100 to 200 tons per day) MRF processing systems are discussed in the following sections. Incoming materials

FIGURE 9.58 Corrugated baling operation.

received at the MRFs are assumed to be a combination of source-separated materials delivered in multicompartment trucks and drop-off containers, and also commingled in refuse packer vehicles which deliver either bagged or unbagged mixed material. The discussions further assume single-shift operation.

Small-Scale, Low-Tech MRFs. As a rule of thumb, when the volume of material to be recycled is less than 25 tons per day, it should be collected in a multicompartment container and tipped as was shown in Fig. 9.4 into roll-off containers and transported to either a scrap or paper broker, or a centralized, regional MRF. On-site processing is not cost-effective and is not recommended for such very small quantities.

A flow diagram for a 25 to 100 tons per day MRF is provided in Fig. 9.60. The system only requires a feed hopper, hand sorting platform, product bunkers under the

FIGURE 9.59 "Chain curtain" separator.

conveyor and/or roll-off storage containers, and a baler. Commingled recylables are first fed into an infeed hopper set into the floor below grade. The feed hopper contains a cleated rubber belt conveyor which inclines up to the picking platform. Rubber cleats avoid the unwanted breakage of glass which can occur with metal cleats. The feed conveyor conveys the commingled material up the incline onto a picking conveyor (may be the same conveyor) with a platform for hand pickers along its side. The conveyor should not be more than 48 in wide. It should be flat and should have side skirts to prevent spillage. Operators stand on one side of the belt and throw hand-picked recylables forward into hoppers on the opposite side of the conveyor. The hoppers have "backstops" behind the opening and are at least 36 in wide. The hoppers may be lined with rubber or wood, at least for glass products, to reduce noise. Each product is positively sorted, i.e., picked off the belt. The residue remains behind as the last "product" which is "negatively sorted" and which will contain nonmarketable materials, broken glass, etc. If the products are very clean coming into the plant, it may be possible to positively pick off the residue and leave another product, say mixed color cullet, as the negatively sorted finished product coming off the end of the conveyor. However, generally, the objective here is to produce high-grade products, and thus each product is "positively" sorted.

HDPE milk bottles and PET are picked off the belt first (i.e., positively sorted). These materials tend to be on top of the commingled stream because they are light in weight. They also represent a high volume of the infeed material. Plastics are stored separately in bunkers or large containers and then fed to the baler or granulator when sufficient quantities have been accumulated to make a whole bale or operate the granulator near capacity for a continuous period of time. HDPE and PET are baled separately. PET is baled as mixed green and clear; HDPE is natural only. A horizontal automatic tie baler is recommended. The ferrous metals are removed magnetically using a cross-belt magnet or can also be separated by hand. Aluminum cans and aluminum food trays are then picked off. Some bimetal cans may be included with the aluminum. The MRF operator should verify that this is acceptable to the product customer. If not,

TABLE 9.9 Existing and Planned MRFs, 1992 Data

Name	Year	City	State	TPD processed	TPD residue	Original cost	Cost year
Construction—low-tech:							
Williamson County		Franklin	TN	100		1900000	92
Hartford		Hartford	VT	25		1300000	91
City of Kenosha		Kenosha	WI	23	2		
Construction—high-tech:							
Rosetto Recycling Center		Toms River	NJ	250	18	6600000	91
Town of Babylon		Babylon	NY	392		20000000	91
Shakedown—low-tech:							
Pacific Rim Recycling		Venicia	CA	100	5	250000	92
Charleston County	91	Charleston	SC	80		1837000	90
Shakedown—high-tech:							
Elk Ridge (BFI)	92	Elk Ridge	MD	150	8	4000000	92
Northern Virginia (BFI)	92	Newington	VA	125			
Monroe County		Rochester	NY	280	8	8200000	91
Buffalo District (BFI)	91	Tonawanda	NY	50	2	2500000	91
Operational—low-tech:							
City of Mobile	90	Mobile	AL	3	0	2692	90
Valley Recycling MRF	90	Chandler	AZ	11	1	125000	90
Phoenix (Why Waste America)	90	Phoenix	AZ	125	6	6000000	90
Tucson (Cutler Recycling)	90	Tucson	AZ	25			
Tucson (Recycle America—Waste Management)	76	Tucson	AZ	100			
San Mateo County (BFI—The Recyclery)	89	Belmont	CA	120	1		
Concord Disposal	85	Concord	CA				
Fresno (BFI)	90	Fresno	CA	13	0		
Western Waste Industries	91	Chino	CA	30	1		
Western Waste Industries	89	Redondo Beach	CA	65	2		
South Valley Refuse Disposal	91	San Martin	CA	5			
San Diego Recycling	90	San Diego	CA	225	2		
San Jose (BFI—Newby Island)	91	Milpitas	CA	200	70	11000000	91
Monterey City Disposal Service	89	Monterey	CA	10	1		

	Company	City	State				
90	Napa Garbage Service	Napa	CA	45	2	1000000	90
90	Upper Valley Disposal	St. Helena	CA	3	0		
89	East Bay Disposal (Durham Road Landfill)	Fremont	CA	55	2	375000	89
91	Fresno (City)	Fresno	CA	210	8	1200000	91
86	Waste Management of Santa Clara	San Jose	CA	201	2	900000	86
89	Pleasant Hill–Bayshore Disposal (BFI)	Pacheco	CA	47	6	340000	89
91	Richmond Sanitary Service	Richmond	CA	30	2		
78	Empire Waste Management	Santa Rosa	CA	120			
90	G.I. Industries	Simi Valley	CA	70	1		
88	Pacific Rim Recycling	Walnut Creek	CA	21	0	75000	89
90	Gold Coast Recycling	Ventura	CA	100	18	3500000	90
90	North Ft. Myers	North Ft. Myers	FL	200	12	1200000	90
90	Jacksonville (BFI)	Jacksonville	FL	110	7	2200000	90
90	Dade County (Attwoods)	Hialeah	FL	200	8	8000000	90
90	Orange County (Waste Management)	Orlando	FL	100	1	3500000	90
90	Pinellas Park (Recycle America—Waste Management)	Pinellas Park	FL	175	14		
83	Pinellas County (BFI)	Pinellas Park	FL	160		875000	83
90	Leon County (Capitol Recycling)	Tallahassee	FL	22	1	250000	91
91	Recycle America (Waste Management)	Atlanta	GA	70	4		
91	Macon (Industrial Recovered Materials)	Macon	GA	400	120	700000	90
90	Carroll County Solid Waste Management Commission	Carroll County	IA	9	1		
89	Northwest Iowa Solid Waste Agency	Sheldon	IA	6	0		
88	Garden City Disposal	Bensenville	IL	34	0	100000	89
89	Meyer Brothers Services (Waste Management)	Chicago Ridge	IL	28	0	250000	88
90	Elgin-Wayne Recycling & Waste Services	Elgin	IL	17	1		
88	Waste Management of McHenry County	McHenry County	IL	9	0		
89	Fox Valley Disposal (Waste Management)	Batavia	IL	30	2		
88	Best Recycling & Waste Services	Wheeling	IL	65	3		
91	Southeast Louisiana (Waste Management)	Walker	LA	70	6	6000000	90
91	SEMASS Recycling Facility	Avon	MA	120	12	5000000	91
91	Lowell	Lowell	MA	25	3	175000	91
91	Waukesha County	Waukesha	WI	40	1	1520000	91
91	Onondaga County (Laidlaw)	Syracuse	NY	300	15	1000000	91
90	Phoenix Recycling	Finksburg	MD	100			

TABLE 9.9 Existing and Planned MRFs, 1992 Data (*Continued*)

Name	Year	City	State	TPD processed	TPD residue	Original cost	Cost year
Operational—low-tech (*Continued*):							
Wilmington (BFI)	91	Wilmington	DE	40			
Hagerstown (BFI)	90	Hagerstown	MD	40			
Salisbury (BFI)	91	Salisbury	MD	40			
Anne Arundel County	90	Severn	MD	37	4	500000	86
Town of Bowdoinham	89	Bowdoinham	ME	2	0	25000	89
Waste Management of Michigan (Southwest)	91	Battle Creek	MI	35	4	1500000	91
Michigan Recycling	89	Kalamazoo	MI	68	7		
Mr. Rubbish Recycling & S.W. Processing Facility	91	Whitmore Lake	MI	400	1	3000000	91
Dakota County	89	Burnsville	MN	60		300000	89
Swift County	90	Benson	MN	6	0	350000	90
Invergrove Heights (BFI)	90	Invergrove Heights	MN	70	14	2800000	89
Jackson County (Fallenstein Recycling)	90	Jackson	MN	1	0	600000	89
Fillmore County	87	Preston	MN	3	1	730000	87
Ramsey County (Super Cycle MRF)	87	St. Paul	MN	100	2	450000	89
Schaap Sanitation & Recycling	91	Worthington	MN	11	1		
St. Louis (BFI)	90	St. Louis	MO	48	5	1300000	89
Mecklenberg County (FCR Charlotte, Inc.)	90	Charlotte	NC	90	2	2500000	89
GDS	90	Conover	NC	65	2	3000000	90
Turnkey Recycling/Recycle America	91	Rochester	NH	30	4	1300000	91
Hooksett (BFI)	91	Hooksett	NH	50	3	2500000	90
Gloucester County Reclamation & Recovery System	91	Blackwood Terrace	NJ	50	1		
Monmouth County (Automated Recycling Technologies)	87	Long Branch	NJ	175	9	1500000	87
Monmouth County	89	Ocean Township	NJ	290		2000000	88
Camden County	86	Camden	NJ	75		700000	86
Somerset County	87	Bridgewater	NJ	104	8	1000000	87
Silver State Disposal	91	Las Vegas	NV	1000	5	9000000	91
Chemung County	91	Elmira	NY	75		4500000	91
Montgomery Otsego Scoharie S.W. Mgmt. Authority	90	Oneonta	NY	32		1000000	90
Columbia County	89	Hudson	NY	4	6		

Facility		City	State				
Islip (New Facility)	91	Islip	NY	120	4	8400000	89
Town of Smithtown	88	Smithtown	NY	45	8	5500000	88
Lewis County	90	Lowville	NY	5	0	300000	90
Wayne County IPC	91	Arcadia	NY	24		500000	91
Hudson Baylor Corporation	83	Newburgh	NY	75	8		
St. Lawrence County Solid Waste Disposal Authority	89	Ogdenburg	NY	6	0	250000	90
Oswego County	91	Fulton	NY	159	8	4500000	91
St. Lawrence County (Waste Stream Management)	89	Potsdam	NY	15	1	1300000	89
Madison County	90	Lincoln	NY	25	1	1200000	90
Warren County	91	Queensbury	NY	13		1000000	92
Jefferson County	89	Pamelia	NY	12	0	1700000	89
Rumpke Recycling, Inc.	91	Cincinnati	OH	200	10	2000000	91
Beaver Creek (Waste Management—Suburban)	90	Kougler	OH	30	2		
Waste Management of Ohio	91	Lima	OH	14	1		
LAS Recycling	89	Youngstown	OH	300	30	3000000	89
Beaver County (Piccirilli Disposal Service)	90	Ambridge	PA	65	10	1000000	90
Centre County	89	College Township	PA	60	2	2200000	89
York County (Recycle America—Waste Management)	89	York	PA	90	7	1500000	89
Total Recycling	88	Jenner Township	PA	100	0	50000	88
Erie (Waste Management)	91	Erie	PA	50	13	1000000	90
Bucks County Satellite Facility	89	New Britain Township	PA	50	1	500000	88
Beaver County	91	Franklin Township	PA	30	5	900000	90
Recycling Works	90	Lebanon	PA	80	6		
Tri-County Recycling	90	Mars	PA	75			
American Disposal	90	Palmyra	PA	7	1	1000000	90
Grand Central Recycling	83	Pen Argyl	PA	52	5		
Lackawanna County	90	Scranton	PA	85	8	3000000	90
York Waste Disposal	89	York	PA	51	5	150000	89
Harold Pollock Recycling (Ambridge)	90	Harmony Township	PA	21			
Waste Management of North America	90	Kingsport	TN	25	2	2500000	90
Knoxville (BFI)	91	Knoxville	TN	90	5	1200000	90
ACCO Waste Paper (BFI)	86	Austin	TX	67			
Commercial Metals Company	90	Dallas	TX	19			
Houston (BFI)	75	Houston	TX		0		

TABLE 9.9 Existing and Planned MRFs, 1992 Data (*Continued*)

Name	Year	City	State	TPD processed	TPD residue	Original cost	Cost year
Operational—low-tech (*Continued*):							
Waste Reduction Systems	91	Houston	TX	75	4	1500000	91
Plano (BFI)	91	Plano	TX				
San Antonio (BFI)	75	San Antonio	TX				
Richmond (Waste Management)	91	Richmond	VA	125	1		
Seattle (Recycle America—Waste Management)	88	Seattle	WA	200	4	500000	88
City of Olympia (Exceptional Foresters)	88	Tumwater	WA	28	3	40000	88
Wisconsin Intercounty Non-Profit Recycling	84	Baraboo	WI	10	1	90000	84
Waste Management Recycling of Wisconsin	89	Milwaukee	WI	30	1	130000	89
Operational—high-tech:							
Wiregrass Rehabilitation Recycling Center	83	Dothan	AL	60	6	1800000	91
TURF/Bayshore	89	San Francisco	CA	190	8		
Housatonic Resource Recovery Authority	91	Danbury	CT	100	21	3000000	91
Automated Container Recovery, Inc.	91	Berlin	CT	420	20	2000000	91
Hartford	91	Hartford	CT	400	20	13000000	91
Groton (SECRRRA)	82	Groton	CT	34	2	500000	82
Prince George's County (Capitol Heights)	90	Prince George's County	MD	48			
Community Recycling	90	Miami	FL	200	20		
Palm Beach County (North)	91	West Palm Beach	FL	250	13	6200000	90
Wapello County	91	Ottumwa	IA	35	2	1170000	91
Groot Industries	91	McCook	IL	500	200		
Laidlaw Waste Systems	91	Schaumberg	IL	96	5	4200000	91
DuPage County (CRInc.)	91	Carol Stream	IL	155	6	9000000	89
Jet-A-Way	90	Roxbury	MA	375	8	4500000	90
Springfield	90	Springfield	MA	214	24	6650000	89
Pittsburgh (BFI)	91	Carnegie	PA	80			
Montgomery Co. (Shady Grove Road Transfer Station)	91	Gaithersburg	MD	240	18	8500000	91
Ann Arbor (Recycle Ann Arbor)	81	Ann Arbor	MI	45	1		
Rice County	91	Dundas	MN	15	0	1550000	91
Waste Management of Minnesota	89	St. Louis Park	MN	150	8		

Facility	Location	State				
Atlantic County (New Facility)	Egg Harbor Township	NJ	300	1	6700000	91
Cape May County,	Woodbine (Borough of)	NJ	225	7	5100000	89
Cumberland County	Deerfield Township	NJ	37	2	3000000	89
Distributors Recycling	Newark	NJ	250	13	900000	87
Ocean County Materials Processing Facility	Lakewood Township	NJ	350	4	6000000	89
West Paterson (WPAR)	West Paterson	NJ	150	2	1100000	88
Waste Management Recycling, Inc.	Brooklyn	NY	100	5		
Integrated Waste Services	Buffalo	NY	90	5	2100000	91
Brookhaven	Brookhaven	NY	160	16	8200000	89
New York City (East Harlem)	New York	NY	75	11	3600000	88
Karta Container & Recycling	Peekskill	NY	210	35	3000000	89
Dutchess County Resource Recovery	Poughkeepsie	NY	75	8	6000000	90
Onondaga County (RRT)	Syracuse	NY	400	8	3500000	88
Oneida-Herkimer Counties	Utica	NY	200	20	8000000	89
Westbury (New Facility)	Westbury	NY	250	5	1000000	91
Philadelphia Transfer & Recycling Center	Philadelphia	PA	80	8	1400000	89
The Forge	Philadelphia	PA	120	10	1500000	90
Bristol (Otter Recycling)	Bristol	PA	65	6	3500000	88
King of Prussia (BFI)	King of Prussia	PA	125	8	2000000	90
Dixon Recyclers MRF	Lebanon	PA	177	31		
Eagle Recycling	Nanicoke	PA	18	0	2500000	91
Todd Heller, Inc.	Northampton	PA	63	4	200000	90
Chambers Development Corporation	Pittsburgh	PA	100	20	1000000	90
Advanced Environmental Consultants	Wilkinsburg	PA	39	1		
Johnston MRF	Johnston	RI	210	27	5500000	89
Seattle (Rabanco)	Seattle	WA	300	15	6000000	88
Materials Recycling Corp. of America/Waste Mgmt.	Madison	WI	80	3		

Source: Excerpted from *1992–1993 Materials Recovery and Recycling Yearbook, Directory and Guide.* Government Advisory Associates, Inc., New York, 1992.

TABLE 9.10 MRF Owners and Operators, 1992 Data

Name	Owner	Operator
Construction—low-tech:		
Williamson County	Williamson County	Waste Management of North America
Hartford	Town of Hartford	Town of Hartford
City of Kenosha	BFI (Town & Country)	BFI (Town & Country)
Construction—high-tech:		
Rosetto Recycling Center	Rosetto Recycling Corporation	Rosetto Recycling Corporation
Town of Babylon	Solar International Trading Corp.	Solar International Trading Corp.
Shakedown—low-tech:		
Pacific Rim Recycling	Pacific Rim Recycling	Pacific Rim Recycling
Charleston County	Charleston County	Charleston County
Shakedown—high-tech:		
Elk Ridge (BFI)	BFI Recycling Systems	BFI Recycling Systems
Northern Virginia (BFI)	Browning-Ferris Industries, Inc.	Browning-Ferris Industries, Inc.
Monroe County	Monroe County	RRT/Empire Returns Corporation
Buffalo District (BFI)	BFI, Buffalo District	BFI, Buffalo District
Operational—low-tech:		
City of Mobile	City of Mobile	Goodwill Industries
Valley Recycling MRF	Valley Recycling Works, Inc.	Valley Recycling Works, Inc.
Phoenix (Why Waste America)	Why Waste America, Inc.	Why Waste America, Inc.
Tucson (Cutler Recycling)	Cutler Recycling	Cutler Recycling
Tucson (Recycle America—Waste Management)	Waste Mgmt. (Recycle America)	Waste Mgmt. (Recycle America)
San Mateo County (BFI—The Recyclery)	Browning-Ferris Industries, Inc.	Browning-Ferris Industries, Inc.
Concord Disposal	Concord Disposal	Concord Disposal
Fresno (BFI)	Browning-Ferris Industries, Inc.	Browning-Ferris Industries, Inc.
Western Waste Industries	Western Waste Industries, Inc.	Western Waste Industries, Inc.
South Valley Refuse Disposal	South Valley Refuse Disposal	San Martin Transfer Station
San Diego Recycling	EDCO Disposal Corporation	San Diego Recycling
San Jose (BFI—Newby Island)	Browning-Ferris Industries, Inc.	Browning-Ferris Industries, Inc.
Monterey City Disposal Service	Monterey City Disposal Service	Monterey City Disposal Service
Napa Garbage Service	Napa Garbage, Inc.	Napa Garbage, Inc.

Upper Valley Disposal	Upper Valley Disposal	Upper Valley Disposal
East Bay Disposal (Durham Road Landfill)	Waste Mgmt. (Oakland Scavenger)	Waste Mgmt. (Oakland Scavenger)
Fresno (City)	Waste Mgmt. of Fresno County	Waste Mgmt. of Fresno County
Waste Management of Santa Clara	Waste Management of Santa Clara	Waste Management of Santa Clara
Pleasant Hill–Bayshore Disposal (BFI)	BFI/Pleasant Hill–Bayshore Disposal	BFI/Pleasant Hill–Bayshore Disposal
Richmond Sanitary Service	Richmond Sanitary Service	Richmond Sanitary Service
Empire Waste Management	Waste Management of North America	Waste Management of North America
G.I. Industries	G.I. Industries	G.I. Industries
Pacific Rim Recycling	Pacific Rim Recycling	Pacific Rim Recycling
Gold Coast Recycling	Gold Coast Recycling	Gold Coast Recycling
North Ft. Myers	Lee County Division of S.W. Mgmt.	Lee County/Goodwill Industries
Jacksonville (BFI)	Browning-Ferris Industries, Inc.	Browning-Ferris Industries, Inc.
Dade County (Attwoods)	Attwoods (Community Recycling)	Attwoods (Community Recycling)
Orange County (Waste Management)	Waste Management of Florida	Waste Mgmt. (Recycle America)
Pinellas Park (Recycle America—Waste Management)	Waste Management of North America	Waste Management of Pinellas County
Pinellas County (BFI)	Browning-Ferris Industries, Inc.	Browning-Ferris Industries, Inc.
Leon County (Capitol Recycling)	Capitol Recycling	Capitol Recycling
Recycle America (Waste Management)	Waste Management of North America	Waste Mgmt. (Recycle America)
Macon (Industrial Recovered Materials)	Industrial Recovered Materials	Industrial Recovered Materials
Carroll County Solid Waste Management Commission	Carroll County S.W.M. Commission	Carroll County S.W.M. Commission
Northwest Iowa Solid Waste Agency	Northwest Iowa Solid Waste Agency	Northwest Iowa Solid Waste Agency
Garden City Disposal	Waste Management of North America	Waste Management of North America
Meyer Brothers Services (Waste Management)	Waste Management of North America	Meyer Brothers Services
Elgin-Wayne Recycling & Waste Services	Waste Mgmt. (Elgin-Wayne Recycling)	Waste Mgmt. (Elgin-Wayne Recycling)
Waste Management of McHenry County	Waste Management of North America	Waste Management of McHenry County
Fox Valley Disposal (Waste Management)	Waste Mgmt. (Fox Valley Disposal)	Waste Mgmt. (Fox Valley Disposal)
Best Recycling & Waste Services	Waste Management of North America	Best Recycling & Waste Services
Southeast Louisiana (Waste Management)	Waste Management of North America	Waste Management of North America
SEMASS Recycling Facility	SEMASS Recycling Management Corp.	SEMASS Recycling Management Corp.
Lowell	CRInc./Resource Control	CRInc./Resource Control
Waukesha County	Waukesha County	CRInc.
Onondaga County (Laidlaw)	Laidlaw Waste Systems, Inc.	Laidlaw Waste Systems, Inc.
Phoenix Recycling	Phoenix Recycling, Inc.	Phoenix Recycling, Inc.
Wilmington (BFI)	Browning-Ferris Industries, Inc.	Browning-Ferris Industries, Inc.

TABLE 9.10 MRF Owners and Operators, 1992 Data (*Continued*)

Name	Owner	Operator
Operational—low-tech (*Continued*):		
Hagerstown (BFI)	Browning-Ferris Industries, Inc.	Browning-Ferris Industries, Inc.
Salisbury (BFI)	Browning-Ferris Industries, Inc.	Browning-Ferris Industries, Inc.
Anne Arundel County	Anne Arundel County	Browning-Ferris Industries, Inc.
Town of Bowdoinham	Town of Bowdoinham	Town of Bowdoinham
Waste Management of Michigan (Southwest)	Waste Management of Michigan	Waste Management of Michigan
Michigan Recycling	Michigan Disposal Services	Michigan Disposal Services
Mr. Rubbish Recycling & S.W. Processing Facility	Mr. Rubbish	Mr. Rubbish
Dakota County	Dakota County	Recycle Minnesota's Resources (RMR)
Swift County	Swift County	Swift County
Invergrove Heights (BFI)	Browning-Ferris Industries, Inc.	Browning-Ferris Industries, Inc.
Jackson County (Fallenstein Recycling)	Fallenstein Refuse & Recycling	Fallenstein Refuse & Recycling
Fillmore County	Fillmore County	Fillmore County
Ramsey County (Super Cycle MRF)	Ramsey County/Super Cycle	Super Cycle
Schaap Sanitation & Recycling	Schaap Sanitation & Recycling	Schaap Sanitation & Recycling
St. Louis (BFI)	Browning-Ferris Industries, Inc.	Browning-Ferris Industries, Inc.
Mecklenberg County (FCR Charlotte, Inc.)	Fairfield County Redemption, Inc.	Fairfield County Redemption, Inc.
GDS	Garbage Disposal Service, Inc.	GDS Recycling Services
Turnkey Recycling/Recycle America	Waste Management of North America	Recycle America/Turnkey Recycling
Hooksett (BFI)	Browning-Ferris Industries, Inc.	Browning-Ferris Industries, Inc.
Gloucester County Reclamation & Recovery System	McKay Disposal	Gloucester County R & R System
Monmouth County (Automated Recycling Technologies)	Automated Recycling Technologies	Automated Recycling Technologies
Monmouth County	Monmouth Processing	Monmouth Processing
Camden County	Camden County (equipment)	Resource Recovery Systems, Inc.
Somerset County	Somerset County	Somerset County
Silver State Disposal	Silver State Disposal	Silver State Disposal
Chemung County	Chemung Co. Solid Waste Mgmt. Dist.	Chemung Co. Solid Waste Mgmt. Dist.
Montgomery Otsego Scoharie S.W. Mgmt. Authority	Montgomery Otsego Scoharie S.W.M.A.	Montgomery Otsego Scoharie S.W.M.A.
Columbia County	Columbia County	Columbia County
Islip (New Facility)	Town of Islip	Town of Islip
Town of Smithtown	Town of Smithtown	Town of Smithtown

Lewis County	Lewis County	Lewis County
Wayne County IPC	Western Finger Lakes S.W.M. Auth.	Western Finger Lakes S.W.M. Auth.
Hudson Baylor Corporation	Hudson Baylor Corporation	Hudson Baylor Corporation
St. Lawrence County Solid Waste Disposal Authority	St. Lawrence Co. S.W.D. Authority	St. Lawrence Co. S.W.D. Authority
Oswego County	Oswego County	Oswego Industries
St. Lawrence County (Waste Stream Management)	Waste Stream Management	Waste Stream Management
Madison County	Association for Retarded Citizens	Association for Retarded Citizens
Warren County	Warren County	Warren County
Jefferson County	Jefferson County	Jefferson County
Rumpke Recycling, Inc.	Rumpke Recycling, Inc.	Rumpke Recycling, Inc.
Beaver Creek (Waste Management—Suburban)	Waste Management of North America	Waste Management of Ohio
Waste Management of Ohio	Waste Management of North America	Waste Management of North America
LAS Recycling	LAS Recycling	LAS Recycling
Beaver County (Piccirilli Disposal Service)	Waste Management of North America	Piccirilli Disposal Service
Centre County	Centre County Solid Waste Authority	Centre County Solid Waste Authority
York County (Recycle America—Waste Management)	Waste Management of North America	Waste Management of North America
Total Recycling	Total Recycling	Total Recycling
Erie (Waste Management)	Waste Management of Erie	Waste Management of Erie
Bucks County Satellite Facility	Bucks County	RRT/Empire Returns Corporation
Beaver County	Franklin Township	Franklin Township
Recycling Works	Recycling Works	Recycling Works
Tri-County Recycling	Tri-County Recycling	Tri-County Recycling
American Disposal	American Disposal	American Disposal
Grand Central Recycling	Grand Central Sanitation, Inc.	Grand Central Recycling
Lackawanna County	Lackawanna County	Lackawanna County
York Waste Disposal	York Waste Disposal, Inc.	York Waste Disposal, Inc.
Harold Pollock Recycling (Ambridge)	Joseph Brunner, Inc.	Harold Pollock Recycling
Waste Management of North America	Waste Management of North America	Waste Management of North America
Knoxville (BFI)	Browning-Ferris Industries, Inc.	Browning-Ferris Industries, Inc.
ACCO Waste Paper (BFI)	BFI (ACCO Waste Paper)	BFI (ACCO Waste Paper)
Commercial Metals Company	Commercial Metals Company	Commercial Metals Company
Houston (BFI)	Browning-Ferris Industries, Inc.	Browning-Ferris Industries, Inc.
Waste Reduction Systems	Waste Reduction Systems, Inc.	Waste Reduction Systems, Inc.
Plano (BFI)	Browning-Ferris Industries, Inc.	Browning-Ferris Industries, Inc.

TABLE 9.10 MRF Owners and Operators, 1992 Data (*Continued*)

Name	Owner	Operator
Operational—low-tech (*Continued*):		
San Antonio (BFI)	Browning-Ferris Industries, Inc.	Browning-Ferris Industries, Inc.
Richmond (Waste Management)	Waste Management of Richmond	Waste Management of Richmond
Seattle (Recycle America—Waste Management)	Waste Management of North America	Waste Management of North America
City of Olympia (Exceptional Foresters)	Exceptional Foresters	Exceptional Foresters
Wisconsin Intercounty Non-Profit Recycling	WI Intercounty Non-Profit Recycling	WI Intercounty Non-Profit Recycling
Waste Management Recycling of Wisconsin	Waste Management of North America	Waste Management of North America
Operational—high-tech:		
Wiregrass Rehabilitation Recycling Center	Waste Recycling	Waste Recycling
TURF/Bayshore	Norcal Solid Waste, Inc.	West Coast Salvage & Recycling Co.
Housatonic Resource Recovery Authority	Recycling Tech, Inc.	Recycling Tech, Inc.
Automated Container Recovery, Inc.	Automated Container Recovery, Inc.	Automated Container Recovery, Inc.
Hartford	Conn. Resources Recovery Authority	RRT/Empire Returns (Cap. Recycling)
Groton (SECRRRA)	S.E. CT R.R.R. Auth./Town of Groton	Resource Recovery System, Inc.
Prince George's County (Capitol Heights)	Eagle Management	Eagle Management
Community Recycling	Attwoods, Inc.	Community Recycling
Palm Beach County (North)	S.W. Authority of Palm Beach County	RRT/Empire Returns Corporation
Wapello County	Ottumwa-Wapello Landfill Commission	Ottumwa-Wapello Landfill Commission
Groot Industries	Groot Industries	Groot Industries
Laidlaw Waste Systems	Laidlaw Waste Systems, Inc.	Laidlaw Waste Systems, Inc.
DuPage County (CRInc.)	DuPage County	CRInc.
Jet-A-Way	Jet-A-Way	Jet-A-Way
Springfield	R.R.S./State of Massachusetts	Resource Recovery Systems, Inc.
Pittsburgh (BFI)	BFI of Pennsylvania	BFI of Pennsylvania
Montgomery Co. (Shady Grove Road Transfer Station)	Montgomery County	CRInc./Wellman, Inc.
Ann Arbor (Recycle Ann Arbor)	City of Ann Arbor/Recycle Ann Arbor	Recycle Ann Arbor
Rice County	Rice County	Rice County

Waste Management of Minnesota	Waste Management of Minnesota	Waste Management of Minnesota
Atlantic County (New Facility)	Atlantic County Utilities Authority	Atlantic County Utilities Authority
Cape May County	Cape May Co. Mun. Utilities Auth.	RRT/Empire Returns Corporation
Cumberland County	Cumberland Co. Improvement Auth.	Cumberland Co. Improvement Auth.
Distributors Recycling	REI Distributors, Inc.	REI Distributors, Inc.
Ocean County Materials Processing Facility	Ocean County	RRT/Empire Returns Corporation
West Paterson (WPAR)	WPAR	WPAR
Waste Management Recycling, Inc.	Waste Management Recycling, Inc.	Waste Management Recycling, Inc.
Integrated Waste Services	Integrated Waste Services, Inc.	Integrated Waste Services, Inc.
Brookhaven	Town of Brookhaven	CRInc. (Materials Recovery of N.Y.)
New York City (East Harlem)	City of New York	Resource Recovery Systems, Inc.
Karta Container & Recycling	Karta Container & Recycling	Karta Container & Recycling
Dutchess County Resource Recovery	Dutchess Co. Res. Recovery Agency	CRInc.
Onondaga County (RRI)	RRT/Empire Returns Corporation	RRT/Empire Returns Corporation
Oneida-Herkimer Counties	Oneida-Herkimer S.W.M. Authority	Oneida-Herkimer S.W.M. Authority
Westbury (New Facility)	OMNI Recycling of Westbury, Inc.	OMNI Recycling of Westbury, Inc.
Philadelphia Transfer & Recycling Center	Waste Management of North America	Waste Management of North America
The Forge	Waste Management of North America	Waste Management of North America
Bristol (Otter Recycling)	Otter Recycling	Otter Recycling
King of Prussia (BFI)	Browning-Ferris Industries, Inc.	Browning-Ferris Industries, Inc.
Dixon Recyclers MRF	Dixon Recyclers	Dixon Recyclers
Eagle Recycling	Eagle Recycling	Eagle Recycling
Todd Heller, Inc.	Todd Heller, Inc.	Todd Heller, Inc.
Chambers Development Corporation	Chambers Development Corporation	Chambers Development Corporation
Advanced Environmental Consultants	Advanced Environmental Consultants	Advanced Environmental Consultants
Johnston MRF	R.I. Solid Waste Management Corp.	CRInc.
Seattle (Rabanco)	Rabanco, Ltd.	Rabanco, Ltd.
Materials Recycling Corp. of America/Waste Mgmt.	Waste Management of North America	Waste Management of North America

Source: Excerpted from *1992–1993 Materials Recovery and Recycling Yearbook, Directory and Guide.* Government Advisory Associates, Inc.. New York. 1992.

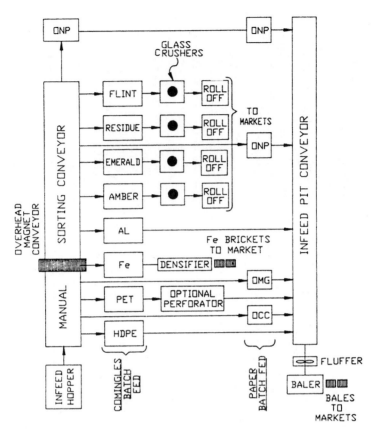

FIGURE 9.60 Typical flow diagram—small scale (25 to 100 tons per day) MRF. (*Source: Mathews, J.A., and T. C. Forbes, "A Comparison of Intermediate Processing Center Reference Designs and Costs," Proceedings of the 1992 National Waste Processing Conference, American Society of Mechanical Engineers.*)

a small magnet should be placed under the aluminum chute to drop bimetal cans into a separate hopper from the all-aluminum cans. If a ferrous magnet is utilized, the bimetal cans will automatically be sorted to go with the ferrous metal, since they will behave as magnetics. Finally, glass is sorted by color into flint, amber, and emerald green. The glass is stored in separate roll-off containers for each color. The glass may optionally be pulverized, one color at a time, followed by screening before final loading onto roll-offs. The screen oversize, consisting mostly of caps and labels, is discarded as residue. The residue is discarded in a roll-off container or is placed in a stationary compactor if the quantities are large.

Old newspaper and corrugated containers should be tipped on a separate area on the tipping floor. The tipping floor should be large enough to leave adequate room for storage, working area, and interim storage while waiting for baling of each separate paper product as well as baling of other containers once separated. Walls in the tipping area should have backstops to push and pile the paper and corrugated cardboard against so that these materials can be piled up to 12 ft high (Fig. 9.61). Sorters, standing on the floor, pull the corrugated from the other ONP and throw it aside temporari-

FIGURE 9.61 Tipping floor area dedicated to paper and corrugated containers. (*Courtesy of Mayfran International and Silver State Disposal.*)

ly. Paper bags are removed from the ONP as are any plastic bags and string. These materials are discarded as residue. The corrugated cardboard is thrown to one side while the heavier newspaper is fed into the baler. The baler is fed from a loading conveyor which is slightly, about 1 or 2 ft, below floor grade. If possible a paper picker is stationed on a platform next to the feed conveyor to positively sort out residue and contaminants from the baler feedstock just before material is fed into the baler feed hopper. It is critical that unwanted material is removed from the bale. Whole bales of paper are produced and tied off. When enough corrugated cardboard has been accumulated for a whole bale, feeding of newspaper ceases and a bale of corrugated cardboard is produced. Intermittently, the baler can be utilized to make a bale of PET, HDPE, aluminum, or ferrous metals.

Bales are picked up by a forklift (Fig. 9.62), weighed, marked with weights and number, and stacked for later loading into transfer trailers. A digital weigh scale is recommended. The bale size should allow full utilization of standard-sized enclosed trailers. Adequate space should be provided in the plant or under shelter to store between one and two trailers' worth of each product. Truck drivers delivering each product to market will not be kept waiting while the operator tries to produce the product needed to fill the truck, and it is too costly to ship a partially filled truck. Shipping mixed loads is not practical in most instances unless both products are going to the same broker or mill.

For the smaller range facilities, closer to 25 to 50 tons per day, a "bobcat" loader may be adequate for both product and infeed materials; however, it is wise to either maintain a second standby unit or have rapid access to a rental unit. It is also wise to have a backup forklift. For facilities in the 50- to 100-ton range, a small articulated frame loader such as a small Caterpillar is a better option than a bobcat because it is

FIGURE 9.62   Forklift transports and stacks bales of natural HDPE plastics.

capable of heavier service, has higher lifting capacity, and can push larger loads, especially of corrugated cardboard and paper, on the floor. It has a frame which provides greater clearance from the floor, thus avoiding wrapping of wire and strapping around the axles and various other undercarriage parts of the machine. Further, the radiator is designed to withstand heavier-duty service in high-dust areas. A swing-out radiator which can be periodically blown out with an air gun is highly desirable. Foam-filled tires are also advantageous.

High-Tech MRFs.   MRFs employing a highly mechanized process line have been developed for processing large quantities of recyclables from commingled feed streams. An example of an automated MRF is the Johnston, R.I., facility, owned by the Rhode Island Solid Waste Management Corporation (RISWMC) and designed and operated by New England CRInc (CRInc). The process is shown in Fig. 9.63. This facility was designed to process 130 tons per day of commingled recyclables received in cocollected, separate fractions of mixed paper (ONP and OCC) and mixed containers (ferrous, HDPE, PET, three colors of glass, and aluminum).

As of 1990, the facility throughput was increased to approximately 200 tons per day by operating a second shift. Mixed paper is removed from the tip floor and manually sorted on conveyors prior to baling into its constituent fractions. Commingled containers are loaded onto a computer-regulated conveyor that senses the quantity of materials fed per lineal foot in order to maintain a steady feed rate. Ascending to elevated separation stations, material is initially visually inspected for gross contaminants and hazardous materials, which are removed manually. After magnetic belts separate ferrous materials, the remaining fraction cascades downward on the conveyor and through a series of suspended metal bars that, relying on the weight, particle size, and aerodynamic differences of aluminum and plastic containers, separate them from glass. Also, because of gravity, glass continues down the line with other containers

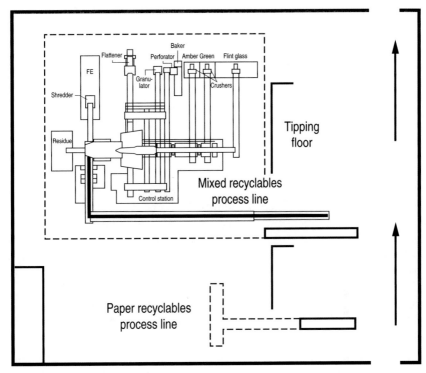

FIGURE 9.63   Johnston, R.I., MRF process plan.

diverted to either side. Glass is screened, with the overs manually sorted by color and the unders remaining as mixed cullet. Clear glass overs are negatively sorted and visually inspected to assure high quality of this most valuable glass color. Containers on the diverted line pass through an eddy-current separator to remove aluminum, and plastics are manually sorted by resin type.

Materials are prepared for market as follows. Ferrous is shredded in a flail mill (which also removes and separates the aluminum tops of bimetal cans) and is containerized in loose form. Aluminum is shipped similarly after passing through a can flattener. Glass is crushed and boxed or shipped loose in truckload quantities. PET is perforated and baled, while HDPE is shredded and shipped in gaylord-style boxes. Papers are baled. The RISWMC facility has experienced a high on-line performance. Residue, primarily mixed glass cullet from the screen unders, is estimated to be 10 percent of the daily throughput.

Large-scale, high-tech MRFs can be quite complex, as shown by the process flow diagrams for Ocean County, N.J. (Figs. 9.64 and 9.65). This type of system requires a complex, capital-intensive processing plant having substantial annual operating and maintenance costs. The proper design and ultimate success of an MRF is highly dependent on integrating the key components—facility, process equipment, and market, to cost-effectively convert solid waste into marketable products. To allow for composition fluctuations and material surges, the design capacity of a facility should be greater than its rated throughput while integrating a redundant systems approach to

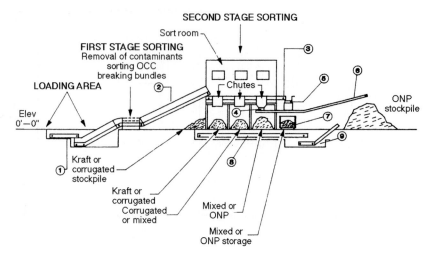

FIGURE 9.64   Paper process system, Ocean County, New Jersey. (1) Material infeed conveyor. (2) inclined-feed conveyor. (3) Paper sort conveyor. (4) Reversing transfer conveyor. (5) Reversing transfer conveyor. (6) Loose news stockpiling conveyor. (7) Paper storage conveyor. (8) Baler feed conveyor. (9) Baler incline conveyor. (*Source: Romeo, E. J., Jr., "Material Recovery Design of Ocean County, New Jersey," Proceedings of the 1992 Waste Processing Conference, American Society of Mechanical Engineers, 1992.*)

minimize downtime. Cost-effective designs must utilize space effectively while providing adequate room for maintenance, employee access, safety features, material storage, and future flexibility.[20] The proper selection of equipment to perform reliably and maintain quality levels is a key design feature in MRFs of the future. As equipment technologies and markets evolve, so will MRF designs with enhanced flexibility to process a variety of incoming materials.

Processing of Mixed Waste for Recovery of Recyclables

In a mixed waste processing program, only one trash can per collection is required. The recyclables are mixed in with all the other municipal solid waste and delivered to a mixed waste processing facility. Mixed waste processing systems are often termed "front-end processing systems" or "refuse-derived fuel (RDF) processing systems." The processing technology can involve either mechanically assisted hand separation of mixed MSW or fully automated processing of mixed MSW.

An example of a mixed waste or front-end processing system colocated with a waste-to-energy facility is shown in Fig. 9.66. This facility features a relatively low technology process that relies on manual inspection and picking of recyclable products from conveyors. It is supplemented by two-stage screening for size classification and magnetic separation of ferrous metals. Materials to be recovered from the solid waste stream include ferrous metals and aluminum; HDPE, PET, and mixed film plastics; amber, green, and flint glass; and corrugated, newsprint, and fine paper.[21]

Figure 9.67 depicts the operations at a mixed waste processing facility where, through a combination of mechanical systems and hand separation, color-sorted glass, ferrous metal, aluminum, and plastics (PET, HDPE, and film) were recovered. At this

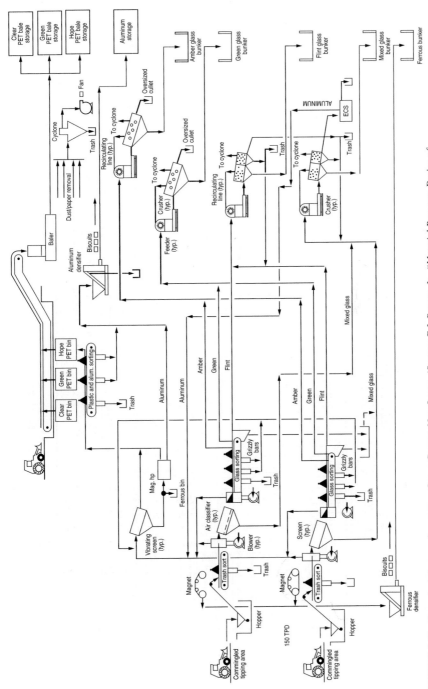

FIGURE 9.65 Commingled processing system, Ocean County, New Jersey. (*Source: E.J. Romeo, Jr., "Material Recovery Design of Ocean County, New Jersey," "Proceeding of the 1992 Waste Processing Conference, American Society of Mechanical Engineers, 1992.*)

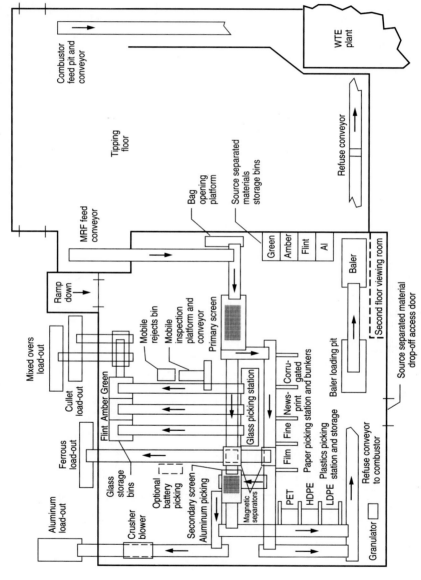

FIGURE 9.66 Mixed waste processing center (3500 tons per week).

(a)

(b)

FIGURE 9.67   Front-end processing for materials recovery. (*a*) Ferrous metal removal. (*b*) Handpicking containers. (*c*) Interior plant view, product chutes, and containers.

(c)

FIGURE 9.67   (*Continued*)

facility, mixed waste is brought in by packer truckers and dumped on the tipping floor. Conveyors carry the waste to a bag breaker. This is followed by screening of the waste on disc screens which separate the larger paper and corrugated materials from bottles, cans, and other finer material. Cardboard boxes and paper are sorted from the waste stream unless they are too contaminated by food waste, in which case they become part of the refuse-derived fuel stream. The tin cans and other ferrous metals are removed magnetically. Aluminum containers are removed either by handpicking or by eddy-current separation. Glass bottles are removed and color sorted by handpicking. Plastics are also removed by handpicking. This type of processing system requires moderate capital cost and is labor-intensive, but there are substantial savings in the collection costs since there is no need for separate collection of recyclables.

In fully automated processing of MSW for recovery of recyclables, the capital costs are quite high, but so are the processing rates. Although recently (1993) shut down after 10 years of operation, the Delaware Reclamation Project (DRP) serves as an example of a fully automated 1000 ton per day facility.

Although these types of systems can be used to recover paper, they are normally used to produce a refuse-derived fuel. In RDF plants, unlike mass burn facilities, materials such as glass, aluminum, and steel are separated from the waste, leaving only a combustible fuel to be burned. The RDF, if properly prepared, can be burned in existing boilers along with pulverized coal or alternatively fired by itself in RDF boilers as is done in both Akron and Columbus, Ohio and elsewhere.

Figure 9.68 depicts process operations. From the tipping floor, waste is conveyed to one of two vertical shredders where bags are opened and contents liberated. An air classifier then separates heavier glass and metals from the lighter paper and plastic films. The heavier material from the air classifiers is magnetically separated to remove the ferrous product. A rotary trommel is used to screen the remaining waste to separate out glass and aluminum. Aluminum and other nonferrous metals are then removed

(a)

(b)

FIGURE 9.68 Fully automated mixed waste processing. (*a*) Tipping floor. (*b*) Shredder. (*c*) Belt magnet. (*d*) Froth flotation cells.

(c)

(d)

FIGURE 9.68   (*Continued*)

using eddy-current separators. The glass, which is very fine in size, is screened out of the MSW, and contamination such as grass and food waste is floated away using a mineral jig. Froth flotation is then used to separate pure glass away from bricks, rocks, and other contamination. A reagent is added to the glass which selectively changes its surface tension. Bubbles are blown into the mixer. The glass attaches itself to the bubbles and floats away from the other contaminants. The final product is a mixed-glass cullet which can be used to produce glass wool insulation or in the manufacture of amber or green glass bottles. The materials which remain after the production of RDF and the removal of aluminum, ferrous metals, and glass are mixed with sewage sludge to make compost. Such fully automated projects are highly capital-intensive and require a large volume of waste. Further, they are expensive to operate and maintain. They should be considered only on a large-scale or regional basis.

9.5   PRODUCTS AND MARKETS

Large piles of collected and processed recyclables are merely solid waste unless someone wants to buy and use them. Thus true "recycling" occurs only at a mill or factory where secondary materials are used to make new products. The key, then, is to market recyclables to consuming companies.

Recycling Rates

The recycling market in the United States exceeds $15 billion in sales annually. There are nearly 100 glass container production facilities that want old bottles. Ferrous scrap can go to steel mills in nearly every state. More than 200 U.S. paper and paperboard mills rely solely on waste paper for their feedstock.[22]

We are using more and more recyclables every year. In 1992, over 32 percent of the fiber needs of American paper and paperboard producers came from waste paper. Sixty-eight percent of aluminum beverage cans weighing 1.07 million tons were recycled in the United States. Glass container recycling is reaching 33 percent of production, according to the Glass Packaging Institute. That rate, up 2 percent from the previous year, includes containers recycled back into containers, reused bottles, and glass used in secondary markets such as roadbed.[23]

More than 1 million tons of steel cans (food, beverage, and other) were recycled in 1992 for a national recycling rate of 40.9 percent, up 34 percent from 1991, according to the Steel Can Recycling Institute. The higher recycling rate was attributed to curbside recycling programs primarily, and also magnetic separation at waste-to-energy plants. Twenty-seven percent of all postconsumer PET plastic containers, including soft drink bottles, was recycled in 1992, a 3 percent increase from 1991.[23] In addition, we ship more recyclables to foreign consumers than any other country. For instance, nearly 20 percent of the waste paper collected each year in the United States (about 4 million tons) goes offshore to foreign mills, particularly those located in Japan, South Korea, and Taiwan.[22]

Even though the recycling industry is a major economic sector here and abroad, the recent rush to collect recyclables has had disastrous market results. With the mistaken belief that collected materials could be sold easily, programs were launched without secure market arrangements. For example, a massive glut of old newspapers from curbside recycling programs caused the worldwide market to crash in 1989. Many communities had to pay to get rid of their waste paper and some abandoned their collection efforts, waiting for the market to rebound. While we should not give up in achieving high recycling rates, we should proceed cautiously and be sure that added increases in recycling collection levels actually achieve increases in tons recycled.

A mixture of newspaper, plastic-coated corrugated boxboard, envelopes containing cellophane windows, magazines containing high-gloss clay fillers, and laser printed kraft paper may be useless for anything other than the liner of a pizza box or the backing of a writing tablet. We certainly cannot supply nearly 50 percent of our entire waste stream, which paper represents, in this limited application for recycled paperboard. Thus separation and segregation may be important criteria necessary to meet aggressive recycling objectives (e.g., 50 percent of all paper contained in our waste).

Relationship between Recyclable Composition and Market Demand

The materials which are in our waste stream are multicomponent materials. A juice container may be plastic-lined and contain layers of aluminum sandwiched between paper fiber, making recovery of any one element of its overall composition quite complex. Further, there are so many grades and types of paper and plastic that, when combined in a mixture, the mixture may be nonrecyclable even though each component individually may be recyclable. For example, there is no market demand for "paper" as such. The demand is for a particular grade of paper. According to The Paper Stock Institute, there are 51 *grades* of paper,[24] with some of the major recyclable categories being old newspaper (ONP), old corrugated containers (OCC), and brown kraft grocery bags. If all of these grades are mixed together as "paper," the product will not be marketable even though there may be an unsaturated demand for the individual components.

From this example, we may infer a recycling mixing principle. One part of recyclable material No. 1 plus one part of recyclable material No. 2 produces 2 parts of lower-value recyclable material or alternative equals municipal solid waste. It is the exception to the rule, and very unlikely, that the combined value of the mixture will equal the combined value of the parts. In the case of plastics, a strong market exists for PET bottles. A market also exists for polyvinyl chloride (PVC) bottles. But no market exists for an intimate mixture of PET and PVC. A consistent, though weaker market exists for natural HDPE fractional melt index bottles (e.g., milk bottles). A still weaker market exists for mixed color, high melt index HDPE bottles (e.g., laundry detergent, bleach). Even though both plastics are HDPE, little or no market exists for an intimate mixture of high and low melt index HDPE.

There is a strong market for clear (flint) bottle glass, and for postconsumer green (emerald) bottles. A strong market exists for recycled window or plate glass. Virtually no market exists for a combination of these types of glass. Even though they are all glass, they must be segregated into their various grades in order for demand to exist. Why do purchasers want bottles delivered unbroken? So that they can tell whether there is contamination with other grades of glass before melting.

The key to securing strong, high-value markets for recyclable materials is quality, not quantity. It is necessary to segregate the materials into their various components or grades with minimal contamination and to produce products of consistent quality. Whenever industry representatives speak to citizens about obstacles to recycling, they constantly caution that it is necessary to keep supply and demand in balance and that the key to recycling is developing sufficient markets and demand for the finished products. However, when comparing supply and demand, an inherent assumption usually exists regarding the product meeting *existing* material specifications designed around supply from virgin feedstock. These specifications assume a level of quality which frequently is not achieved in actual recycling practice. Normally, an industry representative thinks of recycled materials not in terms of a broad general material type, but rather as a single grade of material meeting a predefined product specification.

> The greatest danger to the success of recycling is the inappropriate generation of raw materials. There must be, in place, sufficient end-use demand and conversion capacity to absorb the raw materials. To provide the proper basic incentives to invest tens of millions of dollars in a high capacity project, there must be dependable sources of quality raw materials available in appropriate quantities at the same time that end-use demand is sufficient to support conversion. The absence of this balance has limited the general availability of recycled material.[25]

When we talk about demand for recycled materials, too often there is little or no discussion regarding the particular grade of product under consideration and the relationship between product purity and industry demand. However, this market differentiation is a *critical factor* in assessing the demand and value of the product.

Sample Product Specifications

The following are sample product specifications.[21] They are considered to be typical of that required by end users.

Newsprint. Newsprint shall be separated from all nonpaper products and baled so as to be suitable for overseas export. The density of the bales shall be approximately 25 lb/ft³, yielding an average weight of 1100 lb per bale. Nonnewsprint contamination is limited to a maximum of 2 percent "outthrows paper" and "prohibitive material" as defined by the Paper Stock Institute of America (PS-86) *Special News* No. 7. The newsprint bale should consist of baled, sorted, fresh, dry newspaper, not sunburned and free from paper other than news, containing not more than the percentage of rotogravure and colored sections normally contained in newspaper delivered to the household.

Glass Cullet. The glass product shall be separated from noncontainer-type glass material with the exception of paper labels. The glass shall be segregated by color (amber, flint, and green) prior to crushing. The cullet size shall be greater than 0.25 in in diameter and less than 2.0 in in diameter. Flint cullet shall contain not more than 5 percent other glass colors by weight, amber cullet shall contain not more than 5 percent other glass colors by weight, and green cullet shall contain not more than 5 percent other glass colors by weight. No stones, ceramics, or noncontainer glass shall be contained in the outbound product. Nonglass contaminants shall not exceed 1 percent of the total product weight.

Aluminum. Aluminum used beverage containers (UBCs) shall be separated from all nonaluminum and other aluminum material and baled. All nonaluminum contamination, including moisture, shall be less than 1.5 percent of total product weight. Minimum bale density shall be 20 lb/ft³. The other aluminum should be separated into cast and foils fractions and shipped loose in palletized gaylords (as a minimum).

Tin-Plated Steel Cans. Tin-plated steel cans shall be separated from all other material and shredded. The cans, initially up to 1 gal in size, shall be shredded to a maximum dimension of 2 in and a minimum density of 65 lb/ft³. Nontin plated steel can contamination (including foil, food, aluminum, labels, and plastic) shall be less than 2 percent of total product weight.

PET Plastic.   PET plastic shall be separated from all nonplastic material and further sorted from high-density polyethylene prior to perforation and baling. All PET beverage bottles shall be perforated and baled to a minimum density of 20 lb/ft^3. Contamination of all non-PET beverage bottle material shall be less than 3 percent by weight of total product weight.

HDPE Plastic.   HDPE plastic translucent "milk jug-type" containers shall be separated from all nonplastic material and further sorted from other plastic prior to baling. Colored HDPE content shall not exceed 10 percent by weight. Non-HDPE and nonplastic contamination shall not exceed 1.0 percent by weight.

Mixed Rigid Plastic.   Mixed rigid plastic containers shall be separated from all nonplastic material and from PET and translucent HDPE prior to perforation and baling. Contamination of all nonplastic material shall be less than 3 percent by total product weight.

Product Recovery and Markets

Table 9.11 presents data on the primary materials which are being recovered or are planned to be recovered at operating and planned U.S. MRFs.[26] These materials and the percentages of the facilities reported to recover them are summarized in Table 9.12. Successful MRFs are highly responsive to location-specific needs and especially to the requirements of the markets. Recognizing the lack of design standardization and the material-specific, end-use specifications, the following description of recovery techniques is presented on a material-specific basis.

Paper.   Old newspaper (ONP), old corrugated cardboard (OCC), high-grade office paper, mixed paper, and specialty cellulosic materials can be recycled for a variety of uses. To make use of recycled paper, manufacturers usually must employ specialized equipment to repulp, remove ink and other contaminants, screen, and otherwise refine fiber for mixing with virgin feedstock.[27] Certain high-grade office papers can be remixed directly and therefore command a higher secondary market price than commodity-grade ONP and OCC. Specifications for grades of waste paper are well developed with guidelines for numerous grades of used paper stock. These specifications focus on percentages of "prohibitive materials" and "outthrows" for any contaminants that render the recyclable paper unusable in reprocessing. Depending on the reprocessor's needs, paper is sold baled or loose.

Paper Recovery.   To prevent contamination from glass, moisture, and beverage and food residue, source separation of paper is the preferred method. Even in source separation of commingled recyclables, paper is best recycled if separated from the remaining fraction. An MRF processing capability affords a program the opportunity to collect more than one grade of paper in its paper fraction. Incoming mixed papers would typically be isolated on the MRF tipping floor and pushed onto a box conveyor for manual picking by paper grade. Paper grades then would be baled or containerized (e.g., truckload, container, shrinkwrap) for shipment. Typical problems encountered in mixed paper separation include cross contamination or moisture in the material from exposure to precipitation at the curbside. Separation of paper grades from totally commingled recycling streams conceptually is less effective owing to the risk of residue contamination. If necessary from a collection standpoint, manual sorting on a conveyor is the preferred method.

Issues Affecting the Market for Newspaper.   While it is recognized that many finished grades of secondary (recycled) fiber are the result of hand sorting raw materials, a distinct paper stock grade must be composed primarily of a particular type of paper feedstock and cannot contain more than the specified maximum quantity of "prohibitive materials" (materials which would make the paper unusable or which would damage the processing equipment) and "outthrows" (materials which are unsuitable for that grade of paper). There are actually four different grades of newspaper: news, grade 6; special news, grade 7; special news deink quality, grade 8; and overissue news, grade 9.

News consists of baled newspapers containing less than 5 percent of other *papers.* Prohibitive materials may not exceed 0.5 percent and total outthrows may not exceed 2 percent.

Special news consists of baled, sorted, fresh dry newspapers, not sunburned, free from paper other than news, containing not more than the normal percentage of rotogravure and colored sections. No prohibitive materials may be present and total outthrows may not exceed 2 percent.

Special news deink quality consists of baled, sorted, fresh dry newspapers, not sunburned, free from magazines, white blank, pressroom overissues, and paper other than news, containing not more than the normal percentage of rotogravure and colored sections. The packing must be free from tare. No prohibitive materials are permitted and total outthrows may not exceed 0.25 percent.

Overissue news consists of unused, overrun regular newspapers printed on newsprint, baled or securely tied in bundles, containing not more than the normal percentage of rotogravure and colored sections. No prohibitive materials and no outthrows are permitted.

The actual contamination levels which are acceptable is a function of the unique requirements of each particular customer for the recycled paper product. But generally each purchaser adheres fairly closely to the industry guidelines. Putting $\frac{1}{2}$ lb of a quick-service restaurant's "recyclable" unbleached bags into every 100 lb of the highest two grades of "news" would probably be cause for rejection of the entire load. It would certainly be cause for downgrading the value. By including with these bags the occasional paper napkin and sandwich wrapper that might be left inside the bag because they too were considered recyclable could be cause for rejection for any grade of news, even the lowest grade, "mixed news." Thus, while the bag may contain recycled material and itself be recyclable, is it really recycled? What is the grade of paper which will take the unbleached bag, and are there collection and processing systems through which the consumer can place the bag to ensure that it is actually recycled?

Let us look at the brown grocery bags. Grade 14, used brown kraft, consists of baled brown kraft bags free of objectionable liners or contents. No prohibited materials can be present and outthrows must be less than 0.5 percent. Grocery bag waste, grade 19, consists of baled, new brown kraft bag cuttings, sheets and misprinted bags; no prohibitive materials permitted. There does not appear to be a place for the unbleached take-out bag in these grades either. Of course, a few bags here and there would be no problem. It might be counted with the outthrows, but as the effort of recycling becomes more successful and more people participate, the level of contamination rises, and the quantity of brown bags in the news category becomes greater. The contamination problem becomes more serious and the potential for rejection or downgrading of the feedstock becomes more likely.

The issue here is that there may be substantial demand for clean, segregated product, but there is little or no demand for unsegregated, dirty product. If all 51 grades were blended together, the paper would be unmarketable. If one grade, say No. 1 bleached cup stock, were contaminated with recycled cups of the same grade, but

TABLE 9.11 MRF Products

Name	TPD processed	News-paper	Computer paper	Office paper	Card-board
Construction—low-tech:					
Williamson County	100	Y		Y	Y
Hartford	25	Y		Y	Y
City of Kenosha	23	Y			Y
Construction—high-tech:					
Rosetto Recycling Center	250	Y	Y	Y	Y
Town of Babylon	392		Y	Y	Y
Shakedown—low-tech:					
Pacific Rim Recycling	100	Y	Y	Y	Y
Charleston County	80	Y			
Shakedown—low-tech:					
Elk Ridge (BFI)	150	Y	Y	Y	Y
Northern Virginia (BFI)	125	Y	Y	Y	Y
Monroe County	280	Y			Y
Buffalo District (BFI)	50	Y			Y
Operational—low-tech:					
City of Mobile	3	Y			Y
Valley Recycling MRF	11	Y			Y
Phoenix (Why Waste America)	125	Y	Y	Y	Y
Tucson (Cutler Recycling)	25	Y			Y
Tucson (Recycle America—					
Waste Management)	100	Y		Y	Y
San Mateo County (BFI—The Recyclery)	120	Y	Y	Y	Y
Concord Disposal		Y		Y	Y
Fresno (BFI)	13	Y		Y	
Western Waste Industries	30	Y			
Western Waste Industries	65	Y			
South Valley Refuse Disposal	5	Y			
San Diego Recycling	225	Y	Y	Y	Y
San Jose (BFI—Newby Island)	200	Y	Y	Y	Y
Monterey City Disposal Service	10	Y	Y	Y	Y
Napa Garbage Service	45	Y	Y	Y	Y
Upper Valley Disposal	3	Y		Y	Y
East Bay Disposal (Durham Road Landfill)	55	Y			
Fresno (City)	210	Y			
Waste Management of Santa Clara	120	Y		Y	Y
Pleasant Hill–Bayshore Disposal (BFI)	47	Y			Y
Richmond Sanitary Service	30	Y	Y	Y	Y
Empire Waste Management	120	Y	Y	Y	Y
G.I. Industries	70	Y		Y	Y
Pacific Rim Recycling	21	Y	Y	Y	Y
Gold Coast Recycling	100	Y	Y	Y	Y
North Ft. Myers	200	Y			
Jacksonville (BFI)	110	Y		Y	Y
Dade County (Attwoods)	200	Y	Y	Y	Y
Orange County (Waste Management)	100	Y			Y
Pinellas Park (Recycle America—Waste					
Management)	175	Y	Y	Y	Y
Pinellas County (BFI)	160	Y		Y	Y
Leon County (Capitol Recycling)	22	Y			

Clear glass	Brown glass	Green glass	Mixed glass	Tin	Aluminum	Bi-metal	PET	HDPE	Other plastic
Y	Y	Y		Y	Y	Y	Y	Y	Y
Y	Y	Y		Y	Y	Y	Y	Y	
Y	Y	Y	Y	Y	Y		Y	Y	
Y	Y	Y		Y	Y	Y	Y	Y	Y
Y	Y	Y		Y	Y	Y	Y	Y	
Y	Y	Y	Y	Y	Y	Y	Y	Y	
Y	Y	Y		Y	Y		Y	Y	
Y	Y	Y		Y	Y	Y	Y	Y	
Y	Y	Y		Y	Y	Y	Y	Y	
Y	Y	Y	Y	Y	Y	Y	Y	Y	
Y	Y	Y	Y	Y	Y	Y	Y	Y	Y
					Y		Y	Y	
Y	Y	Y	Y	Y	Y		Y	Y	
Y	Y	Y		Y	Y	Y	Y	Y	Y
Y	Y	Y		Y	Y		Y	Y	
Y	Y	Y		Y	Y	Y	Y	Y	
Y	Y	Y		Y	Y	Y	Y	Y	
Y	Y	Y		Y	Y	Y	Y	Y	
Y	Y	Y		Y	Y	Y	Y	Y	Y
Y	Y	Y	Y	Y	Y	Y	Y	Y	
Y	Y	Y	Y	Y	Y	Y	Y	Y	
Y	Y	Y		Y	Y	Y	Y	Y	Y
Y	Y	Y	Y	Y	Y	Y	Y	Y	Y
Y	Y	Y	Y	Y	Y	Y	Y	Y	Y
Y	Y	Y		Y	Y	Y	Y	Y	Y
Y	Y	Y		Y	Y	Y	Y	Y	
Y	Y	Y		Y	Y	Y	Y	Y	Y
Y	Y	Y		Y	Y	Y	Y	Y	
Y	Y	Y		Y	Y		Y	Y	
Y	Y	Y		Y	Y		Y	Y	Y
Y	Y	Y		Y	Y	Y	Y	Y	
Y	Y	Y	Y	Y	Y	Y	Y	Y	
Y	Y	Y	Y	Y	Y	Y	Y	Y	Y
Y	Y	Y	Y	Y	Y	Y	Y	Y	Y
Y	Y	Y	Y	Y	Y	Y	Y	Y	
Y	Y	Y	Y	Y	Y	Y	Y	Y	
Y	Y	Y		Ti	Y		Y	Y	
Y	Y	Y	Y	Y	Y		Y	Y	
Y	Y	Y		Y	Y	Y	Y	Y	Y
Y	Y	Y		Y	Y	Y	Y	Y	
Y	Y	Y		Y	Y		Y	Y	
Y	Y	Y		Y	Y	Y	Y	Y	Y
Y	Y	Y	Y	Y	Y	Y	Y	Y	

TABLE 9.11 MRF Products (*Continued*)

Name	TPD processed	News-paper	Computer paper	Office paper	Card-board
Operational—low-tech (*Continued*):					
Recycle America (Waste Management)	70	Y		Y	Y
Macon (Industrial Recovered Materials	400	Y	Y	Y	Y
Carroll County Solid Waste Management					
Commission	9	Y			Y
Northwest Iowa Solid Waste Agency	6	Y		Y	Y
Garden City Disposal	34	Y	Y	Y	Y
Meyer Brothers Services (Waste Management)	28	Y			
Elgin-Wayne Recycling & Waste Services	17	Y	Y	Y	Y
Waste Management of McHenry County	9	Y			Y
Fox Valley Disposal (Waste Management)	30	Y		Y	
Best Recycling & Waste Services	65	Y			
Southeast Louisiana (Waste Management)	70	Y	Y	Y	Y
SEMASS Recycling Facility	120	Y		Y	Y
Lowell	25				
Waukesha County	40	Y		Y	Y
Onondaga County (Laidlaw)	300	Y	Y	Y	Y
Phoenix Recycling	100	Y			Y
Wilmington (BFI)	40	Y	Y	Y	Y
Hagerstown (BFI)	40	Y			Y
Salisbury (BFI)	40	Y	Y	Y	Y
Anne Arundel County	37	Y			Y
Town of Bowdoinham	2	Y			Y
Waste Management of Michigan (Southwest)	35	Y			Y
Michigan Recycling	68	Y		Y	Y
Mr. Rubbish Recycling & S.W. Processing					
Facility	400	Y			Y
Dakota County	60	Y		Y	Y
Swift County	6	Y		Y	Y
Invergrove Heights (BFI)	70	Y	Y	Y	Y
Jackson County (Fallenstein Recycling)	1	Y		Y	Y
Fillmore County	3	Y		Y	Y
Ramsey County (Super Cycle MRF)	100	Y	Y	Y	Y
Schaap Sanitation & Recycling	11	Y		Y	Y
St. Louis (BFI)	48	Y	Y		Y
Mecklenberg County (FCR Charlotte, Inc.)	90	Y			
GDS	65	Y	Y	Y	Y
Turnkey Recycling/Recycle America	30	Y	Y	Y	Y
Hooksett (BFI)	50	Y	Y	Y	Y
Gloucester County Reclamation & Recovery					
System	50	Y		Y	Y
Monmouth County (Automated Recycling					
Technologies)	175				
Monmouth County	290	Y		Y	Y
Camden County	75				
Somerset County	104	Y		Y	Y
Silver State Disposal	1000	Y		Y	Y
Chemung County	75	Y		Y	Y
Montgomery Otsego Scoharie S.W. Mgmt.					
Authority	32	Y	Y	Y	Y
Columbia County	4	Y			Y

Clear glass	Brown glass	Green glass	Mixed glass	Tin	Aluminum	Bi-metal	PET	HDPE	Other plastic
Y	Y	Y	Y	Y	Y	Y	Y	Y	
Y	Y	Y	Y	Y	Y	Y	Y	Y	
Y	Y	Y		Y	Y	Y	Y	Y	
Y	Y	Y		Y	Y	Y	Y	Y	
Y	Y	Y		Y	Y	Y	Y	Y	
Y	Y	Y	Y	Y	Y	Y	Y	Y	
Y	Y	Y		Y	Y	Y	Y	Y	
Y	Y	Y		Y	Y	Y	Y	Y	
Y	Y	Y		Y	Y	Y		Y	
Y	Y	Y		Y	Y	Y	Y	Y	
Y	Y	Y		Y	Y	Y	Y	Y	Y
Y	Y	Y	Y	Y	Y	Y	Y	Y	Y
Y	Y	Y		Y	Y	Y	Y	Y	
Y	Y	Y		Y	Y	Y	Y	Y	
Y	Y	Y		Y	Y	Y	Y	Y	Y
Y	Y	Y	Y	Y	Y	Y	Y	Y	
Y	Y	Y		Y	Y	Y	Y	Y	
Y	Y	Y		Y	Y	Y	Y	Y	
Y	Y	Y		Y	Y	Y	Y	Y	
Y	Y	Y		Y	Y	Y	Y	Y	
Y	Y	Y		Y	Y	Y	Y	Y	Y
Y	Y	Y		Y	Y	Y	Y	Y	Y
Y	Y	Y	Y	Y	Y	Y		Y	Y
Y	Y			Y	Y		Y	Y	Y
Y	Y	Y		Y	Y	Y	Y	Y	
Y	Y	Y	Y	Y	Y		Y	Y	Y
Y	Y	Y		Y	Y	Y	Y	Y	Y
Y	Y	Y		Y	Y		Y	Y	Y
Y	Y	Y		Y	Y	Y	Y	Y	
Y	Y	Y		Y	Y	Y	Y	Y	Y
Y	Y	Y	Y	Y	Y	Y	Y	Y	Y
Y	Y	Y		Y	Y		Y	Y	
Y	Y	Y	Y	Y	Y	Y	Y	Y	
Y	Y	Y		Y	Y	Y	Y	Y	Y
Y	Y	Y		Y	Y	Y	Y	Y	
Y	Y	Y		Y	Y	Y	Y	Y	
Y	Y	Y	Y	Y	Y	Y	Y	Y	
Y	Y	Y		Y	Y	Y	Y	Y	
Y	Y	Y		Y	Y	Y	Y	Y	
Y	Y	Y		Y	Y	Y	Y	Y	
Y	Y	Y	Y	Y	Y	Y	Y	Y	Y
Y	Y	Y	Y	Y	Y	Y	Y	Y	
Y	Y	Y		Y	Y	Y	Y	Y	
Y	Y	Y			Y	Y	Y	Y	Y
Y	Y	Y	Y	Y	Y	Y	Y	Y	

TABLE 9.11 MRF Products (*Continued*)

Name	TPD processed	News-paper	Computer paper	Office paper	Card-board
Operational—low-tech (*Continued*):					
Islip (New Facility)	120	Y			Y
Town of Smithtown	45	Y			Y
Lewis County	5	Y			Y
Wayne County IPC	24	Y			Y
Hudson Baylor Corporation	75				Y
St. Lawrence County Solid Waste Disposal Authority	6	Y			Y
Oswego County	159	Y			Y
St. Lawrence County (Waste Stream Management)	15	Y	Y	Y	Y
Madison County	25	Y	Y	Y	Y
Warren County	13	Y			Y
Jefferson County	12	Y		Y	Y
Rumpke Recycling, Inc.	200	Y	Y	Y	Y
Beaver Creek (Waste Management—Suburban)	30	Y		Y	Y
Waste Management of Ohio	14	Y			Y
LAS Recycling	300	Y	Y	Y	Y
Beaver County (Piccirilli Disposal Service)	65		Y	Y	Y
Centre County	60	Y		Y	Y
York County (Recycle America—Waste Management)	90	Y			Y
Total Recycling	100	Y	Y	Y	Y
Erie (Waste Management)	50		Y	Y	Y
Bucks County Satellite Facility	50	Y			
Beaver County	30	Y	Y	Y	Y
Recycling Works	80	Y		Y	Y
Tri-County Recycling	75			Y	Y
American Disposal	7	Y		Y	Y
Grand Central Recycling	52	Y		Y	Y
Lackawanna County	85	Y			Y
York Waste Disposal	51	Y			Y
Harold Pollock Recycling (Ambridge)	21		Y	Y	
Waste Management of North America	25	Y	Y	Y	Y
Knoxville (BFI)	90	Y		Y	Y
ACCO Waste Paper (BFI)	67	Y			
Commercial Metals Company	19	Y			
Houston (BFI)		Y	Y	Y	Y
Waste Reduction Systems	75	Y			Y
Plano (BFI)		Y	Y	Y	Y
San Antonio (BFI)		Y	Y	Y	Y
Richmond (Waste Management)	125	Y	Y	Y	Y
Seattle (Recycle America—Waste Management)	200	Y			
City of Olympia (Exceptional Foresters)	28	Y		Y	
Wisconsin Intercounty Non-Profit Recycling	10	Y		Y	Y
Waste Management Recycling of Wisconsin	30	Y			Y
Operational—high-tech:					
Wiregrass Rehabilitation Recycling Center	60	Y	Y	Y	Y
TURF/Bayshore	190	Y		Y	Y

Clear glass	Brown glass	Green glass	Mixed glass	Tin	Aluminum	Bi-metal	PET	HDPE	Other plastic
Y	Y	Y	Y	Y	Y	Y	Y	Y	
Y	Y	Y	Y	Y	Y	Y	Y	Y	
Y	Y	Y		Y	Y	Y	Y	Y	
Y	Y			Y	Y	Y		Y	
Y	Y	Y	Y	Y	Y	Y	Y	Y	
Y	Y	Y	Y	Y	Y	Y	Y	Y	
Y	Y	Y		Y	Y	Y		Y	
Y	Y	Y		Y	Y	Y	Y	Y	
Y	Y	Y		Y	Y	Y	Y	Y	
Y	Y	Y		Y	Y		Y	Y	
Y	Y	Y		Y	Y	Y	Y	Y	Y
Y	Y	Y	Y	Y		Y	Y	Y	Y
Y	Y	Y	Y	Y	Y	Y	Y	Y	
Y	Y	Y	Y	Y	Y	Y	Y	Y	
Y	Y	Y		Y	Y	Y	Y	Y	Y
Y	Y	Y		Y	Y	Y	Y	Y	
Y	Y	Y		Y	Y	Y	Y	Y	
Y	Y	Y	Y	Y	Y	Y	Y	Y	
Y	Y	Y		Y	Y	Y	Y	Y	
Y	Y	Y	Y	Y	Y	Y	Y	Y	
Y					Y				
Y	Y	Y		Y	Y	Y	Y	Y	Y
Y	Y	Y		Y	Y	Y	Y	Y	
Y	Y	Y		Y	Y	Y	Y	Y	
Y	Y	Y		Y	Y	Y	Y	Y	
Y	Y	Y		Y	Y	Y	Y	Y	Y
Y	Y	Y		Y	Y	Y	Y	Y	Y
Y	Y	Y	Y	Y	Y	Y	Y	Y	
Y	Y	Y		Y	Y	Y	Y	Y	
Y	Y	Y		Y	Y		Y	Y	
Y	Y	Y		Y	Y	Y	Y	Y	
Y	Y	Y	Y	Y	Y				
Y	Y	Y		Y	Y	Y	Y	Y	
				Y	Y	Y	Y	Y	
				Y	Y	Y	Y	Y	
Y	Y	Y		Y	Y	Y	Y	Y	Y
Y	Y	Y	Y	Y	Y	Y	Y	Y	
Y	Y	Y		Y	Y	Y	Y	Y	
Y	Y	Y		Y	Y	Y	Y	Y	
Y	Y	Y	Y	Y	Y				
Y	Y	Y		Y	Y	Y	Y	Y	Y
Y	Y	Y	Y	Y	Y	Y	Y	Y	
Y	Y	Y		Y	Y	Y			
Y	Y	Y		Y	Y	Y	Y		

TABLE 9.11 MRF Products (*Continued*)

Name	TPD processed	News-paper	Computer paper	Office paper	Card-board
Operational—high-tech (*Continued*):					
Housatonic Resource Recovery Authority	100	Y		Y	Y
Automated Container Recovery, Inc.	420	Y		Y	Y
Hartford	400	Y			Y
Groton (SECRRRA)	34				
Prince George's County (Capitol Heights)	48	Y			Y
Community Recycling	200	Y			
Palm Beach County (North)	250	Y			Y
Wapello County	35	Y			Y
Groot Industries	500	Y	Y	Y	Y
Laidlaw Waste Systems	96	Y	Y	Y	Y
DuPage County (CRInc.)	155	Y			Y
Jet-A-Way	375	Y	Y	Y	Y
Springfield	214	Y			Y
Pittsburgh (BFI)	80	Y	Y	Y	Y
Montgomery Co. (Shady Grove Road Transfer Station)	240	Y			
Ann Arbor (Recycle Ann Arbor)	45	Y		Y	Y
Rice County	15	Y		Y	Y
Waste Management of Minnesota	150	Y		Y	Y
Atlantic County (New Facility)	300	Y			Y
Cape May County	225	Y	Y	Y	Y
Cumberland County	37	Y			
Distributors Recycling	250				
Ocean County Materials Processing Facility	350	Y			Y
West Paterson (WPAR)	150				
Waste Management Recycling, Inc.	100				
Integrated Waste Services	90	Y	Y	Y	Y
Brookhaven	160	Y		Y	Y
New York City (East Harlem)	75	Y			
Karta Container & Recycling	210	Y	Y	Y	Y
Dutchess County Resource Recovery	75	Y			
Onondaga County (RRT)	400	Y		Y	Y
Oneida-Herkimer Counties	200	Y	Y	Y	Y
Westbury (New Facility)	250				
Philadelphia Transfer & Recycling Center	80	Y			
The Forge	120	Y			Y
Bristol (Otter Recycling)	65				
King of Prussia (BFI)	125	Y		Y	Y
Dixon Recyclers MRF	177	Y			Y
Eagle Recycling	18	Y	Y	Y	Y
Todd Heller, Inc.	63				
Chambers Development Corporation	100	Y		Y	Y
Advanced Environmental Consultants	39				Y
Johnston MRF	210	Y			
Seattle (Rabanco)	300	Y			Y
Materials Recycling Corp. of America/ Waste Mgmt.	80	Y			Y

Source: Excerpted from *1992–1993 Materials Recovery and Recycling Yearbook, Directory and Guide.* Governmental Advisory Associates, Inc., New York, NY, 1992.

Clear glass	Brown glass	Green glass	Mixed glass	Tin	Aluminum	Bi-metal	PET	HDPE	Other plastic
Y	Y	Y		Y	Y	Y	Y	Y	
Y	Y	Y		Y	Y	Y			
Y	Y	Y	Y	Y	Y	Y	Y	Y	
Y	Y	Y	Y	Y	Y	Y	Y	Y	
Y	Y	Y		Y	Y	Y	Y	Y	
Y	Y	Y	Y	Y	Y	Y	Y	Y	Y
Y	Y	Y		Y	Y	Y	Y	Y	
Y	Y	Y		Y	Y	Y	Y	Y	
Y	Y	Y	Y	Y	Y	Y	Y	Y	
Y	Y	Y		Y	Y	Y	Y	Y	Y
Y	Y	Y	Y	Y	Y	Y	Y	Y	
Y	Y	Y		Y	Y	Y	Y	Y	
Y	Y	Y	Y	Y	Y	Y	Y	Y	Y
Y	Y	Y		Y	Y	Y	Y	Y	
Y	Y	Y		Y	Y	Y	Y	Y	
Y	Y	Y	Y	Y	Y	Y		Y	
Y	Y	Y		Y	Y		Y	Y	
Y	Y	Y		Y	Y	Y	Y	Y	Y
Y	Y	Y	Y	Y	Y	Y	Y	Y	Y
Y	Y	Y		Y	Y	Y	Y	Y	
Y	Y	Y		Y	Y	Y	Y	Y	
Y	Y	Y		Y	Y	Y	Y	Y	
Y	Y	Y	Y	Y	Y	Y	Y	Y	
Y	Y	Y		Y	Y	Y	Y	Y	
Y	Y	Y		Y	Y	Y	Y	Y	
Y	Y	Y		Y	Y	Y	Y	Y	Y
Y	Y	Y		Y	Y	Y	Y	Y	
Y	Y	Y	Y	Y	Y	Y	Y	Y	
Y	Y	Y		Y	Y	Y	Y	Y	
Y	Y	Y	Y	Y	Y	Y	Y	Y	
Y	Y	Y	Y	Y	Y	Y	Y	Y	
Y	Y	Y	Y	Y	Y	Y	Y	Y	Y
Y	Y	Y	Y	Y	Y	Y	Y	Y	
Y	Y	Y	Y	Y	Y	Y	Y	Y	
Y	Y	Y	Y	Y	Y	Y	Y	Y	
Y	Y	Y		Y	Y	Y	Y	Y	
Y	Y	Y		Y	Y	Y	Y	Y	
Y	Y	Y		Y	Y	Y			
Y	Y	Y	Y	Y	Y	Y	Y	Y	Y
Y	Y	Y		Y	Y	Y	PET	Y	
Y	Y	Y		Y	Y	Y	Y	Y	
Y	Y	Y		Y	Y		Y	Y	
Y	Y	Y		Y	Y	Y	Y	Y	
Y	Y	Y		Y	Y	Y			
Y	Y	Y	Y	Y	Y	Y	Y	Y	Y

TABLE 9.12 Product Recovery by MRFs

Material	Low-tech (124 facilities)		High-tech (53 facilities)	
	No. of MRFs	% of total	No. of MRFs	% of total
Newspaper	116	93	44	83
Computer paper	41	33	14	26
Office paper	76	61	24	45
Cardboard	102	82	39	73
Clear glass	121	97	53	100
Brown glass	119	95	53	100
Green glass	118	95	53	100
Mixed glass	41	33	20	37
Tin cans	122	98	53	100
Aluminum	123	99	53	100
Bimetal cans	105	84	51	96
PET	117	94	48	90
HDPE	121	97	48	90
Other plastics	36	29	11	20

containing plastic, wax, or other nonsoluble coatings, there would be no market for the finished product. Thus, while many materials may be theoretically recyclable, as a practical matter, these materials cannot be effectively recycled with the current collection and processing systems. A recycling program which seeks to achieve too high a recycling rate by including all grades of paper could actually hurt the potential to recycle the larger part of our segregated waste paper, namely, old newspaper or old corrugated containers.

Ferrous Metal.   Recovered ferrous metals can be resold to detinning facilities or directly to steel mills for their smelting operations. Detinners are sensitive to contaminants that can impede processing (e.g., aluminum) or exacerbate effluent problems (e.g., labels in sludge). Steel mills are constrained by their basic manufacturing process, metallurgical requirements of end products, and emission and effluent problems. Oxygen furnace mills can usually use up to 30 percent scrap material, but electric arc furnace mills can use up to 100 percent scrap materials.[27]

Figure 9.69 shows an automated continuous detinning facility. In this operation, densified bales are fed into a bale breaker or shredder. The loose bale is then detinned. The finished detinned product is then made into a high-density briquette or bale.

The same metal can sell at $5 per ton or $140 per ton depending on its cleanliness, purity, and density resulting from the degrees of processing varying from collection "as is" from the homeowner to a detinned bale of can scrap. Bimetal cans are the primary source of postconsumer ferrous metal. These materials can be recovered relatively easily from commingled recyclables by stationery or belt magnets. The recovered product can be baled, shredded, or nuggetized in commercially available equipment. According to the Steel Can Recycling Institute,[27] the ferrous product must be free of all nonmetallic, nonferrous materials other than paper labels.

Aluminum.   Aluminum, primarily recovered in the form of used beverage containers (UBCs), can be resold directly to aluminum processors who reprocess it as container flat-rolled stock. Depending on specific alloy specifications, postconsumer aluminum can be reused in amounts up to 100 percent of finished product with substantial energy savings and conservation of the mineral bauxite.[27] Processors prefer that the recovered

(a)

(b)

FIGURE 9.69 Detinning facility operations. (*a*) Facility overview. (*b*) Shredding of bales. (*c*) Shredded product before detinning. (*d*) Detinned product formed into high-density briquettes.

(c)

(d)

FIGURE 9.69 (*Continued*)

aluminum product be densified in bales or biscuits (i.e., nuggets) of specific size and be free of excess moisture and contaminants. Although aluminum comprises only a small fraction of MSW, recovery is highly desirable. Aluminum is easy to recover from commingled recyclables, and its high resale value helps to subsidize the recycling of other materials.

The most common methods of separating aluminum from other recyclables are manual picking from a conveyor belt or use of an eddy-current separator. Air classification also can be used, depending on whether the feed stream also contains plastics, which have aerodynamic characteristics comparable with those of aluminum beverage containers. Small pieces of broken glass can also carry over with the aluminum materials in an air classifier. Other methods for aluminum separation include electrostatic separation and several wet processes (jigging, water elutriation, and heavy media separation).

Repackaging of recovered aluminum for resale involves the flattening of cans in a press or by rollers positioned above a conveyor. Flattened cans can then be baled or compressed into biscuits, or blown into trailers for loose shipment.

Glass. Recovered glass beverage and food containers can be resold to glass container manufacturers for substitution of up to 100 percent for virgin materials or to building material manufacturers for inclusion in road surfacing, glass wool insulation, or aggregate-based products. Substitution of recycled glass enables container manufacturers to operate at lower furnace temperatures and improve emission characteristics. Container manufacturers will accept recycled material in whole container, irregularly broken, or crushed form. Two critical specifications have a direct effect on recycling practices:

- Glass must be sorted by color (i.e., flint, green, and brown) to control the cosmetic appearance of end products.
- Recycled glass must be free of all contaminants, including paper, plastics, metals, textiles, and rocks.

Glass Recovery. As glass containers break during the transition from the point of consumer discard through collection and centralized processing, colors can become mixed and chards of glass can collect the residue of other materials. Metallic tops and paper labels can also remain if not removed by the generator or the centralized processing system. Chards of glass also become embedded in other recyclables with which they come in contact, thereby reducing the marketability of the other material. Glass-impregnated papers, for example, damage rollers and other processing equipment in the manufacture of recycled papers.

If glass is to be separated from mixed waste without subsequent color separation, trommeling, screening, air classification, or combinations thereof are used.[28] Froth flotation also has been demonstrated. These techniques simultaneously break and densify the mixed glass cullet, thereby possibly avoiding the necessity for a discrete densification step. Certain proprietary processes have been developed and are used commercially[29] to beneficiate glass prior to shipment, by removing excessive contaminants through trommeling and wet processes. By contrast, most processes to recover glass by color avoid breakage to facilitate visual recovery. Manual picking of glass colors from a conveyor is the most common method of recovery, although optical scanning and certain proprietary processes have been demonstrated. Densification of the recovered product can occur naturally by handling or by use of a glass crusher and pulverizer.

As more postconsumer glass has become available from MRFs, container manufacturers have become considerably more selective of materials available for sale. More than a phenomenon of increased supply exceeding demand, this has been in response to excessive contamination in postconsumer glass products. Glass recycling can significantly contribute to equipment downtime in an MRF, as the abrasive quality of the material causes accelerated wear of conveyor systems and glass crushers.

Issues Affecting the Market for Glass. Glass makes up only about 2 percent by volume of the municipal solid waste stream. However, because it is heavy for its volume, it makes up 7 to 8 percent of the solid waste by weight. Glass bottles that are recycled and color sorted, and which have not been contaminated, can be easily recycled back into new bottles provided the source of recycled glass is in reasonable geographic proximity to a glass bottling plant. Glass that is not high enough quality to be used in making new bottle glass may be used in lower-value applications such as glass wool, or a substitute for aggregate in bituminous paving mixtures called "glasphalt."[30] Container glass is the only glass that is being recycled in large quantities at the present time. Window panes, light bulbs, mirrors, ceramic dishes and pots, glassware, crystal, ovenware, and fiberglass are not recyclable with container glass and are considered contaminants in container glass recycling.

It is most important that the glass be free of "stones," which are really small pieces of stone, brick, concrete, or ceramic. When consumers throw out glass mirrors, window glass, and ceramic plates along with the bottle glass, these materials form stones in the finished glass product. The stones do not fully melt in the glass furnace, thus creating discontinuities in the finished glass product. These discontinuities in the glass form stress risers or stress concentrators in the bottle and can cause the bottle to break or explode, especially when pressurized with CO_2. Based upon particle counts established by the U.S. Bureau of Mines, in 1 lb of glass there are about 500,000 particles of glass about beach sand size (range 20 to 200 mesh). In that half million particles, the industry standard is that there can be no stones present which are greater than 20 mesh in size. Only two particles of stone or ceramic can be present in the 20- to 40-mesh size range. About 20 much smaller particles in the 40- to 60-mesh size range can be present, since these have a greater tendency to melt during the glass melting operation, which is a batch process. When glass is pushed around on the tipping floor of an MRF by a front-end loader and mixed with other recyclable materials on the tipping floor, some tendency can exist to pick up stones and dirt off the floor. These can make the glass nonrecyclable. Thus, when a MRF operator claims that there is no market for glass, it would be wise to first determine why? The reason could be the presence of stones or ceramics which have contaminated the finished glass product.

While glass is a low-value product and cannot be shipped long distances, the economics of glass manufacture are very favorable for the use of recycled bottle glass because the energy usage is lower for this material, which is essentially "prefluxed." Nine gallons of fuel oil are saved for each ton of glass that is made from recycled cullet rather than from the raw "glass sand" mixtures used to produce new glass. If the market is weak or nonexistent for recycled glass cullet, it probably means there is a problem with meeting specifications on contamination. Thus, while there is a market for flint, green, and amber glass, the market for mixed color glass is somewhat weaker. Further, the presence of stones and contaminants can eliminate any demand for the finished product.

Plastics. All plastics represent only about 7 percent of all MSW by weight. Plastic containers and packaging (those applications found in the MRF stream) represent about 3 percent of all MSW by weight. Industry data on typical plastics products, the resin from which they are produced, and the quantity of product generated in 1992 are given in Table 9.13. The variety of resins and colors often makes it difficult for the generator, curbside collection crew, MRF workers, or MRF mechanical devices to distinguish one type from another. Although of likely resale value, the quantities of cer-

TABLE 9.13 Typical Plastic Products

Plastic	Application	1992 volume, millions of lb
HDPE	Milk and other beverage bottles	1066
HDPE	Pails	613
HDPE	Drums	242
HDPE	Dairy tubs and other food containers	227
HDPE	Ice cream containers	97
HDPE	Paint cans	32
PVC	Cooking oil and beverage bottles	150
PS	Dairy containers	155
PS	Cups and foamed containers	203
PS	Hinged containers	105
PP	Containers	235

Source: *Plastic Recycling Issues,* S. Norwalk & Assoc., July 1993.

tain plastics in the waste stream have precluded recycling at any reasonable net cost. Consequently, plastics recycling technology has been slow to develop.

Primarily because of their high volume and relative ease of identification, containers made from high-density polyethylene (HDPE) and polyethylene terephthalate (PET) are the most commonly recycled plastics. Comprised largely of milk containers and soft drink base cups, HDPE can be sold as is or granulated. The primary source of PET is 2-L soda bottles that can be granulated and shipped loose, shredded and baled, or baled whole. Recycled PET containers can be used in the manufacture of a variety of items such as fiberfill cushioning, geotextile membranes, or industrial strapping. PET can be processed, mixed with virgin resin, and reextruded. Several intermediate plastic processors serve as value-added reprocessors to recycle postconsumer PET in proprietary processes (involving air classification, froth flotation, electrostatic separation, washing, and extrusion) for such reuse applications.

Because of classification difficulty, plastics typically are best separated by primary resin type through manual sorting on a conveyor belt prior to shredding and baling or granulation and packing in gaylord containers for shipment to market. In addition to manual sorting, plastics also can be separated from other materials by air classification or vibration screening. Use of any mechanically assisted separation depends largely on the design approach to glass recycling, its breakage, and cross contamination.

The American Plastics Council recently announced that plastic bottle recycling reached 775 million lb in 1992, or almost 18 percent of the virgin resin sold for bottles. The volume was up 36 percent of 1991 results. The 1990 to 1991 growth rate was 55 percent. Nonbottle recycling of packaging, i.e., other rigid containers and film, lagged behind that for bottles. The APC also announced that 6647 communities were recycling plastics in 1992.[31]

In 1992, the United States recycled 47 percent more postconsumer plastics packaging than in 1991, according to a R. W. Beck and Associates study.[32] The study examined six resins commonly produced and recycled: PET, HDPE, PVC, low-liner low-density polyethylene (LDPE/LLDPE), polypropylene (PP), and polystyrene (PS). Of the six resins, PET postconsumer packaging has the highest recycling rate. More than 402 million lb of PET packaging were estimated to have been recycled in 1992, an increase of 37 percent from 1991 levels. Soft drink bottles contributed significantly to this number. Recycled HDPE packaging consists primarily of milk and water bottles, pigmented bottles, and soft drink bottle base cups. In 1992, approximately 416 million lb of HDPE postconsumer packaging was recycled, up from 277 million lb in 1991. Recycled PVC packaging, which primarily consists of water, food, pharmaceutical, and cosmetic bottles and film, increased from 1.6 million lb in 1991 to 10.2 million lb in 1992. Postconsumer PVC packaging was recycled at a rate of approximately 1.4 percent in 1992, as compared with 0.2 percent in 1991. LLPE/LLDPE packaging such as film packaging and shrink/stretch pallet wrap, and PP packaging, primarily flexible packaging, have shown increases. LDPE/LLDPE recycling has increased 52 percent over 1991 from 41.8 million lb in 1991 to 63.5 million lb in 1992. Only 1 percent of all PP packaging produced domestically has been recycled. Recycling of polystyrene, which is used in food services or as protective packaging, has amounted to approximately 31.6 million lb of resin recycled in 1992.

The success of recycling plastic resins is partially due to the plastics industry and its associations. The increase in PS recycling is primarily due to the commitment from the polystyrene industry, which has built four plants for polystyrene recycling. While recycling of plastics back into reusable polymers will continue to be important, another thrust will be to get pyrolysis of plastics to oils and chemical feedstocks accepted as a form of recycling. The technology is largely in place to accomplish this. The key issue is whether or not legislative bodies will accept pyrolysis in lieu of other forms of recycling.

Market Prices

Although the quantity of recycled products has been increasing, the price paid for those materials has, in most cases, been disappointing during the period of 1992 and 1993. Emphasizing the fact that the product markets are a commodity business, *Recycling Times* provides periodic data on the price paid by processors for recycled materials and the prices paid by end users to the processors for processed products. Prices for old newspaper, computer printout, old corrugated cardboard, and white ledger are provided in Figs. 9.70 through 9.73. Plastic container prices for HDPE and PET are pro-

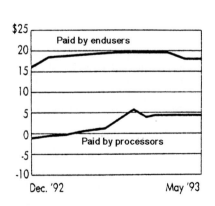

Period	Average Price ($/ton)	
	Paid by Processors	Paid by End-Users
Dec. 3-Dec. 29	$-1.5	$16.3
Dec. 28-Jan. 15	-0.4	19.0
Jan. 13-Feb. 1	-0.2	19.0
Jan. 27-Feb. 13	0.1	19.0
Feb. 10-March 1	0.3	19.3
Feb. 25-March 12	2.0	19.7
March 9-26	3.8	19.7
March 24-April 9	3.8	19.7
April 7-26	4.5	19.7
April 21-May 7	4.0	18.3
May 6-21	4.0	18.3

FIGURE 9.70 ONP prices. Prices are in dollars per ton and represent an average of high-low prices paid for No. 6 old newspapers across seven regions of the United States. Processors are dealers, brokers, recycling centers, etc. End users are paper mills, insulation manufacturers, etc. (*Source: Recycling Times, Mid-Year Market Update 1993 and Waste Age, August 1993.*)

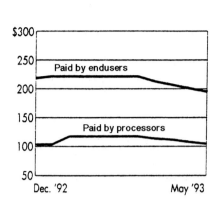

| Period | Average Price ($/ton) | |
	Paid by Processors	Paid by End-Users
Dec. 3-Jan. 29	$106.3	$217.5
Dec. 28-Jan. 15	106.3	219.5
Jan. 13-Feb. 1	113.4	219.5
Jan. 27-Feb. 12	113.4	219.5
Feb. 10-March 1	113.4	219.5
Feb. 25-March 12	113.4	219.5
March 9-26	112.8	219.5
March 24-April 9	112.8	217.9
April 7-26	111.5	216.6
April 21-May 7	109.7	208.3
May 6-21	105.8	195.7

FIGURE 9.71 CPO prices. Prices are in dollars per ton and represent an average of high-low prices paid by processors and end users for baled, laser-free computer printout paper across seven regions of the United States. Processors are dealers, brokers, recycling centers, etc. End users are paper mills. (*Source: Recycling Times, Mid-Year Market Update 1993 and Waste Age, August 1993.*)

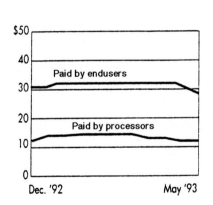

| Period | Average Price ($/ton) | |
	Paid by Processors	Paid by End-Users
Dec. 3-Dec. 29	$14.4	$30.9
Dec. 28-Jan. 15	14.4	30.9
Jan. 13-Feb. 1	14.8	31.6
Jan. 27-Feb. 12	14.8	32.0
Feb. 10-March 1	14.8	32.0
Feb. 25-March 12	14.8	32.0
March 9-26	15.0	32.0
March 24-April 9	13.8	32.0
April 7-26	13.8	32.0
April 21-May 7	12.4	32.0
May 6-21	12.4	29.8

FIGURE 9.72 OCC prices. Prices are in dollars per ton and represent an average of high-low prices for baled old corrugated containers across seven regions of the United States. Many processors pay the same amount for baled or loose paper, because they break the bales anyway to process the paper; few end users purchase any loose paper. Processors are dealers, brokers, recycling centers, etc. End users are paper mills. (*Source: Recycling Times, Mid-Year Market Update 1993 and Waste Age, August 1993.*)

vided in Fig. 9.74; aluminum and steel prices are provided in Fig. 9.75; and glass container prices are provided in Fig. 9.76.

After a disappointing year in 1992, prices for most recyclables slowly began to increase at the beginning of 1993. But, while prices for low grades of paper—old newspaper, old corrugated containers, and colored ledger—began to rise, prices for high grades—computer printout and white ledger—began to fall. A surplus of paper

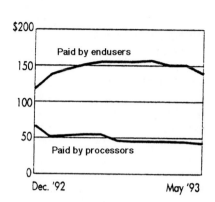

Period	Average Price ($/ton)	
	Paid by Processors	Paid by End-Users
Dec. 3-Dec. 29	$66.5	$129.5
Dec. 28-Jan. 15	51.5	143.5
Jan. 13-Feb. 1	51.5	145.0
Jan. 27-Feb.12	52.7	150.0
Feb. 10-March 1	52.7	150.8
Feb. 25-March 12	52.7	151.6
March 9-26	48.4	151.6
March 23-April 9	48.4	151.6
April 7-26	48.4	150.8
April 21-May 7	47.7	150.8
May 6-21	47.7	145.0

FIGURE 9.73 White ledger prices. Prices are in dollars per ton and represent an average of high-low prices paid by processors and end users for baled, sorted white ledger across seven regions of the United States. Many processors pay the same amount for baled or loose paper, because they break the bales anyway to process the paper; few end users purchase any loose paper. Processors are dealers, brokers, recycling centers, etc. End users are paper mills. (*Source: Recycling Times, Mid-Year Market Update 1993 and Waste Age, August 1993.*)

Period	Average Price (cents/pound)		
	Clear PET	Natural HDPE	Mixed PET/HDPE
Dec. 3-Dec. 29	1.7¢ (6.3)	1.2¢ (7.6)	0.6¢ (3.5)
Dec. 28-Jan. 15	1.7 (6.3)	1.2 (7.6)	0.6 (3.5)
Jan. 13-Feb. 1	1.7 (6.3)	1.2 (7.6)	0.6 (3.5)
Jan. 27-Feb. 12	1.7 (6.4)	1.2 (7.6)	0.6 (3.5)
Feb. 10-March 1	1.7 (6.5)	1.5 (7.7)	0.6 (3.5)
Feb. 25-March 12	1.7 (6.5)	1.5 (7.7)	0.6 (3.5)
March 9-26	1.7 (6.5)	1.5 (7.6)	0.6 (3.5)
March 24-April 9	1.7 (6.5)	1.5 (7.6)	0.6 (3.5)
April 7-26	1.7 (6.5)	1.5 (7.6)	0.6 (3.5)
April 21-May 7	1.7 (6.7)	1.5 (7.5)	0.6 (3.5)
May 6-21	1.8 (6.7)	1.5 (7.5)	0.6 (3.5)

FIGURE 9.74 Plastic container prices. Prices are in cents per pound and represent an average of high-low prices paid by processors for plastic containers across seven regions of the United States. Processors are dealers, brokers, recycling centers, etc. Prices in parentheses represent prices paid by end users. End users are plastic products manufacturers. PET is clear polyethylene terephthalate such as soda bottles. HDPE is natural high-density polyethylene such as milk jugs. (*Source: Recycling Times, Mid-Year Market Update 1993 and Waste Age, August 1993.*)

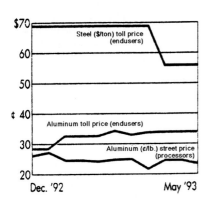

| Period | Aluminum (cents/pound) | | Steel ($/ton) |
| | St. Price | Toll Price | Toll Price |
	(Processors)	(End Users)	(End Users)
Dec. 3-Dec. 29	26.4¢	29.0¢	$69.2
Dec. 28-Jan. 15	27.2	28.8	69.2
Jan. 13-Feb. 1	25.5	33.5	69.2
Jan. 27-Feb. 12	25.5	33.5	69.2
Feb. 10-March 1	25.3	33.5	69.2
Feb. 25-March 12	25.7	34.4	69.2
March 9-26	24.5	32.7	69.2
March 24-April 9	21.3	33.7	NA
April 7-26	25.1	33.7	55.9
April 21-May 7	25.3	33.7	55.9
May 6-21	23.9	33.7	55.9

FIGURE 9.75   Aluminum and bimetal UBC prices. For aluminum used beverage cans, prices are in cents per pound. For steel food cans, prices are in dollars per ton. Aluminum figures represent high-low prices paid by processors (street price) and end users (toll price) across seven regions of the United States. Steel food can figures represent prices paid by end users only, and do not include freight. Processors are dealers, brokers, recycling centers, etc. End users are foundries and smelters. (*Source: Recycling Times, Mid-Year Market Update 1993 and Waste Age, August 1993.*)

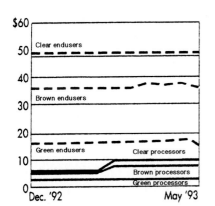

| Period | Average Prices ($/ton) | | |
	Clear	Brown	Green
Dec. 3-Dec. 29	$6.8 (50.4)	$6.0 (37.2)	$4.2 (15.3)
Dec. 28-Jan. 15	6.7 (50.4)	5.9 (37.2)	4.2 (15.3)
Jan. 13-Feb. 1	6.7 (50.4)	5.9 (37.2)	4.2 (15.3)
Jan. 27-Feb. 12	6.7 (50.4)	5.9 (37.2)	4.2 (15.3)
Feb. 10-March 1	6.7 (50.4)	5.9 (37.2)	4.2 (15.3)
Feb. 25-March 12	8.9 (50.4)	7.4 (37.2)	4.2 (15.3)
March 9-26	8.9 (50.4)	7.4 (37.2)	4.2 (15.3)
March 24-April 9	8.9 (50.8)	7.4 (37.6)	4.2 (15.3)
April 7-26	8.9 (50.8)	7.4 (37.6)	4.2 (15.3)
April 21-May 7	8.9 (50.8)	7.4 (37.6)	4.2 (15.3)
May 6-21	8.9 (50.8)	7.4 (36.8)	4.2 (14.9)

FIGURE 9.76   Glass container prices. Prices are in dollars per ton and represent an average of high-low prices paid by processors for clear, brown, and green glass containers across seven regions of the United States. Processors are dealers, brokers, recycling centers, etc. Prices in parentheses represent prices paid by end users. End users are glass products manufacturers. (*Source: Recycling Times, Mid-Year Market Update 1993 and Waste Age, August 1993.*)

on the foreign market caused by Germany's new packaging law and the low price of virgin pulp was blamed for the soft market for high grades of paper. ONP prices peaked in early spring, only to immediately fall. Aluminum prices also faltered. By April, Reynolds Aluminum and Alcoa had dropped their toll prices for aluminum used beverage cans twice. Plastics, particularly natural HDPE, are the only commodities to sustain a steady increase in price since the beginning of the year. Several end users have said it is difficult to get enough quality resin to manufacture their products. Ironically, the lack of available plastics may be related to last year's glut. As communities found it more and more difficult to find markets for plastics, many dropped the material from their curbside collection programs.

PART B

Mary B. Sikora

President, Recyling Research Institute
Suffield, Connecticut

9.6 OPTIONS FOR MANAGING AND MARKETING SCRAP TIRES

Introduction

In the United States today there is great interest in recycling as a solution to the nation's solid waste problem. This interest is driven by a lack of adequate disposal capacity and a growing inability to site new disposal sites because of community opposition. This interest is not restricted to the United States. In the United Kingdom, Canada, France, Italy, Germany, eastern Europe, and almost every other nation, governments and the recycling industry are striving for workable methods to manage wastes.

In the case of tires, decreasing landfill capacity is coupled with the fact that tires won't biodegrade in any foreseeable time in a landfill environment. This has led more and more government agencies worldwide to ban the disposal of whole tires. However, without economical and convenient alternatives to landfilling, illegal stockpiling and disposal will not only continue but is likely to increase.

For recycling and reuse of tires to actually occur, a demand must exist for tire derived materials. Collection and separation alone will not be enough to justify or assure recycling or reuse. This means that the demand for tire derived materials by rubber manufacturers for refabrication into new products, or incorporation into existing products, and by utilities, mills, and kilns for energy recovery must increase. At the same time both industrial and individual consumers must be willing to purchase and use products manufactured with recycled materials. In reality, consumer demand can transform scrap tires into new products.

Today we know more about extending tire life. In addition, during the last decade we learned that the real problem with scrap tires is how to separate the structural components and use them to their best advantage. Ideally, when we can no longer ride on a tire we should be able to reduce it to its wire, fabric, and rubber components—and we can. Scrap tires can be processed for recycling and reuse. However, the questions of cost, overall economic feasibility, and long-term market reliability have not yet been answered. This is the task at hand, if scrap tire recycling is to have a future.

Scope of the Problem

Nationally 200 to 250 million car and truck tires are discarded annually. These tires represent 1 to 1½ percent by weight of all the solid wastes generated in the United States each year. In addition, about 2 billion scrap tires have accumulated in stockpiles or uncontrolled tire dumps across the country, according to the U.S. Environmental Protection Agency (EPA). EPA also estimates that less than 7 percent of the 242 million tires discarded in 1990 were recycled into new products, while about 11 percent were converted into energy. Over 77 percent of the remaining 200 million tires per year were landfilled, stockpiled, or illegally dumped. Another 5 percent were exported. In addition, 33.5 million worn tires were retreaded in 1990 and 10 million were reused.[33]

Tires are chemically complex, precision engineered, and designed to last. Tires maintain their structural integrity under a wide range of driving and road conditions. When the tread rubber wears off, the tire casing with its cords, belting, and metal bead remains tough and flexible.[34] It is not surprising then that scrap tires maintain the basic properties of a tire after they are discarded. This very toughness makes tires both difficult to recycle and difficult to landfill. In a landfill environment tires don't compact well, they consume an inordinate volume of space for their weight, and because of their shape they can fill with methane gas and eventually migrate to the top if not properly compacted and buried.

It is important to note that while many states have banned the landfilling of whole tires and some states have even banned shredded tires, for many communities landfilling is the only legal and/or economical option for tires. (See Table 9.14.) If no markets exist in a region, state and local communities should approach landfill bans with caution since the only alternative to such a ban may be illegal dumping. However, regulators should also be aware that if landfilling is the cheapest option in their community, it will most likely serve as a barrier to both existing markets and future development of additional markets in that community. Usually, if legitimate recycling or reuse markets are available at a reasonable cost, regulators recommend that landfilling of tires be discontinued.

Scrap Tire Sources

A scrap tire is generated whenever a new replacement tire is sold. The major generators of scrap tires are:

- Tire dealers
- New car dealers
- Auto equipment and auto parts shops
- Tire wholesalers
- Tire retread and repair shops
- Cab companies
- Rental car companies
- Fleet owners, including governments
- Auto salvage yards
- Consumers

Tires are discarded when a consumer buys new tires or when a vehicle is decommissioned. Traditionally, when a consumer purchased new replacement tires, the retailer kept the old tires as a trade-in. In fact, the EPA estimates that 95 percent of tires are collected through the commercial waste stream and only 5 percent or less through the household waste stream.[35] In recent years it has become a common practice for the retailer to give the consumer the option of paying a disposal fee ($0.25 to

TABLE 9.14 Scrap Tire Laws and Regulations, January 1993

State	Funding	Regs	Landfill	Market incentives
AL		S P		
AR	$1.50 per tire disposal fee on retail sales $1 per tire imported into state	S P H	Tires must be cut and monofilled	30% equipment tax credit; 10% PP for retreads; grants
AZ	2% sales tax on retail sale	S P H	Bans whole tires	10% equipment tax credit; municipal grants
CA	$0.25 per tire disposal fee	S	Effective 1/93 bans whole tires	40% tax credit for manufacturers using secondary materials; grants and loans; 5% PP for tire materials
CO		S P		Tax credit for manufacturing using secondary materials
CT		S		10% PP
FL	$1 per tire retail sales	S P H	Tires must be cut	Closed-loop purchase contracts; grants; 10% PP innovative technology grants and loans to cities and counties
GA	$1 per tire management fee	S P H G	Bans whole tires, 1/95	
HI			Honolulu only—bans whole tires	
ID	$1 per tire retail sales	S	Bans all tires	Grants to cities and counties; $20 per ton end user rebate; $1 per retread reimbursement
IL	$1 per tire retail sale and $0.50 per vehicle title transfer	S P H	Bans whole tires, 7/94	Rebates, grants, and loans for manufacturing and marketing; procurement goals
IN	Permit fee, tire storage sites	Both proposed S H		10% PP and grants
IA		S P H	Bans whole tires	Grants; recycled content required for state purchases

9.131

TABLE 9.14 Scrap Tire Laws and Regulations, January 1993 (*Continued*)

State	Funding	Regs	Landfill	Market incentives
KS	$0.50 per tire retail sales	S P H	Tires must be cut	Tax credits for equipment, municipal grants
KY	$1 per tire retail sales	S	Tires must be cut	Tax credits for recycling businesses; loans; recycled content preference
LA	$2 per tire retail sales, 2/92	S P H	Tires must be cut	Tax credits; 5% PP
ME	$1 per tire disposal fee	S P H		State should buy recycled; loans and grants
MD	$1 per tire first sale	S P H	Effective 1/94, bans tires	5% PP
MA		S P	Bans whole tires	10% PP
MI	$0.50 per vehicle title transfer	S P H G		Grants and loans; 10% PP
MN	$4 per vehicle title transfer	S P H	Bans whole and cut tires	Grants
MS	$1 per tire retail sales	S P H	Tires must be cut	County and regional grants and loans; 10% PP
MO	$0.50 per tire retail sales	S P H	Bans whole tires	10% PP; grants
MT		S		Tax credits for equipment and products; state required to buy recycled
NE	Business assessment fee $1 per tire retail sale			Grants
NV	$1 per tire on new tire retail sales	To be written		Grants for education and highway projects; 10% PP
NH	Town administration graduated vehicle registration fee	S P	Tires must be cut unless facility is exempt	State should buy recycled
NJ		S P		Grants; tax credits; state should buy recycled
NY		S P H		Grants; DOT specification for crumb rubber
NC	1% sales tax on new tires	S P H	Tires must be cut	Grants; funds county tire collection
ND	$2 per new vehicle sales	S H		
OH		S	Tires must be cut	
OK	$2 per new vehicle sales	S P		Grants; processor credits

9.132

State	Fee/Tax	Categories	Ban	Incentives/Notes
OR		S P H	Tires must be cut	State should buy recycled
PA		S	Operator's option	5% PP on bids; grants and low-interest loans
RI	$0.75 per tire on new tires sales $5 deposit per tire	S P		Funding stockpile cleanup: promotes use of recycled products
SC	$2 per new tire sales	S P H	Bans whole tires	7.5% PP; grants to counties and local governments; state required to buy recycled
SD	$0.25 per tire vehicle registration	S P	Bans tires by 7/1/95 unless allowed by state rules	Grant fund
TN	$1 per tire retail sales		Bans whole tires, 1/95	Grants
TX	$2 per tire retail sales	S P H G D	Bans whole tires	$0.85 per tire processor credit; tax credits; 15% PP on asphalt rubber; low-interest loans
UT	Graduated tax per tire size	S P H		$20 per ton reimbursement to end user
VT		S P	Bans whole tires	State required to buy recycled; 5% PP
VA	$0.50 per tire disposal fee on new tire sales, sunsets 12/31/94	S P	Bans whole tires	Contract services; 10% tax credit for recycled equipment, sunsets 12/31/95
WA	$1 per tire retail sales	S P H		Grants to local governments
WV		S P H	Bans whole tires (1988), bans all tires, 6/1/93	State required to buy recycled
WI	$2 per tire per new vehicle title transfer	S P H	Bans tires, 1/1/95	$20 per ton reimbursement to end user; grants
WY		S		State required to buy recycled

S = storage
P = processor
H = hauler
G = generator
D = disposal
PP = price preference

9.133

$2.00 per tire) or keeping the old tires. In some states, tire dealers are now required by law to take back tires and collect a disposal fee.[36]

Whenever the consumer retains a scrap tire, there is little or no way of tracking how the tire is finally disposed. However, indications are that these tires are rarely retreaded or reused and are most likely placed at the curb for municipal collection or scrapped in an environmentally unacceptable manner. The alleys, vacant lots, and congested roadsides of an urban environment offer ideal dumping areas for consumer discards, as do the open fields, ravines, and wooded areas found in rural locations.

Scrap tires left with the dealer and handled commercially follow a different path which does not necessarily preclude improper disposal. The key players in the commercial handling of scrap tires are the retailer, the broker or jockey, the hauler-transporter-collector and the end users which can be retreaders, used tire dealers, scrap tire processors, energy users, or product manufacturers (see Fig. 9.77).

The regulation of such a diverse network is formidable. Of particular concern in improving proper tire handling and disposal are small operations which can easily avoid identification. Even when commercial tire dealers hire a service to remove their scrap tires, transporters are often not held accountable for the tires they dispose of. Many state and local government agencies are now requiring a manifest system for scrap tire collection, transport, and disposal. It is becoming increasingly important for generators to know the destination of the scrap tires they release to a hauler. Recent cases have shown enforcement authorities will seek to recover cleanup costs or damages from the generators of scrap tires which find their way to unpermitted dump sites.[37]

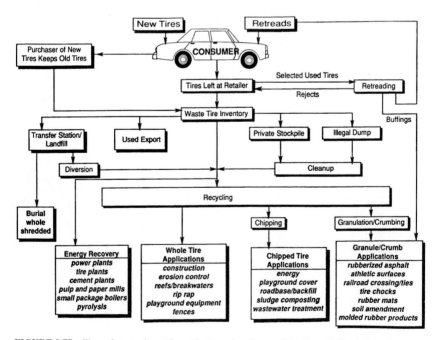

FIGURE 9.77   Flow of waste tires. (*From the Recycling Research Institute, Suffield, Conn.*)

Source Reduction

Basically, the number of tires landfilled, stockpiled, or illegally dumped can be reduced in one of two ways. One is to increase the recovery of scrap tires through expanded market development. The other is to reduce the number of scrap tires generated through source reduction. There are three methods of reducing the number of scrap tires: longer-life tires, reuse of used tires, and retreading.

Technological advances in tire manufacturing have more than doubled the useful wear life of tires in the last half century. Today, 40,000-mile tires are commonplace, with 60,000- to 80,000-mile lifetimes being achieved in increasing numbers. Industry experts say that continuing tire research may yield higher-mileage tires in the future if cost constraints, fuel consumption concerns, and safety requirements can be met.[38]

In the meantime, consumers can extend the present life of their tires by following some basic steps recommended by tire manufacturers and other industry experts. First, buy top-quality tires.[39] Second, maintain tires properly for longer life and better gas mileage. Good maintenance practices include proper inflation and regular tire servicing (rotating, balancing, aligning tires at least twice a year).[40]

Although the reuse of partially worn tires cannot be expected to solve the scrap tire problem, EPA estimates that a minimum of one additional year of tire life can be achieved out of 25 percent of the tires removed from vehicles. EPA offers the following example to estimate the reuse of used tires: "If 25 percent (one tire), on average still has a useful tread of 10,000 miles left, this is equivalent to the set of four tires lasting 42,500 miles instead of 40,000, an increase in life of 6 percent."[41] EPA estimates that 50 percent of the good used tires are being reused and suggests that if the other 50 percent were also used, a 3 percent reduction in tire disposal could be realized.

The third source reduction measure which can extend the useful life of the tire and thus reduce the number of tires scrapped is retreading. Retreading has traditionally been considered the highest and best use for tires once they have completed their useful first life (see Fig. 9.78). In the retread process, new tread rubber is applied to a worn tire that still has a good casing. Bias-ply passenger car worn tire casings in suitable condition can be retreaded four to six times before being discarded. Steel-belted passenger car or light truck radials require a more expensive retreading process and can be retreaded an average of one to two times before being scrapped. Airplane tires are retreaded up to 20 times. Truck tires are retreaded 3 to 5 times.[42] Figure 9.79 summarizes the current state of the retread industry in the United States.

Retreaded tires have always played a role in the replacement tire market, and approximately 10 to 18 percent of all waste tires generated annually are retreaded, with the majority being truck tires. While there has been some consumer concern over safety of passenger tire retreads, reputable dealers can sell as many retreaded tires as they can make. Production is currently limited by the supply of usable, reasonably priced casings. Passenger tire retreading is also limited in today's tire market by the fact that there is virtually no price difference between a new and retreaded automobile tire.

The cost difference between new and retreaded truck tires still favors the retreaded tire, and this market remains strong. Typically, large transport companies, truck fleets in various industries, including waste hauling and collection vehicles, use retreaded tires on the rear axles of their trucks. Bus companies and airlines also use retreads on the rear axle. A retreaded truck or bus tire can cost 25 to 55 percent less than a new tire. A quality retreaded tire, properly maintained, can extend the useful life of a tire casing by 10,000 to 20,000 miles.[43]

While it is true that passenger tire retreading has been on the decline for the reasons cited above, some current legislative and recycling trends are making retreading attractive. In the case of high-performance tires, a new and growing interest in

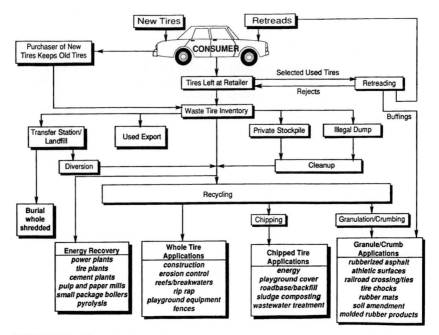

FIGURE 9.78   Hierarchy of scrap tire recycling. (*Source: U.S. Environmental Protection Agency.*)

- Over 30 million retreaded tires were sold in the United States in 1992, with sales totaling approximately 2 billion dollars

- 7.4 million retreaded passenger car tires

- 7.5 million retreaded light truck tires

- 15.2 million retreaded medium and heavy truck tires

- 810,000 other retreaded tires (aircraft, off-the-road vehicle, industrial and lift trucks, motorcycles, farm equipment, specialty, etc.)

- 68 million lb of tread rubber were used by the retread tire industry in 1992

- There are over 1500 retreading plants in the United States and Canada, of which 95 percent are owned and operated by small businesspeople whose collective investment is approximately one billion dollars. The remaining 5 percent are owned and operated by new tire manufacturers

---

FIGURE 9.79   Retread tire market in 1992. (*Source: Retreaders Journal, April 1993.*)

retreads is being driven by the economic incentives similar to those of truck tires. High-performance passenger tires can cost $80 to $200 per tire. New retreading technology for these tires makes retreads a viable replacement option, saving one-third to one-half the cost of some new high-performance tires.[44]

Renewed interest in retreads is also being influenced by legislative measures. In November 1988 the U.S. EPA finalized the retread procurement guideline. The guideline requires that government agencies both purchase retread tires as replacement tires and contract for retreading of used tires if the agency's annual tire purchases exceed $10,000.[45] The U.S. General Services Administration (GSA) is actively developing an affirmative retread procurement program. GSA recently completed revisions to its specifications for pneumatic vehicle tires to include retread tires. This change allows retreaders to qualify their tires for the Qualified Products List (QPL) in the same manner as manufacturers of new tires. In addition, GSA recently changed its policy regarding passenger tire retreads allowing for them to be listed on the Qualified Products List if they qualify in endurance and casing durability.[46] Many states, California and New Jersey among them, are considering or have passed legislation which would require the use of retreads on state vehicles.[47]

Market Status and Potential

Today a number of technically feasible recycling and reuse alternatives exist to utilize scrap tires in ways other than landfilling and stockpiling. In fact, during the last 2 years, markets for scrap tires have more than doubled, according to the *1992 Scrap Tire Use/Disposal Study* published by the Scrap Tire Management Council (STMC). At the end of 1992, about 68 million scrap tires (approximately 28 percent of the annual generation rate) had markets. In 1990, there were markets for about 25 million scrap tires (approximately 10 percent of the annual generation rate).[48]

The largest-volume market reported in the study is fuel—either whole or processed tire-derived fuel (tdf)—which consumed about 58 million scrap tires. The next two largest markets were reported as rubber-modified asphalt and civil engineering applications which consumed approximately 5 million tires each.

The Scrap Tire Management Council predicts that by the end of 1994, some 141 million scrap tires (approximately 55 percent of the annual generation rate) will have markets. The council estimates that the same three major consumers (tdf, rubber-modified asphalt, and civil engineering) will dominate the market with whole or processed tdf commanding 110 million scrap tires, rubber-modified asphalt using some 20 million, and 10 million scrap tires going to meet the demands of civil engineering applications (see Fig. 9.80).

Scrap Tires as Fuel

Each tire has energy potential. The heating value of an average-size passenger tire is between 13,000 and 15,000 Btu/lb, which compares with about 10,000 to 12,000 Btu/lb for coal. According to research by the Goodyear Tire and Rubber Company and by other scientists, tire combustion produces less ash than burning coal, emits less carbon dioxide (coal has a higher carbon-to-hydrogen ratio than tires), and produces lower emissions when replacing high-sulfur coal.[49]

Whole or processed tire-derived fuel (tdf) can be used in cement kilns, pulp and paper mill boilers, and some coal-fired utility or industrial boilers. Scrap tires can also be used in dedicated tires-to-energy facilities. The Scrap Tire Management Council reports that in 1993, 16 cement kilns were using tires as a supplemental fuel. Another 20 to 25 kilns are actively testing or planning to burn tdf. In addition, 11 pulp and paper mill boilers and 7 coal-fired power utility boilers were using tire fuel. Dedicated tire-to-energy facilities in Westley, Calif., and Sterling, Conn., respectively, are currently consuming a total of approximately 15 million whole tires annually. (See Tables 9.15 and 9.16.)

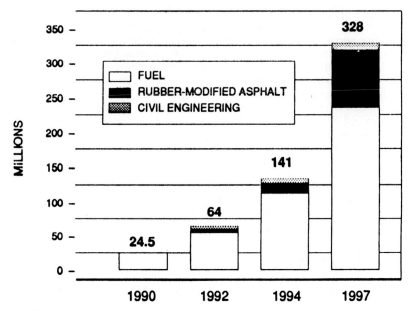

FIGURE 9.80   Uses of scrap tires. (*Source: Scrap Tire Management Council.*)

Shredding scrap tires to produce tdf is a standard materials processing technology. Tire shredding is also the primary method of altering a tire for other secondary uses such as rubber-modified asphalt, civil engineering applications, and the manufacture of rubber products.

Whatever the end product, tdf or crumb rubber, the first step in the process is shredding to reduce the whole tires to smaller pieces. Slow-speed shear shredders are the most commonly used equipment for primary reduction of scrap passenger and truck tires up to 48 in in outside diameter. These shredders are available in two basic designs, a solid cutter shredder and a replaceable cutter tire shredder (see Table 9.17). The most significant difference between these two shredder designs is that the replaceable cutter shredder is made to process tires exclusively while the solid cutter shredder is a multipurpose unit capable of size reducing other materials such as municipal solid waste, pallets, drums, and bulky waste, in addition to tires. The cutters on multipurpose shredders are usually constructed of mild or low-grade steel and do not stay sharp as long as replaceable cutters, which are made from higher alloys. Some users prefer the solid cutter design because, unlike the replaceable cutter shredder, the cutters can be removed and resharpened to a distinct edge several times.[50]

Typically the cost of producing tire fuel-size pieces (2 by 2 in) of rubber is about $20 per ton. (See Table 9.18.) This size of fuel chip is commonly used in pulp and paper mill boilers. Most utilities, however, have stricter fuel tolerance specifications and require chips sized 1-in square or smaller with free wire removed. Some cyclone boilers, common in utilities, require chips under $\frac{1}{2}$ in with $\frac{1}{4}$ in preferred. Producing tdf to these smaller sizes can raise the cost to between $30 to $50 per ton (about 3 to 5 cents per tire).[51]

While cost is a concern in developing fuel markets for scrap tires, other market barriers must also be addressed. Facilities which plan to use the fuel must have the proper

TABLE 9.15 Use of Scrap Tire Material as a Supplemental Fuel for Electric Generating Facilities and Pulp and Paper Mills

Company and location	Status	Potential capacity, million tires per year
Nonutility electricity generating facilities:		
Archer Daniels Midland, Decatur, Ill.	●	10.0
Caterpillar, Mossville, Ill.	◗	0.8
Dow Corning, Midland, Ill.	◗	0.9
Firestone, Decatur, Ill.	●	0.1
Monsanto, Sauget, Ill.	●	1.0
Palm Beach Resource Recovery, West Palm Beach, Fla.	●	0.3
Electric generating facilities:		
Commonwealth Edison, Springfield, Ill.	◗	6.0
Illinois Power, Baldwin, Ill.	◗	3.5
Manitowoc Public Utilities, Manitowoc, Wis.	●	.04
Nebraska Public Power District, Lincoln, Neb.	◗	3.5
New York State Electric & Gas, Binghamton, N.Y.	◗	4.5
Northern Indiana Public Service, Hammond, Ind.	◗	4.5
Northern States Power/Wisconsin, Bayfront, Wis.	◗	1.8
Ohio Edison, Toronto, Ohio	●	3.0
Otter Tail Power, Big Stone, S. Dak.	●	3.0
Traverse City Light & Power, Traverse City, Mich.	○	0.0
United Power Assoc., Elk River, Minn.	○	0.0
Wisconsin Power & Light, Beloit, Wis.	●	1.2
Pulp and paper mills:		
Boise Cascade, Deritter, La.	◗	n/a
Boise Cascade, Wallula, Wash.	○	0.0
Champion International, Bucksport, Me.	●	3.0
Champion International, Sartell, Minn.	◗	0.5
Daishowa America, Port Angeles, Wash.	○	0.0
Fort Howard, Rincon, Ga.	◗	n/a
Fort Howard, Green Bay, Wis.	●	1.1
Georgia Pacific, Cedar Springs, Ga.	●	3.2
Georgia Pacific, Toledo, Ore.	◗	0.4
Georgia Pacific, Big Island, Va.	●	0.7
Georgia Pacific, Bellingham, Wash.	◗	0.5
Inland Rome, Rome, Ga.	●	1.8
James River, Clatskaine, Ore.	○	0.2
Jefferson Smurfit, Newburg, Ore.	●	0.6
Longview Fibre, Longview, Wash.	○	0.0
Louisiana Pacific, Samoa, Calif.	◗	n/a
Mead, Stevenson, Ala.	○	0.0
Packaging Corp. of America, Tomahawk, Wis.	●	1.5
Port Townsend, Port Townsend, Wash.	○	0.0
Potlatch, Lewiston, Idaho	●	1.9
Scott Paper, Everett, Wash.	○	0.0
Sonoco Products, Harleyville, S.C.	n/a	n/a
Union Camp, Franklin, Va.	◗	4.0
Weyerhauser, Longview, Wash.	○	0.0
Westvaco, Covington, Va.	◗	0.3
Willamette, Campti, La.	◗	0.9
Willamette, Albany, Ore.	●	0.3

○ Past use or trial burns—future use unlikely
◗ Past use or trial burns—future use possible
● Currently burning TDF
Source: STMC Scrap Tire Use/Disposal Study, 1992.

TABLE 9.16   Use of Scrap Tire Material as a Supplemental Fuel for Cement Kilns

Company and location	Status	Potential capacity, million tires per year
Arizona Portland, Rillito, Ariz.	●	2.8
Ash Grove Cement West, Durkee, Ore.	●	0.6
Ash Grove Cement West, Inkorn, Idaho	○	0.8
Ash Grove Cement West, Seattle, Wash.	○	n/a
Blue Circle Inc., Atlanta, Ga.	▶	0.0
Blue Circle Inc., Harleyville, S.C.	▶	1.2
Calaveras Cement, Redding, Calif.	●	2.1
California Portland, Mojave, Calif.	○	n/a
Essroc Materials, Fredrick, Md.	●	1.1
Florida Crushed Stone, Brooksille, Fla.	▶	1.0
Giant Cement, Harleyville, S.C.	▶	n/a
Hawaiian Cement, Oahu, Hawaii	●	1.1
Holnam Inc.—Boxcrow Cement, Midlothian, Tex.	▶	1.0
Holnam Inc.—Ideal Basic Industries, Seattle, Wash.	●	1.2
Holnam Inc., Aida, Okla.	○	1.4
Holnam Inc., Morgan, Utah	▶	2.0
Illinois Cement, LaSalle, Ill.	▶	n/a
Independent Cement, Hagerstown, Md.	○	n/a
Kosmos Cement, Kosmosdale, Ky.	▶	1.3
Lafarge Corporation, Davenport, Iowa	○	n/a
Lafarge Corporation, New Braunfels, Tex.	▶	1.5
Lafarge Corporation, Sugar Creek, Mo.	○	n/a
Lafarge Corporation, Whitehall, Pa.	▶	1.5
Lehigh Portland Cement, Leeds, Ala.	●	1.0
Lone Star Industries, Cape Girardeau, Mo.	●	1.5
Lone Star Industries, Oglesby, Ill.	○	0.8
Medusa Cement, Clinchfield, Ga.	●	1.1
Medusa Cement, Charlesvoix, Mich.	○	3.8
Monarch Cement, Humboldt, Kan.	●	2.0
National Cement, Raglin, Ala.	●	0.5
Oldover Corporation, Ashland, Va.	▶	n/a
Phoenix Cement, Clarkdale, Ala.	○	n/a
Rinker Materials, Dade County, Fla.	○	n/a
River Cement, Festus, Mo.	▶	0.0
RMC Lonestar, Davenport, Calif.	▶	5.3
Southdown Inc., Lyons, Colo.	●	1.0
Southwestern Cement, Fairborn, Ohio	●	0.6
Southwestern Cement, Victorville, Calif.	●	2.6
Tarmac Cement, Roanoke, Va.	▶	n/a
Texas Lehigh, Buda, Tex.	▶	n/a

○   Considering use of TDF or whole tires
▶   Test burn conducted
●   Currently burning TDF or whole tires
Source: STMC Scrap Tire Use/Disposal Study, 1992.

pollution control equipment and adequate capacity in their air pollution control systems to handle the tire fuel. In addition, the permitting process is both lengthy and costly.[52]

Another constraint is the energy market. Tdf competes against other fuels, and when fuel prices are low, the price of tdf drops also. To offset these fluctuating fuel prices, tdf producers commonly charge a tipping fee which can range from $55 to $85

TABLE 9.17 Tire Shredding Equipment Capital Costs

Shredder type and manufacturer	Estimated cost* (000s)	System configuration†	Estimated throughput,‡ tons
Replaceable cutter:			
Columbus McKinnon	$500–$525	Portable	12–13 RS
	$435–$460	Stationary	8–10, 2 in
			4–5, 1 in
Triple/S Dynamics	$475–$500 ($575)§	Portable	12–13 RS
	$400–$425	Stationary	8–10, 2 in
			4–5, 1 in
Rotary shear:			
Eidal	$400–$425	Portable	10–12 RS
	$290–$315	Stationary	6–8, 2 in
			3–4, 1 in
ERS	$500–$525	Portable	12–13 RS
	$425–$450	Stationary	8–10, 2 in
			3–4, 1 in
Mac-Saturn	$400–$425§	Portable	10–12 RS
	$340–$365	Stationary	6–8, 2 in
			3–4, 1 in
Mitts & Merrill (Carthage)	$400–$425	Portable	8–10 RS
	$250–$275	Stationary	5–7, 2 in
			2.5–3, 1 in
Shredding systems	$450–$475	Portable	10–12 RS
	$375–$400	Stationary	6–8, 2 in
			3–4, 1 in

General: Portable systems are self-contained with diesel generator; systems include conveyors, sizing device (typically a disc screen) and magnetics for 1 in minus chip production; 1 in minus chip throughputs are estimates based on limited experience.

*Costs are *estimated*; will vary for each application and with shredder model; costs do not include recent price increases that may have occurred.

†"Portable" assumes 1 trailer with diesel generator (unless otherwise noted); "stationary" assumes electric power is available on site.

‡*Estimated* throughput in tons per hour; RS: rough shred throughput, one pass through cutters; 2 in minus chip throughput; 1 in minus chip throughput.

§Two trailer system with trommel as sizing device which requires second trailer.

Source: *Proceedings 1991 Conference on Waste Tires as a Utility Fuel,* Electric Power Research Institute.

per ton. In some states, tdf producers receive a per ton payment from a state recycling fund. Similarly, a few states have offered a direct reimbursement to fuel users (typically 1 cent per lb) to help encourage markets.[53]

As fuel markets have expanded, both tdf processors and energy users have found that one of the key elements to further developing and expediting fuel markets is the ability to demonstrate the value added by the use of the tire fuels. The value added must come in the form of improved performance over existing products and service on the part of the fuel user in the application of the tire fuel. In the case of energy conversion in cement kilns, the wire and compounding ingredients in tires can offset the addition of certain other compounds, such as iron oxide, necessary in the production of cement. For paper mill lime kilns, coal cannot be used and, therefore, natural gas or oil are required. Wire-free tire fuels can be used as partial fuel replacement in lime

TABLE 9.18 Comparative Volume Sensitivity of tdf Production Facilities

| | Processing rate (tires/year) | | | | | |
| | 500,000 | | 1,200,000 | | 2,000,000 | |
Cost component	Basis	$/tire	Basis	$/tire	Basis	$/tire
	Exhibit I: variable costs					
Labor:						
Supervision at $30,000/year	1	$0.06	1	$0.02	2	$0.03
Manual at $18,000/year	2	$0.07	3	$0.05	6	$0.05
Subtotal	$66,000	$0.13	$84,000	$0.07	$168,000	$0.08
Power:						
0.75 kW/hp at $0.07/kW	200 hp	$0.04	300 hp	$0.03	300 hp	$0.03
Subtotal	$22,000	$0.04	$33,000	$0.03	$66,000	$0.03
Maintenance:						
Knives at $5/ton	$25,000	$0.05	$60,000	$0.05	$100,000	$0.03
Other at $9/ton	$45,000	$0.09	$108,000	$0.09	$180,000	$0.09
Subtotal	$70,000	$0.14	$168,000	$0.14	$280,000	$0.14
Subtotal	$158,000	$0.31	$285,000	$0.24	$514,000	$0.25
Wire disposal (if necessary):						
25% by weight at $30/ton	1250 tons		3000 tons		5000 tons	
	$37,500	$0.08	$90,000	$0.08	$150,000	$0.08
Total variable	$195,500	$0.39	$375,000	$0.32	$664,000	$0.33
	Exhibit II: fixed costs					
Administration:						
Manager	$48,000	$0.09	$48,000	$0.04	$48,000	$0.02
Sales and services	By mgr.	$0.00	$36,000	$0.03	$36,000	$0.02
Clerical	$18,000	$0.04	$18,000	$0.01	$18,000	$0.01
Office expense (including travel)	$18,000	$0.04	$21,000	$0.02	$24,000	$0.01
Professional services	$10,000	$0.02	$10,000	$0.01	$10,000	$0.01
Subtotal	$94,000	$0.19	$133,000	$0.11	$136,000	$0.07
Capital charges:						
Depreciation at 20%/year	$100,000	$0.20	$150,000	$0.12	$200,000	$0.10
General expense:						
Insurance at 1% of capital	$5,000	$0.01	$7,500	$0.01	$10,000	$0.01
Property taxes at 2% of capital	$10,000	$0.02	$15,000	$0.01	$20,000	$0.01
Subtotal	$15,000	$0.03	$22,500	$0.02	$30,000	$0.02
Total fixed	$209,000	$0.42	$305,500	$0.25	$366,000	$0.19
Total cost	$404,500	$0.81	$680,500	$0.57	$1,030,000	$0.52
Profit at 25% ROC/BT	$125,000	$0.25	$187,500	$0.16	$250,000	$0.12
Total price	$529,500	$1.06	$868,000	$0.73	$1,280,000	$0.64

Source: T.A.G. Resource Recovery.

kiln applications and can be used to offset these more expensive fuels. Power plants use large quantities of pulverized coal. Tire fuels, if properly processed, blended, and metered with coal, can help improve combustion performance and meet the environmental permits of the energy user.[54]

The most limiting factor to widespread use of tdf today is the availability of a consistent quality supply of tdf. While the tires may be available in an area, the processing systems may not be. In addition, tdf processors must, in some cases, become experts in the application of tire fuels and help assist the energy user in converting whole or processed tire-derived fuels into an environmentally and economically acceptable fuel supply. Successful tdf processors and suppliers must have significant capital, research capabilities, and the technology and equipment in order to reduce the perceived risks associated with these fuels.[55]

Finally, the tdf producer and the fuel user must work hand in hand to address the most persistent barrier to expansion in the tdf market—public perception. The belief still exists on the part of the public that the combustion of tires yields black smoke, noxious odors, and toxic emissions despite EPA reports that the use of tdf actually improves emissions in conventionally fired plants. (See Figs. 9.81 to 9.83.)

- Availability and cost effectiveness of competing supplemental fuels such as solvents and hazardous waste. This procedure may be impacted by stricter BIF regulations
- Significant capital investment needed for installation and modification of feeding system
- High costs of conducting permit required tests
- Reliability of tire and tdf supply in isolated areas
- Local opposition to tire burning

FIGURE 9.81 Principal barriers to tdf use in cement kilns. (*Source: Scrap Tire Use/Disposal Study 1992 Update.*)

- Marginal cost advantage of tdf over typical mill fuels (coal, purchased hog fuel)
- Environmental permit modification requirements; inconsistent regulatory guidance in some states
- Remote location of many mills (higher transportation costs)
- Reliability of tdf supply in remote areas
- Lack of easy access to technical and environmental information regarding the use of tdf in U.S. pulp and paper mills.

FIGURE 9.82 Principal barriers to tdf use in pulp and paper mills. (*Source: Scrap Tire Use/Disposal Study 1992 Update.*)

- Marginal cost advantage of scrap tire material over competing fuels
- Environmental permit modification requirements; inconsistent regulatory guidance in some states
- Reliability of tdf supply in remote locations
- Conservative and risk-averse nature of utility industry

FIGURE 9.83 Principal barriers to tdf use in electric power utilities. (*Source: Scrap Tire Use/ Disposal Study 1992 Update.*)

Crumb Rubber from Scrap Tires

Scrap tire rubber that has been ground into particles ranging in size from $\frac{1}{4}$ in to 100 mesh or finer is generally described as crumb rubber.[56] (Mesh is a term commonly used to measure the size of crumb rubber and is defined by the number of openings found in 1 in^2 of a standard sieve; e.g., 10 mesh means there are 10 holes per square inch, and rubber particles which pass through the openings will be sized as 10 mesh. The larger the mesh designation the smaller the screen aperture.)

Crumb rubber may be processed using either an ambient or a cryogenic grind system. The methods are similar in that each uses a series of size reduction equipment including shredders, granulators, cracker mills, separators, and classifiers to grind rubber particles and separate the fiber and steel. In the cryogenic process liquid nitrogen or other forms of cooling technology are used to embrittle the rubber, allowing for a clean separation of rubber from the wire and cord. This method produces cubical particles of crumb rubber. Ambient grinding is accomplished mechanically and produces irregularly shaped crumb rubber particles.[57]

The primary sources of rubber for making crumb or ground rubber are buffings (fine, elongated particles of rubber generated when the tread area of a worn tire casing is buffed in preparation for applying new tread to the tire), skivings (pieces of wire-free rubber cut from the tread and shoulder area of worn treads that have been removed from truck or passenger tires), tire peels, scrap rubber pieces, and whole passenger and truck tires. It should be noted that it is only in the last few years that processors have begun to produce crumb and ground rubber feedstocks from whole steel belted radial tires. Formerly, nonsteel tires and buffings (as described above) were the primary feedstocks.[58]

Ground or crumb rubber is used to modify asphalt and is incorporated in the manufacture of a variety of rubber products. There are two primary technologies for incorporating scrap tire rubber into asphalt paving materials. The "dry process" incorporates ground tire rubber in the aggregate mix. In the "wet process" finely ground rubber is used to modify the binder. There are several technologies on the market today for using both the wet and dry process. In addition, a generic process which incorporates ground rubber in both the binder and the aggregate is available.

Ground rubber is also used in crack and joint sealants, cape seals and spray applications in stress absorbing membranes (SAMs), and stress-absorbing membrane interlayers (SAMIs). Roofing sealants and insulating coatings, as well as membrane liners for landfills and retention ponds, also use ground rubber.[59]

Although the concept of incorporating scrap tire rubber into asphalt materials has been around for over 25 years, market development efforts for the material were boosted by the passage of the Intermodal Surface Transportation Efficiency Act of 1991 (ISTEA). This law requires state transportation departments, beginning in fiscal 1994, to use increasing amounts of recycled rubber from tires in federally funded road paving projects. The minimum utilization standards set by the law require states to use crumb rubber in 5 percent of the federally funded road projects in their individual states in 1994, 10 percent in 1995, 15 percent in 1996, and 20 percent in 1997 and thereafter.

The Scrap Tire Management Council estimates that when ISTEA is implemented at its minimum rate it will require 17 million tires to supply a sufficient amount of ground rubber to satisfy the 1994 demand. This demand, according to STMC, will increase by 17 million tires each year after until 1997, when the demand to meet the requirements of ISTEA alone will consume about 68 million scrap tires annually. Carried to the year 2000, the ISTEA law will have created a market for nearly 350 million scrap tires over its lifetime.[60]

However, the law is not without opposition. Opponents question the lack of standard performance specifications for crumb rubber modifiers, the potential health effects from the emissions produced when rubber is mixed with hot-mix asphalt, the recyclability of pavements containing crumb rubber modifiers, and the cost to use crumb rubber modifiers in paving applications.[61]

In a joint report issued to Congress in June as required by the ISTEA law, the U.S. Environmental Protection Agency (EPA), U.S. Department of Transportation (DOT), and Federal Highway Administration (FHWA) concluded that there is no marked difference between regular asphalt and asphalt containing rubber.[62]

FHWA is currently evaluating performance specifications for pavements but does not anticipate lowering the specifications because of the addition of rubber.[63] While neither side seems to have definitive answers regarding cost, users of crumb rubber modifiers point out that there is a sufficient body of experience with the material to show that under certain conditions crumb rubber modified asphalt pavements can be applied in reduced thicknesses to conventional asphalt with the same or improved performance.[64] In addition, proponents contend that the cost of crumb rubber modified pavements will decrease as the amounts of roads paved with the material increase and crumb rubber modifiers become more available in the marketplace.[65] (See Fig. 9.84.)

Civil Engineering Applications for Scrap Tires

The use of whole scrap tires and processed scrap tire materials to replace standard construction materials represents one of the newest expanding markets for scrap tires. Civil engineering applications include such uses as:

- Lightweight fill
- Bulking agent in compost applications
- Breakwaters and reefs, shore protection
- Erosion control
- Embankment stabilization
- Beneficial landfill uses (daily cover, leachate collection)
- Crash barriers, side slope stabilizers
- Dock bumpers
- Stone and aggregate replacement septic and other drainage systems

---

- High initial costs
- Inconsistent long-term performance data; long-term technical and economic benefits are difficult to predict
- Lack of standardized long-term environmental testing protocols
- The Federal Department of Transportation and Federal Highway Administration need to provide state DOTs with more details and guidance concerning ISTEA
- Questions regarding the effect on recyclability of used asphalt

---

FIGURE 9.84 Principal barriers to use of scrap tire rubber in asphalt and paving applications. (*Source: Scrap Tire Use/Disposal Study 1992 Update.*)

While these applications offer the significant benefit of using scrap tires in beneficial highway or construction applications, the question of whether tires leach compounds that may adversely affect the environment has not yet been definitively answered. According to the EPA, there has been no evidence to date to support this concern. Tests have been performed by a number of states, industry organizations, and engineering companies, in compliance with EPA's Toxicity Characterization Leaching Procedure (TCLP). Overall the test results showed that none of the cured rubber products tested exceeded TCLP regulatory limits and most compounds detected were found at trace levels ranging from 10 to 100 times less than the TCLP limits and EPA's Drinking Water Standard MCL values[66] (see Fig. 9.85).

A recent project in Virginia as well as several projects in Maine in which processed scrap tires were used as lightweight fill illustrate the potential of this market to consume huge volumes of scrap tires in short sections of highway. In Virginia, about 40,000 tons of wire-containing shredded tire material were graded into 20-ft embankments for an overpass. The project represents the use of about 2.5 million scrap tires. The same length of roadway using crumb rubber additives in the pavement would use the equivalent of only 3500 tires.[67] In Maine, engineers report that they used 20,000 scrap tires in a 600-ft roadbed project—which equals 150,000-plus tires per mile.[68]

Rubber Products Markets on the Rise

Traditionally, manufactured and fabricated rubber goods which incorporated ground or cut rubber from scrap tires have been considered low-volume uses compared with tire-derived fuel, rubberized asphalt, and civil engineering uses. However, scrap tires are finding their way into more and more useful products, conserving energy, reducing pressure on landfills, and saving dollars along the way. From mats to mulch to rubber fenders, the applications for recycling tires and tire-derived materials are increasingly varied, practical, and marketable. Although most of the product applications are not new, consumer interest and market recognition are helping create a new demand for products made of recycled tire rubber.

While no one is saying that recycled rubber products will solve the scrap tire problem, they are making a dent. Perhaps, more importantly, manufacturers of these recy-

---

- Lack of definitive results on the possibility of leachate contamination from scrap tires in certain applications.
- Lack of long-term test data on the suitability of scrap tires for certain applications

---

FIGURE 9.85 Principal barriers to use of scrap tires in civil engineering applications. (*Source: Scrap Tire Use/Disposal Study 1992 Update.*)

- 1992 ground rubber market: 160 million lb
- Total market: over 50% supplied by 6–8 companies
- Industry capacity: 50–100% greater than demand
- Capital-intensive industry
- Majority incapable of processing whole tire material

FIGURE 9.86 Existing ground rubber industry structure. (*Source: Baker Rubber Company.*)

cled rubber products are creating jobs and putting quality products into the marketplace. In 1991, according to the EPA report *Markets for Scrap Tires,* approximately 120 to 150 million lb of ground rubber were converted to new products. (See Fig. 9.86.)

EPA notes, however, that 75 percent of the rubber used came from buffings. In recent years the entry of several companies in the marketplace with the capability to produce ground rubber from whole scrap tires or whole scrap tire materials (i.e., shreds) is rapidly changing this scenario.

Some recent examples illustrate the changing face of the ground rubber markets. Royal Rubber & Manufacturing Co., a California maker of rubber flooring, opened a new plant in Calhoun, Ga., giving jobs to 30 employees. The firm expects to increase the number of jobs at the Georgia plant to 75 after the first year. Royal's flagship plant in South Gate, Calif., already employs about 130 workers to make floor mats from recycled tires. The California plant uses about 15 million lb of recycled tire rubber annually as feedstock for mats and floor tiles. In scrap tire equivalents, that's nearly 4 million tires. Royal uses recycled tire rubber in crumb form to make the mats. Royal, like many other rubber products manufacturers, derives the crumb from buffings but is hoping an affordably priced crumb rubber feedstock will be produced from whole tires in the future.[69]

Increased production has already led to plant expansion for Recovery Technologies, Inc. (RTI), Cambridge, Ontario. Backed by a $1 million grant ($CN) from the provincial government, RTI built a $2.5 million plant to process whole tires into crumb and powdered rubber feedstocks for roofing materials, rubber-plastic compounds, playground composite surfaces, and lay-flat rubber products.

One of the largest volume uses of recycled tire rubber in a molded product is rubber grade crossings. OMNI Rubber Products, the leading U.S. manufacturer of the grade crossings, recycles more than 20 million lb of "off-spec" styrene butadiene rubber each year to make the crossings. As demand for OMNI's rubber grade crossing systems and valve box cushions grows, the company must source additional raw materials. Currently, OMNI purchases tire buffings directly from brokers that accumulate buffings for resale. In addition, OMNI buys crumb rubber produced by whole scrap tire processors and has recently installed a cracker mill to process larger scrap rubber pieces available from other sources.[70] OMNI's large demand for recycled scrap tire rubber has stimulated U.S. and Canadian processors to develop new methods to transform whole scrap tires into recyclable raw products that are price-competitive with tire buffings, according to the company.

The major ground rubber markets are pneumatic tires, friction materials, molded and extruded rubber products, rubber and plastic products, bound rubber products, and athletic and recreational surfaces. (See Fig. 9.87.) The production of the majority of

---

- Pneumatic tire
- Friction materials
- Molded and extruded rubber products
- Rubber and plastic products
- Bound rubber products
- Athletic and recreational surfaces
- Asphalt products

FIGURE 9.87   Major ground rubber markets. (*Source: Baker Rubber Company.*)

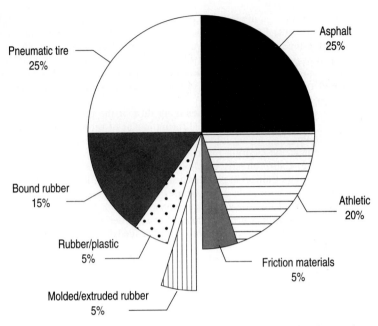

FIGURE 9.88   Estimate of ground rubber market. (*Source: Baker Rubber Company.*)

these products is accomplished through a physical bonding process which uses an adhesive bond rather than a chemical bond to produce the materials.

Among the fastest-growing uses in this category is the use of ground rubber for athletic surfaces (tracks, flooring, playground and recreational surfaces, etc.). (See Fig. 9.88.) In addition, the use of crumb rubber as a cushion covering under playground equipment is becoming popular in many communities. Likewise, crumb rubber used as a soil enhancer has tremendous market potential in sports turfs and professional landscaping applications.

Conclusion

Clearly, there is no shortage of options for managing, reusing, and recycling scrap tires. Continued innovation and development of new recycled rubber products and applications can help eradicate the nation's scrap tires, but not without consumer and market commitment from federal, state, and local governments, businesses, and individuals to buy and use recycled products.

PART C

Gary R. Brenniman
Stephen D. Cosper
William H. Hallenbeck
James M. Lyznicki

Office of Solid Waste Management
Environmental and Occupational Health Sciences Division
School of Public Health, University of Illinois
Chicago, Illinois

9.7 AUTOMOTIVE AND HOUSEHOLD BATTERIES*

Introduction

Americans use considerable quantities of batteries to power a variety of household and industrial products. Batteries are used in motor and marine vehicles, electronics, watches, cameras, calculators, hearing aids, cordless telephones, power tools, and countless other portable household devices. Recently, much attention has been focused on the potential environmental and human health risks associated with the heavy metals present in batteries. Such concern has caused many municipalities to consider programs for recovering the large number of batteries discarded in municipal solid waste (MSW). Historically, residential collection and recycling efforts have been limited to used automobile batteries. In recent years, states and municipalities have begun to focus on the recovery of used household batteries. Such activities have coincided with a number of legislative and industry initiatives to reduce the toxicity of batteries and to promote their safe collection, reclamation, and disposal.

Battery Definitions and Terms

Batteries are complex electrochemical devices, composed of distinct cells, that generate electrical energy from the chemical energy of their cell components. Despite the technical distinction between them, the terms "battery" and "cell" are often used interchangeably. A battery cell consists primarily of a metallic anode (negative electrode), a metallic oxide cathode (positive electrode), and an electrolyte material that facilitates the chemical reaction between the two electrodes. Electric currents are generated as the anode corrodes in the electrolyte and initiates an ionic exchange reaction with the cathode. The electrical energy produced from this reaction is sufficient to power a variety of consumer and industrial devices.

Batteries are classified and distinguished according to their chemical components. Batteries are referred to as wet or dry cells. In wet cell batteries, the electrolyte is a

---

*Section 9.7 contributed by James M. Lyznicki, Gary R. Brenniman, and William H. Hallenbeck.

liquid. In dry cell batteries, the electrolyte is contained in a paste, gel, or other solid matrix within the battery. Primary batteries contain cells in which the chemical reactions are irreversible, and they therefore cannot be recharged. This is in contrast to secondary batteries in which the chemical reactions are reversible and external energy sources can be repeatedly applied to recharge the battery cells.

Batteries are manufactured in a variety of sizes, shapes, and voltages. They are produced in rectangular, cylindrical, button, and coin shapes. In addition, many portable tools and electronic devices utilize rechargeable batteries contained in battery packs. Refer to Table 9.19 for a listing of the common types and general uses of batteries found in MSW.

Composition of Batteries in MSW

Batteries differ in their chemical composition, energy storage capacity, voltage output, and life span. These factors affect their overall performance, utility, and cost. Because of their different intended uses, consumer batteries are usually distinguished as automotive (i.e., lead-acid storage batteries) and household batteries.

Lead-Acid Storage Batteries (Wet Cells). Lead-acid storage batteries are used in automobiles, motorcycles, boats, and a variety of industrial applications. They are primarily used to provide starting, lighting, and ignition for automotive products.[71] These are wet cell batteries consisting of lead electrodes in a liquid sulfuric acid electrolyte. It is estimated that 78 to 80 million automotive and light truck batteries are sold each year in the United States.[72] The average battery weighs approximately 36 lb, one-half of which is composed of the lead anode and lead dioxide cathode.[71] It is estimated that the lead in automobile batteries accounts for approximately two-thirds of the total weight of lead in MSW.[71] In addition to lead, each battery contains approximately 1 gal of sulfuric acid (i.e., 9 lb), almost 3 lb of polypropylene plastic casing, about 3 lb of polyvinyl chloride rubber separators, and about 3 lb of various chemical sulfates and oxides to which the lead is bound.[73] The typical useful lifetime of lead-acid storage batteries is 3 to 4 years.[71]

TABLE 9.19 Types and Uses of Batteries Found in Municipal Solid Waste

Battery type	Shapes	Uses
Wet cells: Lead acid	Rectangular	Cars, motorcycles, boats

Dry cells—primary:
   Zinc-carbon
   Alkaline
	Cylindrical, rectangular,
	   button; AA, AAA,
	   C, D, 9V
		Flashlights, radios,
		   tape recorders, toys

Mercuric oxide
   Silver oxide
   Zinc air
   Lithium
	Button, cylindrical	Hearing aids, watches,
		   calculators, pagers,
		   camcorders, computers,
		   cameras

Dry cells—secondary:
   Nickel-cadmium
   Lead acid
	Cylindrical, button,
	   or in battery packs
		Rechargeable cordless
		   products such as power
		   tools, vacuum cleaners,
		   shavers, phones

Household Batteries (Dry Cells).   Americans use about 8 to 10 household batteries per person each year.[75] In 1992, it is estimated that almost 4 billion dry cell batteries were sold in the United States.[74] Total future sales of household batteries are expected to increase by about 6 percent each year.[74] The types and percentages of household batteries sold in the United States in 1992 are shown in Fig. 9. 89.

Dry cell batteries contain electrodes composed of a variety of potentially hazardous metals including cadmium, mercury, nickel, silver, lead, lithium, and zinc. The electrode materials and electrolytes found in household batteries are listed in Table 9.20. In addition to electrodes and electrolytes, batteries also contain other materials that are added to control or contain the chemical reactions within the battery.[74,76] For example, mercury is added to the zinc anode of primary cells (e.g., alkaline, zinc-carbon) to reduce corrosion and to inhibit the buildup of potentially explosive hydrogen gas.[77] In addition, mercury helps to prevent the batteries from self-discharging and leaking.[78] Other components of batteries include graphite, brass, plastic, paper, cardboard, and steel.

Primary Dry Cell Batteries.   When purchasing batteries, primary dry cell batteries are generally less expensive than secondary or rechargeable batteries. However, when making cost comparisons, consumers should consider that rechargeable batteries are reusable whereas primary batteries must be replaced once they are discharged. Primary dry cells accounted for almost 90 percent of U.S. battery sales in 1992.[75] The majority of batteries purchased were cylindrical and rectangular varieties. Button cells represented only 5 percent of the total battery cell market in 1992.[75]

ALKALINE (MANGANESE) BATTERIES.   Alkaline batteries are the most common household dry cell batteries sold in the United States. It is estimated that they represent over 63 percent of the household battery market and are increasing their market share.[74] Alkaline batteries are manufactured in many sizes and shapes. Their good performance and long shelf life make them appealing for a variety of consumer uses.

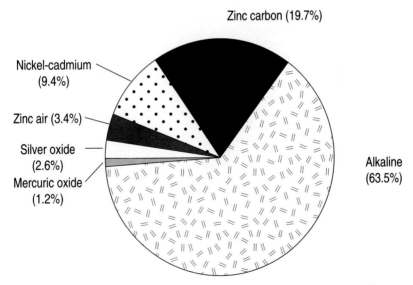

FIGURE 9.89   1992 sales percentage of domestic batteries in the United States.[74] (Note: Lithium batteries accounted for 0.2 percent of battery sales in 1992.)

TABLE 9.20 Primary Chemical Components of Household Batteries in Municipal Solid Waste[71,74]

Battery type	Cathode	Anode	Electrolyte
Alkaline	Manganese dioxide	Zinc	Potassium and/or sodium hydroxide
Zinc-carbon	Manganese dioxide	Zinc	Ammonium and/or zinc chloride
Mercuric oxide	Mercuric oxide	Zinc	Potassium and/or sodium hydroxide
Zinc-air	Oxygen from air	Zinc	Potassium hydroxide
Silver oxide	Silver oxide	Zinc	Potassium and/or sodium hydroxide
Lithium	Various metallic oxides	Lithium	Various organic and/or salt solutions
Nickel-cadmium (rechargeable)	Nickel oxide	Cadmium	Potassium and/or sodium hydroxide
Sealed lead-acid (rechargeable)	Lead oxide	Lead	Sulfuric acid

Recent environmental concerns have resulted in dramatic reductions in the mercury content of alkaline batteries. For example, batteries that contained up to 1 percent mercury by weight in the mid-1980s are now being produced with mercury concentrations of 0.0001 to 0.025 percent.[74] Because of design limitations, such reductions will be more difficult to achieve for button-size batteries than for cylindrical and rectangular batteries.[78] The major alkaline battery manufacturers have established implementation dates for "no mercury added" battery designs for nonbutton cells by 1993.[78] In addition to mercury, alkaline batteries also contain metals such as lead, cadmium, arsenic, chromium, copper, indium, iron, nickel, tin, zinc, and manganese.[71,74]

ZINC-CARBON BATTERIES. Zinc-carbon batteries are the second most commonly used household battery. These batteries represent about 20 percent of the household battery market; however, sales are declining.[74] These batteries are manufactured as inexpensive, general-purpose batteries as well as "heavy-duty" varieties. Zinc-carbon batteries have a shorter shelf life than alkaline batteries, are less powerful, and have a tendency to leak in devices once they are discharged.[76,78] Because of their anode configuration, zinc-carbon batteries require less mercury than alkaline batteries.[76] Reduction and eventual elimination of added mercury to zinc-carbon batteries is anticipated in the near future.[78] In addition to mercury, zinc-carbon batteries also contain metals such as lead, cadmium, arsenic, chromium, copper, iron, manganese, nickel, zinc, and tin.[71,74]

SILVER OXIDE BATTERIES. Silver oxide batteries account for less than 3 percent of all household battery sales and about 5 percent of the button cell market.[74,75] These batteries are manufactured in a variety of button sizes and provide a more constant voltage output than alkaline or zinc-carbon button cells.[78] Silver oxide batteries are interchangeable with mercuric oxide batteries and are increasingly being used to power hearing aids and watches.[77] However, silver oxide batteries are generally more expensive than mercuric oxide cells.[76] Silver oxide batteries contain about 1 percent mercury by battery weight.[78]

MERCURIC OXIDE BATTERIES.   Mercuric oxide batteries account for about 1 percent of annual U.S. battery sales, a percentage that is expected to decline in future years.[74] Most mercuric oxide batteries are manufactured as button cells and represent about 20 percent of that market.[75] Increasingly these batteries have come under scrutiny since more than one-third of their weight is mercury. Suitable alternatives to mercuric oxide cells have been developed (e.g., silver oxide, zinc air) which should reduce consumer dependence on mercuric oxide cells. Despite their decreasing use by household consumers, mercuric oxide batteries continue to be used in a variety of industrial, medical, military, and communications devices.[77]

ZINC AIR BATTERIES.   Zinc air batteries have become increasingly popular in the United States. They represent over 3 percent of the total U.S. household battery sales[74] and over 60 percent of consumer button cell battery purchases.[75] Primary uses are in hearing aids and pagers. Zinc air batteries are advantageous since they have a longer life than silver or mercuric oxide batteries.[78] However, their use is restricted since they require ambient air to provide their oxygen cathode. Consequently, they cannot be used for tightly sealed applications such as watches.[78] These batteries contain about 1 to 2 percent mercury by weight.[74]

LITHIUM BATTERIES.   Lithium batteries represent less than 0.25 percent of the total U.S. household battery market, although their market share is expected to increase in the future.[74] They are manufactured as cylinders, buttons, or coin shapes and may also be contained in battery packs. Despite their high cost, their excellent performance characteristics make them useful in a variety of consumer electronics and computer applications as well as military and medical devices.[78] Lithium is a highly reactive material, especially when mixed with water. Consequently, safety precautions are recommended when collecting, storing, or transporting unspent batteries for disposal or reclamation.[74]

Secondary Dry Cell Batteries.   Secondary dry cells accounted for approximately 10 percent of battery sales in 1992.[74] These rechargeable batteries are preferable to primary cells since they can be used repeatedly. However, their lower performance characteristics may be restrictive for some consumer applications.[74,78] It is estimated that one rechargeable battery can substitute for 100 to 300 single-use batteries.[74,78] Many rechargeable batteries are sealed within consumer products such as cordless telephones, power tools, appliances, personal computers, and other electronics. Recent attention has focused on providing easy access to the rechargeable batteries in consumer products to encourage their recycling or proper disposal. Legislative action by a number of states has resulted in mandates for manufacturers to produce products with removable rechargeable batteries by July 1, 1993.[79]

NICKEL-CADMIUM BATTERIES.   Nickel-cadmium batteries (i.e., Ni-Cd or "ni-cads") are the most common of the secondary or rechargeable household batteries sold in the United States. They represent almost 10 percent of the total U.S. household battery purchases.[79] Future sales are expected to increase.[74] These batteries are available in sizes comparable with alkaline and zinc-carbon batteries. However, they are currently not as powerful as primary cells and tend to discharge more rapidly.[78] Nickel-cadmium batteries are a major consumer of cadmium in the United States.[71] These batteries typically have a cadmium content ranging from 11 to 15 percent of the battery weight.[74]

SEALED LEAD-ACID BATTERIES.   In addition to automotive uses, lead-acid batteries (called sealed lead-acid batteries) are used in a variety of consumer products such as toys, video recorders, portable electronics, tools, appliances, and electric start lawn mowers. These smaller, rechargeable batteries are dry cells since the sulfuric acid electrolyte is contained on a solid separator material or in a gel.[71] These batteries account for less than 1 percent of U.S. household battery purchases.[78]

Environmental Impacts of Batteries in MSW

The disposal of used automobile and household batteries into MSW must be assessed for its potential human and environmental health impacts. An estimated 1.7 million tons of lead-acid batteries were generated in MSW in 1990.[80] This represented less than 1 percent of the total weight of MSW generated during that year. Despite a well-established recycling infrastructure, about 6 percent of lead-acid batteries were landfilled or incinerated in 1990. Household batteries accounted for about 142,000 tons of MSW in 1991 and about 146,000 tons in 1992.[74] This represented less than 0.1 percent of the total weight of MSW generated during those years. Although the tonnage of these materials in MSW may seem small and inconsequential, their potential toxicity must be considered in order to evaluate appropriate and safe disposal practices.[74]

The disposal of used batteries in MSW is problematic for two reasons. First of all, batteries contribute to the total quantity of potentially hazardous waste that is disposed in MSW. Second, and more importantly, batteries contain many potentially toxic chemicals that can have adverse environmental and human health impacts. Assessing the environmental impact of used battery disposal involves evaluating the potential for groundwater contamination due to the leaching of contaminants from MSW landfills; the emission of contaminants into the air from MSW incinerators; the presence of hazardous materials in the residual ash remaining after MSW incineration; and finally, the contamination of composted organic waste with battery components. Such assessments must continually be refined as more stringent regulations are imposed on the design of MSW landfills and incinerators and as the toxicity and types of batteries in MSW continue to change.

Batteries contain a variety of heavy metals that may become toxic contaminants in landfill leachate, incinerator emissions, incinerator ash, and compost.[74,76,78,81] Much concern has been directed to the high percentage of mercury, cadmium, and lead in MSW that is attributed to used batteries.[1,7] Other potentially toxic metals that may be present in batteries include silver, zinc, nickel, manganese, lithium, chromium, and arsenic.[74] Table 9.21 depicts the metallic composition of the typical household batteries found in MSW.

When released, heavy metals persist in the environment. Many of them also have associated human, animal, or plant toxicities.[74,78] Many accumulate in aquatic sedi-

TABLE 9.21 Weight Percentage of Potentially Toxic Heavy Metals in Common Household Batteries[74]

Battery type	Metal, %				
	Cadmium	Mercury	Nickel	Silver	Zinc
Alkaline	0.01	0.025–0.5			8–18
Zinc-carbon	0.03	0.01			12–20
Mercuric oxide		30–43			10–15
Silver oxide		1.0		30–35	30–35
Zinc air		2.0			35–40
Nickel-cadmium	11–15		15–25		

ments and soil and may be metabolized by indigenous microorganisms to more toxic organic forms. Of most concern is the potential for uptake and accumulation of heavy metals or their metabolites in the food chain.[76,78] Recent attention has focused on potential human and environmental exposure risks from the metals present in used batteries, specifically lead, cadmium, and mercury.[74,76,78,81]

Lead. Nearly 1.3 million metric tons of lead are consumed in the United States each year. Of this, over 1 million metric tons (79 percent) are used to manufacture lead-acid batteries; most of this is used to manufacture automobile batteries.[82] Lead-acid storage batteries comprise the largest percentage of the weight of batteries discarded in the United States. In addition, they comprise almost two-thirds of the lead in MSW. Based on 1986 figures, lead-acid batteries contributed about 65 percent of the lead in MSW.[71] Although the total lead discards in MSW are expected to increase yearly, the percentage of lead due to lead-acid batteries is expected to remain fairly constant.[71] Ultimately, the effectiveness of lead-acid battery recycling programs will significantly impact the weight of lead-acid batteries that are landfilled or incinerated.[71]

Cadmium. Based on 1986 figures, household batteries accounted for more than 50 percent of the cadmium in MSW.[71] As nickel-cadmium batteries increase in popularity, the amount of cadmium in the waste stream is expected to increase. It is estimated that by the year 2000, dry cell batteries will account for 76 percent of the cadmium in MSW.[71,74]

Mercury. Based on 1989 figures, household batteries accounted for more than 88 percent of the mercury in MSW.[77] Alkaline batteries accounted for the largest quantity (i.e., 59 percent) of the total weight of mercury in MSW.[74,77] Although the sales of household batteries (including alkaline cells) is projected to increase, the amount of mercury in MSW is expected to decrease in the future. This is due to further reductions in the mercury content of dry cells as well as to the use of alternatives for mercuric oxide batteries. Despite reductions in their mercury content, dry cell batteries are still expected to account for about 56 percent of the mercury in MSW in the year 2000.[74,77]

Used Battery Regulations

Solid waste disposal regulations are based on Subtitles C and D of the Resource Conservation and Recovery Act of 1976 (RCRA) and are codified in Title 40 of the Code of Federal Regulations (CFR). Much of the attention for managing used batteries has centered around whether to classify them as hazardous waste. Such a designation could complicate used battery collection programs by imposing RCRA Subtitle C hazardous waste regulations on the collection, storage, transportation, reclamation, and disposal of used batteries. However, household generated waste, including used batteries, is exempt from RCRA hazardous waste rules and regulations under the "household waste" exclusion rule of 40 CFR, Part 261.4(b)(1).

Waste generated from nonhousehold sources may be declared hazardous waste if it is specifically "listed" (i.e., 40 CFR, Part 261, Subpart D) or if it exhibits one of the following "characteristics" (40 CFR, Part 261, Subpart C): ignitability, corrosivity, reactivity, or toxicity. Although waste batteries are not "listed" hazardous wastes, they may exhibit one of the four "characteristics" of hazardous waste.[74] The toxicity characteristic is of most concern to battery recyclers and is determined by a laboratory procedure called the toxicity characteristic leaching procedure (TCLP). Heavy metals such as lead, mercury, cadmium, and chromium have been detected from batteries at concentrations that exceed the TCLP concentrations. Consequently, some household batteries would be considered characteristic hazardous wastes if it were not for the U.S. EPA exemption.[74]

In addition to RCRA hazardous waste regulations, battery recyclers must adhere to a number of additional federal and state regulations. These include compliance with water and air quality standards, transportation regulations, postal laws, and applicable state hazardous waste laws. Furthermore, potential liability under the Comprehensive Environmental Response Compensation and Liability Act (CERCLA) must also be considered by battery generators and recyclers.[74]

Automobile Batteries. In 1985, the U.S. EPA declared that lead-acid batteries were to be considered a hazardous waste.[73] Spent lead-acid batteries may exhibit the hazardous waste characteristic for toxicity (i.e., lead) as well as for corrosivity (i.e., acidic electrolyte). Regulations in 40 CFR, Part 261.6, (a)(2)(v) and Subpart G of 40 CFR, Part 266 establish a hazardous waste exemption for spent lead-acid batteries. This exemption pertains only to persons who generate, collect, store, and transport spent lead-acid batteries for reclamation but are not directly involved with battery reclamation. Lead-acid battery reclamation facilities must comply with RCRA hazardous waste regulations. If lead-acid batteries are disposed rather than sent to a battery reclamation facility, they are considered hazardous waste and are subject to RCRA Subtitle C hazardous waste regulations.

According to the Battery Council International (BCI), by February 1993, 41 states had passed some form of lead-acid battery legislation.[83] Most states have passed laws following model legislation proposed by the BCI. In addition, states have added more specific and usually more stringent regulatory language. Examples of various state lead-acid regulations include[82,84]

- Prohibition on the disposal of lead-acid batteries in landfills or incinerators
- Mandated delivery of batteries to approved retailers or collection facilities
- Establishment of requirements for retailers and other collection facilities for accepting used lead-acid batteries
- Requirement of posted written notices that inform the public of lead-acid battery recycling
- Imposition of fines to enforce the regulations
- Regulation of spent lead-acid batteries as a hazardous waste
- Regulation of transporters of spent batteries
- Exclusion of lead-acid battery recycling from regulation under hazardous waste provisions
- Deposit fees on the sale of new lead-acid batteries that can be recovered upon return of the used battery
- Requirements for the state to purchase batteries with a minimum specified recycled lead content

Household Batteries. A hazardous waste exemption for used batteries and battery cells that are returned to a manufacturer for "regeneration" is provided in 40 CFR, Part 261.6(a)(3)(ii). However, few companies in the United States are involved with the reclamation of household batteries. Additional regulations for reclaiming precious metals such as silver (i.e., from silver oxide batteries) are provided in 40 CFR, Part 261.6(a)(2)(iv).

To encourage the collection of used household batteries, the U.S. EPA has drafted and proposed a "universal waste rule" to be codified as 40 CFR, Part 273.[85] Ultimately, the rule is intended to encourage the proper collection, treatment, and/or recycling of "postuser" generated hazardous waste. Specific guidelines for used batteries, other than spent lead-acid (e.g., automobile batteries), are in Subpart B of the proposed rule. The rule is designed to streamline the collection process for used batteries by removing current regulatory barriers to their collection. The regulation would apply only to used batteries that exhibit an RCRA hazardous waste characteristic. It would affect their management prior to being received at a permitted hazardous waste treatment, storage, reclamation, or disposal facility.

As of March 1993, 15 states had adopted legislation regarding the management of waste household batteries.[86] Laws range from required battery recycling feasibility studies and plans to mandated collection and disposal programs. A survey of existing state legislation is included in the New York State household battery report.[78] The most significant legislation regarding consumer batteries has been enacted by Connecticut, Minnesota, New Jersey, New York, and Vermont. Essentially, state legislation focuses on reducing the toxicity of dry cell batteries in MSW. It is hoped that such initiatives will create more cost-effective battery reclamation and recycling options than currently exist in the United States. Notable components of state legislation include:[74]

- Manufacturers are directly responsible for the costs of properly disposing of their products. This can include used batteries as well as products powered by rechargeable batteries that may have adverse environmental impacts.

- Mandate maximum mercury content standards for batteries sold in the state. These range from 0.0001 to 0.025 percent by weight for alkaline batteries and 25 mg for alkaline button batteries. These standards also require manufacturers to redesign batteries to minimize or eliminate hazardous components.

- Require accessibility to the rechargeable batteries in consumer products.

- Ban the disposal of recyclable batteries with unregulated MSW.

- Impose regulations to reduce or eliminate various metals (e.g., mercury) from batteries to facilitate reclamation activities in the future.

- Require labeling on all batteries to assist consumers in separating battery types.

Important points for state legislators to consider when developing household battery management programs include:[74] (1) deciding whether to regulate household batteries as hazardous waste if they exhibit any of the RCRA hazardous waste characteristics; (2) deciding whether to enact the RCRA "household waste" exemption rule for household batteries; (3) deciding whether exemptions should be granted for batteries collected for reclamation or for other specified management options in states that choose not to allow the RCRA "household waste" exemption; or (4) deciding whether to adopt specific legislation regarding the composition, collection, transport, processing, and disposal of household batteries.

Collection of Used Batteries for Recycling and Disposal

Battery collection programs are intended to separate batteries from mixed MSW and keep them out of MSW incinerators, MSW landfills, and compost. In addition, such programs are designed to recover certain types of batteries for reclamation and recycling of their components. Used batteries are collected through community-sponsored drop-off locations, residential curbside collection programs, household hazardous waste collection centers, and retailers (e.g., automotive shops, jewelry stores). To encourage battery recovery, some manufacturers now provide prepaid mailers for the return of used rechargeable batteries.[79]

Battery collection programs must ensure convenient facilities for the safe collection and storage of batteries prior to shipment for reclamation or disposal. Program designs should achieve high participation and be cost-effective.[78] Collection and storage procedures will depend on applicable state regulations as they relate to the classification of household batteries as hazardous waste. Ultimately, a successful battery collection program should include ongoing public education to increase consumer participation and awareness of the types of batteries included in the program.

Automobile Batteries. State and federal regulations have established an effective infrastructure for the collection and reclamation of lead-acid batteries. Disposal bans as well as mandatory take-back and deposit programs have created a collection system that includes high retailer and consumer participation rates.[82] The success of automobile lead-acid battery collection programs results from the implementation of centralized and convenient collection locations such as automotive parts stores and service centers. In addition, the collection of lead-acid batteries is supported by the secondary lead industry, which reclaims and markets the battery components.

Household Batteries. Collection of household batteries is hampered by the lack of any centralized recovery and recycling network such as that which exists for spent lead-acid batteries. Deliberation continues over the types and components of batteries in MSW that present the greatest potential environmental and human health risks. Consequently, communities may choose: (1) not to collect any household batteries, (2) to collect only those batteries that can be reclaimed, (3) to collect only the most toxic batteries, or (4) to collect all used household batteries.

Consumers are encouraged to bring their spent household batteries to approved collection facilities for proper handling. Collected batteries should be stored in well-ventilated areas to avoid the buildup of heat as well as mercury and hydrogen gases. Facilities should also have adequate safety and fire-prevention equipment. Batteries should be stored in a dry environment and packed to minimize the potential hazards from short-circuits, leaking cells, and unspent lithium cells.[74,78,81]

Recycling Used Batteries

Used batteries cannot be recycled in the same sense that aluminum or glass containers are recycled into new containers. It is more appropriate to reserve the term "recycling" for those battery components that can be reclaimed and reused. These components include metals (e.g., lead, mercury, silver, nickel, cadmium, steel) and plastic (e.g., the battery case of automobile batteries). Some of the reclaimed materials may then be recycled into new battery components or manufactured into other products.

States have been more aggressive than the federal government in promoting used battery reclamation and recycling. Since automobile lead-acid batteries comprise such a large percentage of the lead consumed in the United States, many states and communities have established programs to collect automobile batteries for reclamation.[82] In contrast, programs for household batteries are not well established and essentially focus primarily on battery collection and safe disposal rather than reclamation.[74] Such

Battery Council International (BCI)   Independent Battery Manufacturers
401 N. Michigan Avenue   Association, Inc. (IBMA)
Chicago, IL 60611-4267   100 Larchwood Drive
(312) 644-6610   Largo, FL 34640
                                             (813) 586-1408
National Electrical Manufacturers
  Association (NEMA)   Association (PRBA)
2101 L Street NW   1000 Parkwood Circle
Suite 300   Suite 430
                                             Atlanta, GA 30339
Washington, DC 20037-1581   (404) 612-8826
(202) 457-8400
Portable Rechargeable Battery

FIGURE 9.90   Sources of information on automobile and household battery recycling.

programs are more concerned with reducing the potential environmental and human health effects from household batteries by disposing of them in approved hazardous waste landfills. Information and technical assistance regarding lead-acid and household battery reclamation is available from the organizations listed in Fig. 9.90.

Automobile Batteries.   Federal and state regulations have raised the costs of battery recycling and have greatly consolidated the lead recycling industry.[73] Recovery of lead-acid batteries for reclamation has varied from 60 percent to well over 90 percent.[80] In 1990, the U.S. EPA estimated that 1.6 million tons (i.e., 94 percent) of the lead-acid batteries in MSW were recovered for recycling. The recycling rates calculated by the BCI for 1990 and 1991 were 98 and 97 percent, respectively.[87] Recovery rates have improved as a result of regulations that ban the landfilling and incineration of lead-acid batteries. Historically, lead-acid battery recycling rates have reflected the market conditions for lead.[82] The goal of the U.S. EPA and state governments is to maintain a high recycling rate despite market fluctuations in lead prices or reductions in processing capacity.

Lead-acid batteries are recycled to reclaim the lead, sulfuric acid, and polypropylene plastic housing. Batteries are processed by secondary smelters who rely on used batteries for more than 70 percent of their lead supply.[72] There are 22 active secondary processors in the United States (Table 9.22). Most of these are independently owned and operated.

At the smelter, the batteries are crushed and then processed to recover the battery components. The sulfuric acid can be reclaimed and used in fertilizer or neutralized for disposal.[89] The plastic battery case can be recycled into new cases or other recycled plastic products. All lead-containing components are loaded into reverberatory furnaces in which the lead is melted and extracted.[89] The furnace residue is further processed in blast furnaces to recover more of the lead. The slag that remains still contains lead and must be tested prior to disposal to determine its hazardous waste characteristics. Alternative lead smelting technologies are now available that significantly reduce the amount of potentially hazardous slag generated during the smelting process.[72,89]

TABLE 9.22 Secondary Lead Smelters in the United States[88]

Secondary lead smelter	Location
ALCO Metals	Los Angeles, Calif.
The Doe Run Co.	Boss, Mo.
East Penn Manufacturing Co.	Lyon Station, Pa.
Exide Corp.	Reading, Pa. (General Battery Corp.); Muncie, Ind.; Dallas, Tex. (Dixie Metals Corp.)
GNB Incorporated	Columbus, Ga.; Frisco, Tex.; Los Angeles, Calif.
General Smelting & Refining	Cottage Grove, Tenn.
Gopher Smelting & Refining	Minneapolis, Minn.
Gulf Coast Lead	Tampa, Fla.
Interstate Lead Co.	Leeds, Ala.
Refined Metals Corp.	Beech Grove, Ind.; Memphis, Tenn.
Ross Metals	Rossville, Tenn.
RSR Corp.	Middletown, N.Y.; Indianapolis, Ind.; Los Angeles, Calif.
Sanders Lead Co.	Troy, Ala.
Schuylkill Metals	Baton Rouge, La.; Cannon Hollow, Mo.

Source: BCI, 1993.

Household Batteries.   Recycling programs for household batteries are not widespread and have been hampered by the limited number of processing facilities available. Such programs may even be misleading, since only certain types of batteries are reclaimed in the United States. As mentioned previously, the largest percentage of household batteries sold in the United States are alkaline and zinc-carbon batteries. However, there is no facility in the U.S. that reclaims these batteries. Instead, they are either shipped overseas for reclamation or disposed in domestic hazardous waste landfills. The mercury present in these batteries complicates the reclamation of other battery components.[74] Expensive mercury recovery systems are required before the zinc, steel, brass, manganese, and carbon can be recovered safely from alkaline and zinc-carbon batteries. Two foreign companies, Sumitomo Heavy Industries (Japan) and Recymet (Switzerland), have developed processes for reclaiming alkaline and carbon-zinc batteries.[74]

Currently, only three U.S. facilities reclaim household batteries. The Mercury Refining Company (MERECO) in New York recovers mercury from mercuric oxide and silver from silver oxide batteries. INMETCO in Pennsylvania recovers nickel and steel from nickel-cadmium batteries. After the batteries are processed, the residue is sent to another U.S. firm to recover cadmium. The Bethlehem Apparatus Company, also in Pennsylvania, recovers mercury from mercuric oxide batteries.[81] Detailed descriptions of U.S. as well as foreign battery reclamation processes have been published previously.[74,81] Refer to Table 9.23 for a listing of U.S. firms that accept waste

TABLE 9.23 U.S. Companies Accepting Waste Household Batteries for Disposal or Reclamation[81,90]

Company and location	Types of batteries processed
BDT 4255 Research Parkway Clarence, NY 14031 (716) 634-6794	Accepts alkaline and lithium batteries for disposal as hazardous waste. Lithium batteries are neutralized prior to disposal
Bethlehem Apparatus Co. 890 Front Street, PO Box Y Hellertown, PA 18055 (215) 838-7034	Accepts mercuric oxide batteries for reclamation of mercury
Environmental Pacific Corp. PO Box 2116 Lake Oswego, OR 97055 (503) 226-7331	Accepts all batteries and provides hazardous waste storage facilities
F. W. Hempel & Co., Inc. 1370 Avenue of the Americas New York, NY 10019 (212) 586-8055	Accepts nickel-cadmium batteries only for shipment to France for processing
INMETCO PO Box 720, Rte 488 Ellwood City, PA 16117 (412) 758-5515	Accepts nickel-cadmium batteries only for reclamation of nickel. Residual sent to Zinc Corporation of America for cadmium reclamation
Kinbursky Brother Supply 1314 N. Lemon Street Anaheim, CA 92801 (714) 738-8516	Accepts lead-acid and nickel-cadmium batteries. Lead plate and nickel sent to a smelter; cadmium sent to France
Mercury Refining Co., Inc. (MERECO) 790 Watervliet-Shaker Road Latham, NY 12110 (518) 785-1703, (800) 833-3505	Accepts all household batteries. Mercuric oxide and silver oxide batteries refined on-site; other cells marketed for disposal or reclamation
NIFE Industrial Boulevard PO Box 7366 Greenville, NC 27835 (919) 830-1600	Accepts nickel-cadmium batteries only. Batteries sent to parent smelting company in Sweden for cadmium reclamation
Quicksilver Products, Inc. 200 Valley Drive, Suite 1 Brisbane, CA 94005 (415) 468-2000	Accepts mercuric oxide batteries only
Universal Metals and Ores Mt. Vernon, NY (914) 664-0200	Accepts nickel-cadmium batteries only. Batteries are marketed overseas for metals reclamation

household batteries for disposal or reclamation. It is recommended that battery reclamation, disposal, and storage facilities be contacted directly regarding their specific guidelines and restrictions for used batteries. In addition, it is essential to carefully evaluate the business practices of these facilities to minimize future RCRA and CERCLA liability.

Summary

- The recovery of used automobile batteries for reclamation continues to be successful as a result of federal and state legislation and cooperative efforts between the battery manufacturers, secondary lead smelters, retail stores, automotive shops, and consumers. Ultimately, these efforts can significantly reduce the amount and potential toxicity of lead in MSW.

- The framework for lead-acid battery collection and reclamation programs can potentially be applied to programs for household batteries. However, consensus legislation and an infrastructure for the collection and reclamation of household batteries is not well established in the United States.

- Debate continues regarding the effectiveness and feasibility of household battery collection and recycling programs. Currently, most collected batteries are disposed in hazardous waste landfills and are not reclaimed, much less recycled. Currently, reclamation of used household batteries is not feasible in the United States and is hindered by the presence of potentially toxic metals (e.g., mercury) in primary cells.

- Source reduction initiatives must continue at both the industrial and municipal levels. Emphasis should be placed on the redesign of batteries to reduce potential toxicity. In addition, battery manufacturers should develop cost-effective reclamation technologies for their products. Source reduction also involves the removal of batteries from mixed MSW prior to composting, burning in MSW incinerators, and disposal in MSW landfills.

- Source reduction can also be promoted by encouraging consumers to use more rechargeable batteries in household products. This would reduce the number of alkaline and zinc-carbon batteries in MSW and also reduce the amount of mercury (and other potentially toxic metals) in MSW. Since nickel-cadmium batteries can be reclaimed in the United States, manufacturers and retailers should more actively promote programs for their collection. The effectiveness of such programs should be monitored to ensure that any increase in nickel-cadmium battery sales is not reflected by an increase in the amount of cadmium in MSW.

- If the toxicity of household batteries in MSW decreases, future assessments should be performed to consider the impact of their disposal in modern, well-designed, and regulated MSW landfills rather than in hazardous waste landfills. Ultimately, the development of cost-effective technologies for the reclamation of used household batteries is needed to ensure that recycling, rather than disposal, becomes the preferred waste management alternative.

9.8   USED OIL*

Introduction

Used oil is a problem waste because its generation is ubiquitous and it can contain hazardous liquid wastes and other contaminants. When used lubricating oil qualifies as a hazardous waste, its disposal becomes complicated and costly. Even when properly handled, re-refined oil carries a misperception of low quality with respect to "new" lubricating oil and often must sell at a lower price than new oil.

---

*Section 9.8 contributed by Stephen D. Casper, William H. Hallenbeck, and Gary R. Brenniman.

The most recent survey of used oil (1988) and its ultimate disposition indicates that of the 1.351 billion gal of used oil (automotive and industrial) generated annually, 901 million gal were reused in some manner such as fuel, secondary industrial use, re-refined lube oil, or road oil.[91,92] Over 400 million gal were disposed of through dumping on the ground or in water, landfilling, or non-energy-recovery incineration. Though inappropriate, the legality of these disposal methods depends on the disposal path, the generator, the type of oil, and the specific state regulations.

Burning as a fuel and re-refining are the two major methods for recycling used oil. By volume, the waste oil fuel industry consumes 58 percent of available used oil while re-refining consumes only 2 percent. Used oil has an excellent heating value (13,000 to 19,000 Btu/lb) and can help meet the growing national energy demand. However, because of the constituents present in used oils, air emission controls may be necessary when burning. A better secondary use of lube oils is its re-refining back into a usable base stock oil. In automotive engines, lube oil becomes dirty and the additives break down, but the base stock oil (roughly 80 percent by volume of marketed product) does not break down, allowing for its redistillation. The re-refining process concentrates contaminant metals into a "bottoms" residue which is reused in asphalt production. Producers of re-refined crankcase oil have shown their product to meet or exceed the engine lubrication testing requirements for "new" oil by the American Petroleum Institute. For these reasons, the re-refining of used oils into a usable product is a preferred energy conservation and pollution prevention alternative.

A lack of regulatory control of used oil disposal and lack of an infrastructure for the collection and processing of used oil from the public has resulted in improper disposal of used oils. To minimize improper disposal, the management system for used oil must be modified:

- To divert the improper disposal of used oil in refuse and the environment and get it into the used oil management system of collectors, handlers, and processors
- To correctly manage used oil after entering the system by including standards which address accountability
- To increase the flow of used oil into re-refineries

This section discusses oil consumption, hazardous contaminants in new and used oil, legislation and regulation of used oil, methods to increase used oil collection, the re-refining process, and recovery of oil and scrap material from oil filters.

Oil Consumption

Methods of Estimating Oil Usage.   There are three methods for estimating oil usage.[93] In the first method, states annually report fuel consumption because of the motor vehicle fuel sales tax. One can base oil consumption estimates on current motor vehicle fuel consumption. The automotive industry recommends changing engine oil after every 3000 miles driven. This translates into approximately 4 or 5 qt of oil generated per 150 gal of gasoline purchased. Second, since motor vehicles use more oil than any other application, motor vehicle registrations may be used to estimate engine-related oil consumption. Third, estimating state oil consumption using population data is a straightforward and commonly used method; although population data may not exactly reflect oil users.

Of all engine oils produced, over 80 percent is multigrade crankcase oil. Engine oils are distributed largely through the retail and commercial sector, which accounts for 88 percent of the sales of engine lubricants.[94] Retail distribution occurs through service stations and other retail outlets such as automotive parts stores or chain stores. Commercial distribution includes sales to commercial truck fleets, governments, railroads, commercial marine, airlines, or industrial plants.

Estimates of Used Oil Generation Factors.   A significant portion of engine oils are nonrecoverable owing to combustion in the engine, residual left in engine components (e.g., oil filters), and inadvertent spilling. Generation factors are defined as the volume of waste oil generated compared with the volume of oil initially purchased. Industrial and engine oil generation factors range from 10 to 80 percent, depending on the application. Engine oil generation factors are higher because of the frequency with which crankcase oil is changed. Various studies have put the engine oil generation rate at about 60 percent of annual purchases.[95–97]

Disposal of Used Oils.   Used oil is improperly disposed of by:

- Direct disposal into the environment by dumping
- Collection with municipal solid waste with subsequent disposal in a landfill or incinerator
- Burning on site
- Disposal to a liquid effluent treatment system
- On-site secondary use such as a machinery lubricant or dust suppressant

The group responsible for the majority of improper disposal is known as do-it-yourselfers (DIY), or people who change the oil on their own vehicles. Only 5 percent of the DIY oil generated in 1988 was channeled into the used oil management system by collection at gas stations, quick lube oil change stations, repair shops, or municipal recyclers.[91] Fifty-one percent of non-DIY used engine oil managed on-site was improperly disposed through dumping. Much of this was from off-road construction and mining sources. This indicates the need to examine regulations and collection methods for used engine oils.

Hazardous Contaminants

Virgin Lube Oil Characteristics.   New and used lubricating oils can contain hazardous constituents such as metals, chlorinated compounds, and polynuclear aromatic hydrocarbons (PAHs). "Finished lubricating oil" is the term used for oil available in the marketplace. It contains a combination of base stock lubricant and various additives. Limited data have shown that virgin base stock oils contain metals such as barium, cadmium, lead, and zinc on the order of 0 to 1 ppm. Lesser amounts of chromium (0 to 0.05 ppm) and benzo(a)pyrene (at <1 ppm), a carcinogenic PAH, have also been identified.[97]

Additive compounds enhance the effectiveness of lubricating oil and greatly influence its composition. They comprise 10 to 30 percent by volume of finished engine oil products. A typical formulation for gasoline engine oil is shown in Table 9.24. Additives inhibit metal corrosion and oxidation of oil, and act as detergents, dispersants, and antiwear compounds. They contain hazardous constituents such as magnesium, zinc, lead, and organics, and they also increase the concentrations of sulfur, chlorine, and nitrogen in lube oil.

Used Oil Contaminants.   During service, lubricating oils become contaminated with metal particles from engine wear, gasoline from incomplete combustion, and rust, dirt, soot, lead compounds, and water vapor from engine blowby (i.e., material that leaks

TABLE 9.24 Typical Formulation of Gasoline Engine Oils[98]

Ingredient	Percent by volume
Base oil (solvent 150 neutral)	86
Detergent inhibitor (ZPDD-zinc dialkyl dithiophosphate)	1
Detergent (barium and calcium sulfonates)	4
Multifunctional additive (polymethyl-methacrylates)	4
Viscosity improver (polyisobutylene)	5

from the engine combustion chamber into the crankcase). Oil additives can oxidize during combustion, forming corrosive acids. Table 9.25 shows the results of used automotive oil analyses when the samples were collected directly from the generator. The table contains 19 constituents, 17 of which are part of the U.S. EPA's list of hazardous constituents listed in Title 40, Code of Federal Regulations, Part 261, Appendix VII (40 CFR 261). Other samples taken from collectors, processors, and refiners which were identified by the respective source as used engine oil showed lower levels of some metals and higher levels of chlorinated solvents, indicating the mixing of different oil types.

A sampling program of curbside collected used crankcase oil was conducted in Oregon in 1986. The state has over 100 curbside used oil programs, affecting a population of 2 million. More than 400 individual samples were used to form 20 composite samples for testing. For the samples, average total halogen level was 357 ppm; lead, 662 ppm; arsenic, 0.21 ppm; cadmium, 1.2 ppm; and chromium, 3 ppm.[99] The lead content exceeds the U.S. EPA lead limit of 100 ppm defining "off-specification" used oil for fuel. This figure should decrease over time owing to the federal phase-down of allowable lead additive levels in leaded gasoline. Overall, the study concluded that used oil collected from households was generally not contaminated with household hazardous material or other inappropriate wastes.

Minimum Testing of Used Oil.   When testing used oil through the services of a laboratory or as part of a waste disposal firm's services, minimum analysis should address metals, halogenated solvents, aromatics, and polychlorinated biphenyls (PCBs). A disposal service will often retain a sample of used oil in order to identify the source. A good practice for the used oil generator is to split the sample collected and retain half in order to serve as the check against the sample taken by a hauler or processor. An initial test which is typically performed on loads of oil at the generator site is a total organic chlorine test. Such a test is only an initial evaluation of whether waste oil may contain chlorinated hazardous waste greater than the 1000 ppm threshold level (see below).

Legislation and Regulations Surrounding Used Oil

Federal Regulation of Used Oil.   Congress and the U.S. EPA have been attempting to deal with used oil regulation since 1976 when the Resource Conservation and Recovery Act (RCRA) was legislated. Recent years have seen a growing debate over whether or not used oil should be classified as a hazardous waste. Table 9.26 gives limits for "specification" used oil fuel. Listing used oil as a hazardous waste, some argue, would discourage the recovery and recycling of used oil and is contrary to the intent of regulation. Essentially, used oil that is not burned for energy recovery is not currently regulated unless it has been mixed with a listed hazardous waste or exhibits a hazardous waste characteristic of ignitability, reactivity, corrosivity, or toxicity.

TABLE 9.25 Concentration of Potentially Hazardous Constituents in Used Automotive Oil Samples Taken Directly from Generators[97]

Constituent	Number tested	Samples above detection limit, %	Mean concentration,* ppm	Median concentration, ppm	Concentration range, ppm Low	Concentration range, ppm High	Regulatory limit for used oil fuel, ppm
Metals:							
Arsenic	24	8	9.9	5	<5	14	5
Barium	113	95	209.5	94	0.78	3,906	
Cadmium	64	93	1.7	1	<0.2	10	2
Chromium	99	97	10.8	8	0.5	50	10
Lead	40	97	2,573.7	1,470	5	21,700	100
Zinc	116	100	982.3	1,000	4.4	3,000	
Chlorinated solvents:							
1,1,1-Trichloroethane	22	18	401.3	6	<1	1,000	
Trichloroethylene	22	9	2.5	5	<1	16	
Tetrachloroethylene	22	36	180.1	9	<2	660	
Total chlorine	36	100	1,200.0	800	<100	4,700	4,000
Other organics:							
Benzene	22	45	589.0	9	1	3,600	
Toluene	22	86	1,010.7	190	1	6,500	
Xylene	22	90	2,005.2	490	2	14,000	
Benzo(a)pyrene	21	100	9.7	10	1.3	17	
PCBs	22	5	39	—	—	—	

*Calculated for detected concentrations only.

†For the purpose of determining median concentration, undetected levels were assumed equal to the detection limit.

‡Federal regulations for burning used oil as fuel under 40 CFR 266, Subpart E.

TABLE 9.26 Limits for "Specification" Used Oil Fuel*

Contaminant and property	Limit
Arsenic	≤5 ppm
Cadmium	≤2 ppm
Chromium	≤10 ppm
Lead	≤100 ppm
Total halogens†	≤4000 ppm
Flash point	≥100°F

*The specification does not apply to used oil fuel mixed with a hazardous waste other than that from a small-quantity generator.

† Used oil containing more than 1000 ppm total halogens is presumed to be a hazardous waste. Such used oil is subject to 40 CFR, Part 279, Subpart G unless rebutted by demonstrating that the used oil does not contain hazardous waste.

At the federal level, household waste is exempt from regulation as a hazardous waste regardless of whether it contains hazardous waste (40 CFR 261.4). Household waste means any material derived from any type of dwelling, and includes waste which has been collected, stored, transported, treated, disposed, recovered, or reused.

The hazardous waste industry and petroleum re-refiners support the listing of used oil as hazardous. They would like to see full RCRA Subtitle C (hazardous waste) regulations for all recycling facilities and transporters, because many of the large companies involved are already permitted for handling RCRA Subtitle C waste and, based on technical grounds, used oil can qualify as a hazardous waste. Service stations and other such entities would be exempt from such regulations. Re-refined oil products and their users would also be exempt.

Oil processors and others, including the American Petroleum Institute (API), are opposed to such a listing. Since processors essentially filter and remove the water from oil for reuse as fuel, the cost to operate is less than a re-refinery. This gives the processor a cost-competitive edge over re-refining. A hazardous listing could force many processors out of business. Additionally, any hazardous listing would require users of such fuel to become a permitted hazardous waste incineration facility.

Current Regulatory Status.   The U.S. EPA has attempted to resolve the above debate by issuing its final ruling on used oil on Sept. 10, 1992.[100] They decided not to list used oil as hazardous waste because existing hazardous waste regulations based on hazardous characteristics (e.g., toxic, corrosive, reactive, ignitable) adequately address the disposal of oil exhibiting hazardous properties. There are also sufficient existing federal and state regulations to control the disposal of nonhazardous used oils.

The U.S. EPA simultaneously promulgated used oil handling standards for generators, transporters, processors, re-refiners, burners, and marketers. These standards apply to DIY generated oil only after it has been collected and aggregated through public or private collection services (e.g., municipal collections or service stations). These standards are located in 40 CFR, Part 279.

The generator regulations apply to facilities that produce more than 25 gal of oil per month (not farmers or DIY). Generators must:

- Maintain storage tanks and containers
- Label storage tanks "Used Oil"

- Clean up any leaks or spills
- Engage a used oil transporter possessing an EPA identification number

Service station owners who comply with the above may accept oil from DIY without liability for subsequent handling mishaps.

There are about 300 used oil processors and re-refiners in the United States. They must follow these management standards:

- Obtain an EPA identification number
- Maintain storage tanks
- Handle or store oil only in areas with impervious floors and secondary containment
- Plan oil testing for halogen content
- Keep records
- Safely manage processing residue
- Plan for proper facility closure

Transporters or collectors deliver oil from one site to another for recycling. Transfer areas (e.g., loading docks, parking areas) must comply with transporter storage requirements, if oil is held for more than 24 h in route to its final destination. Generators who transport less than 55 gal of their own oil are exempt. Transporters must comply with the same requirements as processors (where applicable) and:

- Limit storage at transfer facilities to 35 days
- Test waste in storage tanks that are out of service for hazardous characteristics and, if wastes are hazardous, close them according to existing hazardous waste management requirements

Used oil burners must comply with the same storage requirement as transporters. There is no significant new regulation of used oil marketers.

Used Oil Burned for Energy Recovery.   Federal regulation addresses and controls two of the primary mismanagement activities of used oil: (1) burning contaminated used oils in nonindustrial boilers, and (2) mixing hazardous wastes into used oils. Any oil which meets the specification levels shown in Table 9.26 is subject only to analysis and recordkeeping requirements. Oil which does not meet the requirements of Table 9.26 is "off-specification used oil fuel." Used oil containing more than 1000 ppm total halogens is presumed a hazardous waste because it was mixed with a halogenated hazardous waste (40 CFR, Part 279, Subpart G).

Off-specification used oil is subject to the used oil burning regulations (40 CFR, Part 279, Subpart G). It can be burned in industrial applications such as cement kilns, blast furnaces, manufacturing plant boilers, and utility boilers, but not in nonindustrial boilers such as those located in office buildings, schools, and hospitals. Off-specification used oil may also be burned in space heaters with less than 500,000 Btu/h capacity provided only DIY oil is burned and the heater is vented to the atmosphere. Those who treat off-specification used oil by processing, blending, or other treatment to meet the specification shown in Table 9.26 must document that the used oil meets the specification. Off-specification used oil incurs additional restrictions and requirements on marketers and burners.

Liability Concerns.   Pollution from a leak, spill, or improper disposal of used oil is the primary liability concern which owners and operators of used oil collection and processing facilities should consider. The Comprehensive Environmental Response, Compensation, and Liability Act (CERCLA), passed in 1980 and amended in 1986, allows the courts and government to hold those parties that created dangerous conditions at a hazardous waste disposal site financially responsible for the required cleanup. Under CERCLA's strict liability standard, it is not a defense that the generator exercised "due care" in arranging for disposal through another entity or that the disposal facility complied with all contemporary environmental and safety requirements.[101]

The only specific exclusion from the definition of hazardous substance in CERCLA is for "petroleum," which Congress defined to encompass crude oil or any fraction thereof (including used oil) which is not otherwise specifically listed or designated as a hazardous substance. A follow-up definition of "petroleum" by the U.S. EPA provides guidance on whether used oil products may be required to meet hazardous waste regulations: (1) "Petroleum" must be interpreted to include all hazardous substances, such as benzene, which are indigenous to petroleum substances. Inclusion of hazardous substances which are found naturally in crude oil and its fractions is necessary for the petroleum exclusion to have any meaning. (2) "Petroleum" must be interpreted to also include hazardous substances that are normally mixed with or added to crude oil or crude oil fractions during the refining process. (3) "Petroleum" does not include hazardous substances that are added to petroleum or increase in concentration solely as a result of contamination of petroleum during use.[101] This means that a hazardous substance added to petroleum during use (e.g., as a result of contamination) is not part of the petroleum and cannot be excluded from the requirements of CERCLA.

Regulatory Methods to Increase Used Oil Collection

Methods for increasing collection of used oils should be aimed at impacting do-it-yourself behavior. A number of different methods are available at the municipal, state, and federal level to increase used oil recovery:[92]

- Impose regulations on the generators of used oil
- Provide a deposit-refund system on the purchase and return of new and used oil
- Provide a tax on engine-related purchases to subsidize used oil collection and recycling
- Require sellers of lube oil to maintain collection facilities
- Develop a state-supported infrastructure of public and private collection centers
- Rely on public education and labeling as a method of impacting end use
- Utilize government procurement policies to stimulate the market for re-refined used oil
- Require lube oil producers to reuse a certain amount of used oil (either through re-refining or as a fuel)

Generator Regulation.   Imposing regulations on the generator of used oils can significantly change generator behavior. Such regulations must be limited in their scope because much problem oil is a result of DIY, and environmental compliance monitoring and enforcement of such regulations at the household level is not practical. Semienforceable regulations, such as the banning of used oil disposal in refuse, may have the most impact on DIY generators.

Point-of-Sale Collection.   Requiring retail sellers of lube oil to accept used oil and maintain a collection facility can have both positive and negative impacts. Such a requirement may discourage nonautomotive retailers from selling crankcase oil at all. On the other hand, automotive franchises may be well suited to coordinate a uniform policy for collection, handling, and disposal. Valvoline, through its subsidiary EcoGard, has established a retail used oil collection service in which the store owner receives training materials, consumer education materials and signage, a wheeled collection tank (which is utilized behind the counter by a store employee), generator documentation material, and a collection service.

Government Supported Used Oil Infrastructure. Project ROSE (recycled oil saves energy) is a successful, state-supported program in Alabama which works with private sector businesses such as retail stores and service stations to develop public and private sector collection of used oil. The state's portion of the program, which is funded by the Department of Economic and Community Affairs, has four main objectives:

- Educate citizens about the energy and environmental benefits of used oil recycling
- Create a statewide awareness of the implications of improper disposal of used oil
- Organize and promote used oil collection centers for every county in the state
- Document energy savings for the state

Project ROSE works with retailers, government agencies, and public service groups, and provides citizens up-to-date information concerning used oil recycling in their area. The program also maintains a service for identifying used oil haulers and processors. Thanks to Project ROSE, the state of Alabama has used oil collection in 45 of 67 counties with over 200 collection centers.

Germany uses taxes levied on auto or lube oil purchases to subsidize a used oil recycling infrastructure and to stimulate development of the used oil market. Automobiles are taxed based on engine size and revenues are deposited into a central fund. The recyclers are then reimbursed by the government based on the difference between the cost of re-refining or reprocessing and the market price for the oil.[92] Such a system taxes the consumer to create a relatively new industry, however, this may inhibit competition within the used oil recycling business.

Mandatory Recycling for Lube Oil Producers. Mandatory recycling requires lube oil producers to recycle a percentage of their annual oil production or support the recycling infrastructure through the purchase of recycling credits. Such a regulation would encourage development of the used oil recycling industry (particularly re-refining), provide a market demand for used oil (by viewing it as a resource rather than a waste product), and minimize cost impact because of private sector competition. Development of a "credit system" allows lube oil producers to support recycling activities if they opt not to physically enter into the used oil collection and recycling business. Recycling credits are also meant to reduce the incentives for illegal behavior on the part of the generator by creating the market for collection.

Education and Collection of Used Engine Oil

Participation in Collection Programs. Education and the availability of collection programs are the most important elements in minimizing used oil mismanagement. DIY oil changers tend to be the primary group mismanaging used oil. A statewide survey in Minnesota showed that 58 percent of the population are DIY, and only 37 percent of these disposed of their oil in a responsible manner[102] (Table 9.27). Any used

TABLE 9.27 Used Oil Disposal Practices of Do-It-Yourselfers in Minnesota[102]

Disposition	Statewide, %	Urban, %	Rural, %
Recycled:	37	54	14
Taken to a service station or store	22	31	11
Taken to a recycling center	15	23	3
Thrown away:	24	19	30
In the trash	17	14	20
Taken to a landfill	1	0	1
Dumped into the sewer	1	1	0
Dumped on the ground	6	4	9
On-site reuse and disposal	39	27	56
Road dust control	15	15	14
Reused	10	6	16
Used as fuel	1	1	1
Weed killer	2	0	4
Burned	7	3	14
Kept	4	2	6

oil management program should address these issues. When compared with the rural population, the urban DIY population recycled at a substantially higher rate.

Most of the DIY population would make the effort necessary to recycle if it were marginally convenient or if minor compensation were involved. Shull also showed that convenience in collection is an important factor in recycling DIY oil, particularly in urban areas. Rural residents are less likely to recycle regardless of the options open to them.

Education and Promotion. An educational campaign to promote proper management of used oil should focus on three groups: current DIY, young people in school, and the general public.

In educating consumers about used oil, there should be three goals: educate about the problems raised by mismanagement of used oil; encourage more responsible used oil management; and inform DIY exactly how to recycle oil in their locality. When presenting the problems caused by mismanagement, it is important to note that used oil is a valuable resource.

In order to have a lasting impact in the community, it is necessary to educate young people who will soon be driving. Impressing upon young people that used oil can be re-refined back into a usable product or can be reused in the crude oil production process will show that used oil has value. High schools and driver's education programs are natural places to present short courses on the benefits of used oil recycling and how to change oil properly.

A number of promotional methods can be used to promote a used oil program: program kickoff day, a used oil recycling hotline, handouts, brochures, posters, mailings, windshield service stickers, mailing inserts in utility bills, editorials in newspapers, and broadcast public service announcements. Periodic used oil collection programs should be ideally held during the spring and fall because these are the times when people clean house and may dispose of accumulated used oil.

The U.S. EPA has developed a number of documents on promoting the proper management of used oil. They present clear, simple ways to initiate oil recycling programs and include sample brochures, press releases, signs, and letters to encourage participation. These documents are available from the U.S. EPA, Office of Solid Waste, 401 M Street, SW, Washington, DC 20460, telephone (800) 424-9346.

Used Oil Collection in Rural Areas. It is estimated that an average farmer in Illinois buys 50 gal of motor oil and 40 gal of hydraulic fluid per year.[103] A survey in Minnesota identified that 80 percent of the farmers either burn, use for dust suppression, or lubricate machinery with some of the used oil generated.[102] Nonetheless, the farmers' comments indicated that they thought the best way to handle the used oil would be to have someone collect it.

Used Oil Collection Days. Used oil collection in farming areas is a particular problem because waste oil haulers have to travel long distances between stops and the load collected per stop is minor in relation to the truck tank size. The result is a large collection fee to the farmer. An alternate oil collection method for sparsely populated areas is to organize used oil collection days. This provides a service to the generator, cuts down on the cost of collection, consolidates the collection locations, and eliminates the need for filing a permanent storage permit application. A collection of once or twice per year may be adequate in rural areas because storage space is usually not a great concern for the generator.

The organizer of the collection day typically makes one or more tanks available at a farm cooperative or arranges for a waste oil hauler's truck to be on-site. Using a tank instead of a waste oil hauler truck for collection allows the hauler to perform the normal route for the day and simply make an additional stop at the collection day site. In some cases haulers do not charge for collection because of the oil value and the ease with which it is collected, and in other cases the sponsoring cooperative has paid the waste oil hauling charges so that no cost is passed along to its customers.

To track any potential problems with collected used oil, a small sample of oil should be collected from each generator and a simple form signed by the generator stating, "...the oil is free of contamination such as water, gasoline, anti-freeze, solvents, and farm chemicals...if said product is contaminated, the generator may be held liable for a disposal fee." The cooperatives additionally provide "Waste Oil Only" stickers for its customers to label drums and tanks at home.

Promotional methods for the collection days have included direct mail, notices included in monthly statements, and spots in local newspapers and radio stations.

Two collection days per year in agricultural areas, one in early spring (March) and another in late summer (end of August), may be sufficient. Early spring collection handles the waste oil from the fall and winter, before spring planting, and late summer handles the spring and summer oil when farm machinery is used most.

Employer Collection. Another method for managing used oil in rural areas is to locate a collection station at places of employment. Amana Refrigeration in Amana, Iowa, has opened up a recycling center for employees. Collection includes used oil as well as newspapers and clear HDPE bottles. The recycling center consists of a skid-mounted shed divided into three sections which can be transported with a forklift. Amana provides free oil change recycling tubs with disposable plastic liners or employees simply utilize 1-gal milk jugs and put waste oil on a shelf in the recycling center for disposal. Janitors empty the center daily by putting used oil in 55-gal drums. Once a drum is filled with used oil, a total halogen check is performed on the material and then it is mixed in a large tank with waste process oil from the company's manufacturing process. The waste oil is then picked up by a waste oil hauler at a collection cost of about 10 cents per gallon. Since some of Amana's employees are also farmers, the company has supplied these workers with 55-gal drums to return to the company when full.

Used Oil Collection in Urban Areas. There are a number of methods for collecting used oil in urban areas. Examples include curbside collection, drop-off at recycling stations or in conjunction with local business, drop-off at dedicated used oil collection depots, point-of-purchase collection, door-to-door pickup by appointment on designated days, used oil drop-off collection days, or as part of household hazardous waste collection days.

Curbside Collection of Used Oil. Curbside programs are by far the most successful recycling programs because they make it convenient for the public to participate. An earlier study indicated that 70 to 75 percent of people would save their used engine oil for recycling if it were collected at home.[104] Nationwide, it is estimated there are 170 used oil curbside collection programs, of which 43 are in California.[105] Curbside collection of used oil requires the separation of oil in a sealed container by the generator. As with source separation of any recyclable material, curbside collection promotes attitude change and behavior modification.

Curbside collection of used oil is fairly simple. The most popular method has been to attach a collection tank to the side of a refuse collection or recycling vehicle. The drain funnel for the tank is sized to hold common collection containers such as gallon milk jugs or 1-qt oil bottles. Another method has been to fasten a collection rack to the side of a refuse truck or install an additional compartment on a recycling truck. This allows the operator to collect entire containers of used oil and then empty them at a central facility. Regardless of whether waste oil is emptied on a route or at a central facility, it is imperative that operators be trained to watch for non-waste-oil products being disposed. It is also important to locate the tank or rack in a location that would not cause a problem or spill on the ground if the collection truck is involved in an accident.

The town of Florence, Ala., participates in Project ROSE, the statewide oil collection program mentioned above. The town retrofitted its two recycling vehicles with 75- and 150-gal collection tanks. The tanks are located underneath the recycling truck near the rear with the collection funnel piped to the side of the truck. To keep costs low, the town's sanitation division performed the retrofit. The coordinator for the project, R. Holst of the Northwest Alabama Council of Local Governments, indicates collection amounts are fairly steady at a rate of 100 gal per week for the 5000-household collection area. As expected, initially large amounts were collected that were previously saved up, and there is seasonal variation, with large amounts in the fall and spring. To help resolve this, residents are asked to put out no more than 2 gal per week.

The used oil management program in Florence is noteworthy because it represents a coordination of municipal government and private business to their mutual benefit. The project coordinator arranged for a local franchised quick lube center, Express Oil Change, to accept the used oil collected in the curbside program. Express Oil is not paid to accept the oil but does receive a nominal price from the waste oil hauler. In exchange, the municipality has put Express Oil signs on the recycling truck tanks. Also, an opening day kick-off press conference (with resulting front-page coverage) for the curbside program was held at the Express Oil station where drop-off occurs.

Drop-Off Recycling Stations. Establishing a local drop-off station for used oil is one of the simplest methods for collecting used oil. The basic features include tank- or barrel-type collection above ground with a raised curb, a roof and side walls to prevent water entry, and a fence for security protection. The cost of constructing the Rockford, Ill., used oil collection station was roughly $1500 for the shed and concrete base, $1100 for the two 150-gal tanks, complete with valving, sight glass, drain pan and piping, and $300 for signage.

Preassembled, igloo-shaped collection stations made of fiberglass have been used in Europe for many years, and are gaining popularity in the United States. These should be placed in locations under occasional observation (e.g., in front of a service station).

Point-of-Sale Collection. The solid waste management board of Snohomish County, Washington, has jointly coordinated the countywide point-of-sale collection of used crankcase oil with a retail automotive parts chain.[106] They decided that since most oil is purchased at automotive parts stores, such a location was appropriate for used oil collection. The county has a population of 480,000 and contains urban as well as rural areas.

A local automotive retail subsidiary expressed interest in participating in such a program as a site sponsor. The county placed outdoor collection tanks at each outlet store as well as solid waste handling facilities (a total of 16 tanks). Responsibilities of the retailer were to:

- Provide a site for placing the collection tank (a spot in adjacent parking lots)
- Obtain approval from the actual property owner
- Lock up the collection tank at night
- Clean up minor spills with an oil absorbent or cleanup equipment
- Monitor oil levels in the tanks
- Call for pickup if off schedule
- Maintain a log

The county was responsible for:

- Providing the used oil collection tanks and curbing around the tanks
- Establishing contracts
- Countywide education and promotion of the project
- Cleanup of major spills (e.g., knocking over a collection tank)
- The material collected

The county established three contract agreements:

- With the site sponsor parent company, outlining the responsibilities of the county and the sponsor
- With the waste oil hauler
- With a hazardous waste hauler for handling "hot" loads

A key component of the program was a ruling by the state environmental agency indicating that, since the collection sites are for consumer use, any hazardous waste collected qualifies as household hazardous waste. This ruling exempted the oil from RCRA hazardous waste requirements and relieved the county of generator responsibility.

The county now consistently collects 6000 to 7000 gal per month at its 16 collection sites; less than 2 percent of total volume collected has been contaminated. A key recommendation by the county for keeping hazardous waste disposal in the tanks to a minimum is to ensure alternative disposal means are available for other household hazardous wastes.

The capital cost for the collection tanks, curbing, and site setup was approximately $45,000; annual operating cost is $12,000; and the cost for removing contaminant disposal in 1990 was $6000. Based on a full year of collection, total cost to the county is 28 to 35 cents per gallon of used oil.[106]

Re-refining Used Oil

Re-refining used oil is analogous to recycling an aluminum beverage can, i.e., reman-ufacturing a waste commodity into a new product of the same type. The re-refining process is up to 98 percent efficient in converting used engine and industrial oils into high-quality lubricants for identical applications. One re-refining process consists of four steps: dehydration, defueling, extraction and distillation, and hydrotreating. When waste oil arrives at a re-refining plant, it is first tested for contamination and then bulked and mixed in a storage tank to achieve a uniform feedstock. The first process-ing step, dehydration, is needed because waste oil coming into the plant can contain 12 to 15 percent water by volume. Oil is piped to the dehydration tank and heated to 135°C under atmospheric pressure. This boils off any water and some lighter petrole-um fractions. The wastewater produced is treated on site.[107]

From the dehydrator, the oil is fed to the defueling system and the temperature is raised to 230°C under a vacuum of 100 mmHg. This process removes more light fuel and lube oils, which are then condensed and used as a fuel on site.

In the next process step, the oil is completely vaporized at 400°C under a vacuum of 3 mmHg. It is then condensed into three separate oil fractions and pumped to hold-ing tanks. Of the material not collected, the lightest fraction is marketed as an indus-trial fuel, and the heavy fraction, or "bottoms," is marketed as an asphalt extender. This product contains the additives, polymers, wear metals, contaminants, and oxi-dized materials removed in the distillation process.

In the final re-refining step, each of the three distilled oils is fed into a reactor at high pressure and temperature with hydrogen and a catalyst. This process removes sul-fur, nitrogen, chlorine, oxygenated compounds, heavy metals, and other impurities. The hydrogen is then removed and light distillates stripped.

The end product of the re-refining process is base oil which can be used to formu-late new engine, gear, and hydraulic oils. API has certified many re-refined oil prod-ucts, using the same standards applied to virgin oil products, API SG/CD for 10W30 crankcase oil, as an example. Figure 9.91 lists known U.S. and Canadian oil re-refiners.

Recovery of Used Oil and Scrap Material from Oil Filters

Oil filters and their contents are one component of the waste oil stream which is near-ly always disposed in landfills. A study at the University of Northern Iowa[110] has eval-uated the recovery of used oil from filters as well as the filter material and scrap metal. The evaluation was divided into three phases: methods to reduce residual oil in waste oil filters at the point of generation by simple draining; hydraulic compaction of waste oil filters to extract additional quantities of residual oil; and recyclability poten-tial of the resulting used oil, filter media, and scrap metal.

Oil Filter Recovery Results

Reducing Residual Oil by Draining. Two independent studies found that only about half of the oil contained in a filter can be removed by simple gravity drain-ing.[110,111] Using this method of oil recovery alone would certainly not be practical, especially in a service station setting, owing to the low recovery and time constraints.

Oil Recovery through Mechanical Compaction. Compacting oil filters with a hydraulic press removes about 88 percent of the oil contained in a used filter.[110] The remaining 12 percent cannot be recovered because it is absorbed into the filter media or

Breslube–Safety Kleen
P.O. Box 130
Breslau, Ontario N0B 1M0
519-648-2291

Breslube–Safety Kleen
7001 W. 62nd Street
Chicago, IL 60638
312-229-1500

Cibro Petroleum Products
Bronx, NY
718-824-5000

Consolidated Recycling
8 Commerce Drive
P.O. Box 55
Troy, IN 47588
606-264-7304

Demenno/Kerdoon
2000 N. Alameda St.
Compton, CA 90222
213-537-7100

Ecoguard, Inc.
Promax Division
301 East Main Street
PO Box 14047
Lexington, KY 40512
606-264-7389

Evergreen Oil
5000 Birch Street
Suite 500
Newport Beach, CA 92660
714-757-7770

International Recovery Corp.
Miami Springs, FL
305-884-2001

Lyondell Petrochemical Co.
12000 Lawndale Ave.
PO Box 2451
Houston, TX 77252
713-652-7200

Mid-America Distillations
P.O. Box 2880
Hot Springs, AR 71914
501-767-7776

Mohawk Lubricants
130 Forester St.
N. Vancouver, BC V7112M9
604-929-1285

Motor Oils Refining Co.
7601 W. 47th Street
McCook, IL 60525
708-442-6000

Shannon Environmental
Services
Toronto, ON
416-466-2133

FIGURE 9.91   Used oil re-refiners and marketers.[93,108,109]

Air Boy Sales & Mfg. Co.
PO Box 2649
Santa Rosa, CA 95405
800-221-8333

Danco Development Corp.
10832 Normandale Blvd.
Bloomington, MN 55437
612-888-3255

Graham Resources, Inc.
220 S. Edwards St.
PO Box 15
Pierz, MN 56364
800-228-0901

Morris Enterprises
2393 Teller Road
Newbury Park, CA 91320
800-833-3409

Sun Fire Mfg. Corp.
126 Bonnie Crescent
Elmira, ON N3B 3G2
519-669-1514

United Marketing International
PO Box 989
Everett, WA 98206
800-848-8228

FIGURE 9.92   Used oil filter press manufacturers.[93]

remains as a residue inside the filter. Compaction also reduces the volume of the filter by 73 percent. Figure 9.92 lists North American manufacturers of oil press filters.

Recyclability of Used Oil, Filter Media, and Scrap Metal.   Oil recovered from crushed filters is subject to the same regulations discussed earlier. The crushing could significantly add to the quantity of used oil collected. A maintenance shop that performs 50 oil changes daily could recover an additional 2 to 3 gal per day of used oil that would otherwise be disposed of with the filters.

The compacted filters from this study were processed through a scrap metal shredder which resulted in the separation of canister metal from the filter media. The recovered metal was essentially "oil-free" and acceptable to the existing scrap metal smelting market.

Laboratory analysis of the filter media was inconclusive in determining if the media should be considered a hazardous waste. The U.S. EPA has stated that the TCLP test is not appropriate for oil- and solvent-based waste.

Regulatory Classification of Waste Oil Filters.   The U.S. EPA in 1990 issued a regulatory interpretation regarding the crushing of waste oil filters and subsequent reclamation of contents.[112] Such an interpretation serves as a legal interpretation of U.S. EPA regulations. It basically indicates that if crushed or drained filters are recycled, it is not necessary to determine the hazardous waste status of used filters because of exemption due to recycling of the scrap metal. However, the filter must be drained to the point of having no free-flowing liquid, or crushed. The U.S. EPA recommends that the generator or recycling facility do both. The act of crushing filters is not regulated provided that the oil is collected for recycling. The interpretation makes no specific mention of the oil contained within a filter or the filter media. The standards mentioned earlier for used oil are assumed to apply to the oil contents.

Summary

- The greatest environmental threat from used oil comes from individuals who change the oil on their own vehicles. Many of these people use improper methods to dispose of their oil (e.g., dumping on land or water) which may or may not be legal.

- Used oil generators should retain a sample of each load of oil taken by a waste hauler, to resolve any questions about possible contamination.

- Currently, the U.S. EPA does not regulate used oil as a hazardous waste. However, there are management standards for generators, transporters, processors, re-refiners, burners, and marketers.

- Many different oil collection policies have been tried, ranging from mandated recycling to public drop-off sites. Collection schemes with public and private sector cooperation have proved very effective.

- Oil collection procedures must be tailored to the community. Urban and rural collection systems will be quite different.

- Re-refined motor oil is subject to the same testing and performance standards as virgin oil.

- Mechanical compaction can recover significant amounts of oil from used oil filters.

Other Information Sources

American Petroleum Institute
1220 L Street NW
Washington, DC 20005
202-682-8000

Association of Petroleum Re-refiners
PO Box 605, Ellicott Station
Buffalo, NY 14205
716-855-2757, FAX 716-855-0339

Center for Earth Resources Management
5528 Hempstead Way
Springfield, VA 22151
703-941-4452

Community Coalition for Oil Recycling
PO Box 141255
Dallas, TX 75214
214-821-3000

Hazardous Waste Treatment Council
1440 New York Ave, NW
Washington, DC 20005
202-783-0870

National Institute of Governmental Purchasing
115 Hillwood Avenue
Falls Church, VA 22046
703-533-7715

National Oil Recyclers Association
277 Broadway Avenue
Cleveland, OH 44115
216-791-7316

National Recycling Coalition
1101 30th Street NW
Washington, DC 20007
202-625-6406

Natural Resources Defense Council
40 West 20th St.
New York, NY 10011
212-727-2700

Service Station Dealers Association of America, Inc.
499 S. Capitol St. SW
Suite 1130
Washington, DC 20003-4013
202-479-0196

Sierra Club
408 C Street NE
Washington, DC 20002
202-547-1141

Society of Automotive Engineers
400 Commonwealth Drive
Warrendale, PA 15096
412-776-4841

U.S. EPA, Office of Solid Waste
401 M Street SW
Washington, DC 20460

RCRA Hotline: 800-424-9346
Specific information on used oil rule:
Ms. Rajani D. Joglekar (202-260-3516) or
Ms. Eydie Pines (202-260-3509)

Waste Oil Heating Manufacturers Assoc.
c/o Patton, Boggs, & Blow
2550 M Street NW
Washington, DC 20037
202-457-6420

PART D

Tracy Bone

Environmental Scientist, Municipal and
Industrial Solid Waste Division
Office of Solid Waste, U.S. Environmental Protection Agency
Washington, D.C.

9.9 HOUSEHOLD HAZARDOUS WASTE MANAGEMENT

What Are Household Hazardous Wastes?

The Environmental Protection Agency (EPA) has not officially or completely defined the term "household hazardous waste" in any regulation or publication to date. Materials that are excluded from regulation as hazardous wastes are identified in the federal hazardous waste regulations.[113] One category identified as excluded is "household waste." The regulations define household waste as including any waste materials derived from households, including single and multiple residences, hotels, motels, and other similar sources. In the preamble to the proposal for the federal hazardous waste regulations, EPA stated that "this exemption [from regulation] is based on congressional intent to exempt from the hazardous waste regulations those wastes generated by consumers in their households, and not on the absence of hazard from the waste." In fact, in the preamble to its final hazardous waste regulations, EPA stated that it was not attempting to pass judgment on the health and environmental risks associated with these wastes.

Table 9.28 is a list of generic types of household wastes that may contain hazardous constituents. If a product appears in Table 9.28, it typically exhibits one of the four (federal hazardous waste) characteristic properties; the characteristic is indicated. If a product is considered hazardous because it is composed of a listed compound, that is noted as well. Some specific products within these product types do not contain a listed hazardous compound or exhibit a hazardous characteristic. For example, the majority of oven cleaners are hazardous because of their corrosivity, some do not exhibit the corrosivity characteristic.

Several commonly used household products would be classified as hazardous wastes if they were generated by a commercial operation that exceeded the conditionally exempt small quantity generator limitation (i.e., less than 100 kg per month). In other words, these materials are legally not hazardous because they are generated in small amounts and in homes and in locations similar to homes such as hotels and motels. States can and do have more stringent definitions and regulations than EPA. Quite a few states regulate household hazardous waste (HHW) in one way or another. Contact your state or local solid waste authority to find out the applicable regulations in your area.

TABLE 9.28   Products That May Contain Hazardous Constituents

Automotive products	Air conditioning refrigerants (Listed)
	Body putty (I)
	Carburetor and fuel injection cleaners (I)
	General lubricating fluids (I or E)
	Grease and rust solvents (I)
	Oil and fuel additives (I)
	Radiator fluids and additives (I)
	Starter fluids (I or Listed)
	Transmission additives (I)
	Waxes, polishes, and cleaners (I or C)
Home maintenance products	Adhesives (I)
	Paints (I)
	Paint strippers and removers (I)
	Paint thinners (I)
	Stains, varnishes, and sealants (I)
Household cleaners	Disinfectants (C or I)
	Drain openers (C)
	General-purpose cleaners (C or I)
	Oven cleaners (C)
	Toilet bowl cleaners (C)
	Wood and metal cleaners and polishes (I)
Lawn and garden products	Fungicides or wood preservatives (Listed)
	Herbicides (E or Listed)
	Pesticides (E or listed)
Miscellaneous	Batteries (C or E)
	Electronic items (E)
	Fingernail polish remover (I)
	Photoprocessing chemicals (E, C, or I)
	Pool chemicals (R)

C = corrosive
I = ignitable
Listed = toxic or acutely toxic
E = EP toxic
R = reactive
Source: Reference 113.

How Much HHW Is Generated?

There has never been a statistical national survey on HHW generation. One calculation based on several studies estimates that the average U.S. household generates more than 20 lb of HHW per year.[115] As much as 100 lb can accumulate in the home, often remaining there until the residents move or do an extensive cleanout.[114] Two local government studies[116,117] conducted by Los Angeles County Sanitation District involved sorting and weighing of MSW. One of these estimated that the fraction of HHW in the MSW stream was less than 0.2 percent by weight; the other study estimated 0.00015 percent by weight. The University of Arizona conducted HHW surveys in New Orleans, La., and Marin County, Calif.,[118] and found that the fraction of HHW in the MSW stream was approximately 0.35 to 0.40 percent by weight.

HHW Collection Programs

HHW collection programs (HHWCPs) began out of a concern for the environment and human health. They are locally run and financed. There are many types of programs; the most common are:

> "One-day"—citizens drop their HHW off on a certain day, during certain hours, at a centrally located site.
>
> "Curbside collection"—HHW is collected from each household curb by program staff.
>
> "Door-to-door pickup"—allows housebound individuals to participate by collecting their wastes individually.
>
> "Permanent drop-off"—citizens drop off their HHW at a permanent collection site (defined as a program with at least monthly collections held at a fixed site or at a dedicated mobile facility.[120]

From 1980 through 1993, more than 5,800 collection programs have been conducted in the United States.[120] Most of these collection programs have been 1-day special events, but the use of permanent programs has grown steadily. In 1993, there were 172 communities with a permanent HHW collection facility.

HHWCPs are expensive, averaging $100 per participant.[119] Participation rates usually range from 1 to 3 percent of the eligible households and can be as high as 10 percent.[119] To limit the cost of the HHW program, HHW program planners often exclude certain wastes from collection programs. Frequently excluded wastes include radioactive materials, explosives, banned pesticides, and compressed gas cylinders.

In addition to deciding the type of program and the types of HHW accepted, HHW planners must decide who they want to allow to bring wastes to the collection. Most collections are limited to wastes generated by individuals at home and exclude hazardous waste from commercial and industrial generators. Increasingly, however, HHWCPs are open to small businesses that are "conditionally exempt small-quantity generators" (CESQGs) of hazardous waste. Hazardous waste generators are conditionally exempt from the federal hazardous waste regulations if they generate less than 100 kg (220 lb or about half of a 55-gal drum) of hazardous waste per month. Like HHW, CESQG waste is exempt from most federal hazardous waste requirements; according to federal regulations (state regulations may differ) CESQG waste can be disposed in the municipal solid waste stream. CESQGs include schools, small businesses, farms, government agencies, and other commercial and institutional hazardous waste generators. Unlike HHW generators, HHWCPs generally charge CESQGs a fee to cover the disposal costs for their wastes.

Management Options

Waste management costs are the largest item in an HHW program budget. The overall waste management costs and options will depend on the types of waste collected. For example, programs that accept only recyclable materials or provide a "drop-and-swap" (allowing usable HHW to be taken home) area will have much lower personnel and disposal costs. Reusing or recycling HHW or burning it as a supplemental fuel is less expensive than incinerating the waste at a hazardous waste facility. Pesticides, especially those containing dioxin, and solvent paints and other materials containing PCBs can be very expensive to manage ($850 per 55-gal drum in 1991). Burning used oil and solvent-based paint as supplemental fuel typically costs the sponsor $175 to $250 in management fees. In 1991, the cost of sending most other wastes to a hazardous waste incinerator or land disposal facility ranges from $350 to $500 per drum. These costs can vary and might increase over time. Discussed below are three management options: source reduction, treatment, and recycling.

Source reduction is incorporated into virtually every HHWCP's education information. In addition to safely managing the HHW brought into the collection program, HHWCPs encourage the public to avoid generating HHW. Government and industry also are working to develop consumer products with less or no hazardous constituents. However, no known nonhazardous substitutes exist for some constituents or products, such as batteries and photographic chemicals. Most HHWCPs try to develop some version of the source reduction advice, "Buy only what you need; try to use up what you have or give away any usable product."

Treatment technologies reduce the volume and/or toxicity of HHW after it is generated. These technologies include chemical, physical, biological, and thermal treatment. Common treatment procedures are neutralization of acids and bases, distillation of solvents, and incineration. The methods are dictated by the types of waste, proximity to treatment facilities, cost, and the contractor's access to treatment facilities. These treatment technologies are almost always contracted out to a hazardous waste management firm. As a result of current and pending bans on land disposal of hazardous waste and the efforts of communities to reduce the amount of HHW sent to municipal solid waste landfills, more HHW is being reused, recycled, or treated rather than landfilled.

A significant percentage of HHW can be recycled. For example, used oil can be rerefined for use as a lubricant. It also can be reprocessed for burning as a supplemental fuel (as can some oil-based paint and ignitable liquids). EPA has issued several publications to help communities safely collect and recycle used oil. Other recyclable HHW includes:

- Antifreeze
- Latex paint
- Lead-acid batteries and lead used in dental x-rays
- Some household batteries
- Fluorescent lamps

Some communities sponsor "recyclables-only" days to divert the large-volume materials (motor oil, car batteries, and latex paint) from HHW collections. The results of the state of Florida's "amnesty days" show the great potential for recycling HHW received at 1-day collections.[119] Thirty-six percent of the HHW collected at 107 amnesty days (984,655 lb out of a total of 2.7 million lb) was recycled over a 2-year period. The recycled material consisted of used oil, car batteries, and latex paint. Oil and battery recycling was discussed earlier. Paint and fluorescent lamps are discussed below.

Much of the paint brought to HHWCPs is either usable and/or of low environmental concern as compared with other types of HHW. Paint, by volume, is the largest waste stream brought into household hazardous waste collection programs. By recycling or reusing paint the disposal costs of the HHWCP can be significantly reduced.

Several communities around the country have found ways to recycle the paint they collect. Most paint recycling programs fall into three categories: reuse, low-tech recycling, and high-tech recycling.[122] Paint from a reuse program is not manipulated by the program workers other than sorting by type and color or consolidation into larger volumes. This paint may not be high in quality but is suitable for antigraffiti paint, charity donation, or other uses. Low-tech recycling involves minimal upgrading. This paint may be filtered, screened, improved through addition of ingredients, and then perhaps tested according to performance standards. This paint may be used for any of the uses mentioned above or sold. High-tech recycling upgrades the paint so that it can be sold. A particular level of paint quality is aimed for. More extensive reworking of the paint is required, perhaps even the addition of virgin materials. Low-tech and high-tech recycling may be done by program personnel but more commonly is done by a paint manufacturer.

Perhaps the biggest obstacle to paint recycling is contamination of paint, primarily by mercury. Mercury has been added to paints and coatings to preserve the paint in the can by controlling microbial growth, principally bacteria, and to protect the paint film from mildew attack after it is applied to an outdoor surface. As of Sept. 30, 1991, no residential paint may contain mercury; interior latex paints have not had mercury added since August 1990.[121] Paint manufactured before this date might contain mercury. For this reason, all latex paint in a paint exchange or a "drop and swap" program should be assumed to contain mercury and labeled "For Exterior Use Only."[121]

Alternatively, latex paint may be used as interior paint if mercury levels of less than 200 ppm can be confirmed:

- A commercial laboratory can test the paint for mercury.

- The National Pesticides Telecommunications Network (800-858-7378) provides names of paint brands that contain less than 200 ppm of mercury.

- The date of manufacture might appear on the label; no interior latex paint manufactured after Aug. 20, 1990, contains mercury.

Fluorescent lamps contain small quantities of mercury and other metals that are harmful to the environment and to human health.[123] Small quantities of mercury, cadmium, and antimony are used to manufacture fluorescent lamps and high-intensity discharge (HID) lamps such as metal halide and mercury vapor lamps.

While currently limited to just a few businesses, fluorescent lamp recycling is growing. The reclaimed glass may be remanufactured, the aluminum caps may be sent to an aluminum recycler for remanufacturing, and the mercury extracted in the recycling process may be sent to a mercury distiller, where it can be redistilled for use in thermometers, switches, and other products. It costs approximately 10 cents per foot to recycle a fluorescent lamp (e.g., the standard 4-ft fluorescent lamp used in many commercial installations costs about 40 cents to recycle).[123]

Summary

Any integrated waste management program should consider whether to include HHW collection in their municipal solid waste program. Opportunities to recycle and reuse HHW are increasing every year; however, HHWCPs can be expensive. Therefore, the type of program selected should consider factors such as waste accepted, permanent vs. 1-day programs, inclusion of CESQG waste, and availability of local recycling and disposal facilities.

PART E

David B. Spencer
President, wTe Corporation
Bedford, Massachusetts

9.10   ENVIRONMENTAL IMPACTS

Among solid waste management alternatives, recycling, including only the collection and sortation of recyclables, is believed by many to be environmentally benign. However, there is very little technical data to support this hypothesis and environmental impacts associated with transportation, separation of recyclables from MSW, and from the reformulation of recyclables into new products. In this section, environmental impacts from recycling on groundwater, dust, noise, vector, odors, and the atmosphere are discussed.

Groundwater resources are largely unaffected by recycling. MRFs for curbside separation programs typically are constructed on a concrete pad that prevents seepage of any waste pollutants into the soils. Moreover, these facilities typically handle precleaned, dry, and solid components of the waste stream. Most facilities are new and therefore subject to state-of-the-art design and regulatory scrutiny with respect to surface drainage and run-off. Potential groundwater impacts of a mixed waste processing facility would be similar to an RDF processing plant or a composting plant.

Dust emissions from recycling programs are from two sources: collection operations and processing facilities. Dust emissions are minimal on route. Operations are usually conducted indoors where ventilation and localized dust suppression measures are taken as required for stationary sources. Mixed waste processing results in greater emission of dust, but more sophisticated ventilation and collecting devices, such as cyclones and fabric filters, are typically used.

Significant emission of dust comes from the blowing of paper and plastics from containers, tipping operations, storage, and loading operations. This requires operators and collectors to frequently "police" the area to clean up windblown or spilled materials. If attention is not given to picking up paper and plastics around the area of an operating MRF, significant nuisance and litter will create an environmental "eyesore" even though the environmental damage may be minimal.

Potential noise impacts are from two sources: collection vehicles and machinery. Collection vehicles are equipped with conventional noise abatement devices. Vehicles typically contain loading machinery which is considerably less noisy than conventional packer trucks for MSW. Processing machinery noise is suppressed by restriction of operations to the interior of buildings.

Potential vector impacts are minimal in front-end processing systems in general due to the enclosure of processing operations, ventilation, and pest control. MRFs for curbside source separation programs also process a cleaner fraction of the waste, which often is prewashed by the waste generator of food and other organic residues. The putrescible waste content of the commingled source-separated recyclable stream entering an MRF can be virtually eliminated with a carefully controlled collection program.

Odor emissions are controlled with similar design features for vehicles and machinery as are used to control noise and dust. In addition, in mixed waste processing systems such as front-end systems, the tipping floor areas can be designed to maintain a slightly negative pressure to control odors. Again, due to the minimal putrescible waste content of commingled or source-separated recyclables entering an MRF, odor is typically not a problem.

The most significant air pollution from collection operations and processing facilities, particularly curbside recycling programs that employ dedicated vehicles, is from vehicular emissions to the atmosphere. Atmospheric emissions data from MRFs processing commingled recyclables is largely unavailable, but air pollution (NO_2, SO_2, CO_2, etc.) from trucks is well known. The total pollution depends on the distance traveled per ton of waste collected, but it is exacerbated by frequent stops and starts, particularly in cold weather.

Another source of pollution is the energy necessary to operate MRF facilities. This pollution occurs at the site of energy generation where coal, nuclear reactors, or natural gas is used and waste products are created. This type of pollution can be estimated from available data.

9.11 ENERGY PRODUCTION REQUIREMENTS AND ENERGY SAVINGS

Mixed waste processing facilities have energy requirements comparable to RDF fuel preparation facilities, but MRFs servicing curbside source separation programs require conceptually less energy to operate. Approximately 65 percent of the MRFs in operation in 1993 are "low technology" and require little energy to operate. The other 35 percent use more mechanical equipment and are "high technology" and thus more energy intensive. However, these definitions are somewhat subjective since most facilities employ some combination of mechanical and hand separation.

One appeal of materials recycling is the reported energy savings available in the reprocessing of recycled materials and the avoidance of processing virgin raw materials. Table 9.29 illustrates energy savings claimed for the substitution of recycled feedstock for virgin material in basic manufacturing processes.[124,125]

9.12 INTEGRATION WITH OTHER TECHNOLOGIES

Materials recycling plays an integral role in the overall management of municipal solid wastes:

- *Composting.* Requires materials separation to remove impurities, reduce odor, and remove inorganics.
- *Landfilling.* Landfilling benefits from recycling in the sense that the landfill life is extended when materials are diverted. An MRF can be located at the landfill, reducing residue disposal time and costs.
- *MSW Combustion.* Removal of low Btu materials such as metals and glass improves the fuel quality, whereas removal of high Btu material such as paper and plastic will reduce the fuel yield. The higher heating value (HHV) of the fuel will be affected.

TABLE 9.29 Environmental Benefits Derived from Substituting Recycled Materials for Virgin Resources

Environmental benefit	Aluminum, %	Steel, %	Paper, %	Glass, %
Reduction of energy use	90–97	47–74	23–74	4–32
Reduction of air pollution	95	85	74	20
Reduction of water pollution	97	76	35	—
Reduction of mining wastes	—	97	—	80
Reduction of water use	—	40	58	50

Source: Robinson, W. D. (ed.), *The Solid Waste Handbook, A Practical Guide,* Wiley, New York, 1986.

A study was conducted on the effects of recycling on Massachusetts' solid waste combustion capacity projected to the year 2000.[126] This study considered:

1. The cumulative effects of Massachusetts' goals of 10 percent source reduction and 46 percent recycling by the year 2000
2. A predicted change in the percentage of plastics in the waste steam from 7.3 percent in 1990 to 9.2 percent in the year 2000
3. The diversion to landfill of noncombustible materials such as white goods, street sweepings, and unrecycled metals and glass

The net result of these three factors is an estimated increase in the HHV from 4754 Btu per lb (without recycling) to 5884 Btu per lb, a 24 percent increase.

This increase can largely be attributed to diversion to the landfill of low Btu street sweepings and the recycling of noncombustibles such as metals and glass. It is surprising that the recycling of paper is more than offset by the increase in plastic content heating value (assumes plastic is not recycled) and the diversion of noncombustible materials through recycling.

Most of the combustion facilities in Massachusetts are limited on a heat input basis, and therefore the quantity of fuel that can be burned is a function of its Btu content. Any increase in the energy content of the fuel must be accompanied by a corresponding decrease in the feed rate. The Massachusetts study estimated that for every Btu per lb added to the HHV, the processing capability decreases by approximately 640 tons per year. Thus, Massachusetts will need to provide an additional disposal capacity of 723,000 tons per year to meet the expected disposal requirements in the year 2000 if the recycling goals are not achieved.

9.13 MRF ECONOMICS

A wide range of process and program costs for recycling technologies have been reported that reveal inconsistencies and little or no emerging pattern of costs.[127] This phenomenon can be attributed to a variety of factors.

- Early programs and facilities have had a convoluted history that make expended costs different than replication costs.
- Private vendors have been unwilling to provide proprietary cost information.
- Programs vary widely in target materials, collection methods, and levels of processing.

- Documentation of costs is poor and/or reporting is inconsistent (e.g., exclusion of collection costs, shared overhead, residue disposal charges, material resale credits, etc.).
- It is difficult to tell when the data reported is reflecting sell-price or cost. (Sell prices may not reflect cost of contractor overruns incurred on fixed price contracts.)
- Few reports explain the boundaries for tabulating the costs. Reported literature often suffers from radically different kinds of information, assumptions, and methodologies; the reader has little way of knowing just what is, and what is not included in the costs.
- Public sector cost information is not usually reported in the same manner as private cost information. There are significant differences in the way these two sectors account for and report overhead costs, development costs, legal expenses, capitalized costs, interest expense, and employee benefits, among other items. Further, public sector projects often include significant donated or public service volunteer participation, or state grants.

Because data collection methods have been inconsistent, a comparative analysis by program or technology type is difficult. This section provides an analysis of information available in the literature for various types of MRFs for which costs have been reported.

Table 9.30 provides a compilation and database evaluation of facilities adjusted to 1992 cost levels for 117 existing and 26 planned MRF facilities that process recyclable materials from a variety of curbside programs. Special note should be made of the range of costs and standard deviation on the facilities polled for this survey, which highlights the variations and inconsistencies in the available database.

The data reported in Table 9.31 covers the same facilities reported in Table 9.30, but is detailed by location. This database does incorporate a comprehensive effort to make the information comparable in form or method of reporting. This may contribute in part to the wide variation in the data reported. Original capital costs are provided as well as "added capital costs" where available. Operations and maintenance (O&M) costs are reported; and the year of construction and the year in which added capital costs were expended are noted. There are many holes in the database where no cost information has been reported.

For the same facility population reported in 9.31, Table 9.32 presents plant capital cost ranges as a function of design capacity. "Planned" facilities average 212 tons per day compared to 108 tons per day for "existing" facilities.

Capital cost depends on the degree of mechanization. The effect of degree of mechanization on capital cost is reported in Table 9.33. Operating cost data was available from only 34 existing and 2 planned facilities in this particular data base. This O&M data is reported in Table 9.34.

A more definitive analysis of a much smaller database containing only seven recycling programs is provided in Tables 9.35 and 9.36. This data, collected and analyzed for seven programs, was taken from a sample of 30 cities where the data for these seven was the only data sufficiently detailed to allow an "apples to apples" comparison and where the costs could be reasonably verified.[128]

The data from these seven programs shows the net costs of recycling, including all costs and revenues associated with collection, processing, and marketing ranging from $98 to $138 per ton. In each case, collection costs make up 75 percent of the total costs. Revenue from sale of products averaged between $28 and $39 per ton which covers a significant part of the processing costs, but does not begin to offset the incremental costs of collection.

TABLE 9.30   MRF Capital Costs and Bond Issues

Sample	Mean	Sum	Standard deviation	Minimum	Maximum	N
		Original capital costs				
All facilities	$3,116,851	$452,859,692	3,392,811	$2,692	$20,000, 000	143
Planned	5,748,077	149,450,000	4,843,438	250,000	20,000,000	26
Existing	2,593,245	303,409,692	2,685,188	2,692	13,000,000	117
		Adjusted capital costs (1992 $)				
All facilities	$3,260,774	$466,290,688	3,458,808	$2,781	$20,253,991	143
Planned	5,815,068	151,191,767	4,902,604	250,000	20,253,991	26
Existing	2,693,153	315,098,921	2,770,809	2,781	13,165,094	117
		Additional capital or retrofit costs				
Existing	$701,077	$9,114,000	1,300,408	$50,000	$4,900,000	13
		Bond issues				
All facilities	$31,166,304	$716,825,000	74,583,398	$400,000	$320,000,000	23
Planned	53,295,833	319,775,000	69,686,800	1,875,000	180,000,000	6
Existing	23,355,882	397,050,000	76,698,452	400,000	320,000,000	17
		Ratio: adjusted capital costs/plant capacity (tons per day)				
All facilities	$40,431	—	77,209	$537	$847,684	142
Planned	35,807	—	23,706	6,667	103,294	25
Existing	41,419	—	84,405	537	847,684	117
Low-Tech	$44,670	—	95,930	537	$847,684	88
High-Tech	33,523	—	25,977	3,305	140,653	54

Note: No capital cost data were available from 24 planned and 55 existing MRFs. Only minimal information was available on additional capital costs and the size of bond issues and these data have been presented for illustrative purposes only.

Source: Governmental Advisory Associates, Inc., *1992–1993 Materials Recovery and Recycling Yearbook,* 1992.

Note that set-out rates ranged from 33 to 45 percent but were relatively constant, not seasonal, for each individual program over the course of the year. However, the pounds picked up per household varied by 30 to 40 percent over time. ONP overwhelmingly predominated as the largest tonnage collected representing 67 percent to 75 percent of all recyclable materials collected. By volume, ONP represented about 43 percent.

A breakdown of collection costs (the majority of total costs) indicates that labor including wages, benefits, and workers' compensation insurance makes up 30 to 33 percent of the total collection costs while equipment including trucks, containers, and scales makes up 22 to 26 percent. The remainder of collection costs is comprised of general administration (17 to 20 percent), repairs and maintenance (11 to 18 percent), fuel (3 to 5 percent), and other (1 to 7 percent).

Perhaps the most comprehensive cost analysis conducted on MRFs to date is included in a special report by the National Solid Wastes Management Association (NSWMA).[129] Ten facilities were evaluated. All of the facilities received, sorted, and processed recyclables in a variety of ways including different combinations of auto-mated and hand-sorting. Four of the ten MRFs were located in the Northeast, three

TABLE 9.31 Detailed Cost Data, Existing and Planned MRFs

Name	TPD	Original capital cost	Year	1992 adjusted cost	Added capital cost	Year	O&M cost per ton	Year
Construction—low tech:								
City of Kenosha	23	$ 0		$ 0	$0		$0	
Hartford	25	1,300,000	91	1,316,509	0		0	
Williamson County	100	1,900,000	92	1,900,000	0		0	
Construction—high tech:								
Rosetto Recycling Center	250	$6,600,000	91	$ 6,683,817	$0		$0	
Town of Babylon	392	20,000,000	91	20,253,991	0		0	
Shakedown—low tech:								
Charleston county	80	$1,837,000	90	$1,897,508	$0		$0	
Pacific Rim Recycling	100	250,000	92	250,000	0		0	
Shakedown—high tech:								
Buffalo District (BFI)	50	$2,500,000	91	$2,531,749	$0		$0	
Elk Ridge (BFI)	150	4,000,000	92	4,000,000	0		0	
Monroe County	280	8,200,000	91	8,304,136	0		0	
Northern Virginia (BFI)	125	0		0	0		0	
Operational—low tech:								
ACCO Waste Paper (BFI)	67	$ 0		$ 0	$0		$0	
American Disposal	7	1,000,000	90	1,032,939	1,000,000	92	0	
Anne Arundel County	36	500,000	86	562,022	139,000	90	0	
Beaver County	30	900,000	90	929,645	0		0	
Beaver County (Piccirilli Disposal Service)	65	1,000,000	90	1,032,939	0		0	
Beaver Creek (Waste Management—Suburban)	30	0		0	0		0	
Best Recycling & Waste Services	65	0		0	0		0	

Facility								
Bucks County Satellite Facility	50	500,000	88	537,144	0		7	91
Camden County	75	700,000	86	786,830	0		68	89
Carroll County Solid Waste Management Commission	9	700,000	90	723,057	0		60	91
Centre County	60	2,200,000	89	2,331,131	0		0	
Chemung County	75	4,500,000	91	4,557,148	0		0	
City of Mobile	3	2,692	90	2,781	0		24	91
City of Olympia (Exceptional Foresters)	28	40,000	88	42,972	0		0	
Columbia County	4	0		0	0		0	
Commercial Metals Company	19	0		0	0		0	
Concord Disposal	0	0		0	0		0	
Dade County (Attwoods)	200	8,000,000	90	8,263,509	0		21	91
Dakota County	60	300,000	89	317,882	0		0	
East Bay Disposal (Durham Road Landfill)	55	375,000	89	397,352	50,000	90	0	
Elgin-Wayne Recycling & Waste Services	16	0		0	0		0	
Empire Waste Management	120	0		0	0		0	
Erie (Waste Management)	50	1,000,000	90	1,032,939	0		0	
Fillmore County	3	730,000	87	801,822	0		0	
Fox Valley Disposal (Waste Management)	30	0		0	0		0	
Fresno (BFI)	13	0		0	0		0	
Fresno (City)	210	1,200,000	91	1,215,239	0		0	
G.I. Industries	70	0		0	0		0	
GDS	65	3,000,000	90	3,098,816	450,000	90	0	
Garden City Disposal	34	100,000	89	105,961	0		70	91
Gloucester County Reclamation & Recovery System	50	0		0	0		0	
Gold Coast Recycling	100	3,500,000	90	3,615,285	0		0	
Grand Central Recycling	52	0		0	0		0	
Hagerstown (BFI)	40	0		0	0		0	
Harold Pollock Recycling (Ambridge)	21	0		0	0		0	
Hooksett (BFI)	50	2,500,000	90	2,582,346	0		0	
Houston (BFI)	0	0		0	0		0	
Hudson Baylor Corporation	75	0		0	0		0	
Invergrove Heights (BFI)	70	2,800,000	89	2,966,894	0		40	91
Islip (New Facility)	120	8,400,000	89	8,900,683	0		66	90
Jackson County (Fallenstein Recycling)	0	600,000	89	635,763	0		0	
Jacksonville (BFI)	110	2,200,000	90	2,272,465	0		0	

TABLE 9.31 Detailed Cost Data, Existing and Planned MRFs (*Continued*)

Name	TPD	Original capital cost	Year	1992 adjusted cost	Added capital cost	Year	O&M cost per ton	Year
Operational—low tech (*Continued*):								
Jefferson County	12	$1,700,000	89	$1,801,329	$		$130	91
Knoxville (BFI)	90	1,200,000	90	1,239,526	0		0	
LAS Recycling	300	3,000,000	89	3,178,815	0		0	
Lackawanna County	85	3,000,000	90	3,098,816	0		47	91
Leon County (Capitol Recycling)	22	250,000	91	253,175	0		0	
Lewis County	4	300,000	90	309,882	0		0	
Lowell	25	175,000	91	177,222	0		0	
Macon (Industrial Recovered Materials)	400	0		0	0		0	
Madison County	25	1,200,000	90	1,239,526	0		82	91
Mecklenberg County (FCR Charlotte, Inc.)	90	2,500,000	89	2,649,013	0		0	
Meyer Brothers Services (Waste Management)	28	250,000	88	268,572	50,000	91	26	89
Michigan Recycling	68	0		0	0		0	
Monmouth County	290	2,000,000	88	2,148,576	0		23	89
Monmouth County (Automated Recycling Technologies)	175	1,500,000	87	1,647,580	0		20	90
Monterey City Disposal Service	9	0		0	0		0	
Montgomery Otsego Scoharie S.W. Mgmt. Authority	32	1,000,000	90	1,032,939	0		0	
Mr. Rubbish Recycling & S.W. Processing Facility	400	3,000,000	91	3,038,099	0		0	
Napa Garbage Service	45	1,000,000	90	1,032,939	0		0	
North Ft. Myers	200	1,200,000	90	1,239,526	0		7	90
Northwest Iowa Solid Waste Agency	6	0		0	0		34	90
Onondaga County (Laidlaw)	300	1,000,000	91	1,012,700	0		0	
Orange County (Waste Management)	100	3,500,000	90	3,615,285	0		0	
Oswego County	159	4,500,000	91	4,557,148	0		0	
Pacific Rim Recycling	21	75,000	89	79,470	0		0	
Phoenix (Why Waste America)	125	6,000,000	90	6,197,631	0		0	
Phoenix Recycling	100	0		0	0		0	
Pinellas County (BFI)	160	875,000	83	1,024,381	1,000,000	89	7	90
Pinellas Park (Recycle America—Waste Management)	175	0		0	0		0	
Plano (BFI)	0	0		0	0		0	

Pleasant Hill–Bayshore Disposal (BFI)	47	340,000	89	360,266	180,000	92	0	
Ramsey County (Super Cycle MRF)	100	450,000	89	476,822	0	0	80	91
Recycle America (Waste Management)	70	0		0	0	0	0	
Recycling Works	80	0		0	0	0	0	
Richmond (Waste Management)	125	0		0	0	0	0	
Richmond Sanitary Service	30	0		0	0	0	0	
Rumpke Recycling, Inc.	200	2,000,000	91	2,025,399	0	0	0	
SEMASS Recycling Facility	120	5,000,000	91	5,063,498	0	0	0	
Salisbury (BFI)	40	0		0	0	0	0	
San Antonio (BFI)	0	0		0	0	0	0	
San Diego Recycling	225	0		0	0	0	15	
San Jose (BFI—Newby Island)	200	11,000,000	91	11,139,695	0	0	0	91
San Mateo County (BFI—The Recyclery)	120	0		0	0	0	0	
Schaap Sanitation & Recycling	10	0		0	0	0	0	
Seattle (Recycle America—Waste Management)	200	500,000	88	537,144	0	0	0	
Silver State Disposal	1000	9,000,000	91	9,114,296	4,900,000	90	189	90
Somerset County	104	1,000,000	87	1,098,386	0	0	0	90
South Valley Refuse Disposal	5	0		0	0	0	0	
Southeast Louisiana (Waste Management)	70	6,000,000	90	6,197,631	0	0	90	
St. Lawrence County (Waste Stream Management)	15	1,300,000	89	1,377,487	0	0	150	91
St. Lawrence County Solid Waste Disposal Authority	6	250,000	90	258,235	0	0	0	91
St. Louis (BFI)	48	1,300,000	89	1,377,487	0	0	0	
Swift County	6	350,000	90	361,528	0	0	0	
Total Recycling	100	50,000	88	53,714	0	0	0	
Town of Bowdoinham	1	25,000	89	26,490	0	0	169	91
Town of Smithtown	45	5,500,000	88	5,908,584	0	0	110	91
Tri-County Recycling	75	0		0	0	0	0	
Tucson (Cutler Recycling)	25	0		0	0	0	0	
Tucson (Recycle America—Waste Management)	100	0		0	0	0	0	
Turnkey Recycling/Recycle America	30	1,300,000	91	1,316,509	0	0	0	
Upper Valley Disposal	2	0		0	0	0	0	
Valley Recycling MRF	11	125,000	90	129,117	0	0	0	
Warren County	13	1,000,000	92	1,000,000	0	0	0	
Waste Management Recycling of Wisconsin	30	130,000	89	137,749	0	0	0	
Waste Management of McHenry County	9	0		0	0	0	0	

TABLE 9.31 Detailed Cost Data, Existing and Planned MRFs (*Continued*)

Name	TPD	Original capital cost	Year	1992 adjusted cost	Added capital cost	Year	O&M cost per ton	Year
Operational—low tech (*Continued*):								
Waste Management of Michigan (Southwest)	35	$1,500,000	91	$1,519,049	$		$ 0	
Waste Management of North America	25	2,500,000	90	2,582,346	0	0	0	
Waste Management of Ohio	14	0		0	0		0	
Waste Management of Santa Clara	120	900,000	86	1,011,639	0		0	
Waste Reduction Systems	75	1,500,000	91	1,519,049	0		0	
Waukesha County	40	1,520,000	91	1,539,303	0		0	
Wayne County IPC	24	500,000	91	506,168	0		0	
Western Waste Industries	30	0		0	0		0	
Western Waste Industries	65	0		0	0		0	
Wilmington (BFI)	40	0		0	0		0	
Wisconsin Intercounty Non-Profit Recycling	10	90,000	84	103,926	0		99	90
York County (Recycle America—Waste Management)	90	1,500,000	89	1,589,408	0		0	
York Waste Disposal	51	150,000	89	158,941	0		0	
Operational—high tech:								
Advanced Environmental Consultants	39	$0		$0	0		0	
Ann Arbor (Recycle Ann Arbor)	45	0		0	0		0	
Atlantic County (New Facility)	300	6,700,000	91	6,785,087	0		0	
Automated Container Recovery, Inc.	420	2,000,000	91	2,025,399	0		0	
Bristol (Otter Recycling)	65	3,500,000	88	3,760,008	0		66	89
Brookhaven	160	8,200,000	89	8,688,762	0		31	91
Cape May County	225	5,100,000	89	5,403,986	0		41	91
Chambers Development Corporation	100	1,000,000	90	1,032,939	155,000	91	0	
Community Recycling	200	0		0	0		0	
Cumberland County	37	3,000,000	89	3,178,815	0		25	90
Distributors Recycling	250	900,000	87	988,548	450,000	90	0	
Dixon Recyclers MRF	177	0		0	0		0	
DuPage County (CRInc.)	155	9,000,000	89	9,536,446	0		0	

Facility								
Dutchess County Resource Recovery	75	6,000,000	90	6,197,631	0		0	
Eagle Recycling	18	2,500,000	91	2,531,749	0		0	
Groot Industries	500	0		0	0		0	
Groton (SECRRRA)	34	500,000	82	627,191	290,000	87	8	89
Hartford	400	13,000,000	91	13,165,094	0		0	
Housatonic Resource Recovery Authority	100	3,000,000	91	3,038,099	0		0	
Integrated Waste Services	90	2,100,000	91	2,126,669	0		0	
Jet-A-Way	375	4,500,000	90	4,648,224	0		0	
Johnston MRF	210	5,500,000	89	5,827,828	0		35	91
Karta Container & Recycling	210	3,000,000	89	3,178,815	0		0	
King of Prussia (BFI)	125	2,000,000	90	2,065,877	0		0	
Laidlaw Waste Systems	96	4,200,000	91	4,253,338	0		0	
Materials Recycling Corp. of America/Waste Mgmt.	80	0		0	0		0	
Montgomery Co. (Shady Grove Road Transfer Station)	240	8,500,000	91	8,607,946	0		0	
New York City (East Harlem)	75	3,600,000	88	3,867,436	0		71	91
Ocean County Materials Processing Facility	350	6,000,000	89	6,357,631	0		13	91
Oneida-Herkimer Counties	200	8,000,000	89	8,476,841	0		0	
Onondaga County (RRT)	400	3,500,000	88	3,760,008	0		0	
Palm Beach County (North)	250	6,200,000	90	6,404,219	0		0	
Philadelphia Transfer & Recycling Center	80	1,400,000	89	1,483,447	0		0	
Pittsburgh (BFI)	80	0		0	0		0	
Prince George's County (Capitol Heights)	48	0		0	0		0	
Rice County	15	1,550,000	91	1,569,684	0		0	
Seattle (Rabanco)	300	6,000,000	88	6,445,727	0		0	
Springfield	214	6,650,000	89	7,046,374	200,000	91	21	90
TURF/Bayshore	190	0		0	0		0	
The Forge	120	1,500,000	90	1,549,408	0		0	
Todd Heller, Inc.	62	200,000	90	206,588	0		0	
Wapello County	35	1,170,000	91	1,184,858	0		0	
Waste Management Recycling, Inc.	100	0		0	0		0	
Waste Management of Minnesota	150	0		0	0		0	
West Paterson (WPAR)	150	1,100,000	88	1,181,717	250,000	90	0	
Westbury (New Facility)	250	1,000,000	91	1,012,700	0		0	
Wiregrass Rehabilitation Recycling Center	60	1,800,000	91	1,822,859	0		0	

Source: Governmental Advisory Associates, Inc., *1992–1993 Materials Recovery and Recycling Yearbook*, 1992.

TABLE 9.32 Adjusted Capital Costs by Plant Capacity (Percentages)

Adjusted capital costs (1992 $)	Plant capacity (tons per day)			
	Less than 100	100 to 199	200+	All facilities
Less than $1,000,000	40.3	10.3	4.9	23.9 (34)
$1,000,001 to $5,000,000	55.6	62.1	39.0	52.1 (74)
More than $5,000,000	4.2	27.6	56.1	23.9 (34)
Total %	100.0	100.0	100.0	100.0
Total no. of plants	(72)	(29)	(41)	(142)

Note: No information was available from 80 MRFs with respect to plant capacity or adjusted capital costs.

Source: Governmental Advisory Associates, Inc., *1992–1993 Materials Recovery and Recycling Yearbook,* 1992.

TABLE 9.33 Adjusted Capital Costs by Degree of Mechanization (Percentages)

Adjusted capital costs (1992 $)	Degree of mechanization		
	Low-tech	High-tech	All facilities
Less than $1,000,000	36.0	5.6	24.5 (35)
$1,000,001 to $5,000,000	53.9	48.1	51.7 (74)
More than $5,000,000	10.1	46.3	23.8 (34)
Total %	100.0	100.0	100.0
Total no. of plants	(89)	(54)	(143)

Note: No information was available from 79 MRFs with respect to adjusted capital costs.

Source: Governmental Advisory Associates, *1992–1993 Materials Recovery and Recycling Yearbook,* 1992.

were in Florida, one was in the Midwest, and two were on the West Coast. Four facilities were partially or completely publicly owned. All were privately operated. Five were sited near a landfill and the other five were located at a commercial or industrial site. Average throughput was 162 tons per day.

This study looked only at processing costs and did not take into account recycling collection costs, revenues from sale of recyclables, taxes, profit/loss calculations, or avoided costs of disposal. The study looked at the costs to process each individual recyclable material and the average costs for processing all the materials at the MRF. The following conclusions were reached:

- The average cost to process recyclables at an MRF, before revenues from the sale of the recyclables are considered, is $50.30 per ton with a range of costs from $28.11 to $72.06 per ton (Fig. 9.93). The median cost (half the facilities are higher, half lower) is $48.54 per ton.

TABLE 9.34 Operating Costs

Sample	Mean	Standard deviation	Minimum	Maximum	N
Annual operating costs					
All facilities	$917,513	1,020,712	$17,536	$5,652,632	36
Planned	1,282,100	732,421	764,200	1,800,000	2
Existing	896,066	1,039,331	17,536	5,652,632	34
O&M costs per ton processed					
All facilities	$55.76	46.96	$7.00	$189.15	36
Planned	27.56	11.94	19.11	36.00	2
Existing	57.42	47.78	7.00	189.15	34
Residue disposal costs ($ per ton)					
All facilities	$52.10	27.37	$6.25	$125.00	94
Planned	38.10	20.28	6.25	75.00	13
Existing	54.35	27.78	10.00	125.00	81

Note: No O&M cost data were available from 48 planned and 138 existing MRFs. An additional 13 existing MRFs provided information on residue costs in the unit of dollars per cubic yard: mean = $9.31; minimum = $4.50; maximum = $15.00.

Source: Governmental Advisory Associates, Inc., *1992–1993 Materials Recovery and Recycling Yearbook,* 1992.

- The average cost to process commingled recyclables is $83.36 per ton. Processing costs for commingled bottles and cans can range from $40.76 per ton to $146.29 per ton.
- The average cost to process paper is $33.55 per ton. The range for paper processing varies from $20.43 per ton to $55.93 per ton. Paper accounts for approximately two-thirds of the throughput.
- Newspaper is the least expensive recyclable to process per ton of infeed to the plant, averaging $33.59 per ton (Fig. 9.94). Mixed paper which has a processing cost averaging $36.76 per ton and corrugated boxes having a processing cost of $42.99 per ton are also relatively inexpensive to process per unit of weight.
- Plastic is the most expensive material to process per ton of infeed to the plant averaging $183.84 per ton for polyethylene terephthalate (PET) and $187.95 for high density polyethylene (HDPE).
- Aluminum cans have an average processing cost of $143.41 per ton, nearly as high as plastic. Aluminum also has the greatest range of costs varying from $72.88 to $362.59 per ton.
- The cost of processing clear and mixed glass is generally lower than the cost for processing amber and green glass. For most MRFs, this is because clear and mixed glass are more abundant in the waste stream and typically are being removed by negative sorting or mechanical means.
- Residue, on average, amounts to 4.5 percent.
- Worker productivity, defined in terms of tons per employee per day, is provided in Fig. 9.95. The MRF average is 5.04 tons per employee per day. For each individual material recovered at the MRF, the results are: paper, 7.42 tons average with a

TABLE 9.35 Comparison of Seven Recycling Programs

City (no. of households)	Materials collected (% total)*		Container	Bin replacement	Crew size/ truck capacity	Pick-ups daily	Pounds per household (per pick-up)*	Set-out rate weekly, %	Residential waste diverted, %	Total MSW diverted, %	Distance to transfer station
City A (15,000–20,000)	Alum. Tin/steel Glass HDPE PET ONP	.8 3.4 25.3 2.4 1.0 67.2	3 11 gal. (each)	5% year	1-person 30 cu. yd.	750	14	36%	4.5%	2.9%	15 miles
City B (<10,000)	Alum. Tin/steel Glass Plastic ONP	<.5 3.8 20.2 4.0 72.0	3 11 gal. (each)	<5%	1 person 28 cu. yd.	650	11.3–15.2	33–40%	4.2%	3%	10 miles
City C (10,000–15,000)	Metals Glass Plastic ONP	3.8 18.0 3.6 74.9	2 1—18 gal 1—14 gal	6%	1 person 28 cu. yd.	650–700	8.7–12	42%	5%	3%	N/A

City	Material	%									
City D (15,000–30,000)	Alum.	<.5	3	<5%	1 person 28 cu. yd.	700	10–18	43%	4.8%	4%	15 miles
	Tin/steel	4.2									
	Glass	22.2									
	HDPE	3.5									
	PET	<.5									
	ONP	70.0									
City E (30,000–50,000)	Alum.	N/A	1 18 gal.	N/A	1–2 persons 2–6 in truck	480	17–18	45%	12% (projected)	3% (projected)	N/A
	Tin/steel	N/A									
	Glass	N/A									
	HDPE	N/A									
	PET	N/A									
	PS	N/A									
	ONP/OCC	N/A									
City F (Drop-off program)	ONP		Drop-off bin	N/A	N/A	N/A	32 tons/month (avg.)	N/A	1.9%	1.0%	N/A
	Alum.										
	Glass										
City G (Drop-off program)	ONP	67.1	Drop-off bin	1 Drop-off bin/month	N/A	N/A	1,022 tons/month (avg.)	35% regular particip.	3.8%	1.9%	N/A
	Alum.	4.6									
	Bi-metal										
	Plastic	5.3									
	Glass	22.9									
	Other	<1.0									

*Information regarding amounts of materials collected were largely drawn from direct sorts (by weight) and household counts, rather than from program estimates or extrapolations from other data. N/A = not applicable or not available.

Source: Scarlett, L., "Recycling Costs: Clearing Away Some Smoke," *Solid Waste & Power*, July/August 1993, p. 13.

TABLE 9.36 Recycling Program Costs ($/ton)

City	Collection	Processing	Revenue[a]	Net cost
A	112.78[b]	40.00	(28.60)	124.18
		40.00	(52.00)c	100.78[c]
B	123.00[b]	42.00	(27.60)	137.40
			(53.96)[c]	111.04[c]
C	115.38	24.00	(30.00)	109.38
			(37.96)c	101.42[c]
D	110.26	26.03	(38.45)	97.84
E	—[d]	—	—	138.50
F	Drop-off[e]			136.25
				99.74[f]
G	Drop-off[e]			129.00

[a]Revenues are based on actual receipts, averaged over a minimum of five months. An average was used since receipts for materials collected in a given month are not always received in that month.

[b]Collection figures exclude public department costs, thus understating total costs.

[c]Revenues include subsidies paid for rigid containers by California as part of its "bottle bill" law.

[d]All collection, processing, and marketing costs, plus revenues are built into a comprehensive service fee. Contractor's fee plus city's in-house costs are reflected in net cost column.

[e]Drop-off program costs include cost of processing materials, but these are a very small percentage, since most materials delivered are clean, separated products.

[f]Direct costs only, as reported by city. Includes equipment, labor, cost of drop-off boxes, transportation, sorting, baling, etc. Excludes full department costs; e.g., overhead from public sanitation department such as allocation of building space, supervisor, etc.

Source: Scarlett, L., "Recycling Costs: Clearing Away Some Smoke," *Solid Waste & Power,* July/August 1993, p. 14.

range of 4.29 to 12.85 tons; metals, 5.96 tons average with a range of 1.19 to 17.25 tons; glass, 4.21 tons average with a range of 1.59 to 9.97 tons; and plastics, 1.57 tons average with a range of 0.59 to 3.16 tons.

- The overall costs of MRF processing (Fig. 9.96) are: labor, 33.4 percent of total costs ranging from 27.1 to 43.3 percent; building rental/amortization, 16.7 percent ranging from 7.5 to 34.4 percent; equipment amortization, 13.5 percent ranging from 6 to 25.2 percent; general administration, 13.0 percent ranging from 2.1 to 28.2 percent); residue hauling/disposal, 7.7 percent ranging from 1.5 to 16.6 percent; maintenance/repairs, 6.1 percent ranging from 2.0 to 14.8 percent; and insurance, taxes and miscellaneous, 9.7 percent ranging from 5.2 to 12.4 percent.

Although the NSWMA report focused on MRF processing costs, it noted that, based upon market prices reported in *Recycling Times,* July 14, 1992 (Table 9.37), market values for products recoverable from a ton of recyclables estimated to be approximately $30 did not offset the average cost of processing which was $50.30 per ton.

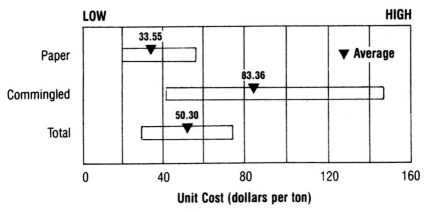

FIGURE 9.93   MRF process costs. (*Source: National Solid Wastes Management Association, Special Report: The Cost to Recycle at a Materials Recovery Facility, 1992.*)

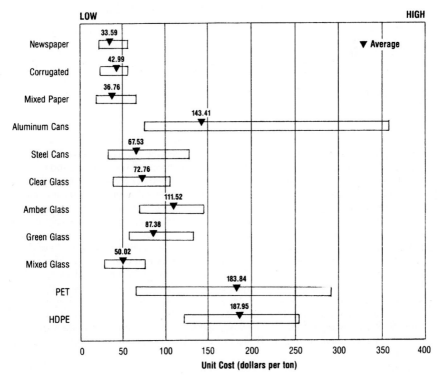

FIGURE 9.94   Recyclables processing costs. (*Source: National Solid Wastes Management Association, Special Report: The Cost to Recycle at a Materials Recovery Facility, 1992.*)

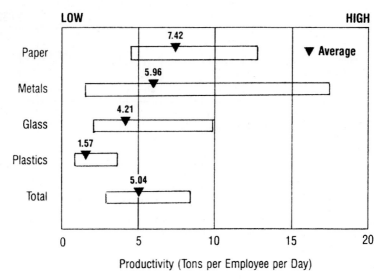

FIGURE 9.95   MRF employee productivities. (*Source: National Solid Wastes Management Association, Special Report: The Cost to Recycle at a Materials Recovery Facility, 1992.*)

9.14 OVERALL COMPARATIVE RECYCLING ECONOMICS

The need to understand the economics of different recycling methods is obvious. A recycling program is no different than any other service; everyone wants to get the best value for their money. In some instances, the choice of recycling method can be heavily influenced by the population that is to be served. There are not enough recyclable materials generated by an individual village or small town to justify a complex centralized processing system. Unless a large group of towns can consolidate their recyclables for processing, a central processing facility will be prohibitively expensive. Large cities or regional waste districts have more options open to them, but does that mean they have significant cost advantages over the smaller population bases?

Unfortunately, there are no concrete answers. As in any cost analysis of services, coming up with "the real numbers" is difficult. This is complicated in recycling because it is an emerging business, and costs are not well defined. At best, only generalizations and gross trends can be discerned. Table 9.38 compares recycling methods that are based on curbside collection of the recyclables. The numbers used to generate Table 9.38 are given in Tables 9.39, 9.40, and 9.41. They are based on data from the author's own evaluations, publications, bid awards, and operations reports of existing facilities.

Collection costs, processing or consolidation costs, and net revenues vary widely from method to method, but their estimated "average" net costs of recycling are surprisingly close. The cost advantages appear to be in favor of the recycling methods which are technically and mechanically more complex. For example, the net cost of processing commingled recyclables at a mechanized MRF appears to be less than collecting and separating at curbside. Front-end processing of the entire waste stream is more difficult, both technically and mechanically than an MRF, but with an appropri-

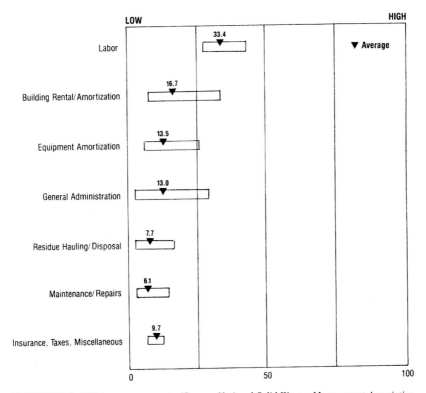

FIGURE 9.96   MRF cost components. (*Source: National Solid Wastes Management Association, Special Report: The Cost to Recycle at a Materials Recovery Facility, 1992.*)

ately sized population base, it can be less expensive. However, there is enough varia- tions in the numbers that an individual program's particular circumstances could easi- ly tip the scales in favor of any of the recycling methods. As the industry matures, these costs will be better defined.

The lowest collection cost is associated with front-end processing. In this case, the recyclables are collected with the remainder of the MSW. There are no extra trucks, extra manpower, or any of the other collection costs. The cost of collecting the recy- clables is the same as the cost to collect MSW. This lower collection cost is offset by the cost of processing since the recyclables now need to be recovered from a very complex mix of difficult-to-handle materials.

A separate collection for recyclables is inevitably more expensive than collecting them with the rest of the MSW. Collection of commingled recyclables as a group for subsequent separation and processing in a centralized facility reduces collection costs to an intermediate level. But, this "savings" is offset by the cost of building and oper- ating an MRF.

Recycling programs that do not do any processing can save on the costs of central- ized facilities. Such programs merely consolidate individual materials for shipment to other processors and markets, such as scrap dealers. Capital costs are limited to the cost of consolidation containers and a simple method for loading into them. However,

TABLE 9.37   Regional 1992 Prices for Selected Recyclables ($ per ton)

Old newspaper:		Old corrugated:	
Northeast	10–25	Northeast	15–25
East Central	0–15	East Central	20–25
South	0–25	South	35–40
West	12–20	West	40–50
Clear (flint) glass:		Amber (brown) glass:	
Northeast	26–30	Northeast	15–26
East Central	40–50	East Central	22–25
South	50	South	50
West	40	West	20
Green glass:		Steel cans:	
Northeast	0–15	Northeast	65
East Central	0–15	East Central	30–60
South	10–20	South	40–60
West	10	West	40–60
PET:		HDPE natural:	
Northeast	120–140	Northeast	100–200
East Central	120–200	East Central	80–120
South	80–100	South	120–160
West	120–160	West	0–100
HDPE mixed color:		Aluminum cans:	
Northeast	80	Northeast	780–800
East Central	120–140	East Central	700–780
South	80	South	780–800
West	60	West	780–840

Note: All prices are per ton, end-market prices. Prices are intended as general guides and are not intended to make a market or as a firm quote for recyclables.

Source: As shown in Markets Page of *Recycling Times,* July 14, 1992. From NSWMA, *Special Report: The Cost to Recycle at a Materials Recovery Facility,* 1992.

TABLE 9.38   Comparison of Recycling Costs

Type of recycling	Average collection costs*	Average processing or consolidation costs†	Estimated net revenues‡	Average net cost	Potential range of net cost
Source-separated recyclables	$90	$10	$7	$93	$68 to $118
Curbside separation of commingled recyclables	100	10	13	97	$72 to $122
Central processing of commingled recyclables	75	60	40	95	$60 to $130
Front-end processing	40	90	50	90	$60 to $120

*See Table 9.24.

†See Table 9.25; includes capital costs.

‡See Table 9.26.

TABLE 9.39   Collection Costs

Type of recycling	$/ton-cost low end	$/ton-cost high end	$/ton-cost average
Source-separated recyclables	$70	$110	$90
Curbside separation of commingled recyclables	80	120	100
Central processing of commingled recyclables	60	90	75
Front-end processing	30	50	40

TABLE 9.40   Processing/Consolidation Costs

Type of recycling	$/ton-cost low end	$/ton-cost high end	$/ton-cost average
Source-separated recyclables	$5	$15	$10
Curbside separation of commingled recyclables	5	15	10
Central processing of commingled recyclables	30	80	55
Front-end processing	70	110	90

TABLE 9.41   Estimated Net Revenues

Typical materials	Percent of total recyclables	Consolidated materials		Processed materials	
		Individual item value, $/ton	Item value × percent, $/ton	Individual item value, $/ton	Item value $/ton
Newspaper	37	($30)	($11)	$ 0	$0
Glass	32	30	10	40	13
Ferrous	18	0	0	25	5
Aluminum	2	800	16	900	18
Plastics	11	(10)	(1)	40	4
	100	Total	$14	Total	$40

such a simple system does not reap the same revenues as the programs that produce a high purity product and that densify recyclables for economical shipping over long distances. On the collection side of these types of programs, the homeowner or the collector can do the separations. Obviously paying someone to do the separations costs more than having the homeowner do it. However, participation can suffer if the homeowner is asked to do too much separation or follow a complicated collection schedule. The net result is a lower amount of material recycled and lower revenues.

In summary, the cost advantage currently appears to be with the larger, more complex, front-end processing facilities if the population base is sufficiently large. However, it is not a large advantage and costs vary widely for any given method of recycling.

9.15   SUMMARY AND CONCLUSIONS

Generally speaking, there are three different types of processing systems depending on the form in which the recycled materials are delivered to a central processing facility.

1. *Central bottle bill processing plants.*   These processing facilities handle clean bottle bill material or source-separated material collected at curbside. The material is sorted prior to its arrival at the processing plant. The only need for processing is to densify the material so that it can be shipped to market economically. Technology involves balers, pulverizers, screens, and magnetic separators. Some inspection by hand-picking is usually required as a quality control step.

2. *Materials recycling facilities [MRFs] or intermediate processing centers [IPCs].* These processing facilities receive commingled recyclables which have not been segregated by the homeowner or at curbside and which must be separated and processed at a central processing plant. The only materials delivered to these plants are recyclables. The balance of the waste stream is collected separately and delivered to an incinerator or landfill. The plants involve separation systems such as magnetic separators, shredders or pulverizers, screens, air classifiers, eddy current separators, balers, and other devices. Additional modules to recover and produce high grade plastic materials may be present. Modules which recover plastic may also be called plastic recycling facilities [PRFs].

3. *Front end processing systems or RDF processing systems.*   These very complex regional waste processing systems accept raw municipal solid waste and recycle glass, aluminum, ferrous metals, as well as producing RDF and compost. These facilities may also include modules for recovery of plastics and possibly incineration systems for combustion of the RDF which may be produced in several different forms from a fluff to a densified refuse derived fuel. Technology includes hammermills, shear shredders, conventional shredders, flail mills, magnetic separators, disc screens, rotary trommels, eddy current separators, froth flotation cells, mineral jigs, digesters, rod mills, dryers, and many other specific items of equipment depending on the particular system employed. Larger systems of 1000 tons per day or more are fully automated and employ only a limited amount of hand-picking. Smaller systems may employ hand-picking which is mechanically assisted to reduce labor cost.

As the type of system becomes more complex, moving from method 1 to method 3, the amount of participation will increase because it is easier for the homeowner to participate. Further, collection costs will decrease. However, the capital and operating costs of the processing facility will increase.

The economic tradeoffs for one system over the other would depend on the size of the system. Clearly, for a smaller community or one just starting out in recycling, system 1 would be preferred because of the lower capital investment. A large regional authority or state may want to consider the more complex system 3. Given the problems which have been experienced historically with RDF systems, system 2, the MRF or IPC may offer a good compromise of capital cost and complexity with ease of operation. However, the collection costs for system 2 will be considerably higher than for system 3.

REFERENCES

1. Melosi, Martin V., "Down in the Dumps Is: There a Garbage Crisis in America?" Draft Paper for delivery to the Urban Affairs Association Meeting, Vancouver, Apr. 17, 1991, p. 12.

2. Porter, J. W., "Municipal Solid Waste Recycling: The Big Picture," U.S. Conference of Mayors Recycling Conference, Mar. 29, 1990.

3. U.S. Environmental Protection Agency, Franklin Associates, *Characterization of Municipal Solid Waste in the United States: 1990 Update*, EPA/530-SW-90-042, Washington, D.C., June 1990.

4. deKadt, M., "Evaluating Recycling Programs: Do You Have the Data?" *Resource Recycling,* June 1992, p. 28.

5. McMillen, A. P., "Separation and Collection Systems Performance Monitoring," chap. 5, H. F. Lund (ed.), *The McGraw-Hill Recycling Handbook,* McGraw-Hill, New York, 1993.

6. Powell, J., "Keeping It Separate or Commingling It: The Latest Numbers," *Resource Recycling,* March 1991, p. 68.

7. Bagby, J., T. Diangson, and G. Patterson, "Participation in Seattle's Curbside Recycling Collection Program," *Resource Recycling,* December 1992, p. 64

8. Pieters, Riek, and Theo Verhallen, "Participation in Source Separation Projects: Design Characteristics and Perceived Costs and Benefits," *Resources and Conservation,* vol. 12, pp. 95–111, 1986.

9. Apotheker, S., "Finding a Formula for Successful Recycling Collection," *Resource Recycling,* October 1992, p. 28.

10. Hatfield, M. O., and E. J. Markey, "In Our Opinion…Beverage Container Deposit Laws: Central to a Comprehensive Solution to our Solid Waste Crisis," *Resource Recycling,* April 1992, p. 64.

11. Rankin, S., "Recycling Plastics in Municipal Solid Wastes I: Myths and Realities," *Journal of Resource Management and Technology,* October 1989, pp. 143–148.

12. Sanderlin, G. H., "Delaware Does Drop Off," *Resource Recycling,* January 1992, pp. 43ff.

13. "New Age Drop-off Programs," *The BioCycle Guide to Collecting, Processing and Marketing Recyclables,* The JG Press, Inc., Emmaus, Pa., 1990.

14. "Collecting Bagged Recyclables," *BioCycle,* November 1992, p. 37.

15. Woods, R., *Automated Plastic Sorting Industry Finds Its Legs, Recycling Times,* July 27, 1993, p. 7.

16. Gottesman, R. T., *Vinyl Recycling in the United States.* The Vinyl Institute.

17. Schut, J. H., "Automatic Resin and Color Sorting Proves a Boon to Recyclers," *Technology News,* September 1992, pp. 15–19.

18. Powell, J., "Automated Plastic Bottle Sorting: An Emerging Technology," *Resource Recycling,* August 1992, pp. 62–69.

19. National Solid Wastes Management Association, *Recycling in the States, Mid-Year Update 1990,* Washington, D.C., 1990.

20. Mathews, J. A. and T. C. Forbes, "A Comparison of Intermediate Processing Center Reference Designs and Costs," *Proceedings of the 1992 National Waste Processing Conference,* American Society of Mechanical Engineers, 1992, pp. 385–393.

21. National Renewable Energy Laboratory, *Data Summary of Municipal Solid Waste Management Alternatives,* vol. VII: Appendix E—Material Recovery/Material Recycling Technologies, NREL/TP-431-4988E, prepared by wTe Corporation, October 1992.

22. Spencer, D. B., and J. Powell, "Recycling," chap. 5, F. Kreith (ed.), *Integrated Solid Waste Management, Options for Legislative Action,* Genium Publishing Corporation, 1990. (Based upon a seminar on managing solid waste sponsored by National Conference of State Legislatures, U.S. Environmental Protection Agency and American Society of Mechanical Engineers, Breckenridge, Colo., June 1989.)

23. *Solid Waste & Power,* July/August 1993.

24. The Paper Stock Institute, Institute of Scrap Recycling Industries, *Guidelines for Paper Stock.*

25. Friar, Byron L., and Lisa Max, "Papers," chap. 11, Herbert F. Lund (ed.), *The McGraw-Hill Recycling Handbook,* McGraw-Hill, New York, 1993, p. 11.22.

26. Governmental Advisory Associates, Inc., *1992–1993 Materials Recovery and Recycling Yearbook, Directory and Guide,* New York, 1992.

27. Finelli, A., "Secondary Materials Markets: A Primer," *Solid Waste & Power,* August 1990, 5 pp.

28. Savage, G. M., and L. F. Diaz, "Processing of Solid Waste for Material Recovery," *Proceedings of 1990 National Waste Processing Conference,* The American Society of Mechanical Engineers, June 1990, pp. 417–426.

29. Egosi, N. G., and E. J. Romeo, "Meeting High Expectations through MRF Design," *Solid Waste & Power,* June 1991, p. 48.

30. Gilmore, Michael W., and Tammy L. Hayes, chap. 13, "Glass Beverage Bottles," Herbert F. Lund (ed.), *The McGraw-Hill Recycling Handbook,* McGraw-Hill, New York, 1993, p. 13.1

31. *Plastic Recycling Issues,* S. Norwalk & Assoc., July 1993.

32. *World Wastes,* 1993.

33. U.S. Environmental Protection Agency, *Markets for Scrap Tires,* EPA/530-SW-90-074B.

34. Snyder, Robert H., Market-Oriented Processing Systems, Presented at the International Scrap Tire Seminar. Toronto, Canada, April 1990.

35. U.S. Environmental Protection Agency, Office of Solid Waste, *The Solid Waste Dilemma: An Agenda for Action,* February 1989.

36. Conversation with Donald Wilson, Government Affairs Director, National Tire Dealers and Retreaders Association.

37. Conversation with Jim Miller, President, North Central (Minn.) Tire Dealers and Suppliers Association.

38. Conversation with Jack Zimmer, Goodyear Tire and Rubber Company.

39. Center for Auto Safety, 2001 S Street, N.W., Washington, D.C. 20009 and the National Highway Traffic Safety Administration, Washington, D.C.

40. Consumer Tire Guide, Tire Industry Safety Council, Washington, D.C.

41. U.S. EPA, *Markets for Scrap Tires,* EPA/530-SW-90-074A.

42. Tire Retread Information Bureau Informational Packet. 900 Weldon Grove, Pacific Grove, Calif. 93950.

43. American Retreaders Association, Louisville, Ky.

44. Conversation with Richard Gust, President, Lakin General Corp. Chicago, Ill.

45. U.S. EPA, Office of Recycling Fact Sheet on Procurement Guidelines.

46. Conversation with Kenneth L. Collings, Federal Tire Program Manager, GSA, Crystal City, Va.

47. *Scrap Tire Management in the States,* 1992 Legislative and Regulatory Review, Recycling Research Institute.

48. Scrap Tire Management Council informational materials. Washington, D.C.

49. Drabeck, John, and Jay Willenberg, Measurement of Polynuclear Aromatic Hydrocarbons and Metals from Burning Tire Chips for Supplementary Fuel, 1987 TAPPI Environmental Conference, Portland, Ore., April 1987.

50. Bakkom, Trygve, and Martin R. Felker, Tire Shredding Equipment, 1991 Conference on Waste Tires as a Utility Fuel. San Jose, Calif., January 1991.

51. Mayo, Matt, and Jim Sullivan, "Processing Scrap Tires for Multiple Markets," *Solid Waste and Power Magazine,* March/April 1992, pp. 18–25.

52. Ohio Air Quality Emissions Development Authority, Air Emissions Associated with the Combustion of Scrap Tires for Energy Recovery, Malcolm Pirnie, Inc., May 1991.

53. *Scrap Tire Management in the States,* 1992 Legislative and Regulatory Review, Recycling Research Institute.

54. Rouse, Michael, Scrap Tire Perspective. Waste Management, Inc., Recycling Meeting, Chicago, Ill. September 1989.

55. Gray, Terry A., "Scrap Tire Processing in the 1990s," *Scrap Tire News,* vol. 6, no. 3, March 1992.

56. Crumb Rubber Modifier Manual, *Design Procedures and Construction Practices,* Federal Highway Administration, Washington, D.C., February 1993.

57. Baker, Tim, and Michael Rouse, Production of Crumb Rubber, Presented at Crumb Rubber Modifier Workshop, Atlanta, Ga., February 1993.

58. U.S. EPA, *Markets for Scrap Tires.*

59. Conversation with Sean Reed, Rubber Pavements Association. Washington, D.C.

60. Blumenthal, Michael, *Scrap Tire Connection,* Scrap Tire Management Council, spring 1993.

61. Conversation with John Rugg, National Asphalt Pavement Association, Lanham, Md.

62. Federal Highway Administration Implementation Guidance, June 1993.

63. Conversation with Michael Heitzman, Federal Highway Administration.

64. California Transportation Authority (CALTRANS), Research Reports, March 1993.

65. Baker, Tim, testifying before the House Appropriations Subcommittee on Transportation, May 1993.

66. A Report on the RMA TCLP Assessment Project, Radian Corporation, Austin, Tx., September 1989.

67. Conversation with Chris Kuhn, Virginia Tire Recycling Corporation, Providence Forge, Va.

68. Conversation with Dana Humphrey, professor of civil engineering, University of Maine.

69. Conversation with Bruce Crenshaw, Royal Rubber & Manufacturing, South Gate, Calif.

70. Conversation with Ronald Nutting, OMNI Rubber Products, Portland, Ore.

71. U.S. Environmental Protection Agency, *Characterization of Products Containing Lead and Cadmium in Municipal Solid Waste in the United States, 1970 to 2000, Office of Solid Waste,* Washington, D.C., EPA 530-SW-89-015A, January 1989.

72. Apotheker, S., "Get the Lead Out," *Resource Recycling,* vol. 10, pp. 58–63, 1991.

73. Apotheker, S., "Does Battery Recycling Need a Jump?" *Resource Recycling,* vol. 9, pp. 21–23, 1990.

74. Hurd, D., D. Muchnick, M. Schedler, and T. Mele, *Feasibility Study for the Implementation of Consumer Dry Cell Battery Recycling as an Alternative to Disposal,* Recoverable Resources/Boro Bronx 2000, Inc., New York, April 1992.

75. Reutlinger, N., and D. de Grassi, "Household Battery Recycling: Numerous Obstacles, Few Solutions," *Resource Recycling,* vol. 10, pp. 24–29, 1991.

76. Eutrotech Inc., *Used Batteries and the Environment: A Study on the Feasibility of Their Recovery,* Environment Canada, Montreal, EPS 4/CE/1, May 1991.

77. U.S. Environmental Protection Agency, *Characterization of Products Containing Mercury in Municipal Solid Waste in the United States, 1970 to 2000,* Office of Solid Waste and Emergency Response, Washington, D.C., EPA 530-R-92-013, April 1992.

78. New York State Departments of Environmental Conservation and Economic Development, *Report on Dry Cell Batteries in New York State,* December 1992.

79. Cohen, S., "Recycling Nickel-Cadmium Batteries," *Resource Recycling,* vol. 12, pp. 47–54, 1993.

80. U.S. Environmental Protection Agency, *Characterization of Municipal Solid Waste in the United States: 1992 Update,* Office of Solid Waste and Emergency Response, Washington, D.C., EPA 530-R-92-019, July 1992.

81. Arnold, K., *Household Battery Recycling and Disposal Study,* Minnesota Pollution Control Agency, St. Paul, June 1991.

82. U.S. Environmental Protection Agency, *States' Efforts to Promote Lead-Acid Battery Recycling,* Office of Solid Waste and Emergency Response, Washington, D.C., EPA 530-SW-91-029, January 1992.

83. Battery Council International, "List of States without Lead-Acid Recycling Laws," memorandum from Saskia Mooney of Weinberg, Bergeson, and Neuman, Washington, D.C., Feb. 1, 1993.

84. Gaba, J., and D. Stever, *Law of Solid Waste, Pollution Prevention and Recycling,* Clark, Boardman, and Callaghan, Deerfield, Ill., 1992.

85. U.S. Environmental Protection Agency, "Modification of the Hazardous Waste Recycling Regulatory Program: Proposed Rule," *Federal Register,* Government Printing Office, Washington, D.C., vol. 58, pp. 8101–8133, Feb. 11, 1993.

86. Portable Rechargeable Battery Association Newsletter, *The Recharger,* Atlanta, March 1993.

87. Smith, Bucklin and Associates, Inc., *Battery Council International 1991 National Recycling Rate Study,* Chicago, April 1993.

88. Battery Council International, Chicago, Ill., personal communication.

89. Apotheker, S., "Batteries Power Secondary Lead Smelter Growth," *Resource Recycling,* vol. 9, pp. 46–47, 1990.

90. Adams, A., and C. Amos, Jr., "Batteries," in H. F. Lund (ed.), *The McGraw-Hill Recycling Handbook,* McGraw-Hill, New York, 1993.

91. Temple, Barker & Sloane, Inc., "Generation and Flow of Used Oil in the United States in 1988," presented at 5th Conference on Used Oil Recovery and Reuse, Baltimore, Md., Association of Petroleum Re-refiners, Buffalo, N.Y., 1989.

92. McHugh, R., "Incentives for Recycling: A World Review," presented at 6th International Conference on Used Oil Recovery and Reuse, San Francisco, Calif., Association of Petroleum Re-refiners, Buffalo, N.Y., 1991.

93. Hegberg, B., W. Hallenbeck, and G. Brenniman, *Used Oil Management in Illinois* (OTT-10), University of Illinois, School of Public Health, Office of Technology Transfer, Chicago, Ill., 1991.

94. National Petroleum Refiners Association, *1989 Report on U.S. Lubricating Oil Sales,* Washington, D.C., 1990.

95. Temple, Barker & Sloane, Inc., Memorandum to U.S. EPA, Flow Estimate Documentation, Washington, D.C., 1989.

96. Franklin Associates, Ltd., *Background Report: [Michigan] Used Motor Oil Development Study,* Michigan Department of Natural Resources, Lansing, Mich., 1987.

97. Franklin Associates, Ltd., *Composition and Management of Used Oil Generated in the United States* (EPA/530-SW-013), Washington, D.C., 1985.

98. Weinstein, K., *Waste Oil Recycling and Disposal* (EPA-670/2-74-052), Washington, D.C., 1974.

99. Spendelow, P., Recycling Specialist, Used Oil Recycling Program, Oregon Department of Environmental Quality. "Analysis of Used Oil Collected through Curbside Recycling," presented at 5th Conference on Used Oil Recovery and Reuse, Baltimore, Md., Association of Petroleum Re-refiners, Buffalo, N.Y., 1989.

100. *Federal Register,* Sept. 10, 1992 (57 FR 41566), Hazardous Waste Management System; Identification and Listing of Hazardous Waste; Recycled Used Oil Management Standards, U.S. EPA, Final rule, 1992.

101. Nolan, J., C. Harris, and P. Cavanaugh, *Used Oil: Disposal Options, Management Practices and Potential Liability,* 3d ed., Rockville, Md., Government Institutes, Inc., 1990.

102. Shull, H., M. Barnes, V. Leak, J. Powers, M. Rouse, and T. Van Hale, *Feasibility Study on Long-Term Management Options for Used Oil in Minnesota.* Minnesota Waste Management Board, Crystal, Minn., 1987.

103. Peterson, G., Manager, Fuels and Lubricants Marketing Division, Growmark, Inc., Bloomington, Ill., personal communication, 1991.

104. Market Facts, *Analysis of Potential Used Oil Recovery from Individuals, Final Report* (DOE-AC19-79BC10053), U.S. Department of Energy, Washington, D.C., 1981.

105. Arner, R., "Curbside Programs: Success or Failure?" presented at 6th International Conference on Used Oil Recovery and Reuse, San Francisco, Calif., Association of Petroleum Re-refiners, Buffalo, N.Y., 1991.

106. Wolfin, J., Senior Planner, Snohomish County Solid Waste Management Division. "The Snohomish County Used Oil Collection and Recycling Project," presentation at 6th International Conference on Used Oil Recovery and Reuse, San Francisco, Calif., Association of Petroleum Re-refiners, Buffalo, N.Y., 1991.

107. Safety-Kleen, *Safety-Kleen Re-Refined Base Oils/Products,* Company publication, Chicago, Ill., 1993.

108. Wolfe, P., "Economics of Used Oil Recycling Still Slippery," *Resource Recycling,* September 1992, p. 28.

109. Arner, R., *Used Oil Recycling Markets and Best Management Practices in the United States,* Northern Virginia Planning District Commission, Annandale, Va., 1992.

110. Konefes, J., and J. Olson, *Motor Vehicle Oil Filter Recycling Demonstration Project,* University of Northern Iowa, Iowa Waste Reduction Center, Cedar Falls, Iowa, 1991.

111. Minnesota Technical Assistance Program, *Management Options for Motor Vehicle Oil Filters (Draft),* Minneapolis, Minn., 1991.

112. U.S. Environmental Protection Agency, "Regulatory Determination on Used Oil Filters" (Memorandum, S. Lowrance, Dir. Office of Solid Waste to R. Duprey, Dir. Hazardous Waste Management Div., Region VIII, 10/30/90), Washington, D.C., 1990.

113. U.S. EPA, 45 *Federal Register* 33119, Washington, D.C., Government Printing Office, May 19, 1980.

114. SCS Engineers, "A Survey of Household Hazardous Wastes and Related Collection Programs," U.S. EPA, Washington, DC, EPA/530-SW-86-038, October 1986.

115. Rassbach, K., of Brown, Vence & Associates, "Data," Proceedings of the Third National Conference on Household Hazardous Waste Management, The Center for Environmental Management, Tufts University, Medford Mass., Nov. 2–4, 1988.

116. Los Angeles County Sanitation District, *Hand Sorting Fact Sheet,* Solid Waste Management Department, Whittier, Calif., 1979.

117. Los Angeles County Sanitation District, Unannounced Search, summer 1984, Solid Waste Management Department, Whittier, Calif., 1984.

118. University of Arizona, Bureau of Applied Research in Anthropology, and Florida State University, Center for Biomedical and Toxicological Research and Hazardous Waste Management, *Characterization of Household Hazardous Waste from Marin County, California, and New Orleans, Louisiana,* U.S. Environmental Protection Agency, Las Vegas, Nev., 1987.

119. U.S. EPA, *Household Hazardous Waste Management: A Manual for 1-Day Community Collection Programs,* EPA/530-R-93-026, U.S. EPA, Washington, D.C., 1993.

120. Duxbury, D., *Proceedings of the Eight National U.S. EPA Conference on Household Hazardous Waste Management, Nov. 6–10, 1993,* The Waste Watch Center, Andover, Mass., pp. 613–618.

121. Duxbury, D., *Managing Unwanted Paint Continued..., Household Hazardous Waste Management News,* vol. II, no. 6, The Waste Watch Center, Andover, Mass., summer 1990.

122. Waste Watch Center, *Paint Reuse and Recycling Consensus Meeting Summary,* unpublished. The Waste Watch Center, Andover, Mass., November 1992.

123. U.S. EPA, *Light Brief,* EPA Green Lights Program, U.S. EPA, Washington, D.C., January 1992.

124. Thurner, C., and D. Ashley, "Developing Recycling Markets and Industries," National Conference of State Legislatures, Washington D.C., July 1990.

125. Robinson, W. D. (ed.), *The Solid Waste Handbook, A Practical Guide,* Wiley, New York, 1986.

126. Neal, D., "Effects of Recycling on Solid Waste Combustion Capacity in Massachusetts," New England Environmental Expo, Boston, Mass., May 21, 1991.

127. *The BioCycle Guide to Collecting, Processing and Marketing Recyclables,* The JG Press, Inc., Emmaus, Pa., 1990.

128. Scarlett, Lynn, "Recycling Costs: Clearing Away Some Smoke," *Solid Waste & Power,* July/August 1993.

129. National Solid Wastes Management Association, *Special Report: The Cost to Recycle at a Materials Recovery Facility,* 1992.

BIBLIOGRAPHY

U.S. EPA: "Used Dry Cell Batteries: Is a Collection Program Right for Your Community?" EPA/530-K-92-006, U.S. EPA, Washington, D.C., December 1992.

U.S. EPA: "Household Hazardous Waste: Steps to Safe Management," EPA/530-F-92-031, Washington, D.C., April 1993.

CHAPTER 10

COMPOSTING OF MUNICIPAL SOLID WASTES

L. F. Diaz, G. M. Savage, and C. G. Golueke

CalRecovery, Inc.
Hercules, California

Composting is one element of an integrated solid waste management strategy that can be applied to mixed MSW or to separately collected leaves, yard wastes, and food wastes. The four basic functions of composting are preparation, decomposition, postprocessing, and marketing. This chapter treats the last three functions. Preparation or preprocessing is described in Chap. 9. MSW composting results in a volume reduction of up to 50% and consumes about 50% of the organic mass on a dry weight basis, by releasing mainly CO_2 and water. Composting breaks down easily degradable plant and animal tissue but does not produce appreciable changes in difficult-to-degrade organics (cellulose, leather, polymers) or in inorganics (dirt, glass, ceramics, and metals). A typical composting process flow diagram is shown in Fig. 10.1. The most important preprocessing steps are receiving, removal of recyclable material, size reduction, and possibly some adjustment of the waste properties (e.g., carbon-to-nitrogen ratio). Three basic systems used for the decomposition steps are static windrows (piles), turned windrows, and in-vessel composting.

Yard waste composting is a relatively simple open air process. The first step is to "chip" the yard waste to reduce its size and promote the breakdown of organic matter. It is then set out in long piles or windrows that are periodically turned over to expose all the material to air. Alternatively, the piles can be placed on a porous pad that is connected to a blower to supply air. The processing of MSW to make a commercially valuable compost, however, is a complex process. MSW composting begins with separating the organic materials from the rest of the waste and then shredding or grinding the organics (the remaining MSW is usually landfilled). In some cases, the organics are then initially composted inside a vessel that provides mechanical agitation and forced aeration; in other cases, composting takes place entirely in the open. Enclosed composting can help to control odors through better control of aeration and temperature, but composting in a vessel is generally followed by additional open air composting.

The number of U.S. composting programs has increased steadily to more than 1500 in 1993. But at least 500 of these programs compost only leaves on a seasonal basis. For composting mixed MSW, the United States had only 16 operating plants with a total combined design capacity of over 2000 tons per day in 1992, but another 11 plants with similar combined capacity were under construction. As recycling goals are

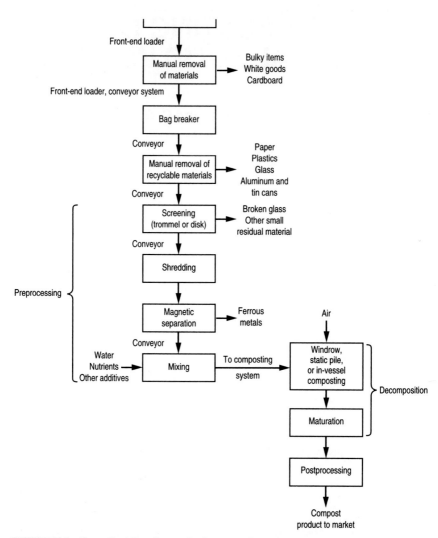

FIGURE 10.1   Generalized flow diagram for the composting process.

set to go up in many states, composting is expected to grow as an important facet of an integrated waste management program. Theoretically, composting could be used to process the 18 to 20 percent of MSW in the waste stream (see Chap. 3) that is yard waste. This quantity can be increased if some of the paper fraction (i.e., soiled paper) as well as kitchen residues are included in the feedstock to the composting facility. Because the level of technology and mechanization for composting varies widely, the costs of composting systems also vary as shown in this chapter. But, for an order-of-magnitude estimate, the empirical relations shown in Fig. 10.2 can be used to estimate the capital cost in 1991 dollars as a function of capacity in tons per day.

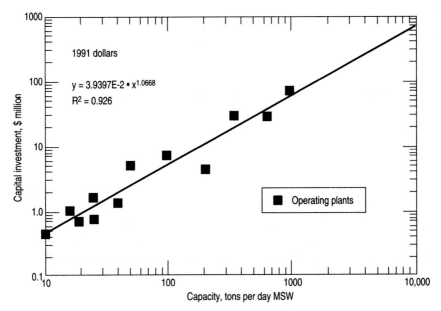

FIGURE 10.2 Capital investment for composting plants as a function of capacity (excluding cost associated with collection, e.g., trucks). (*From SRI International, Data Summary of Municipal Solid Waste Management Alternatives, NREL, 1992.*)

10.1 PRINCIPLES

Definition

A definition of composting as applied to MSW management is as follows: "Composting is the biological decomposition of the biodegradable organic fraction of MSW under controlled conditions to a state sufficiently stable for nuisance-free storage and handling and for safe use in land applications."[1-3] Definitive, i.e., distinguishing, terms in the definition are "biological decomposition," "biodegradable organic fraction of MSW," and "sufficiently stable."

The specification "biological decomposition" confines composting to the treatment and disposal of the biologically originated organic fraction of MSW. The specification "under controlled conditions" distinguishes composting from the simple decomposition that takes place in open dumps, landfills, and feedlots. The specification "sufficiently stable" is a prerequisite for nuisance-free storage and handling and for safe use in land applications.

Biology

The organisms actively involved in composting can be classified into six broad groups. Named in order of decreasing abundance, the groups are bacteria, actinomycetes, fungi, protozoa, worms, and some larvae. The bacteria include a wide spectrum of classes, families, genera, and species. For example, pseudomonads have been isolated and classified down to the genus level. Although actinomycetes are bacteria, they are named separately because of their particular role in the curing stage of the process.[1] Two genera of actinomycetes have been isolated and identified, *Actinomyces* and *Streptomyces*.[1] The fungi rival the bacteria in terms of number and importance in the later stages of the process. The worms include nematodes and some earthworms (species of annelids). The larvae are of various types of flies.

Attempts to identify a hierarchy of microbes down to the species level on the basis of number and activity in the compost process have met with little success, because of the inevitable local differences in the gamut of environmental and operational situations. An even greater uncertainty arises from the limitations of analytical procedures and techniques presently available. However, the following very broad generalizations have proved to be adequate for routine composting, particularly of MSW. In terms of number and activity, the predominant organisms are bacteria and fungi and, to a far lesser extent, protozoa. However, some higher organisms such as earthworms and various larvae may appear in the later stages of the composting process.

Of great practical and economic importance is the fact that the presence of all these organisms is a characteristic of all wastes—particularly of yard waste and MSW. Hence, as has been confirmed by carefully conducted scientific research, the use of inoculums (including enzymes, growth factors, etc.) not only would be unnecessary, it would also be an economic handicap.

Classification

The compost process can be classified in terms of distinguishing cultural condition and in terms of technology. This section deals with classification in terms of cultural conditions, i.e., aerobic vs. anaerobic and mesophylic vs. thermophylic. Classification on the basis of technology is reserved for Sec. 10.2.

Aerobic vs. Anaerobic.   Originally, composting was classified into aerobic vs. anaerobic,* and many arguments were offered in favor of one or the other. However, in time, the aerobic approach became the usual one, and anaerobic composting fell into disfavor. In fact, a tendency has developed in recent years to define composting as "aerobic decomposition," thereby invalidating the terms "anaerobic composting." Nevertheless, many practitioners, especially those well versed in composting, are not following the trend. Regardless of one's views on the matter, maintenance of completely aerobic conditions in a composting mass would be exceedingly difficult and certainly impractically expensive. In recognition of the foul odors associated with anaerobiosis, a more realistic approach is to design the composting system such that aerobiosis is promoted and anaerobiosis is minimized as much as is feasible.

Mesophylic vs. Thermophylic.   In modern compost practice, the question of relative advantages of an all-thermophylic vs. an all-mesophylic process is moot because with few exceptions, modern composting incorporates the rise and fall of temperature levels that normally occurs unless positive measures are taken to circumvent it. Mesophilic is the temperature range from about 5 to 45°C. Thermophilic is the temperature range from about 45 to 75°C.

---

*Aerobic processes are those carried out in the presence of oxygen. Anaerobic processes are those carried out in the absence of oxygen.

Compost Phases

Composting characteristically is an ecological succession of microbial populations almost invariably present in wastes. The succession begins with the establishment of composting conditions. "Resident" (indigenous) microbes capable of utilizing nutrients in the raw waste immediately begin to proliferate. Owing to the activity of this group, conditions in the composting mass become favorable for other indigenous populations to proliferate. Plotting the effect of the succession of total bacterial content of the mass would result in a curve the shape of which would roughly mirror those of the normal microbial growth curves and of the rise and fall of temperature during composting (cf. Fig. 10.3). Judging from the curve, composting proceeds in three stages, namely, an initial lag period ("lag phase") and a period of exponential growth and accompanying intensification of activity ("active phase") that eventually tapers into one of final decline, which continues until ambient levels are reached ("curing phase" or "maturation phase"). In practice, this progression of phases is manifested by a rise and fall of temperature in the composting mass. A plot of the temperature rise and fall would result in a curve the shape of which would be roughly identical with that of the growth curve.

The course of the process in all its aspects and the characteristics of its product are all determined by the environmental factors to which the process is exposed, by the operational parameters being followed, and by the technology employed. An abrupt deviation during any of the phases (e.g., sharp drop in temperature) betokens a malfunction. The phase resumes upon the elimination of the malfunction.

Lag Phase. The lag phase begins as soon as composting conditions are established. It is a period of adaptation of the microbes characteristically present in the waste.

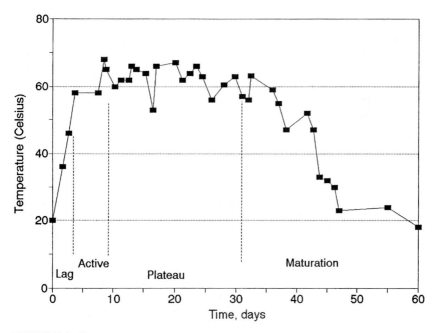

FIGURE 10.3 Temperature curve.

Microbes begin to proliferate, by using sugars, starches, simple celluloses, and amino acids present in the raw waste. Breakdown of waste to release nutrients begins. Because of the accelerating activity, temperature begins to rise in the mass. Pseudomonads have been routinely identified as being among the more numerous types of bacteria. Protozoa and fungi, if present, are not discernible. The lag period is very brief when highly putrescible materials and/or herbaceous yard wastes are involved. It is somewhat longer with mixed MSW and woody yard waste, and is very protracted with dry leaves and resistant wastes such as dry hay, straw, rice hulls, and sawdust.

Active Phase.   The transition from lag phase to active phase is marked by an exponential increase in microbial numbers and a corresponding intensification of microbial activity. This activity is manifested by a precipitous and uninterrupted rise in the temperature of the composting mass. The rise continues until the concentration of easily decomposable waste remains great enough to support the microbial expansion and intense activity. Unless countermeasures are taken, the temperature may peak at 70°C or higher.

The activity remains at peak level until the supply of readily available nutrients and easily decomposed materials begins to dwindle. In a plot of the temperature curve, this period of peak activity is indicated by a flattening of the curve (i.e., by a plateau). This "plateau" phase may be as brief as a few days or, if the concentration of resistant material is high, as long as a few weeks.

The duration of the entire active stage (exponential plus plateau) varies with substrate and with environmental and operational conditions. Thus it may be as brief as 5 or 6 days or as long as 2 to 5 weeks. It should be pointed out that a sudden drop in temperature during the active stage is an indication of some malfunction that requires immediate attention (e.g., insufficiency of oxygen supply, excess moisture). Temperature drop due to turning is of brief duration.

Maturation or Curing Phase.   Eventually, the supply of easily decomposable material is depleted, and the maturation stage begins. In the maturation phase, the proportion of material that is resistant steadily rises and microbial proliferation correspondingly declines. Temperature begins an inexorable decline which persists until ambient temperature is reached. The time involved in maturation is a function of substrate and environmental and operational conditions, i.e., as brief as a few weeks to as long as a year or two.

Environmental Factors and Parameters

Nutrients and Substrate.   In composting, "substrate" and "nutrient supply" are synonymous because the substrate is the source of nutrients. In the composting of yard waste and MSW, the biologically originated organic fraction of the wastes is the substrate. The specification "biologically originated" eliminates synthetic organic wastes. The exclusion of synthetic organics has a very practical significance because it eliminates many types of plastics. Wastes of biological origin differ from synthetic organic wastes in terms of molecular structure and arrangement. Examples of organic wastes of biological origin are wood, paper, and plant and crop debris. Plastics are examples of synthetic organic materials.

There are exceptions to the biological origin requirement. In fact, the number of exceptions is growing because of strides being made in microbial genetics, gene manipulation, and molecular engineering. This is particularly true with toxic organics and chemical pesticides. However, much remains to be done before the exceptions become generalities.

Although the ideal waste would contain all necessary nutrients, in practice it may be necessary at times to add a chemical nutrient to remedy a nutrient deficiency.

Chemical Elements. The major nutrient elements ("macronutrients") are carbon (C), nitrogen (N), phosphorus (P), and potassium (K). Among the nutrient elements used in minute amounts ("micronutrients" or "trace elements") are cobalt (Co), manganese (Mn), magnesium (Mg), and copper (Cu). Calcium (Ca) falls between macro and micronutrients. Carbon is oxidized (respired) to produce energy and metabolized to synthesize cellular constituents. Nitrogen is an important constituent of protoplasm, proteins, and amino acids. An organism can neither grow nor multiply in the absence of nitrogen in a form that is accessible to it. Although microbes continue to be active without having a nitrogen source, the activity rapidly dwindles as cells age and die. The principal use of calcium is as a buffer (resists change in pH). Phosphorus is involved in energy storage and to some extent in the synthesis of protoplasm.

Availability of Nutrients. An aspect of nutrition is that the mere presence of a nutrient element in a substrate does not suffice. To be utilized, the element must be in a form that can be assimilated by the organism. In short, the element must be "available" to the organism. This even applies to sugars and starches, most of which are readily decomposed.

Availability to a microbe is a function of the organism's enzymatic makeup. Thus members of certain groups of microbes have an enzymatic complex that permits them to attack, degrade, and utilize organic matter present in a raw waste. Groups that lack the needed complex can utilize as a nutrient source only the decomposition products (intermediates) produced by enzymatically endowed organisms.

The restriction regarding availability is particularly significant. The significance is in the fact that it makes the composting of a waste the result of the activities of a dynamic succession of groups of microorganisms. In this succession, groups "prepare" the way for their successor. In this context, succession does not necessarily imply that groups "come and go" in series. On the contrary, some or even most groups may persist. However, some persistent groups may become less prominent in the ongoing activity.

Certain organic substances are not readily decomposed even by microbes that have the required enzymatic complex. Such resistant materials are broken down slowly despite the maintenance of conditions at optimum levels. Examples are lignin and chitin. Lignin is the principal constituent of wood, whereas chitin is a major constituent of feathers and shellfish exoskeletons. Although the cellulose C in wood, straw, and pith is readily available to many fungi, it is resistant to most microbes.

Nitrogen is readily available when it is in the proteinaceous, peptide, or amino acid forms. On the other hand, because of the resistance of the chitin and lignin molecules to microbial attack, the small amount of nitrogen in them is released too slowly for practical composting.

Carbon-to-Nitrogen Ratio. The available carbon to available nitrogen ratio (C/N) is the most important of the nutritional factors, inasmuch as experience shows that most organic wastes contain the other nutrients in the required amounts and ratios for composting. The ideal ratio is about 20 to 25 parts of available carbon to 1 of available nitrogen. The nitrogen content and C/N of several wastes are given in Table 10.1.

A C/N higher than 20/1 or 30/1 can slow the compost process. A C/N that is too low (less than 15/1 to 20/1) leads to loss of nitrogen as ammonium N. The addition of a nitrogenous waste can lower an unfavorably high C/N, whereas the addition of a carbonaceous waste can raise an undesirably low C/N. Examples of nitrogenous wastes are grass clippings, green vegetation, food wastes, sewage sludge, and commercial

TABLE 10.1 Percent N, C/N, and Moisture of Selected Materials

Material	% N (dry wt.)	C/N (weight to weight)	% moisture (wet wt.)
Corncob	0.4–0.8	56–123	9–18
Cornstalks	0.6–0.8	60–73	12
Fruit wastes	0.9–2.6	20–49	62–88
Rice hulls	0.0–0.4	113–1120	7–12
Vegetable wastes	2.5–4.0	11–13	*
Poultry litter (broiler)	1.6–3.9	12–15	22–46
Cattle manure	1.5–4.2	11–30	67–87
Horse manure	1.4–2.3	22–50	59–79
Garbage (food wastes)	1.9–2.9	14–16	69
Paper (from domestic refuse)	0.2–0.25	127–178	18–20
Refuse	0.6–1.3	34–80	
Sewage sludge	2.0–6.9	5–16	72–84
Grass clippings	2.0–6.0	9–25	
Leaves	0.5–1.3	40–80	
Shrub trimmings	1.0	53	15
Tree trimmings	3.1	16	70
Sawdust	0.06–0.8	200–750	19–65

*Not reported.
Source: From Ref. 9.

chemical fertilizers. Examples of carbonaceous wastes are hay, dry leaves, paper, and chopped twigs.

The requirement that the carbon be in an *available* form minimizes or even eliminates wood and woody materials as a carbon source in the composting of sewage sludge. Thus sawdust, wood chips, or woody shavings used to bulk sewage sludge should not be regarded as a carbon source. Although the carbon in paper, dry leaves, and chopped twigs is relatively slowly available, the materials can serve as carbon sources only to a limited extent. Furthermore, because the carbon in the latter materials is only slowly available, their use can raise the permissible upper C/N to as high as 35 to 40/1.

Animal manures, sewage sludge, and commercial (agricultural) chemical fertilizers are adequate sources of nitrogen and any other element that may be needed. For example, experience indicates that the unfavorably high C/N ratio of the organic fraction of refuse can be advantageously lowered through the addition of digested sludge.[4]

Particle Size. Theoretically the smaller the particle size, the more rapid the rate of microbial attack. In practical composting, however, there is a minimum size below which it is exceedingly difficult to maintain an adequate porosity in a composting mass. This size is the "minimum particle size" of the waste material. In composting, the practical "optimum" is a function of the physical nature of the waste material. With a rigid or not readily compacted material such as fibrous waste, twigs, prunings, and corn stover, the suitable size is from 1/2 in (13 mm) to about 2 in (50 mm). The particle size of the greater part of a fresh green plant mass such as vegetable wastes, fruits, and lawn clippings should be no less than 2 in (50 mm). On the other hand, depending upon their overall decomposability, their maximum particle size can be as large as 6 in (0.15 m) or even larger.

Oxygen. Oxygen availability is a prime environmental factor in composting, inasmuch as composting is an aerobic process. Oxygen is a key element in the respiratory and metabolic activities of microbes. Interruption in the availability leads to a shunt metabolism, the products of which are reduced intermediates, which characteristically are malodorous. The microbes involved in the composting process obtain their oxygen from the air with which they come in contact, i.e., the air that impinges upon them. Consequently, the oxygen content of this air must be continually replenished or the air itself must be continually replaced. The interstitial oxygen content in a windrow can be estimated by use of an oxygen probe inserted into the windrow. The oxygen content of the airstream into and out of a static windrow (forced aeration) and in-vessel systems can be directly measured. For convenience, the amount of oxygen required by the microbes is termed "oxygen demand."

Attempts to establish a universally applicable numerical rate of oxygen uptake for use as a design parameter have been unsuccessful. The underlying reason for the lack of success is the variability of key factors that influence oxygen demand. Among such factors are temperature, moisture content, size of bacterial population, and availability of nutrients. Therefore, determination of the amount of aeration that would meet a specific demand adds another level of complexity, because the capacity and performance of the aeration equipment and the physical nature of the composting mass must be taken into account. The straightforward methods (procedures) used for determining oxygen demand in wastewater treatment (e.g., COD, BOD) are poorly or not at all applicable to composting.

The variability of oxygen demand is demonstrated by the diversity of results reported in the literature. One of the earlier reports described a study in which air was passed at a known rate through composting material enclosed in a drum and the oxygen content of the influent and effluent airstream were measured.[7,8] Oxygen uptake rose from 1 mg/g of volatile matter at 30°C to 5 mg/g at 63°C. In a later study, Chrometska[10] observed oxygen requirements that ranged from 9 mm^3/g•h for ripe compost to 284 mm^3/g•h for raw substrate. "Fresh" compost (7 days old) required 176 mm^3/g•h. Lossin[11] reports average chemical oxygen demands that range from almost 900 mg/g on the first day of composting to about 325 mg/g on the twenty-fourth day. In a review of the decomposition of cellulose and refuse, Regan and Jeris[12] observed that oxygen uptake was lowest (1.0 mg oxygen/g volatile matter per hour) when the temperature of the mass was 30°C and the moisture content was 45 percent. The highest uptake (13.6 mg/g volatile matter per hour) occurred when the temperature was 45°C and the moisture content 56 percent.

With respect to the design of airflow through an in-vessel reactor and to a lesser extent through a static pile, the indicated procedure would be to estimate the carbon content and to determine the amount of oxygen consumed in oxidizing the carbon. The flaw in such an approach is that it would result in an overdesign, because normally only a fraction of the carbon is available to the microbial population. Nevertheless, some overdesign is advisable because of the impossibility of aerating a mass such that all microorganisms simultaneously have access to sufficient oxygen. Perhaps the airflow requirement estimated by Schulze (562 to 623.4 m^3/Mton volatile matter per day) could be of some use. However, his estimates are based upon the use of his particular equipment and on a laboratory-scale experiment.

Moisture Content

Maximum. Theoretically, the optimum moisture content of the wastes is one that approaches saturation, provided that the material can be sufficiently aerated to meet the oxygen demand. Although meeting the demand is technologically feasible, it also is economically unfeasible. Hence the term "permissible maximum" is introduced. It is the moisture content above which oxygen availability becomes inadequate and anaerobiosis ensues. The maximum permissible moisture content usually is also the optimum content.

Because the air entrapped in interstices between particles is the primary source of oxygen for the microbial population, interstitial ("pore") volume is a decisive factor; i.e., the more numerous the pores the greater is the interstitial volume. Hence porosity is a key consideration. The relation to moisture stems from the fact that the greater the fraction of the pore volume occupied by water, the less is the volume available for air and hence for oxygen.

Interstitial volume, also known as porosity, is determined by (1) the size of individual particles, (2) the configuration of the particles, and (3) the extent to which individual particles maintain their respective configuration. Maintenance of configuration depends upon the structural strength of the individual particle, i.e., resistance to flattening. Because these characteristics vary, maximum permissible moisture content varies from substrate to substrate. Thus the maximum permissible moisture content is higher with wastes in which straw and wood chips rather than paper are the bulking materials. Several wastes and their respective permissible moisture contents are listed in Table 10.2.

With in-vessel systems, the microbial oxygen demand is more or less met by direct exposure to air brought about by agitating the composting mass. Reliance upon interstitial air is correspondingly reduced. Although the reliance upon interstitial air may be reduced, it is never eliminated because agitation is neither complete nor uninterrupted. Therefore, moisture content continues to be a decisive factor. Moreover, excessive moisture adversely affects materials handling characteristics.

Minimum.   Moisture inadequacy is a common operational factor because the combination of relatively high temperatures and intense aeration is conducive to evaporation. Minimum moisture content becomes a consideration at moisture levels lower than those at which oxygen availability is a limiting factor. At such levels, microbial biological requirement for water becomes the determinant.

The penalty for moisture shortage is inhibition of microbial activity. Because almost all biological activity ceases at moisture contents lower than about 12 percent, the more closely the moisture content of a composting mass approaches that level, the less is the intensity of the microbial activity. The consensus is that efficient composting requires that the moisture content of the composting mass be maintained at or above 45 to 50 percent.

The effect of moisture insufficiency is illustrated by its effect on oxygen demand. In a report, Lossin states that moisture content is a determinant of oxygen needs. For

TABLE 10.2   Maximum Permissible Moisture Contents

Type of waste	Moisture content, % of total weight
Theoretical	100
Straw	70–85
Rice hulls	70–85
Wood (sawdust, small chips)	80–90
Manure with bedding	60–65
Wet wastes (vegetable trimmings, lawn clippings, kitchen wastes, etc.)	50–55
MSW (refuse)	55–60

example, fresh compost having a moisture content of 45 percent required 263 mm^3/g•h, whereas at a moisture content of 60 percent the demand was 306 mm^3/g•h.[5] This increase in demand indicates that moisture insufficiency had inhibited bacterial activity. It would have taken another determination at a higher moisture content to determine the level at which moisture would no longer be limiting.

pH Level.   The optimum pH range for most bacteria is between 6.0 and 7.5, whereas the optimum for fungi is 5.5 to 8.0. Precipitation of essential nutrients out of solution rather than inhibition due to pH per se establishes the upper pH limit for many fungi.

In practice, little should be done to adjust the pH level of the composting mass. Owing to the activity of acid-forming bacteria, the pH level generally begins to drop during the initial stages of the compost process. These bacteria break down complex carbonaceous materials (polysaccharides and cellulose) to organic acid intermediates. Some acid formation may also occur in localized anaerobic zones. Some may be due to the accumulation of intermediates formed by shunt metabolisms. Shunt metabolism may be triggered by an abundance of carbonaceous substrate and/or perhaps by interfering environmental conditions. Whatever the cause may be, the early pH drop in composting MSW may be to 4.5 or 5.0. The drop could well be lower with other wastes.

Organic acid synthesis is paralleled by the development of a microbial population for which the acids serve as a substrate. The consequence is a rise in pH level to as high as 8.0 to 9.0. The mass becomes alkaline in reaction.

Buffer against the initial pH drop through the addition of lime is unnecessary. Moreover, it promotes a loss of nitrogen. The loss can be particularly serious during the active stage of the compost process. For example, in research conducted at the University of California at Berkeley in the 1950s, nitrogen loss always was greater from piles to which lime [Ca(OH)$_2$] had been added to raise the pH.[1]

Despite the potential promotion of nitrogen loss, the addition of lime might be beneficial in cases in which the raw waste is rich in sugars or other readily decomposed carbohydrates (e.g., fruit and cannery waste). Acid formation in such wastes is more extensive than in MSW and yard waste. For example, it was found in studies on the composting of fruit waste bulked with sawdust, rice hulls, or composted refuse[6] that the 3 to 4 days' lag in temperature rise characteristic of unbuffered fruit waste could be eliminated by adding lime. However, nitrogen loss also was greater.

Occasionally, the addition of lime may lessen offensive odors because of the effect of pH. Lime addition also improves the handling characteristics of some wastes.

Temperature.   In the consideration of temperature as an environmental factor, the interest is in the effect of temperature on the well-being and activities of the microbial population, rather than in the effect of microbial well-being and activity on temperature level. In short, environmentally oriented interest is in the effect of temperature on microbial well-being and activity; whereas operationally oriented interest is in the effect of microbes on temperature.

As an Environmental Factor.   In the section on classification, brief mention was made of the relation between temperature and rate of composting. The question is not so much one of the effect of temperature within either the mesophylic or the thermophylic ranges as it is the relative advantages of one range over the other, i.e., mesophylic vs. thermophylic. Thus each group of mesophyles has an optimum range specific to it in the mesophylic range. Similarly, each group of thermophyles has its specific optimum level in the thermophylic range. The result is that because of the diverse population and variation in temperature in the composting mass, chances are that in any given instant in time, temperature will be optimum for some group. Conversely, chances of its being optimum for all groups at any single instant in time would be nil.

For example, the optimum for the mesophyle *Pseudomonas delphinium* is 25°C; whereas for *Clostridium acetobutylicum,* another mesophyle, it is 37°C. The high degree of activity is indicative of a satisfactory temperature for most of the microbes.

A straight-line relationship exists in terms of increase in process efficiency and speed and rise in temperature because of the overlapping of optimum temperatures at levels lower than 30°C. The slope of a curve showing efficiency or speed of the process as a function of temperature would flatten somewhat between 35 and 55°C, perhaps with some decline between 50 and 55°C. The existence of an activity plateau at the transition from the mesophylic range to the thermophylic range is due not only to the involvement of many types of organisms but also to adaptation of organisms or enrichment for organisms adapted to a given range. As the temperature rises above 55°C, efficiency and speed drop and are negligible at temperatures above 70°C. At temperatures higher than 65°C, spore formers rapidly enter the spore stage, and as such are dormant. Most non–spore formers die off.

Mesophylic vs. Thermophylic Composting. In the 1950s there was some debate on the relative merits of mesophylic vs. thermophylic composting regarding nature, extent, and rate of decomposition.[13] The opinion in favor of thermophylic composting largely rested on the fact that experimental evidence shows that up to a certain point, chemical and enzymatic reactions are accelerated by each increment in temperature. For enzymes the acceleration continues up to the point above which they are inactivated.

Experience shows that the upper limit for most thermophyles involved in composting is between 55 and 60°C, and accordingly the process is adversely affected if temperatures rise above this range.[12,14] In practice the question is moot not only because of the costs involved in establishing and maintaining a thermophylic environment but also because the heat generated in a reasonably large or insulated mass inevitably brings the internal temperature to thermophylic levels unless something is drastically amiss with the operation. In fact, measures should be taken to avoid the inhibitory range.[14] Therefore, if the question of mesophylic vs. thermophylic has any significance, it would mainly concern in-vessel systems because with them, heat is dissipated by the continued agitation and ventilation of the mass in the compost unit. Of course, temperature can rise in the units if the mass of composting material is sufficiently large and the degree of agitation of the content is kept below a critical level.

Operation and Performance Parameters

Commonly used operational parameters include these eight: oxygen uptake, temperature, moisture content, pH, odor, color, destruction of volatile matter, and stability. With respect to the first four, the distinction between their status as environmental factor and that as operational parameter is very difficult to define because the two overlap in that operational parameters evolve from environmental factors.

Oxygen Uptake. Oxygen uptake is a very useful parameter, because it is a direct manifestation of oxygen consumption by the microbial population, and hence of microbial activity. Microbes use oxygen to obtain the energy to carry on their activities.

A very effective means of monitoring for adequacy of oxygen supply is by way of the olfactory sense, namely, detection of odors. The emanation of putrefactive odors from a composting mass is a positive indication of anaerobiosis. The intensity of the odors is an indication of the extent of anaerobiosis. Attempts to measure odoriferous constituents (e.g., H_2S) have been only indifferently successful. Because of their anaerobic origin, the malodors soon decrease after aeration is intensified. Although reliance upon the detection of objectionable odors may seem to be rather primitive, nevertheless it is a useful supplement in routine monitoring. It does have the disadvantage of being an "after-the-fact" indicator. Therefore, in operations in which an oxygen probe can be used or the oxygen of input and output airstreams can be measured a direct monitoring of oxygen is advisable.

An important operational consideration is that although the input airstream may be sufficiently great to meet the theoretical microbial oxygen demand and the discharge airstream may contain some oxygen, localized anaerobic zones may be present. The zones may be due to inadequate mixing or to shortcircuiting of air through the mass. In practice, the complete prevention or elimination of these zones would be economically, if not technologically, unfeasible. Fortunately, the complete elimination is not essential for a nuisance-free operation, provided the number and size of the zones does not become excessively large.

According to Ref. 15, four generalizations can be made despite the many uncertainties mentioned or implied in the preceding paragraphs. The generalizations are:

1. An oxygen pressure greater than 14 percent of the total indicates that not more than one-third of the oxygen in the air has been consumed.
2. The optimum oxygen level is 14 to 17 percent.
3. Aerobic composting supposedly ceases if the oxygen concentration drops to 10 percent.
4. If CO_2 concentration in the exhaust gas is used as a parameter for oxygen concentration, then the CO_2 in the exhaust gas should be between 3 and 6 percent by volume.

Temperature.   Temperature is a very useful parameter because it is a direct indicator of microbial activity. However, in the application of temperature as an operational parameter, it must be remembered that in a practical operation, the desired temperature range should include thermophylic temperatures. The reasons are: (1) some of the organisms involved in the process have their optimum level in the thermophylic range; (2) weed seeds and most microbes of pathogen significance cannot survive exposure to thermophylic temperatures; and (3) unless definite countermeasures are taken, thermophylic levels will be reached during the active stage.

In general, any abrupt and unexplained deviation from the normal course of temperature rise and fall is an indication of an environmental or operational deficiency that requires attention. An exception to this general rule is the need to prevent the temperature from exceeding 55 to 60°C, i.e., reaching a level that is inhibitory to most microbes. Probably the most effective remedial measure is ventilation.

Moisture.   The numerical value of the operational parameter, moisture, is the maximum permissible moisture content. As stated earlier, this value varies from substrate to substrate. Table 10.2 lists several maximum permissible moisture contents. The relatively low value for MSW reflects the high paper content of the waste.

Regardless of substrate, the lowest permissible moisture content for efficient composting is about 45 percent. An unfavorably low moisture content is a common problem in compost practice, because conditions in a composting mass are conducive to evaporation, i.e., water loss. Unless this water is replaced, moisture is likely to become limiting.

pH.   Unless the substrate is unusually acidic, which rarely is the case with MSW, pH level has little value as an operational parameter. If the pH level is lower than 4.5, some buffering may be indicated, e.g., adding lime. Liming may also be indicated for certain cannery wastes.

Odor.   Odor as an operational parameter received some attention in the discussion of aeration. Attempts to develop a quantitative standard for odor based on hydrogen sulfide concentration have met with little if any success, because the olfactory nerve senses H_2S concentrations lower than the detection level of H_2S analytical tests. In waste treatment practice, all odors are regarded as being objectionable to the public.

Color.   Although the color of the composting mass progressively darkens, it is a crude parameter and at best is roughly qualitative and highly subjective.

Destruction of Volatile Solids.   Inasmuch as composting is a decomposition process, it is characterized by some destruction of volatile solids. Complete destruction is neither desirable nor necessary because the value of the compost product, particularly as a soil conditioner, is mostly due to its volatile (i.e., organic) solids content. Hence rate rather than extent of destruction would be the useful parameter. The problem is in the establishment of a standard rate. Rates vary with several important factors. The best that is presently available is to the effect that volatile matter *is* being destroyed.

Stability.   "Stability" is a broad term that may refer to chemical and physical stability and/or to biological stability. As applied in composting, the composting mass is judged "stable" when it has reached a state of decomposition at which it can be stored without giving rise to health or nuisance problems. This excludes the temporary stability due to dehydration or other condition that inhibits microbial activity. Despite many claims to the contrary, a satisfactory quantitative method for determining degree of stability has yet to be developed, at least one that can be used as a "universally" applicable standard.

   The search for a method of determining stability that can be sufficiently standardized is almost as old as the compost practice. The list of proposed methods is correspondingly lengthy. It includes final drop in temperature,[1] degree of self-heating capacity,[16] amount of decomposable and resistant organic matter in the material,[17] rise in the redox potential,[18] oxygen uptake,[7] growth response of the fungus *Chaetomium gracillis,*[19] and the starch test.[20] Of this array of tests, the final drop in temperature is the most reliable, because it is a direct consequence of the entire microbial activity as well as of the intensity of the activity. The weakness of temperature decline as a parameter is its time element. Because the decline represents a trend, it involves a succession of readings taken over a period of days. The other tests lack the necessary universality. For example, a redox potential that characterizes stability under one set of compost conditions does not necessarily do so under another set. With certain tests, lack of universality is aggravated by the difficulty of conducting them, e.g., the *Chaetomium* test.

   Phytoxicity frequently is regarded as being an indication of stability, although it is true that in the early stages of maturation, composting material often contains a substance that is inhibitory to plants (phytoxic), and which almost invariably disappears as maturation progresses. However, the disappearance does not always coincide with the attainment of the required degree of stability.

10.2 TECHNOLOGY

Facility Site Selection Considerations

Buffer Zone.   Measures taken with any and all waste treatment facilities regarding site selection and preparation to protect air, water, and soil resources must also be taken with a compost facility. In addition to the usual topographical, hydrological, economic, political, and sociological considerations involved in the selection process, provision of an adequate "buffer zone" between the facility and residential areas is particularly essential to the continued survival of a compost facility. Of the factors responsible for the need of a buffer zone (e.g., vehicular traffic, noise), odors rank among the highest. Moreover, because of the likelihood of offensive odors, the size of the buffer zone must be substantial. Nevertheless, the actual dimensions of the zone required for a particular facility depend upon the magnitude of the operation, the nature of the waste, the type of compost system employed, and the degree of enclosure and control of emissions. Obviously, the buffer zone indicated for a few tons per day of yard waste compost operation is far smaller than that for a full-scale MSW, manure, or sewage sludge compost facility.

Odors.   In a compost facility, offensive odors usually have two main origins: (1) the raw wastes that serve as substrate; and (2) operational shortcomings and mishaps. (Raw yard wastes have very little, if any, odor unless they include high concentrations of grass or food waste.) Although operational odors can be kept at a minimum, those due to odoriferous raw wastes are inevitable. However, their intensity can be substantially reduced by proper storage and prompt processing.

The impact of the odor problem can be considerably reduced by enclosing the raw waste receiving and storage areas and the active and early maturation stages of the composting process. The ventilation of the housing structure can be designed such that air is exhausted through an air scrubber[21] or an odor filter. Unfortunately, doing so is costly and is not failproof.

Compost Systems

The rationale underlying compost system design is twofold: (1) provide optimum conditions for composting in an environmentally and economically acceptable manner; and (2) determine the type and size of the compost system and other aspects of technology by the type, volume, nature of waste, and the size of the available buffer zone.

Classification of Compost Systems.   Compost systems fall into two very broad groups, windrow and in-vessel. Reflecting their mechanisms of aeration, windrows may be the turned type, forced aeration (static pile) type, or a combination of turned and forced aeration. A typical windrow is presented in Fig. 10.4. Windrows may be sheltered, i.e., contained within a structure, or they may be unsheltered. Shelters must be provided with a ventilation system such that emissions can be satisfactorily conditioned. Also, in winter, consideration must be given to the control of condensation of moisture released by actively composting material. Reactors in in-vessel systems have one of the following configurations: horizontal drum, which is rotated slowly and which may be compartmentalized; vertical silo; and an open tank equipped with a stirring or an agitation device. All designs of in-vessel systems have provisions for forced aeration. Because of economic constraints, the usual procedure with in-vessel systems is to use the reactor for the lag and active phases and to rely upon windrowing for the maturation phase. A photograph of an in-vessel system is given in Fig. 10.5.

Aeration Mechanisms.   The provision of satisfactory aeration is an essential feature of almost all existing compost systems. Aeration mechanisms involved in providing atmospheric oxygen fall into three broad groups, namely, agitation, forced aeration, and turning. A particular system may incorporate one or a combination of mechanisms. Agitation is accomplished by tumbling, stirring, and/or the act of mixing the composting mass. In forced aeration, air is either pushed or pulled through the com-

FIGURE 10.4 Photograph of a typical windrow.

FIGURE 10.5 Photograph of an in-vessel system (*Courtesy of OTVD, Inc.*)

posting mass. Most in-vessel systems rely on a combination of the three mechanisms. In windrow composting, milling and stacking the raw waste accomplishes the initial aeration. As stated in the section on moisture content, most of the microbial oxygen requirement in windrowed material is met by the air entrapped in the windrows, i.e., interstitial air. Interstitial air is renewed by turning the windrow or by forcing air through the windrow, i.e., by ventilating the pile. Very little oxygen comes by way of diffusion of ambient air into the outer layer of the windrow.

Windrow Systems

Site Preparation. The preparation of concern in this discussion is that of the working area, i.e., the area in which the windrows are constructed and the associated maintenance equipment is maneuvered. For convenience, the working site is also referred to as the "compost pad" in this discussion. Not included is the preparation of the facility site as a whole (access roads, grading, construction of structures, and provision of utilities).

Pad Specifications. Pad specifications cover a wide spectrum and are influenced by the size of the operation, the nature of the wastes to be composted, and the dictates of circumstances specific to it (e.g., proximity to residential areas, land use, financial capacity). Among the most applicable specifications are availability of essential utilities, surface and construction geared to all-weather accessibility and use regardless of whether the pad is sheltered or is exposed to the elements, appropriate aereal dimensions, prevention of water intrusion, collection and disposal (treatment) of runoff from pad, and leachate collection and disposal.

Utilities. Access should be had to water and electricity. Water occasionally must be added to the composting mass to keep the moisture content from dropping to inhibitory levels. Water also should be at hand for fire and dust control. Although access to power is not as important as access to water, power has many useful applications (e.g., powering blowers and illumination).

Pad Surface and Construction. All working areas should be paved and be ready for use regardless of weather. The pad should be sufficiently rugged to support the combined weight of the composting mass and associated materials handling equipment, as well as the maneuvering of the latter. The required degree of conformity with the specifications (i.e., flexibility of application) regarding surface and construction is closely related to the stage of the compost process. Pad conformity with specifications is most necessary for the lag, active, and early maturation stages, decreases as maturation progresses, and is least necessary during storage.

Calculation of Total Area of Windrow Pad. A variety of factors combine to determine the dimensions of the area requirement. Among them are total volume of material to be accommodated during all stages of the compost process, i.e., from the construction of the windrows through disposal of the stored product, the configuration of the windrows, space required for the associated materials handling equipment and the maneuvering thereof, and the aeration system (forced or turning).

The following is a summary of the steps involved in calculating the pad area. For convenience, the steps are arranged in four main groups: (A) total volume of feedstock to be composted, (B) area occupied solely by windrows, (C) maneuvering area, and (D) total pad area.

A. TOTAL VOLUME FOR FEEDSTOCK

$$\text{Total volume of feedstock (ft}^3 \text{ or m}^3) = \frac{\text{retention time (days)} \times \text{rate of feedstock delivery (lb/day or kg/day)}}{\text{bulk density (lb/ft}^3 \text{ or kg/m}^3)} \quad (10.1)$$

B. AREA OCCUPIED SOLELY BY WINDROWS

Step 1. Determine the volume of each windrow.

$$\text{Volume (ft}^3 \text{ or m}^3) = \text{cross-sectional area (ft}^2 \text{ or m}^2) \times \text{length of windrow (ft or m)} \quad (10.2)$$

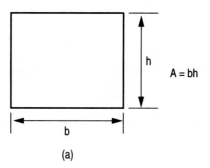

$A = bh$

(a)

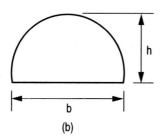

$A = \dfrac{\pi\,bh}{4}$, where $b = 2h$

(b)

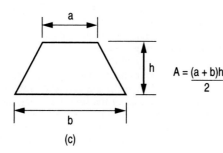

$A = \dfrac{(a + b)h}{2}$

(c)

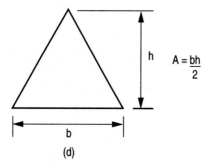

$A = \dfrac{bh}{2}$

(d)

FIGURE 10.6   Areas of various potential cross sections for compost piles.

Cross-sectional area is a function of cross-sectional configuration. Figure 10.6 illustrates four types of cross-sectional configurations. The cross-sectional area of a pile having a square or rectangular cross section (Fig. 10.6a) is given by Eq. (10.3).

$$\text{Cross-sectional area (ft}^2 \text{ or m}^2) = \text{base (ft or m)} \times \text{height (ft or m)} \qquad (10.3)$$

On the other hand, the cross-sectional area of the configuration in Fig. 10.6b is defined by Eq. (10.4).

$$\text{Cross-sectional area (ft}^2 \text{ or m}^2) = \pi/4 \times b \text{ (ft or m)} \times h \text{ (ft or m)} \qquad (10.4)$$

The cross-sectional area of the configuration in Fig. 10.6c is given in Eq. (10.5).

$$\text{Cross-sectional area (ft}^2 \text{ or m}^2) = 1/2 \, (a + b) \text{ (ft or m)} \times h \text{ (ft or m)} \qquad (10.5)$$

The area of the configuration in Fig. 10.6d is defined by Eq. (10.6)

$$\text{Cross-sectional area (ft}^2 \text{ or m}^2) = 1/2 \text{ (ft or m)} \times \text{h (ft or m)} \qquad (10.6)$$

Step 2. Determine the number of windrows:

$$\text{Number of windrows} = \frac{\text{total volume of feedstock (ft}^3 \text{ or m}^3)}{\text{volume per windrow (ft}^3 \text{ or m}^3 \text{ per windrow)}} \qquad (10.7)$$

Step 3. Determine the area solely occupied by windrows:

$$\text{Total windrow area (ft}^2 \text{ or m}^2) = \text{number of windrows}$$
$$\times \text{ area per windrow (ft}^2 \text{ or m}^2 \text{ per windrow)} \quad (10.8)$$

C. MANEUVERING AREA.   Maneuvering area is the space required to maneuver the turning and other equipment. Two such spaces must be provided for each windrow (one on each side of a windrow). The area of each space is given by Eq. (10.9).

$$\text{Area of space (ft}^2 \text{ or m}^2) = \text{windrow/length (ft or m)}$$
$$\times \text{ width of space (ft or m)} \qquad (10.9)$$

The width of a space depends upon the type of turning machine. The following widths are times width of the space. The width of a space depends upon the type of turning machine. The following widths are approximate estimates: If turned with a bucket loader, the width may be as little as 4 ft (1.22 m). If a self-propelled turner is used, the width may be from 3 to 5 ft (0.9 to 1.5 m). If a tractor-assisted turner is used (two passes), a 6- to 8-ft (1.8 to 2.4 m) space is indicated. The space between two individually aerated piles is about 20 ft (6.1 m).

D. TOTAL AREA OF PAD.   The total pad area is the sum of the area required for the windrows plus that needed for maneuvering the material (e.g., constructing windrows, turning the composting mass, water trucks, force aeration equipment, etc.).

It is emphasized that the calculations do not allow for the shrinking of the piles that will occur due to destruction of volatile matter and loss of moisture. Because of the variation in percentage loss due to differences in nature of feedstock and compost system applied, it is impractical to apply a single shrinkage value to all situations. Therefore, the value calculated for the total area will be the maximum value, i.e., the maximum requirement.

A sample calculation to determine the area for a composting pad follows:

1. Volume of material to be composted = 24 m³/day
2. Composting period (detention time) = 50 days
3. Total volume of material on pad = 50 days × 24 m³/day = 1200 m³
4. Dimensions of windrow: length = 50 m, height = 3 m, and width = 4 m
5. Volume of windrow: $V = 2/3 \times (4 \times 3) \times 50$ m³
6. Number of windrows = total volume of material/volume of windrow = 1200/400 = 3
7. Distance between windrows = 4 m
8. Space around perimeter of composting area = 3 m
9. Length of composting area = windrow length and perimeter space = 50 m + 2(3) = 56 m
10. Width of composting area: width of windrows + distances between windrows + perimeter space = (4 × 3) + (2 × 4) + (2 × 3) = 12 + 8 + 6 = 26 m
11. Area required = length × width = 56 × 26 = 1456 m²

These calculations do not include required setbacks or buffer zones.

Windrow Construction. A windrow is constructed by stacking the prepared feedstock in the form of an elongated pile. The procedure involved in stacking the material is influenced by the volume and nature of the feedstock, the design and capacity of the available materials handling equipment, and the physical layout of the windrow pad. If more than one feedstock is involved (e.g., cocomposting sewage sludge and MSW, or yard waste and food waste), or an additive is to be employed, the incorporation would take place at this time. If cocomposting or additives are not involved, the windrows are set up directly after preprocessing is completed. If cocomposting is involved, one approach is to build up the windrow by alternating layers of one of the feedstocks with layers of the other feedstock or doses of the additive. The first and subsequent turning accomplish the necessary mixing of the components. If turning is not the method of aeration, necessary mixing is done immediately prior to constructing the windrow.

Conventional materials handling equipment such as a bulldozer or a bucket loader can be used for windrow construction. An alternative approach involves the use of a conveyor belt as follows: Directly after having been preprocessed, the feedstock is transferred to the windrow pad by way of a conveyor belt, the discharge end of which has been adjusted to the height intended for the completed windrow.

Windrow Dimensions. Three key factors enter into the determination of windrow dimensions, namely, aeration requirements, efficient utilization of land area, and the structural strength and size of the feedstock particles. Structural strength, in turn, is a key factor in the maintenance of the interstitial integrity needed to ensure a sufficient oxygen supply.

All dimensions could be expanded during winter to enhance self-insulation. In windy and arid regions, the dimensions can also be expanded so as to minimize moisture loss through evaporation.

HEIGHT. Interstitial integrity and height of the windrow are closely related, because the higher the pile, the greater is the compressive weight on the particles. Hence the greater the structural strength, the higher is the permissible height. With the organic fraction of municipal refuse, the height is on the order of 5 to 6 ft (1.5 to 1.8 m). Depending upon the size of the shrubbery trimmings fraction, it can be slightly higher with yard waste. In practice, the actual height is determined by the type of equipment used to aerate the composting mass. It generally is lower than the maximum permissible height.

WIDTH. Other than its determination of the ratio of surface area exposed to inward diffusion of air, width has little effect on aeration. However, the amount of inward air diffusion usually is negligible. In summary, width is dictated by convenience. If other factors do not intervene, a width of about 8 to 9 ft (2.4 to 2.7 m) is suitable.

Windrow Geometry. Windrow geometry should be geared to climatic conditions and efficient use of pad area. However, in practice the determinant is the type of windrow construction and turning equipment, principally the latter. For example, as was indicated in the section on calculation of pad dimensions, cross-sectional configuration exerts a significant impact on the ratio of windrow volume to area—and hence on efficient use of land area. However, in regions in which rains are frequent or heavy and the windrows are not sheltered, the cross-sectional configuration should be conical so as to shed water. On the other hand, a flattened top (square or rectangular configuration) is appropriate where rainfall is not a problem. With such a configuration, heat loss is less and windrow volume per unit pad area is greatest.

Turned Windrow Aeration. As stated earlier, windrows can be aerated by turning, by forced aeration, or by a combination of the two. A long record of successful experience has demonstrated the efficacy of turning.

Turning is accomplished by tearing down and then reconstructing the windrow. The windrow can be reconstructed either in its original position or immediately adjacent or somewhat removed from its prior position. Tearing down and reconstructing the windrow exposes the composting material to the ambient air and replenishes the interstitial oxygen supply. The resulting mixing renews microbial access to nutrients and disperses metabolic intermediates. The cooling effect of turning can be used for lowering a pile temperature that has reached inhibitory levels.

To avoid defeating its purpose, turning should be done in a manner that does not compact the composting mass.

Although ideally the turning should be such that the outer layers of the original pile become the inner layers of the reconstituted pile, limitations of turning equipment make it unfeasible to do so. However, the benefits that accrue from the reversal of positions can be gained by increasing the frequency and number of turnings. Benefits to be gained are twofold: (1) It promotes uniform decomposition. (2) It subjects all material to an eventual exposure of all material to the high temperatures characteristic of the interior of an actively composting windrow. The temperatures are high enough to be lethal to disease-causing organisms and to most weed seeds.

Frequency of Turning. Because required frequency of turning exerts a strong influence on design and size of equipment used in accomplishing turning, a few words on frequency are appropriate at this point. Ideally, frequency should be a function of rate of oxygen uptake by the active microbial population. For example, judging from past experience,[1-3] turning every third day is sufficient to meet the oxygen uptake in actively composting MSW. Of course, this assumes that the MSW is neither waterlogged nor compacted. Incidentally, waterlogged conditions and compaction can be remedied by increasing the frequency of turning.

It should be noted that a turning frequency of three times per week may not be sufficient to kill off all pathogens.[22] Undoubtedly, the incomplete die-off is a consequence of only about 40 to 50 percent of a typical windrow being exposed at a given interval to lethal temperatures. Each subsequent turning brings about a recontamination of "sterilized" material. One solution is to resort to attrition by increasing the frequency such that intervals between turning become too brief for appreciable regrowth.

Equipment. The simplest, but also the least satisfactory, method of turning involves the use of a bulldozer for tearing down and reforming a windrow. With such an approach, mixing and aeration are minimal and the material is compacted instead of being fluffed. The situation is somewhat improved when a bucket loader is used. However, owing to materials handling limitations, both approaches become increasingly inefficient when the volume involved exceeds a few tons per day. Nevertheless, it may be that economic circumstances render the use of a more complex turner unfeasible. Such a situation may not be unusual, which perhaps explains why the use of a bulldozer or bucket loader for turning continues to be a fairly widespread practice. If a bucket loader is used, it should be operated such that the bucket contents are discharged in a cascading manner, rather than dropped as a single mass.

Among the first of the automatic turners was one used in the mushroom industry in the 1950s. In the succeeding years other mechanical turners began to appear in increasing numbers and design variation. Consequently, several types of turners are now available. A good idea of the diversity may be gained from the lists in Ref. 9 and by consulting the advertisements in publications such as *BioCycle.*[25] The several types of turners presently on the market fit one or the other of three general groups divided on the basis of the design of the turner mechanism. They are the auger turner, the elevating face conveyor, and the rotary drum with flails. Some types of turners are designed to be towed and others are self-propelled. As is to be expected, the self-propelled types are more expensive than the towed types. An advantage of the towed type is the fact that the tow vehicle (tractor) can be used for other purposes between turnings. In addition to convenience, the self-propelled type requires much less space for maneuvering and therefore the windrows can be closer to each other (Fig. 10.7). The turning capacity of the machines ranges from about 800 tons per h (727 Mtons/h) with the smaller models to as much as 3000 tons per h (2727 Mtons/h) with the larger self-propelled versions. Similarly, the dimensions and configuration of the windrows vary with type of machine, e.g., 9 to 15 ft (4.6 m) wide and 4 to 10 ft (1.2 to 3.0 m) high.

To allow for an increased frequency dictated by emergency situations (e.g., excessive moisture) and to ensure hygienic safety, the equipment capacity should be sufficient to permit daily turning.

Forced Aeration (Static Pile). The substitution of forced aeration for turning as an effective means of aeration has long intrigued compost practitioners.[26,27] A major factor, if not the decisive factor, in favor of forced aeration is the fact that it would be less expensive than turning. Supposedly, a prime saving would be the elimination of the need for expensive turning equipment. However, in practice this saving may not always materialize because some turning inevitably is necessary for the satisfactory remedying of localized problem zones, for ensuring uniformity of decomposition, and for the adequate destruction of pathogens.

Windrow Construction. The construction of a windrow for forced aeration begins with the installation of a loop of perforated pipe on the compost pad. The perforations are evenly spaced in a long row slightly off-center at the top of the pipe. The pipe diameter is 4 to 5 in (10.2 to 12.7 cm). The loop is oriented longitudinally and is centered under what is to be the ridge of the windrow. Short circuiting of air is avoided by not extending the piping the full length of the windrow. The perforated pipe is connected to a blower by way of a nonperforated pipe. After the pipe is in place, it is cov-

FIGURE 10.7 Self-propelled windrow-straddling composting machine. A rotary drum, not seen within the machine, picks up the material, throws it over the drum, and redeposits it into a windrow ready for subsequent turning. (*Courtesy of Resource Recovery Systems of Nebraska, Inc.*)

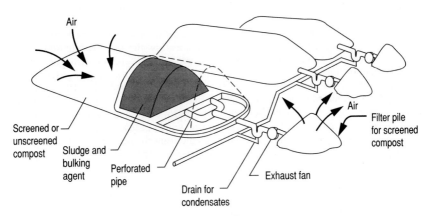

FIGURE 10.8 Windrow arrangement for forced aeration.

ered with a layer of bulking material or finished compost that extends over the area to be covered by the windrow. This base layer (bed) is intended to serve as a means of facilitating the movement and uniform distribution of air during composting. Additionally, the bed absorbs excess moisture and thereby minimizes seepage from the windrow. The compost feedstock is then stacked upon the piping and bed of bulking material to form a windrow which has the configuration diagrammed in Fig. 10.8. The finished windrow is of indeterminate length, about 13 ft (3.9 m) wide, and 8 ft (2.4 m) high.

The completed windrow is entirely covered with a 12- to 18-in (30.5- to 47.7-cm) layer of wood chips, finished compost, or similar material. The covering absorbs objectionable odors. It also results in the occurrence of high temperatures throughout the composting mass, and in that way leads to a more complete pathogen kill as well as uniform decomposition.

Process Management.   Experience indicates that intermittent forcing of air into the windrows serves to maintain aerobiosis at an adequate level. This was confirmed by results obtained in a study that involved a 50-ft (15.2-m) windrow which contained about 73 tons (66.4 Mtons) of sludge. In the study it was found that a timing sequence that forced air into the windrow at 16 m³/h for 5- to 10-min intervals was fully adequate. This particular rate was based upon a need of about 4 L/s/Mton of dry sludge solids. It should be noted that these numbers are only indicative. For a given situation, the required rate of air input should be determined experimentally, inasmuch as it will depend upon a number of variable factors.[23,24]

An innovation introduced during the past decade calls for tying airflow rate and timing with temperature control. The underlying rationale is to use windrow ventilation as a means of cooling the interior of the windrow.[14] The temperature control approach attempts to maintain optimum windrow temperatures (e.g., 130 to 140°F or 54.4 to 60°C). Because temperature directly indicates the status of the process, electronic temperature sensors, such as thermocouples or thermistors, provide a means to control airflow as well as monitor the temperature. An electronic signal from the sensor causes a control circuit to switch the blowers on or off when the windrow temperature reaches set limits. Similarly, blowers are shut off when the temperature drops below a set level. Another innovation is to use an electronic oxygen-sensing device to activate the blowers when the oxygen level drops below a predetermined level (e.g., 5 percent). From the standpoint of process management, temperature control is better aeration strategy, because it prevents the attainment of inhibitory temperature levels. However, it involves greater airflow rates, larger blowers, and more expensive and sophisticated temperature-based control systems than do timer sequenced systems.

Direction of Air Flow.   The direction of the airflow through the windrow may or may not be reversed during the course of the process. A common arrangement is to initially pull air through the windrow (suction) and pass the discharged gaseous emissions through an emission conditioning filter (e.g., odor control). The filter may consist of fully composted material, organically rich soil, or other materials. The rationale is that the suction arrangement facilitates the control of gaseous emissions during the initial phases (lag and active), i.e., phases during which gaseous emissions are particularly troublesome. Airflow direction is reversed during the maturing and curing phases, inasmuch as objectionable emission characteristics are likely to be at an acceptable level.

In-Vessel Systems

Currently, there are several in-vessel systems in the market. The primary objective of the design is to provide the best environmental conditions, particularly aeration, temperature, and moisture. Nearly all in-vessel systems use forced aeration in combination with stirring, tumbling, or both.

Past and current experience indicates in-vessel composting does not guarantee a nuisance-free, specifically odor-free operation. All recent forced closures were occasioned, in part, by complaints about odors. With very few exceptions, modern "in-vessel" compost systems are in reality combinations of in-vessel and windrow composting in which the vessel (reactor) is reserved for the active stage of the composting process and the windrowing is reserved for curing and maturation. Relatively high capital and operation and maintenance costs of most in-vessel composting systems together with the long residence times required to achieve stabilization make such a hybridization of reactor and windrow mandatory. Most of the problems leading to closure of the facilities can be traced to the discharge of the composting mass from the reactor before the completion of the active stage and the failure to make the necessary compensation in the windrowing phase.

The four basic configurations of reactor are the vertical silo, the horizontal silo, the horizontal drum (usually rotating), and the horizontally oriented open tank (rectangular or circular). The method of aerating the composting mass varies with type of reactor. Modes of aeration are forced and agitation. Agitation may be accomplished by stirring or tumbling the composting mass.

Representative Systems

1. *Plug-flow vertical reactor.* Pertinent features are illustrated in Fig. 10.9. Apparently, experience with the use of the plug-flow vertical reactor has been the difficulty of adequately aerating the contents throughout the column.

Another version of forced aeration stirring involves the use of three completely enclosed vessels. A special feature of this version is a rotating screw device installed at the bottom of the vessel for discharging the compost. One of the three tanks serves

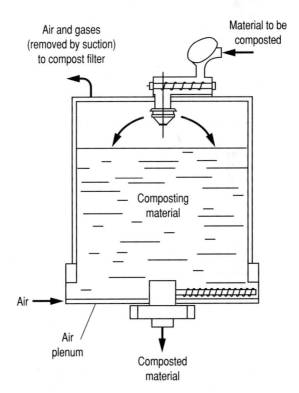

FIGURE 10.9 Schematic diagram of plug-flow vertical reactor.

as a storage container for carbonaceous material intended for use as a bulking agent and for correcting the C/N ratio. The composting process takes place in the second and third vessels, the "bioreactor" and the "cure reactor." Air is fed continuously into the bottom of the bioreactor, with positive control maintained by pulling air off the top. Composted material from the bioreactor is transferred into a cure reactor, in which further stabilization takes place. Air is fed continuously into the cure reactor to maintain aerobic conditions and to remove moisture due to evaporative cooling. Since the retention period in the bioreactor is 14 days, the normal daily operating sequence begins with the bioreactor outfeed discharging approximately one-fourteenth of the contents into a conveyor. The conveyor transports the material to the top of the cure reactor. At the same time, the outfeed device in the cure reactor is started and final compost product is discharged. Retention time in the cure reactor is on the order of 20 days. Problems frequently encountered in the operation of the units are: (1) a tendency of the material to "bridge" over the discharge screw, (2) failure of the rotating screw, and (3) excessive condensation of the upper layer of the bioreactor. The system originally was designed for sewage sludge and manure composting. Up to now, most of the experience has been with sewage sludge. The extent of the experience with MSW has been limited.

2. *Rotating, horizontal drum.* One of the earliest in-vessel systems to utilize the tumbling mode of aeration is the rotating horizontal drum. In most versions, the major piece of equipment is a long, slightly inclined drum at least 9 ft (2.7 m) in diameter that is rotated at about 2 r/min. According to the promotional literature, the retention periods in the drum may range from 1 to 6 days. However, the degree of stability acquired in such a time is not sufficient. Consequently, it is necessary to windrow the partially composted material for periods of 1 to 3 months in order to produce a properly matured (stabilized) product. The windrows should receive some aeration during the maturation period. MSW should be size reduced and sorted before it is introduced into the drum. Liquid or dewatered sewage sludge may be added to the MSW.

The relatively high capital, operational, and maintenance costs involved would seriously detract from the use, if not the economic feasibility, of a drum system for yard waste composting.

A schematic diagram indicating a design of a composting facility using a rotating drum is shown in Fig. 10.10.

3. *Open, horizontal, rectangular tank.* An in-vessel system that has much in its favor also is based upon a combination of forced aeration and tumbling like that shown in Figs. 10.5 and 10.11. It involves the use of a long, horizontal bin. In the operation of the system, properly prepared waste is placed in the bin. Tumbling is accomplished by way of a traveling endless belt, and air is forced through the perforated plates that make up the bottom of the bin and into the composting mass in the bin. The belt is passed through the composting material periodically. After a 6- to 12-day retention in the bin, the material is windrowed over a 1- to 2-month period. This system is well suited for MSW, sewage sludge, mixture of MSW and sludge, and properly bulked high-moisture food (cannery) wastes. Times involved (in bin plus required windrowing) range from 1 to 2 months. Currently there are several variations of the bin system. The open, horizontal rectangular tank is one of the most successful in-vessel units. This is probably related to the combination of aeration and adequate detention times. Gaseous emissions from the system must, however, be properly collected and treated.

4. *Vertical, mixed reactor.* A system that is based on a combination of forced aeration and stirring involves the use of a cylindrical tank. The tank is equipped with a set of augers supported by a bridge attached to a central pivoting structure. The bridge

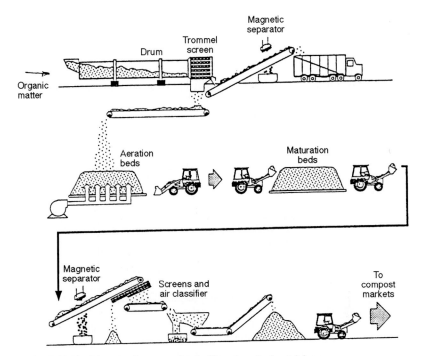

FIGURE 10.10 Diagram of a composting facility using a horizontal drum.

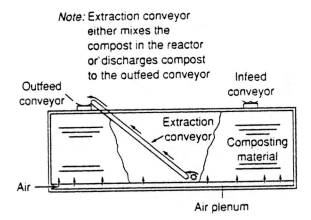

FIGURE 10.11 Schematic diagram of open, horizontal, rectangular tank.

with its set of hollow augers is slowly rotated. The augers are turned as the arm rotates. The hollow augers are perforated at their edges. Air is forced through the perforations and into the composting material. Retention time varies. If it is less than 3 weeks or so, the discharged material must be windrowed until stability is reached. A schematic diagram of the reactor is shown in Fig. 10.12. In addition, the diagram in Fig. 10.13 describes the position of vertical mixed reactor in an overall resource recovery operation. The system is that of the Delaware Reclamation Project in which

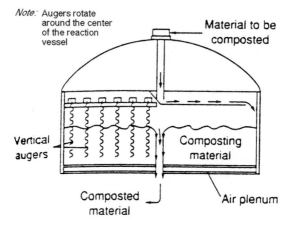

FIGURE 10.12   Schematic diagram of a vertical mixed reactor.

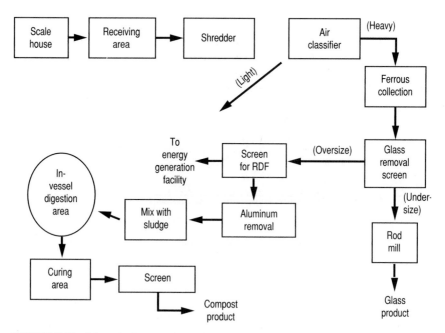

FIGURE 10.13   Schematic diagram of the Delaware Reclamation Project.

several inorganic materials, as well as RDF, are separated mechanically from mixed MSW. The process leaves a highly organic fraction which is mixed with sludge and introduced into the reactor. As of this writing, the composting portion of the facility had been closed because of odor complaints.

5. *Plug-flow, horizontal tank.* A diagram of this system is presented in Fig. 10.14. As the figure shows, the material to be composted is introduced into a rectangular tank. The material is forced into the tank by means of a hydraulic ram. After a certain detention time, the material exits the unit. As shown in Fig. 10.14, the material is aerated while in the tank. Problems that may be encountered with this reactor are due to inadequate aeration, mixing, and moisture control throughout the composting mass. In this case, the inadequacy is a result of the compaction exerted by the ram used to move the mass through the reactor.

Anaerobic Processes

As stated earlier, anaerobic composting was once ranked with aerobic composting by some authorities on composting. Because of a record of environmental problems (e.g., foul odors, strong leachates), anaerobic composting fell into disfavor. However, in keeping with the cyclic nature of the record of composting, anaerobic composting is again regaining some consideration.

It is true that the difference between high-solids anaerobic digestion and anaerobic composting is rather insignificant and is primarily a matter of objective. If the objective is recovery of energy through biogasification, the accepted classification is as anaerobic digestion or biogasification. If the objective is simply stabilization of a solid waste (e.g., refuse), the process is called anaerobic composting. However, it must be admitted that the distinction between the two objectives entails considerable differences in approach. With biogasification, process parameters are aimed at methane production; whereas in anaerobic composting, methane production is incidental and parameters center on stabilization.

In essence, the principles discussed in the preceding sections are directed to aerobic composting. To ensure anaerobiosis, the technology would involve the use of enclosed reactors. The main difference would be the absence of aeration. In fact, the presence of oxygen would inhibit the process. Finally, in order to produce an acceptable compost product, it would be necessary to follow the anaerobic phase by an aerobic phase.

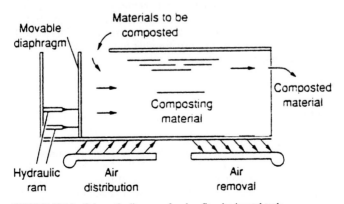

FIGURE 10.14   Schematic diagram of a plug-flow horizontal tank.

Equipment and System Vendors

The number of enterprises dealing with composting has expanded substantially during the last few years. A partial listing of equipment and system vendors is given in Appendix 10.A.

10.3 ECONOMICS

Introduction

Economics is a key element in all phases of a compost undertaking. Thus it is one of the considerations that enter into the decision to select composting as a waste management and treatment option. Economics enters into all subsequent decisions involved in the implementation of the compost option. Accordingly, it influences the selection of a particular compost system and of the associated equipment. Most importantly, it acts as a constraint on the continuity of the operation, in that an operation can survive only as long as its economic condition permits. This fact underlines the need to have a financial base sufficiently large to meet anticipated and unanticipated problems.

As would be true regarding the economics of all waste management alternatives, the status of the economics of composting is measured by the extent to which costs are balanced by returns. Costs include those related to the facility (capital, operation, and maintenance) and those involved in the disposal and/or marketing of the product. Returns include (1) the extensive list of benefits associated with composting and its product, (2) avoided costs, (3) income from sale of the product, and (4) collected tipping fees.

The item of interest in making an economic analysis of composting and decisions regarding the choice of the composting option is the net cost. Net cost can be expressed arithmetically as

$$\text{Net cost} = \text{gross cost} - (\text{income} + \text{avoided costs} + \text{other benefits}) \quad (10.10)$$

The principal element of uncertainty in the use of the equation is the assigning of a monetary value to the benefits.

The cost of composting is discussed under three main headings, namely, general composting costs, mixed MSW composting, and yard waste composting. In keeping with its heading, the first subsection deals with composting in general, i.e., cost of the compost process as applied to all wastes. Its main emphasis is on the problems attending the interpretation and utilization of the data reported in the literature. The second and third subsections concentrate on costs peculiar to municipal solid waste composting and yard waste composting, respectively.

General Composting Costs

Gross Costs. Costs cited in the literature cover a wide range of values. This situation is a reflection of the bearing exerted on cost by technology, climate, topography, demography, and availability of skilled and unskilled personnel. Nevertheless, accumulated data presently available permit the making of reasonably realistic projections regarding the economics of existing and of proposed operations.

The range of reported costs is illustrated by data cited in two references. One reference[28] presents estimated capital and operation and maintenance (O&M) costs for two 400 TPD* MSW compost facilities, one of which is a windrow facility and the other an in-vessel MSW facility. The other reference[29] cites data on capital and O&M costs for eight compost facilities in the United States at the time the report was written. Designed capacities ranged from 10 to 800 TPD. In 1988 dollars, the capital costs reported for the eight facilities ranged from about $30,000 to $75,000 per TPD of installed capacity. Estimates for the 400 TPD facilities ranged from $49,000 to about $156,000 TPD of installed capacity. Annual O&M costs ranged from about $150,000 to $6,000,000.[28,29] A summary of reported costs for some MSW composting facilities is presented in Table 10.3.

Assigning realistic costs to yard waste composting is difficult to do because of the paucity of reported economic information. The few published costs for operation and maintenance range from approximately as little as $4 to as much as $56 per ton of yard waste.

Factors Underlying Uncertainty of Reported Costs.   The factors that account for the wide range of the reported costs are important contributors to the uncertainty in assessing compost economics. Therefore, they should serve as constraints upon the use of the reported information in decision making. For example, variations in size of the facility and type and nature of the technology employed and failure to report costs in constant dollars are responsible for the wide range of reported capital and O&M costs.

An important contributor to the uncertainty is the difficulty of making accurate economic comparisons relative to operating facilities and compost technologies. The difficulty can be traced to the exceedingly broad collection of factors that not only are unique to each facility but also determine its economics. The collection includes everything from climate, labor, and equipment to cost accounting practices. Regarding cost accounting practice, in some cases, costs associated with a compost project are not segregated from existing solid waste operations. Cost accounting practice as applied to a yard waste compost operation may take the form of cost sharing the use and costs of items such as land, labor, and equipment with other ongoing operations. In this case, costs attributed to composting reflect estimated incremental costs. Yet another complication is the absence of consistent and precise definitions of operations and maintenance costs. The many obstacles to arriving at an accurate prediction of costs can be countered by defining proposed project requirements and performing detailed cost analyses.

Personnel and Training.   The extent of the personnel and training expenditure depends upon the size of the staff. Staffing requirements for composting operations vary with facility size and with the relative allocation of capital equipment vs. labor. The number of personnel range from part-time employment for small, seasonal leaf composting operations to approximately 30 full-time employees for large MSW compost operations. Labor requirements for manual removal of recyclables and inerts vary roughly in proportion to plant throughput. Mechanical separation reduces the need for sorters.

Personnel training requirements and costs usually do not vary markedly as facility size increases, although more people are involved with the larger facilities. However, the range of skills required is similar to that of some smaller facilities.

---

*All costs are given in TPD (U.S. tons per day). Multiply TPD by 0.907 to obtain Mtons per day.

TABLE 10.3 Reported Costs for Some MSW Composting Facilities

Facility	Year opened	System	Capacity, TPD	Capital cost		O&M costs[a]		Tipping fee, $/ton	Reference
				Total, $	$/TPD	Annual, $	$/ton		
Lake of the Woods County, Minn.	1989	TW	10	500,000	50,000	150,000	60	Zero[b]	45
Fillmore County, Minn.	1987	A-SP	15–20	1,310,000	75,000	NA		40	46
Swift County, Minn.	1990	A-SP	30	1,400,000	46,667	266,000	51	69	47
Portage, Wis.[c]	1986	Drum-TW	30	850,000[d]	28,300[d]	NA		Zero[b]	45
St. Cloud, Minn.	1988	Drum-TW	100	NA	NA	NA		50	45
Portland, Ore.	[e]	Drum-A-SP	600	20,000,000	33,300	5,000,000	27	42	48
Pembroke Pines, Fla.	[f]	A-SP?	667	48,500,000	72,700	NA		NA	45
Dade County, Fla.	1990[a]	TW	800	25,000,000	31,250	NA		24[g]	49

All costs are given in U.S. tons. Multiply U.S. tons by 0.907 to obtain Mtons.

TW = turned windrow

A-SP = aerated static pile

NA = not available

[a]Projected.

[b]Funded through taxes.

[c]Cocomposting facility.

[d]Built with used equipment.

[e]Closed as of mid-1993.

[f]Undergoing modifications.

[g]Tipping fee mandated by county.

Impact of Technology on Costs. In this section, the emphasis is on the impact exerted on costs by the type of technology employed, by system residence time, by equipment redundancy, and by plant utilization. The section closes with a few words on the impact of feedstock characteristics.

In windrow composting, the composting mass often is sheltered in a structure during the active and early curing stages. This practice is prevalent in regions characterized by seasonal changes, especially in winter, and in regions subjected to heavy rainfall.

Most in-vessel systems reserve the reactor for the active stage of the compost process and rely upon windrowing for the curing and maturation stages. The rationale is twofold, namely, to maintain conditions at optimum levels during the active stage, thereby accelerating the rate of microbial activity and correspondingly shortening the active phase. The economic gain in shortening the active stage is the reduction of residence time in the reactor and hence increase in its processing capacity, i.e., reduction in size of the reactor.

The approach to hastening the active stage has been to increase the sophistication of the reactor. However, acceleration made possible by sophistication of the reactor ultimately is limited by the genetic makeup of the microbial population. Obviously, the more expensive the reactor, the greater the required degree of process acceleration. This requirement often is underestimated, which in turn leads to disastrous consequences. The trench type of reactor is a good compromise with respect to both economy and processing efficiency.

It is difficult to make a comparison between the impact of windrow technology and in-vessel technology. At a minimum, it is assumed in making such a comparison that the products of the two are competitive with each other. All other process variables being the same, particularly the active stage, the capital cost of structures and equipment for the active phase of mixed MSW composting would be higher for the various in-vessel systems than for a conventional windrow system. Theoretically, however, this disadvantage is balanced by lower operating costs. In-vessel proponents claim that their systems have the potential for lowering operating costs by virtue of increased automation. Many in-vessel process configurations also require less land area per unit of throughput than is required for windrow composting. An important point is that regardless of the sophistication of the system, prudence demands that a generous buffer zone be provided.

The present consensus is that because of the factors mentioned above, high labor and land costs tend to favor the selection of labor- and land-efficient in-vessel systems. Conversely, low labor and land costs tend to favor the selection of windrow composting.

Feedstock. Feedstock differences can also have a significant effect on the cost of composting MSW, in particular if undesirable materials are removed by source separation. Virtually any MSW feedstock rich in paper and yard waste would require some shredding in order to produce a marketable compost in a reasonable amount of time. However, if glass, metal, and plastics were to be removed from the waste stream via source separation, some of the cost of preprocessing could be avoided, by eliminating the labor and equipment required for processing the recyclables. Eliminating glass, plastic, and metal from the compost plant's feedstock could reduce the unit cost of composting by 10 to 20 percent. The effect is relatively small owing to the elimination of the offsetting revenue from recyclables.

The level of plant utilization also has a significant influence on the unit cost of composting. Throughput per unit of capital cost increases markedly as facility utilization moves from 40 h per week to near continuous operation. In anticipation of growth in the waste stream, because of noise and traffic requirements, or for other reasons, some facilities are designed for 8 h per day, 5 day per week operation. A related plant utilization issue is the optimum level of equipment redundancy. In general, a process facility functions at or near its lowest unit cost when the process operates continuously except for interruptions for necessary repairs and maintenance.

Income. Among the available sources of income are tipping fees, taxation, state grants, sale of collected recyclables, and sale of compost product. In some respects, taxation and state grants could be interpreted as monetary expressions of benefits.

The collection of recyclables is a part of preprocessing in most MSW compost facilities. The range of reported prices received for MSW compost is from "no charge" (i.e., zero) to $5/yd^3 ($6.6/m^3) at the site. Unsold compost can be put to a variety of beneficial uses such as weed abatement, improvement of marginal land, or landfill cover.

Avoided Costs. A commonly used measure of costs avoided through the use of composting is the cost of disposing of the waste by landfilling. On that basis, the avoided costs are the equivalent of landfill tipping fees plus collection and hauling costs. Obviously, the degree of the equivalence is the extent to which the tipping fees in force reflect the true costs of the landfill, including amortization of future costs for the development of MSW landfills.

Inasmuch as composting MSW inevitably leaves some residue that must be disposed of by landfilling, the cost of landfilling this residue must be taken into consideration.

Mixed MSW Composting

In this section, the subject matter becomes more specific. Instead of dealing with the cost of composting as a waste treatment system, it deals with the cost of composting mixed MSW.

In addition to technology and size, a number of ancillary factors determine system costs. The nature of the ancillary factors and degree of their influence ultimately depend upon type of the selected system and its technology and upon size. Among the ancillary factors specific to the selected system are site preparation costs, land costs, climate, dust and odor-level stipulations, restrictions on waste receiving and operating hours, and the amount of process redundancy included. The diversity of these factors and their variability make a simple estimate of costs extremely difficult.

The authors of an EPA manual[29] approached the cost projection problem by considering two facilities that rely upon windrow systems but are widely divergent as to size in that one is a 100 TPD (91 MT/D) facility, whereas the other is a 1000 TPD (909 MT/D) facility. The smaller facility involves a system in which forced aeration is the method of aeration. Turning through the use of mechanical turners is the method of aeration in the larger facility. Both have preprocessing systems. In both cases, the active stage takes place inside a building. Odor control is more expensive in the smaller facility because the degree of control is greater owing to location of the site. Moreover, in both cases, it is assumed that all compost can be sold at a net revenue of $2.50/ton ($2.27/Mton) of compost. It should be noted that revenue from sales is minor in comparison with the cost of production.

The EPA manual[29] presents detailed cost analyses of the two systems by way of three tables. Table 10.4 is a summary of the data presented in these three tables. As the data in Table 10.4 show, the net daily unit cost with the smaller facility is $63/ton ($57.3/Mton) and for the larger facility it is $48/ton ($43.6/Mton). Additional information is presented in Appendix 10.B.

TABLE 10.4 Summary of Unit Costs for Two Facilities (1990 $)

Cost item (TPD of capacity)	100 TPD (91 MT/D)† facility	1000 TPD (909 MT/D) facility
Capital ($)	67,300	55,300
O&M* ($/ton, $/Mton)	45 (41)	35 (31.8)
Net unit cost ($/ton, $/Mton)	63 (57.3)	48 (43.6)

*Based on 312 operating days per year.
†U.S. ton × 0.907 = 1 Mton.
Source: Data adapted from Ref. 29.

Yard Waste Composting

Introduction. Basing estimates of the costs of a planned yard waste facility on those of existing operations or those reported in the literature is an exceedingly difficult task. The reasons are similar to those encountered with MSW undertakings, namely, the wide range of cited costs that reflect variations in facility size, design, and ancillary conditions. Thus, collectively, existing yard waste compost facilities include very simple facilities, some fairly complex and fully equipped facilities, and a full gamut of intermediate facilities.

The nature and breadth of the gap between a fully equipped facility and a simple, minimally equipped facility can be indicated by calling attention to their similarities and differences regarding key construction and equipment features. Typically, the entire site of a fully equipped facility is graded, and the receiving, processing, and about a fourth of composting areas are paved. A fully equipped facility has a shredder, a mechanical turner, a front-end loader, screens, and necessary conveyors. "Minimal" refers to the lowest level at which environmental and public health requirements can be met. The minimal construction requirements are a graded site, and paved and fenced-in processing and active compost stage areas. The minimal equipment requirements are a shredder and a turning device (e.g., front-end loader). A shredder provides a very essential function, namely, size reduction of woody yard waste (brush, branches, etc.). Its final product is inferior to that from the fully equipped facility, because the product has not been screened.

Capital Costs. An excellent detailed analysis of the likely investment requirements for two widely divergent hypothetical yard waste compost facilities is presented in Ref. 29. One of the two facilities is highly capital-intensive and has the characteristics of the fully equipped facility. The second facility is much less, perhaps even minimally, capital-intensive. It has the characteristics of the "minimally" equipped facility described above. The daily throughput capacity of each of the two facilities is 70,000 yd³ (53,200 m³) of input waste, from which each produces 10,000 yd³ (7600 m³) of compost. Each facility has a total area requirement of 12 acres (4.9 ha).

Table 10.5 is a summary of the estimated initial investment costs of the two facilities as listed in the reference. (The reference treats land cost as an operational expense rather than as a capital cost item, because it assumes that the land is leased.) Although relative, the dollar values listed in the table are indicative of the magnitude of the monetary outlay involved in implementing a decision to establish a yard waste compost facility, and a measure of its economic feasibility. Additional information is given in Appendix 10.B.

TABLE 10.5 Estimated Costs of Yard Waste*

Item	Fully equipped	Minimally equipped
Initial investment costs:		
Construction	392,500	116,750
Engineering	84,700	48,600
Utility hookup	40,000	40,000
Equipment	405,000	260,000
Total investment costs	922,200	465,350
Net annual costs		
Annual costs		
Amortized investment	150,100	75,730
Annual O&M	244,450	256,000
Total annual cost	394,550	331,730
Annual revenues		
Sale of compost	90,000	70,000
Net annual cost	304,550	261,730
Net unit costs ($ per ton yard waste)		
Unit cost	32	28
Avoided cost	37	37

*Throughput 70,000 yd^3 per day (53,200 m^3 per day); composted product, 10,000 yd^3 per day (7,600 m^3 per day). Area of each, 12 acres (4.9 ha).

Source: Adapted from Ref. 29, in 1990 dollars.

Site-Specific Factors.   The extrapolation of many reported costs from one facility to another is constrained by the site specificity of the costs. The cost of bringing utilities to a facility (i.e., utility hookup) is an example. The hookup cost is largely a function of the location of the nearest service and the distance between it and the facility. Construction costs serve as another example, inasmuch as they are dependent upon conditions unique to the site. Thus differences between sites with respect to soil conditions can introduce a variability factor of 200 percent for cost of grading and paving.

The equipment costs shown in the example presented previously are typical costs for adequately sized machines. Actual equipment costs can be determined after particular pieces have been chosen in the design stage of the operation.

Operation and Maintenance Expenses.   Among the many items to be considered in arriving at an estimate of the O&M cost are insurance, fuel, labor, lease on land, trailer rental, water, and power. Judging from the analysis reported in Ref. 29 and summarized in Table 10.5, the estimated annual O&M cost of the fully equipped hypothetical facility is on the order of $244,450; and of the minimally equipped facility, $256,000. The higher costs of the minimally equipped facility are due to higher estimated fuel and labor costs. However, the facility's maintenance, insurance, and power costs are somewhat lower.

Income and Avoided Cost.   The estimated income listed in Table 10.5 is based upon a 60 percent product yield and the existence of a market for the entire yield at a net return equal to $9/yd^3 ($11.8/m^3) of product from the fully equipped facility and $7/yd^3 ($9.2/m^3) from the minimally equipped facility. The estimate of the avoided cost is $5/yd^3 ($6.6/m^3) and is the same for both facilities.

The projected negative unit cost that results from the magnitude of avoided unit cost being higher than unit costs probably is overly optimistic.

Public vs. Private Ownership

Much can be said for and against public ownership and operation. A compromise that is fairly common is public ownership and private operation. Three considerations rank high in the decision regarding ownership and operation. They are project control, risk allocation, and project costs.

Project Control. A municipality can exercise a large measure of control over a privately operated large facility by way of a series of contracts negotiated prior to the actual construction of the facility. A relatively small privately operated facility can be controlled through a contract which specifies areas subject to municipal control. However, implementation of major operational changes according to a timetable desired by the municipality is more easily done if the facility is municipally rather than privately operated.

Risk Assumption and Allocation. Major risks that can adversely affect the cost of operations are increases in costs associated with system reliability, management, labor productivity, inflation, landfill, and insurance. For a privately operated facility, the allocation of these risks is a subject of contract negotiations. It is up to the municipality to develop contracts to reduce risks that can best be borne by the private sector.

Project Costs. Public or private operation should matter little with respect to project costs unless a municipality has surplus labor and/or equipment that could be effectively utilized through implementation of the project. The issue could become important if labor, fuel, utilities, or material costs are substantially different or can be more easily met by one of the two.

The need for the private owner or operator to show a profit can tip project cost in favor of public ownership or operation. Determination of which type of operation will be most cost-effective must be made on a case-by-case basis.

10.4 MARKETING PRINCIPLES AND METHODS

The principles involved in marketing compost are basically the same as those in marketing any commodity—whether it be a feedstuff, chemical fertilizer, or compost. Application of these principles takes the form of the following sequence of steps: (1) determine the possible uses of the product and its application; (2) identify potential users (market analysis); (3) make the potential user aware of the product characteristics and its utility, as well as of the benefits from using the product; (4) persuade the potential customer to procure and use the product; and (5) establish a satisfactory distribution program.[3] In connection with step 3, the producer and/or vendor should be able to assure the customer that the product will be unfailingly available and that its specifications remain constant. Each of these steps is discussed separately in this section.

In this section, "compost product" and "compost" are used synonymously. Unless otherwise indicated, the primary emphasis is on MSW compost, although the information is readily applicable to other types of compost.

Uses of the Compost Product and Its Application

The characteristics and utility of compost were discussed in detail earlier in this chapter. In summary, the primary use of compost is as a soil amendment. A distant secondary use is as a source of fertilizer elements, especially of nitrogen, phosphorus, and potassium (NPK). Applications include landfill covers, as well as the entire gambit of agricultural activities.

Market Analysis

The objective of a market analysis is to arrive at an estimate of the full size of the potential market. It is best begun with a survey designed to obtain the information on which a realistic estimate of the full size of the market potential can be made. Preferably, the survey is conducted by way of interviews or questionnaires or by a combination of the two to identify the needs of prospective customers and establish the dimensions of their potential demands. An important fact to keep in mind when designing or conducting an interview is that successful selling presupposes knowledge of the customer's values and motivations.

Parties to be interviewed are representative members of major agricultural sectors, of government agencies, of the general public, of the landscapers, of soil vendors and distributors, of nurseries, and of any organization that may have a use for the product. If the feasibility of conducting personal interviews becomes a limiting factor, they can be supplemented by mailed questionnaires. Questions posed in the interviews and questionnaires should be relevant to the targeted market.

Acquainting the Customer

"Acquainting the customer" is used in the sense of ways and means of imparting a knowledge of the utility of compost to all sectors of the actual and potential market. Imparting this knowledge involves the identification and description of compost characteristics and benefits associated with its use. It includes explaining and illustrating ways of obtaining the benefits and utility. In short, the objective is "to educate the market."

The need for education is emphasized by the fact that, particularly with MSW, marketing compost is seriously encumbered by inertia and bias. The encumbrance is largely due to potential users being unaware of the true worth of compost. Obviously, the best means of removing or minimizing the obstacle is to instill in potential users an awareness of the real worth of MSW compost. This can be done through a program of education and salesmanship. The task is made easier by the fact that the product does indeed have value and genuine utility.

Public education can be accomplished through presentations in the media. The presentations may deal with the advantages and disadvantages of compost utilization, with methods of producing compost, with information on obtaining compost and on how compost can and should be used. The presentations should be backed by carefully orchestrated demonstrations.

An important aspect of the process is the dispelling of troubling doubts. The removal of doubts should be accompanied by explanations regarding the best utilization of compost for particular applications.

The line of demarcation between education and advertising is not clear-cut. Probably, education becomes advertising when a product of a particular producer is promoted. Given the current situation, the logical course would be to precede the advertising phase by a carefully planned and conducted program of educating the largest group of prospective customers, namely, farmers from the field crop, row crop, and orchard sectors.

The participation of governmental and educational professionals who specialize in advising and guiding farmers on a local level is very helpful. Farmers tend to take the advice of such specialists seriously, because of the close association of the specialists with them. Moreover, the specialists have a better understanding of the problems that beset farmers locally. The specialists could furnish practical advice and assistance in designing, implementing, and publicizing demonstrations. The demonstrations could range in magnitude from small plots involving two or three types of plants to a large undertaking comparable with one conducted in Johnson City.[30]

Product Sales and Salesmanship

Having established a receptive climate through education and demonstration, the next step is to narrow the focus to particular potential users, i.e., to advance from sectors to the individuals in the sectors.

Importance.   Disposal of the waste depends upon persuading potential users to acquire and use the product. Failure to accomplish this step leaves the compost producer with the responsibility of satisfactorily disposing of the product. Failure might even reduce the value of composting as a feasible disposal option. In short, failure to find appropriate uses would defeat one of the principal objectives of MSW composting.

The position of composted MSW in the hierarchy of uses is immaterial, provided the waste generator and compost producer are divested of the disposal responsibility. This assumes that the compost is used in an environmentally sound manner, from use as landfill cover to use as a soil amendment in food crop production.

Compost marketing specialists stress the importance of analyzing the needs and the potential of each sector, including delivery requirements, storage capabilities, and pricing policies.[31] It is necessary for the seller to understand the needs and requirements of the targeted market.

Methods.   The first step in persuading a customer is to emphasize the benefits that the user will gain from using the compost product in preference to a competitive product. This can be done by calling attention to the benefits described in the section on uses of compost. Some benefits that are deemed especially important by potential compost users are once again mentioned at this time. It should be pointed out that the cash value of a crop strongly influences the purchaser's decision whether or not to buy and use compost. If the cash value of a crop is high, a potential user is more likely to purchase and use compost. The willingness to purchase declines with decline in dollar value of the increase in crop yield resulting from compost use.

Particularly important to most potential users are the compost properties that facilitate and improve crop production through remedying soil deficiencies and improving soil characteristics. For example, because it is predominantly organic and is an excellent medium for soil bacteria, compost improves the tilth of soil, thereby enhancing the soil's productivity. Moreover, tilth is a key consideration in soil management. Another benefit that should be stressed is that compost incorporated into the soil lowers fertilizer expenditures by lessening nutrient loss through leaching. For example, as much as 30 percent of applied chemical nitrogen may be dissolved and leached to the groundwater. In addition to minimizing nutrient loss through leaching, compost also increases the ability of plants to utilize nutrients efficiently. Increase in efficiency results in higher yields—and likelihood of greater financial return. One property of compost will be of particular interest to customers in arid regions or in any region where water is in limited supply either seasonally or year-round. The property is the high water-retention capacity of compost that is second only to that of peat. (Peat is more expensive than compost, and its plant and soil bacteria nutrient content is much less than that of compost.)

Particularly attractive not only to agriculturalists but also to soil conservationists is the reduction in loss of topsoil through wind and water erosion. Loss of topsoil reduces crop yield and increases cost of soil management because the lower strata in the soil profile are more difficult to till.

In terms of percentage of the soil amendment market, steer manure offers the greatest competition at present. However, composted and noncomposted steer and other animal manures are handicapped by having a higher sodium (Na) and chloride (Cl) content than do MSW and yard waste composts. The higher Na and Cl contents are due to the presence of urine in mammalian manures. (Sewage sludge is an exception, because being of human origin, it also includes urine.)

Advertising. The means of communicating the message presented in the preceding paragraphs is through conventional advertising practices. The efficacy of conventional advertising has been amply demonstrated. With regard to compost, the task of advertising is considerably lightened by the public educational campaign that should both precede and accompany it.

Not to be ignored is word-of-mouth advertising. This form of advertising can be initiated and facilitated by carefully planned demonstrations, of which the "first user" is the most venerable. Moreover, the approach not only is venerable, it is effective. The demonstration involves persuading a representative potential user to try the product. Persuasion is strongly facilitated by providing the compost either free or at a very low price. It is imperative that the participant be carefully guided and supervised during the demonstration. A successful demonstration will attract favorable attention on the part of neighbors and onlookers. The demonstration may be expanded to include several neighborhoods and participants.

A key requisite for a successful advertising campaign is the ability to convincingly assure targeted customers that: (1) the compost will be unfailing available; (2) there will be no large deviations in quality other than improvement; (3) no unwelcome deviations in product characteristics and specifications will take place; and (4) the price always will be "right."

Market Continuity. Four factors also are applicable to market continuity: product quality, availability, constant specifications, and pricing.

Product Quality. Most users of compost base their evaluation of compost quality as a soil amendment and organic fertilizer on certain characteristics of the product. Among the characteristics of importance are NPK, moisture content, extent of contamination, odors, and particle size. The concentration of NPK should be high enough to justify the application of compost. The higher the NPK, the less is the amount of product at which the crop's NPK requirements can be met. Moisture content is a factor because of its influence on ease of handling and on cost of transport. Handling is more difficult at high moisture contents. Furthermore, water adds to the weight of the product and increases the cost of transport. Pathogens and toxic and nontoxic contaminants adversely affect product quality and act as constraints on use of the product. Foul odors, even in very low intensity, adversely affect quality perception. Particle size relates to visual quality, ease of handling, and applicability.

Grading is generally recommended as a means of coping with the many variations in products with respect to visual and nutritive quality. Grading assures the most effective utilization of the product. A facility may produce only one type of compost. On the other hand, the output may be separated into different products on the basis of quality. Effective use is ensured by matching type of application with appropriate quality of compost. For example, a relatively low grade of compost would be adequate for the reclamation of excavations and denuded forests. On the other hand, a high grade of compost is required in row crop production or use by homeowners.

As previously stated, aside from the grading demanded by the market, serious efforts to establish a formal system of grading did not occur until the late 1980s. However, guidelines have been and continue to be proposed by federal and state agencies. Guidelines for grading generally are based on constraints on application of the product. A sample of grades proposed by the authors for the state of Washington is presented in Table 10.6.[32] In this grading system, there are no constraints on the use of grade 1 composts. They can be used on food chain and row crops and for all other uses. Grade 2 composts cannot be used in food chain and row crop production but can be used in orchards, viticulture, landscaping, etc. The third and lowest grade can be used only for all other applications. Assignment to a grade usually is based on toxic substance concentration (e.g., Cd, Pb, PCBs), number of viable pathogens, weed seed concentration, contaminant content (e.g., plastic, glass), plant nutrient concentration, degree of maturity, and major physical characteristics (e.g., particle size distribution, moisture content).

Maintaining product consistency is an overriding requirement for market continuity. The customer demands product consistency because efficient utilization, particularly in crop production, depends upon the use of a soil amendment of known composition and physical characteristics. Variation in consistency lessens the usefulness of the product, resulting in a loss of customer confidence and interest. Therefore, it is extremely important that the compost meet a fixed set of specifications.

Unfailingly Available.　As far as market continuity is concerned, availability implies that production not only must be sufficiently large to permit adequate introduction of the product into the market but it also must be consistently available in the future. The market could not long survive sporadic availability.

TABLE 10.6　Marketability Standards

	Unit	Grade A	Grade B
Bulk density	lb/yd^3 (kg/m^3)	600–800 (356–475)	400–1000 (238–594)
CEC*	meq/100 g	>100	>100
Foreign matter	Maximum %	2	5
Moisture content	%	40–60	30–70
Odor		Earthy	Minimal
Organic matter	Minimum %	50	40
pH		5.5–6.5	5–8
Size distribution	Nominal, in	<1/2	<7/8
Water-holding capacity	Minimum %	150	100
C/N ratio	Maximum	15	20
Nitrogen	Minimum %	1	0.5
Conductivity (soluble salts)	mmhos/cm	<2	<3
Seed germination	Minimum %	95	90
Viable weed seeds		None	None

*CEC = cationic exchange capacity, expressed in milliequivalents (meq) exchangeable cations per 100 g of dry soil.

Source:　Reference 32.

Price.   Unlike conventional marketing, the pricing policy for MSW compost, and to a lesser degree for yard waste compost, is not to make a monetary profit from the operation but rather to defray as much of the cost as is possible. The main objective for composting is usually treatment and disposal. If the selling price is too high, potential users will turn to less expensive competing products. Compost must compete with other organic products for a share of the organic fertilizer market. If the spread between the price of compost and a competing product is too wide, the user buys the less expensive product despite being aware of the benefits to be gained from use of compost. Inasmuch as the primary reason for composting MSW and yard waste is waste disposal, unsold compost must be disposed of either by landfilling or by incineration.

An upper limit on the selling price for compost also is determined by the potential user's ability to pay. If the greater share of the compost market is the agricultural sector, the limit would be relatively low because the profit margin characteristic of agricultural enterprises is small. The average farmer can afford to make a relatively small expenditure for fertilizer, chemical or organic. Compost produced specifically for landscaping and cultivation of ornamentals should be priced according to the buyer's ability and willingness to pay as well as by the price of competitive products.

At the time of this writing, the prices for composts produced from MSW, yard waste, and sludge range from zero (i.e., is given away free) to about $5/yd^3 ($6.6/m^3).

Public versus Private Marketing.   Generally, public vs. private marketing of the compost is a question only when the public entity (community, district, etc.) has sole ownership of the MSW, yard waste, or sewage sludge compost. Moreover, the question applies only at the wholesale (bulk sales) level; because the consensus is that with few exceptions, private enterprise is better qualified at the retail level. Therefore, the discussion that follows is not concerned with retail selling.

If a facility is privately owned and operated, it is to be expected that the entrepreneur owns the product and hence is responsible for marketing or disposing of it—unless the contract with the community states differently. If a community owns a facility, but by way of a contract has a private party operate the facility, ownership of the product is specified in the contract. It follows that if a public entity owns and operates a facility, the entity owns the product and is responsible for marketing or otherwise disposing of it. The entity has two choices: (1) it can sell the entire compost output to a single entrepreneur, who thereupon is responsible for the disposition of the product; or (2) the entity can do the marketing.

If the entity opts to do the marketing, its success will depend upon meeting certain requirements. First and foremost, the entity must be prepared to unreservedly do everything that is needed. Selling requires the full-time input of highly qualified, knowledgeable, and dedicated professional staff. Such a staff can give the task the necessary effort and attention. The difficulty is that most entities either cannot afford such a staff or are unwilling to make the necessary expenditure. Short of these specifications, the selling will be less than adequate.

Distribution

The significance of distribution is readily apparent because it is the link between the production facility and users. Methods range from free transport in bulk form to end users, to bagging and distribution through existing channels established for other soil amendments. Despite its significance and a long record of compost production and use, progress in the distribution of compost has been very limited. Distribution channels are still not well defined.

The rate established for shipping secondary (recovered) materials is a major element in the economy of a resource recovery operation. In general, regulated freight rates are based on cost and value of service. In turn, several factors have an impact on the rate structure established by freight carriers (motor, railroad, ship, barge). These factors are too numerous and varied to be adequately discussed in this chapter. However, of particular importance is the fact that the combination of relatively low monetary value of MSW compost and its low bulk density exacerbates the cost of long-distance transport, and consequently sharply limits the distance at which haul is economically feasible.

Motor Freight.   The National Motor Freight Classification is followed in the classification of commodities transported via motor freight. According to the National Motor Freight Classifications, compost would be classified as soil and would be considered to be either in class 50 LTL (less than truckload) or in class 35 TL (truckload).

Criteria have been established for special rate classifications. To qualify for special rates, the total load should be ready to be loaded at one time. The particular classification to be applied depends primarily upon the total weight of the shipment, loading and unloading restrictions, and the limits of liability. The curve plotted in Fig. 10.15[28] exemplifies the relation between motor freight rates and distance of the haul.

Railroad Freight.   Rates established by railroads are based on several factors. An important factor is the type of commodity being shipped. Commodities are classified according to the "Unified Freight Classification—12." The rule of analogy applies to commodities not specifically named in the "Classification." According to this rule, compost is classified as soil or agricultural mulch. Another key factor is "the minimum weight shipment per railroad car." For compost, the minimum weight per car in California is about 50 tons (45.5 Mtons).

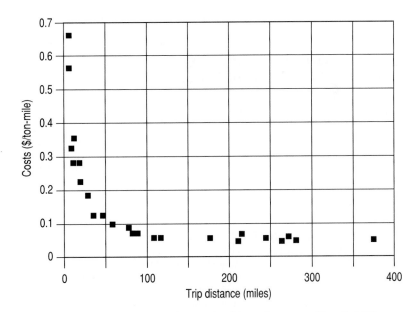

FIGURE 10.15   Intrastate motor carrier rates for all bagged composts. (*From Ref. 28.*)

Water Transport

Ship Transport.   Rates are highly variable inasmuch as rates are not governmentally regulated and are dependent upon the individual shipping lines. However, shipping lines have banded together to set up cartels to establish mutually binding guidelines (e.g., "Pacific Westbound Conference"). Not surprisingly, independent ship lines impose rates lower than those established by cartels.

At present and in the near future, it is very likely that transporting compost by ship will be almost negligible. The exception might be transport between coastal ports or between ports in the Great Lakes states.

Barge Transport.   Compost transport via barge could be practical in certain regions of the United States, particularly those that have access to the Ohio and Mississippi Rivers and a few canals designed for barge traffic.

Cost estimates made by the authors of this chapter on the basis of information gained in an informal survey[28] can serve as guidelines for estimating barging costs. In general, barging rates apparently are independent of the type of commodity being transported. On the other hand, weight of the load being transported and the crew time involved are primary factors.

The five considerations upon which the authors based their estimates can serve as a model for estimating barging service costs per ton of compost (dollar values are for 1989): (1) Barges (i.e., barge train and tug) move at about 5.6 nautical miles/h. (2) The barges are "bare-boat" charter at $2300 per day. The one who charters the barge is responsible for insurance and maintenance costs, which usually amount to about $0.25 to $0.35/ton ($0.23 to $0.32/Mton). (3) Based on 1989 labor and fuel costs, the daily rate for a 1000-hp tug for the operation would be about $7200, assuming a single rate for a 30-day operation is arrived at by combining straight time and weekend overtime. (4) The estimated costs for one tug and four barges moving 60,000 tons per month (54,500 Mtons per month) of material is about $8.10/ton ($7.4/Mton). One tug plus three barges (15 trips per month) could haul only 24,000 tons per month (21,800 Mtons per month), in which case the estimated cost would be $13/ton or $11.8/Mton.

10.5 ENVIRONMENTAL, PUBLIC, AND INDUSTRIAL HEALTH CONSIDERATIONS

Major potential negative impacts of a compost operation would be lowering the quality of water and air resources, and compromising the public health and well-being. It should be emphasized that these are potential negative impacts, and that they become *actual* impacts *only* when an inadequate technology is used, a normally adequate technology is improperly applied, or preventive or corrective measures are not taken.

Water and Air Resources

Water Resource.   The quality of the water resource can only be adversely affected through contamination with either leachate from raw, composting, or composted refuse or with runoff from the compost operation. Leachate is formed only when the moisture content of the material is higher than the optimum for composting. Aside from maintaining the moisture content of the material at or below the optimum level, chances of uncontrolled increases in moisture content from rain or snow can be minimized by protecting the material from the elements. As a precautionary measure, provision should be made to keep leachate from reaching ground and/or surface water resources by conducting all phases of the operation on an impermeable surface. The surface should be equipped to collect all leachate for treatment or discharge into a public sewer. Runoff can be avoided by selecting a site where it would not be likely to occur. If this is not possible, runoff can be prevented from entering the operation site by constructing ditches to divert the runoff around the site. Runoff from the site can be intercepted and channeled to a treatment facility (e.g., conventional stabilization lagoon). It is important that leachate does not reach a body of water, since leachate from raw waste is similar to raw sewage sludge in terms of pollutant concentration. Although the compost process sharply reduces the pollutant concentration, leachate from a properly matured compost mass would still reduce the quality of ground and surface water.[4,22]

Air Resource.   Biological and nonbiological agents from a compost operation most likely would enter the environment by way of dust particles and aerosols generated during the various stages of the operation and subsequently be discharged into the air. Some of the microbes transported in this manner could pose a health hazard to a susceptible individual who by chance might ingest the dust particle or aerosol. Aerosols, in particular, are vehicles for a wide variety of microorganisms. These microorganisms may occur as single entities, as clumps of organisms, or as adhering to dust particles. Two types of infections may be acquired from such contaminants, namely, those that are limited to the respiratory tract and those that may affect another part of the body. Both are taken in by way of the respiratory tract. The existence of a hazard from the spores of *Aspergillus fumigatus* is yet to be demonstrated. The infectivity of the spores is low. Consequently, any danger posed by it would be significant only to an unusually susceptible individual. Nevertheless, prudence indicates that an open-air compost plant should not be sited in close proximity to human habitation.[33,34]

Dust suppression at all stages of a compost operation can be accomplished through the use of conventional dust control measures. Some of these measures include the use of mist sprayers in the working area, and the installation of air collection and particulate control devices such as cyclones and fabric filters.

Odors.   Although the generation of objectionable odors lowers the quality of the air resource in terms of human well-being, it does not become a health hazard until the odors become particularly foul. Some odors are inescapable, e.g., those from raw wastes. Odor control during preprocessing can be accomplished by enclosing all the operations in a building; conditioning the feed; and treating exhaust gases through absorption, adsorption, or oxidation methods. Foul odors are generated during the composting stages principally through improper management of the composting process, e.g., failure to maintain aerobic conditions. The use of an in-vessel system does not always ensure odor-free operation.

In the absence of proper management, all materials can become sources of foul odors until they are adequately matured. However, conventional techniques are available for treating foul odors. Control and containment are effective approaches to preventing the development of odor problems or contending with those that may escape prevention. Means of control involve trapping the odors through ventilation or containment of the composting process. Exhaust air can be treated by passing it through a chemical scrubbing system or by way of biofiltration. Difficulties with biofiltration are its large filter areas and relatively sophisticated management requirements.[35,36]

Vectors

Fly and rodent attraction is almost inevitable because of the nature of organic residues and the long time interval between reception of the raw material and the stage of the compost process in which conditions become lethal to flies and intolerable to rodents.

Most likely, flies and rodents would not constitute a serious problem in a yard waste compost operation. However, food wastes or sewage sludge cocomposted with the yard wastes could serve as strong attractants for flies and rodents. Although a fly and rodent problem could be almost completely eliminated by enclosing the entire facility, it can be considerably eliminated by using certain measures. For example, an important mitigating measure would be careful "housekeeping" throughout all stages of the operation. Storage of raw wastes should be as brief as possible. Preprocessing, particularly size reduction, wreaks a substantial destruction on fly eggs and fly larvae, and lowers the value of the refuse as a feedstuff for rodents. Migration of fly larvae that survive the preprocessing and the early compost stages can be prevented by the use of a paved surface.

Industrial Health and Safety

The greatest potential for accidents in a composting facility is in the preprocessing stage. Chances of injuries from accidents are greatest in this stage because of the extensive exposure of the workers to machinery. Standard measures for minimizing such hazards are well developed and easily available.

The greatest hazard to the health of the workers comes from the dusts suspended in the air in a compost plant. Hazards associated with dust are greatest in the preprocessing phase and become much less in the subsequent stages of the compost process. In addition to the biological burden generated by the dust particles and aerosols, there is a fibrous fraction of dust that may have a health significance. Health problems associated with dusts can be substantially controlled by the use of face masks and protective clothing, and the installation of adequate sanitation facilities.

The highest levels of noise that would occur in a preprocessing plant would be from about 95 dBA to about 105 dBA (slow response). These levels can be generated by a shredder or a front-end loader. With the present state of the technology, some type of ear protection is needed for exposures longer than about 2 h.

Constraints on Use of the Compost

Constraints on the use of the compost with respect to the health and safety of humans arise from the harmful substances that may be in the compost. Examples of such harmful substances are heavy metals, toxic organic compounds, glass shards, and pathogenic organisms. PCBs are beginning to appear in troublesome concentrations in municipal wastes.[37] As yet, however, reports of findings are too fragmentary to indicate trends. Because composting of wastes has been almost entirely limited to sewage sludge, most of the existing information deals with that compost.[38] The sources of harmful substances obviously are the wastes used as feedstock for the process. Concentrations of harmful substances usually are higher in sludges than in the organic fraction of municipal solid wastes, as shown by the data in Table 10.7. When considering the data, it should be kept in mind that the concentrations vary widely from operation to operation (e.g., cadmium ranges from 0 to 1100 μg/g dry sludge[39]), because of a variety of site-specific differences (e.g., mostly residential vs. mostly industrial generators). Reliable information on concentrations of toxic substances and pathogens in composted yard debris is extremely scarce. The few data that are available indicate that the concentrations of heavy metals in compost from yard wastes are relatively low (<0.1 ppm to about 10 ppm for Hg, Cd, Cu, and Ni; and from 50 ppm to about 200 ppm for Pb and Zn). Similarly, concentrations of pesticides, PCBs, and

TABLE 10.7 Composition and Characteristics of Sludge and of the Light Fraction of Air Classified Refuse

Item	Units	Air classified light fraction		Sludge cake	Refuse-sludge mixture
		As received	As analyzed		
Carbon (C) (total organic)	%	15.9	16.8	15.6	15.8–18.0
Nitrogen (N)	mg/kg dry	7080	7500	41,000	11,000–13,000
(total Kjeldahl)	%	0.7	0.75	4.1	1.1–1.3
Zinc (Zn)	mg/kg dry	226.6	240.0	2000	680–840
Cadmium (Cd)	mg/kg dry	1.0	1.1	93.0	3.7–21.0
Lead (Pb)	mg/kg dry	29.3	31.0	1000	47–110
Nickel (Ni)	mg/kg dry	6.1	6.5	150.0	10–35
Copper (Cu)	mg/kg dry	22.7	24.0	8900	

Source: From Ref. 50.

pathogens are quite low.[37,40] Average concentrations of pesticides in composts produced from yard wastes in Illinois were found to be between 1 and about 10 ppm Carbaryl, Atrazine, and 2,4,5-T. All other pesticides analyzed for were found at levels between about 0.01 and 1 ppm.[40] Therefore, compost products, especially those from MSW and sewage sludge, should be routinely analyzed as a precautionary measure.

The harmful effect on humans and animals may be exerted directly by eating food crops grown on soil that has been amended with compost. The effect could be exerted indirectly through the consumption of meat and other products involving animals fed on such food crops. The effects are due to persistence of the inorganic contaminants and survival of certain pathogens through the food chain.

Cadmium can be used to demonstrate how a heavy metal passes through the food chain. A certain fraction of the cadmium in a compost incorporated into soil is assimilated by plants grown on that soil. The amount of cadmium assimilated by the plant depends upon a number of factors, such as the availability of the metal, plant species, and the particular part of the plant. Availability depends upon the concentration of the metal in the soil, the pH of the soil, concentration of organic matter in the soil, ion exchange capacity of the soil, and several other factors. Generally, availability decreases as the soil pH changes from acidic to alkaline. As a rule, leafy vegetables assimilate more than cereal crops. In cereal crops, concentration is greater in the root and leafy portions than in the grain. If those plants are eaten by humans, a fraction of the cadmium in the plants is assimilated in the tissues of the persons who eat the plants. If the plants are consumed by animals, some of the cadmium is assimilated by the animals and remains in their meat and in products (e.g., eggs) produced by them. This cadmium awaits assimilation by humans who consume the meat and the products. The distribution of cadmium in the soil, plant, and animal as it passes through the food chain, and the contribution of sewage sludge are described in detail in Refs. 38, 39, and 41. The incidence of other metals and chemicals in the food chain is summarized in Ref. 34.

Restraints due to the presence of pathogens in compost range from negligible to substantial, depending upon the waste composted and the conditions under which it was composted. Such constraints can be eliminated by rendering the product free of pathogens through pasteurization. Pasteurization can be accomplished by way of composting or through the application of an external source of heat. Except through contamination by contact (e.g., adhering compost particles), direct transfer of pathogenic organisms between members of the food chain even without disinfection is either nonexistent or very minor.

Types and concentrations of pathogens that might be in the product *prior* to pasteurization depend upon the feedstock. Yard wastes are not likely to contain human pathogens because human body wastes are not involved. However, they may contain organisms pathogenic to pets or to plants. Sewage sludge, on the other hand, has a wide range of human pathogens.[34] MSW may contain some human pathogens because of contamination by body wastes. Because the indicators are of pet rather than human origin,[4,22] the extent, if any, of such contamination cannot be measured by concentration of "indicator organisms." (The concentration rivals that in sewage sludge.) Improperly composted food wastes could contain zoonotic organisms (trichina, ascaris, taenia) by way of meat scraps. In summary, composted yard waste is not likely to contain human pathogens, whereas inadequately composted sewage sludge and food waste could.

Health constraints are receiving legal backing in the form of "Classifications" proposed or actually promulgated by the U.S. EPA and various state regulatory bodies, although generally yard waste composts have been and are being less tightly regulated. Tables 10.8 and 10.9 are examples of such classifications based on heavy metal content. It should be noted that these tables are only examples because at present classification development is rapidly changing.

Among the legal constraints other than those that are health-oriented is an important one pertaining to labeling. It prohibits labeling a compost product as a "fertilizer" when the product's NPK (nitrogen, phosphorus, potassium) concentration is less than a total of 6 percent. (The required total may vary from one state to another.) Permitted are labels of "soil amendment," "soil conditioner," or simply "compost." The NPK of a compost product depends on the NPK of the wastes from which the compost is produced. Because of the wide variations between products in terms of nutrient content, it would be misleading to list particular concentrations as being "typical." Other conventional designations are named and described in Refs. 42 and 43. Convincing arguments and a plea for the setting of labeling requirements and regulations on potting soil are given in Ref. 44.

10.6 CONCLUSIONS

The application of composting to the management of municipal solid wastes (MSW) in the United States has undergone extraordinary changes during the last 40 years.

TABLE 10.8 Classification for the State of Minnesota (Maximum Allowable Limits)

Metal	Class I, ppm dry wt.	Class II
Cadmium	10	
Chromium	1000	
Copper	500	All composts that do not meet class I
Lead	500	standards are placed in class II.
Mercury	5	Therefore, there are only two classes
Nickel	100	
Zinc	1000	
PCB	1	

Source: From Ref. 37.

TABLE 10.9 Regulations for the States of New York and Massachusetts (Maximum Allowable Limits)

State	Class I food-chain crops	Class II non-food-chain crops
New York (sludge, MSW compost):*		
Mercury	10	10
Cadmium	10	25
Nickel	200	200
Lead	250	1000
Chromium	1000	1000
Copper	1000	1000
Zinc	2500	2500
PCB	1	10
Particle size	<10 mm	<25 mm
Massachusetts (sludge, MSW, yard waste compost):†		
Cadmium	2	25
Mercury	10	10
Molybdenum	10	10
Nickel	200	200
Lead	300	1000
Boron	300	300
Chromium	1000	1000
Copper	1000	1000
Zinc	2500	2500
PCB	2	10

*From Ref. 52.
†From Ref. 51.

Recent changes are reflected by the data presented in Table 10.10. The data clearly show the substantial increase in the number of compost facilities in the United States—particularly those used for treating yard waste. In many instances, the increase has been a direct response to regulatory constraints (e.g., ban on the disposal of yard wastes in landfills). In other cases, growth has been primarily due to a desire to apply an appropriate, environmentally benign technology to the recycling of organic wastes.

The expansion of composting practice was not entirely advantageous, however. The problem was that the rapidity and magnitude of the expansion were such that the waste management industry could not adequately meet the substantial demand. The compost bonanza attracted a wide diversity of industries, equipment and system ven-

TABLE 10.10 Change in the Number of Composting Facilities in the United States (1988–1991)

Year	Number of facilities	
	MSW	Yard waste
1988	5	651
1989	7	986
1990	9	1407
1991	18	2201

dors, financial institutions, and a sizable number of companies and individuals to promote and develop compost programs. This development brought to light another problem, namely, the absence of a matching "infrastructure" to satisfy program demands. Here, we use "infrastructure" in the sense of collection of human resources, equipment, markets, material specifications, guidelines, and other factors. The large demand coupled with the inadequacy of the infrastructure led to several costly and painful errors.

Among the major causes of failed programs are (1) a tendency to oversimplify the compost process, (2) underestimation of the complexity of large-scale compost plants, and (3) insufficient understanding of mechanical and biological processes.

A fourth cause that often is cited is the apparent frequency of incidents of generation of malodors. The desire to minimize malodor generation has been made the reason for the modification of some facilities and for the closure of others. When evaluating this problem, it should be remembered that there is no such a thing as an "odorless" waste treatment facility. Of itself, the delivery of feedstock entails the generation of odors that differ from customary background odors. Odors due to handling and delivering the feedstock will be present even though the entire compost process is conducted properly. Furthermore, malodors are symptoms of a variety of problems, most of which have been identified in this chapter. Among these problems are (1) an inordinately long storage time (on the order of 1 to 2 months) of the raw feedstock (e.g., yard wastes); (2) inadequate mechanical processing (i.e., insufficient size reduction); (3) unrealistically short detention times during the composting process (a few days rather that weeks); (4) abbreviated maturation time; and (5) shortage of dedicated land area. Any one of these situations can be an occasion for the generation of malodors.

The majority of these problems can be prevented and even completely avoided by complying with the basic principles of feedstock preparation and composting outlined in this chapter. Furthermore, it is emphasized that prudence and experience dictate that the entire process of system development, i.e., identification of need, procurement, system selection, and monitoring, must not be delegated to individuals who are novices in the waste management business. The seriousness of this admonition has been learned the hard way by several hapless entities.

Currently, the future of composting in waste management appears to be favorable, especially in the United States. However, it certainly can be substantially improved. For example, improvement can be brought about by inculcating a greater awareness and appreciation of the limitations inexorably imposed by the very nature of the compost process. This awareness ensures the rational application of the basic principles of composting in the design and implementation of an integrated solid waste management program that involves composting.

The future can be further ensured by enriching the technical literature and popular media on composting with well-balanced, objective reports of activities and programs that are in progress or as yet in the proposal stage. The literature and popular media on composting are important elements in the success or failure of compost ventures. All too often, the literature and the media are the sole source of information on which decision makers, officialdom, and the public base their compost programs. Unfortunately, the quality of current and past literature and media accounts leaves much to be desired. The shortcoming is a consequence of undue reliance upon promotional publications and insufficiently substantiated reports as sources of information. For example, not infrequently the accounts fail to clearly or even confusedly delineate the status of the program, i.e., whether it has been implemented or even has failed. Fortunately, the reporting is steadily improving with gain in experience and expertise.

APPENDIX 10.A. PARTIAL LISTING OF VENDORS OF EQUIPMENT AND SYSTEMS FOR COMPOSTING MSW AND OTHER ORGANIC

Composting Systems: Biosolids, MSW, Yard Waste

Ag-Renu, Inc.
256 McCullough
Cincinnati, OH 45226

Agrisystems Engineering and Construction
3100 AM Schiedam
The Netherlands

Airite Environmental, Inc.
511 Woodland Acres
Crescent, RR2
Maple, Ontario
Canada K6A 1G2

Amerecycle
P.O. Box 338
Sumterville, FL 33585

American Bio Tech, Inc.
P.O. Box 19769
Jacksonville, FL 32245

American Compost Technologies, Inc.
17 W. Union St.
Wilkes Barre, PA 18701

Badger Composting Systems
15444 Clayton Road, Suite 302
St. Louis, MO 63011

Bedminister Bioconversion
52 Haddonfield-Berlin Road, Suite 1000
Cherry Hill, NJ 08034

Bio Gro Systems
180 Admiral Cochrane Dr., Suite 305
Annapolis, MD 21401

Bio Mac Composting Systems
Rt. 1, Box 324
Presque Isle, ME 04769

Buhler, Inc.
P.O. Box 9497
Minneapolis, MN 55440

Ceres, Inc.
2504 West County Road B
St. Paul, MN 55113

Compost Management, Inc.
4140 Skyron Drive
Doylestown, PA 18901

Compost Systems Co.
P.O. Box 354
Marion, OH 43302

ComTech Environmental, Inc.
372 Englishtown Road
Jamesburg, NJ 08831

Comtek Products, Inc.
14 Home Place
Topsham, ME 04086

Daneco, Inc.
119 N. Fourth St., Suite 508
Minneapolis, MN 55401

Double T Equipment
P.O. Box 3637
Airdrie, Alberta
Canada T4B 2B8

Earthcare Systems, Inc.
P.O. Box 998
Lincoln, AR 72744

Earthgro
Route 207
Lebanon, CT 06249

Ebara Environmental
R.D. 6, Box 516
Greensburg, PA 15601

Ecology Systems & Design
P.O. Box 1314
Murfreesboro, TN 37133

EKO Systems
P.O. Box 14026
Lakewood, CO 80214

Engineered Material Handling, Inc.
1451 Wolters Blvd.
St. Paul, MN 55110

Environmental Recovery Systems, Inc.
1625 Broadway, No. 2600
Denver, CO 80202

Fairfield Service Co.
240 Boone Avenue
Marion, OH 43302

Farmer Automatic
P.O. Box 39
Register, GA 30452

Green Mountain Technologies
237 East 10th St., Suite 6B
New York, NY 10003

International Process Systems, Inc.
Liberty Lane
Hampton, NH 03842

Keith Manufacturing
P.O. Box 1
Madras, OR 97741

LH Resource Management
RR 1
Walton, Ontario
Canada NOK 1Z0

Naturizer International, Inc.
P.O. Box 755
Norman, OK 73070-0755

Omni Technical Services
50 Charles Lindbergh Blvd.
Uniondale, NY 11553

OTVD
135 E. 57th St., 23rd Floor
New York, NY 10022

PWT Waste Solutions, Inc.
500 Southland Dr., Suite 124
Birmingham, AL 35226

Resource Conservation Services
Fort Andross
14 Main St., Suite 205
Brunswick, ME 04011

Reuter, Inc.
410 11th Avenue, South
Hopkins, MN 55343-7878

Ryan Companies
700 International Centre
900 Second Avenue South
Minneapolis, MN 55402

SPM Systems, Inc.
500 Market Tower
10 West Market St.
Indianapolis, IN 46204

Stinnes Enerco
2905 S. Sheridan Way
Oakville, Ontario
Canada L6J 7C7

Taulman, Inc.
415 E. Paces Ferry Road, NE
Atlanta, GA 30305

Universal Entech
44-E Inverness Dr., East
Englewood, CO 80112

USA DANO
1312 East Burnside
Portland, OR 97214

WPF Corporation
P.O. Box 381
Bellevue, OH 44811

Compost Turning and Mixing Equipment

Allied Steel & Tractor
5800 Harper Road
Solon, OH 44139

Brown Bear Corp.
P.O. Box 29
Corning, IA 50841

Cobey Composter
(Eagle Crusher Co., Inc.)
4250 S.R. 309
Galion, OH 44833

Double T Equipment
P.O. Box 3637
Airdrie, Alberta
Canada T4B 2B8

Eagle Crusher Co., Inc.
c/o Cobey Composter
4250 S.R. 309
Galion, OH 44833

Frontier Manufacturing Co.
P.O. Box 9176
Brooks, OR 97305

Fuel Harvesters Equipment
2501 Commerce Drive
Midland, TX 79703

JWI, Inc.
2155 112th Avenue
Holland, MI 49424

Knight Industrial Division
P.O. Box 167
Brodhead, WI 53520

Littleford Day, Inc.
7451 Empire Drive
Florence, KY 41042

McLanahan Corp.
200 Wall Street
Hollidaysburg, PA 16648

Olathe Manufacturing, Inc.
201 Leawood
Industrial Airport, KS 66031

Pike Lab Supplies, Inc.
RR 2, Box 710
Strong, ME 04983

Processall
10596 Springfield Pike
Cincinnati, OH 45215

Product Development Industries, Inc.
15444 Clayton Road
St. Louis, MO 63011

Resource Recovery Systems of Nebraska, Inc.
511 Pawnee Drive, Suite 4
Sterling, CO 80751-8969

Re-Tech
222 S. Market Street
Elizabethtown, PA 17022-0106

Scarab Manufacturing and Leasing, Inc.
P.O. Box 1047
White Deer, TX 79097

Scat Engineering
P.O. Box 266
Delhi, IA 52223

Scott Equipment
605 4th Ave., NW
New Prague, MN 56071

Simcorp, Inc.
Rt. 1, Box 202
Canyon, TX 79015

Sludge Systems, International
1125 Starr Avenue
Eau Claire, WI 54703

Valoraction, Inc.
Box 892
Sherbrooke, Quebec
Canada J1H 5L1

Wildcat Manufacturing Co., Inc.
Box 523
Freeman, SD 57029

Dewatering Systems

Andritz-Ruthner, Inc.
1010 Commercial Blvd. S
Arlington, TX 76017

Andritz-Sprout-Bauer, Inc.
Sherman Street
Muncy, PA 17756

Andy River, Inc.
23 Hatch Road
Lisbon, ME 04250

Arus-Andritz
Arlington S. Industrial Park
1010 Commercial Blvd.
Arlington, TX 76017

Bio Gro Systems
180 Adm. Cochrane Drive, Suite 305
Annapolis, MD 21401

Bio-Nomic Services
516 Roundtree Road
Charlotte, NC 28217-2133

Brown Bear Corp.
P.O. Box 29
Corning, IA 50841

Carrier Vibrating Equipment, Inc.
P.O. Box 37070
Louisville, KY 40233

The Carylon Corp.
2500 W. Arthington Street
Chicago, IL 60612-4108

D. R. Sperry & Co.
112 N. Grant Street
N. Aurora, IL 60542

Envirex
P.O. Box 1604
Waukesha, WI 53187

F. B. Leopold Co., Inc.
227 S. Division Street
Zelienople, PA 16063

Hydropress, Inc.
59 Dwight St.
Hatfield, MA 01038

JWI, Inc.
2155 - 112th Ave.
Holland, MI 49423

K-F Environmental Technologies, Inc.
210 W. Parkway, Unit 5
Pompton Plains, NJ 07444

Komile-Sanderson Engineering Corp.
12 Holland Ave.
Peapack, NJ 07977

Laidlaw Environmental Services (FS)
2202 Genoa-Red Bluff Rd.
Houston, TX 77034

Midwest Auger-Aerator
P.O. Box 445
Pontiac, IL 61764

Mobile Dredging & Pumping Co.
3100 Bethel Road
Chester, PA 19013

Olson Manufacturing Co.
620 S. Broadway
P.O. Box 289
Albert Lea, MN 56007

Parkson Corp.
P.O. Box 408399
Ft. Lauderdale, FL 33340

Rayfo, Inc.
4180 160th St., E.
Rosemount, MN 55068

Roediger Pittsburgh, Inc.
3812 Route 8
Allison Park, PA 15101

SP Industries, Inc.
2982 Jefferson Road
Hopkins, MI 49328

Sebright Products
127 N. Water St.
Hopkins, MI 49328

Simon Waste Solutions
820 Gessner, Suite 240
Houston, TX 77024

Somat Corporation
855 Fox Chase Road
Coatesville, PA 19320

Stord, Inc.
309 Regional Road, S.
Greensboro, SC 27409

Tetra Technologies, Inc.
P.O. Box 73087
Houston, TX 77273

Transportable Treatment Services, Inc.
1461 Harbor Ave.
Long Beach, CA 90813-2741

Magnetic Separation

Dings Magnetic Co.
4740 W. Electric Ave.
Milwaukee, WI 53219

Eriez Magnetics
P.O. Box 10608
Erie, PA 16514

Newell Industries, Inc.
P.O. Box 10629
San Antonio, TX 78210

Stearns Magnetics, Inc.
5400 Dunham Road
Maple Heights, OH 44137

Walker Magnetics
Rockdale St.
Worcester, MA 01606

Maintenance, Spare Parts

Accessory Sales
P.O. Box 160
Kimball, NE 69145

Applied Compost Technology and
  Manufacturing
45540 Desert Fox Drive
La Quinta, CA 92253

Arkin Sales, Inc.
P.O. Box 297
Clarendon Hills, IL 60514

Beall Manufacturing, Inc.
P.O. Box 70
East Alton, IL 62024

Cast Hammers, Inc.
P.O. Box 126
Waukegan, IL

CBT Wear Parts
223 N. Cedar Lake Rd.
Round Lake, IL 60073

Kenco Engineering
P.O. Box 1467
Roseville, CA 95661

Power Transmission Technology, Inc.
P.O. Box 305
Sharon Center, OH 44274

Marketing

Allgro
Liberty Lane
Hampton, NH 03842

American Soil Products
2222 Third St.
Berkeley, CA 94710

Compost Management, Inc.
4140 Skyron Drive
Doylestown, PA 18901

Earthgro
Route 207
Lebanon, CT 06249

Kellogg Supply Co.
350 W. Sepulveda Blvd.
Carson, CA 90745

Kenetech Resource Recovery
6447 33rd St. East
Sarasota, FL 34243

Kurtz Brothers
P.O. Box 31179
Independence, OH 44131

PRSM
882 S. Matlack St., Unit E
West Chester, PA 19382

The O. M. Scott & Sons Co.
14111 Scottslawn Rd.
Marysville, OH 43041

Soil Products, Inc.
P.O. Box 145
Hermitage, TN 37076

Materials Handling

Ag-Bag Corp.
P.O. Box 418
Astoria, OR 97103

Aggregates Equipment, Inc.
9 Horseshoe Road, Box 39
Leola, PA 17540-0039

Bio Gro Systems
180 Adm. Cochrane Drive, Suite 305
Annapolis, MD 21401

Bouldin & Lawson, Inc.
Rt. 10, Box 208
McMinnville, TN 37110

Buhler, Inc.
P.O. Box 9497
Minneapolis, MN 55440

Consilium Bulk Babcock
P.O. Box 47129
Atlanta, GA 30362

Converto Manufacturing, Inc.
P.O. Box 287
Cambridge City, IN 47327

Duraquip, Inc.
P.O. Box 948
Tualatin, OR 97062

East Manufacturing Co.
P.O. Box 277
Randolph, OH 44265

Engineered Material Handling, Inc.
1451 Wolters Blvd.
St. Paul, MN 55110

Gensco America, Inc.
2372 S. Stone Mountain
Lithonia Road
Lithonia, GA 30058

Hallco Manufacturing Co., Inc.
P.O. Box 505
Tillamook, OR 97141

J. C. Steele & Sons, Inc.
P.O. Box 1834
Statesville, NC 28687

Keith Manufacturing
P.O. Box 1
Madras, OR 97741

Laidig Industrial Sales
14535 Dragoon Trail
Mishawaka, IN 46544

Phelps Industries, Inc.
P.O. Box 1093
Little Rock, AR 72203

Royer Industries, Inc.
P.O. Box 1232
Kingston, PA 18704-0232

Tink, Inc.
2361 Durham-Dayton Hwy.
Durham, CA 95938

Walluski Western Ltd.
P.O. Box 642
Astoria, OR 97103

Yargus Manufacturing, Inc.
Box 238
Marshall, IL 62441

Mixing

Aggregates Equipment, Inc.
9 Horseshoe Road, Box 39
Leola, PA 17540-0039

Amadas Industries
1100 Holland Road
Suffolk, VA 23434

American Recycling Equipment
3200 Bordentown Ave.
P.O. Box 125
Parlin, NJ 08859

Blue Planet Ltd.
P.O. Box 1500
Princeton, NJ 08542

Brown Bear Corp.
P.O. Box 29
Corning, IA 50841

Cemen Tech, Inc.
1100 N. 14th Street
Indianola, IA 50125

Detcon
P.O. Box 2249
Farmingdale, NJ 07727

Knight Industrial Division
P.O. Box 167
Brodhead, WI 53520

Koch Engineering
P.O. Box 8127
Wichita, KS 67208

F. B. Leopold Co., Inc.
227 S. Division Street
Zelienople, PA 16063

Littleford Day, Inc.
7451 Empire Drive
Florence, KY 41042

McLanahan Corp.
P.O. Box 229
Hollidaysburg, PA 16648

Midwest Auger-Aerator
P.O. Box 445
Pontiac, IL 61764

Multitek, Inc.
P.O. Box 170
Prentice, WI 54556

Processall
10596 Springfield Pike
Cincinnati, OH 45215

Scat Engineering
P.O. Box 266
Delhi, IA 52223

Scott Equipment
605 4th Ave., NW
New Prague, MN 56071

Sludge Systems, International
1125 Starr Avenue
Eau Claire, WI 54703

Odor Control Aeration Systems

ABB Air Preheater, Inc.
Andover Road or
P.O. Box 372
Wellsville, NY 14895

Aero Tech Labs
(chemicals only)
728 N.W. 7 Terrace
Ft. Lauderdale, FL 33311

AirEactor
6619 13th Avenue
Brooklyn, NY 11219

Air Scent International
215 8th Ave.
P.O. Box 1000
Braddock, PA 15104

American Air Filter
P.O. Box 35690
Louisville, KY 40232

BioFiltration, Inc.
P.O. Box 15268
Gainesville, FL 32604

BioThermal Associates
40½ Paul Gore St.
Jamaica Plan, MA 02130

Calvert Environmental
5985 Santa Fe St.
San Diego, CA 92109

Duall Division
700 S. McMillan Street
Owosso, MI 48867

Ecolo Odor Control Systems, Inc.
1222 Fewster Dr., Unit 9
Mississauga, Ontario
Canada L4W 1A1

Howe-Baker Engineers
3102 E. 5th Street
P.O. Box 956
Tyler, TX 75710

M&C Manufacturing and Distribution
216 W. Falls Road
Grafton, WI 53024

Met-Pro Corp.
Duall Division
1550 Industrial Drive
Owosso, MI 48867

Monsanto-Enviro-Chem Systems, Inc.
14522 So. Outer Forty
Chesterfield, MO 63017

NuTech Environmental Corp.
5350 N. Washington St.
Denver, CO 80216

Odomaster Canada
994 W. Port Crescent, Unit A7
Mississauga, Ontario
Canada L5T 1G1

PEPCON Systems, Inc.
3770 Howard Hughes Pkwy., Suite 340
Las Vegas, NV 89109

Purafil, Inc.
P.O. Box 287
Oconomowoc, WI 53066

Quad Environmental Technologies Corp.
3605 Woodhead Dr., No. 103
Northbrook, IL 60062

Wheatec
2 S. 076 Orchard Road
Wheaton, IL 60187

Screens

Aggregates Equipment, Inc.
P.O. Box 39
Leola, PA 17540-0039

Amadas Industries
1100 Holland Road
Suffolk, VA 23434

Banner Welder, Inc.
N117 W18200 Fulton Drive
Germantown, WI 53022

Bulk Handling Systems, Inc.
1040 Arrowsmith
Eugene, OR 97402

Central Manufacturing, Inc.
P.O. Box 1900
Peoria, IL 61656

Construction Steel, Inc.
1772 Corn Road
Smyrna, GA 30080

Excel Recycling & Manufacturing, Inc.
P.O. Box 31118
Amarillo, TX 79120

Fuel Harvesters Equipment
2501 Commerce Drive
Midland, TX 79703

The Heil Company
Engineers Systems Division
Arbor Terrace II
205 Bishops Way, Suite 201
Brookfield, WI 53005

Jiffy Sift
P.O. Box 891
Yelm, WA 98597

Lindemann Recycling Equipment
42 W. 38th St., Suite 1102
New York, NY 10018

Lindig Manufacturing Corp.
P.O. Box 130130
Roseville, MN 55113

Ohio Central Steel Co.
7001 Americana Pkwy.
Reynoldsburg, OH 43068

Powerscreen of America
11001 Electron Drive
Louisville, KY 40299

Rader Companies, Inc.
P.O. Box 181048
Memphis, TN 38181

Read Corporation
P.O. Box 1298
Middleboro, MA 02346

Re-Tech
499 Houtztown Road
Myerstown, PA 17067

Royer Industries, Inc.
P.O. Box 1232
Kingston, PA 18704-0232

SPM Group, Inc.
317 Tall Grass Ct.
Lawrence, KS 66049

Triple/S Dynamics, Inc.
1031 S. Haskell Ave.
Dallas, TX 75223

West Salem Machinery
P.O. Box 5288
Salem, OR 97304

Whirl-Air-Flow
P.O. Box 18190
Minneapolis, MN 55418

Wildcat Manufacturing Co., Inc.
Box 523
Freeman, SD 57029

Size Reduction: Shredders, Chippers, Tub Grinders

Allegheny Paper Shredders Corp.
Old Wm. Penn Highway E
Delmont, PA 15626

Amadas Industries
1100 Holland Road
Suffolk, VA 23434

American Pulverizer Co.
5540 W. Park Ave.
St. Louis, MO 63110

Bandit Industries
6750 Millbrook Road
Remus, MI 49340

Banner Welder, Inc.
N117 W18200 Fulton Drive
Germantown, WI 53022

Blower Application Co.
W19125 Clinton Drive
Germantown, WI 53022

Conair/Wor-Tex Corp.
Harry S. Truman Pkwy.
Bay City, MI 48706

Construction Steel, Inc.
1772 Corn Road
Smyrna, GA 30080

Continental Biomass Industries
P.O. Box 656
Salem, NH 03079

CW Manufacturing, Inc.
P.O. Box 246
Sabetha, KS 66534

Diamond Z
1102 Franklin Blvd.
Nampa, ID 83653

EAC Engineering
332-C Edwardia Dr.
Greensboro, NC 27409

EnVnet Corporation
P.O. Box 321
Akron, OH 44309-0321

Excel Recycling & Manufacturing Co., Inc.
P.O. Box 31118
Amarillo, TX 79120

Farmhand/Ag Equipment Group, Lpn.
RR 1, Box 55
Grinnell, IA 50112

Fecon, Inc.
9281 Le Saint Drive
Fairfield, OH 45014

Fuel Harvesters Equipment
2501 Commerce Drive
Midland, TX 79703

Gensco America, Inc.
2372 S. Stone Mountain
Lithonia Road
Lithonia, GA 30058

Gruendler Crusher
212 S. Oak Street
Durand, MI 48429

Haybuster Manufacturing, Inc.
P.O. Box 1940
Jamestown, ND 58402-1940

The Heil Company
Engineers Systems Division
Arbor Terrace II
205 Bishops Way, Suite 201
Brookfield, WI 53005

W. J. Heinrichs, Inc.
21013 E Dinuba Ave.
Reedley, CA 93654

Hi-Torque Shredder Co.
230 Sherman Avenue
Berkeley Heights, NJ 07922

Hobart
401 West Market Street
Troy, WI 45374

Iggesund Recycling
Aitkin Industrial Park
P.O. Box 387
Aitkin, MN 56431

Industrial Paper Shredders, Inc.
P.O. Box 180
Salem, OH 44460

Innovator Manufacturing, Inc.
120 Weston Street
London, Ontario
Canada N6C 1R4

Jacobson Companies
2445 Nevada Ave., N.
Minneapolis, MN 55427

Jeffrey Division
P.O. Box 387
Woodruff, SC 29388

Jones Manufacturing Co.
P.O. Box 38
Beemer, NE 68716

JWC Environmental
16802 Aston St., Suite 200
Irvine, CA 92714

La Bounty Manufacturing
100 State Road 2
Two Harbors, MN 55616

Lindemann Recycling Equipment
42 W. 38th St., Suite 1102
New York, NY 10018

Lindig Manufacturing Corp.
P.O. Box 130130
Roseville, MN 55113

Mac/Saturn Corp.
201 E. Shady Grove Road
Grand Prairie, TX 75050

Micron Powder Systems
10 Chatham Road
Summit, NJ 07901

Miller Manufacturing
P.O. Box 336
Turlock, CA 95381-0336

Montgomery Industries
P.O. Box 3687
Jacksonville, FL 32206

Morbark Sales Corp.
P.O. Box 1000
Winn, MI 48896

Nelmore Co., Inc.
Rivulet Street
North Uxbridge, MA 01538

Newell Industries, Inc.
P.O. Box 10629
San Antonio, TX 78210

Norcia
RD 4, Box 451
N. Brunswick, NJ 08902

O&E Machine Corp.
P.O. Box 1836
Green Bay, WI 54305

Ohio Central Steel Co.
7001 Americana Pkwy.
Reynoldsburg, OH 43068

Olathe Manufacturing, Inc.
201 Leawood Drive
Ind. Airport, KS 66031

Old Dominion Brush
(ODB)
5118 Glen Alden Dr.
Richmond, VA 23231

PCR, Inc.
Route 1
Coon Valley, WI 54623

Prodeva, Inc.
100 Jerry Drive
Jackson Center, OH 45334

Reduction Technology, Inc.
P.O. Box 297
Leeds, AL 35094-0297

Rexworks, Inc.
P.O. Box 2037
Milwaukee, WI 53201

Rome Waste Management Systems
P.O. Box 48
Cedartown, GA 30125

Seppi Division, C.T & E. Co.
P.O. Box 69
Rosemount, MN 55068

Shred Pax Corp.
136 W. Commercial Ave.
Wood Dale, IL 60191-1304

Shred-Tech Limited
P.O. Box 1508
Cambridge, Ontario
Canada N1R 7G8

Somat Corp.
855 Fox Chase
Coatesville, PA 19320

SSI Shredding Systems
9760 S. W. Freeman Drive
Wilsonville, OR 97070-9286

Stumpmaster, Inc.
P.O. Box 103
Rising Fawn, GA 30738

Sundance
P.O. Box 2437
Greeley, CO 80632

Triple/S Dynamics, Inc.
1031 S. Haskell Ave.
Dallas, TX 75223

Universal Engineering
800 First Ave., NW
Cedar Rapids, IA 52405

Universal Refiner Corp.
P.O. Box 151
Montesano, WA 98563

Valby Woodchippers
Northeast Implement Corp.
Box 402
Spencer, NY 14883

West Salem Machinery
P.O. Box 5288
Salem, OR 97304

W-H-O Manufacturing Co., Inc.
P.O. Box 1153
Lamar, CO 81052

R.M. Wilson, Co.
P.O. Box 6274
Wheeling, WV 26003

Spreaders and Applicators

Ag-Chem Equipment Co.
5720 Smetana Drive, Suite 100
Minnetonka, MN 55343

Amadas Industries
P.O. Box 1833
Suffolk, VA 23434

Field Gymmy
Box 121
Glandorf, OH 45848

Gehl Co.
143 Water St.
West Bend, WI 53095

Henderson Manufacturing
P.O. Box 40
Manchester, IA 52057

Highway Equipment Co.
616 D Ave. NW
Cedar Rapids, IA 52405

Knight Industrial Division
P.O. Box 167
Brodhead, WI 53520

L.W.T., Inc.
Box 250
Somerset, WI 54025

Millcreek Mfg. Co.
112 S. Railroad Ave.
New Holland, PA 17557

Thermometers and Monitoring

Arthur Technologies/Tech-Line Instruments
P.O. Box 1236
Fond du Lac, WI 54935

Concord, Inc.
2800 7th Ave. North
Fargo, ND 58102

Hanna Instruments, Inc.
Highland Industrial Park
584 Park East Drive
Woonsocket, RI 02895

Meriden Cooper Corp.
Box 692
Meriden, CT 06450

Omega Engineering, Inc.
P.O. Box 4047
Stamford, CT 06907

Reotemp Instrument Corp.
11568 Sorrento Valley Road, Suite 10
San Diego, CA 92121

Soil Control Lab
42 Hangar Way
Watsonville, CA 95076

Spectrum Technologies
12010 S. Aero Drive
Plainfield, IL 60544

Tel Tru, Inc.
408 St. Paul St.
Rochester, NY 14605

Trend Instruments
887 S. Matlack St.
West Chester, PA 19380

Walden Instrument Supply Co.
910 Main Street
Wakefield, MA 01880

Yard Waste Debaggers

Dover Conveyor Co.
P.O. Box 300
Midvale, OH 44653

Empire Organic Greenhouse
15325 Babcock Ave.
Rosemount, MN 55068

Lande Environmental Equipment
11615 N. Shore Road
Whitmore Lake, MI 48189

Magnificent Machinery
2366 Woodhill Road
Cleveland, OH 44106

Scarab Manufacturing and Leasing, Inc.
P.O. Box 1047
White Deer, TX 79097

Scat Engineering
P.O. Box 266
Delhi, IA 52223

SSI Shredding Systems
9760 S.W. Freeman Drive
Wilsonville, OR 97070-9286

Superior Tech, Inc.
P.O. Box 5177
Lancaster, PA 17601

Wayne Engineering Corp.
P.O. Box 648
Cedar Falls, IA 50613

Whirl-Air-Flow
P.O. Box 18190
Minneapolis, MN 55418

Wildcat Manufacturing Co., Inc.
P.O. Box 523
Freeman, SD 57029

APPENDIX 10.B.   COSTS FOR COMPOSTING MSW AND YARD WASTES

TABLE 10.11   Projected Capital Construction Cost Estimate for Two Hypothetical MSW Composting Facilities (1990 $)

Category	100 TPD aerated windrow	1000 TPD turned windrow
Site preparation (mobilization, earthwork, paving, utilities, connections, landscape, fencing, etc.)	$400,000	$5,000,000
Building, structures, foundations (receiving floor, equipment areas, office building, scale, scale house, etc.)	3,000,000	26,000,000
In-plant mobile equipment	450,000	2,500,000
Composting equipment (conveyors, shredders, screens, preprocessing, postprocessing, etc.)	1,300,000	11,000,000
Miscellaneous (supplies, office, furnishing, insurance, etc.)	100,000	800,000
Engineering, permits, construction management	470,000	4,500,000
Haul and package equipment	Not included	Not included
Land purchase	650,000	2,500,000
Working capital (start-up)	360,000	3,000,000
Total	$6,730,000	$55,300,000
$ per TPD of capacity	67,300	55,300

Quantities given in TPD (U.S. tons per day). Multiply TPD by 0.907 to obtain Mtons per day.

TABLE 10.12 Projected Annual Operation and Maintenance Cost Estimate for Two Hypothetical MSW Composting Facilities (1990 $)

Category	100 TPD aerated windrow	1000 TPD turned windrow
Labor (O&M personnel, scale operations, supervisory and office personnel including fringes)	$580,000	$4,500,000
Maintenance and materials	300,000	2,500,000
Utilities (water, sewer, electric)	90,000	900,000
Administration and insurance	150,000	750,000
Regulatory compliance	40,000	100,000
Miscellaneous (contract services)	55,000	350,000
Haul and residue disposal*	175,000	1,747,000
Total	$1,390,000	$10,847,000
Unit O&M cost, $ per ton MSW†	$45	$35

Quantities given in TPD (U.S. tons per day). Multiply TPD by 0.907 to obtain Mtons per day.

*11.2 percent of MSW input at $50 per ton.

†Based on 312 operating days per year.

TABLE 10.13 Cost Summary for Hypothetical MSW Composting Facilities (1990 $)

Cost item	100 TPD aerated windrow	1000 TPD turned windrow
Net annual costs:		
A. Amortized investment cost*	$811,000	$6,608,000
B. Annual operating and maintenance cost	1,390,000	10,847,000
Total annual cost	2,201,000	17,455,000
C. Recyclables revenue	214,000	2,140,000
Compost revenue	31,200	312,000
Total annual revenue	245,200	2,452,000
D. Net annual cost	1,955,800	15,003,000
E. Avoided annual cost†	1,248,000	12,480,000
F. Annual net, minus avoided cost	$707,800	$2,523,000
Net unit costs‡,§ ($ per ton)		
A. Amortized investment cost*	$26	$21
B. Operating and maintenance cost	45	35
Total unit cost	71	56
C. Revenue	8	8
D. Net unit cost	63	48
E. Avoided cost	40	40
F. Unit cost minus avoided cost	$23	$8

Quantities given in TPD (U.S. tons per day). Multiply TPD by 0.907 to obtain Mtons per day.

*Interest rate = 10 percent per year. Building and most stationary equipment are assumed to have a life of 20 years. Mobile equipment is assumed to have a life of 10 years.

†Avoided cost = $40 per ton of MSW.

‡Unit costs are given in dollars per ton of MSW feedstock. Based on 312 operating days per year.

§Round-off affects numerical values presented.

TABLE 10.14 Variations between the Yard Waste Compost Plant Designs Considered for the Economic Analysis (1990 $)

	Option A	Option B
Construction:		
Grading	Entire site	Entire site
Paving	Processing area, receiving area, and 1/4 of composting area	Processing area
Fencing	Entire site	Processing area
Equipment:		
Grinder	1	1
Compost turner	1	0
Front-end loader	1	1
Screen	1	0
Conveyors	4	2

Quantities given in TPD (U.S. tons per day). Multiply TPD by 0.907 to obtain Mtons per day.

TABLE 10.15 Estimated Initial Investment Costs for Options A and B in Table 10.14 for Composting Yard Wastes (1990 $)

	Option A	Option B
Construction:		
Grading site	$50,000	$50,000
Paving site	300,000	46,000
Road in	5,000	5,000
Fencing	30,000	8,250
Water distribution system	7,500	7,500
Engineering:		
Design	72,700	39,100
Construction supervision	5,000	2,500
Training of site operators	7,000	7,000
Utility hookups:		
Water	20,000	20,000
Power	20,000	20,000
Equipment:		
Compost turner	125,000	0
Front-end loader	75,000	125,000
Grinder	100,000	100,000
Screen	50,000	0
Conveyors	40,000	20,000
Miscellaneous equipment	15,000	15,000
Total investment costs	$922,200	$465,350

Quantities given in TPD (U.S. tons per day). Multiply TPD by 0.907 to obtain Mtons per day.

TABLE 10.16 Estimated Annual Operating and Maintenance Costs for Options A and B in Table 10.14 for Composting Yard Wastes (1990 $)

	Option A	Option B
Maintenance	$20,250	$13,000
Insurance	8,100	5,200
Fuel	16,800	21,900
Labor	70,000	87,600
Lease on land ($800 per acre per month)	115,200	115,200
Trailer rental	2,100	2,100
Water	2,000	2,000
Power	10,000	9,000
Total operating and maintenance costs	$244,450	$256,000

Quantities given in TPD (U.S. tons per day). Multiply TPD by 0.907 to obtain Mtons per day.

TABLE 10.17 Cost Summary for Hypothetical Yard Waste Composting Facilities in Table 10.14 (1990 $)

	Option A	Option B
Net annual costs		
Annual costs:		
Amortized investment cost*	150,100	75,730
Annual operating and maintenance costs	244,450	256,000
Total annual cost ($)	394,550	331,730
Annual revenues:		
Sale of compost†	90,000	70,000
Total revenues	90,000	70,000
Net annual cost	304,550	261,730
Annual avoided cost‡	350,000	350,000
Annual net minus avoided cost	(54,450)	(88,270)
Net unit costs ($ per ton of yard waste)§		
Amortized investment cost	16	8
Operating and maintenance cost	26	27
Total unit cost	42	35
Revenue	10	7
Net unit cost	32	28
Avoided cost	37	37
Net unit cost minus avoided cost	(5)	(9)

Quantities given in TPD (U.S. tons per day). Multiply TPD by 0.907 to obtain Mtons per day.

*Based on a 10 percent discount rate, 10-year term, and no salvage value.

†10,000 yd^3 at $9/yd^3 for option A and $7/yd^3 for option B.

‡Based on $5/yd^3 of yard waste.

§Based on 70,000 yd^3 of annual yard waste at an average bulk density of 270 lb/yd^3.

CHAPTER TEN

REFERENCES

1. Golueke, C. G., and P. H. McGauhey, "Reclamation of Municipal Refuse by Composting," *Technical Bulletin* 9, Sanitary Engineering Research Laboratory, University of California, Berkeley, June 1955.
2. Golueke, C. G., *Composting: A Study of the Process and Its Principles,* Rodale Press, Inc., Emmaus, PA, 1972.
3. Diaz, L. F, G. M. Savage, L. L. Eggerth, and C. G. Golueke, *Composting and Recycling· Municipal Solid Waste,* Lewis Publishers, Inc., Ann Arbor, Mich., 1993.
4. Diaz, L. F., et al, *Public Health Aspects of Composting Combined Refuse and Sludge and of the Leachates Therefrom,* University of California, Berkeley, September 1977.
5. Lossin, R. D., "Compost Studies. Part III. Measurements of Chemical Oxygen Demand of Compost," *Compost Science,* vol. 12, no. 2, pp. 31–32, March 1971.
6. "Composting Fruit and Vegetable Refuse. Part II. Investigation of Composting as a Means for Disposing of Fruit Waste Solids," Progress Report, National Canners Association Research Foundation, Washington, D.C., August 1964.
7. Schulze, K. F., "Rate of Oxygen Consumption and Respiratory Rate Quotients during the Aerobic Composting of Synthetic Garbage," *Compost Science,* vol. 1, no. 36, spring 1960.
8. Schulze, K. F., "Relationship between Moisture Content and Activity of Finished Compost," *Compost Science,* vol. 2, no. 32, summer 1964.
9. Rynk, R. (ed.), *On-Farm Composting Handbook,* Northeastern Regional Agricultural Service, 152 Riley-Robb Hall, Cooperative Extension, Ithaca, N.Y. 1992.
10. Chrometska, P., "Determination of the Oxygen Requirements of Maturing Composts," International Research Group on Refuse Disposal, *Information Bulletin* 33, August 1968.
11. Lossin, R. D., "Compost Studies, Part II, *Compost Science,* vol. 12, no. 1, pp. 12–13, January–February 1971.
12. Regan, R. W., and J. S. Jeris, "A Review of the Decomposition of Cellulose and Refuse," *Compost Science,* vol. 11, no. 17, January–February 1970.
13. Wiley, J. S., "Progress Report on High-Rate Composting Studies," *Engineering Bulletin, Proceedings of the 12th Industrial Waste Conference,* Series 94, pp. 590–595, May 13–15, 1957.
14. Finstein, M., Composting in the Context of Municipal Solid Waste Management, chap. 14 in *Environmental Microbiology,* pp. 355–374, Wiley-Liss, Inc., New York, 1992.
15. Diaz, L. F., G. M. Savage, and C. G. Golueke, *Resource Recovery from Municipal Solid Wastes,* chap. II, Final Processing, CRC Press, Boca Raton, Fla., 1982.
16. Niese, G., "Experiments to Determine the Degree of Decomposition of Refuse by Itself—Heating Capability," International Research Group on Refuse Disposal, *Information Bulletin* 17, May 1963.
17. Rolle, G. and E. Organic, "A New Method for Determining Decomposable and Resistant Organic Matter in Refuse and Refuse Compost," International Research Group on Refuse Disposal, *Information Bulletin* 21, August 1964.
18. Moller, F., "Oxidation-Reduction Potential and Hygienic State of Compost from Urban Refuse," International Research Group on Refuse Disposal, *Information Bulletin* 32, August 1968.
19. Obrist, W., "Enzymatic Activity and Degradation of Matter in Refuse Digestion: Suggested New Method for Microbiological Determination of Degree of Digestion," International Research Group on Refuse Disposal, *Information Bulletin* 24, September 1965.
20. Lossin, R., "Compost Studies," *Compost Science,* vol. 11, no. 6, 1970.
21. "Odor Control at Honolulu," *Water Environment Research,* March–April, 1992, in "Environmental Waste Control Digest," *Public Works,* vol. 132, no. 9, pp. 132–133, August 1992.
22. Cooper, R. C., S. A. Klein, C. J. Leong, J. L. Potter, and C. G. Golueke, *Effect of Disposable Diapers on the Composition of Leachate from a Landfill,* Final Report, SERL Report 74-3, Sanitary Engineering Research Laboratory, University of California, Berkeley, February 1974.
23. Epstein, E., G. B. Willson, W. D. Burge, D. C. Mullen, and N. K. Enkiri, "A Forced Aeration System for Composting Wastewater Sludge," *Journal of the Water Pollution Control Federation,* vol. 38, no. 4, p. 688, 1976.
24. Willson, G. B., J. F. Parr, E. Epstein, P. B. Marsh, R. L. Chaney, D. Colacicco, W. D. Burge, L. J. Sikora, C. F. Tester, and S. Hornick, *Manual for Composting Sewage Sludge by the*

Beltsville Aeration Pile Method, EPA 600/S-80-022, Office of Research and Development, U.S. EPA, Cincinnati, Ohio, May 1980.

25. *BioCycle,* The JG Press, Inc., Emmaus, PA, 18049.
26. Wylie, J. S., "Progress Report on High-Rate Composting Studies," *Engineering Bulletin, Proceedings of the 12th Industrial Waste Conference,* Series 94, May 12–15, 1957.
27. Senn, C. L., "Role of Composting in Waste Utilization," *Compost Science,* vol. 15, no. 4, pp. 24–28, September–October 1974.
28. *Composting Technologies, Costs, Programs, and Markets,* prepared by Cal Recovery Systems, Inc. (now CalRecovery, Inc.), for the Congress of the United States, Office of Technology Assessment, Washington, D.C., 1989.
29. *Municipal Solid Waste Composting Programs: A Guidebook for Officials,* prepared by Eastern Research Group and CalRecovery, Inc. for the U.S. Environmental Protection Agency, Cincinnati, Ohio, May 1991.
30. *Composting at Johnson City,* Final Report on Joint USEPA-TVA Composting Project, Vols. I, II, U.S. EPA, EPA/530/SW-31s.2, 333 pp., 1975.
31. Snyser, S., "Taking the Sludge to Market," *BioCycle,* vol. 23, no. 1, pp. 21–24, January–February 1982.
32. *Compost Classification/Quality Standards for the State of Washington,* prepared by CalRecovery, Inc., for the State of Washington, Department of Ecology, September 1990.
33. "Aspergillus—Implications on the Health and Legal Implications of Sewage Sludge Composting," *Workshop Proceedings,* vol. 1, prepared by Energy Resources, Inc., for Division Problems—Focused Research Applications, National Science Foundation, December 1978.
34. Golueke, C. G., "Epidemiological Aspects of Sewage Sludge Handling and Management," *Proceedings, International Symposium on Land Application of Sewage Sludge,* October 1982.
35. Miller, F. C., and B. J. Macauley, "Odors Arising from Mushroom Composting: A Review," *Australian Journal of Experimental Agriculture,* vol. 28, no. 19, pp. 553–560, 1988.
36. Hay, J., J. H. Ahn, S. Chang, R. Caballero, and H. Kellogg, "Alternative Bulking Agent for Sludge Composting, Part II," *BioCycle,* vol. 29, no. 10, pp. 46–49, November–December 1988.
37. *Portland Area Compost Products Market Study,* prepared by Cal Recovery Systems, Inc., for Metropolitan Service District, Portland, Ore., 100 pp., 1988.
38. Chaney, R. L., "The Establishment of Guidelines and Monitoring System for Disposal of Sewage Sludge to Land," *Proceedings, International Symposium on Land Application of Sewage Sludge,* Tokyo, October 1982.
39. Sharma, R. P., "Plant-Animal Distribution of Cadmium in the Environment," chap. 14 in Nriagu, J. O. (ed.), *Cadmium in the Environment. Part 1, Ecological Cycling,* Wiley, New York, 682 pp., 1980.
40. Miller, T. L., R. R. Swager, S. G. Wood, and A. D. Atkins, *Sampling Compost Sites for Metals and Pesticides, Resource Recycling,* December 1992.
41. Mennear, J. H. (ed.), *Cadmium Toxicity,* Marcel Dekker, New York, 224 pp., 1978.
42. Verdonck, O., M. DeBrodt, and R. Gabriel, "Compost as a Growing Medium for Horticultural Plants," in *Compost: Production, Quality, and Use,* Elsevier Applied Science, New York, 1987, pp. 399–414.
43. Anon., "Mulch: 1988 Garden Hero," *Sunset,* August 1988, pp. 56–59.
44. Pittenger, "Potting Soil Label Information Is Inadequate," *California Agriculture,* vol. 40, pp. 6–8, November–December 1986.
45. Goldstein, N., "Solid Waste Composting in the U.S.," *BioCycle,* vol. 30, no. 11, pp. 32–37, November 1989.
46. Craig, N., "Solid Waste Composting Underway in Minnesota," *BioCycle,* vol. 29, no. 7, pp. 30–33, August 1988.
47. *Swift County Solid Waste Compost/Recycling Project,* prepared by CalRecovery, Inc., for Swift County Auditor, Benson, Minnesota, March 1988, updated November 1988.
48. Gorham, D., and R. Zier, "Portland Moves to Solid Waste Composting," *BioCycle,* vol. 29, no. 7, pp. 34–37, August 1988.
49. "Agripost News Construction of Compost Facility," *BioCycle,* vol. 29, no. 10, pp. 13–14, November–December 1988.

50. Golueke, C. G., D. Lafrenz, B. Chaser, and L. F. Diaz, *Benefits and Problems of Refuse-Sludge Composting,* prepared for the National Science Foundation (PFR-791707), 1980.

51. Trubiano, R. P., M. Thayer, S. J. Kruger, and F. Senske, "Boston Project Tests Methods and Markets," *BioCycle,* vol. 28, no. 7, p. 25, August 1987.

52. *Solid Waste Management Facilities, Revised 6 NYCRR, Part 360,* New York State, Department of Environmental Conservation, August 1988.

CHAPTER 11
WASTE-TO-ENERGY COMBUSTION

Calvin R. Brunner

PART A

11.1 Incineration

Floyd Hasselriis

PART B

11.2 Ash Disposal

Aaron Teller

PART C

11.3 Emission Control

T. Randall Curlee

PART D

11.4 The Socioeconomics of Waste-to-Energy in the United States

*INTRODUCTION**

Waste-to-energy combustion is handling an increasing percentage of the municipal solid waste. But its growth has recently slowed while communities wrestle with issues that range from flow control, impact on recycling, cost effectiveness, and political acceptability. Nevertheless, indications are that waste-to-energy combustion will be an increasingly important factor in the overall integrated solid waste management strategy. The traditional term "incineration" has acquired a bad connotation in the mind of the public due to the poor operation of some waste combustors in the past. Therefore, the term waste-to-energy combustion is now widely used in its place. The term incineration, as used in this chapter, refers to the modern practice of incineration of waste that cannot be recycled economically. The technology offers great opportunities for reducing the volume of waste to be landfilled, as well as for generating heat and power. Raw solid waste has a heating value between 4000 and 7000 Btu/lb compared to coal, which releases about 10,000 Btu/lb. Hence, a large amount of heat can be released by burning municipal waste, and that heat can be used to generate power. It has been estimated that waste-to-energy facilities could supply as much as 2 percent of the electrical power needed in this country. But, more importantly, incineration reduces the volume of waste dramatically, up to tenfold. Thus, incineration is becoming particularly attractive for large metropolitan areas where landfills are a long distance from the population center.

The major constraints on waste-to-energy combustion facilities are their cost, the level of sophistication needed to operate them safely, and the fact that the American

---

*Introduction by Frank Kreith.

public lacks confidence in their safety. The public is concerned about stack emissions of dioxins and the toxicity of ash residues. This concern exists, despite the assurance of experts that incineration in a modern plant with proper air pollution control equipment does not pose any dangers to health and environment. A panel of experts at the 1990 U.S. Conference of Mayors, evaluating the health and environmental impact of waste incineration, concluded that "Inclusive of ash residue management, properly designed, operated and maintained incinerators equipped with state-of-the-art pollution devices can be used [to burn solid waste] in a manner that maintains associated risks below levels set by regulatory bodies for the protection of human health."

Incineration has been used widely in Europe and Japan without any adverse health impacts. Switzerland, a country with high environmental standards, incinerates about 75 percent and Japan more than 50 percent of its solid waste, according to a survey by the Integrated Waste Services Association in the spring of 1993. Sweden incinerates 60 percent and composts up to 25 percent. Waste-to-energy combustion is slowly gaining public acceptance also in the U.S., according to a survey conducted by Cambridge Reports/Research International. This poll found that in 1990, 50 percent of the general public (up from 44 percent in July of 1989) said that they would not object to a local waste-to-energy plant. The big jump in public support came from the "don't know" respondents who accounted for 20 percent in the 1989 sample, but dropped to 8 percent in March of 1990. The increase in public support for waste-to-energy combustion suggests that, as more information on this technology becomes available, political support for siting new facilities will increase and pave the way for full integration of this technology in waste management schemes. In any successful integrated waste management systems in the United States designed for the next century, waste-to-energy combustion is bound to perform an important role.

This chapter is organized to present technical information needed, as well as the political and social challenges to be considered in designing and siting a waste-to-energy combustion facility. The main part of the chapter has been written by Calvin R. Brunner, a consulting engineer who has previously published *The Handbook of Incineration Systems*. Part A of this chapter has been adapted from this large and extensive work to reflect the current technology and cost of municipal solid waste systems. The chapter also deals with the incineration of medical wastes, which is becoming increasingly important in many metropolitan areas.

Significant contributions have been made in Part B of the chapter by Floyd Hasselriis, who has covered the disposal of ash residues from waste-to-energy incineration. In addition to addressing concerns about ash toxicity, he also presents novel ways of converting ash to useful products, thereby completely recycling the solid waste. Dr. Aaron Teller deals in Part C with the problem of mitigating emissions from the stack by means of various control devices. The data in his section conclusively show that the emissions of dioxins and furans can be reduced to levels that will not pose a health hazard, as the panel of experts for the U.S. mayors concluded. Dr. Teller's results also show that, on a comparative basis, power can be generated from municipal waste with less pollution than from a coal-fired power plant. The final contribution in Part D deals with the socioeonomic aspects of solid waste management and has been written by T. Randall Curlee of the Oak Ridge National Laboratory. This section has been extracted from a book that has recently been published by the staff of ORNL, which provides help on siting incineration facilities and describes various case studies and their implications.

There are over 150 waste-to-energy plants in operation today and more than 50 are either planned or under construction. The technology is proven and reliable and the information presented in this chapter should assist in integrating waste-to-energy combustion systems in a state-of-the-art municipal solid waste management system.

PART A

Calvin R. Brunner

Consulting Engineer
Reston, Virginia

11.1 INCINERATION*

One of the most effective means of dealing with many wastes, to reduce their harmful potential and often to convert them to an energy form, is incineration. In comparing incineration (the destruction of a waste material by the application of heat) to other disposal options such as land burial, the advantages of incineration are:

- The volume and weight of the waste are reduced to a fraction of its original size.
- Waste reduction is immediate; it does not require long-term residence in a landfill or holding pond.
- Waste can be incinerated on-site, without having to be carted to a distant area.
- Air discharges can be effectively controlled for minimal impact on the atmospheric environment.
- The ash residue is usually nonputrescible, or sterile.
- Technology exists to completely destroy even the most hazardous of materials in a complete and effective manner.
- Incineration requires a relatively small disposal area, compared to the acres required for land burial.
- By using heat-recovery techniques the cost of operation can often be reduced or offset through the use or sale of energy.

Incineration will not solve all waste problems. Some disadvantages include:

- The capital cost is high.
- Skilled operators are required.
- Not all materials are incinerable, e.g., construction and demolition wastes.
- Supplemental fuel is required to initiate and at times to maintain the incineration process.

Incinerable Waste.   The Incinerator Institute of America was a national organization attempting to quantify and standardize incinerator design parameters. It went out of business over 20 years ago; however, a number of its standards are still in use. One such standard, given in Table 11.1, is used by manufacturers of small and packaged

---

*Adapted from *Handbook of Incineration Systems,* published by McGraw-Hill in 1991.

TABLE 11.1 Classification of Wastes to Be Incinerated

Type	Description	Principal components	Approximate composition, % by weight	Moisture content, %	Incombustible solids, %	Refuse as fired, Btu/lb	Btu of auxiliary fuel per lb of waste to be included in combustion calculations	Recommended min Btu/h burner input per lb waste
"0	Trash	Highly combustible waste, paper, wood, cardboard cartons, including up to 10% treated papers, plastic or rubber scraps; commercial and industrial sources	Trash 100%	10	5	8500	0	0
"1	Rubbish	Combustible waste, paper, cartons, rags, wood scraps, combustible floor sweepings; domestic, commercial, and industrial sources	Rubbish 80% Garbage 20%	25	10	6500	0	0
"2	Refuse	Rubbish and garbage; residential sources	Rubbish 50% Garbage 50%	50	7	4300	0	1500
"3	Garbage	Animal and vegetable wastes, restaurants, hotels, markets; institutional, commercial, and club sources	Garbage 65% Rubbish 35%	70	5	2500	1500	3000
4	Animal solids and organic wastes	Carcasses, organs, solid organic wastes; hospital, laboratory, abattoirs, animal pounds, and similar sources	100% Animal and human tissue	85	5	1000	3000	8000 (5000 Primary) (3000 Secondary)
5	Gaseous, liquid or semi-liquid wastes	Industrial process wastes	Variable	Dependent on predominant components	Variable according to wastes survey	Variable according to wastes survey	Variable according to wastes survey	Variable according to wastes survey
6	Semi-solid and solid wastes	Combustibles requiring hearth, retort, or grate burning equipment	Variable	Dependent on predominant components	Variable according to wastes survey	Variable according to wastes survey	Variable according to wastes survey	Variable according to wastes survey

[a]The above figures on moisture content, ash, and Btu as fired have been determined by analysis of many samples. They are recommended for use in computing heat release, burning rate, velocity, and other details of incinerator designs. Any design based on these calculations can accommodate minor variations.

Source: Ref. 1.

incinerators in rating their equipment. The classifications in the table represent incinerable wastes, wastes which are combustible and are viable candidates for incineration.

Incinerability can be defined more specifically by consideration of the following factors:

Waste moisture content. The greater the moisture content, the more fuel is required to destroy the waste. An aqueous waste with a moisture content greater than 95 percent or a sludge waste with less than 15 percent solids content would be considered poor candidates for incineration.

Heating value. Incineration is a thermal destruction process where the waste is degraded to nonputrescible form by the application and maintenance of a source of heat. With no significant heating value, incineration would not be a practical disposal method. Generally, a waste with a heating value less than 1000 Btu/lb as received, such as concrete blocks or stone, is not applicable for incineration. There are instances, however, where an essentially inert material has a relatively small content (or coating) of combustibles and incineration would be a viable option even with a small heating value. Two such cases are incineration of empty drums with a residual coating of organic material on their inner surfaces and incineration of grit from wastewater treatment plants. The grit adsorbs grease from within the wastewater flow which results in a slight heating value to the grit material, normally less than 500 Btu/lb.

Inorganic salts. Wastes rich in inorganic, alkaline salts are troublesome to dispose of in a conventional incineration system. A significant fraction of the salt can become airborne. It will collect on furnace surfaces, creating a slag, or cake, which severely reduces the ability of an incinerator to function properly.

High sulfur or halogen content. The presence of chlorides or sulfides in a waste will normally result in the generation of acid-forming compounds in the off-gas. The cost of protecting equipment from acid attack must be balanced against the cost of alternative disposal methods for the waste in question.

Radioactive waste. Incinerators have been developed specifically for the destruction of radioactive waste materials. Unless designed specifically for radioactive waste disposal, however, an incinerator should not be used for the firing of a radioactive waste.

Load Estimating. The quantity of solid waste generated in the United States, industrial and municipal, is approximately 300 million tons/year. Of this figure approximately 2000 lb of household refuse is produced per year per capita.

The estimation of incinerator loading, where the waste quantity is not known, usually requires a survey of the area in question including a study of past records, demographic trends, etc. Table 11.2 can be used as a guide in determining the solid waste generated from various sources.

Table 11.3 lists the average weight of various solid wastes, and Table 11.4 lists per capita waste generation in the United States.

Another major waste, sewage sludge, can be estimated to be generated at the rate of 0.2 lb/day of sludge solids per capita.

Estimating Solid Waste Quality. While a general figure for waste generation can be obtained as noted in the previous sections, a more accurate means of determining the quality of a solid waste stream is by use of Table 11.5 and/or Table 11.6. By a visual

TABLE 11.2   Incinerator Capacity Chart

Classification	Building types	Quantities of waste produced
Industrial buildings	Factories Warehouses	Survey must be made 2 lb/(100 ft^2 · day)
Commercial buildings	Office buildings Department stores Shopping centers Supermarkets Restaurants Drugstores Banks	1 lb/(100 ft^2 · day) 4 lb/(100 ft^2 · day) Study of plans or survey required 9 lb/(100 ft^2 · day) 2 lb per meal per day 5 lb/(100 ft^2 · day) Study of plans or survey required
Residential	Private homes Apartment buildings	5 lb basic & 1 lb per bedroom 4 lb per sleeping room per day
Schools	Grade schools High schools Universities	10 lb per room & ½ lb per pupil per day 8 lb per room & ½ lb per pupil per day Survey required
Institutions	Hospitals Nurses' or interns' homes Homes for aged Rest homes	15 lb per bed per day 3 lb per person per day 3 lb per person per day 3 lb per person per day
Hotels, etc.	Hotels—1st class Hotels—Medium class Motels Trailer camps	3 lb per room and 2 lb per meal per day 1½ lb per room & 1 lb per meal per day 2 lb per room per day 6 to 10 lb per trailer per day
Miscellaneous	Veterinary hospitals Industrial plants Municipalities	Study of plans or survey required

Do not estimate more than 7-h operation per shift of industrial installations.

Do not estimate more than 6-h operation per day for commercial buildings, institutions, and hotels.

Do not estimate more than 4-h operation per day for schools.

Do not estimate more than 3-h operation per day for apartment buildings.

Whenever possible an actual survey of the amount and nature of refuse to be burned should be carefully taken. The data herein are of value in estimating capacity of the incinerator where no survey is possible and also to double-check against an actual survey.

Source: Ref. 1.

TABLE 11.3   Average Weight of Solid Waste

Type	lb/ft^3
Type 0 waste	8 to 10
Type 1 waste	8 to 10
Type 2 waste	15 to 20
Type 3 waste	30 to 35
Type 4 waste	45 to 55
Garbage (70% H_2O)	40 to 45
Magazines and packaged paper	35 to 50
Loose paper	5 to 7
Scrap wood and sawdust	12 to 15
Wood shavings	6 to 8
Wood sawdust	10 to 12

Source: Ref. 1.

TABLE 11.4   Average Solid Waste Collected (lb per person per day)

Solid Wastes	Urban	Rural	National
Household	1.26	0.72	1.14
Commercial	0.46	0.11	0.38
Combined	2.63	2.60	2.63
Industrial	0.65	0.37	0.59
Demolition, construction	0.23	0.02	0.18
Street and alley	0.11	0.03	0.09
Miscellaneous	0.38	0.08	0.31
Totals	5.72	3.93	5.32

Source: Ref. 2.

TABLE 11.5   Typical Moisture Content of Municipal Solid Waste (MSW) Components

Component	Moisture, percent	
	Range	Typical
Food wastes	50–80	70
Paper	4–10	6
Cardboard	4–8	5
Plastics	1–4	2
Textiles	6–15	10
Rubber	1–4	2
Leather	8–12	10
Garden trimmings	30–80	60
Wood	15–40	20
Glass	1–4	2
Tin cans	2–4	3
Nonferrous metals	2–4	2
Ferrous metals	2–6	3
Dirt, ashes, brick, etc.	6–12	8
Municipal solid waste	15–40	20

Source: Ref. 3.

inspection of the waste, a percentage of each waste component as listed in these tables can be established. By multiplying the moisture percentage or heating value or density of each of these components by the indicated moisture, heating value, or density, a more accurate figure for the total waste quality can be estimated. (A more detailed analysis of heating value of wastes is included in this chapter.)

As an example, to estimate the heating value of a particular municipal solid waste, with the waste components as listed below, using the heating value listed in Table 11.6, the total waste heating value is calculated as follows:

Component	Solid wastes, %	Inherent energy, Btu/lb	Total energy contribution, Btu/lb
Food wastes	15	2,000	300
Paper	40	7,200	2880
Cardboard	5	7,000	350
Plastics	5	14,000	700
Wood	15	8,000	1200
Glass	10	60	6
Tin cans	10	300	30
Total	100		5466

Total energy content is therefore 5466 Btu/lb.

TABLE 11.6   Typical Heating Value of MSW Components

Component	Energy, Btu/lb	
	Range	Typical
Food wastes	1500–3000	2000
Paper	5000–8000	7200
Cardboard	6000–7500	7000
Plastics	12000–16000	14000
Textiles	6500–8000	7500
Rubber	9000–12000	10000
Leather	6500–8500	7500
Garden trimmings	1000–8000	2800
Wood	7500–8500	8000
Glass	50–100	60
Tin cans	100–500	300
Nonferrous metals	—	—
Ferrous metals	100–500	300
Dirt, ashes, brick, etc.	1000–5000	3000
Municipal solid wastes	4000–6500	4500

Source: Ref. 3.

Solid Waste Incineration.   Solid waste incinerators are usually categorized according to the nature of the material which they are designed to burn, i.e., refuse or industrial waste. However, more than one waste type can often be burned in a given unit.

Incinerators for destruction of solid waste are the most difficult class of incinerators to design and operate, primarily because of the nature of the waste material. Solid waste can vary widely in composition and physical characteristics, making the effects of feed rates and parameters of combustion very difficult to predict. Solid waste incinerators most often burn wastes over a range of low and high heat values, i.e., from wet garbage with an as-received heat value as low as 2500 Btu/lb, to plastic wastes, over 19,000 Btu/lb. Materials handling, firing, and residue removal equipment are more critical, cumbersome, expensive, and difficult to control with these than with other types of incinerators.

Types of Solid Waste Incinerators.   Waste incineration includes the following techniques:

1. Open burning
2. Single-chamber incinerators
3. Teepee burners
4. Open-pit incinerators
5. Multiple-chamber incinerators
6. Controlled air incinerators
7. Central-station disposal
8. Rotary kiln incinerators

Open Burning.   Open burning is the oldest technique for incineration of wastes. Basically it consists of placing or piling waste materials on the ground and burning them without the aid of specialty combustion equipment.

This type of system is found in most parts of the United States. It results in excessive smoking and high particulate emission, and it presents a fire hazard.

Open burning has been utilized to dispose of high-energy explosives such as dynamite or TNT. For proper incineration, the waste is placed on a refractory pad which is in turn placed over gravel, in a cleared location, remote from populated areas.

Single-Chamber Incinerators.   Single-chamber incinerators will, in general, not meet the air pollution emission standards that have been developed over the past 10 to 15 years. A typical single-chamber incinerator is shown in Fig. 11.1. Solid waste is placed on the grate and fired. These incinerators have also been manufactured in top-loading (flue loading) configuration for apartment house waste disposal, firing waste in 55-gal drums or wire baskets or in a concrete or refractory-lined structure with a cast-iron grate, etc. This equipment may or may not have a firing system to ignite the waste. As with open burning, smoking and excessive air pollution emissions can occur.

Attempts have been made to control emissions to reasonable levels by the addition of an afterburner. Normally a temperature of 1400°F is required, at a retention time of 0.5 s, and the afterburner is used to obtain these combustion parameters in the exiting off-gas.

A *jug incinerator* is another type of single-chamber unit. A typical jug incinerator is shown in Fig. 11.2. This is a specialty incinerator used for the destruction of cotton waste and other waste agricultural products. It is a brick-lined vertical cylindrical or conical structure. Waste is fed through the top section of the incinerator and falls to its floor, which may or may not be provided with grates. Waste is pneumatically conveyed to the incinerator charging system, and the transfer air is the only combustion air supplied to the incinerator. Afterburners are provided in the stack to control air emissions, although many such incinerators discharge from their conical top, without provision of a stack or afterburning equipment.

Open-Pit Incinerators.   Open-pit incinerators have been developed for controlled incineration of explosive wastes, wastes which would create an explosion hazard or high heat release in a conventional, enclosed incinerator. They are constructed as shown in Fig. 11.3 with an open top and a number of closely spaced nozzles blowing air from the open top down into the incinerator chamber. Air is blown at high velocity, creating a rolling action, i.e., a high degree of turbulence. Burning rates within the incinerator provide temperature in excess of 2000°F with low smoke and relatively low particulate emissions discharges.

Incinerators of this type may be built either above or below ground. They are constructed with refractory walls and floor or as earthen trenches. The width of an open-

ON-SITE SOLID WASTE DISPOSAL

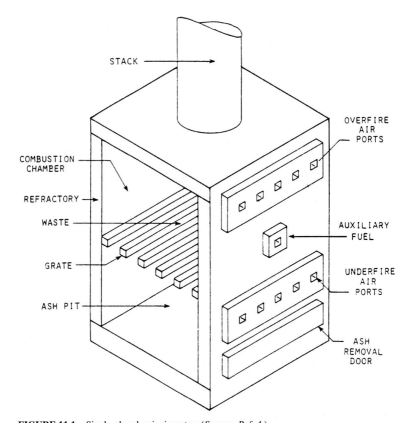

FIGURE 11.1   Single-chamber incinerator. (*Source: Ref. 4.*)

pit incinerator is normally on the order of 8 ft, with a depth of approximately 10 ft. The length varies from 8 to 16 ft.

Overfire air nozzles are 2 to 3 in in diameter, located above one edge of the pit. They fire down at an angle of 25° to 35° from the horizontal. The incinerator is normally charged from a top-loading ramp on the edge opposite the air nozzles. Some units have a mesh placed on their top to contain larger particles of fly ash. Residue cleanout doors are often provided on aboveground incinerators.

For a waste with a heating value of 5000 Btu/lb note the following typical parameters of design for open-pit incinerators:

- Heat release of 3.4 MBtu/h per foot of length.
- Provision of 100 to 300 percent excess air.
- Overfire air of 850 st ft³/min per foot of pit length at 11 in water column (WC).

Particulate emissions are normally below 0.25 gr/dry st ft³, corrected to 12 percent CO_2, which is unacceptable with regard to current air pollution control standards.

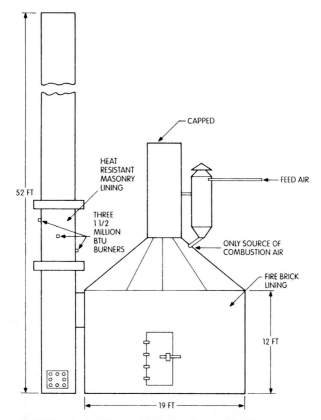

FIGURE 11.2 Modified jug incinerator. (*Source: Ref. 4.*)

(Most current statutes limit air pollution emissions from burning refuse to 0.08 gr/dry st ft^3 corrected to 12 percent CO_2.) Other than combustion control by control of over-fire air, there is no mechanism practicable for control of exhaust emissions. This incinerator, while effective in destruction of some waste, cannot normally be used without relaxation of local air pollution emission requirements.

Multiple-Chamber Incinerators.   In an attempt to provide complete burnout of combustion products and decrease the airborne particulate loading in the exiting flue gas, multiple-chamber incinerators have been developed. A first, or primary, chamber is used for combustion of solid waste. The secondary chamber provides the residence time, and supplementary fuel, for combustion of the unburned gaseous products and airborne combustible solids (soot) discharged from the primary chamber. There are two basic types of multiple-chamber incinerators: the retort and the in-line systems.

   Retort Incinerator.   This unit is a compact cubic-type incinerator with multiple internal baffles. The baffles are positioned to guide the combustion gases through 90° turns in both lateral (horizontal) and vertical directions. At each turn ash drops out of the flue gas flow. The primary chamber has elevated grates for discharge of waste and an ash pit for collection of ash residual. A cutaway view of a typical retort-type incinerator is shown in Fig. 11.4. Figure 11.5 gives dimensional data for typical retort units.

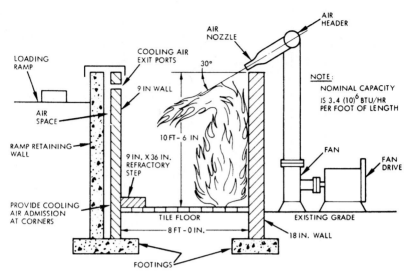

FIGURE 11.3   Open-pit incinerator.

Overfire air and underfire air are provided above and below the primary chamber grate. This air is normally supplied by forced-air fans at a controlled rate. Flue gas exits the primary chamber through an opening, termed a *flame port,* which discharges to the secondary chamber or to a smaller *mixing chamber* immediately before the secondary chamber. The flame port is actually an opening atop the bridge wall, separating the primary from the secondary chamber.

Air ports are provided in the secondary combustion chamber and, when present, in the mixing chamber. Supplemental fuel is provided in the secondary and primary chambers. Depending on the nature of waste charged, the fuel supply in the primary chamber may be unnecessary after start-up, i.e., after bringing the chamber temperature to a level high enough for the waste to ignite and sustain its own combustion. The secondary chamber normally requires a continuous supplemental fuel supply.

As the flue gas enters and exits the secondary combustion chamber, larger airborne particles settle out of the gas stream. Temperatures in the secondary chamber are high enough (in the range of 1400°F for refuse and other carbonaceous waste) to destroy unburned airborne particles. This equipment therefore has relatively low particulate emissions and in many cases can meet an emission standard of 0.08 gr/dry st ft^3 corrected to 12 percent CO_2, without additional air pollution control equipment.

In-Line Incinerator.   This is a larger unit than the retort incinerator. Flow of combustion gases is straight through the incinerator, axially, with abrupt changes in the direction of flow only in the vertical direction, as shown in Fig. 11.6, an in-line incinerator using natural gas as supplemental fuel. Waste is charged on the grate, which can be stationary or moving. A moving grate lends itself to continuous burning whereas stationary grates, as with the retort incinerator, are used for batch or semicontinuous operation. In-line incinerators are often provided with automatic ash removal equipment or ash discharge conveyers which also contribute to continuous operation of the incinerator.

As with the retort type, changes in the flow path and flow restrictions in an in-line incinerator provide settling out of larger airborne particles and increase turbulence for more effective burning. Supplemental fuel burners in the primary chamber ignite the

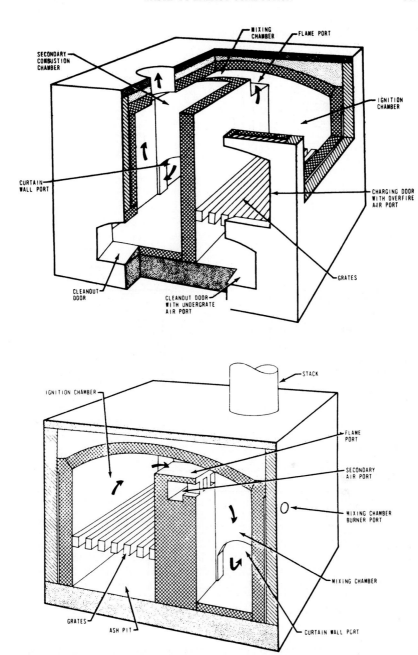

FIGURE 11.4 Cutaway of a retort multiple-chamber incinerator. (*Source: Ref. 5.*)

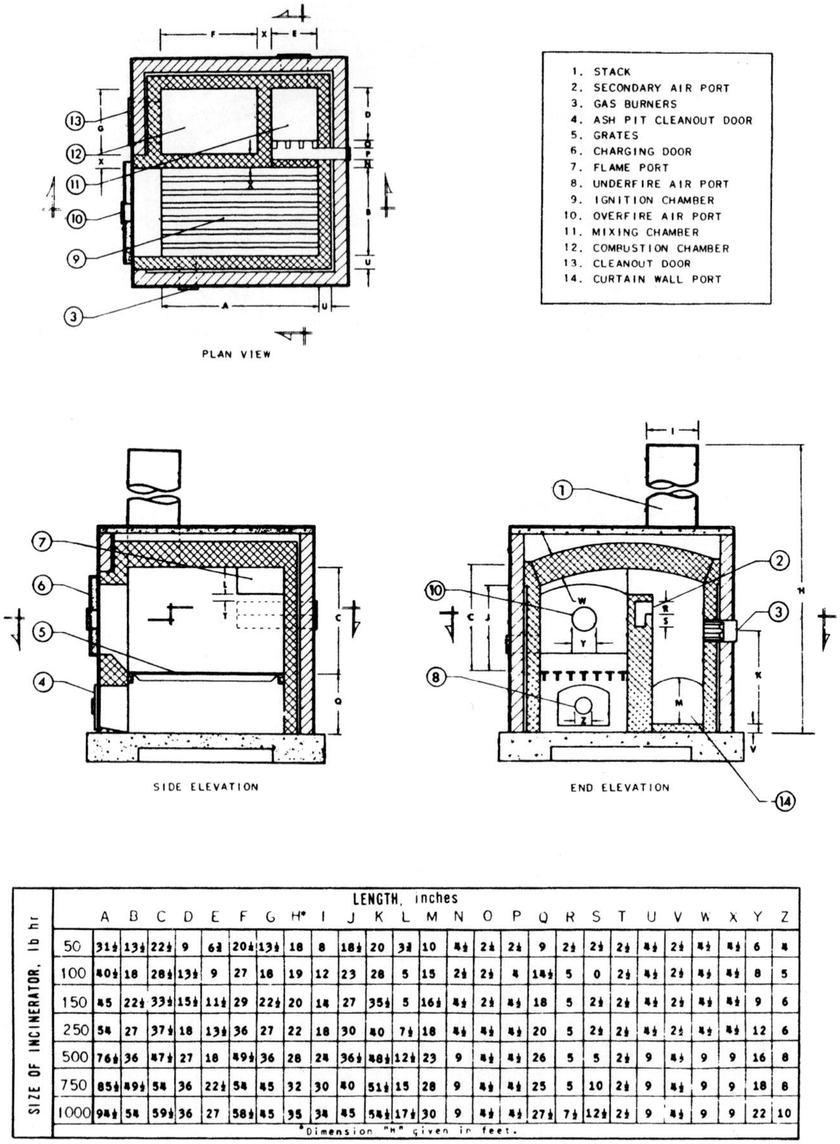

PLAN VIEW
SIDE ELEVATION
END ELEVATION
1. STACK
2. SECONDARY AIR PORT
3. GAS BURNERS
4. ASH PIT CLEANOUT DOOR
5. GRATES
6. CHARGING DOOR
7. FLAME PORT
8. UNDERFIRE AIR PORT
9. IGNITION CHAMBER
10. OVERFIRE AIR PORT
11. MIXING CHAMBER
12. COMBUSTION CHAMBER
13. CLEANOUT DOOR
14. CURTAIN WALL PORT

SIZE OF INCINERATOR, lb hr	LENGTH, inches																									
	A	B	C	D	E	F	G	H*	I	J	K	L	M	N	O	P	Q	R	S	T	U	V	W	X	Y	Z
50	31½	13½	22½	9	6¾	20¼	13½	18	8	18½	20	3¾	10	4½	2¼	2¼	9	2½	2½	2½	4½	2½	4½	4½	6	4
100	40½	18	28½	13½	9	27	18	19	12	23	28	5	15	2½	2½	4	14½	5	0	2½	4½	2½	4½	4½	8	5
150	45	22½	33½	15½	11½	29	22½	20	14	27	35½	5	16½	4½	2½	4½	18	5	2½	2½	4½	2½	4½	4½	9	6
250	54	27	37½	18	13½	36	27	22	18	30	40	7½	18	4½	4½	4½	20	5	2½	2½	4½	2½	4½	4½	12	6
500	76½	36	47½	27	18	49½	36	28	24	36½	48½	12½	23	9	4½	4½	26	5	5	2½	9	4½	9	9	16	8
750	85½	49½	54	36	22½	54	45	32	30	40	51½	15	28	9	4½	4½	25	5	10	2½	9	4½	9	9	18	8
1000	94½	54	59½	36	27	58½	45	35	34	45	54½	17½	30	9	4½	4½	27½	7½	12½	2½	9	4½	9	9	22	10

*Dimension "H" given in feet.

FIGURE 11.5 Design standards for multiple-chamber retort incinerators. (*Source: Ref. 5.*)

waste whereas secondary-chamber supplementary fuel burners provide heat to maintain complete combustion of the burnable components of the exhaust gas.

The retort incinerator is used in the range of 20 lb/h to approximately 750 lb/h. In-line incinerators are normally provided in the range of 500 to 2000 lb/h and greater with automatic charging and/or ash removal equipment not ususally provided for units smaller than 1000 lb/h in capacity. Figure 11.6 shows an in-line incinerator with mov-

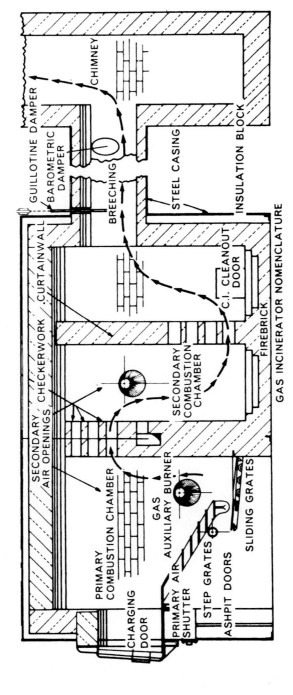

FIGURE 11.6 Schematic diagram of a gas incinerator. (*Source: Incinerator Committee, Industrial & Commercial Gas Section, American Gas Association, New York.*)

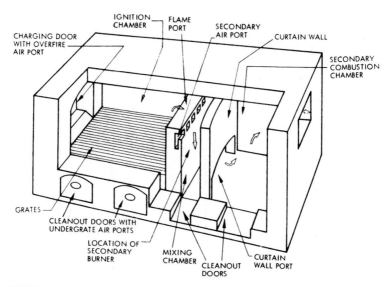

FIGURE 11.7 Cutaway of an in-line multiple-chamber incinerator. (*Source: Ref. 6.*)

ing grates for charging and ash disposal. Figure 11.7 is another type of in-line inciner-ator utilizing manual charging, i.e., fixed grates. Typical in-line unit dimensions are shown in Fig. 11.8.

Combustion air requirements are the same for either of these incinerators: approxi-mately 300 percent excess air. Approximately half the required air enters as leakage through the charging port and other areas of the incinerator. Of the remaining air requirement, 70 percent should be provided in the primary combustion chamber as overfire air, 10 percent as underfire air, and 20 percent in the mixing chamber or in the secondary combustion chamber.

Multiple-chamber incinerators will produce significantly lower emissions than sin-gle-chamber incinerators, as illustrated in Table 11.7. Water curtains across the path of the flue gases exiting the secondary combustion chamber will decrease emissions even further. Multiple-chamber incinerators have been designed for specialty wastes. Typical are pathological waste incinerators such as that shown in Fig. 11.9. Table 11.8 lists chemical composition and combustion data for pathological waste. Design factors and gas velocities for pathological waste incinerators are listed in Tables 11.9 and 11.10, respectively. Air emissions from pathological incinerators based on two test runs with and two runs without afterburner firing are listed in Tables 11.11 and 11.12, respective-ly. Note the significant decrease in emissions with the afterburner in operation.

A crematory retort is shown in Fig. 11.10. Its operating parameters for typical cre-matory waste are listed in Table 11.13.

Rotary Kiln Technology

The rotary kiln incinerator is the most universal of thermal waste disposal systems. It can be used for the disposal of a wide variety of solid and sludge wastes and for the incineration of liquid and gaseous waste. The rotary kiln system has found application in both municipal and industrial waste incineration.

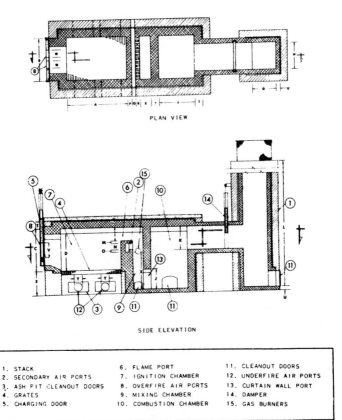

PLAN VIEW

SIDE ELEVATION

1. STACK	6. FLAME PORT	11. CLEANOUT DOORS
2. SECONDARY AIR PORTS	7. IGNITION CHAMBER	12. UNDERFIRE AIR PORTS
3. ASH PIT CLEANOUT DOORS	8. OVERFIRE AIR PORTS	13. CURTAIN WALL PORT
4. GRATES	9. MIXING CHAMBER	14. DAMPER
5. CHARGING DOOR	10. COMBUSTION CHAMBER	15. GAS BURNERS

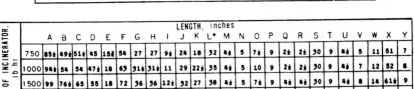

SIZE OF INCINERATOR, lb hr	LENGTH, inches																								
	A	B	C	D	E	F	G	H	I	J	K	L*	M	N	O	P	Q	R	S	T	U	V	W	X	Y
750	85½	49½	51½	45	15½	54	27	27	9½	24	18	32	4½	5	7½	9	2½	2½	30	9	4½	5	11	51	7
1000	94½	54	54	47½	18	63	31½	31½	11	29	22½	35	4½	5	10	9	2½	2½	30	9	4½	7	12	52	8
1500	99	76½	65	55	18	72	36	36	12½	32	27	38	4½	5	7½	9	4½	4½	30	9	4½	8	14	61½	9
2000	108	90	69½	57½	22½	79½	40½	40½	15	36	31½	40	4½	5	10	9	4½	4½	30	9	4½	9	15	63½	10

*Dimension "L" given in feet.

FIGURE 11.8   Design standards for multiple-chamber, in-line incinerators. (*Source: Ref. 7.*)

Kiln System.   A rotary kiln system used for waste incineration is shown in Fig. 11.11. It includes provisions for feeding, air injection, the kiln itself, an afterburner, and an ash collection system. The gas discharge from the afterburner is directed to an air emissions control system. An induced-draft (ID) or exhaust fan is provided within the emission control system to draw gases from the kiln through the equipment line and discharges through a stack to the atmosphere.

As shown in Fig. 11.11, a rotary kiln system may include a waste heat boiler between the afterburner and the scrubber for energy recovery. The waste heat boiler reduces the temperature of the gas stream sufficiently to allow the use of a fabric filter, or baghouse, for particulate control. The scrubber in this illustration utilizes water

TABLE 11.7 Comparison between Amounts of Emissions from Single- and Multiple-Chamber Incinerators

Item	Multiple chamber	Single chamber
Particulate matter, gr/st ft^3 at 12% CO_2	0.11	0.9
Volatile matter, gr/st ft^3 at 12% CO_2	0.07	0.5
Total, gr/st ft^3 at 12% CO_2	0.18	1.4
Total, lb/ton refuse burned	3.50	23.8
Carbon monoxide, lb/ton of refuse burned	2.90	197–991
Ammonia, lb/ton of refuse burned	0	0.9–4
Organic acid (acetic), lb/ton of refuse burned	0.22	<3
Aldehydes (formaldehyde), lb/ton of refuse burned	0.22	5–64
Nitrogen oxides, lb/ton of refuse burned	2.50	1
Hydrocarbons (hexane), lb/ton of refuse burned	<1	—

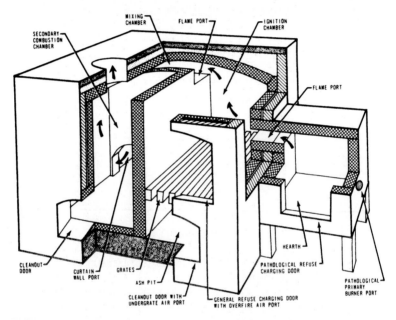

FIGURE 11.9 Multiple-chamber incinerator with a pathological waste retort. (*Source: Ref. 5.*)

and alkali injection. It is used for acid gas control. Dry scrubbing may be used in lieu of the wet scrubbing system shown.

There are a number of areas within the kiln system where leakage can occur, as can be seen in each of the figures cited above. The feeding ports cannot be completely sealed, and the kiln seals are areas of potential leakage. The ash system is normally provided with a water seal, but for dry ash collection there will usually be some leakage. To ensure that the leakage is into the system, that no hot, dirty gases leak out of the kiln to the surrounding areas, the kiln is maintained with a negative draft. The ID fan is sized to maintain a negative pressure throughout the system so that leakage is always into, not out of, the kiln system.

TABLE 11.8 Chemical Composition of Pathological Waste and Combustion Data

	Ultimate analysis (whole dead animal)	
	As charged	Ash-free combustible
Constituent	% by weight	% by weight
Carbon	14.7	50.80
Hydrogen	2.7	9.35
Oxygen	11.5	39.85
Water	62.1	–
Nitrogen	Trace	–
Mineral (ash)	9	–

Dry combustible empirical formula - $C_5H_{10}O_3$

	Combustion data (based on 1 lb of dry ash-free combustible)	
	Quantity	Volume
Constituent	lb	scf
Theoretical air	7.028	92.40
40% sat at 60°F	7.069	93
Flue gas with theoretical air 40% saturated CO_2	1.858	16.06
N_2	5.402	73.24
H_2O formed	0.763	15.99
H_2O air	0.031	0.63
Products of combustion total	8.054	105.92
Gross heat of combustion	8,820 Btu per lb	

Source: Ref. 7.

TABLE 11.9 Design Factors for Pathological Ignition Chamber (Incinerator Cavity, 25 to 250 lb/h)

Item	Recommended value	Allowable deviation, %
Hearth loading	10 lb/(h · ft²)	±10
Hearth length-to-width ratio	2	±20
Primary burner design	$\dfrac{10 \text{ ft}^3 \text{ natural gas}}{\text{lb waste burned}}$	±10

Source: Ref. 7.

TABLE 11.10 Gas Velocities and Draft (Pathological Incinerators with Hot-Gas Passage below a Solid Hearth)

Item	Recommended values	Allowable deviation, %
Gas velocities		
Flame port at 1600°F, ft/s	20	±20
Mixing chamber at 1600°F, ft/s	20	±20
Port at bottom of mixing chamber at 1550°F, ft/s	20	±20
Chamber below hearth at 1500°F, ft/s	10	±100
Port at bottom of combustion chamber at 1500°F, ft/s	20	±20
Combustion chamber at 1400°F, ft/s	5	±100
Stack at 1400°F, ft/s	20	±25
Draft		
Combustion chamber, in WC	0.25[a]	$\begin{cases} -0 \\ +25 \end{cases}$
Ignition chamber, in WC	0.05–0.10	± 0

[a]Draft can be 0.20 in WC for incinerators with a cold hearth.
Source: Ref. 7.

TABLE 11.11 Emissions from Two Pathological-Waste Incinerators with Secondary Burners

Test no.	A	B
Rate of destruction to powdery ash, lb/h	Mixing chamber burner operating	Mixing chamber burner operating
	19.2	99
Type of waste	Placental tissue in newspaper at 40°F	Dogs freshly killed
Combustion contaminants		
gr/st ft³ [a] at 12% CO_2	0.200	0.300
gr/st ft³	0.014	0.936
lb/h	0.030	0.360
lb/ton charged	3.120	7.260
Organic acids		
gr/st ft³	0.006	0.013
lb/h	0.010	0.050
lb/ton charged	1.040	1.010
Aldehydes		
gr/st ft³	N.A.[b]	0.006
lb/h	N.A.[b]	0.020
lb/ton charged	N.A.[b]	0.400
Nitrogen oxides		
ppm	42.70	131
lb/h	0.08	0.099
lb/ton charged	8.84	2
Hydrocarbons	Nil	Nil

[a]CO_2 from burning of waste used only to convert to basis of 12% CO_2.
[b]Not available.
Source: Ref. 7, p. 462.

TABLE 11.12 Emissions from Two Pathological-Waste Incinerators without Secondary Burners (Source Tests of Two Pathological-Waste Incinerators)

Test no.	A	B
	Mixing chamber burner not operating	Mixing chamber burner not operating
Rate of destruction to powdery ash, lb/h	26.4	107
Type of waste	Placental tissue in newspaper at 40°F	Dogs freshly killed
Combustion contaminants		
gr/st ft^3 [a] at 12% CO_2	0.500	0.300
gr/st ft^3	0.017	0.128
lb/h	0.030	0.430
lb/ton charged	2.270	8.040
Organic acids		
gr/st ft^3	0.010	0.034
lb/h	0.020	0.110
lb/ton charged	1.514	2.050
Aldehydes		
gr/st ft^3	0.007	0.010
lb/h	0.013	0.033
lb/ton charged	0.985	0.617
Nitrogen oxides		
ppm	14.700	95
lb/h	0.016	0.082
lb/ton charged	1.210	1.550
Hydrocarbons	Nil	Nil

[a]CO_2 from burning of waste only used to convert to basis of 12% CO_2.
Source: Ref. 7

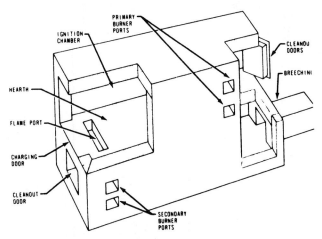

FIGURE 11.10 Crematory retort. (*Source: Ref. 7.*)

TABLE 11.13 Operating Procedures for Crematory

Phase	Duration, 1½ h operation, min	Burner settings	Casket	Body Moisture	Tissue	Bone calcin
Charging[a]	—	Secondary zone on				
Ignition	15	All on	20% burns	—	10% burns	—
Full combustion	30	All on	80% burns	20% evap.	90% burns	—
Final combustion	45	All on	—	80% evap.	—	50%
Calcining	1 to 12 h	All off (or small primary on)	—	—	—	50%

Phase	Duration, 2½ h operation, min	Burner settings	Casket	Moisture	Tissue	Bone calcin
Ignition	15	All on	20% burns	—	—	—
Full combustion	30	Primary off	60% burns	20% evap.	—	—
Final combustion	15	All on	20% burns	20% evap.	20% burns	50%
	90	All on	—	60% evap.	80% burns	—
Calcining	1 to 12 h	All off (small primary may be on)				

[a]Charge Casket 75 lb wood
     Body 180 lb
     Moisture: 108 lb
     Tissue: 50 lb
     Bone: 22 lb

Source: Ref. 7.

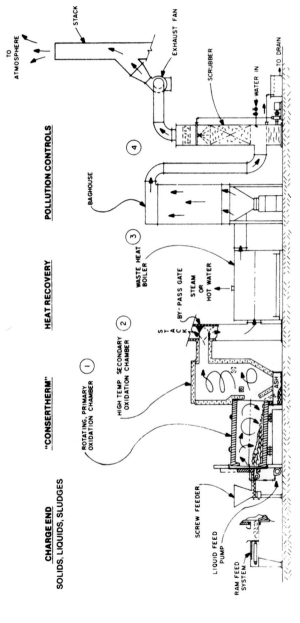

FIGURE 11.11 Rotary kiln with waste heat boiler. (*Source: R. Rayve, Consertherm, East Hartford, Conn.*)

11.23

The Primary Combustion Chamber (Rotary Kiln).   The conventional rotary kiln is a horizontal cylinder, lined with refractory, which turns about its horizontal axis. Waste is deposited in the kiln at one end, and the waste burns out to an ash by the time it reaches the other end. Kiln rotational speed is variable, in the range of $\frac{3}{4}$ to $2\frac{1}{2}$ r/min. The ratio of length to diameter of a kiln used for waste disposal is normally in the range of 2:1 to 5:1.

Most kiln designs utilize smooth refractory on the kiln interior. Some designs, particularly those for the processing of granular material (dirt or powders), may have internal vanes or paddles to encourage motion along the kiln length and to promote turbulence of the feed. Care must be taken in the provision of internal baffles of any kind. With certain material consistencies, such as soil of from 10 to 20 percent moisture content, baffles may tend to retard the movement of material through the kiln.

The kiln is supported by at least two trunnions. One or more sets of trunnion rollers are idlers. Kiln rotation can be achieved by a set of powered trunnion rollers, by a gear drive around the kiln periphery, or through a chain driving a large sprocket around the body of the kiln. The kiln trunnion supports are adjustable in the vertical direction. The kiln is normally supported at an angle to the horizontal, or rake. The rake will normally vary from 2 to 4 percent ($\frac{1}{4}$ to $\frac{1}{2}$ in/ft of length), with the higher end at the feed end of the kiln. Other kiln designs have a zero or slightly negative rake, with lips at the input and discharge ends. These kilns are operated in the slagging mode, as discussed subsequently, with the internal kiln geometry designed to maintain a pool of molten slag between the kiln lips.

A source of heat is required to bring the kiln up to operating temperature and to maintain its temperature during incineration of the waste feed. Supplemental fuel is normally injected into the kiln through a conventional burner or a ring burner when gas fuel is used.

There are a number of variations in kiln design, including the following:

- Parallel flow or counterflow
- Slagging or nonslagging mode.
- Refractory or bare wall

The more commonly used kiln design, referred to as the *conventional kiln,* is a parallel-flow system, nonslagging, lined with refractory.

Kiln Exhaust Gas Flow.   When gas flow through the kiln is in the same direction as the waste flow, the kiln is said to have *parallel* or *cocurrent flow.* With countercurrent flow, the gas flows opposite to the flow of waste. The burner(s) is(are) placed at the front of the kiln, the face of the kiln from which air or gas originates.

Generally, a countercurrent kiln is used when an aqueous waste, one with at least 30 percent water content, is to be incinerated. Waste is introduced at the end of the kiln far from the burner. The gases exiting the kiln will dry the aqueous waste, and its temperature will drop. If aqueous waste were dropped into a kiln with cocurrent flow, water would be evaporated at the feed end of the kiln. The feed end would be the end of the kiln at the lowest temperature, and a much longer kiln would be required for burnout of the waste.

Wastes with a light volatile fraction (containing greases, for instance) should utilize a kiln with cocurrent flow. These volatiles will likely be released from the feed immediately upon entering the kiln. Use of a cocurrent kiln provides a higher residence time than use of a countercurrent kiln for the effective burnout of these volatiles.

TABLE 11.14　Slagging versus Nonslagging Kiln

Factor	Effect
Construction	More complex with slagging kiln
Duty	Slagging kiln can accept drums, salt-laden wastes; nonslagging kiln is limited
Temperature	Higher with slagging kiln
Retention time	Greater residence required in nonslagging kiln
Process control	Thermal inertia or forgiveness in slagging kiln
Emissions	Less particulate, greater NO_x in slagging kiln
Slag	Slagging kiln may require CaO, Al_2O_3, SiO_2 additives; dissolves drums, salts
Ash	Wet, less leachable with slagging; wet or dry with nonslagging kiln
Maintenance	Higher with slagging kiln
Refractory	More critical with slagging kiln

Slagging Mode.　At temperatures in the range of 2000 to 2200°F, ash will start to deform for many waste streams; and as the temperature increases, the ash will fluidize. The actual temperatures of initial deformation and subsequent physical changes to the ash are a function of the chemical constituents present in the waste residual. They are also a function of the presence of oxygen in the furnace. The ash deformation temperatures will vary with reducing vs. oxidation atmospheres, as noted in Table 11.18, Ash Fusion Temperatures, which lists deformation temperatures for coal and a typical refuse mix. Eutectic properties can be controlled by the use of additives to the molten material.

A kiln can be designed to generate and maintain molten ash during operation. Operation in a slagging mode provides a number of advantages over nonslagging operation. When a kiln is operating in a nonslagging mode, however, and slagging occurs, slagging is undesirable and must be eliminated.

Differences in slagging vs. nonslagging kilns are outlined in Table 11.14. As noted, the construction of a slagging kiln is more complex than that of a nonslagging kiln, requiring provision of a lip at the kiln exit to contain the molten material. A nonslagging kiln will normally have a smooth transition with no impediments to the smooth discharge of ash.

Slagging kilns have been designed and operated with a negative rake; e.g., the outer surface of the kiln at the feed end is lower than the kiln surface at the discharge end. This will permit the accumulation of more slag in the kiln than with zero or positive rake. The kiln internal surface must be designed for this operating mode. For instance, as noted previously, an internal refractory lip is required on the kiln feed end.

The slagging kiln can accept metal drums. The ash eutectic properties at the molten slag temperatures will tend to dissolve a ferrous metal drum placed in the kiln. The placing of drums containing waste in a kiln may be undesirable from a safety and maintenance standpoint (even with the tops of the drums removed, localized heating of the drum surface may occur, causing an explosion, and the impact of a dropping drum will eventually damage kiln refractory). However, if drums are to be placed in a kiln (they should be quartered), slagging kilns are able to absorb the drum into a homogeneous residue discharge. The nonslagging kiln can only move the drum through the unit and must include specialty equipment for handling the drum body as it exits the kiln.

Salt-laden wastes will tend to melt in the range of 1300 to 1600°F and can produce severe caking, or deposits, in a nonslagging kiln. Often salt-bearing wastes are prohibited from kilns because they will produce an unacceptable buildup on the kiln surface which can eventually choke off the kiln. In a slagging kiln, however, the temperature is kept high enough to keep the salts in a molten state. The salts combine with the molten ash in the pool at the bottom of the kiln and are maintained in their molten state until quenched. The temperature in a slagging kiln must be sufficiently high to maintain the ash as a molten slag. Temperatures as high as 2600 to 2800°F are not uncommon. A nonslagging kiln will normally operate at temperatures below 2000°F.

The destruction of organic compounds is achieved by a combination of high temperature and residence time. Generally, the higher the temperature, the shorter the residence time required for destruction. Conversely, the higher the residence time, the lower the required temperature. The use of higher temperatures in the slagging kiln reduces the residence time requirements for the off-gas. The afterburner associated with a slagging kiln can often be smaller than that required for a nonslagging kiln.

The molten slag can weigh hundreds or even thousands of pounds. As a concentrated material, a liquid, it represents a significant thermal inertia within the kiln. The molten slag tends to act as a heat sink which provides thermal stability to the system. The slagging kiln is much less subject to temperature extremes than the nonslagging kiln because of the presence of this massive melt. It will maintain a relatively constant-temperature profile under rapid changes in kiln loading. This stability leads to more predictable system behavior. Safety factors employed in the design and operation of downstream equipment (such as an exhaust gas scrubber or the induced-draft fan) can be reduced when a slagging kiln is used.

The tumbling action of a rotating kiln encourages the release of particulate to the gas stream. From 5 to 25 percent of the nonvolatile solids in a feed stream may become airborne with the use of a conventional nonslagging kiln. The presence of the molten slag in a slagging kiln acts much like the fluid ash in a pulverized-coal (PC) burner. The slag will absorb particulate matter from the gas stream and can reduce particulate emissions from the kiln to 25 to 75 percent of the emissions from a nonslagging kiln. However, emissions of NO_x are greater with a slagging than with a nonslagging kiln. The generation of NO_x is generally not significant until the temperature of the process increases above 2000°F. Above this temperature the formation of NO_x will increase substantially. At 2600°F the generation of NO_x is almost 10 times as great as at 1800°F.

A danger in slagging kiln operation is that the melt will solidify. When this happens the kiln will be off-balance. With an eccentric-turning kiln, if rotation of the kiln is not stopped, damage to kiln supports and to the kiln drive may occur. In addition, the incineration process will degrade under a melt freeze. Operating stability will be lost, and demands on downstream equipment (the gas scrubbing system, for instance) may be too severe. One reason for the loss of a molten slag, besides a drop in temperature, is a change in the feed quality. To ensure the maintenance of an adequate melt, additives may have to be employed. These additives may include CaO, Al_2O_3, SiO_2, or another compound or set of compounds, depending on the nature of the waste. Additives will help maintain the eutectic, to ensure that the melt will remain in a molten state.

The molten slag from a slagging kiln is dropped into a wet sump. (The hot slag can "pop" or explode as it contacts the cooling water in the sump.) The slag immediately hardens into a granular material (termed *frit*) with the appearance of gravel or dark glass. The ash from a nonslagging kiln can be collected wet or dry.

Refractory for a slagging kiln will experience more severe duty than that for non-slagging kiln service. The higher operating temperatures will directly affect refractory life, as will the corrosive effect of the melt. In addition, if steel drums are dropped into the kiln, the physical impact of the drum on the kiln surface will be damaging. The molten slag will absorb the steel and ferrous metals, as well as other metals, which are highly corrosive to the refractory. The refractory must resist this corrosive attack, high temperatures, and impact loading. The resulting refractory system will be expensive and will require frequent maintenance.

Operation. The waste retention time in a kiln can be varied. It is a function of kiln geometry and kiln speed, as shown in the following equation:

$$t = \frac{2.28\ L/D}{SN}$$

where

t = mean residence time, min
L/D = internal length-to-diameter ratio
S = kiln rake slope, in/ft of length
N = rotational speed, r/min

For a given L/D ratio and rake, the solids residence time within the unit is inversely proportional to the kiln speed. By doubling the speed, the residence time will halve. An example of this calculation is as follows:

Calculating the residence time for a kiln rotating at $N = 0.75$ r/min with a 1 percent slope ($S = 0.12$ in/ft of length), with a 4-ft inside diameter and 12-ft length L, we get

$$t = \frac{2.28(12/4)}{0.12 \times 0.75} = 76\ \text{min}$$

By inspection note that a doubling of rotation N would halve the retention time, and halving the rake S would double the retention time.

The above calculation was for the residence time of solids or other materials within the kiln, not the kiln exhaust gas. The off-gas residence time can be determined by the application of the heat balance and flue gas analyses developed later in this text.

Kiln Seal. Sealing a kiln is a difficult task. Efficient kiln operation requires that kiln seals be provided and maintained to control the infiltration of unwanted airflow into the system. With too much air, fuel usage increases and process control deteriorates.

The kiln turns between two stationary yokes. The kiln diameter, which can vary from 4 to 20 ft, will have a periphery of from 12 to 60 ft. At 1 ft/min velocity, the kiln surface is moving at a rate of up to 60 ft/min. A seal must close this gap between the yoke and the kiln surface while the kiln is moving at this surface velocity. The kiln surface is not a machined surface and will have variations in texture and dimension, making the task of sealing very difficult. A further problem is that the kiln interior is normally at relatively high temperatures, which tend to encourage wear of the kiln surface.

Two types of seals are illustrated in Fig. 11.12. The rotating portion of this seal, a T-ring in this illustration, is mounted on the kiln surface. There are as many variations in kiln seal designs as there are kiln and kiln seal manufacturers.

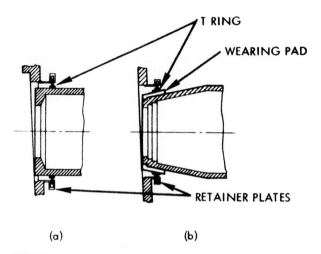

FIGURE 11.12   Kiln seal arrangements. (*a*) Single floating-type feed-end air seal. (*b*) Single floating-type air seal on air-cooled tapered feed end. (*Source: Ref. 8.*)

Design Variations.   In an effort to control the air distribution and temperature profile along the length of the kiln, a rotary kiln was developed with air injection ports. This kiln, developed by Universal Energy International, Inc., has a combustion air plenum inserted high throughout its length. Combustion air will cool the plenum as well as provide air for feed volatiles. The airflow within the plenum can be directed to any of a number of zones within the kiln. The control of air has been found to allow low- or substoichiometric operation in portions of the kiln or throughout the entire length of the kiln.

The rotary kiln system illustrated in Fig. 11.13 has been developed for municipal solid waste application. The kiln, or rotary combustor, has no refractory. Without refractory, it is believed that the system maintenance cost is reduced. It is constructed of water tubes which absorb from 25 to 35 percent of the heat generated by the burning waste. Burning begins in the kiln, with air injected through openings in the tube wall construction. Burning of the off-gas is completed within the boiler, which is also constructed of water tubes, with no refractory. A rotary joint in the kiln hot-water circulation system maintains a water seal under the physical motion of the tubes and the relative high-pressure demands of the hot fluid.

Pyrolysis and Controlled Air Incineration

General Description.   Pyrolysis is the destructive distillation of a solid, carbonaceous, material in the presence of heat and in the absence of stoichiometric oxygen. It is an exothermic reaction; i.e., heat must be applied for the reaction to occur.

Ideally a pyrolytic reaction will occur as follows, using cellulose:

$$C_6H_{10}O_5 \xrightarrow{\text{heat}} CH_4 + 2CO + 3H_2O + 3C$$

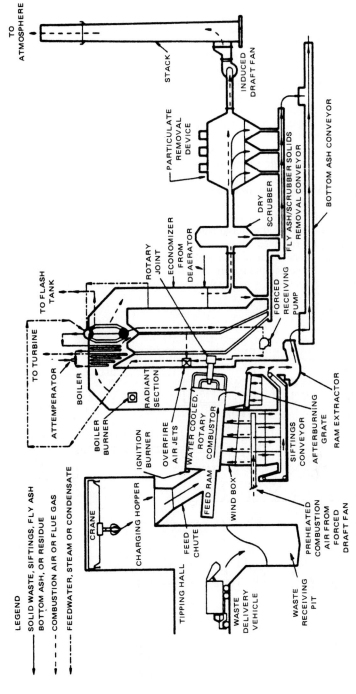

FIGURE 11.13 Rotary combustor. (*Source: Westinghouse/O'Connor Combustion Corp., Pittsburgh.*)

A gas is produced containing methane, CH_4, carbon monoxide, CO, and moisture. The carbon monoxide and methane components are combustible, providing heating value to the off-gas. The carbon residual, a char, also has heating value. This is an idealized reaction. No oxygen is added, and the original material is pure cellulose, $C_6H_{10}O_5$. In general the initial material is not pure and contains additional components, both organic and inorganic. The off-gas is a mixture of many simple and complex organic compounds. The char is often a liquid which contains minerals, ash, and other inorganics as well as residual carbon or tars.

Pyrolysis as an industrial process has been in use for years, and, although attempts to apply this process to municipal solid waste disposal have been made since the 1960s, it has not met with success in this area in the United States. The pyrolysis process produces charcoal from wood chips, coke and coke gas from coal, fuel gas and pitch from heavy-hydrocarbon still bottoms, etc.

The Pyrolysis System.   An idealized pyrolysis system for disposal of mixed waste is shown in Fig. 11.14. The waste received is sorted for removal of glass, metal, and cardboard, all of which has possible resale value. The waste stream enters a shredder (grinder), and the shredded material passes through a magnetic separator where residual ferrous metal is removed, for resale.

The balance of the waste stream will be fed into the reactor from a feed hopper. The hopper discharge and the feeder must be provided with air locks to minimize the infiltration of air (oxygen), which will degrade the pyrolysis reaction. Shredding is a

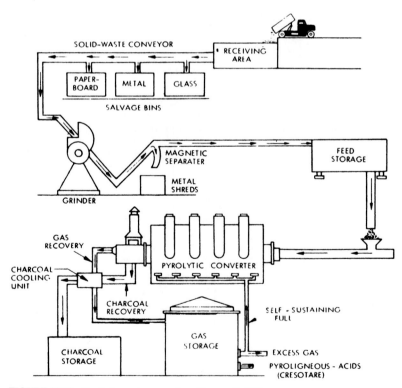

FIGURE 11.14   Pyrolytic waste conversion. (*Source: Ref. 7.*)

necessary step, not only to allow metal removal but also to provide a uniform-size feed of relatively small particles to the reactor. The converter is heated externally, as shown. Other types of pyrolytic reactors are designed to allow sufficient air infiltration to provide some burning within the reactor, generating enough heat internally to sustain the process.

Gas, exiting the reactor, is collected in a storage tank where organic acids and other organic compounds condense and are eventually discharged. Between 30 and 40 percent of the gas is required to heat the pyrolytic reactor; the balance of the gas stream can be used for other processes. In this generalized scheme a significant portion of the heating value of the off-gas is contained within the condensables. If the gas is heated and the discharged char residue is cooled, the condensables will remain in the gaseous state. As the gas cools and the condensables leave the gas stream, the gas heating value will decrease. Therefore, for maximum energy reclamation from the gas, it is important that the gas be kept in a heated state as long as possible—at least long enough for the gas to reach the farthest gas burner. Storage should be minimized because the condensables will leave the gas stream relatively readily in any quiescent area. The residual solid material is termed *charcoal*. This is, ideally, a desired by-product of this reaction.

A stack is shown immediately downstream of the converter. Upon start-up of the process, when an outside source of heat is required to initiate the reaction (not shown), the initial off-gas is basically composed of steam, carbon dioxide, entrapped air, and trace amounts of carbon monoxide. These components can be vented through the stack until the process stabilizes and pyrolysis gas is produced.

Severe problems have occurred in commercial attempts to develop this technology. It has been found difficult, if not impossible, to clean up the gas exiting the reactor. Although the gas is passed through a secondary combustion chamber and is subject to high-energy scrubbing systems, the organics within the gas stream have not been effectively controlled. Another severe problem lies within the reactor itself. As the waste is heated in the reducing atmosphere of the reactor, the ash, metals, glass, and other noncombustible materials tend to liquefy and form a slag. It is impossible to remove all of the glass or metals from the waste stream, and even the relatively small amount of these materials that may be present in the waste will contribute to the formation of a slag. In a number of the designs of commercial pyrolysis systems, the movement of the slag and the quantity of the slag generated has been impossible to control. Slagging has reached burner ports, and has risen in the reactor to interfere with the pyrolysis reaction itself.

Interest in the development of pyrolysis as an effective means of treatment of municipal solid waste is reviving in Europe, but American firms have abandoned the marketing of this technology.

Related Systems. There are many studies in progress not far distant from those of the traditional alchemists. Instead of yellow gold the goal is black gold—petroleum and petroleum-derived products. Currently processes and systems are exemplary if they just dispose of waste, generating innocuous residual materials without creating nuisance odor or budget overruns. But new processes will undoubtedly follow the perfectly sound theory of oil from waste (one hydrocarbon from another), and perhaps a workable system will be found to generate the potential of 1 barrel of oil per ton of waste by pyrolysis. A related process, starved air or controlled air combustion has been successfully developed.

Starved Air Incineration. In the early 1960s a new type of incinerator started gaining in popularity. The *modular combustion unit* has become an economical and efficient system for on-site and central destruction of waste. These incinerators are also

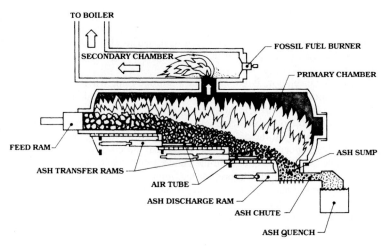

FIGURE 11.15   Starved air incinerator. (*Source: Ref. 7.*)

known as *controlled air* units. They can be operated as excess air units (EAU) or starved air units (SAU).

Theory of Operation.   The SAU consists of two major furnace components, as shown in Fig. 11.15, a primary and a secondary combustion chamber. Waste is charged into the primary chamber, and a carefully controlled flow of air is introduced. Only enough air is provided to allow sufficient burning for heating to occur. Typically 70 to 80 percent of the stoichiometric air requirement is introduced into the primary chamber.

The off-gas generated by this starved air reaction will contain combustibles, and this gas is burned in the secondary chamber, which is sized for sufficient residence time to totally destroy organics in the off-gas. As in the primary chamber, a carefully controlled quantity of air is introduced into the secondary chamber, but in this case excess air, 140 to 200 percent of the off-gas stoichiometric requirement, is maintained to effect complete combustion. Gas-cleaning devices such as wet scrubbers or electrostatic precipitators may not be required. The burnout of the off-gas in the secondary chamber is usually sufficient to clean the gas to 0.08 grains/st ft^3.

Figure 11.16 illustrates the variety of configurations currently marketed for controlled air incineration. They all have a starved air primary section and a secondary or afterburner chamber.

Control.   As can be seen in Fig. 11.17, the temperature is directly related to the excess air provided. Temperature, therefore, is normally utilized to control airflow in both primary and secondary chambers. Below stoichiometric the temperature of the reaction increases with an increase in airflow. As more air is provided, more combustion will occur, so more heat will be released. This heat release will result in higher temperatures produced. Control of the primary-chamber operation, therefore, where less than complete oxidation is provided, is as follows:

• With higher temperatures, decrease airflow.

• With lower temperatures, increase airflow.

The secondary chamber is designed for complete combustion, greater than stoichiometric air is supplied. At stoichiometric conditions all the combustible material present

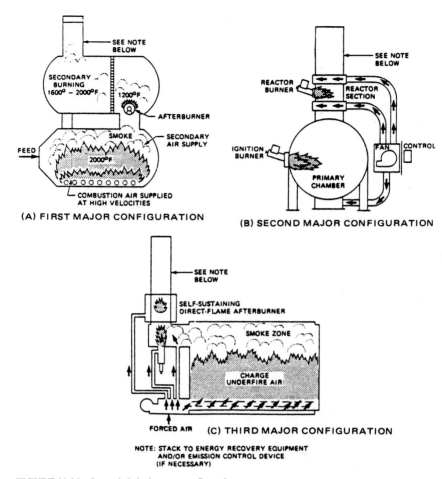

FIGURE 11.16 Starved air incinerator configurations.

will combust completely. Additional air will act to quench the off-gas, i.e., will lower the resulting exhaust gas temperature. Therefore, control of the secondary-chamber operation is as follows:

- With higher temperatures, increase airflow.
- With lower temperatures, decrease airflow.

SAUs are often provided with temperature detectors which automatically control fan damper positioning to provide the required chamber airflow.

Incinerable Wastes.   The SAUs were originally developed for the destruction of trash. They are applicable for other solid waste destruction, and their secondary chamber can be used for destruction of gaseous or liquid waste in suspension. It is not applicable for incineration of endothermic materials. The nature of the SAU process is such that turbulence of the waste feed is minimal. Materials requiring turbulence for

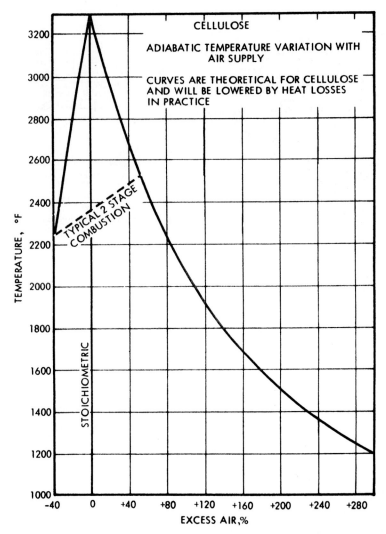

FIGURE 11.17   Adiabatic temperature variation with air supply. (*Source: Ref. 7.*)

effective combustion such as powdered carbon or pulp wastes are not appropriate candidates for starved air incineration.

Air Emissions.   Compared to other incineration methods, the airflow in the primary chamber, firing the waste, is low in quantity and is low in velocity. The low velocity and near absence of turbulence of the waste result in minimal amounts of particulate carried along in the gas stream. Complete burning is accomplished in the secondary chamber, and the resulting exhaust gas is clean and practically free of particulate matter, i.e., smoke and soot. The SAU can usually comply with exhaust emissions standards to 0.08 g/st ft^3 without the use of supplemental gas-cleaning equipment such as scrubbers or baghouses.

Waste Charging. Smaller units, under 750 lb/h, are normally batch-fed. Waste is charged over a period of hours, and after a full load has been placed in the chamber, the chamber is sealed and the waste fired.

Figure 11.18 illustrates a typical hopper-ram assembly designed to minimize the quantity of air infiltration into the primary chamber when charging. Figure 11.19 illustrates a double-ram charging system which allows a more continuous feed than the single-ram. Note that the furnace charging door is not opened until the hopper is sealed by the upper ram, preventing air infiltration from the hopper. Larger units are usually provided with a continuous waste charging system, a screw feeder, or a series of moving grates.

Ash Disposal. As with waste charging, SAUs are provided with both manual and automatic discharge systems. With smaller units, after burnout the chamber is opened and ash residue is manually raked out. With continuous operating units such as that shown in Fig. 11.15, ash is continually discharged, normally into a wet well, where it is transferred to a container or truck by means of a drag conveyer.

Energy Reclamation. Waste heat utilization is a viable option provided with SAUs. The hot gas exiting the secondary chamber is relatively clean. Boiler or heat-exchanger surfaces placed within this gas stream will therefore be subject to minimal particulate matter carryover and attendant problems of erosion and plugging.

Typical Systems. Dimensional data of a typical SAU (Morse Boulger, Inc.) system are shown in Fig. 11.20. Its internal configuration is similar to that of the unit in the upper left of Fig. 11.16. Its charging system is similar to that of Fig. 11.18.

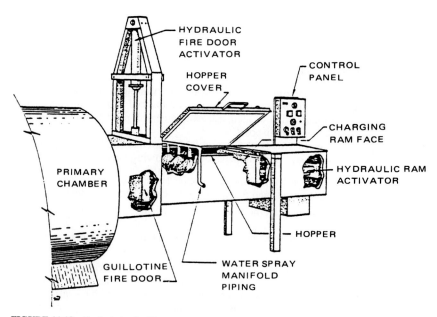

FIGURE 11.18 Typical standard hopper-ram assembly.

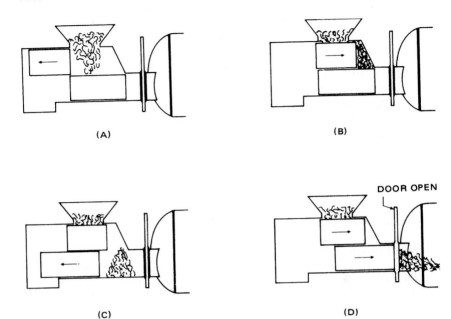

FIGURE 11.19   Double-ram type of charging system.

Table 11.15 lists typical SAU systems which are provided with energy generation systems. These are basically "standard" models which are normally modified to the customers' specific waste, heat recovery mode, or other needs.

Central Disposal Systems

The Need for Central Disposal.   Central disposal of waste is a consideration in the disposal of municipal solid waste. In recent years industrial firms with many plants have looked toward incineration of their wastes in a central plant location as an efficient and economical means of disposal.

In Europe, after World War II, the disposal of municipal solid waste at a central location, by incineration, was given impetus by the following factors:

1. Population concentrations and increases required the use of more land for housing and for farming. The use of land for burying refuse was becoming impractical.

2. Technology developed to the point where it became economical to generate energy, i.e., steam and/or electric power, from incineration. Economics of scale dictated that the larger the facility, the more efficient would be its energy generation potential.

3. In general all utilities, including refuse disposal and electric or steam power generating industries, were state-owned. The interests of the electric utility and the refuse disposal authority were therefore common. This conflicts with conditions in the United States, where refuse collection is a public or government function and electric power generation is generally a private-sector function. Cooperation between these agencies in Europe has promoted the development of energy-producing incinerators.

The power utility readily purchases energy from an incineration facility, providing revenues for the incinerator operation.

4. The higher cost of fossil fuel, particularly fuel oil, has helped promote energy generation, hence, central disposal facilities.

Only in the past decade, when the United States came to the realization that the cost and availability of energy were unreliable and out of its control, has a serious attempt been made to generate energy from waste in central collection and incineration facilities. Table 11.16 lists information on selected mass burning central disposal facilities in the United States. Cost data are included for reference only.

Municipal Solid Waste.   Central-station incineration is usually applied to municipal solid waste destruction. The average characteristics of refuse and other wastes are listed earlier in this chapter. The actual variation in average waste composition from one country to another is listed in Table 11.17. The Ash column represents the residual from coal or wood burning for domestic heat in the winter months. For instance, 43 percent of the composition of refuse in the United States was ash due to household coal burning in 1939, whereas 30 years later this component, i.e., ash from coal burning, was absent from the refuse.

When burning refuse, the generation of dry gas and moisture from combustion can be estimated as follows:

Dry gas	7.5 lb/10,000 Btu fired
Moisture	0.51 lb/10,000 Btu fired

Grate System.   The grate system is one of the most crucial systems within the mass burning incinerator. The grate must transport refuse through the furnace and, at the same time, promote combustion by adequate agitation and good mixing with combustion air. Abrupt tumbling caused by the dropping of burning solid waste from one tier to another will promote combustion. This action, however, may contribute to excessive carryover of particulate matter in the exiting flue gas. A gentle agitation will decrease particulate emissions.

Combustion is largely achieved by injection of combustion air below the grates, i.e., underfire air. Underfire air is also necessary to cool the grates. It is normally provided at a rate of approximately 40 to 60 percent of the total air entering the furnace. Too low a flow of underfire air will inhibit the burning process and will result in high grate temperatures.

Note the ash fusion temperatures listed in Table 11.18. These temperatures limit the operating temperatures of the grate areas. With insufficient air a reducing atmosphere will result, and the ash deformation temperature can be as low as 1800°F. If the refuse reaches this temperature, slagging will begin, further reducing the air supply by clogging the grates and forming large, unwieldy clinkers. The ash properties of coal are listed for comparison.

Overfire air is injected above the grates. Its main purpose is to provide sufficient air to completely combust the flue gas and flue gas particulate rising from the grates. Numerous injection points are located on the furnace walls above the grates to provide a turbulent overfire air supply along the furnace length.

Ash and other particles dropping through the grates are termed *siftings,* and they must be effectively removed from the system. Siftings can readily clog grate mechanisms, generate fires, and create housekeeping problems if not attended to. Siftings, due to their small particle size, have been found to be more dense than incinerator ash, approximately 1780 versus 1040 lb/yd^3 for typical incinerator ash.

Grate Design.   A number of different types of grate designs are used in central waste burning facilities. Each grate system manufacturer provides a unique grate feature,

attempting to obtain a competitive edge in the marketplace. The grate system manufacturer should be contacted for design and sizing information for a particular grate design. The following listing describes typical grate systems, both generic grate types and grates specific to certain manufacturers:

Traveling Grate. This type is no longer in common usage. As shown in Fig. 11.21, it is normally not a single grate but a series of grates which are placed in a manner that separates the drying and burning functions of the incinerator.

Rocking Grate. As shown in Fig. 11.22, these grate sections are placed across the width of the furnace. Alternate rows are mechanically pivoted or rocked to produce an upward and forward motion, advancing and agitating the waste. The stroke of the grate sections is 5 to 6 in. This grate will handle refuse on a continuous basis.

Reciprocating Grate. As shown in Fig. 11.23, this grate consists of sections stacked above each other similar to overlapping roof shingles. Alternate grate sections slide back and forth while adjacent sections remain fixed. Drying and burning are accomplished on single, short but wide grates. The moving grates are basically bars, stoking bars, which move the waste along and help agitate it.

Rotary Kiln. As shown in Fig. 11.24, two traveling grates are initially used for

CAPACITY lbs./hr.	A	B	C	CC*	D	E	F	FF*	G	H	J	JJ*	CHUTE CAP. cu. yd.
100-160	8'-6"	4'-6"	5'-0"	9'-0"	6'-4"	3'-6"	9'0"	13'-0"	1'-4"	11'-0"	2'-0"	6'-0"	0.5
220-350	9'-6"	5'-6"	6'-0"	10'0"	7'-0"	4'-6"	11'-0"	15'-0"	1'-7½"	11'-0"	2'-0"	6'-0"	0.5
320-525	9'-6"	6'-0"	6'-6"	10'-6"	7'-0"	5'-6"	12'-6"	16'-6"	1'-9"	11'-0"	2'-6"	6'-6"	1.0
430-700	10'-0"	6'-6"	7'-0"	11'-0"	8'-0"	5'-6"	13'-0"	17'-0"	2'-2½"	11'-0"	2'-6"	6'-6"	1.0
640-1050	11'-6"	7'-6"	8'-0"	12'-0"	8'-0"	6'-0"	15'-0"	19'-0"	2'-6"	13'-0"	2'-6"	6'-6"	2.0
870-1400	12'-0"	8'-0"	8'-6"	12'-6"	9'-0"	6'-6"	16'-0"	20'-0"	2'-9½"	13'-0"	3'-6"	8'-0"	2.0
1300-2100	12'-6"	9'-6"	10'-0"	14'-0"	9'-6"	8'-6"	19'6"	23'-6"	3'-5"	13'-0"	3'-6"	8'-6"	3.0
1950-3200	14'-6"	10'-6"	11'-0"	15'-0"	10'-0"	9'-6"	21'-6"	25'-6"	3'-11"	13'-0"	4'-0"	9'-0"	3.0
2400-3900	16'-0"	11'-0"	11'-6"	15'-6"	12'-0"	9'-6"	22'-0"	26'-0"	4'-5"	14'-0"	5'-6"	9'-6"	4.0
2900-4700	18'-0"	11'-6"	12'-0"	16'-0"	14'-0"	9'-6"	22'-6"	26'-6"	4'-9"	14'-0"	5'-6"	9'-6"	4.0

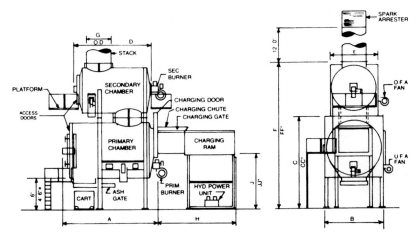

FIGURE 11.20 Typical SAU system. (*Source: Morse Boulger, Inc., Queens, N.Y.*)

TABLE 11.15 SAU Energy Generation

Burning rate, lb/h	Waste feed, Btu/lb	Energy generation mode	Energy generation rate	Manufacturer
700	6,000	Steam, 100 lb/in^2 gauge	2,550 lb/h	Morse Boulger
1,000	6,000	Steam, 100 lb/in^2 gauge	3,850 lb/h	Morse Boulger
1,400	6,000	Steam, 100 lb/in^2 gauge	5,100 lb/h	Morse Boulger
3,200	6,000	Steam, 100 lb/in^2 gauge	11,600 lb/h	Morse Boulger
4,700	6,000	Steam, 100 lb/in^2 gauge	17,000 lb/h	Morse Boulger
1,280	6,285	Steam, 160 lb/in^2 gauge	2,025 lb/h	George L. Simonds
1,650	8,500	Steam, 150 lb/in^2 gauge	3,500 lb/h	George L. Simonds
1,050	6,240	Steam, 125 lb/in^2 gauge	1,900 lb/h	George L. Simonds
650	6,500	Steam, 150 lb/in^2 gauge	2,500 lb/h	George L. Simonds
1,800/15 gal/h	8,500 trash/ waste oil	Steam, 150 lb/in^2 gauge	8,000 lb/h	George L. Simonds
1,000	6,500	Steam, 100 lb/in^2 gauge Hot water, 105° Δt	3,458 lb/h 66 gal/min	Smokatrol
1,500	6,500	Steam, 100 lb/in^2 gauge Hot water, 105° Δt	5,187 lb/h 86 gal/min	Smokatrol
2,000	6,500	Steam, 100 lb/in^2 gauge Hot water, 105° Δt	6,916 lb/h 132 gal/min	Smokatrol
2,500	6,500	Steam, 100 lb/in^2 gauge Hot water, 105° Δt	8,645 lb/h 165 gal/min	Smokatrol
1,250	4,500 7,000	Steam, 150 lb/in^2 gauge	3,200 lb/h 4,950 lb/h	Consumat
2,100	4,500 7,000	Steam, 150 lb/in^2 gauge	5,400 lb/h 8,400 lb/h	Consumat
6,250	4,500 7,000	Steam, 150 lb/in^2 gauge	16,100 lb/h 25,000 lb/h	Consumat
8,400	4,500 7,000	Steam, 150 lb/in^2 gauge	21,600 lb/h 33,600 lb/h	Consumat

Source: Selected manufacturers' data.

drying the incoming refuse and for initial ignition. The kiln is at the heart of this system. By varying the kiln rotational speed, burnout of the refuse is accurately controlled. The refuse burns out in the kiln, and ash is discharged from the end of the kiln to residue conveyers. Some of the flue gases are diverted for drying the incoming refuse. Flue gas can be passed through a waste heat boiler for energy recovery.

Martin System.   As shown in Fig. 11.25, this system utilizes reverse reciprocating grates. As the grates move forward and then reverse, there is continuous agitation of the waste. The bars making up the grates are hollow, allowing air to circulate within them and keep them relatively cool.

Von Roll System.   As shown in Fig. 11.26, a series of reciprocating grates is used to move refuse through the furnace. The first grate section dries the refuse, the second is a burning grate, and burnout to ash takes place on the third grate.

VKW System.   Shown in Fig. 11.27 is a variation of the traveling grate concept. A series of drums is utilized as grates. The drums rotate slowly, agitating the waste and moving it along to subsequent drums. Air passes through openings in these drums, or roller grates, as underfire air. Both speed of rotation on the roller grates and quantity of underfire air per roller grate are variable.

Alberti System.   As illustrated in Fig. 11.28, this grate system has a single-section grate constructed of fixed and moving elements arranged, as shown, in a series of steps. Feed is rammed from the feed hopper to the grates. The fixed grate contains the refuse while the moving elements agitate the waste, driving it down to the next grate.

Esslingen System.   As shown in Fig. 11.29, a traveling grate is used to feed a

TABLE 11.16 Field-Erected Mass Burn Facilities

Name	Status*	Design tons/day	Original capital costs, $	Year	Capital cost, 1990 $	Capital cost per ton, 1990 $	Additional capital cost	Year	Additional cost, 1990 $	Additional capital cost per ton, 1990 $
				Process: MB—refractory						
Energy type—steam										
Betts Avenue	05	1000	5,000,000	65	21,594,896	21,595	36,500,000	89	37,440,152	37,440
City of Waukesha (old plant)	05	175	1,700,000	71	4,856,118	27,749	3,900,000	79	5,806,047	33,177
Davis County	05	400	40,000,000	88	41,693,608	104,234	0		0	0
Average		525	15,566,667		22,714,874	51,193	20,200,000		21,623,100	35,309
Standard deviation		348	17,329,423		15,059,680	37,590	16,300,000		15,817,053	2,132
Energy type—electricity										
McKay Bay Refuse-to-Energy Facility	05	1000	72,700,000	85	81,083,856	81,084	500,000	87	532,861	533
Average		1000	72,700,000		81,083,856	81,084	500,000		532,861	533
Standard deviation		0	0		0	0	0		0	0
Energy type—steam and electricity										
Muscoda	05	125	8,250,000	87	8,792,208	70,338	0		0	0
Average		125	8,250,000		8,792,208	70,338	0		0	0
Standard deviation		0	0		0	0	0		0	0
				Process: MB—waterwall						
Energy type—steam										
Brooklyn Navy Yard	02	3000	426,000,000	90	426,000,000	142,000	0		0	0
Hampton/NASA Project Recoup	05	200	10,400,000	78	16,823,896	84,119	2,450,000	87	2,611,020	13,055
Norfolk Naval Station	08	360	3,220,000	67	12,995,170	36,098	5,400,000	87	5,754,899	15,986
Savannah	05	500	35,000,000	85	39,036,240	78,072	0		0	0
Average		1015	118,655,000		123,713,827	85,072	3,925,000		4,182,960	14,521
Standard deviation		1151	177,836,647		174,807,968	37,713	1,475,000		1,571,940	1,466
Energy type—electricity										
Albany (American Ref-Fuel)	02	1500	200,000,000	89	205,151,520	136,768	0		0	0
Alexandria/Arlington R.R. Facility	05	975	75,900,000	85	84,652,896	86,823	2,000,000	89	2,051,515	2,104
Babylon Resource Recovery Project	05	750	85,520,000	85	95,382,272	127,176	0		0	0

Bergen County	02	3000	335,000,000	91	335,000,000	111,667	0	0	0	0
Bridgeport RESCO	05	2250	211,000,000	85	235,332,768	104,592	0	0	0	0
Bristol	05	650	58,800,000	85	65,580,888	100,894	0	0	0	0
Broome County	02	571	77,000,000	90	77,000,000	134,851	0	0	0	0
Broward County (Northern Facility)	03	2250	216,007,000	90	216,007,000	96,003	0	0	0	0
Broward County (Southern Facility)	03	2250	277,816,000	90	277,816,000	123,474	0	0	0	0
Camden County (Foster Wheeler)	03	1050	96,000,000	87	102,309,328	97,437	0	0	0	0
Camden County (Pennsauken)	03	500	88,000,000	90	88,000,000	176,000	0	0	0	0
Central Mass. Resource Recovery Project	05	1500	140,000,000	86	152,686,272	101,791	0	0	0	0
City of Commerce	05	400	35,010,000	86	38,182,472	95,456	1,000,000	89	1,025,758	2,564
Concord Regional S.W. Recovery Facility	05	500	53,500,000	90	59,669,688	119,339	0	0	0	0
Dakota County	02	800	108,852,000	90	108,852,000	136,065	0	0	0	0
East Bridgewater (American Ref-Fuel)	02	1500	150,000,000	90	150,000,000	100,000	0	0	0	0
Eastern-Central Project	02	550	78,000,000	89	80,009,088	145,471	0	0	0	0
Essex County	03	2277	252,500,000	89	259,003,776	113,748	0	0	0	0
Fairfax County	05	3000	195,500,000	88	203,777,536	67,926	0	0	0	0
Falls Township (Wheelabrator)	02	2250	200,000,000	91	200,000,000	88,889	0	0	0	0
Glendon	02	500	63,500,000	90	63,500,000	127,000	0	0	0	0
Gloucester County	05	575	60,000,000	90	60,000,000	104,348	0	0	0	0
Haverhill (Mass Burn)	05	1650	120,000,000	87	127,886,656	77,507	0	0	0	0
Hempstead (American Ref-Fuel)	05	2505	255,000,000	85	284,406,912	113,536	0	0	0	0
Hennepin County (Blount)	05	1200	80,000,000	88	83,387,216	69,489	0	0	0	0
Hillsborough County S.W.E.R. Facility	05	1200	80,500,000	87	85,790,640	71,492	0	0	0	0
Hudson County	02	1500	179,000,000	89	183,610,592	122,407	0	0	0	0
Huntington	03	750	153,500,000	90	153,500,000	204,667	0	0	0	0
Johnston (Central Landfill)	02	750	80,000,000	90	80,000,000	106,667	0	0	0	0
Lake County	04	528	60,000,000	90	60,000,000	113,636	0	0	0	0
Lancaster County	03	1200	102,000,000	89	104,627,280	87,189	0	0	0	0
Lee County	02	1800	146,964,600	90	146,964,600	81,647	0	0	0	0
Lisbon	02	500	100,000,000	90	100,000,000	200,000	0	0	0	0
Marion County Solid W-T-E. Facility	05	550	47,500,000	86	51,804,272	94,190	0	0	0	0
Montgomery County	02	1800	280,000,000	89	287,212,096	159,562	0	0	0	0
Montgomery County	03	1200	115,000,000	89	117,962,112	98,302	0	0	0	0
Morris County	02	1340	141,900,000	89	145,555,008	108,623	0	0	0	0

TABLE 11.16 Field-Erected Mass Burn Facilities (*Continued*)

Process: MB—waterwall (*Continued*)

Name	Status*	Design tons/day	Original capital costs, $	Year	Capital cost, 1990 $	Capital cost per ton, 1990 $	Additional capital cost	Year	Additional cost, 1990 $	Additional capital cost per ton, 1990 $
Energy type—electricity										
New Hampshire/Vermont S.W. Project	05	200	26,500,000	85	29,556,012	147,780	0		0	0
North Andover	05	1500	185,000,000	85	206,334,432	137,556	0		0	0
North Hempstead	02	990	135,000,000	89	138,477,280	139,876	0		0	0
Oklahoma City	08	820	35,000,000	87	37,300,272	45,488	0		0	0
Onondaga County	02	990	132,000,000	90	132,000,000	133,333	0		0	0
Oyster Bay	02	1000	135,000,000	90	135,000,000	135,000	0		0	0
Pasco County	03	1050	90,600,000	89	92,933,632	88,508	0		0	0
Passaic County	02	1434	142,000,000	90	142,000,000	99,024	0		0	0
Pinellas County (Wheelabrator)	05	3150	83,000,000	83	94,280,208	29,930	60,000,000	86	65,436,968	20,774
Portland	05	500	45,500,000	87	48,490,360	96,981	20,600,000	90	20,600,000	41,200
Preston (Southeastern Connecticut)	03	600	83,000,000	87	88,454,944	147,425	0		0	0
S.E. Resource Recovery Facility (SERRF)	04	1380	106,000,000	87	112,966,544	81,860	0		0	0
Saugus	05	1500	33,000,000	74	74,160,992	49,441	95,000,000	90	95,000,000	63,333
Spokane	03	800	82,149,000	87	87,548,016	109,435	0		0	0
Stanislaus County Res. Recovery Facility	05	800	82,200,000	85	91,679,408	114,599	0		0	0
Sturgis	02	560	0	0	0	0	0	0	0	0
Union County	02	1440	150,000,000	90	150,000,000	104,167	0		0	0
Warren County	05	400	50,300,000	89	51,595,608	128,989	0		0	0
Washington-Warren Counties	03	400	50,000,000	90	50,000,000	125,000	0		0	0
West Pottsgrove Recycling/R.R. Facility	02	1500	150,000,000	91	150,000,000	100,000	0		0	0
Westchester	05	2250	179,000,000	83	203,327,168	90,368	0		0	0
Average		1230	122,359,975		127,837,294	110,691	35,720,000		36,822,848	25,995
Standard deviation		723	69,637,080		70,966,036	32,515	36,537,017		37,301,493	23,547
Energy type—steam and electricity										
Charleston County	05	644	59,000,000	89	60,519,688	93,975	0		0	0
Davidson County	03	210	7,000,000	87	7,460,056	35,524	0		0	0
Harrisburg	05	720	8,300,000	71	23,709,280	32,930	21,300,000	86	23,230,124	32,264

Jackson County/Southern Michigan State Prison	05	200	28,000,000	86	30,537,256	152,686	0		0	0
Kent County	05	625	62,200,000	89	63,802,128	102,083	0		0	0
Nashville Thermal Transfer Corp. (NTTC)	05	1120	24,500,000	74	55,058,920	49,160	36,500,000	85	40,709,224	36,348
Northwest Waste-To-Energy Facility	05	1600	23,000,000	68	86,385,568	53,991	5,000,000	88	5,211,701	3,257
Olmstead County	05	200	30,000,000	87	31,971,664	159,858	0		0	0
Quonset Point	02	710	83,000,000	90	83,000,000	116,901	0		0	0
S.W. Resource Recovery Facility (BRESCO)	05	2250	185,000,000	83	210,142,624	93,397	0		0	0
University City Res. Recovery Facility	05	235	27,000,000	87	28,774,496	122,445	0		0	0
Walter B. Hall Res. Recovery Facility	05	1125	114,000,000	87	121,492,320	107,993	0		0	0
Wayne County	02	300	27,000,000	90	27,000,000	90,000	0		0	0
Average		765	52,153,846		63,834,923	93,149	20,933,333		23,050,350	23,956
Standard deviation		597	48,428,117		52,029,873	39,364	12,862,435		14,492,361	14,731

Process: MB—rotary combustor

Energy type—steam										
Galax	05	56	2,100,000	85	2,342,175	41,825	160,000	88	166,774	2,978
Average		56	2,100,000		2,342,175	41,825	160,000		166,774	2,978
Standard deviation		0	0		0	0	0		0	0
Energy type—electricity										
Auburn (New Plant)	03	200	26,500,000	90	26,500,000	132,500	0		0	0
Delaware County Regional R.R. Project	03	2688	276,000,000	90	276,000,000	102,679	0		0	0
Gaston County/Westinghouse R.R. Center	02	440	42,000,000	89	43,081,824	97,913	0		0	0
MacArthur Energy Recovery Facility	05	518	38,700,000	85	43,162,936	83,326	2,500,000	91	2,500,000	4,826
Mercer County	02	975	117,500,000	88	122,474,976	125,615	0		0	0
Monmouth County	02	1700	220,000,000	90	220,000,000	129,412	0		0	0
Montgomery County (North)	05	300	7,494,000	69	25,688,292	85,628	9,700,000	87	10,337,504	34,458
Montgomery County (South)	02	900	6,150,000	69	21,081,268	23,424	5,000,000	85	5,576,606	6,196
Oakland County	02	2000	172,000,000	90	172,000,000	86,000	0		0	0
San Juan Resource Recovery Facility	02	1040	91,400,000	89	93,754,240	90,148	0		0	0
Skagit County	05	178	14,000,000	87	14,920,112	83,821	0		0	0
Westinghouse/Bay Resource Mgmt. Center	05	510	38,000,000	86	41,443,416	81,262	0		0	0
York County	05	1344	91,200,000	88	95,061,424	70,730	0		0	0
Average		984	87,764,923		91,936,038	91,728	5,733,333		6,138,037	15,160
Standard deviation		737	83,297,756		80,657,588	27,471	2,984,776		3,224,182	13,657

TABLE 11.16 Field-Erected Mass Burn Facilities (*Continued*)

Name	Status*	Design tons/day	Original capital costs, $	Year	Capital cost, 1990 $	Capital cost per ton, 1990 $	Additional capital cost	Year	Additional cost, 1990 $	Additional capital cost per ton, 1990 $
Process: MB—rotary combustor (*Continued*)										
Energy type—steam and electricity										
Dutchess County	05	506	35,000,000	84	39,213,904	77,498	0		0	0
Falls Township (Technochem)	02	70	7,000,000	90	7,000,000	100,000	0		0	0
Monroe County	02	500	100,000,000	91	100,000,000	200,000	0		0	0
Sangamon County	02	450	38,160,000	90	38,160,000	84,800	0		0	0
Sumner County	05	200	9,800,000	81	12,655,414	63,277	5,340,000	90	5,340,000	26,700
Waukesha County (New Plant)	02	600	100,000,000	90	100,000,000	166,667	0		0	0
Average		388	48,326,667		49,504,886	115,374	5,340,000		5,340,000	26,700
Standard deviation		188	38,326,286		37,635,694	50,187	0		0	0

*Facility status: advanced planning 02, construction 03, shakedown 04, operation 05 through 07, and temporarily shutdown 08.

Source: *Data Summary of Municipal Solid Waste Management Alternatives*, vol. III: *Appendix A—Mass Burn Technologies*, National Technical Information Service, October 1992, NTIS accession number DE92016433.

TABLE 11.17 A Summary of International Refuse Composition (in percentages)

	Ash	Paper	Organic matter	Metals	Glass	Miscella-neous
United States (1939)	43.0	21.9	17.0	6.8	5.5	5.8
United States (1970)	0	44.0	26.5	8.6	8.8	12.1
Canada	5	70	10	5	5	5
United Kingdom	40–40	25–30	10–15	5–8	5–8	5–10
France	24.3	29.6	24	4.2	3.9	14
West Germany	30	18.7	21.2	5.1	9.8	15.2
Sweden	0	55	12	6	15	12
Spain	22	21	45	3	4	5
Switzerland	20	40–50	15–25	5	5	—
Netherlands	9.1	45.2	14	4.8	4.9	22
Norway (summer)	0	56.6	34.7	3.2	2.1	8.4
Norway (winter)	12.4	24.2	55.7	2.6	5.1	0
Israel	1.9	23.9	71.3	1.1	0.9	1.9
Belgium	48	20.5	23	2.5	3	3
Czechoslovakia (summer)	6	14	39	2	11	28
Czechoslovakia (winter)	65	7	22	1	3	2
Finland	—	65	10	5	5	15
Poland	10–21	2.7–6.2	35.3–43.8	0.8–0.9	0.8–2.4	—
Japan (1963)	19.3	24.8	36.9	2.8	3.3	12.9

Source: Ref. 7.

TABLE 11.18 Ash Fusion Temperatures

	Reducing Atmosphere, °F	Oxidizing Atmosphere, °F
Refuse		
Initial deformation	1880–2060	2030–2100
Softening	2190–2370	2260–2410
Fluid	2400–2560	2480–2700
Coal		
Initial deformation	1940–2010	2020–2270
Softening	1980–2200	2120–2450
Fluid	2250–2600	2390–2610

Source: Ref. 7.

rocking grate system. It is normally provided with a single-grate section composed of semicircular rocking elements. Each movement of these elements promotes transport and agitation of the waste. Underfire air passes through the rocking elements, keeping them cool and providing for combustion of the waste.

The Heenan Nichol System.   As shown in Fig. 11.30, this system utilizes grates composed of three or more sections which are arranged in steps. Each pair of elements moves in a rocking manner so that at any moment half of the elements are moving. All odd-numbered elements are linked to each other, as are all the even-numbered elements. The rocking action moves and agitates the waste.

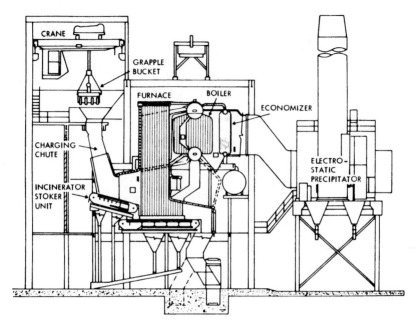

FIGURE 11.21 Traveling grate system. (*Source: Ref. 7.*)

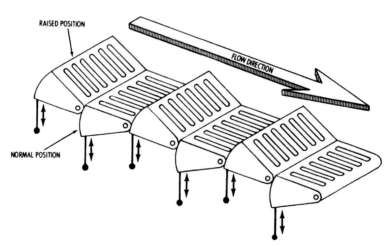

FIGURE 11.22 Rocking grates.

CEC System. This system, illustrated in Fig. 11.31, utilizes a single-section grate. Two or more grate sections can be arranged in parallel. The grate is constructed of successive sliding, rocking, and fixed elements. The sliding and rocking elements are synchronized so that the sliding elements move over the rocking elements when the rocking elements are retracted. The sliding elements, therefore, promote transport of the waste while the rocking elements provide the required agitation.

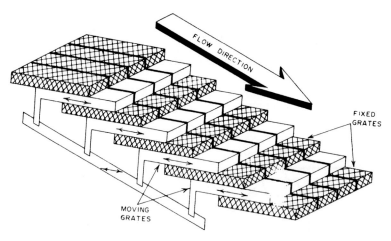

FIGURE 11.23   Reciprocating grates.

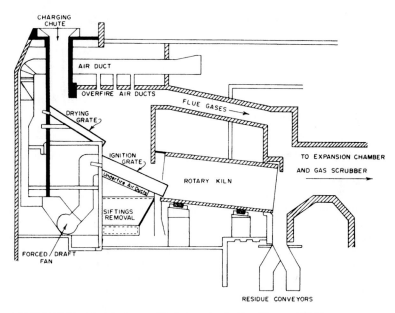

FIGURE 11.24   Municipal rotary kiln incineration facility. (*Source: Ref. 7.*)

Bruun and Sorensen System.   As shown in Fig. 11.32, this system utilizes a series of rollers, up to six in each of its three sections. Odd-numbered rollers turn clockwise while even-numbered rollers turn counterclockwise. Underfire air passes through the rotary grates, or drums. The action of the drums provides good agitation, and the slope of the grate to the horizontal promotes the transport of waste along the grate.

Volund System.   This system is shown in Fig. 11.33. It utilizes a rotary kiln for controlled burning of waste. Reciprocating grates are used for waste drying and initial combustion. Burnout takes place within the kiln.

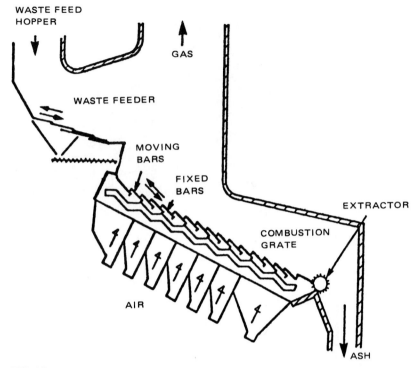

FIGURE 11.25 Martin system. (*Source: Ref. 7.*)

Associated Disposal Systems. There are refuse burning systems in use which cannot strictly be classified as grate systems.

Suspension Burning. An incinerator coupled to a refuse processing system is pictured in Fig. 11.34. Refuse is shredded and air classified into light and heavy fractions. The light fraction is blown into the boiler through a pneumatic charging system. Figure 11.35 shows the air distribution within the furnace and around the waste feed. Waste that is not burned in suspension will drop onto the shredder stoker, a variation of the traveling grate, and will burn out. Ash not airborne, produced by suspension burning, will also drop onto the spreader stoker. The stoker moves slowly, discharging its ash load to an ash hopper for ultimate disposal. Heavier components of the refuse that are not incinerated are composed mainly of metals and glass. These materials can, in certain instances, be marketed.

Fluid Bed Incineration. There have been some attempts to adapt limited European experience with fluid bed incineration of municipal solid waste to the United States. This technology requires that glass and low-melting-point metals (such as aluminum) be removed from the waste stream. These components, in even relatively small quantities, will slag the furnace bed. In addition, the feed must be reduced to uniform size, no larger than 1- to $1\frac{1}{2}$-in mean particle size.

The advantage of this type of incineration system is the ability to add limestone (or other alkali) to the bed, which will capture halogens (chlorides and fluorides) and other compounds, significantly reducing the discharge of acid gases. The effort and

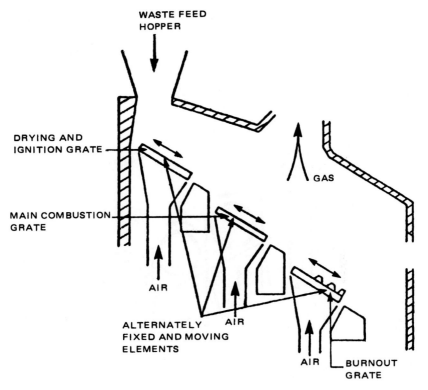

FIGURE 11.26 Von Roll system. (*Source: Ref. 7.*)

resultant high cost required to remove the aluminum and glass from the waste stream, however, restrict the use of this technology.

Sludge Burning in an MSW Incinerator.   Moisture content is the single most important parameter in determining the burning characteristics of a material. The higher the moisture content, the longer it will take that material to burn. When materials of different moisture content are placed in an incinerator at the same time, the lower-moisture-content material will burn off first, while moisture is evaporated from the second material. It will exit the incinerator as it came into the incinerator, but with less moisture.

Municipal solid waste has a moisture content in the range of 20 to 30 percent. The moisture content of sewage sludge is normally in the range of 70 to 80 percent. The only effective method of firing these two waste streams has been by a reduction of the sludge moisture content to that of the MSW.

A refuse incinerator is located at the site of a sewage treatment plant at a New York facility. Sewage sludge, having approximately 75 percent moisture content, is sprayed on top of the refuse as the refuse was entering the incinerator. The refuse has a moisture content of between 20 and 30 percent. At the incinerator ash discharge the refuse is burned out, however, the sludge is unburned. It is found exiting the incinerator smoldering and identifiable as sludge.

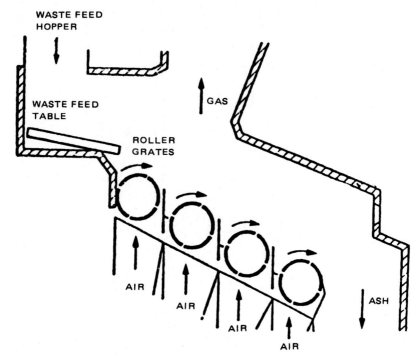

FIGURE 11.27   VKW system. (*Source: Ref. 7.*)

In contrast, at a Connecticut facility a refuse incinerator is also located on the site of a wastewater treatment plant. Sludge, as above, is generated at approximately 75 percent moisture content. An extensive sludge drying and conveying system has been employed at this site, however, to reduce its moisture content to from 15 to 20 percent. This allows the sludge to burn in no longer a period of time than it takes the refuse to burn. There have been no problems with burnout at this plant. Sludge is completely fired, with no sludge residual in the ash discharge of the incinerator.

A major reason for the absence of sewage sludge burning facilities at refuse incinerator plants is in the problems inherent in the burning process when materials of these dissimilar moisture fractions are present. The cost of drying equipment to reduce the moisture content of the sludge to that of the refuse is usually prohibitive and the sludge is either taken to its own dedicated incinerator or an alternative sludge disposal method is found.

Incinerator Corrosion Problems.   Severe corrosion has been found in three major areas of the mass burning incinerator system.

Scrubber Corrosion.   The acidic components of the flue gas present corrosion problems in wet scrubbing equipment, which will be discussed later in this chapter.

Corrosion of Grates.   With insufficient airflow, high temperatures and a reducing atmosphere can occur and ash can soften or fluidize. Fluid ash can be exceedingly corrosive, readily attacking cast iron or steel.

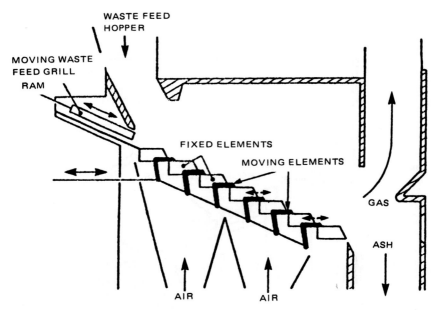

FIGURE 11.28   Alberti system. (*Source: Ref. 7.*)

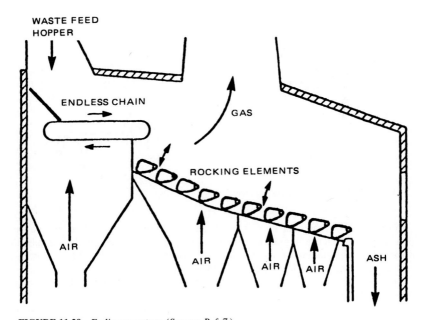

FIGURE 11.29   Esslingen system. (*Source: Ref. 7.*)

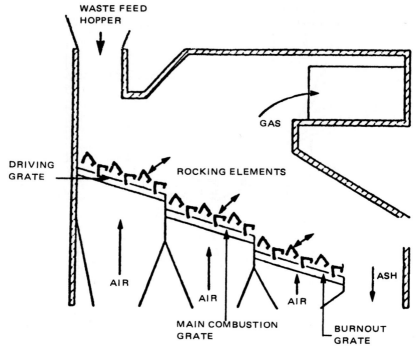

FIGURE 11.30   Heenan Nichol system. (*Source: Ref. 7.*)

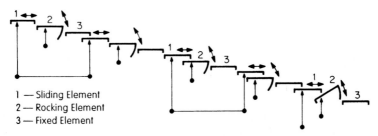

1 — Sliding Element
2 — Rocking Element
3 — Fixed Element

FIGURE 11.31   CEC system. (*Source: Ref. 7.*)

   Table 11.19 illustrates the corrosion rate as a function of temperature for two steel alloys in widespread use in grate construction. Temperature alone greatly increases the corrosive rate, the amount of material lost per month of service. At 1200°F over $\frac{3}{8}$ in of material is "wasted" or lost from steel components, for instance.

   Fireside Corrosion.   There are two modes of corrosion that affect boiler tubes. Low-temperature or dewpoint corrosion is metal wastage caused by sulfuric or hydrochloric acid condensation. Chlorides and sulfides within the refuse (chlorides are present in plastics) will partially convert to hydrogen chloride, sulfur dioxide, and sulfur trioxide in the exhaust gas stream. These gases will condense at temperatures below 300°F, and their condensate or liquid phase will be hydrochloric and sulfuric acid, both of which will attack steel. It is important, therefore, that the temperature of the boiler

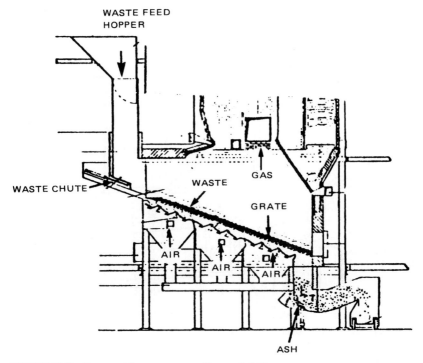

FIGURE 11.32   Bruun and Sorensen system. (*Source: Ref. 7.*)

tubes, constructed of steel, be kept above 300°F. This is a function of the temperature of the steam or hot water generated. The boiler tubes will be at a temperature close to that of the circulating fluid, and this 300°F rule therefore limits the minimum temperature of the steam or hot water generated.

High-temperature corrosion is a more complex problem. Table 11.20 lists the steam pressure, temperature, and external tube temperatures for an assortment of incinerator systems burning municipal solid waste. At temperatures exceeding 700°F a complex reaction takes place between the sulfide and chlorine/chloride-bearing flue gas and the steel boiler tube, as illustrated in Fig. 11.36. Chlorine reacts with the iron in the tube wall to produce ferrous chloride which, upon contact with oxygen in the flue gas, converts to iron oxide. The iron oxide (rust) will leave the surface of the steel, causing wastage of the steel surface. Other components of the refuse which become airborne, such as alkali salts, will promote this corrosion. For incinerators operating above 700°F metal temperatures (most of the incinerator tubes noted in Table 11.20 operate at temperatures in excess of 700°F), special refractory-lined water walls must be utilized to protect the tubes from this metal wastage.

Table 11.21 indicates the relative performance of various alloys in incinerator fireside areas. Although stainless steels appear to have favorable corrosion resistance, the danger of stainless-steel stress corrosion cracking prohibits its use in pressure vessels such as boilers and high-temperature hot-water heaters. Figure 11.37 further illustrates the rate of corrosion of carbon steel by chloride attack as a function of metal temperature.

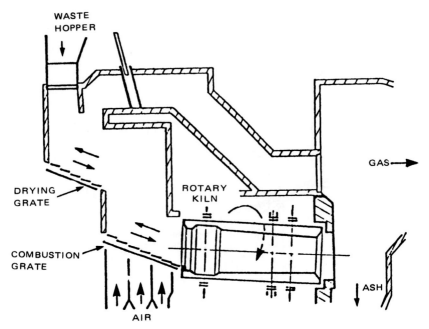

WASTE
HOPPER

GAS→

DRYING
GRATE

ROTARY
KILN

COMBUSTION
GRATE

ASH

AIR

FIGURE 11.33 Volund System. (*Source: Ref. 7.*)

Incinerator Refractory Selection. Table 11.22 illustrates the types of problems that can be expected within the various areas of a large incinerator system associated with refractory selection. The nature of the hot gas stream within the incinerator can create significant detrimental effects on grates, walls, ceilings, and other areas within the furnace enclosure. Some of the more common concerns in refractory selection are discussed here.

Abrasion. This is the effect of impact of moving solids within the gas stream or of heavy pieces of materials charged into the furnace upon refractory surfaces. Fly ash also causes abrasive effects. Abrasion is the wearing away of refractory, or any surface, under direct contact with another material with relative motion to its surface.

Slagging. When a portion of charged material, usually ash, metals, or glass within an incinerator, reaches a high enough temperature, deformation of that material will occur. The material will physically change to a more amorphous state and may begin to flow, as a heavy liquid. When the temperature of the material is then reduced below that required for deformation, the material will solidify into a hard slag. This process can take place when high temperatures are experienced on a grate. When the grate section moves through a lower-temperature zone, a slag may form. Molten ash may become airborne, then attach to a refractory or a metal surface within the furnace which is cooler than the air stream. Slag will then form on this surface. This slag can be acidic (as a result of silicon, aluminum, or titanium oxides released from the burning waste), or it can be basic (due to the generation of oxides of iron, calcium, magnesium, potassium, sodium, or chromium), and the selection of refractory must be compatible with these materials to help ensure long refractory life. (For acidic slag, fireclay or high-alumina and/or silica firebrick would be used. Chrome, magnesite, or forsterite brick would be used for basic slags.)

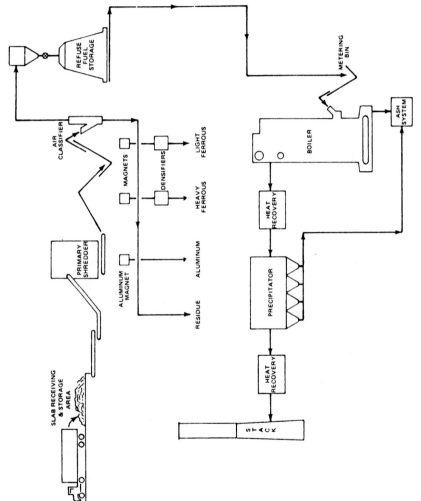

FIGURE 11.34 Akron recycle energy system. (*Source: Akron Recycle Energy System, Teledyne National, Akron, Ohio.*)

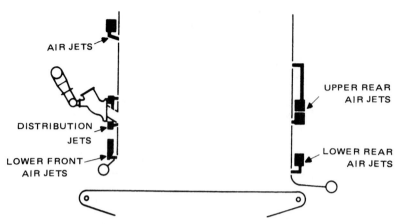

FIGURE 11.35   Air and feed distribution. (*Source: Babcock & Wilcox, North Canton, Ohio.*)

TABLE 11.19   Gas-Phase Corrosion at Elevated Temperatures

Alloy	Temperature, °F	Corrosion rate, mils/month
A106	800	0.9
A106	1000	8
A106	1200	36
T11	800	0.8
T11	1000	6
T11	1200	29

1 mil = 2.54×10^{-2} mm.
Source: Ref. 7.

Mechanical Shock.   The impact of falling refuse can cause mechanical shock, as can constant vibration caused by grate bushings or supports and vibration set up by turbulent flow adjacent to air inlet ports.

Spalling.   The flaking away of the refractory surface, or spalling, is most commonly caused by thermal stresses or mechanical action. Uneven temperature gradients can cause local thermal stresses in brick which will degrade the refractory surface, causing a spalling condition. The more common type of mechanical spalling is caused by rapid drying of wet brickwork. The steamed water does not have an opportunity to escape the brick surface through the natural porosity of the refractory, but expands rapidly, causing cracking and spalling of the brick.

Fly Ash Adherence.   As noted above, fly ash can have fluid properties within the hot gas stream and adhere to refractory or other cooler surfaces within the furnace chamber. Fly ash accumulation can result in corrosive attack on these surfaces. Heavy accumulations will interfere with the normal surface cooling effects within the furnace, resulting in a decrease of furnace refractory life.

As noted above, Table 11.22 lists various areas within an incinerator and describes the severity of the above problems at each of these locations. Normal temperature ranges are noted as well as recommended refractory types.

TABLE 11.20   Nominal Operating Conditions of Waterwall Incinerators

Location	Steam pressure, lb/in² gauge	Steam temp., °F	Metal temp., °F (approx.)
Milan, Italy	500	840	890
Mannheim, Germany	1800	980	1030
Frankfurt, Germany	960	930	980
Munster, Germany	1100	980	1030
Moulineaux, France	930	770	820
Essen Karnap, Germany	--	930	980
Stuttgart, Germany	1100	980	1030
Munich, Germany	2650	1000	1050
Rotterdam, Netherlands	400	680	730
Edmonton, England	625	850	900
Coventry, England	275	415	465
Amsterdam, Netherlands	600	770	820
Montreal, Canada	225	395	445
Chicago (N. W.), Illinois	265	410	460
Oceanside, New York	460	465	515
Norfolk, Virginia	175	375	425
Braintree, Massachusetts	265	410	460
Harrisburg, Pennsylvania	275	460	510
Hamilton, Ontario	250	400	450

Source: Ref. 7

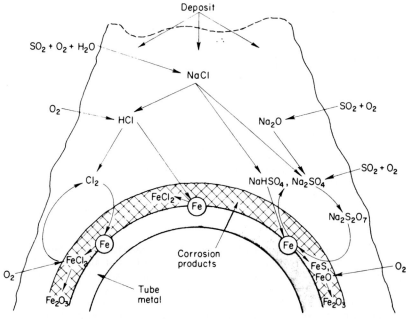

FIGURE 11.36 Sequence of chemical reactions explaining corrosion on incinerator boiler tube. (*Source: Ref. 7.*)

TABLE 11.21 Performance of Alloys in Fireside Areas[a]

| | Resistance to wastage | | |
Alloy	300–600°F	600–1200°F	Moist deposit
Incoloy 825	Good	Fair	Good
Type 446	Good	Fair	Pits
Type 310	Good	Fair	SCC[b]
Type 316L	Good	Fair	SCC
Type 304	Good	Fair	SCC
Type 321	Good	Fair	SCC
Inconel 600	Good	Poor	Pits
Inconel 601	Good	Poor	Pits
Type 416	Fair	Fair	Pits
A106-Grade B (carbon steel)	Fair	Poor	Fair
A213-Grade T11 (carbon steel)	Fair	Poor	Fair

[a]Arranged in approximate decreasing order.
[b]Stress-corrosion cracking.
Source: Ref. 7.

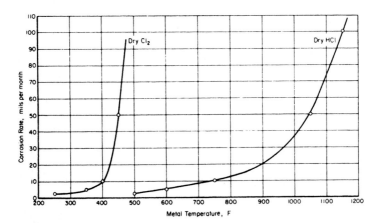

FIGURE 11.37 Corrosion of carbon steel in chlorine and hydrogen chloride. (*Source: Ref. 7.*)

Heat Recovery or Wasting of Heat.   Steam can be generated by utilization of incinerator waste heat, as shown in Table 11.23, which lists steam production rates. Variations in waste heating value will produce variations in steam generation, as shown in Table 11.24. But these quantities must be weighed against the overall implications of a boiler installation for waste burning. Note the following comparison between an energy recovery system and an incinerator without provision for heat recovery:

TABLE 11.22 Suggested Refractory Selection for Incinerators

Incinerator part	Temperature (°F) range	Abrasion	Slagging	Mechanical shock	Spalling	Fly ash adherence	Recommended refractory
Charging gate	70–2600	Severe, very important	Slight	Severe	Severe	None	Superduty
Furnace walls, grate to 48 in. above	70–2600	Severe	Severe, very important	Severe	Severe	None	Silicon carbide or superduty
Furnace walls, upper portion	70–2600	Slight	Severe	Moderate	Severe	None	Superduty
Stoking doors	70–2600	Severe, very important	Severe	Severe	Severe	None	Superduty
Furnace ceiling	700–2600	Slight	Moderate	Slight	Severe	Moderate	Superduty
Flue to combustion chamber	1200–2600	Slight	Severe, very important	None	Moderate	Moderate	Silicon carbide or superduty
Combustion chamber walls	1200–2600	Slight	Moderate	None	Moderate	Moderate	Superduty
Combustion chamber ceiling	1200–2600	Slight	Moderate	None	Moderate	Moderate	Superduty
Breeching walls	1200–3000	Slight	Slight	None	Moderate	Moderate	Superduty
Breeching ceiling	1200–3000	Slight	Slight	None	Moderate	Moderate	Superduty
Subsidence chamber walls	1200–1600	Slight	Slight	None	Slight	Moderate	Medium duty
Subsidence chamber ceiling	1200–1600	Slight	Slight	None	Slight	Moderate	Medium duty
Stack	500–1000	Slight	None	None	Slight	Slight	Medium duty

Source: Ref. 7.

With heat recovery	Without heat recovery
Reduced gas temperatures and volumes due to absorption of heat by heat recovery system	Hotter gas temperatures
Moderate excess air	High excess air required to control furnace temperatures
Moderate-size combustion chamber	Large refractory-lined combustion chamber to handle high gas flow
Smaller air and induced-draft fans required for smaller gas volume	Higher gas volumes due to higher temperatures, requiring larger air and gas flow equipment
Steam facilities including integral water wall, boiler drums, and boiler auxiliary equipment required	No steam facilities required
Operations involve boiler system monitoring, adjustments for steam demand, etc.	Relatively simple operating procedures
Steam tube corrosion is possible as well as corrosion within exhaust gas train	Corrosion possible in the exhaust gas train
Licensed boiler operators are required to operate incinerator	Conventional operators satisfactory
Considerable steam credits possible, including in-plant energy savings in addition to salvage	Only credits are possible salvage of the equipment after its useful life

TABLE 11.23 Typical Steam Generation

Solid waste type	MSW	MSW	MSW	RDF
Steam temperature, °F	620	500	465	400
Steam pressure, lb/in^2 gauge	400	225	260	250
Steam production, tons/ton refuse	3.6	1.4–3.0	1.5–4.3	4.2

Note:
MSW: Municipal solid waste.
RDF: Refuse-derived fuel, i.e., shredded MSW less metals.
Source: Ref. 7.

TABLE 11.24 Steam Production Related to MSW Characteristics

	As-received heating value, Btu/lb				
	6500	6000	5000	4000	3000
Refuse:					
% moisture	15	18	25	32	39
% noncombustible	14	16	20	24	28
% combustible	71	66	55	44	33
Steam generated, tons/ton refuse	4.3	3.9	3.2	2.3	1.5

Source: Ref. 7.

Resource Recovery Plant Emissions. A number of states have established regulations specifically governing emissions of central disposal incinerators firing municipal solid waste (resource recovery facilities). Table 11.25 lists these regulations for six states and the EPA. Most of these criteria are noted as guidelines. A guideline is not necessarily established by statute, but it is used by the regulator as a criterion for the permitting of these facilities.

Biomedical Waste Incineration

Biomedical wastes are generated by hospitals, laboratories, animal research facilities, and other institutional sources. The disposal of these wastes is coming under severe public scrutiny, and regulations are being promulgated to control their disposal. Incineration is a favored method of treating these wastes because it is the only commercially available method of treatment which destroys the organisms associated with this waste completely and effectively.

The Waste Stream. *Biomedical waste* is a term coming into common usage to replace what had been referred to as *pathological* or *infectious* wastes and to include additional related waste streams. Where the term *pathological waste* is used here, it refers to anatomical wastes, carcasses, and similar wastes. Table 11.26 is a listing of wastes classified as biomedical and includes a description of each waste as well as typical characteristics. The bag designations (red, orange, yellow, blue) are used in Canada. In the United States, generally, most of these wastes are classified as "red bag."

The hospital waste stream has changed significantly in the last few years. Disposable plastics have been replacing glass and clothing in what appears to be, at first look, a means of cutting costs. They represent a greater cost in their disposal, however, since many of these plastics contain chlorine and with the increase in the use of plastics, the increase in chlorine creates the need for additional equipment in the incineration process. The plastics content of the hospital waste stream has grown from 10 percent to over 30 percent in the past 10 years.

It is rare to find an incinerator designated for biomedical waste destruction to be fired solely on this type of waste. Generally, particularly in hospitals, installation of an incinerator encourages the disposal of other wastes in the unit. Besides the cost savings this represents in not having to cart away this trash, there is the potential for heat recovery. For example, hospitals generally require steam throughout the year for their laundry, sterilizers, autoclaves, and kitchens. As more waste is fired, more heat is produced and more steam is generated.

Another set of wastes includes those generated in hospital laboratories that are hazardous wastes under the Resource Conservation and Recovery Act (RCRA). Table 11.27 lists some of these wastes. If more than 200 lb/month of these wastes is generated, the incinerator must be permitted under the provisions of RCRA, which are additional to state requirements.

Waste generation rates will vary from one hospital to another, as a function of the number of hospital beds, the number of intensive-care beds, and the presence of other specialty facilities. In the absence of specific generation data, the figures in Table 11.28 can be used as an estimate of waste generation rates.

Regulatory Issues. Biomedical waste incinerators are generally small, much smaller than the central disposal incinerators that have been in the public eye in many of the densely populated areas of the country. Regulations have addressed the larger municipal solid waste incinerators. Smaller units, such as 2 to 10 tons/day biomedical waste

TABLE 11.25 Emissions Limitations for Municipal Waste Incinerators

Pollutant	California guidelines 7% O_2	Illinois guidelines 12% CO_2	New Jersey guidelines 7% O_2	New York guidelines 12% CO_2	Pennsylvania BAT criteria 7% O_2	USEPA guidelines 12% CO	Wisconsin guidelines 12% CO_2
Particulate	0.01	0.01	0.015	0.01	0.015	—	0.015
(below 2 μm), gr/dry st ft³	0.008						
HCl	30 ppm	30 ppm/1 h or 90% reduction	50 ppm/1 h or 90% reduction	50 ppm/8 h or 90% reduction	30 ppm/1 h or 90% reduction	—	50 ppm
SO_2	30 ppm	50 ppm/1 h or 70% reduction	50 ppm/1 h or 80% reduction	0.2–2.5 lb/(MBtu/h)	50 ppm/1 h or 70% reduction	—	—
NO_x	140–200 ppm	100 ppm/1 h	350 ppm/1 h	BACT	—	—	—
Hydrocarbons	70 ppm	—	70 ppm/1 h	—	—	—	—
CO	400 ppm	100 ppm/1 h	400 ppm/1 h 100 ppm/4 days	—	400 ppm/8 h 100 ppm/4 days	50 ppm/4 h	—
Dioxin	—	—	—	2 ng/Nm³	—	—	3 ng/Nm³
Furnace temp., design, °F	1800 ±200	1800	1800	1800	—	—	—
Furnace temp., min., °F		1500	1500	1500 for 15 min	1800	1800	1500
Residence time, min., s	1	1.2	1	1	1	1	1
Lime injection, min., lb/h	—	100	—	—	—	—	—
Baghouse temp., max., °F	—	—	—	300	—	—	250
Combustion efficiency, %	—	99.9/2 h	—	99.9/8 h 99.95/7 days	99.9/4 days	—	—
Minimum O_2, %	—	6	6	—	—	6-12	—
Opacity, max., %	—	10	20	10/6 min	30/3 min/h	—	20

TABLE 11.26 Characterization of Biomedical Waste

Waste class	Component description	Typical component weight, % (as fired)	HHV dry basis, Btu/lb	Bulk density as fired, lb/ft	Moisture content of component, wt %	Weighted heat value range of waste component, Btu/lb	Typical component heat value of waste as fired, Btu/lb
A1 (red bag)	Human anatomical	95–100	8,000–12,000	50–75	70–90	760–2,600	1,200
	Plastics	0–5	14,000–20,000	5–144	0–1	0–1,000	180
	Swabs, absorbents	0–5	8,000–12,000	5–62	0–30	0–600	80
	Alcohol, disinfectants	0–0.2	11,000–14,000	48–62	0–0.2	0–28	20
	Total bag						1,480
A2 (orange bag)	Animal infected anatomical	80–100	9,000–16,000	30–80	60–90	720–6,400	1,500
	Plastics	0–15	14,000–20,000	5–144	0–1	0–3,000	420
	Glass	0–5	0	175–225	0	0	0
	Beddings, shavings, paper, fecal matter	0–10	8,000–9,000	20–46	10–50	0–810	600
	Total bag						2,520
A3a (yellow bag)	Gauze, pads, swabs, garments, paper, cellulose	60–90	8,000–12,000	5–62	0–30	3,360–10,800	6,400
	Plastics, PVC, syringes	15–30	9,700–20,000	5–144	0–1	1,440–6,000	3,250
	Sharps, needles	4–8	60	450–500	0–1	3–5	5
	Fluids, residuals	2–5	0–10,000	62–63	80–100	0–11	30
	Alcohols, disinfectants	0–0.2	7,000–14,000	48–62	0–50	0–28	15
	Total bag						9,700
A3b (yellow bag) Lab waste	Plastics	50–60	14,000–20,000	5–144	0–1	6,930–12,000	9,000
	Sharps	0–5	60	450–500	0–1	0–3	0
	Cellulosic materials	5–10	8,000–12,000	5–62	0–15	340–1,200	650
	Fluids, residuals	1–20	0–10,000	62–63	95–100	0–100	30
	Alcohols, disinfectants	0–0.2	11,000–14,000	48–62	0–50	0–28	20
	Glass	15–25	0	175–225	0	0	0
	Total bag						9,700
A3c (yellow bag) R&D	Gauze, pads, swabs	5–20	8,000–12,000	5–62	0–30	280–3,600	1,000
	Plastics, petri dishes	50–60	14,000–20,000	5–144	0–1	6,930–12,000	9,000
	Sharps, glass	0–10	60	450–500	0–1	0–6	0
	Fluids	1–10	0–10,000	62–63	80–100	0–200	100
	Total bag						10,100
B1 (blue bag)	Noninfected Animal anatomical	90–100	9,000–16,000	30–80	60–90	810–6,400	1,400
	Plastics	0–1	14,000–20,000	5–144	0–1	0–20,000	1,000
	Glass	0–3	0	175–225	0	0	0
	Beddings, shavings, fecal matter	0–10	8,000–9,000	20–46	10–50	0–810	600
	Total bag						3,000

Source: Ref. 9

incinerators, have generally not been subject to rigorous regulatory attention in the past. The only restriction on their operation in many parts of the country is that they not create a public nuisance. That has meant that no odors are to be generated and that the opacity is to be low, i.e., no greater than Ringleman no. 1 for more than, for instance, 5 min/h. Incinerators have been designed to this standard, which is virtually no standard at all. As public attention is starting to focus on hazardous, dangerous, and toxic wastes, the regulatory attitude toward biomedical waste incinerators is starting to change. These incinerators are not addressed by the federal government yet, but many states are moving in the direction of regulation. In some states these wastes are classified as hazardous; in others they are regulated as a unique waste stream with its own set of regulations; and in still others there is still no regulation of biomedical wastes per se.

Hazardous Waste Incineration.   Where hazardous regulations must be complied with, the incinerator design and operation must be subject to the RCRA regulations for handling and disposal. Incineration regulations under RCRA require an extensive analytical and compliance process. In addition to operating requirements, the RCRA incinerator regulations mandate extensive record-keeping and reporting procedures. A detailed, comprehensive operator training program must also be implemented.

Combined Hazardous Waste Systems.   Hazardous waste incineration systems require an RCRA permit and strict operating controls and reporting standards. Another significant issue associated with hazardous waste incinerators is that the ash is always considered hazardous. Procedures exist for delisting ash (declaring ash nonhazardous), but this requires extensive testing and administrative activity (filings and petitions) which represent at least 18 months of reporting and review.

If, for example, 1000 lb of biomedical waste were incinerated, approximately 200 lb of ash would be generated. In a state not classifying such waste as hazardous, the ash could be deposited directly in a nonhazardous (municipal waste) landfill. If 100 lb of a hazardous waste were fired in the secondary chamber of this same incinerator, all the ash would be considered hazardous and would have to be deposited in a hazardous waste landfill (unless it were delisted). Where 100 lb of hazardous waste was originally present, now at least 200 lb of hazardous waste must be disposed of. As a general

TABLE 11.27   Hazardous Wastes under RCRA Typically Generated by In-Hospital Laboratories

Acetone	Methyl alcohol
Antineoplastics	Methyl cellosolve
Butyl alcohol	Pentane
Cyclohexane	Petroleum ether
Diethyl ether	Tetrahydrofuran
Ethyl alcohol	Xylene

Source: Ref. 10.

TABLE 11.28   Estimated Waste Generation Rates

Hospital	13 lb/(occupied bed • day)
Rest home	3 lb/(person • day)
Laboratory	0.5 lb/(patient • day)
Cafeteria	2 lb/(meal • day)

Source: Ref. 11.

rule, it is impractical and uneconomical to incinerate hazardous and nonhazardous waste in the same incinerator.

Waste Combustion. Hospital wastes will contain paper and cardboard, plastics, aqueous and nonaqueous fluids, anatomical parts, animal carcasses, and bedding, glass bottles, clothing, and many other materials. Much of this waste is combustible. Lighting a match to a mixed assortment of hospital waste will generally result in a sustained flame, unless it contains a high proportion of liquids, anatomical, or other pathological waste materials.

Thermal treatment technologies include starved air and excess air combustion processes, as described previously in this text. The main advantage of starved air is a low air requirement in the primary chamber. With little air passing through the waste there is less turbulence within the system and less particulate carryover from the burning chamber. This low airflow also results in very low nitrogen oxide generation, although this is not normally a concern in hospital incinerators. Less supplemental fuel is required than with excess air systems, where the entire airflow must be brought to the operating temperature of the incinerator. With the lower airflow, fans, ducts, flues, and air emissions control equipment can be sized smaller than in excess air systems.

In small systems, with less than 100 lb/h throughput, starved air operation is difficult, if not impossible, to achieve for two reasons. It is difficult to control air leakage into the system, and it is not possible to determine an accurate waste heating value on which to base a definition of the stoichiometric air requirement (see Table 11.29). As listed in Table 11.29, the stoichiometric air requirement can vary by a factor of 4 depending on the types of materials normally found in a biomedical waste stream. With larger systems the significance of air leakage decreases and the variations in feed characteristics will tend to even out. With units above 1500 lb/h, starved air can be sustained for many medical waste streams.

Waste Destruction Criteria. Generally, paper waste (cellulosic materials) requires that a temperature of 1400°F be maintained for a minimum of 0.5 s for complete burnout. The temperature-residence time requirement for biomedical waste destruction must be at least equal to the requirements for paper waste; however, the specific relationship between temperature and residence time must be determined for the specific waste. Many states require that a temperature of 1800°F be maintained for a minimum of 1 or 2 s, and some states require a 2000°F off-gas temperature.

On the high-temperature side, it is necessary to consider the general nature of much of the biomedical waste stream. It has a high proportion of organic material, including cellulosic waste. The noncombustible portion of this waste (ash) will begin to melt, or at least desolidify, as the temperature increases. Table 11.18 lists the ash fusion temperature of refuse, which represents the same constituents as much of the

TABLE 11.29 Waste Combustion Characteristics

Waste constituent	Btu/lb	lb air/lb waste*
Polyethylene	19,687	16
Polystyrene	16,419	13
Polyurethane	11,203	9
PVC	9,754	8
Paper	5,000	4
Pathological	Will not support combustion	

*Stoichiometric requirement.
Source: Ref. 12.

biomedical waste stream. Above 1800°F, the ash produced will begin to deform in a reducing atmosphere, i.e., where there is a lack of oxygen in the furnace. When ash starts to deform and then is moved to a cooler portion of the incinerator, or to an area of the furnace where additional oxygen is present, the ash will harden into slag or clinker. This hardened ash can clog air ports, disable burners, corrode refractory, and interfere with the normal flow of material through the furnace. To prevent slagging, the temperature within the incinerator should never be allowed to rise above 1800°F. Higher temperatures also encourage the discharge of heavy metals to the gas stream, which is another reason not to impose an arbitrarily high temperature on the process.

Past Practices.   Modular units have been popular in the past because of their relatively low cost. A major factor contributing to their cost advantage over rotary kilns and other equipment is that they require no external air emissions control equipment to produce a fairly clean stack discharge. When properly designed and operated, they can achieve a particulate emissions rate of 0.08 gr/dry st ft^3 (corrected to 50 percent excess air). As new regulations are promulgated, however, lower emissions limitations will make the use of baghouses or electrostatic precipitators mandatory, and the modular incinerator (with inclusion of control equipment in the system package) will likely lose its price advantage over other systems.

The higher cost of rotary kilns was due to their need for external air emissions control equipment and to inclusion of the drive mechanism necessary for its operation.

Starved Air Process Limitations.   The most important issue associated with starved air combustion is the nature of the waste. (Note that there may be references to pyrolysis in some literature, including sales brochures for such equipment, but this term is usually used in error. The process is starved air combustion.)

For any starved air reaction to occur, the waste must be basically organic and able to sustain combustion without the addition of supplemental fuel, a definition of *autogenous* combustion. Without sufficient heat content to sustain combustion, the concept of substoichiometric burning has no meaning.

A waste with a moisture content in excess of 60 percent will not burn autogenously at 1600°F. As noted in Table 11.26, the moisture contents of red bag, orange bag, and blue bag wastes are generally in excess of this figure. Starved air combustion will not work when wastes with this moisture fraction are placed in the furnace.

Usually a starved air incinerator is designed to fire a paper waste, and the burners in both the primary and secondary chambers of the incinerator are sized appropriately: for a relatively small supplemental-fuel requirement. When a bag of pathological waste is placed in the primary combustion chamber, the waste will not burn autogenously and supplemental fuel must be added. In most present starved air incinerator designs, the burners have too low a capacity to provide the heat required when organics released by a starved air process are not present. Without unburned organics in the secondary chamber (when starved air combustion has not occurred in the primary combustion chamber), the sizing of the secondary burner is generally inadequate (the burner is too small) to provide the fuel required for complete burnout of the gas stream.

On the other end of the operating range, when a waste with a very high heating value is introduced (a plastic material such as polyurethane or polystyrene, for instance, as noted in Table 11.29), a good deal of air is required to generate even the substoichiometric requirement necessary to generate heat for the process to advance. This air quantity is often much greater than the airflow present from the fans provided with the unit. If a high-quality paper waste (waste paper, boxes, cartons, cardboard, etc.) is introduced into an incinerator, starved air combustion will generally work, assuming there is good control of infiltration air. The incinerator should be designed, however, not for paper waste, but for the firing of mixed paper and pathological waste, which requires a relatively large heat input and a coordinated increase in air-flow (primary and secondary combustion air and burner air fans) in both the primary and the secondary combustion chambers.

Starved air operation of an operating incinerator can be easily checked. Increase the airflow in the primary combustion chamber, and if its temperature increases, the incin-erator is operating in the starved air mode. If its temperature decreases, the incinerator is operating as an excess air unit; i.e., starved air operation is not occurring.

One must expect, in the design of biomedical waste incinerators, that although the incinerator charge may be sufficiently large to preclude swings from very high to very low heat value wastes, this is not always the case. It is likely that a single incinerator charge can contain a polystyrene mattress (and very little else) that will start to burn almost at once upon insertion into the incinerator.

Likewise, since much of the waste charged into an incinerator is in opaque bags, and the waste cannot be identified, a charge can consist almost wholly of anatomical waste, or animal carcasses, or liquid (aqueous) materials which have a very low heat content. An incinerator must be designed for this certain variation in waste stream quality. Of the incinerators on the market today, the starved air unit is least able to adapt to changes in waste constituents. As noted previously, this is of particular con-cern with smaller systems.

Incinerator Analysis

Combustion Properties.   The effectiveness of combustion is related to the combina-tion of three factors, the "three T's": temperature in the furnace, time of residence of the combustion products at the furnace temperature, and turbulence within the furnace.

In general a solid or liquid must be converted to a gaseous phase before burning will occur. (Examine a lit match or a burning log. The flame does not rise directly from the solid. There is a zone immediately above the match, or log, where the gaseous fuel phase has been generated and is mixing with combustion air prior to burning.) The three T's are factors which control the rapidity of conversion of solid and liquid fuel to the gaseous phase.

Furnace Temperature.   Furnace temperature is a function of fuel heating value, furnace design, air admission, and combustion control. The minimum temperature must be higher than the ignition temperature of the waste. The upper temperature limit is normally a function of the enclosure materials and the ash melting temperature. Over 2400°F operation requires use of special refractory materials. The rates of com-bustion reactions increase rapidly with increased temperatures. Of the three T's (tem-perature, time, and turbulence), only temperature can be significantly controlled after a furnace is constructed. Time and turbulence are fixed by furnace design and airflow rate and can normally be controlled only over a limited range.

Furnace Temperature Control.   This can be achieved as follows:

Excess air control, i.e., control of the air-fuel ratio. Temperature produced is a direct function of the fuel properties and excess air introduced. Excess air control requires either automatic control or close manual supervision.

Direct heat transfer by the addition of heat-absorbing material within the furnace, such as water-cooled furnace walls. The addition of water sprays in the combustion zone (1000 Btu/lb water evaporated is equivalent to $\frac{1}{2}$ Btu/°F sensible heat in the flue gas) will reduce the furnace temperature. Use of water sprays must be carefully con-trolled to avoid thermal shock to the furnace refractory.

Furnace Gas Turbulence. *Turbulence* is an expression relating the physical relationship of fuel and combustion air in a furnace. A high degree of turbulence, intimate mixing of air and fuel, is desirable. Burning efficiency is enhanced with increased surface area of fuel particles exposed to the air. Fuel atomization maximizes the exposed particle surface. Turbulence helps to increase particle surface area by promoting fuel vaporization. In addition, good turbulence exposes the fuel to air in a rapid manner, helping to promote rapid combustion and maximizing fuel release.

A burner requiring no excess air and producing no smoke is said to have perfect turbulence (a *turbulence factor* of 100 percent). If, for instance, 15 percent excess air is required to achieve a no-smoke condition, the turbulence factor is calculated as follows:

$$\text{Turbulence factor} = \frac{\text{stoichiometric air}}{\text{total air} \times 100\%}$$

$$= \frac{1.00}{1.15} \times 100\% = 87\%$$

Fuel gas burners can be designed to produce a turbulence factor close to 100 percent.

Retention Time. Combustion does not occur instantaneously. Sufficient space must be provided within a furnace chamber to allow fuel and combustible gases the time required to fully burn. This factor, termed *dwell time, residence time,* or *retention time,* is a function of furnace temperature, degree of turbulence, and fuel particle size.

Retention time required may be a fraction of a second, as when gaseous waste is burned, or many minutes, as when solid granular waste such as powdered carbon is burned.

Waste Combustion. The characteristics of an incineration process depend upon the characteristics of the waste incinerated (Table 11.30). Waste characteristics of interest include the heat release, the air required for combustion of the waste, and the dry gas and moisture generated from waste burning.

Municipal waste will have three components: moisture, ash, and combustibles. Moisture will generally be in the range of 18 to 25 percent, ash (or noncombustibles) will normally be from 10 to 20 percent of the waste, and the balance of the waste will be combustible (or volatiles).

TABLE 11.30 Waste Burning Characteristics for Stoichiometric Combustion

Q, Btu/lb	Air, lb/10 kB	Dry gas, lb/10 kB	H_2O, lb/10 kB	C, %	H_2, %	O_2, %	S, %	N_2, %	Cl_2, %
7,000	7.13	8.00	0.56	47.62	4.41	46.07	0.25	0.82	0.83
8,000	7.01	7.67	0.59	49.36	5.29	43.45	0.25	0.82	0.83
9,000	6.91	7.41	0.61	51.11	6.16	40.84	0.25	0.82	0.83
10,000	6.83	7.20	0.62	52.85	7.03	38.22	0.25	0.82	0.83
11,000	6.76	7.03	0.64	54.59	7.90	35.60	0.25	0.82	0.83
12,000	6.72	6.90	0.65	56.38	8.79	32.93	0.25	0.82	0.83
13,000	6.67	6.77	0.66	58.08	9.65	30.37	0.25	0.82	0.83
14,000	6.63	6.67	0.67	59.83	10.52	27.75	0.25	0.82	0.83
15,000	6.59	6.58	0.68	61.57	11.39	25.14	0.25	0.82	0.83

Note: kB = 1000 Btu (10 kB = 10 kBtu = 10,000 Btu).

The combustible content of a waste is that portion of the waste that will burn. An equation to estimate of heat release Q in Btu/lb, is as follows:

$$Q = 14406C + 67276H_2 - 6187O_2 + 4142S + 2433Cl_2 - 1082N_2$$

where C, H_2, O_2, S, Cl_2 and N_2 are the fractions of carbon, hydrogen, oxygen, sulfur, chlorine, and nitrogen, respectively, in the combustible content of the waste. Table 11.30 is based on this equation. The heat content of the volatile fraction of municipal waste is normally in the range of 8000 to 12,000 Btu/lb.

Gas Properties.   The gases generated by incineration will include nitrogen, carbon monoxide, oxygen, sulfur dioxide, hydrogen chloride, water vapor, and trace amounts of other gases. These gases can be considered to have a dry component and moisture. In most municipal waste incineration systems from 150 to 250 percent of the stoichiometric (ideal) air requirement is injected into the incinerator as combustion air. This means that from 50 to 150 percent of the air injected into the incinerator remains in the exhaust gas after burning is complete.

The dry gas component of the incinerator off-gas stream will have characteristics similar to those of air because of this relatively large air carryover through the process. In the following calculations, the characteristics of the incinerator off-gas are considered to be the same as that of dry air and moisture. This introduces a slight error, less than 3 percent, in the calculation of incinerator temperature, fuel requirement, scrubber performance, etc. This error is insignificant compared to the error inherent in other waste parameters. For instance, the waste feed rate is not normally known to within 3 percent and the heat content of the waste is generally an estimate. It cannot be determined with an accuracy of less than 5 percent, not just because of the heterogeneous nature of the waste, but because of the difficulty of obtaining a truly representative waste sample for analysis. By assuming that the dry gas characteristics are those of dry air rather than of the individual gases, the calculations are made more expeditious.

Table 11.31 lists the enthalpy of dry gas and moisture relative to 60°F and 80°F. The properties of a saturated mixture of moisture vapor in air are listed in Table 11.32.

The Mass Balance.   The flow weight into an incinerator must equal the flow of products leaving the incinerator. Input includes waste fuel, air (including humidity entrained within the air), and supplementary fuel. The flow exiting the incinerator includes moisture and dry gas in the exhaust as well as ash, both in the exhaust as fly ash and exiting as bottom ash.

Table 11.33, the mass flow table, provides an orderly method of establishing a mass balance surrounding a combustion system. Of initial interest is waste quality: its moisture content, ash content (noncombustible fraction), and heat content. Of prime importance is the generation of moisture and dry flue gas from the combustion process.

For the purpose of this example, the following waste will be assumed, at the indicated firing rate and other combustion parameters:

8000 lb/h of municipal solid waste

5000 Btu/lb as fired

25 percent moisture as fired

20 percent ash as fired

Fired with 100 percent excess air

TABLE 11.31 Enthalpy of Air, and Moisture

Relative to 60°F			Relative to 80°F	
H_{Air}, Btu/lb	H_{H_2O}, Btu/lb	Temp., °F	H_{Air}, Btu/lb	H_{H_2O}, Btu/lb
21.61	1091.92	150	16.82	1071.91
33.65	1116.62	200	28.86	1096.61
45.71	1140.72	250	40.92	1120.71
57.81	1164.52	300	53.02	1144.51
69.98	1188.22	350	65.19	1168.21
82.19	1211.82	400	77.40	1191.81
94.45	1235.47	450	89.66	1215.46
106.79	1259.22	500	102.00	1239.21
119.21	1283.07	550	114.42	1263.06
131.69	1307.12	600	126.90	1287.11
144.25	1331.27	650	139.46	1311.26
156.87	1355.72	700	152.08	1335.71
169.59	1380.27	750	164.80	1360.26
187.38	1405.02	800	177.59	1385.01
195.26	1430.02	850	190.47	1410.01
208.21	1455.32	900	203.42	1435.31
221.25	1480.72	950	216.46	1460.71
234.36	1506.42	1000	229.57	1486.41
247.55	1532.40	1050	242.76	1512.40
260.81	1558.32	1100	256.02	1538.31
274.15	1584.80	1150	264.36	1564.80
287.55	1611.22	1200	282.76	1591.21
301.02	1638.26	1250	296.23	1618.20
314.56	1665.12	1300	309.77	1645.11
328.17	1692.15	1350	323.38	1672.15
341.85	1719.82	1400	337.06	1699.81
355.58	1747.70	1450	350.82	1727.70
369.37	1775.52	1500	364.58	1755.51
397.17	1832.12	1600	392.33	1812.11
425.08	1890.11	1700	420.29	1870.10
453.24	1948.02	1800	448.45	1928.01
481.57	2007.17	1900	476.78	1987.70
510.07	2067.42	2000	505.28	2047.41
538.72	2128.70	2100	533.93	2108.70
567.52	2189.92	2200	562.73	2169.91
596.45	2252.60	2300	591.66	2232.60
625.52	2315.32	2400	620.73	2295.31
654.70	2377.80	2500	649.91	2357.80
684.01	2443.30	2600	679.22	2423.30
713.42	2511.88	2700	708.63	2491.80

Source: Ref. 7.

TABLE 11.32 Saturation Properties of Dry Air with Moisture

Temp., °F	Humidity, lb H_2O/lb DA	Enthalpy, Btu mixture/lb DA	Volume, ft^3 mixture/lb DA
60	0.01108	0.000	13.329
61	0.01149	0.690	13.363
62	0.01191	1.393	13.398
63	0.01234	2.114	13.433
64	0.01279	2.849	13.468
65	0.01326	3.602	13.504
66	0.01374	4.369	13.540
67	0.01423	5.155	13.680
68	0.01474	5.958	13.613
69	0.01527	6.781	13.650
70	0.01581	7.621	13.688
71	0.01638	8.484	13.762
72	0.01700	9.366	13.764
73	0.01755	10.266	13.803
74	0.01817	11.191	13.842
75	0.01881	12.136	13.882
76	0.01946	13.094	13.922
77	0.02014	14.094	13.963
78	0.02084	15.109	14.004
79	0.02156	16.149	14.046
80	0.02231	17.214	14.088
81	0.02308	18.306	14.131
82	0.02387	19.426	14.175
83	0.02468	20.571	14.219
84	0.02552	21.744	14.264
85	0.02639	22.947	14.309
86	0.02728	24.181	14.355
87	0.02821	25.445	14.402
88	0.02916	26.744	14.449
89	0.03014	28.074	14.497
90	0.03115	29.441	14.547
91	0.03219	30.841	14.597
92	0.03326	32.279	14.647
93	0.03437	33.751	14.699
94	0.03551	35.266	14.751
95	0.03668	36.815	14.804
96	0.03789	38.408	14.854
97	0.03914	40.039	14.913
98	0.04043	41.715	14.968
99	0.04175	43.438	15.025
100	0.04312	45.209	15.083
101	0.04453	47.023	15.142
102	0.04498	48.886	15.202
103	0.04748	50.806	15.263
104	0.04902	52.770	15.325
105	0.05061	54.792	15.389
106	0.05225	56.868	15.453
107	0.05394	59.000	15.519
108	0.05568	61.190	15.587

TABLE 11.32 Saturation Properties of Dry Air with Moisture (*Continued*)

Temp., °F	Humidity, lb H$_2$O/lb DA	Enthalpy, Btu mixture/lb DA	Volume, ft^3 mixture/lb DA
109	0.05747	63.444	15.655
110	0.05932	65.764	15.725
111	0.06123	68.144	15.796
112	0.06319	70.589	15.869
113	0.06522	73.102	15.944
114	0.06731	75.690	16.020
115	0.06946	78.357	16.098
116	0.07168	81.095	16.178
117	0.07397	83.197	16.259
118	0.07633	86.815	16.343
119	0.07877	89.800	16.428
120	0.08128	92.880	16.515
121	0.08388	96.840	16.603
122	0.08655	99.300	16.695
123	0.08931	102.65	16.789
124	0.09216	106.11	16.885
125	0.09511	109.67	16.983
126	0.09815	113.34	17.084
127	0.10129	117.14	17.187
128	0.10453	121.04	17.293
129	0.10788	125.06	17.402
130	0.11130	129.22	17.514
131	0.11490	133.50	17.628
132	0.11860	137.87	17.746
133	0.12240	142.46	17.867
134	0.12640	147.18	17.991
135	0.13040	152.02	18.119
136	0.13460	157.04	18.251
137	0.13900	162.24	18.386
138	0.14350	167.58	18.525
139	0.14820	173.11	18.669
140	0.15300	178.82	18.816
141	0.15800	184.73	18.969
142	0.16320	190.85	19.126
143	0.16850	197.17	19.288
144	0.17410	203.72	19.454
145	0.17980	210.49	19.626
146	0.18580	217.52	19.804
147	0.19200	224.82	19.987
148	0.19840	232.35	20.176
149	0.20510	240.17	20.374
150	0.22120	248.29	20.576
151	0.21920	256.71	20.786
152	0.22670	265.45	21.004
153	0.23440	274.53	21.229
154	0.24250	283.96	21.462
155	0.25090	293.78	21.704
156	0.25960	303.98	21.955
157	0.26880	314.64	22.216

TABLE 11.32 Saturation Properties of Dry Air with Moisture (*Continued*)

Temp., °F	Humidity, lb H$_2$O/lb DA	Enthalpy, Btu mixture/lb DA	Volume, ft^3 mixture/lb DA
158	0.27820	325.69	22.487
159	0.28810	337.23	22.769
160	0.29850	349.36	23.063
161	0.30920	361.79	23.368
162	0.32050	374.90	23.685
163	0.33230	388.58	24.017
164	0.34460	402.89	24.365
165	0.35750	417.84	24.725
166	0.37100	433.53	25.102
167	0.38510	449.95	25.492
168	0.40000	467.18	25.914
169	0.41560	485.28	26.347
170	0.43200	504.29	26.804
171	0.44930	524.31	27.272
172	0.46750	545.38	27.787
173	0.48670	567.61	28.315
174	0.50700	591.07	28.876
175	0.52840	615.86	29.465
176	0.55110	642.11	30.089
177	0.57520	669.92	30.749
178	0.60080	699.48	31.449
179	0.62790	730.86	32.193
180	0.65600	764.31	32.984
181	0.68780	800.01	33.829
182	0.72090	838.20	34.731
183	0.75630	879.02	35.694
184	0.79430	922.91	36.728
185	0.83520	970.11	37.839
186	0.87940	1021.0	39.037
187	0.92710	1076.1	40.332
188	0.97900	1135.9	41.737
189	1.03550	1201.0	43.265
190	1.09700	1272.2	44.935
191	1.16500	1350.7	46.764
192	1.24000	1436.3	48.780
193	1.32200	1531.6	51.011
194	1.41400	1637.4	53.488
195	1.51700	1756.2	56.265
196	1.63300	1889.5	59.381
197	1.76500	2040.8	62.918
198	1.91500	2214.0	66.963
199	2.08900	2413.9	71.630
200	2.29200	2648.4	77.102
201	2.53200	2924.2	83.543
202	2.82000	3255.4	91.270
203	3.17300	3660.8	100.750
204	3.61400	4169.8	112.590
205	4.18100	4821.7	127.800
206	4.93900	5694.0	147.600

TABLE 11.32   Saturation Properties of Dry Air with
Moisture (*Continued*)

Temp., °F	Humidity, lb H₂O/lb DA	Enthalpy, Btu mixture/lb DA	Volume, ft³ mixture/lb DA
207	6.00000	6913.0	176.560
208	7.59400	8748.0	219.300
209	10.248	11802	290.440
210	15.54	17887	432.250
211	31.49	36241	859.820

Note: DA = dry air.

Source: From C. R. Brunner, *Handbook of Hazardous Waste Incineration,* TAB Books Inc., Blue Ridge Summit, Pa., 1989. Derived from O. Zimmerman, I. Lavine, *Psychrometric Tables and Charts,* 1st ed., Industrial Research Service, Dover, N.H.

For the example in Table 11.33:

Wet feed, as received charging rate is 8000 lb/h.

Moisture, by weight, of the wet feed is 25 percent. The moisture rate is 25 percent of the wet feed rate, 0.25×8000 lb/h = 2000 lb/h.

Dry feed equals total wet feed less moisture, 8000 lb/h − 2000 lb/h = 6000 lb/h.

Ash is the percentage of total feed that remains after combustion. From the data provided, 20 percent of the total wet feed, 0.20×8000 lb/h = 1600 lb/h ash.

Volatile is that portion of feed that is combusted. It is found by subtracting ash from dry feed: 6000 lb/h − 1600 lb/h = 4400 lb/h.

Volatile heating value is the Btu value of the waste per pound of volatile matter. The total heating value as charged is 8000 lb/h $\times$ 4500 Btu/lb = 36,000,000 Btu/h. With M representing 1 million, the total waste heat value is 36.00 MBtu/h. On a unit volatile basis, with 4400 lb/h volatile charged, the heating value per pound volatile is 36 MBtu/h ÷ 4400 lb/h = 8200 Btu/h.

Dry gas produced from combustion of the waste is 7.60 lb/10 kBtu from Table 11.30. The dry gas flow is this figure multiplied by the Btu released, or (7.60 ÷ 10,000) lb/Btu $\times$ 36.00 MBtu/h = 27,360 lb/h dry gas.

Combustion H₂O is the moisture generated from burning the waste, 0.60 lb/10 kB from Table 11.30. The combustion moisture flow rate is (0.60 ÷ 10,000) lb/Btu $\times$ 36.00 MBtu/h = 2160 lb/h moisture.

Dry gas + combustion H₂O is the sum of the dry gas and the moisture products of combustion, 27,360 + 2160 = 29,520 lb/h. This figure is convenient for obtaining the amount of air required for combustion.

100 percent air is the dry gas and moisture weights that are produced by combustion of the volatile component, which equals the weight of the volatile component plus the weight of the air provided. Likewise, the air requirement is equal to the sum of the dry gas and moisture of combustion less the volatile component. This air requirement is the stoichiometric air requirement, that amount of air necessary for complete combustion of the volatile component of the waste. Using the above figures, the value for 100 percent air is as follows:

TABLE 11.33 Mass Flow

	Example
Wet feed, lb/h	8,000
Moisture, %	25
lb/h	2,000
Dry feed, lb/h	6,000
Ash, %	20
lb/h	1,600
Volatile	4,400
Volatile htg. value, Btu/lb	8,200
MBtu/h	36.00
Dry gas, lb/10 kBtu	7.60
lb/h	27,360
Comb. H_2O, lb/10 kBtu	0.60
lb/h	2,160
Dry gas + comb. H_2O, lb/h	29,520
100% Air, lb/h	25,120
Total air fraction	2.00
Total air, lb/h	50,240
Excess air, lb/h	25,120
Humid/dry gas (air), lb/lb	0.01
Humidity, lb/h	502
Total H_2O, lb/h	4,662
Total dry gas, lb/h	52,480

29520 lb/h (dry gas + H_2O) − 4400 lb/h (volatile) = 25120 lb/h

Total air fraction is the air required for effective combustion. This is basically a function of the physical state of the fuel (gas, liquid, or solid) and the nature of the burning equipment. In this example, 100 percent air is required, providing 1.00 + 1.00 = 2.00 total air fraction.

Total air is the stoichiometric requirement multiplied by the total air fraction, 25,120 lb/h × 2.00 = 50,240 lb/h.

Excess air provided to the system is the total air less the stoichiometric air requirement, 50,240 lb/h − 25,120 lb/h = 25,120 lb/h.

Humidity/dry gas (air) The humidity of the air entering the incinerator may have a significant effect on the heat balance to be calculated subsequently. In this case, assume a humidity of 0.01 lb of moisture per pound of dry air.

Humidity is the flow of moisture into the system with the air supply. Given the air-flow and the fractional humidity, the humidity can be calculated as 50,240 lb/h air × 0.01 lb H_2O/lb air = 502 lb/h moisture.

Total H_2O is that moisture exiting the system. It is equal to the sum of three moisture components: moisture in the feed plus moisture of combustion plus humidity, or 2000 lb/h + 2160 lb/h + 502 lb/h = 4662 lb/h total H_2O.

Total dry gas exiting the system is equal to the sum of the dry gas generated by the combustion of the volatiles, with stoichiometric air, plus the flow of excess air into the system. Thus, 27,360 lb/h dry gas + 25,120 lb/h excess air = 52,480 lb/h total dry gas.

Heat Balance. Heat, like mass, is conserved within a system. The heat exiting a system is equal to the amount of heat entering that system. Table 11.34 presents a quantitative means of establishing a heat balance for an incinerator. The heat balance must be preceded by a mass flow balance, as discussed previously. The result of a heat balance is a determination of the incinerator outlet temperature, outlet gas flow, supplemental fuel requirement, and total air requirement. The total heat into the incinerator was calculated previously in the mass flow computations. By determining how much of the total heat produced is present in the exhaust gas, the exhaust gas temperature can be calculated. If the calculated exhaust gas temperature is equal to the desired exhaust gas outlet temperature, the process is autogenous and supplemental fuel is not required. If the desired temperature is lower than the actual outlet temperature, additional air must be added (or additional water, if this is possible) to lower the outlet temperature to that desired. This condition is also autogenous burning since supplemental fuel is not added to the system. For the case where the actual outlet temperature is less than the desired outlet temperature, supplemental fuel must be added. The products of combustion must include the products of combustion of the supplemental fuel. Dry flue gas properties are assumed to be identical to those of dry air.

As with Table 11.33, Table 11.34 will be described on a step-by-step basis.

Cooling air wasted: assume the incinerator shell is cooled by a flow of 2000 st ft³/min of air. A standard cubic foot of air weighs 0.075 lb/ft³, therefore the mass airflow is 2000 ft³/min × 60 min/h × 0.075 lb/ft³ = 9000 lb/h. It is further assumed that this flow is wasted, i.e., discharged to the atmosphere. The temperature of the air at the discharge point is assumed to be 450°F. From Table 11.31 the enthalpy of air at 450°F, the cooling air discharge temperature, is 94 Btu/lb. The total heat loss due to the wasted cooling air is the quantity of cooling air discharged multiplied by its enthalpy, 9000 lb/h × 94 Btu/lb = 0.85 MBtu/h.

Ash generated is 1600 lb/h, from the mass flow table. The heating value of ash can be assumed to be 130 Btu/lb based on the equation $Q = c_p(T - 60)$, where Q is the heat value, c_p is the ash specific heat (taken as 0.24 Btu/lb • °F), and T is the ash discharge temperature, at 600°F [0.24(600 − 60) = 130 Btu/lb]. The heat loss through ash discharge is therefore 1600 lb/h ash × 130 Btu/lb = 0.21 MBtu/h.

Radiation: the heat lost by radiation from the incinerator shell can be approximated as a percentage of the total heat of combustion. Table 11.35 lists typical values of radiation loss. For this case, with a heat release of 36.00 MBtu/h, a radiation loss of 1.5 percent is used. The total loss by radiation is equal to 36.00 MBtu/h released × 1.5 percent, or 0.54 MBtu/h.

Humidity is water vapor within the air. The humidity component of the air, 502 lb/h, is taken from the mass flow sheet.

Correction: when one is considering the heat absorbed by the moisture or water within the incinerator, exhaust stream humidity has released its heat of vaporization because it is in the vapor phase. The other moisture components, moisture in the feed and moisture of combustion, enter the reaction as liquid, and the heat of vaporization is released by the reaction. To simplify these moisture calculations, the heat of vaporization of humidity moisture at 60°F, 970 Btu/lb, is added to the total heat capacity of the flue gas. Therefore the correction factor is 502 lb/h humidity × 970 Btu/lb = 0.49 MBtu/h.

Total losses of an incinerator are the sum of the heat discharged as cooling air and the heat lost in the ash discharge and the radiation loss. To this is added the correction for humidity. In this case the total loss is 0.85 MBtu/h + 0.21 MBtu/h + 0.54 MBtu/h − 0.49 Btu/h = 1.11 MBtu/h.

TABLE 11.34 Heat Balance

	Example w/fuel oil	Example w/gas
Cooling air wasted, lb/h	9,000	
°F	450	
Btu/lb	94	
MBtu/h	0.85	
Ash, lb/h	1,600	
Btu/lb	130	
MBtu/h	0.21	
Radiation, %	1.5	
MBtu/h	0.54	
Humidity, lb/h	502	
Correction (@970 Btu/lb), MBtu/h	− 0.49	
Losses, total, MBtu/h	1.11	
Input, MBtu/h	36.00	
Outlet, MBtu/h	34.89	
Dry gas, lb/h	53,480	
H_2O, lb/h	4,662	
Temperature, °F	1,915	
Desired temp., °F	2,000	
MBtu/h	36.41	
Net MBtu/h	1.53	
Fuel oil, air fraction	1.20	1.10
Net Btu/gal	57,578	406
gal/h	26.57	3,768
Air, lb/gal	125.06	0.791
lb/h	3,323	2,980
Dry gas, lb/gal	125.54	0.748
lb/h	3,336	2,818
H_2O, lb/gal	8.75	0.103
lb/h	232	388
Dry gas w/fuel oil, lb/h	55,816	56,298
H_2O w/fuel oil, lb/h	4,894	5,050
Air w/fuel oil, lb/h	53,563	53,220
Outlet MBtu/h	38.61	38.66
Reference t, °F	60	60

TABLE 11.35 Furnace Radiation Loss Estimates

Furnace rate, MBtu/h	Radiation loss, % of furnace rate
< 10	3
15	2.75
20	2.50
25	2
30	1.75
> 35	1.50

Input to the system is the heat generated from the combustible, or volatile, portion of the feed. From the mass flow table this figure is 36.00 MBtu/h.

Outlet heat content is that amount of heat exiting in the flue gas. The heat left in the flue gas is the heat generated by the feed (input) less the total heat loss, 36.00 MBtu/h − 1.11 MBtu/h = 34.89 MBtu/h.

Dry gas is 52,480 lb/h, from the mass flow sheet.

H_2O is 4662 lb/h, from the mass flow sheet.

Temperature is the temperature of the exhaust gas where the heat content of the dry gas flow plus the heat content of the moisture flow exiting the incinerator equals the outlet MBtu/h. With 52,480 lb/h dry gas and 4662 lb/h moisture at 1900°F and 2000°F, the dry gas enthalpy is 481.57 and 510.07, respectively, and the moisture enthalpy is 2007.17 and 2067.42, respectively (from Table 11.31). Therefore, the total calculated enthalpy is:

1900°F	34.63 MBtu/h
X°F	34.89 MBtu/h
2000°F	36.41 MBtu/h

By interpolation, the exhaust-temperature X is

$$X = 1900 + (2000 - 1900) \times (34.89 - 34.63) \div (36.41 - 34.63)$$

$$= 1915°F$$

Therefore the exhaust gas temperature is 1915°F.

Desired temperature is the temperature to which it is desired to bring the products of combustion. For this example a temperature of 2000°F will be used. At this temperature the heat content of the wet gas stream is that of the dry gas and that of the moisture at 2000°F or 52,480 lb/h dry gas × 510.07 Btu/lb + 4662 lb/h moisture × 2067.42 Btu/lb = 36.41 MBtu/h.

Net MBtu/h is the amount of heat that must be added to the flue gas to raise its heat content to the desired level. (In this case the desired heat level is evaluated at 2000°F.) The net MBtu/h, therefore, is the desired less the outlet MBtu/h, 36.42 MBtu/h − 34.89 MBtu/h = 1.53 MBtu/h.

Note: this analysis will continue on the basis of using No. 2 fuel oil as supplemental fuel. Table 11.36 lists fuel oil combustion parameters. Table 11.37 is a list of combustion parameters for natural gas. The incinerator analysis of Tables 11.38 and 11.42 includes listings for the use of natural gas as supplemental fuel. The values for natural gas are calculated like those for fuel oil.

Fuel oil, air fraction is the total air fraction required for combustion of supplementary fuel. The total air normally required for efficient combustion of gas fuel is from 1.05 to 1.15 and for light fuel oil is 1.10 to 1.30. In this case assume No. 2 fuel oil is used for supplemental heat, with a total air stoichiometric ratio of 1.20.

Net Btu/gal: when fuel is combusted, the products of combustion must be heated to the desired flue gas temperature. The amount of heat required to heat these combustion products must be subtracted from the total heat of combustion to obtain the effective heating value of the fuel. As the temperature to which the fuel products must be raised increases, the net heat available from the fuel decreases. From Table 11.36, by bringing the products of combustion of fuel oil, with 1.2 total air, to 2000°F, a net heating value of 57,578 Btu/gal is available. The gallons of fuel oil required to provide the heat required to bring the exhaust gas temperature from its actual temperature, 1915°F, to the desired temperature, 2000°F, are equal to the net

MBtu/h required divided by the net Btu/gal available: 1.53 MBtu/h ÷ 57,578 Btu/gal = 26.57 gal/h.

Air required for combustion of fuel oil, with 1.2 total air, from Table 11.36, is 125.06 lb/gal of fuel oil. The fuel combustion airflow is the unit flow multiplied by the fuel quantity, 125.06 lb/gal × 26.57 gal/h fuel oil = 3323 lb/h air.

Dry gas. From Table 11.36 the dry gas produced from combustion of fuel oil with 1.2 total air is 125.54 lb/gal of fuel oil. The dry gas flow rate is 125.54 lb dry gas/gal × 26.57 gal/h fuel oil = 3336 lb/h.

H_2O produced from combustion of fuel oil with 1.2 total air and 0.013 humidity is 8.75 lb/gal fuel oil from Table 11.36. The moisture flow rate from combustion of fuel oil is 8.75 lb H_2O/gal fuel oil × 26.57 gal fuel oil/h = 232 lb/h.

TABLE 11.36   No. 2 Fuel Oil (139,703 Btu/gal, 7.6 lb/gal)

Total Air:	1.1	1.2	1.3
lb air/gal	114.640	125.062	135.483
lb dry gas/gal	115.115	125.537	135.958
lb H$_2$O/gal	8.615	8.751	8.886

Temp., °F	Heat available, Btu/gal		
200	126,210	125,707	125,206
300	123,016	122,255	121,495
400	119,802	118,780	117,760
500	116,562	115,277	113,995
600	113,283	111,732	110,184
700	109,965	108,146	106,328
800	106,029	103,885	101,742
900	103,197	100,829	98,463
1,000	99,747	97,099	99,747
1,100	96,255	93,325	90,397
1,200	92,721	89,505	86,291
1,300	89,147	85,643	82,140
1,400	85,535	81,738	77,943
1,500	81,887	77,796	73,707
1,600	78,205	73,817	69,431
1,700	74,487	69,799	65,115
1,800	70,746	65,757	60,771
1,832	69,538	64,452	59,369
1,900	66,971	61,679	56,389
2,000	63,175	57,578	51,984
2,100	59,341	53,445	59,349
2,192	55,813	49,628	44,385
2,200	55,507	49,294	43,084
2,300	41,637	45,114	38,594
2,400	47,750	40,916	34,085
2,500	43,852	36,706	29,562
2,600	39,914	32,453	24,995
2,700	35,938	28,162	20,388

TABLE 11.37 Natural Gas (1000 Btu/st ft^3, 0.050 lb/st ft^3)

Total Air:	1.05	1.10	1.15
lb air/st ft^3	0.755	0.791	0.827
lb dry gas/st ft^3	0.712	0.748	0.784
lb H$_2$O/st ft^3	0.103	0.103	0.104
Temp., °F	Heat available, Btu/st ft^3		
200	861	860	857
300	839	837	834
400	817	814	810
500	794	790	785
600	772	767	761
700	749	743	736
800	722	715	707
900	702	694	685
1000	678	670	660
1100	654	644	633
1200	629	619	607
1300	605	593	580
1400	579	567	553
1500	554	541	526
1600	529	514	498
1700	503	487	470
1800	477	460	442
1832	468	451	433
1900	450	433	414
2000	424	406	385
2100	397	378	356
2192	372	352	329
2200	370	350	327
2300	343	322	298
2400	316	294	269
2500	289	265	239
2600	261	237	210
2700	233	208	179

Dry gas with fuel oil is the total quantity of dry gas exiting the system. It is equal to the dry gas produced from combustion of the waste plus the dry gas produced from fuel combustion: 52,480 lb/h + 3336 lb/h = 55,816 lb/h dry gas.

H$_2$O with fuel oil is the total quantity of moisture exiting the system, that calculated in the mass flow sheet plus the contribution from combustion of supplementary fuel, 4662 lb/h + 232 lb/h = 4894 lb/h.

Air with fuel oil is the total amount of air entering the incinerator, calculated from the mass flow sheet, plus that needed for supplemental fuel combustion, 50,240 lb/h + 3323 lb/h = 53,563 lb/h.

Outlet MBtu/h, the total heat value of the flue gas exiting the incinerator, is the sum of the heat content of the gas prior to adding supplemental fuel and the heat addition of the supplemental fuel. The supplemental fuel adds 26.57 gal/h ×

TABLE 11.38 Flue Gas Discharge

	Example w/fuel oil	Example w/gas
Inlet, °F	2,000	2,000
Dry gas, lb/h	55,816	56,298
Heat, MBtu/h	38.61	38.66
Btu/lb dry gas	692	687
Adiabatic t, °F	178	178
H_2O saturation, lb/lb dry gas	0.6008	0.6008
lb/h	33,534	33,824
H_2O inlet, lb/h	4,894	5050
Quench H_2O, lb/h	28,640	28,774
gal/min	58	58
Outlet temp., °F	120	120
Raw H_2O temp., °F	60	60
Sump temp., °F	148	148
Temp. diff., °F	88	88
Outlet, Btu/lb dry gas	92.880	92.880
MBtu/h	5.18	5.23
Req'd. cooling, MBtu/h	33.43	33.43
H_2O, lb/h	379,886	379,886
gal/min	760	760
Outlet, ft^3/lb dry gas	16.515	16.515
ft^3/min	15,363	15,496
Fan press., in WC	30	30
Outlet, actual ft^3/min	16,586	16,729
Outlet, H_2O/lb dry gas	0.08128	0.08128
H_2O, lb/h	4,537	4,576
Recirc. (ideal), gal/min	58	58
Recirc. (actual), gal/min	464	464
Cooling H_2O, gal/min	760	760

139,703 Btu/gal = 3.72 MBtu/h to the flue gas. Therefore the flue gas outlet contains 34.89 MBtu/h + 3.72 MBtu/h = 38.61 MBtu/h. As a check on this figure, the outlet temperature will be calculated by using the flue gas flow (55,816 lb/h dry gas and 4894 lb/h moisture):

2000°F	38.59 MBtu/h
X°F	38.61 MBtu/h
2100°F	40.49 MBtu/h

By interpolation,

$$X = 2000 + (2100 - 2000) \times (38.61 - 38.59) \div (40.49 - 38.59)$$

$$= 2001°F$$

This calculation of flue gas temperature, 2001°F, is in good agreement with the desired gas outlet temperature, 2000°F.

Reference t is the datum temperature for enthalpy. It is the temperature at which feed, supplemental fuel, and air enter the system, 60°F for this example.

Flue Gas Discharge. To meet the rigorous air pollution codes in effect today, gas scrubbing equipment is often necessary. Table 11.38, the flue gas discharge table, provides a method of calculating gas flow volumes exiting a wet scrubbing system as well as scrubber flow quantities. This table can also be used when calculating volumetric flow from a dry flue gas system.

Table 11.38 entries are as follows:

Inlet: insert the incinerator outlet temperature, 2000°F, from the heat balance table. This temperature is the inlet temperature of the flue gas processing system.

Dry gas is the flow of dry gas exiting the incinerator. From the heat balance table, this figure is 55,816 lb/h.

Heat is the total heat exiting the incinerator in the flue gas, MBtu/h. From the heat balance table this figure is 38.61 MBtu/h. The heat is calculated in terms of the dry gas component of the flue gas, as Btu/lb dry gas. From the entries for total heat and dry gas flow, the heat is 38.61 MBtu/h ÷ 55,816 lb/h = 692 Btu/lb dry gas.

Adiabatic t: when 1 lb of water evaporates, it absorbs approximately 1000 Btu, without a change in temperature. This heat adsorption is called the *heat of vaporization* or *latent heat.* Latent heat is opposed to sensible heat, which is the heat required for a change in temperature without a change in phase. For evaporated water (steam) the sensible heat is approximately 0.5 Btu/lb steam for every rise of 1°F.

The adiabatic temperature t of the flue gas is the quench temperature. Quenching of a gas is defined as the use of latent heat of water (or other fluid) to decrease the gas temperature. The process does not involve the addition or removal of heat, only the use of the heat of vaporization of the quench liquid, i.e., water. The term *adiabatic* defines a process where heat is neither added nor removed from a system. Considering the properties of dry flue gas equal to the properties of dry air, listed in Table 11.32, note that a maximum amount of moisture can be held in dry air at a particular temperature. The table lists saturation moisture quantities, volumes, and enthalpy as a function of temperature. The temperature at which the enthalpy of the saturated flue gas (dry air) is equal to the enthalpy calculated above, Btu/lb dry gas, is the adiabatic temperature of the system. There has been no transfer of heat from the system, only the conversion of latent heat in the quench water to sensible heat in the dry flue gas. In this example the adiabatic temperature is found in Table 11.32 as that temperature where the dry flue gas (saturated mixture) will have an enthalpy of approximately 692 Btu/lb, 178°F.

H_2O saturation: The quenched flue gas, at the adiabatic temperature (178°F in this example) will contain an amount of moisture equal to the maximum amount of moisture that it can hold, saturation. From Table 11.32, the saturation moisture, in lb H_2O/lb dry gas (air), is read opposite the adiabatic temperature. In this case, for 178°F adiabatic temperature, the saturation moisture is 0.6008 lb H_2O/lb dry gas. The moisture flow is the saturation moisture multiplied by the dry gas flow, 0.6008 lb H_2O/lb dry gas × 55,816 lb dry gas/h = 33,534 lb H_2O/h.

H_2O inlet: The moisture component of the flue gas exiting the incinerator is inserted here from the heat balance table: 4894 lb/h.

Quench H_2O is the moisture required for quenching the incoming flue gas to its adiabatic temperature. This is equal to the saturated moisture content of the flue gas less the moisture initially carried into the system with the flue gas. This figure is equal to H_2O saturation (33,534 lb/h) less H_2O inlet (4894 lb/h), which in this example is equal to 28,640 lb/h. The conversion factor from lb/h of water to gal/min (8.34 lb/gal × 60 min/h) is 500. The quench water required is equal to 28,640 lb/h ÷ 500 = 58 gal/min.

Outlet temperature is that temperature entering the low (negative) pressure side of the induced-draft (ID) fan, or, with no ID fan, the temperature within the stack. This temperature is normally selected in the range of 120 to 160°F. The lower this temperature, the smaller the size of the outlet plume and the lower the volumetric flow rate of flue gas. For this example 120°F was chosen as the outlet temperature. As can be seen below, a lower outlet temperature would require additional amounts of cooling water.

Raw H₂O temperature: this entry is the temperature of the water available for cooling the flue gas from the adiabatic temperature to the outlet temperature. In this example a raw water temperature of 60°F was chosen.

Sump temperature: normally a quantity of water in excess of that calculated for cooling the flue gas is provided for particulate removal. The excess water is generally collected in a sump where a quiescent period is allowed to permit larger particles within the spent water to settle to the sump floor, eventually to be drained. The temperature of the water in the sump must be ascertained to determine the effective cooling rate of the water flow. The sump temperature is a practical impossibility to forecast accurately through detailed calculations, but an empirical relationship has been established. The sump temperature is assumed equal to the adiabatic temperature divided by 1.2. In this example the sump temperature is estimated at 178 ÷ 1.2 = 148°F.

Temperature differential: the temperature differential of note is the difference in temperature between the raw water entering the cooling tower (or scrubber) and the water temperature exiting the tower, the sump temperature. Sump temperature less raw H_2O temperature is, in this example, $148 - 60 = 88$°F, the temperature differential.

Outlet: the gas exiting the scrubber system is designed to be at the outlet temperature, 120°F in this case, saturated with moisture. The outlet enthalpy is inserted from Table 11.32 for the outlet temperature chosen: 92.880 Btu/lb dry gas. The total heat in the outlet flue gas is its enthalpy multiplied by its flow, that is, 92.880 Btu/lb × 55,816 lb dry gas/h = 5.18 MBtu/h.

Required cooling: as noted previously, the flue gas is initially quenched, without heat addition or removal, to its adiabatic temperature. To reduce the temperature to the desired outlet temperature, a supply of cooling water is required. This cooling water must remove the heat content at adiabatic conditions relative to the heat content at outlet conditions. The required cooling is therefore the heat inlet less the outlet MBtu/h. For this example 38.61 MBtu/h heat inlet less 5.18 MBtu/h outlet equals 33.43 MBtu/h required cooling; i.e., with removal of 33.43 MBtu/h from the saturated flue gas stream the flue gas temperature will fall from 178°F to 120°F.

H₂O: the moisture flow referred to is that required to achieve the desired cooling effect. With $Q = WC \Delta t$ or $W = Q/C \Delta t$, where W = cooling water in lb/h, Q = cooling load in Btu/h, C = specific heat of water [1 Btu/(lb • °F)], and Δt = temperature difference of the cooling water across the flue gas stream (sump water temperature less raw water temperature), the required cooling water flow rate can be calculated. For this example:

$$W = 33.43 \text{ MBtu/h} \div 1 \text{ Btu(lb • °F)} \div 88°F = 379,886 \text{ lb/h}$$

The flow in gallons per minute is that in lb/h divided by 500, 379,886 ÷ 500 = 760 gal/min.

Outlet: the outlet volumetric flow is obtained with use of Table 11.32. The specific volume, ft^3 mixture/lb dry gas (air), is found in this table for the outlet temperature. For this example, with an outlet temperature of 120°F, the specific volume is 16.515 ft^3/lb dry gas. The volumetric flow is equal to the specific volume multiplied by the dry gas flow divided by 60 min/h; that is, 16.515 ft^3/lb dry gas × 55,816 lb/h ÷ 60 min/h = 15,363 ft^3/min.

Fan pressure: induced draft fans used to clean the incinerator off-gases may require a relatively high pressure. The actual differential pressure value across the fan, inserted here, will be used to modify the volumetric flow rate. For this example the fan pressure is 30 in. WC.

Outlet actual ft^3/min. The volumetric flow immediately prior to entering the induced draft fan will experience an expansion because of the fan suction. The volumetric flow correction is as follows: Multiply the value in ft^3/min determined above by the ration 407 ÷ (407 − p), where p is the pressure across the fan in inches of water column. The figure 407 is atmospheric pressure (14.7 lb/in^2 absolute) expressed in inches of water column (14.7 lb/in^2 absolute ÷ 62.4 lb H_2O/ft^3 × 1728 in^3/ft^3 = 407 inches WC). In this example the corrected, or actual, volumetric flow entering the ID fan is [407 ÷ (407 − 30)] × 15,363 ft^3/min = 16,586 ft^3/min actual flow.

Outlet: From Table 11.32, insert the saturation humidity, lb H_2O/lb dry gas (dry air), corresponding to the outlet temperature. For this example, with an outlet temperature of 120°F, the humidity is 0.08128 lb H_2O/lb dry gas. The total moisture exiting the stack is the saturation humidity multiplied by the dry gas flow. For this case, 0.08128 lb H_2O/lb dry gas × 55,816 lb dry gas/h = 4537 lb H_2O/h.

Recirculation (ideal): the recirculation flow is that amount of flow required for quenching, as calculated above (58 gal/min for this example). The temperature of this water flow is not critical. The flow is used adiabatically, where the latent heat (not a temperature change) in the water flow reduces the temperature of the flue gas. It is termed a *recirculation flow* because spent scrubber water from the scrubber sump can be recirculated to the venturi for use at 140°F or greater, instead of a cooler flow of water.

Recirculation (actual): In practice the ideal flow is inadequate to fully clean the gas stream of particulate. Ideal quenching requires intimate contact of each molecule of water with gas, instantaneous evaporation, and instantaneous heat transfer between the moisture and the gas, none of which occurs. To compensate for actual versus ideal conditions, an empirical factor is used. In this case this factor is 8. Therefore, for actual recirculation flow use the ideal flow multiplied by 8, 58 gal/min × 8 = 464 gal/min.

Cooling H_2O: Insert the flow of cooling water calculated above, in this case, 760 gal/min.

Computer Program.   The mass flow, heat balance, and flue gas discharge analyses presented in this chapter have been developed into a series of computer programs. The programs accept a number of different waste streams, consider individual gas components, and have afterburner, waste heat boiler, and air emission control systems options, as well as the ability to operate in the starved air or excess air modes. These programs display and print out more comprehensive information than is immediately available from the analysis sheets in this chapter. The programs are available from Incinerator Consultants Incorporated, 11204 Longwood Grove Drive, Reston, VA 22094 [phone: (703) 437-1790, fax: (703) 437-9048].

Energy Recovery

Recovering Heat.   Steam is used for incinerator heat recovery far more frequently than hot water or hot air (gas) generation. Steam is more versatile in its application, and 1 lb of steam contains significantly more energy than 1 lb of water or air. In general, while hot water is normally of use only for building heat during winter months or can be used in limited quantities for feedwater heating, steam can be used for process requirements and for equipment loads, which are year-round loads. Further, steam can be converted to hot water or used for air heating when these needs arise. The calculations presented herein are for steam generation.

Approach Temperature.   With t the temperature of the heated medium (steam or hot water), t_i the temperature of the entering flue gas, and t_o the temperature of the flue gas exiting the boiler (see Fig. 11.38), the heat available can be calculated. For any heat exchanger there is an approach temperature t_x. This temperature is the difference between the temperatures of the heated medium (t) and of the exiting flue gas (t_o). Therefore,

$$t_x = t_o - t$$

The more efficient the heat exchanger, the lower the approach temperature. The larger the heat exchanger, the lower t_x until, in the extreme case, with an infinitely large heat exchanger, t_x will be zero and the steam (or hot water) will be at the same temperature as the exiting flue gas. In practice, the approach temperature of a waste heat boiler is on the order of 100°F for efficient and 150°F for standard, economical construction.

Available Heat.   The heat available in exhaust or flue gas is equal to that heat at the boiler inlet less the gas heat content at the boiler outlet. With Q the heat available

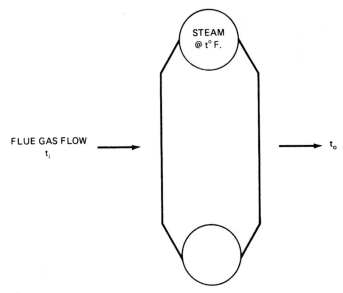

FIGURE 11.38   Waste heat boiler.

from the flue gas stream, in Btu per pound, note the following:

$$Q = W(h \, @ \, t_i - h \, @ \, t_o)$$

The flue gas will have a dry and a wet component. Considering the dry gas component to have the properties of air (W_{dg}, h_a) and W_m the moisture component, this equation becomes

$$Q = W_{dg}(h_{ai} - h_{ao}) + W_m(h_{mi} - h_{mo})$$

The inlet temperature t_i is the temperature of the incinerator outlet. The outlet temperature of the heat exchanger (t_o) is defined by the approach temperature (t_x) and the temperature of the heated medium (t):

$$t_o = t_x + t$$

The enthalpy at the outlet of the heat exchanger must be evaluated at t_o.

Example.   Consider a gas flow at 1400°F of 15,000 lb/h of dry gas plus 2000 lb/h of moisture. Let the available heat be calculated for the generation of saturated steam at 100, 200, and 400 lb/in² absolute with a 150°F approach temperature (note Table 11.31 for enthalpy values):

Inlet condition:

$$t_i = 1400°F$$

$$h_a = 341.85 \text{ Btu/lb}$$

$$h_{mi} = 1719.82 \text{ Btu/lb}$$

$$W_{dg} = 15,000 \text{ lb/h}$$

$$W_m = 2000 \text{ lb/h}$$

Outlet condition:
With $p = 100$ lb/in² absolute,

$$t = 328°F$$

$$t_o = 150 + 328 = 478°F$$

By interpolation,

$$h_{ao} = 101 \text{ Btu/lb}$$

$$h_{mo} = 1249 \text{ Btu/lb}$$

$$Q = 15,000(341.85 - 101) + 2000(1719.82 - 1249)$$

$$= 4.554 \text{ MBtu/h}$$

With $p = 200$ lb/in² absolute,

$$t = 382°F$$

$$t_o = 150 + 382 = 532°F$$

By interpolation,

$$h_{ao} = 115 \text{ Btu/lb}$$

$$h_{mo} = 1274 \text{ Btu/lb}$$

$$Q = 15,000(341.85 - 115) + 2000(1719.82 - 1274)$$

$$= 4.294 \text{ MBtu/h}$$

With $p = 400 \text{ lb/in}^2$ absolute,

$$t = 445°F$$

$$t_o = 150 + 445 = 595°F$$

By interpolation,

$$h_{ao} = 130 \text{ Btu/lb}$$

$$h_{mo} = 1304 \text{ Btu/lb}$$

$$Q = 15,000(341.85 - 130) + 2000(1719.82 - 1305)$$

$$= 4.007 \text{ MBtu/h}$$

These calculations are summarized in Table 11.39. Steam temperature is listed in Table 11.40. The inlet is the total heat in the flue gas, related to 60°F:

$$15,000 \times 341.85 + 2000 \times 1714.82 = 8.567 \text{ MBtu/h}$$

The column Δt is the difference in flue gas temperatures entering and leaving the boiler. The efficiency noted is the available heat divided by the total heat in the flue gas entering the boiler.

Of significance is the relationship between available heat and Δt, the temperature difference of the flue gas across the boiler. The available heat is proportional to Δt.

For example, Q at $\Delta t = 805$ °F versus $\Delta t = 922$ °F:

$$Q @ 805 = \frac{805}{922} \times 4.554 = 4.0 \text{ MBtu/h}$$

Q at $\Delta t = 868$ versus $\Delta t = 922$:

$$Q @ 868 = \frac{868}{922} \times 4.554 = 4.3 \text{ MBtu/h}$$

By comparing these values to the calculated values for Q in Table 11.39, it is clear that the available heat Q is directly proportional to Δt, the temperature loss in the flue gas.

Steam Generation.   Given the heat availability, the amount of steam that can be generated can be calculated. Figure 11.39 shows typical flow through a waste heat boiler producing steam. Makeup water temperature is raised to feedwater temperature by steam flow from the boiler and by the heat contained in return condensate. Condensate is returned to the deaerator.

Besides raising the feedwater temperature prior to injection into the boiler, the deaerator acts to help release dissolved oxygen from feedwater. Additional feedwater treatment is usually employed to reduce, or prevent, scaling and corrosion of boiler

TABLE 11.39

Inlet, MBtu/h	°F	Steam pressure, psia	Steam temperature, °F	Flue gas temperature, °F	Δt, °F	Available heat, MBtu/h	Efficiency, %
8.567	1400	100	328	478	922	4.554	53
8.567	1400	200	382	532	868	4.294	50
8.567	1400	400	445	595	805	4.007	47

TABLE 11.40 Saturated Steam Properties

Pressure, lb/in² absolute	Temperature, °F	Pressure, lb/in² absolute	Temperature, °F
14.7	212	45	274
15	213	50	281
16	216	55	287
17	219	60	293
18	222	65	298
19	225	70	303
20	228	75	308
21	231	80	312
22	233	90	320
23	235	100	328
24	238	125	344
25	240	150	358
26	242	175	371
27	244	200	382
28	246	250	401
29	248	300	417
30	250	350	432
32	254	400	445
34	258	450	456
36	261	500	467
38	264	600	486
40	267	700	503

Source: Ref. 13.

surfaces. Water softeners are used to remove most of the calcium and magnesium hardness from raw water. Chemical addition is also used, typically as follows:

- *Sodium sulfite.* This is an oxygen-scavenging chemical that chemically removes the dissolved-oxygen component not removed in the deaerator. Hydrazine is another oxygen scavenger that is used in high-pressure (over 1200 lb/in² absolute) boiler applications.

- *Amine.* There are a number of amines in use for feedwater treatment. They are used for boiler pH or alkalinity control. Excess alkalinity (pH greater than 11) will result in accelerated scale buildup while low pH (below 6) can cause excessive boiler tube corrosion. Normally boiler water pH is maintained in the range of 8.0 to 9.5, slightly alkaline.

- *Phosphates.* This treatment is used to precipitate residual calcium and magnesium hardness remaining in feedwater after softening. Certain phosphates will act as dispersants, preventing adhesion of the precipitate to tube walls.

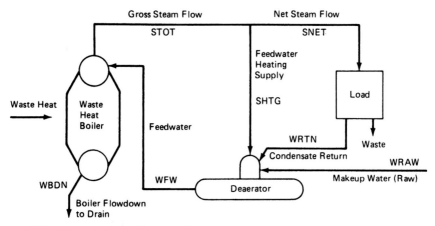

FIGURE 11.39 Waste heat boiler, steam flow.

These chemicals will form a sludge, or mud, which will accumulate in the lower drum of a boiler. The boiler water must have a blowdown on a regular basis to prevent a buildup of mud within the boiler. This blowdown will normally represent from 2 to 5 percent of the boiler steam generation.

Calculating Steam Generation. By using the steam tables (Table 11.40), calculations will be performed for obtaining steam, makeup, blowdown, and feedwater flows.

With an approach temperature of 150°F, generating 100 lb/in^2 absolute steam, dry and saturated ($h_{stm} = 1187$ Btu/lb, $h_{bdn} = 298$ Btu/lb), the heat available (from Table 11.39) is 4.554 MBtu/h. For this illustration, blowdown is 4 percent of the feedwater flow, feedwater is provided to the boiler at 220°F ($h_{fw} = 188$ Btu/lb), and 20 percent of the steam generation is returned as condensate at 170°F ($h_{ret} = 138$ Btu/lb). In addition, raw water enters the deaerator at 60°F ($h_{mu} = 27$ Btu/lb), and radiation loss from the boiler is 1 percent of the total boiler input.

The heat available for generating steam is the heat lost by the flue gas less the heat lost by boiler radiation:

$$\text{Waste heat} = 4.554 \text{ MBtu/h} - 0.01 \times 4.554 \text{ MBtu/h} = 4.508 \text{ MBtu/h}$$

Referring to Fig. 11.39,

$$\text{Waste heat} = \text{heat in steam} + \text{heat in blowdown} - \text{heat in feedwater}$$

$$\text{Heat in blowdown} = \text{WBDN} \times h_{bdn}$$

$$\text{WBDN} = 0.04 \times \text{WFW}$$

$$\text{Heat in blowdown} = 0.04 \times 298 \times \text{WFW} = 11.92\text{WFW}$$

$$\text{Heat in feedwater} = \text{WFW} \times h_{fw}$$

$$= 188\text{WFW}$$

$$\text{Heat in steam} = \text{STOT} \times h_{stm} = \text{STOT} \times 1187 \text{ Btu/lb}$$

$$\text{STOT} = \text{WFW} - \text{WBDN} = \text{WFW} - 0.04\text{WFW} = 0.96\text{WFW}$$

$$\text{Heat in steam} = 0.96\text{WFW} \times 1187 = 1139.52\text{WFW}$$

$$\text{Waste heat} = 4.508 \text{ MBtu/h} = 11.92\text{WFW} + 1139.52\text{WFW} - 188\text{WFW}$$

$$= 963.44\text{WFW}$$

Therefore,

$$\text{WFW} = 4679 \text{ lb/h}$$

and

$$\text{STOT} = 0.96 \times 4679 = 4492 \text{ lb/h}$$

also

$$\text{WBDN} = 0.04 \times 4679 = 187 \text{ lb/h}$$

To calculate steam required for feedwater heating, makeup, and condensate return flows, a material balance and a heat balance must be performed around the deaerator. Material (flow) balance:

$$\text{WFW} = \text{SHTG} + \text{WRTN} + \text{WRAW}$$

From above:

$$\text{WRTN} = 0.2 \times \text{STOT} = 0.2 \times 4492 = 898 \text{ lb/h}$$

$$\text{WFW} = 4679 \text{ lb/h}$$

Therefore:

$$\text{SHTG} = \text{WFW} - \text{WRTN} - \text{WRAW} = 4679 - 898 - \text{WRAW} = 3781 - \text{WRAW}$$

Heat balance:

$$\text{WFW} \times h_{\text{fw}} = \text{SHTG} \times h_{\text{stm}} + \text{WRTN} \times h_{\text{ret}} + \text{WRAW} \times h_{\text{mu}}$$

Therefore,

$$4679 \times 188 = (3781 - \text{WRAW}) \times 1187 + 898 \times 138 + \text{WRAW} \times 27$$

$$\text{WRAW} = 3218 \text{ lb/h}$$

$$\text{SHTG} = 3781 - 3218 = 563 \text{ lb/h}$$

Where condensate is not returned to the deaerator, i.e., to the boiler system, the steam required for feedwater heating will increase:
Material balance:

$$\text{SHTG} = \text{WFW} - \text{WRAW} = 4679 - \text{WRAW}$$

Heat balance:

$$4679 \times 188 = (4679 - \text{WRAW}) \times 1187 + \text{WRAW} \times 27$$

Therefore,

$$\text{WRAW} = 4030 \text{ lb/h}$$

$$\text{SHTG} = 649 \text{ lb/h}$$

In general, with no separate source of heat for feedwater heating (such as returned condensate), 12 to 15 percent of generated steam is required.

Table 11.41 summarizes the above calculations.

The net flow of steam (SNET) is that quantity of steam available for useful work. As can be seen, use of condensate return for feedwater heating increases the quantity of steam available for a load.

Waterwall Systems. Larger incinerators dedicated to destruction of refuse or other paper-type waste materials are often designed with "waterwall" construction, as described previously.

These installations can be provided with a variety of features including the following:

- *Convection boiler section.* Boiler tubes are placed perpendicular to the flow of gas as it exits the incinerator. A major portion of available heat is captured by these tubes, producing saturated steam.
- *Economizer.* This is used to heat feedwater by extracting heat from gases as they leave the convection boiler section.
- *Superheater.* A tubular section is normally placed upstream of the convection section. Hot incinerator gases superheat steam generated from the convection section of the boiler.
- *Air preheater.* This is used in lieu of, or directly downstream of, the economizer. It produces heated combustion air from the relatively low-temperature gas flow at this location.

Calculations of steam generation from each of these sections are a complex task and will not be detailed here. Tables 11.23 and 11.24 indicate steam generation for typical waterwall incinerators for a variety of waste quality.

To calculate available heat by the methods of this chapter, the exit gas temperature (the temperature of flue gas exiting the boiler sections and entering the air emissions control system) can be assumed to be in the range of 350 to 550°F.

Electric Power Generation. As discussed above, steam is a useful by-product of the incineration process. The generation of steam from typical facilities is listed in Table 11.42. Steam rates follow the cost of energy, and can range from $10 to $15 per 1000 lb,

TABLE 11.41 Calculated Steam Generation

Avail. heat, MBtu/h	STOT, lb/h	WBDN, lb/h	WFW, lb/h	WRET, lb/h	WRAW, lb/h	SHTG, lb/h	WNET, lb/h
4.554	4492	187	4679	898	3218	563	3929
4.554	4492	187	4679	0	4030	649	3843

TABLE 11.42 Typical Steam Generation Rates

MSW quality*	6500	6000	5000	4000	3000
Moisture, %	15	18	25	32	39
Noncombustible, %	14	16	20	24	28
Combustible, %	71	66	55	44	33
Steam generation, lb/ton MSW	8600	7800	6400	4600	3000

*Heat value, Btu/lb as received.

which represents a revenue of $70 to $105 per ton, based on a steaming rate of 7000 lb per ton of MSW.

Revenue from steam sales can be generated only if there is a market for steam and if the user is relatively close to the incinerator, where pipeline losses are not significant. In many areas there is no local steam customer, or the potential user does not have a year-round need for steam. If the main use for steam is for winter heating, the sale of steam will go wanting for half the year.

Electric power has a universal market. It is salable practically anywhere in the country. The conversion of steam to electric power, however, results in a loss of energy. Electric power generation requires that steam pass through a turbine, and that the turbine drive a generator to feed a power grid. While 7000 lb of steam per ton of MSW converts to 2050 kWh, this same ton of MSW will generate less than 600 kWh of electrical energy. At a cost of electric power of from $0.09 to $0.15 per kWh, the electric power revenue will bring in from $54 to $90 per ton of MSW. This is less than the revenue generated by steam.

It is usually preferable to find a steam customer. In some areas of the country a crucial factor in the location of an incinerator is its physical proximity to a potential user of steam.

Table 11.43 lists energy production rates of incinerators throughout the United States. The net kWh listed is the gross generation of electric power less the amount of power that is used to operate the facility. The gross power generation rate will vary from 577 kWh/ton MSW for waterwall units to 350 kWh/ton MSW for smaller, modular systems.

TABLE 11.43 Energy Production at Mass Burn Facilities in the United States

Facility	State	Design capacity, tons/day	Net power output, MW	Gross power output, MW	Ratio net/gross power output	Net kWh per ton processed	Gross kWh per ton processed	Ratio net/gross kWh/ton	Steam, lb/h	Btu/lb	Start-up year
					Process: Mass burn—waterwall						
Albany (American Ref-Fuel)	NY	1500	40	50	0.80	N/A	N/A	N/A	400,000	5500	88
Alexandria/Arlington R.R. Facility	VA	975	20	22	0.90	470	520	0.90	255,000	4800	89
Babylon Resource Recovery Project	NY	750	14	17	0.82	410	N/A	N/A	185,000	5000	
Bergen County	NJ	3000	80	88	0.91	482	N/A	N/A	808,000	4500	
Bridgeport RESCO	CT	2250	60	67	0.90	640	720	0.89	576,000	5300	88
Bristol	CT	650	14	16	0.84	535	620	0.86	148,000	5000	88
Brooklyn Navy Yard	NY	3000	N/A	N/A	N/A	N/A	N/A	N/A	847,000	N/A	
Broome County	NY	571	15	18	0.83	467	560	0.83	184,000	5200	
Broward County (Northern Facility)	FL	2250	60	67	0.90	638	709	0.90	573,500	5200	
Broward County (Southern Facility)	FL	2250	57	63	0.90	608	676	0.90	576,700	5200	
Camden County (Foster Wheeler)	NJ	1050	21	30	0.70	482	N/A	N/A	260,400	4500	
Camden County (Pennsauken)	NJ	500	10	13	0.78	425	N/A	N/A	110,000	5200	
Central Mass. Resource Recovery Project	MA	1500	36	40	0.90	600	N/A	N/A	336,000	5000	88
Charleston County	SC	644	11	13	0.84	N/A	N/A	N/A	164,000	5000	89
City of Commerce	CA	400	10	12	0.87	630	725	0.87	115,000	5600	87
Concord Regional S.W. Recovery Facility	NH	500	12	13	0.92	470	550	0.85	135,400	5000	89
Dakota County	MN	800	20	23	0.87	550	N/A	N/A	410,000	5000	
Davidson County	TN	210	3	4	0.81	N/A	N/A	N/A	34,000	6000	
East Bridgewater (American Ref-Fuel)	MA	1500	40	50	0.80	N/A	N/A	N/A	400,000	5500	
Eastern-Central Project	CT	550	12	15	0.83	560	N/A	N/A	155,500	5300	
Essex County	NJ	2277	72	76	0.95	N/A	501	N/A	633,000	4500	
Fairfax County	VA	3000	73	85	0.86	540	610	0.89	822,504	4400	90
Falls Township (Wheelabrator)	PA	2250	65	72	0.90	600	N/A	N/A	570,000	5200	
Glendon	PA	500	13	14	0.89	525	N/A	N/A	130,000	5200	
Gloucester County	NJ	575	12	14	0.86	425	475	0.89	135,400	4500	90
Hampton/NASA Project Recoup	VA	200	N/A	N/A	N/A	N/A	N/A	N/A	66,000	N/A	80
Harrisburg	PA	720	5	8	0.64	500	N/A	N/A	170,000	4500	72
Haverhill (Mass Burn)	MA	1650	41	46	0.89	572	N/A	N/A	396,000	5081	89

TABLE 11.43 Energy Production at Mass Burn Facilities in the United States (*Continued*)

Facility	State	Design capacity, tons/day	Net power output, MW	Gross power output, MW	Ratio net/ gross power output	Net kWh per ton processed	Gross kWh per ton processed	Ratio net/gross kWh/ton	Steam, lb/h	Btu/lb	Start-up year
					Process: Mass burn—waterwall (*Continued*)						
Hempstead (American Ref-Fuel)	NY	2505	64	72	0.89	570	N/A	N/A	604,000	4500	90
Hennepin County (Blount)	MN	1200	33	38	0.88	540	700	0.77	350,000	5800	90
Hillsborough County SWER Facility	FL	1200	28	30	0.92	492	N/A	N/A	270,000	4500	87
Hudson County	NJ	1500	38	45	0.85	455	N/A	N/A	410,000	4500	
Huntington	NY	750	21	25	0.84	627	736	0.85	225,000	6000	
Jackson County/Southern MI State Prison	MI	200	2	2	0.85	N/A	N/A	N/A	49,600	4900	87
Johnston (Central Landfill)	RI	750	17	21	0.81	543	N/A	N/A	150,000	5200	
Kent County	MI	625	16	18	0.86	410	N/A	N/A	158,000	5350	
Lake County	FL	528	10	15	0.69	N/A	525	N/A	120,000	5000	90
Lancaster County	PA	1200	30	36	0.83	560	N/A	N/A	291,000	5000	
Lee County	FL	1800	47	50	0.94	630	N/A	N/A	506,250	5000	
Lisbon	CT	500	13	15	0.87	550	600	0.92	135,400	4500	
Marion County Solid W-T-E Facility	OR	550	11	13	0.84	450	N/A	N/A	133,446	4700	86
Montgomery County	MD	1800	69	84	0.83	644	N/A	N/A	512,000	5500	
Montgomery County	PA	1200	29	34	0.85	N/A	460	N/A	269,082	4500	
Morris County	NJ	1340	34	40	0.85	N/A	535	N/A	433,300	5500	
Nashville Thermal Transfer Corp. (NTTC)	TN	1120	3	7	0.40	N/A	N/A	N/A	308,000	4900	74
New Hampshire/Vermont S.W. Project	NH	200	4	5	0.84	N/A	440	N/A	46,200	5400	87
Norfolk Naval Station	VA	360	N/A	N/A	N/A	N/A	N/A	N/A	40,000	N/A	67
North Andover	MA	1500	32	38	0.84	550	N/A	N/A	344,000	5500	85
North Hempstead	NY	990	17	21	0.81	N/A	N/A	N/A	N/A	N/A	
Northwest Waste-To-Energy Facility	IL	1600	N/A	N/A	N/A	N/A	N/A	N/A	330,000	N/A	70
Oklahoma City	OK	820	10	22	0.46	N/A	N/A	N/A	240,000	5200	85
Olmstead County	MN	200	2	3	0.75	N/A	293	N/A	50,000	5500	87
Onondaga County	NY	990	32	38	0.84	640	N/A	N/A	311,646	6000	
Oyster Bay	NY	1000	27	31	0.87	N/A	N/A	N/A	248,000	6000	
Pasco County	FL	1050	29	31	0.94	550	650	0.85	270,900	4800	
Passaic County	NJ	1434	37	45	0.83	625	753	0.83	445,620	5500	
Pinellas County (Wheelabrator)	FL	3150	56	62	0.90	430	N/A	N/A	750,000	4000	83

Facility	State										
Portland	ME	500	10	14	0.74	N/A	500	N/A	120,000	5000	88
Preston (Southeastern Connecticut)	CT	600	16	18	0.89	520	N/A	N/A	144,000	5000	
Quonset Point	RI	710	18	21	0.86	455	N/A	N/A	182,000	4750	88
S.E. Resource Recovery Facility (SERRF)	CA	1380	30	36	0.83	540	400	0.88	351,000	4800	85
S.W. Resource Recovery Facility (BRESCO)	MD	2250	34	60	0.57	350	N/A	N/A	441,000	5100	75
Saugus	MA	1500	40	50	0.80	550	N/A	N/A	340,000	4500	87
Savannah	GA	500	N/A	N/A	N/A	N/A	N/A	N/A	120,000	N/A	
Spokane	WA	800	22	26	0.85	497	N/A	N/A	222,600	N/A	89
Stanislaus County Res. Recovery Facility	CA	800	17	23	0.76	450	N/A	N/A	201,000	4750	
Sturgis	MI	560	11	13	0.85	N/A	N/A	N/A	100,000	6000	
Union County	NJ	1440	39	44	0.89	567	670	0.85	360,000	5400	89
University City Res. Recovery Facility	NC	235	4	5	0.75	395	476	0.83	50,000	4500	86
Walter B. Hall Res. Recovery Facility	OK	1125	15	17	0.88	530	600	0.88	240,000	5000	88
Warren County	NJ	400	11	14	0.78	482	N/A	N/A	112,000	4650	
Washington/Warren Counties	NY	400	11	13	0.85	N/A	N/A	N/A	115,000	5500	
Wayne County	NC	300	4	5	0.85	N/A	N/A	N/A	36,000	N/A	
West Pottsgrove Recycling/R.R. Facility	PA	1500	40	45	0.89	N/A	N/A	N/A	336,000	5200	
Westchester	NY	2250	56	60	0.93	590	N/A	N/A	504,000	4800	84
Numerical average of nonzero values		1138	27	32	0.83	526	577	0.87	291,520	5065	
Standard deviation		754	20	22	0.10	74	115	0.03	199,429	450	

Process: Mass burn—modular (*Continued*)

Facility	State										
Agawam/Springfield	MA	360	7	9	0.83	390	N/A	N/A	85,500	4200	88
Barron County	WI	80	0	0	0.26	N/A	N/A	N/A	16,500	4750	86
Batesville	AR	100	N/A	N/A	N/A	N/A	N/A	N/A	6,200	N/A	81
Bellingham	WA	100	1	2	0.67	350	N/A	N/A	23,000	4500	86
Beto 1 Unit (Texas Dept. of Corrections)	TX	25	N/A	N/A	N/A	N/A	N/A	N/A	7,000	N/A	80
Cassia County	ID	50	N/A	N/A	N/A	N/A	N/A	N/A	9,000	N/A	82
Cattaraugus County R-T-E Facility	NY	112	N/A	N/A	N/A	N/A	N/A	N/A	26,000	N/A	83
Center	TX	40	N/A	N/A	N/A	N/A	N/A	N/A	9,000	N/A	86
City of Carthage/Panola County	TX	40	N/A	N/A	N/A	N/A	N/A	N/A	2,500	N/A	86
Cleburne	TX	115	1	1	0.77	N/A	N/A	N/A	18,000	4500	86
Collegeville	MN	50	N/A	N/A	N/A	N/A	N/A	N/A	11,000	N/A	81
Dyersburg	TN	100	N/A	N/A	N/A	N/A	N/A	N/A	20,000	N/A	80
Eau Claire County	WI	150	3	3	0.91	263	323	0.81	37,000	5000	
Elk River R.R. Authority (TERRA)	TN	200	N/A	N/A	N/A	N/A	N/A	N/A	50,000	N/A	

TABLE 11.43 Energy Production at Mass Burn Facilities in the United States (*Continued*)

Facility	State	Design capacity, tons/day	Net power output, MW	Gross power output, MW	Ratio net/gross power output	Net kWh per ton processed	Gross kWh per ton processed	Ratio net/gross kWh/ton	Steam, lb/h	Btu/lb	Start-up year
					Process: Mass burn—modular (*Continued*)						
Energy Gen. Facility at Pigeon Point	DE	600	11	13	0.79	532	N/A	N/A	152,000	5500	87
Fergus Falls	MN	94	N/A	N/A	N/A	N/A	N/A	N/A	30,000	N/A	88
Fort Dix	NJ	80	N/A	N/A	N/A	N/A	N/A	N/A	12,000	N/A	86
Fort Leonard Wood	MO	75	N/A	N/A	N/A	N/A	N/A	N/A	8,740	N/A	82
Fort Lewis (U.S. Army)	WA	120	N/A	N/A	N/A	N/A	N/A	N/A	42,000	N/A	
Gatesville (Texas Dept. of Corrections)	TX	13	N/A	N/A	N/A	N/A	N/A	N/A	3,000	N/A	80
Hampton	SC	270	N/A	N/A	N/A	N/A	N/A	N/A	45,000	N/A	85
Harford County	MD	360	N/A	N/A	N/A	N/A	N/A	N/A	75,000	N/A	88
Harrisonburg	VA	100	N/A	N/A	N/A	N/A	N/A	N/A	17,000	N/A	82
Key West	FL	150	2	3	0.85	300	N/A	N/A	42,740	5000	86
Lamprey Regional Solid Waste Cooperative	NH	108	N/A	N/A	N/A	N/A	N/A	N/A	20,000	N/A	80
Lassen Community College	CA	100	1	2	0.78	N/A	N/A	N/A	24,000	6500	84
Lewis County	TN	50	N/A	N/A	N/A	N/A	N/A	N/A	14,000	N/A	88
Long Beach	NY	200	3	5	0.67	N/A	N/A	N/A	58,000	5000	88
Manchester	NH	560	13	14	0.89	425	N/A	N/A	20,000	4500	
Mayport Naval Station	FL	50	N/A	N/A	N/A	N/A	N/A	N/A	N/A	N/A	79
Miami	OK	108	N/A	N/A	N/A	N/A	N/A	N/A	23,000	N/A	82
Miami International Airport	FL	60	N/A	N/A	N/A	N/A	N/A	N/A	15,000	N/A	83
Muskegon County	MI	180	2	3	0.82	373	N/A	N/A	34,000	N/A	
New Hanover County	NC	100	2	4	0.50	N/A	N/A	N/A	54,000	N/A	84
North Slope Borough/Prudhoe Bay	AK	100	N/A	N/A	N/A	N/A	N/A	N/A	N/A	N/A	81
Oneida County	NY	200	1	2	0.55	N/A	N/A	N/A	26,000	N/A	85
Osceola	AR	50	N/A	N/A	N/A	N/A	N/A	N/A	10,000	N/A	80
Oswego County	NY	200	1	4	0.28	275	N/A	N/A	45,000	5000	86
Park County	MT	75	N/A	N/A	N/A	N/A	N/A	N/A	13,000	N/A	82
Pascagoula	MS	150	N/A	N/A	N/A	N/A	N/A	N/A	24,000	N/A	85
Perham	MN	116	N/A	N/A	N/A	N/A	N/A	N/A	23,000	N/A	86
Pittsfield	MA	240	N/A	N/A	N/A	N/A	N/A	N/A	50,000	N/A	81
Polk County	MN	80	N/A	N/A	N/A	N/A	N/A	N/A	21,000	N/A	88
Pope-Douglas W-T-E Facility	MN	80	N/A	N/A	N/A	N/A	N/A	N/A	11,000	N/A	87
Red Wing	MN	72	N/A	N/A	N/A	N/A	N/A	N/A	15,000	N/A	82

Facility	State										
Richard Asphalt	MN	57	N/A	N/A	N/A	N/A	N/A	N/A	13,500	N/A	82
Rutland	VT	240	6	7	470	N/A	0.86	N/A	40,000	N/A	88
Salem	VA	100	N/A	N/A	N/A	N/A	N/A	N/A	14,000	N/A	78
St. Croix County	WI	115	1	1	85	110	0.58	0.77	23,500	5000	89
Tuscaloosa Energy Recovery Facility	AL	300	N/A	N/A	N/A	N/A	N/A	N/A	55,880	N/A	84
Wallingford	CT	420	9	11	384	500	0.85	0.77	105,000	4850	89
Waxahachie	TX	50	N/A	N/A	N/A	N/A	N/A	N/A	15,000	N/A	82
Westmoreland County	PA	50	N/A	N/A	N/A	N/A	N/A	N/A	10,000	4500	88
Windham	CT	108	2	2	350	150	0.86	0.78	16,800	5000	81
Numerical average of nonzero values		143	4	5	322	271	0.71	0.78	29,651	4920	
Standard deviation		122	4	4	166	155	0.19	0.02	27,108	525	
Process: Rotary Combustor											
Auburn	ME	200	4	5	N/A	N/A	0.76	N/A	113,800	5200	N/A
Delaware County Regional	PA	2688	80	90	600	N/A	0.79	N/A	664,972	5200	N/A
Dutchess County	NY	506	9	10	140	320	0.92	0.44	110,000	N/A	88
Falls Township (Technochem)	PA	70	0	1	130	275	0.47	0.47	16,000	4500	
Galax	VA	56	N/A	N/A	N/A	N/A	N/A	N/A	12,000	N/A	86
Gaston County/Westinghouse R.R. Center	NC	440	6	7	550	N/A	0.81	N/A	N/A	4450	89
MacArthur Energy Recovery Facility	NY	518	8	12	370	655	0.70	0.85	118,000	5000	
Mercer County	NJ	975	32	36	560	N/A	0.89	N/A	314,500	5000	
Monmouth County	NJ	1700	57	63	N/A	N/A	0.90	N/A	N/A	4950	
Monroe County	IN	500	9	11	N/A	N/A	0.85	N/A	110,000	N/A	
Montgomery County (North)	OH	300	6	6	523	550	0.95	0.95	72,000	5000	88
Montgomery County (South)	OH	900	18	19	482	507	0.95	0.95	240,000	5000	
Oakland County	MI	2000	54	62	645	N/A	0.87	N/A	600,000	5200	
San Juan Resource Recovery Facility	PR	1040	22	27	510	N/A	0.81	N/A	254,000	4500	
Sangamon County	IL	450	6	8	380	N/A	0.75	N/A	90,000	N/A	88
Skagit County	WA	178	2	2	345	N/A	0.85	N/A	40,000	4500	81
Sumner County	TN	200	0	1	N/A	N/A	0.86	N/A	50,000	N/A	
Waukesha County (New Plant)	WI	600	N/A	N/A	N/A	480	N/A	0.90	200,000	5500	
Westinghouse/Bay Resource Mgmt. Center	FL	510	10	12	432	N/A	0.83	N/A	136,000	4600	87
York County	PA	1344	30	35	540	465	0.86	0.76	330,000	4500	89
Numerical average of nonzero values		759	20	25	443	465	0.83	0.76	192,848	4864	
Standard deviation		680	22	25	151	131	0.11	0.22	181,052	336	

N/A = not available.

Source: *Data Summary of Municipal Solid Waste Management Alternatives,* vol. III: *Appendix A—Mass Burn Technologies,* National Technical Information Service, October 1992, NTIS accession number DE92016433.

PART B

Floyd Hasselriis
Consulting Engineer
Forest Hills, New York

11.2 ASH DISPOSAL

Ash residues from combustion of MSW need to be disposed of in an environmentally sound and economical manner. Whether placed in contained landfills or beneficially used, account must be taken of their characteristics and the effect of ash management procedures on their properties.

Sources and Types of Ash Residues

Ash residues are discharged at various locations from the combustion and emission control equipment.

Bottom ash, discharged after the waste has progressed down the stoker, consists of inert residues, glass and metallic objects, and 2 to 10 percent carbon. Bottom ash is usually quenched with water, although it can also be collected in a dry state.

Stoker grate siftings fall through clearances in the grates, and are collected with bottom ash. These may include unburned organic matter.

Boiler ash, carried by combustion gases, consists of flying particles and condensible metal vapors which may attach to refractory and water-cooled walls and be caught by boiler tube surfaces. It may fall onto the stoker into the bottom ash, or it may be collected in hoppers and discharged into the bottom ash.

Fly ash, carried by the combustion gases through the furnace, boiler, and scrubber, is collected by the particulate control device. If a wet scrubber is employed, it will be discharged with the scrubber blowdown. Fly ash collected by an electrostatic precipitator (ESP) or fabric filter may be discharged into the bottom ash or collected separately. Fly ash can be conditioned (moistened) to prevent dusting and fugitive emissions.

Scrubber reaction products, collected at the bottom of spray-dry or dry lime-injection acid gas scrubbers, include fly ash and reacted or partially reacted alkaline reagent (such as lime) and some carbon.

Mixed ash may contain siftings, bottom ash, boiler deposits, scrubber residues, fly ash, and scrubber products.

Properties of Ash Residues

Ash residue properties depend on the municipal solid waste (MSW) burned, the combustion and emission control systems, and the methods of residue collection. What goes in must come out somewhere and in some form.

Composition of Municipal Solid Waste.   Unprocessed municipal solid waste generally contains about 52 percent combustible matter, 26 percent moisture, and 22 percent ash and noncombustible (inert) materials, as shown in Fig. 11.40.[14] Recycling of metals and glass reduces but does not eliminate the noncombustible fraction.

Chemical Composition of MSW.   The chemical composition of raw MSW is determined after removing (and separately accounting for) the large inert materials (metals, glass, and ceramics), and shredding and performing laboratory analysis on the remaining fraction. Table 11.44 shows a typical ultimate analysis and analysis of noncombustibles for major and trace metals as performed by ASTM procedures.[15] The combustible remainder contains 6.86 percent inherent ash. Major metals constitute 99 percent of this, mainly aluminum, calcium, sodium, and potassium, as shown in Fig. 11.41.[16] Trace metals, totaling about 1 percent of the combustibles, are mainly zinc, tin, and lead (Fig. 11.42).

Composition and Quantities of Ash Residues.   The composition of bottom ash residues remaining after combustion of unprocessed MSW, determined after separation through a 2-in mesh screen, is described in Table 11.45.[17]

The typical range in total quantities of residues generated from combustion of MSW and associated emission controls, shown in Table 11.46, indicates that dry bottom ash and fly ash may be 27 to 39 percent of the weight of MSW, and residues from various types of acid gas emission controls may add 1 to 5 percent to the residues requiring disposal.[18]

Potentially recoverable are a fraction of the 84 percent mixed metals in the 19 percent oversize fraction and the 22.7 percent ferrous metal and 3.4 percent nonferrous metals in the $-$ 2-in screened fraction. The fine fly ash component of the bottom ash has been found to contain relatively high levels of toxic metals, and thus contaminates the bottom ash.

Density of Ash Residues.   Ash residues from combustion of MSW generally have densities ranging from 65 to 75 lb/ft^3 at a water content of 15 to 25 percent. They can be compacted in the ash fill to 135 lb/ft^3 (2180 kg/m^3) at about 20 percent moisture.[19]

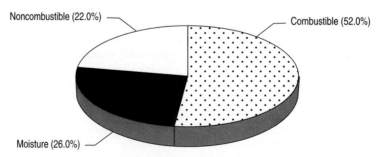

FIGURE 11.40   Composition of municipal solid waste. (*Source: Ref. 14.*)

TABLE 11.44 Chemical Analysis of Residential and Commercial Waste

(From analysis of Portland, Oregon Waste)

Ultimate analysis	Percent	Minerals	Percent
		Major metals:	
Carbon	35.02	Aluminum	33.21
Oxygen	30.16	Calcium	30.20
Hydrogen	6.67	Sodium	17.50
Sulfur	1.88	Potassium	13.12
Chlorine	0.25	Silicon	5.00
Moisture	19.16		
Ash of Combustibles	6.86	Subtotal:	99.04
Total	100.00	Trace metals:	
		Zinc	0.40
		Lead	0.31
		Tin	0.15
		Chromium	0.047
		Nickel	0.020
		Cadmium	0.009
		Copper	0.007
		Subtotal:	0.96
		Total:	100.00

Source: Gershman, Brickner & Bratton, Inc.[16]

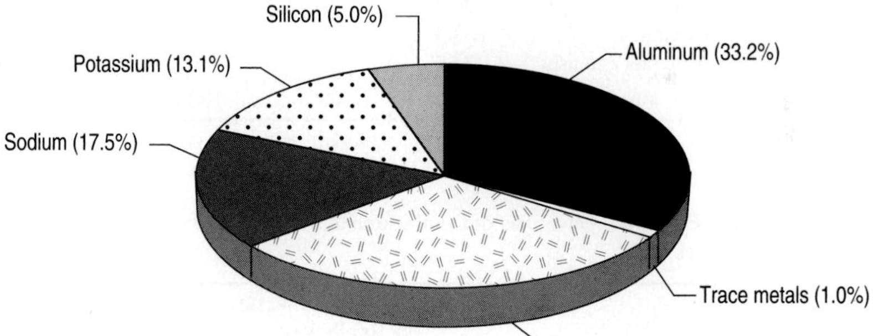

FIGURE 11.41 Major minerals in MSW ash residue. (*Source: Ref. 16.*)

This may be compared with portland cement weighing about 94 lb/ft^3 and gravel or aggregate at 105 lb/ft^3. (1 lb/ft^3 = 16.2 kg/m^3.) Specific gravities of fine and coarse residues from a facility employing a dry lime-injection scrubber and baghouse ranged from 1.9 to 2.5 (120 to 160 lb/ft^3).

Moisture Content. Moisture content of ash residues generally ranges widely, from 15 to 57 percent, depending on whether a semidry ash discharger or a water quench tank is used. Excessive moisture content increases disposal costs. Maintaining minimum but sufficient moisture essentially eliminates dust liberation and fugitive dust

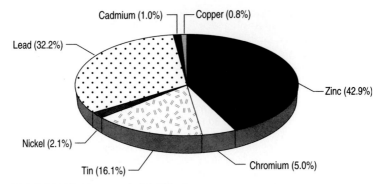

FIGURE 11.42   Trace metals in MSW ash.

TABLE 11.45   Material Fractions of Municipal Solid
Waste Ash Residues

	Total (100%)	Separation	
		+ 2-in (19.2%)	− 2-in (80.8%)
Metal	16.1	84.0	
Other	1.1	6.4	
Combustibles	4.0	9.7	2.7
Ferrous metal	18.3		22.7
Nonferrous metal	2.7		3.4
Glass	26.2		32.4
Ceramics	8.3		10.3
Minerals and ash	23.0		28.5
	100%	100%	100%

Source: Ref. 17.

TABLE 11.46   Quantities of Combustion and Emission Control
Residues

	Quantity, % of waste (lb/100 lb waste)	% of total
Bottom ash (slag)	25.0–35.0	90
Filter dust (fly ash)	2.0–4.0	10
Total:	27.0–39.0	100
Additional residues:		
Wet scrubber residue	0.8–1.5	3–4
Spray-dry scrubber residue	1.6–3.5	6–9
Dry injection residue	2.5–4.5	9–12

Source: Ref. 18.

TABLE 11.47   Total Metals in Combined Ash, Fly Ash, and Fly Ash Scrubber Residues

(Parts per million parts of ash by weight)

	Combined ash	Fly ash	Fly ash/scrubber
Aluminum	17,857	27,500	12,714
Calcium	33,642	64,857	176,428
Sodium	3,828	26,928	1,292
Potassium	3,071	36,928	8,450
Iron	20,428	10,857	3,314
Chloride	928	65,428	164,285
Sulfate	7	33	8
Lead	3,142	22,143	3,257
Cadmium	35	642	160
Zinc	4,107	53,500	9,143
Manganese	534	649	365
Mercury	ND	3	73

ND = not determined.
Source: Ref. 20.

problems. Landfill density can be optimized by controlling moisture. Dry fly ash can be conditioned with water to eliminate dusting.

Chemical Composition of Ash Residues.   The noncombustible components of the MSW appear in the bottom ash and fly ash residues in different fractions, as can be seen in Table 11.47.[20] Combined bottom ash (including fly ash) consists mainly of the mineral metals found in common earth. Fly ash has greater proportions of the earth metals, the trace metals, lead, cadmium, zinc, and the highly soluble chloride and sulfate salts.

Alkaline materials used to scrub acids from the combustion products increase the calcium (or sodium) compounds in the ash residues. Fly ash/scrubber residues from lime-injection acid gas controls contain predominantly calcium and chlorine, mainly calcium chloride salt resulting from the reaction of calcium hydroxide with the hydrogen chloride arising from combustion of chlorine-bearing components in the waste. This high salt content may be the main component which must be considered in the disposal of scrubber reaction products, dry or wet.

TABLE 11.48   Comparison of MSW APC Waste with Portland Cement and Slag*

Component	MSW/APC* residue	Portland cement†	Utility FGD waste†	MSW slag‡
CaO	62–67	24	37	14
SiO_2	18–24	28	24	61
Al_2O_3	4–8	14	6	5
Fe_2O_3	1.5–4.5	5	3	

*MSW/APC residue is fly ash plus spray-dry acid gas scrubber residue.
†Ref. 21.
‡Slag obtained by fusing MSW ash residues.[22]

The chemical forms of fly ash/scrubber residues, shown in Table 11.48, are mainly calcium, silica, alumina, and iron oxides. This composition is similar to that of portland cement and flue gas desulfurization (FGD) waste.[21] Vitrified slag from MSW consists primarily of silica, iron, calcium, sodium, and aluminum.[22]

Metals Found in Bottom Ash and Fly Ash.   The trace (or minor) metals found in combined bottom ash/fly ash, fly ash, and combined fly ash/scrubber residues are shown in Table 11.47.[20] Lead, cadmium, and zinc have high concentrations in the fly ash since they are volatilized at normal combustion temperatures.

The potentially toxic metals in the fly ash, such as lead, cadmium, chromium, and nickel, are mainly derived from pigments, fillers, and inks used in or on paper and plastic products. Reduction of these sources can reduce the quantities of these metals in the ash residues from combustion of MSW.

Particle (Grain) Size Distribution.   Size distributions of typical bottom ash and fly ash, an important factor in disposal and beneficial use of ash residues, are shown in Fig. 11.43. Typically these distributions plot as straight lines on log-log paper. The knee in the fly ash curve indicates the presence of agglomerated fly ash. The size distribution of bottom ash corresponds well with specifications for aggregate materials.[17]

Acidity and Alkalinity.   The acidity or alkalinity of ash residues in the presence of water, as measured by pH, generally ranges from neutral (pH = 7) to alkaline (pH = 9). When large amounts of unreacted lime are present, higher alkalinities, up to pH = 11, may be encountered.

Solubility of Metals in Water.   Water containing high concentrations of dissolved metal compounds has the potential for contaminating the environment and drinking

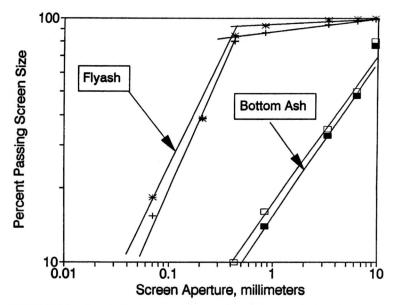

FIGURE 11.43   Size distribution of particles in boiler ash, bottom ash, fly ash, and combined fly ash and boiler ash. (*Source: Ref. 17.*)

water supplies. Metals in solid form are not readily dissolved by pure neutral water. The solubility of the trace toxic metals depends on their chemical form and the acidity or alkalinity (pH) of the water with which they are in contact. Hydroxides, sulfates, and chlorides are more soluble and more readily available for leaching than oxides, silicates, and carbonates. The sulfate and chloride forms of lead and cadmium, found mainly in fly ash as the result of reaction with the sulfur and chlorine in the MSW, are highly soluble.

Most metals are only slightly soluble in water at a pH range from 6 to 9. However, under acidic conditions, such as pH of 5 or less, solubility increases rapidly. While most metals are not soluble under alkaline conditions, lead becomes highly soluble at a pH greater than 10. Since lime has a pH of 12, large quantities of unreacted lime in the fly ash from acid gas control devices can increase the concentration of lead in the solution.

Figure 11.44 shows the solubility of lead and cadmium under the range from acidity to alkalinity. The limiting concentrations of lead and cadmium prescribed by the U.S. Environmental Protection Agency (EPA) are shown: 5 mg/L for lead and 1 mg/L for cadmium. It is apparent that the concentrations of lead and cadmium are less than these limits within a wide range of pH values.[23]

Lead carbonate is relatively insoluble. The carbonic acid which forms when carbon dioxide in the atmosphere is dissolved in rain may account for the low concentrations of lead found in leachates from ash residues which have been exposed to the atmosphere.[24]

The solubility of metals and the pH—the total acid content of the leaching water and the alkaline buffering content of the ash—are crucial factors in the management of ash residues. If the water is neither highly alkaline nor acidic, the soluble metals will be leached out relatively slowly, and the leachate will contain only low concen-

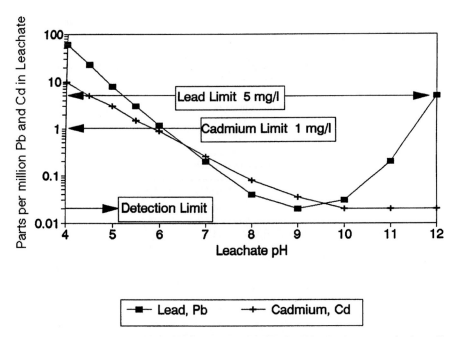

FIGURE 11.44 Lead and cadmium concentrations in MSW ash residue leachate versus leachate pH, showing U.S. EPA toxic limits and detection limit.

trations of the metals. When leached with simulated acid rain, the concentrations of soluble metals in the leachate falls rapidly to a minimum at a liquid/solids ratio of less than 1:1, indicating that most of the leachable metals are on the surface of the fly ash particles.[25]

Soluble Salts in Fly Ash.   In general, fly ash contains a large fraction of soluble salts, because of the sulfur and chlorine in the waste. Alkaline reagents such as caustic soda and lime, used as scrubbing agents for acid gas control, react with the acid gases to form soluble salts such as sodium and calcium chloride, and various sulfates. Table 11.49 shows an analysis of the fly ash and scrubber residues from a spray-dry scrubber, and the soluble fraction, which totals 56 percent of the fly ash, primarily chlorides and sulfates.[26]

Permeability of Ash Residues.   Permeability is the property of a material which measures the velocity at which water will pass through the material, usually reported as centimeters per second (cm/s). The permeability of compacted ash has been measured to be from 1×10^{-6} to 1×10^{-9} cm/s. For comparison, landfill liners are typically required to have a permeability of 1×10^{-7} cm/s. It is possible to prepare ash for use as a landfill liner, thus meeting this specification, by proper compaction, and by adding portland cement and/or lime.[20]

Ash Handling Equipment

Mechanical devices are used to handle and process ash residues.

Wet quench systems cool the residues and permit them to be removed from the quench tank by means of drag conveyers. Ash residues commonly have a moisture content of about 50 percent.

Semiwet systems quench the residues with water, but employ mechanical dischargers to push the residues out with a minimum amount of moisture. The heat remaining in the residues serves to drive off moisture so that the discharged residues may have a moisture content close to 25 percent.

TABLE 11.49   Solubility of Fly Ash Residues of Spray-Dry Scrubber

	Weight %	Soluble %
Fly ash	40	
$CaCl_2$	27	27
$CaSO_3 \cdot 0.5 H_2O$	20	20
$CaSO_4 \cdot 2 H_2O$	5	5
CaF_2	2	2
Lime inerts	2	
$CaCO_3$	2	
$MnCl_2$	1	1
Heavy metals	1–2	1
	100	56

Source: Ref. 23.

Dry removal of the ash residues makes it possible to remove the fine ash component by screening, leaving a useful granular aggregate and a relatively clean ferrous product.

Drag conveyers extract ash residues from water-filled quench tanks. They consist of flytes attached to moving chains carried over sprockets, configured to carry the ash residues up a slope so that they can be conveniently discharged.

Screw conveyers move powder-like dry ash from fly ash hoppers to other locations where they are dropped into water tanks, conditioning (wetting) devices for damp discharge, or into dry collecting containers.

Vibrating conveyers carry damp ash residues from ash dischargers to processing devices or receiving containers.

Vibrating screens separate ash residues into various size fractions, to remove rejects and produce useful fractions.

Grizzley screens remove oversize objects such as wood, bulky appliances, wheels, and miscellaneous metal objects from the ash residue. Vibrating rails separate out the oversize objects.

Trommels are rotary screens which remove oversize objects and clean ash residues to obtain more uniform products.

Magnetic separators, usually placed after grizzley screens to reduce interference by oversize objects, recover ferrous metal.

Pneumatic conveyers transfer fly ash from ash hoppers to remotely located containers. Blower air and fabric filters are used to separate the fly ash from the transport air.

Ash Handling and Storage

Handling and disposal of ash residues must not cause contamination of the environment by fugitive dust or by leaching into the environment or water supplies. Dusting is minimized by keeping the residues in a moist condition.[27]

Stored ash residues must be properly managed to prevent unacceptable discharge of dust or leachate. Runoff and leachate must be collected, supervised and properly disposed of so that soluble metal compounds and salts do not contaminate the environment. Storing of ash allows chemical reactions to take place which bind the metals, reducing leaching potential. Rainwater percolating through the pile cleans the ash by slowly leaching out the available (soluble) metals.

Ash containers must be drainable to recover leachate and watertight to prevent leachate from running off into the environment.

Transportation of ash residues must be in covered containers to prevent dust from escaping into the environment. Containers, vehicles, and roads must be washed if contaminated with ash. The wash water must be disposed of properly or recycled back to the ash quench tank.

Ash Processing

Ash residues can be processed at the waste-to-energy facility to reduce the rate of release of contaminants to the environment, facilitate disposal, improve the quality of the residues, remove valuable, useful, or harmful materials, and prepare portions of the ash for beneficial use. The residues can be treated by washing, chemical treatment, or use of additives and specific chemicals in order to retain, remove, or immobilize potentially toxic compounds.

Ferrous metal can be separated from ash residues by solid magnets or electromagnets. As much as 15 percent of the bottom ash from mass burn facilities is ferrous material which can be extracted magnetically. The quality of the ferrous metal is measured by the amount of contamination with combustibles and fine ash materials. Tumbling of the ferrous product in a trommel can separate contaminants to improve quality, as can washing with water.

Screening processes remove unwanted oversize and undersize components and separate the ash into usable products, including aggregate for use in construction.

Washing processes can provide clean aggregate materials and ferrous metals for beneficial use, and remove the cementitious fly ash. The wash water can be processed and recirculated. The blowdown stream must be treated or evaporated to remove/recover and render harmless the dissolved metals and salts.[28,29]

Treatment of Ash Residues

Various methods of treatment may be used to reduce the amount of leachable metal and salt concentrations, and thus render the ash more environmentally acceptable, as well as improve the chemical and physical stability and durability of product so that it can be used for a variety of purposes. Treatment methods include: ferrous separation and compaction, and various methods which modify release rates by chemical and physical changes, including solidification, stabilization and encapsulation; addition of portland cement, phosphates, waste pozzolans and bituminous materials; washing and chemical treatment, thermal treatment and vitrification. After ferrous separation and screening, bottom ash residues have been used for fill and road base under certain conditions.

Encapsulation in asphalt, for use in bituminous paving mixtures and in cement, serves to minimize the leaching of metals and salts from the product. Cement blocks and special forms can be made from aggregate processed from ash residues.

Lime and/or portland cement can be mixed with fly ash to encapsulate the toxic metals and/or render them insoluble.[30]

Phosphate treatment converts soluble lead compounds to insoluble phosphates, thus reducing the leachable lead to acceptable levels.[31]

Carbonic acid absorbed from stack gases can convert soluble lead compounds into relatively insoluble lead carbonates.[24,32] Heavy metals can be removed from fly ash by using carbonic acid recovered from flue gases, producing insoluble carbonates.

Washing ash residues can produce clean metals and aggregate suitable for use in concrete and road base, at the expense of physical and chemical processing of the wash water.

Chemical processing systems can process the ash residues while treating the flue gases and minimizing or eliminating water discharges. Calcium hydroxide (CaOH) specifically removes hydrogen chloride (HCl) and mercury; sodium hydroxide [$Na_2(OH)_3$] removes sulfur dioxide (SO_2); and ammonia (NH_3) removes nitrogen oxides (NO_x).[33] Hydrochloric acid used in a primary gas scrubber can be used to remove most of the cadmium and a major portion of the lead content of combined ash. Washing may be especially justified for treatment of fly ash, which can contain as much as 50 percent soluble salts in facilities employing acid gas controls.[32]

Vitrification of bottom ash, mixed ash, or fly ash can be employed to obtain a high-quality glassy frit from which the trace toxic metals such as lead and cadmium do not leach out.[34] Thermal treatment can produce glassy materials or sintered and ceramic materials.

Landfill Disposal

Landfilling of untreated ash is perhaps the simplest means of disposal. Ash residues have been codisposed with municipal solid waste, but for various reasons it may be preferable to place ash residues in separate or dedicated cells. Ash residues have been used to cover MSW as daily cover or as final cover. Placing ash residues in ash fills has the advantage that a solid, relatively impervious mass is created, over which trucks can drive as soon as it is placed. Removing ferrous metal from the ash residues improves the density of the ashfill as well as its stability.

Efficient management of ashfills can increase the density of the ash to as high as 3300 lb/yd^3 (122 lb/ft^3), as compared with about 1800 lb/yd^3 of uncompacted ash. Another benefit of compaction is the potential for reducing the permeability to as low as 1×10^{-6} to 1×10^{-9} cm/s. Ashfills are so impervious to water that only a small fraction of rain falling on the top surface is able to penetrate the fill; 90 percent or more will run off, without leaching much of the soluble material in the ash. It is important to provide effective runoff collection systems, since the water may be significantly contaminated, especially when the ashfill is in the process of being filled, prior to capping.[19]

Ash residues have been used as lining materials for landfills, in lieu of costly clay liners. To prepare the ash for use as a landfill liner, portland cement can be added at the landfill at 6 to 10 percent plus lime at 6 to 7 percent by weight. If the fly ash component of the ash residues contains excess lime from the acid gas scrubber, less lime will be needed.

Codisposal with MSW. There has been concern that, when ash residues are codisposed in landfills with raw MSW, the acids generated by decomposing MSW would increase concentrations of soluble toxic metals in the collected leachate, requiring more stringent containment, leachate treatment, and groundwater monitoring.[35] On the other hand, the alkalinity of the ash can neutralize (buffer) the acids, reducing acid leaching.

Ashfills. Ashfills provide dedicated disposal of ash residues in cells separate from MSW. Ash residues have a high density and low permeability which minimize the need for leachate collection and treatment.[20,36]

Liners and Containment. Liners are provided in landfills and ashfills to contain the ash residues, to minimize or eliminate leachate penetration into the surroundings, and to provide means for collection and removal of leachate and monitoring for indications of leakage. Only a small fraction of the rain falling on an ashfill can percolate through to the leachate collection system. The remainder is runoff which must be collected and properly managed.[20]

Ash residues containing mixed bottom ash, fly ash, and acid gas scrubber residues have cementitious and compaction properties which make them relatively impervious to the penetration of leachate, especially if moisture and lime content are optimized. Completed cells can be covered with plastic liners, ash residues, or other relatively impervious materials to essentially eliminate the generation of leachate after the ashfill cell is closed.[36]

TABLE 11.50   Ranges of Leachate Concentrations of Inorganic Constituents from Monofills

Constituent	Concentration (CORRE study), mg/L	EP toxicity maximum allowable limit, mg/L	Primary drinking water standard, mg/L
pH	5.2–7.4		
Arsenic	ND–0.400	5.0	0.05
Barium	ND–9.22	100	1.00
Cadmium	ND–0.004	1.0	0.01
Chromium	ND–0.032	5.0	0.05
Copper			1.00
Iron			0.30
Lead	ND–0.054	5.0	0.05
Manganese			0.50
Mercury	ND	0.2	0.002
Selenium	ND–0.340	1.0	0.01
Silver	ND	5.0	0.05
Zinc			5.00
Chloride			250

ND = not determined.
Source: Ref. 38.

Leachate Disposal and Treatment.   The composition of leachate must be known before it can be disposed of or treated. Table 11.50 shows that actual leachate concentrations measured at various ashfill sites had concentrations of the potentially toxic metals which were far below the EPA toxic limits. In most cases the concentrations of the regulated toxic metals were close to the EPA drinking water limit.[37–39]

The leachate may be discharged or trucked to wastewater disposal plants if found to be acceptable, or it may require treatment before such disposal. The leachate from fly ash and from mixed bottom ash and fly ash contains substantial amounts (roughly 50 percent) of soluble salts resulting from the removal of acid gases by the emission controls which are now required. It has been described as similar to saltwater.[25] The salt content may be more likely to require attention than the low concentrations of soluble metals.

Salinity of Leachate from Ash Residues.   The salinity of leachate is measured by electrical conductivity. The effect of highly salty leachate on the environment has been studied. Soils producing leachates which have an electrical conductivity greater than 16 mho/cm are classified as saline, causing interference in the uptake of water by plants. Column simulations have shown that ash residues having an initial leachate conductivity of 21 mho/cm were reduced by simulated annual acid rainfall to 8 mho/cm after 20 years of leaching, a level having relatively little effect on most plants.[40]

Neutralizing Capacity—Ash/Acid Deposition Mass Balance.   The soluble toxic metals are only slowly released because of the presence of alkaline materials which provide powerful buffering against MSW-produced acids and the low quantity of weak acids in acid rain. It has been estimated that acid rain would be resisted for over 1000 years. Long before this time the leachable materials would presumably have been removed.[41]

Regulatory Aspects

The management, disposal, and beneficial use of ash residues and their products is subject to federal and state regulations which require sampling and analysis of the leaching characteristics of residues from combustion of municipal and other types of wastes.

Federal Regulations.   Federal regulations broadly classify wastes into hazardous and nonhazardous. The Resource Recovery and Conservation Act of 1976 (RCRA) empowered the U.S. EPA to regulate residues from solid waste incinerators. In 1992 the EPA administrator sent a memorandum to EPA regional administrators stating that the Resource Conservation and Recovery Act completely excludes ash from municipal waste combustors from regulation as a hazardous waste under Subtitle C as long as it is not characterized as toxic, since ash can be managed safely in solid waste landfills under Subtitle D, Section 3001(i) of RCRA. Prior to this statement, states developed various requirements, many stipulating that MWC ash be disposed in monofills for ash only (ashfills), employing single liners as compared with the double liners and stricter operating procedures required for hazardous waste landfills. On May 2, 1994, the U.S. Supreme Court ruled that the exemption of MSW from hazardous waste regulations does not contain any exclusion of the ash itself. Under this ruling, ash will have to be regularly tested: if the ash exceeds the Federal standards, it will have to be sent to the special more costly landfills built, operated, and permitted for hazardous wastes.

The U.S. EPA has developed test procedures designed to screen wastes to determine their classification. Various states had developed different regulations as to whether or not ash must be sampled and analyzed prior to disposal. Under the new ruling it is the responsibility of the producer of the ash to test the ash in order to determine whether or not it exhibits the characteristic of toxicity in accordance with EPA procedures.

Ash Residue Extraction Leaching Procedures.   The following leaching procedures have been applied to determine the characteristics of ash residues under a wide range of conditions to which they might be exposed, and also to discover which procedures might more closely simulate actual conditions of disposal or beneficial use:

* Extraction procedure toxicity (EP-Tox) test using acid no. 1 (acetic acid extraction fluid at pH of 4.87 to 5.2)
* Threshold characteristic leaching procedure (TCLP) using TCLP fluid no. 1 (acid no. 2) or TCLP fluid no. 2 (acid no. 3); similar to EP-Tox
* California waste extraction test (WET) using citric acid
* Deionized water (method SW-924), also known as the monofill waste extraction procedure (MWEP)
* CO_2-saturated deionized water
* Simulated acid rain (SAR)
* ASTM shake extraction procedure using distilled water
* Leaching column tests using simulated acid rain

The original extraction procedure (EP) toxicity test produced erratic results when testing municipal waste combustion ash residues, and hence was replaced by the threshold characteristic leaching procedure.[42] These screening procedures use an acid leaching medium intended to simulate leaching of ash codisposed in a landfill in a proportion of 15 percent ash to 85 percent MSW. In the TCLP test, a unit sample of ash residue is immersed in 20 units of a specified acetic acid solution. The acidity of the solution is maintained at a pH value between 4.87 to 5.2 and stirred for a 24-h period. The extract (leachate) is analyzed, and the results are compared with the EPA-established limits shown in Table 11.50. The sample would be characterized as hazardous if any of these limits were exceeded.[42,43] These limits assume that a 100-time dilution would occur before the leachates could reach drinking water, hence are 100 times the drinking water standard.

Actual leachate from MSW landfills as well as codisposal and ash residue landfills (ashfills) do not generally have pH values as low as 5.0. Extensive testing of actual leachate from these various types of landfills shows that they do not contain significant amounts of the toxic metals lead and cadmium, contrary to the results of laboratory tests employing the TCLP method. In other words, the TCLP test does not simulate actual disposal conditions.[38]

A comparison of the effects of various laboratory leaching procedures on the cadmium and lead concentrations is shown in Table 11.51.[43] Actual leachate concentrations from tests sponsored by CORRE/EPA are compared with the EPA toxic limits in Table 11.50. The carbonic acid test was in closest agreement, while the EP toxicity procedure overestimated leaching by over 100 times.[38]

In general, it may be concluded that actual leachates from landfills are more closely simulated by leaching column tests and tests using simulated acid rain or carbon dioxide–saturated water and/or deionized water. More aggressive leaching tests, such as the California WET and the EP and TCLP tests, which represent conditions not likely to occur in the environment, serve as the basis for classifying the wastes as potentially able to produce toxic levels of metals in the leachate.

Leaching column tests show that the soluble metals and salts are gradually removed from the ash as the leachate absorbs and removes them. The larger the quantity of acid in the water, the faster the rate of removal. In some cases two to four quantitative washes will have removed essentially all of the lead and cadmium which was soluble at the leaving pH level. Test borings of one ash pile showed that after several years of natural acid rainfall the leachable metals remaining in the ash pile had been reduced to the nondetectable level.

State Regulations.   State regulations must be at least as stringent as federal regulations, but may be more detailed and suited to specific state environments. Many states

TABLE 11.51   Metal Concentrations in Extracts of MSW Ash Residues, mg/L (parts per million)

	Facility			
	Chicago	Sumner	Hampton	Auburn
Cadmium:				
WET test	1.6	0.81	1.52	0.18
EP toxicity	0.71	0.24	0.50	0.02
Acetate	0.19	0.52	0.33	0.03
Carbonic acid	0.016	0.012	0.07	0.005
Water	<0.0005	<0.005	<0.005	<0.005
Lead:				
WET test	29.0	35.0	46.0	29.0
EP toxicity	5.8	6.4	10.3	3.15
Acetate	0.5	0.28	1.62	4.20
Carbonic acid	0.025	0.004	0.095	0.012
Water	<0.002	<0.002	<0.002	<0.002

Source: Ref. 43.

require the collection and analysis of ash residue samples on a periodic basis, and require that these samples, on average, pass the prescribed toxicity tests.

Testing of ash residues may be required in order to obtain confidence that there will be no harmful effects on the environment after the residues are disposed of or used beneficially.

Samples of ash residues should represent the stream of ash residues from which they are taken. Mixing the fly ash properly with bottom ash avoids unrepresentative hot spots. Aging samples with normal moisture allows the chemical reactions to take place which would occur under the conditions of disposal, such as conversion of soluble lead chlorides to insoluble lead carbonates.

Frequency of testing is generally regulated, including extensive testing after start-up of the plant, followed by one or more tests per year in order to assure consistent operation.[44] If some of the analytical results are found to be critically close to acceptable limits, more samples may be taken to obtain confidence in the average values.

Regularly time-spaced samples will represent the true average characteristic of the ash residues for the period of time over which they are produced, landfilled, or otherwise used. Daily average samples are collected at uniform time intervals over the entire day. Daily samples are well mixed and coned and quartered to reduce the sample size and provide several identical samples. Weekly samples should include the entire week; monthly samples, each week; and annual averages, all months. As analyses are accumulated over long periods, greater confidence is established as to the true mean, and fewer analyses are required to assure representation of the residue stream.

Statistical analysis may be needed, especially if a high degree of variability is observed in leaching characteristics of ash residues. While cadmium data are fairly consistent, greater variations have often been found in lead concentrations.[45] Analyses of fly ash samples generally show that the fly ash contains enough soluble lead and cadmium to exceed the toxic limit according to the EP or TCLP test. On the other hand, bottom ash generally passes the test, although individual samples of mixed bottom and fly ash samples may occasionally fail.

Figure 11.45 shows the data from extraction procedure tests of fly ash and mixed bottom ash plotted to reveal the distribution of lead concentration data of samples collected daily on two separate weeks.[46] In spite of the wide range, each week's data generally falls on straight lines in this log-log plot, which may be described as log normal. Eighty percent of the bottom ash (#1) and 50 percent of bottom ash (#2) passed the limit of 5 mg/L.

One of the two sets of fly ash samples, both of which fail the EP test, shows a multimodal distribution which may depart from the definition of log normal. To obtain lognormal distributions the two or more components must be separated into their separate distributions. The wide range during and between the two weeks illustrates why a large number of samples is needed to obtain confidence in the true value of the average.

Beneficial Use of Residues

Ash residues represent approximately 20 percent of the municipal waste stream. For this reason, reduction in the amount of ash which must be transported and disposed of in ashfills can offer a substantial saving in landfill space and cost.[47]

Ash Landfill Operations. The pozzolanic behavior of MSW residues from facilities with acid gas control, because of their high free-lime content and cement-like mineralogy, is beneficial for disposal site management practices. These properties allow the residues to be disposed of as a liner/cap over lifts of MSW and as a final capping

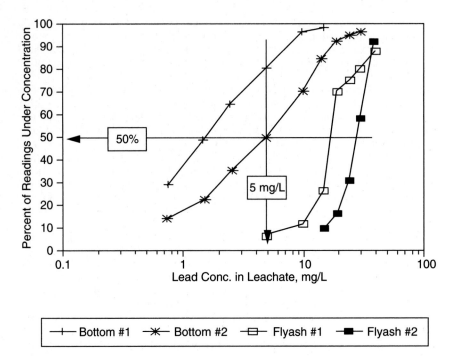

FIGURE 11.45 Distributions of lead concentrations in leachate from two sets of MSW bottom ash and fly ash samples collected during different sampling weeks. (*Source: Ref. 46.*)

material over MSW or other materials. In addition to providing high densities, using more lime or portland cement makes it possible to achieve permeabilities below the liner requirement of 1×10^{-7} cm/s.[20]

Use of Ash Residues for Construction. Ash residues from combustion of MSW have been used as roadway fill and subbase for parking lots, stabilized road base, bituminous paving mixtures, concrete masonry block, and portland cement concrete. Concern about leaching of metals into the environment has led to extensive research into the characteristics and environmental impact of these practices. In Denmark, laws and regulations define how ash residues may be used, and in what quantities, so that environmental impact will be insignificant.[48] In general, fly ash generally is not mixed with bottom ash in order to produce an environmentally benign bottom ash material by minimizing the quantity of soluble salts.

Table 11.52 describes the physical and chemical tests which may be used to evaluate the waste material for reuse.[49]

Percent Available as Aggregate. Recent tests of ash residues from a waste-to-energy facility in Concord, N.H., analyzed residues which combined bottom ash and residues from the dry lime-injection scrubber/baghouse system. The average percent of material smaller than ¾ in, suitable for asphaltic base course, was 65 percent, with a standard deviation of 13 percent. A portion of the remaining 40 percent also has the potential for use in road construction.

TABLE 11.52   Tests Recommended for Waste Reuse of Granular and Asphaltic Materials

Chemical tests	Physical tests
Elemental composition	Moisture content (ASTM D2216)
Mineralogy	Percent rejected ($>\frac{3}{4}$ in)
Acid neutralizing capacity	Organic content (loss on ignition)
Distilled water leach test	Ferrous content
Bioavailability leach test	Particle size distribution (ASTM C136)
Toxicity characteristics leaching procedure (TCLP)	Absorption and specific gravity (ASTM C127 and C128)
Lysimeter leach test	Unit weight and voids (ASTM C29)
	Moisture density test (ASTM D1557)
	CBR (ASTM D1863)
	Sodium sulfate soundness of aggregates (ASTM C88)
	Los Angeles Abrasion Test (ASTM C131)
	Unconfined Compressive Strength (ASTM D2166)
	Marshall Stability of Asphaltic Material (ASTM D1559)

Source: Ref. 49.

Use of Ash Residues in Asphaltic Mixtures.   Asphalt has been found to encapsulate ash residues effectively, reducing leaching potential to acceptable levels so that the asphalt can be safely used for road construction.[50]

Ash residue used as an aggregate in bituminous base course construction has been studied since the 1970s. Test sections have been continuously evaluated over periods from 1 to 5 years. The surface pavement and binder courses using combined bottom ash and fly ash from the WTE facility in Lynn, Mass., constructed in 1980, used 50 percent ash, 2 percent lime, 50 percent natural aggregate, and 6.5 percent asphalt, and were assessed in 1991 to be still performing well. A section built in Washington, D.C., contained 68.5 percent residue, 1.5 percent hydrated lime, 15 percent sand, 15 percent limestone, and 9 percent asphalt. Surface asphalt road sections at the SEMASS facility in Rochester, Mass., and in New Jersey, and ash use in road base and subbase structures at SEMASS and in New Hampshire are being evaluated for soil contamination, runoff, and leachate.[51]

Portland Cement Treatment.   A patented portland cement–based ash aggregate, McKaynite, using combined ash and bottom ash, has been tested in Florida as landfill cover and as aggregate in road projects, and found to meet physical standards for road construction materials, showing no adverse effects on groundwater, soil, or ambient air quality. The State of Tennessee Highway Department has developed a standard for acceptable use of ash residues as aggregate in road-base construction. Progress continues in many other states toward acceptance of ash residues for beneficial use.

At the Commerce WTE facility, in Los Angeles, fly ash from the spray-dry scrubber/baghouse is mixed with portland cement, then blended with bottom ash in a cement mixer truck. The treated ash-concrete is poured into roll-off containers and stored for 24 h before transport to the landfill where it is crushed for use as a subbase for roads at the landfill.[51]

Building Blocks and Other Uses. Use of ash residue to produce aggregate material for use in concrete both encapsulates the heavy metals and converts them to chemical forms which are essentially insoluble. Leaching tests of concrete blocks made from bottom ash only and mixed bottom ash/fly ash have shown that the rate of leaching of metals is insignificant in both underwater marine environments and in aboveground applications.[52]

Cement blocks made from MSW combustion in Montgomery County, Ohio, have been used to build buildings on the county landfill, after testing according to structural testing protocols, including ASTM tests for strength, UL tests for fire resistance, and TCLP tests. The blocks exhibited somewhat higher than normal shrinkage indices. The blocks made from coarser bottom ash released more easily from the molds than blocks which included the fly ash.[51]

Vitrification of Ash Residues. Vitrification of residues from thermal processes produces a dense, grainless, amorphous, glass-like material which contains no organic material and from which inorganic mineral matter does not leach significantly, and which has many beneficial uses.

In Japan, where landfills are especially scarce, nonexistent, or costly, vitrification has been applied at many facilities, employing fossil fuels, electric heating, and electric arc as the heat source. Due to the relatively low cost of landfill in the United States, there has been less incentive to pursue this course.

Electric Arc Furnace Vitrification. An extensive investigation of vitrification was carried out by the American Society of Mechanical Engineers (ASME) and the U.S. Bureau of Mines with the support of governmental agencies and industry.[34] The results of this program provide information on operating parameters and potential constraints of the technology; identify and quantify process residuals, effluents, or emissions; identify beneficial uses for the vitrified products; and develop economic data.

Tests were performed on combined furnace bottom ash and fly ash from three mass burn waste-to-energy (WTE) plants, fly ash from an RDF-fired WTE plant with acid gas scrubber, and combined ash from a multiple-hearth waste water treatment sludge incinerator. The tests were performed in a sealed submerged-arc electric furnace having a capacity of 1 ton/h. Fumes from the furnace were cooled and the deposits analyzed. A fabric filter was used to collect particulate for analysis, and an afterburner was provided to burn off any possible organic matter. The stack gases were analyzed for trace organics and metals.

The composition of the furnace feed, vitrified product, metal, matte, and fumes for a typical mass burn WTE furnace is shown in Table 11.53. Although the mass balance is not perfect, it is evident that most of the nonvolatile elements reported to the vitreous product and only traces to the fume.

Approximately 83 percent of the feed weight was converted to vitrified product, and 5 percent metal was withdrawn, consisting mostly of about 75 percent Fe, and 8 percent copper, 6 percent phosphorus, and 7 percent silica. The matte, which collects on the top of the charge, contained about 48 percent copper, 15 percent iron and about 27 percent sulfur. The principal components of the fume solids were NaCl, ZnS, and KCl. They contained 70 percent of the zinc and 37 percent of the lead in the furnace feed, and from 7 to 26 percent silica, varying with the carryover which depended on the nature of the ash residues. Most of the chlorine in the feed material left the system in gaseous form, since essentially none was found in the solid products.

The vitrified product was tested in accordance with TCLP procedures which showed that the vitrified residues were environmentally benign. The levels of leaching from the samples were in all cases 10 to 50 times less than the EPA maximum limit. The cost of vitrification was estimated to range from $200 per ton of dry ash residue

TABLE 11.53 Composition of Vitrification Furnace Feed and Products

Element	Feed	Vitrified product	Metal	Matte	Fume
Percent	93	83	5	1	2
Si	186,860	154,018	24	0.09	19
Fe	122,484	85,040	2028	38	11
Ca	40,409	67,257			21
Al	31,649	48,312	0.28	0.02	8
Na	22,259	14,811			38
Cl	13,089	51			69
C	9,568	372	2.85	0.04	1
Mg	8,458	10,471			2
S	8,343	1,764	93	25	20
K	6,392	4,035			25
Cu	5,842	1,020	202	31	3
Ti	4,131	6,483			1
Pb	3,287	187	4	2	30
Zn	2,718	1,353	2	0.33	59
P	2,399	1,387	75	0.11	7
Bi	1,013	1,766			0.26
Mn	963	1,542	0.06	0.10	0.19
Ba	631	737	0.05		0.11
Sr	306	391			0.04
Cr	280	1,216	1.05	0.03	0.06
Ni	222	53	13	0.22	0.04
Sn	182	138			1.30
B	167	48			0.24
Sb	91	8	3.73	0.14	0.34
Ce	86	69			0.04
Br	63	1			0.97
V	48	59			0.01
Y	43	34			0.02
Mo	43	34	1.02	0.01	0.02
Co	35	21			0.00
As	28	2	0.89	0.01	0.07
Li	19	16			0.02
Cd	18	6	0.12		0.18
Ag	9.51	1.38	0.33	0.16	0.03
Tl	4.32	3.44	0.43		0.00
Hg	2.59	0.69	0.00		0.01
Be	0.86	0.69	0.00		
Se	0.86	0.69	0.53		0.01
Au			0.17		
Total	472,145	402,709	2449	98	317

for a 50-ton/day facility to $115 per ton for a 300 ton/day facility, based on electric power costing $0.051 per kWh. The components of the cost of vitrification can be roughly estimated to be 5 percent for drying, 45 percent for electricity, water, and gas, 30 percent for labor and maintenance, and about 20 percent for the cost of capital.

Cold Crown Glass Furnace Vitrification. Glass furnace technology was investigated as a means of vitrifying WTE facility air pollution control (APC) fly ash residues. In this process, the ash must first be washed to extract chlorine, since the maximum tolerable level is about 5 percent. About 22 percent of the APC residue was dissolved and removed in two extraction steps, during which 25 to 30 percent of the lead was also washed out. Washing removed 96 percent of the chloride, 30 percent of the calcium, 60 percent of the potassium, 20 percent of the lead, and 19 percent of the sulfur, but 99 percent of the volatile heavy metals were retained in the glassy product. The melting temperature ranged from 2200 to 2800°F (1200 to 1550°C). The wash water can be treated to precipitate the metals, especially lead, by raising the pH to about 9. In the second step, the dechlorinated APC residue is blended with glass-forming additives, forming a mixture of about 52 percent residue, 36 percent silica, 9 percent sodium carbonate, and 4 percent sodium nitrate. The volume of the resulting product was reduced by 40 percent from the original APC volume. The product was subjected to the TCLP and the accelerated strong acid durability test, neither of which tests showed appreciable leaching of elements of concern, most being below detectable limits of analysis.[53]

Statistical Analysis of Ash Residue Test Data

The variability of analytical test data of ash residues has created serious problems. Individual samples have shown heavy-metal concentrations, 5 to 10 times the average, occasionally exceeding the EPA toxic limit values. Early sporadic analyses of ash residues showed that more than half the samples exceeded the limits, causing ash to acquire the name *toxic ash*. As more data became available, and procedures improved, it has become apparent that average ash quality is not as highly variable, and that bottom ash and mixed bottom and fly ash generally do not exhibit the characteristic of toxicity as determined by the TCLP leaching procedure.

The variability of the leaching characteristics of ash residues must be investigated and understood in order to obtain confidence in its quality. The following paragraphs apply statistical methods to an unusually extensive database in order to illustrate the principles of statistical analysis.

Basic Elements of Ash Sampling and Testing.   Ash sampling is carried out in order to obtain sufficient sample to estimate whether specific characteristics of the ash material meet required specifications. The ASTM *Standard Guide for General Planning of Waste Sampling* (D4687) contains guidelines for developing a sampling plan, including sampling procedures, safety plans, quality assurance, general considerations, preservation and containerization, labeling and shipping, and chain of custody (ASTM, 1989).[47]

Specification.   A material specification generally contains a target range of values to be met, the test methods to be used, and the desired confidence level. The confidence level is the selected degree of confidence that the difference between the mean of the sample and the mean of the population of all possible samples of the material being tested is less than some allowable error. For instance, the U.S. EPA has proposed a 90 percent confidence level in its SW846.[54] Specifically, an ash residue may be classified as exhibiting the characteristic of a hazardous waste if extracted leachate obtained from TCLP leaching procedures exceeds the target of 5 mg/L, on average, for lead and 1 mg/L for cadmium, the most likely metals to exceed the limits. The method of obtaining the average becomes more critical as the analytical results approach this target.

Selecting Physical Sampling Procedures.   Sampling procedures should be related to the purpose of the test program. Performance tests require determination of the average moisture and heating value of ash residues which are representative of the period of days or weeks of testing. Procedures used to determine the leaching characteristics of ash residues destined for landfilling or for constructive use should simulate the ongoing production of ash over the time periods during which the residues are produced, such as monthly or annually.

To obtain an appropriate test-sized sample: (1) take samples or increments for composited samples, (2) combine increments into composite samples, (3) blend or process (screen or crush) the composite, and (4) take subsamples from the composite for laboratory analysis and reference. Representative ash samples can be taken from the full width of a conveyer, or from the conveyer drop-off.

Figure 11.46 shows a procedure for collecting and dividing ash residue samples in order to obtain laboratory and reference samples over a period of 8 h, during a facility performance test. In this case 40- to 60-lb samples are collected from the drop-off of a conveyer belt every 10 min. Alternatively, a sample could be taken once per hour over a 24-h period. Samples weighing about 10 lb each are sent to the laboratory and the client, and kept for reference in case any samples are lost or damaged.

Coning and quartering reduces the quantity of the sample while retaining representativeness. The composite sample is mixed in a pile then split into four quarters. Two opposite quarters are retained and two are set aside. The retained quarters are combined into a pile. The process is repeated until the desired sample quantity is obtained.

The average characteristics of the ash residue stream leaving the facility can be obtained by taking daily samples for a month, to determine the statistical properties of the residue, after which weekly samples might be taken to sustain a moving average. Unusual results might provoke a return to daily samples until confidence was again restored.

At least 8 samples, and preferably 16 or more samples, should be analyzed to obtain the statistical properties of the ash. The variability of individual samples of ash is illustrated in Fig. 11.47, which graphs chronologically the results of EP toxicity analysis for lead and cadmium of 48 samples taken hourly for 2 days. These data are listed in Table 11.54.[46] It is apparent that a series of spikes occurred in the lead analyses, and to a smaller extent in the cadmium analyses, at roughly 6-h intervals. These spikes were attributed to boiler cleaning, in this case by tube rapping. The deposits which form on the tubes fell onto the stoker, or were conveyed externally to the ash quench tank. Analysis of the tube deposits showed high concentrations of lead and cadmium which, volatilized in the combustion process, condensed on the particulate that adhered to the tubes. The lead spikes were 2 to 3 times the EP limit of 5 mg/L, although the average of 48 analyses was only 2.59 mg/L. Cadmium concentrations were far less than the EPA limit of 1 mg/L.

Graphs help visualize data. A set of data can be sorted in ascending order and plotted versus the number of samples as shown in Fig. 11.48. The lead and cadmium EP data plotted on a logarithmic scale fall on a fairly straight line for 41 points, representing the log-normal distribution which is typical of natural variability. The upper seven lead data points, which exceeded 5 mg/L, represent the spikes due to tube rapping. They must be considered to be a different population, which distorts the normal distribution. Three high cadmium points also represent spikes.

Histograms of data show how the data are distributed. The average concentration of lead in the EP extract is not easily determined from the linear graph shown in Fig. 11.49, whereas the histogram of the logarithm of the data exhibits the typical bell curve of the probability distribution, wherein all points have an equal probability of happening (see Fig. 11.50). The point which has the same weight on either side appears to be about 1.1, which translates back to 3.0 mg/L. This graph, called a log-normal distribution, is characteristic of ash properties, because it covers a wide range of natural phenomena, such as the breakage of coal.

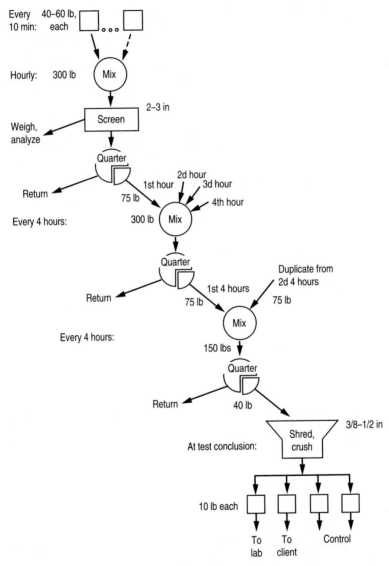

FIGURE 11.46   Schematic of ash sampling procedure.

To plot the points, sort the data, increasing or decreasing, and plot at equal intervals on probability or Rosin-Rammler paper. For instance, place four data points at the 20, 40, 60, and 80 percent "less than" locations. With normally or log-normally distributed data, most of the points will fall on a straight line.

Numerical methods can be used to analyze data. A pocket calculator can be used to determine the mean $\bar{x}$ and the standard deviation s of a set of data. In general, about 85 percent of the data points will have values less than the mean plus 1 standard devia-

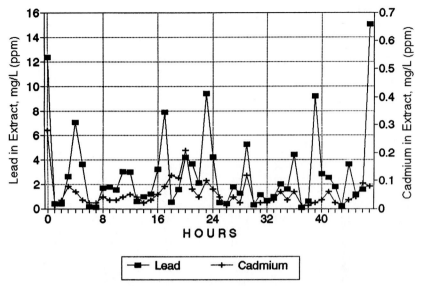

FIGURE 11.47 Time plot of hourly samples—EP toxicity data for lead and cadmium. (*Source: Ref. 46.*)

tion ($\bar{x} + s$), 95 percent will be less than $\bar{x} + 2s$, and 98 percent will be less than $\bar{x} + 3s$. The arithmetic mean obtained in this manner should represent the true average of the properties of the samples collected uniformly over a period of time.

In general, if the standard deviation of a set of data exceeds the average or mean, the data are not normally distributed. Transformation using the logarithm of the data often solves the problem, so that the standard deviation does not exceed the mean. When transformed back, the log-transformed mean is called the *geometric mean.* The geometric mean and standard deviations are obtained by obtaining the mean and standard deviations of the logarithms of the data, then converting back. In general it is found that the geometric mean is substantially less than the arithmetic mean, but the value of the geometric mean plus 2 standard deviations is close to the arithmetic mean plus 2 standard deviations. Using the geometric mean avoids the exaggerated estimate of the mean which the arithmetic mean produces when there is a wide range in data values. When the standard deviation exceeds the arithmetic mean, the outlying data should be inspected to see if it belongs in the set. If a valid reason for the outliers can be discovered, the outlying data points may be removed from consideration.

Determining the Number of Samples Needed. The number of samples analyzed influences the accuracy of the average or mean value obtained, as can be seen in Table 11.54. The relationship between the number of samples and the accuracy and confidence in the estimate of the mean value of the whole population can be calculated once the mean and standard deviation of the properties in question have been determined.[14]

For a normal distribution, the number of samples required is determined as follows:

$$n_r = t^2 s^2 / e^2$$

where n_r = number of samples required to be analyzed to show that the specified target range of values has been met at the desired confidence level

TABLE 11.54 EP Extracts of Lead and Cadmium in Bottom Ash Collected from June 3 to June 5, 1982

Sample number	Lead, mg/L	Cadmium, mg/L	Sample number	Lead, mg/L	Cadmium, mg/L
1	12.40	0.28	25	4.27	0.07
2	0.41	0.01	26	0.51	0.04
3	0.38	0.03	27	0.40	0.01
4	2.62	0.08	28	1.79	0.04
5	7.11	0.06	29	1.26	0.02
6	3.65	0.03	30	5.28	0.12
7	0.18	0.02	31	0.34	0.02
8	0.15	0.02	32	1.10	0.02
9	1.70	0.04	33	0.65	0.02
10	1.81	0.03	34	0.98	0.03
11	1.52	0.03	35	2.04	0.06
12	3.04	0.04	36	1.60	0.03
13	3.01	0.05	37	4.43	0.06
14	0.57	0.04	38	0.08	0.01
15	0.98	0.02	39	0.59	0.01
16	1.19	0.03	40	9.21	0.02
17	3.23	0.05	41	2.87	0.03
18	7.93	0.08	42	2.54	0.06
19	0.53	0.12	43	1.81	0.02
20	1.56	0.11	44	0.21	0.01
21	4.22	0.21	45	3.66	0.03
22	3.69	0.07	46	1.15	0.04
23	2.11	0.04	47	1.58	0.09
24	9.44	0.10	48	15.08	0.08
Avg. of 48	2.59	0.05	Avg. of 16	2.55	0.05
Std. dev.	2.69	0.05	Std. dev.	3.07	0.06
90 % UCL	3.46	0.05	90% UCL	3.21	0.05
Avg. of 24	3.06	0.06	Avg. of 8	3.18	0.06
Std. dev.	3.10	0.06	Std. dev.	3.90	0.08
90 % UCL	3.68	0.06	90% UCL	4.04	0.06

Source: Ref. 46.

t = the confidence factor for a designed confidence level: Student's t values are used, based on the desired confidence level and the actual or planned number of samples analyzed; for a confidence limit of 90 percent, t is 1.31 for 30 samples, 1.38 for 10 samples, and 1.64 for 4 samples

s^2 = the variance (square of the standard deviation), calculated for the data set, or estimated from prior sampling and testing

e^2 = the allowable error, or absolute value of the difference from the specified target value which defines the maximum acceptable target value and the mean of the corresponding data set

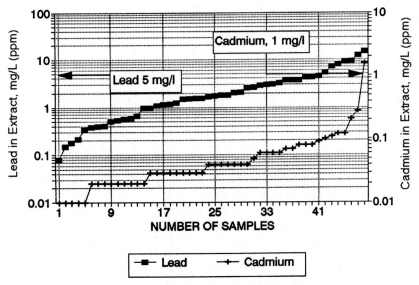

FIGURE 11.48　Distribution of EP test hourly samples for lead and cadmium. (*Source: Ref. 46.*)

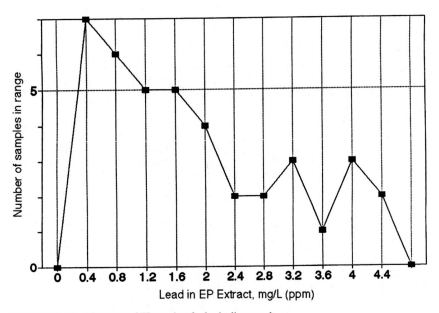

FIGURE 11.49　Histogram of EP test data for lead—linear scale.

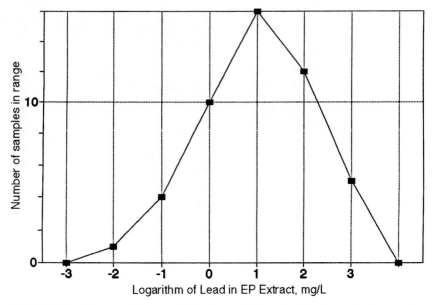

FIGURE 11.50   Histogram of EP test data for lead—logarithmic scale; 48 hourly samples, bottom ash.

Confidence Limits.   The upper confidence limit (UCL), from Ref. 54, is

$$\text{UCL} = \bar{s} + t_n s_{\bar{x}} = \bar{x} + t_n s/(n)^{0.5}$$

where UCL = upper confidence limit below which the actual mean of the character-
             istic being tested will be found at the specified confidence level
      $\bar{x}$ = the mean of the data set (or of the logarithms of the data)
      t_n = the probability factor (Student's t) corresponding to the desired level
             of confidence in the conclusions made about the actual mean of the
             data
      $s_{\bar{x}}$ = the standard error of the mean, calculated by dividing the standard
             deviation s of the data set (or of the logarithms) by the square root of
             the number of analyses in the data set ($= s/\sqrt{n}$)

TABLE 11.55   Calculation of Sampling Error versus Number of Samples

Number of samples, n	Degrees of freedom, $n-1$	Student's t	Standard deviation s	Average (mean) $\bar{x}$	Error e	UCL = mean + standard error
4	3	1.68	4.46	3.95	4.17	8.12
8	7	1.415	4.10	3.36	2.19	5.55
16	15	1.341	3.07	2.55	1.55	4.10
30	29	1.31	2.91	2.90	0.91	3.81
48	47	1.31	3.21	2.85	0.61	3.46

Example.   The number of samples required can be found by trial-and-error procedures, sharpening the numbers by stages. Data from other sources or from preliminary
tests may be used to obtain a first estimate of the number of samples needed to obtain
the desired confidence that the estimate of the mean will be safely below the target.

Table 11.55 is the result of a series of trials to determine how many samples must
be taken to be certain that the estimate of the mean is safely below the target, for lead,
of 5 mg/L. The data set in Table 11.54 has been used as the basis. Starting with the
first sample, the mean and standard deviation were calculated for different numbers of
consecutive samples, as if the sampling were stopped at different points. For each
number of samples analyzed, n, the degrees of freedom, $n - 1$, is used to select the
value of Student's t. The error is then calculated, using the mean and the number of
samples. The resulting error is added to the mean to obtain the estimate of the mean.
With 4 and 8 samples, the error was greater than the mean. With 16 samples, the mean
plus error was 4.1. Compared with the target of 5.0, this represents a safety margin of
about 10 percent. With 30 samples, the mean plus error was only slightly lower, 3.81,
and the safety margin increased to about 25 percent, while 48 samples increased it to
40 percent. It is evident that 16 samples would be sufficient in this case, but if the
mean were higher, more samples would have to be analyzed.

The UCL for 16 samples would be

$$UCL = 2.55 + 1.34[3.07/(16)(0.5)]$$

$$= 3.58 \text{ mg/L}$$

This is within about 28 percent of the target of 5.0 mg/L. With 30 and 48 samples, the
UCLs are almost the same, due to the effects of an outlying data point. If the UCL
were closer to the target of 5.0 mg/L, more samples would be needed.

When there are cyclical factors in the flow of ash, as seen in Figure 11.47, the time
at which samples are collected can severely affect the data. The UCL calculated using
the 6 samples taken during the first hour of each 8-h shift over the 48-h test period
(after cleaning of the boiler tubes) is 4.33 mg/L, whereas the UCL calculated from 6
samples collected during the seventh hour (prior to cleaning the tubes) is 2.12 mg/L.
The average of these two UCL values is close to the average of the UCL found for 48
hourly samples, 3.46 mg/L. Taking consecutive hourly samples evidently produced
much more reliable results than taking one sample per shift at the same hour. This
would also be true for samples which are composited and one analysis performed per
composite sample.

PART C

Aaron Teller

Senior Vice President, Air & Water Technologies, Inc.
Research Cottrell/Metcalf & Eddy, Sunrise, Florida

11.3   EMISSION CONTROL

Background

The utilization of incineration as a disposal mode of municipal waste has varied widely. To a high degree, the utilization is reflective of social recognition of pollution problems resulting from incineration, rejection of incineration for other options, excessive expectations from the alternative methods, technical solutions for pollution problems, recognition of energy benefits, and renewal of selection of incineration, followed by discovery of a new emission. In the year 1993, incineration was at a low ebb but there was evidence of its return because of the unanticipated costs and waste of excessive attempts at recycling.

The concern with incineration lies predominantly with the emissions, and the problem has been

1. Slow emergence of knowledge regarding the compositions, concentrations, and effects of the pollutants
2. Wide variation in emission standards and degree of enforcement of standards
3. Increasing sophistication of the public with regard to pollution

At this writing, it is believed that the study of the emissions has matured to the level where knowledge of composition and concentrations of the emissions are well-established. And in spite of the controversy about the effect of these pollutions, the emission standards that have emerged in Europe, Japan, and several of the states afford, within the context of epidemiological information, excellent protection for the public. Emission control technology in the United States has evolved to achieve greater reductions in emissions than required in Europe, and with high reliability and consistency.

The major technology now specified is the semidry reactor–fabric filter system. It has evolved since the first installation in Massachusetts in 1978 to a highly efficient, reliable system that provides effective control for acid gases, particulates, respirable particulates, heavy metals including mercury, and organic compounds including dioxins.

This section establishes a basis for evaluation and analysis of application to specific problems by presenting data and information in the following order:

Emissions and regulatory limits.   The contaminants in both the gas and solid discharge, the range of concentrations and the regulatory requirements in both Europe and the United States.

Control system requirements. The variation and rapidity of variation in the emissions. The limited effect of "good combustion practice." The availability of technology to modulate variability.

Types of systems. The technologies that have been, are being, and will be stipulated for future installation by the regulatory authorities. Descriptions of the systems and their components, capabilities, and behavior.

System performance. The performance and comparison of performance of generic operating systems for reduction of

- Particulate
- Sulfur dioxide
- Hydrochloric acid
- Heavy metals
- Organic components

inclusive of specific system data for pressure drop, reagent consumption, and contaminant reduction. Specific system behavior:

- Wet systems
- Dry and semidry—temperature effect on performance for various contaminants; reaction time effect of utilization of fabric filter as backup reactor and separator for various contaminants
- Specific contaminants—problems, modes of control, performance with respect to heavy metals, mercury, dioxins and furans, and nitrogen oxides
- Capital and operating costs

Emissions and Regulatory Limits

The incineration of municipal solid waste (MSW) is conducted to (1) reduce the mass and volume of the waste and (2) provide energy from the combustion of the waste. The incineration or combustion process produces a flue gas and a solid residue of the noncombustible components of the waste. Both require treatment to remove or fix, in insoluble form, the toxic or noxious components present in the flue gas or solid.

Municipal waste consists of a wide range of materials and concentrations of these materials. The product combustion gases contain inorganic and organic gases, volatilized metals, and particulates incorporating heavy metals and organic compounds, in concentrations varying as much as 50 to 1 (Table 11.56). It has also been established that even under the good combustion practice guidelines, organic contaminants such as dioxins can be synthesized downstream of the combustion zone.

The solid discharges, consisting of bottom ash or product falling through the grates and the fly ash or the particulate entrained in the flue gas, contain heavy metals that are subject to leaching by water when deposited in landfills. Reflecting the heterogeneity of municipal waste, the concentrations of these metals vary as much as 1000 fold (Table 11.57).

The emissions from the incineration process have been subject to regulatory limits. These limits are still evolving as the quality of measurements improves and as ecological impacts are evaluated (Table 11.58). In general, the European and Japanese regulatory limits are more severe than those required by the U.S. EPA. However, some states, notably California, Florida, and New Jersey, have established some emission limits significantly lower than proposed by the U.S. EPA (Table 11.58).

The emission limits for flue gas contaminants are normally established for a specific oxygen or carbon dioxide content in the gas stream so that dilution cannot be used

TABLE 11.56 Range of Emissions from Incineration of MSW

In flue gas component	Range of concentrations observed	
	mg/Nm^3	ng/Nm^3
Hydrogen chloride	100–3300	
Sulfur oxides	50–2900	
Hydrogen fluoride	0.5–45	
Nitrogen oxides	100–1000	
Carbon monoxide	10–200	
Total particulate	1000–10,000	
Particulate <10 μm (PM10)	50–2000	
Arsenic	0.02–0.2	
Beryllium	0.01–0.1	
Cadmium	0.05–0.3	
Chromium	0.5–2.5	
Copper	0.5–2.5	
Lead	1–50	
Manganese	1–5	
Mercury	0.2–5	
Molybdenum	0.05–0.25	
Nickel	5–30	
Selenium	0.01–0.1	
Tin	0.05–0.3	
Vanadium	0.05–0.5	
Zinc	1.5–100	
Dioxins and furans (TEQ)		0.2–10
PAH		4–30

TABLE 11.57 Range of Concentrations of Heavy Metals in Ash Discharge

In ash (bottom ash, fly ash)

Component	Concentration, μg/g
Arsenic	3–56
Boron	24–174
Cadmium	0–152
Chromium	12–1,500
Copper	40–9,300
Lead	31–36,600
Manganese	14–3,150
Mercury	0–25
Molybdenum	2–290
Nickel	13–12,900
Tin	13–380
Vanadium	13–150
Zinc	2,120–46,000

TABLE 11.58 Regulations on Emissions from MSW Incineration Concentrations, mg/Nm^3

Component	Most restrictive (1989)[2,5,6,7]	U.S. EPA recommended, new facilities >250 tons/day[3]
Hydrogen chloride	10 (11% O$_2$)[1,2]	40.7 (7% O$_2$) = 29 (11% O$_2$), or 95% reduction
Sulfur oxides	40 (11% O$_2$)[2]	85.7 (7% O$_2$) = 61 (11% O$_2$), or 80% reduction
Hydrogen fluoride	1 (11% O$_2$)[1,2]	
Nitrogen oxides	70 (11% O$_2$)[2]	139 (7% O$_2$) = 99.5 (11% O$_2$)
Carbon monoxide	50 (11% O$_2$)[1,2]	125 (7% O$_2$) = 89 (11% O$_2$)
Organics	5 (11% O$_2$)[3]	30 (7% O$_2$) = 21.4 (11% O$_2$)
Arsenic	0.007 (11% O$_2$)[6]	
Mercury	0.05 (11% O$_2$),[1,2] 0.027[7]	
Cadmium + mercury	0.05 (11% O$_2$)[1,2]	
Nickel + arsenic	1.0 (11% O$_2$)[3]	
Lead + chromium + copper + manganese	1.0 (11% O$_2$)[2]	
Arsenic + cobalt + nickel + selenium + tellurium	1.0 (11% O$_2$)[1,3]	
Antimony + lead + chromium + cyanide + fluorine + copper + manganese + vanadium	5.0 (11% O$_2$)[1]	
Total heavy metals	0.5 (11% O$_2$)[1,3]	34 (7% O$_2$) = 24.3 (11% O$_2$)
Total particulate	10 (11% O$_2$)[1]	34 (7% O$_2$) = 24 (11% O$_2$)
Particulate <10μm (PM10)	5 (3% O$_2$)[5]	
Dioxins and furans		5–30 (7% O$_2$)
Dioxins and furans (TEQ)	0.1 (11% O$_2$)[1,2]	[approximately 0.04–0.32 (11% O$_2$)]

[1]Germany [2]Netherlands [3]European Community [4]Sweden
[5]California SCAQMD [6]Florida [7]New Jersey

as a method for reduction of concentration. The permissible effluent concentrations are expressed for the carbon dioxide content of 12 percent and for oxygen contents of 3, 7, and 11 percent. The conversion to these standard concentrations from the conditions of measurement are for carbon dioxide correction

$$C_{12} = C_M \frac{12}{(CO_2)_M}$$

where C_{12} = concentration of contaminant at 12 percent CO_2 content
C_M = measured concentration of contaminant under sampling conditions
$(CO_2)_M$ = concentration of CO_2 in the gas at sampling conditions, percent volume

For oxygen correction:

$$C_{11} = \frac{21 - 11}{21 - (O_2)_M} C_M$$

$$C_7 = \frac{21 - 7}{21 - (O_2)_M} C_M$$

TABLE 11.59 Limits of
Concentration of Heavy Metals in
Leachate

Metal	TCLP, mg/L
Arsenic	5
Barium	10
Cadmium	1
Chromium	5
Lead	5
Mercury	0
Selenium	1
Silver	5

TCLP = toxicity characteristics
leaching procedure test 3.

$$C_3 = \frac{21 - 3}{21 - (O_2)_M} C_M$$

where C_3, C_7, C_{11} = corrected concentration of contaminant at 3, 7, 11 percent oxygen
content.

The solids discharged from the incineration process and the flue gas treatment system are subject to restrictions on the concentrations of heavy metals present in the leachate when the solids are landfilled (Table 11.59).

Emission Control System Requirements

The control system for gaseous and gasborne emissions from the incineration process must be able to modulate the rapid variation in inlet composition and concentrations. As great as a twofold change in inlet concentrations of contaminants can occur in a 1-min interval (Figs. 11.51 and 11.52). The causes of this wide variation are

1. Heterogeneity of the feed material to the combustion zone
2. Heterogeneity of the combustion process, reflecting the variation in water content and bulk density of the feed material
3. Variation in composition of the particulate components, affecting postcombustion synthesis of dioxins and furans and adsorption of contaminants such as mercury

Review of boiler outlet compositions and concentrations indicate the variation in emissions that can be anticipated (Table 11.60). The ranges of the components of the incinerator feed and the pollutants released by their combustion are listed in Table 11.61.

Source separation, while reducing the emissions of some of the compounds, does not achieve modulation of the variation in concentrations. Combustion control, while reducing some of the concentrations of organic compounds, is often counteracted by resynthesis of these compounds, such as dioxins from combustion fragments, downstream of the combustion zone.

The requirements for good combustion practice (GCP) apply primarily to effective thermal destruction of organic compounds, such as polychlorinated biphenyls (PCBs) and the limitation of carbon monoxide emissions. GCP does not have effects on acid

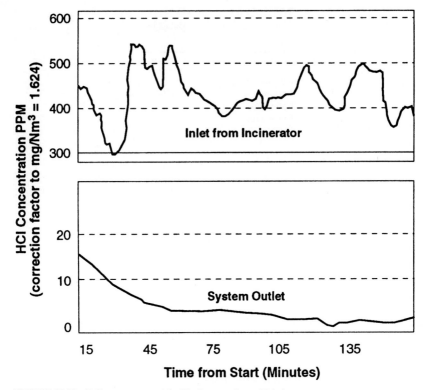

FIGURE 11.51   Teller system municipal incinerator, Isogo-Yokohama.

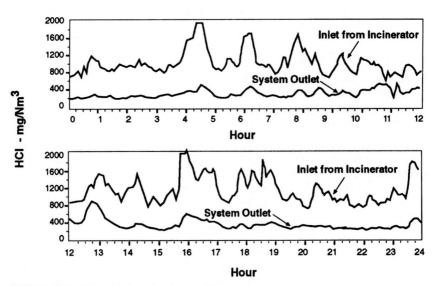

FIGURE 11.52   Hydrochloric acid emissions, Malmo system.

TABLE 11.60 Variation in Incinerator Emissions

Component	Ratio of concentrations	Range of concentration observed
Hydrochloric acid	3	100–3,000 mg/Nm^3
Sulfur oxides	40	50–2,000 mg/Nm^3
Dioxins, furans	50	0.2–10 mg/Nm^3 TEQ
Lead	50	1–50 mg/Nm^3
Chromium	5	0.5–2.5 mg/Nm^3
Mercury	25	0.2–5 mg/Nm^3
Particulate	10	1,000–10,000 mg/Nm^3
Humid content	4	7–30% v/v

TABLE 11.61 Components of MSW and Combustion Products

Component	Range of concentration, %	Pollutants formed
Paper and paperboard	30–50	Organics, heavy metals, HCl, SO$_x$
Food and yard waste	15–30	Organics
Glass	5–15	
Metals	6–12	Heavy metals
Plastics	3–9	HCl, SO$_x$, organics, heavy metals
Others, including building material waste, batteries, paints	6–12	SO$_x$, heavy metals, Hg
Batteries	Variable	SO$_x$, heavy metals, Hg

gas emissions and has only minor effects on particulate emissions. Thus, the emission control system must have the capability, by design and/or by instrument response, to modulate the rapid variation in system inlet concentrations. This modulation capability is indicated in Table 11.62.

Types of Systems

The requirements imposed on the performance of emission control systems employed in municipal waste incineration are now directed at the emissions of:

Particulates—total

Particulates—PM 10 (less than 10 μm)

Heavy metals

Mercury

Acid gases—HCl, SO$_x$, HF

Nitrogen oxides

Dioxins, furans

PAH

The types of systems now in operation can be classified into three generic forms (Table 11.63): dry, semidry, and wet. As regulations for additional substances were

TABLE 11.62   Tsushima Daily Printout of Outlet Concentrations, September 11, 1985

Inlet HCl 350–550 ppm, SO$_2$ 30–100 ppm

Time	HCl conc., ppm	SO$_2$ conc., ppm	NO$_x$ conc., ppm
00:00	3	1	102
01:00	5	0	101
02:00	5	0	105
03:00	5	0	98
04:00	5	0	98
05:00	5	0	98
06:00	4	0	94
07:00	4	0	99
08:00	3	0	105
09:00	4	0	91
10:00	4	0	112
11:00	5	0	97
12:00	5	0	110
13:00	5	0	106
14:00	5	1	101
15:00	3	1	83
16:00	2	1	89
17:00	4	2	90
18:00	4	2	117
19:00	4	1	103
20:00	5	2	120
21:00	5	1	122
22:00	4	1	108
23:00	4	1	109

promulgated, some systems were adapted by modification of internal features, reagents, and operating conditions to achieve compliance. Often, however, new components were added, thus both increasing the complexity of the systems and blurring the generic classifications.

The disposal of the unreacted reagents, reaction products, and captured materials is a function of the collection system design. The dry and semidry systems produce a solid product. In some jurisdictions it has been designated as hazardous and, if disposed of, is directed to hazardous substance landfills. Alternatively, the solid product has been mixed with pozzolanic materials or has been vitrified. With vitrification, leaching of heavy metals is insignificant and the glass-like product can be used as a construction material.

Wet system products must be treated for removal of the heavy metals by precipitation or adsorption and in some cases for removal of some anions, i.e., fluorides before disposal to the sewage system. As noted in Table 11.63, the fact that a system design is capable of recovering certain materials is not an indication of its efficiency in the recovery of those materials.

Dry Systems.   These systems are the simplest from an operation aspect. They consist of a solid-gas reactor or dry venturi and a particulate collection component, either an electrostatic precipitator or fabric filter (Fig. 11.53). The reagents must be fed in solid form and the neutralization reaction occurs in the reactor and in the downstream particulate collector.

TABLE 11.63 Emission Control System Types

Dry	Semidary	Wet	Materials addressed*					
			Particulate	Acid gases	Heavy metals	Mercury	Organics	NO$_x$
Electrostatic precipitator			X		X			
Fabric filter			X		X			
Furnace injection + FGD		Tray or packed scrubber		X	X			
		Condensing scrubber		X	X		X	
Dry reactor or dry venturi + ESP	Quench reactor + ESP	ESP, wet scrubber	X	X	X		X	
+ FF	+ FF	FF, wet scrubber	X	X	X		X	
Dry reactor or dry venturi + ESP + carbon injection	Quench reactor (+ DV) + ESP + carbon injection	ESP or FF + wet scrubber inclusive of oxidation	X	X	X	X	X	
+ FF + carbon injection	+ FF + carbon injection		X	X	X	X	X	
SNCR or SCR with dry reactor or dry venturi + ESP + carbon injection	SNCR or SCR with quench reactor (+ DV) + ESP + carbon injection	SNCR or SCR with wet scrubber inclusive of oxidation	X	X	X	X	X	X
+ FF + carbon injection	+ FF + carbon injection		X	X	X	X	X	X

*Merely addressing specific materials is not reflective of the efficiency achieved.

ESP = electrostatic precipitator

FF = fabric filter

DV = dry scrubber

SNCR = selective noncatalytic reduction

SCR = selection catalytic reduction

FGD =

Dry System

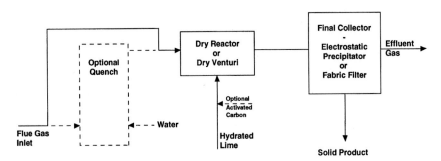

FIGURE 11.53   Dry system.

Where the temperature of the inlet flue gas exceeds 200°C, the gas is generally quenched with water to the range of 120 to 160°C. Not only is the stress on materials of construction reduced but, with the cooler gas and the humidity increase, the efficiency of neutralization and adsorption is enhanced.

Two general types of contacting devices are used for dry reaction. A typical gas-dispersed system (Fig. 11.54) uses the flue gas to entrain the fresh hydrated lime, recycled solids, and activated carbon, if desired, into the reactor zone, the pressure drop necessary for dispersion being supplied by the total gas stream. After contact, an initial gravitational separation of solids and gas can be used to provide for recycling. External cyclones or partial recycling of the final separator solids can also be used for recycling.

The dry venturi (Fig. 11.55) utilizes a small air or recycle gas stream to disperse the reagents and Tesisorb into the flue gas stream. The Tesisorb is used to collect fine particulates[56] by inertial impact and to modify the cake characteristics in the downstream fabric filter. The change to a higher porosity filter cake enhances the fabric filter's contribution as a neutralization reactor and adsorber by increasing the residence time of the reagents.

Dry sorbent injection (DSI), the introduction of hydrated lime or calcium carbonate into the zone of the furnace prior to the boiler, has been utilized in Japan since 1981. It also has been used at the Davis Co. Utah installation and has been tested at the Commerce California facility. The only external installation is the particulate collector, an electrostatic precipitator or fabric filter.

Semidry Systems.   Semidry systems (Figs. 11.56 and 11.57) use a slurry of lime to both cool the inlet gas via evaporation and to achieve partial neutralization, simultaneously. The equipment utilized to achieve this is either a spray drier[57] (Fig. 11.58) or quench reactor (Fig. 11.59). Both provide a dried solid in the effluent gas.

The gas proceeds to a final collector (Figs. 11.56 and 11.57), either a fabric filter or electrostatic precipitator, where the solid products of reaction, the particulate, and unreacted reagents are collected or through a dry venturi. The gas then proceeds to the final collector. Where the dry venturi is used, components such as activated carbon and lime reagent may be added to the gas stream along with Tesisorb material, which modifies the fabric filter cake to increase the time for reaction and neutralization.

Dry Reactor
Flue Gas Dispersed

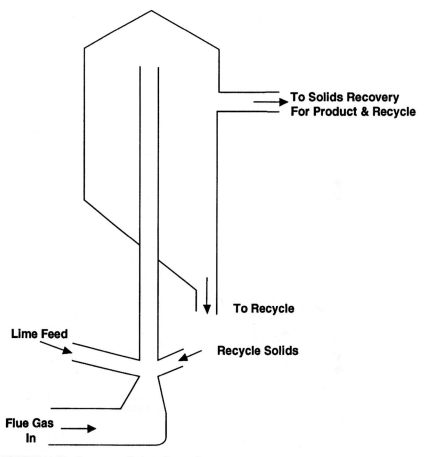

FIGURE 11.54 Dry reactor, flue gas dispersed.

The spray drier adsorber[57] (Fig. 11.58) uses a rotary atomizer to disperse the slurry into the gas stream with a resulting concurrent gas-dispersed slurry downward flow. A portion of the entrained solids is separated from the gas stream by centrifugal action as the gas leaves the system.

The upflow quench reactor[58] (Fig. 11.59) uses two fluid spray nozzles to disperse the slurry into the gas stream with a resulting concurrent gas-dispersed slurry upward flow. The effect of the upward flow is to effectively double the time of exposure of the spray particles to the hot gas stream compared to an equal size vessel with the downward flow pattern, thus providing more complete drying. This phenomenon, uti-

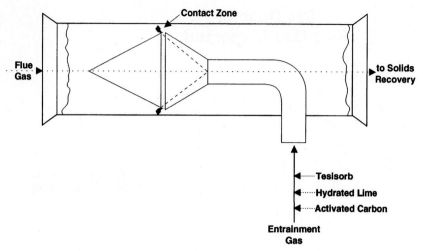

FIGURE 11.55 Dry venturi, solids dispersed.

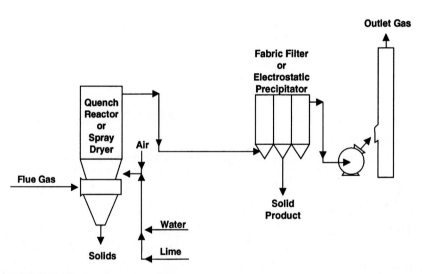

FIGURE 11.56 Quench reactor–collector.

lizing the gravitational effect on the dispersed slurry, requires that particles larger than 400 μm be removed from the gas stream prior to entering the spray zone. A cyclonic section at the inlet provides this separation.

Wet Systems. Wet scrubbers may be used in several configurations:

- Tray type (Fig. 11.60)
- Packed type (Fig. 11.61)
- Venturi type (Fig. 11.62)

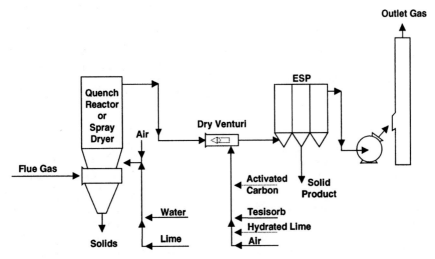

FIGURE 11.57   Quench reactor–dry venturi–collector.

If preparticulate separation is part of the system, utilizing an electrostatic precipitator or fabric filter, then any scrubber type may be used. If no particulate removal precedes the scrubbing process, the potential exists for plugging a packed tower unless the tower is preceded by a venturi. The venturi scrubber is an effective particulate remover at pressure drops exceeding 15 in H$_2$O gauge for particle sizes in the range of 1 to 2 μm, but has limited adsorption capability. The tray type or packed type can reduce acid gases by more than 95 percent, provided an adequate number of trays or depth of packing is used.

The reagents for neutralization of the acid gas that create minimum operating problems are sodium salts, carbonates, and hydroxides inasmuch as the neutralized acid gases form soluble sodium salts. Lime, however, will form relatively insoluble calcium sulfate and calcium sulfite that can deposit and cause plugging of scrubber internals. Several design variations splitting the adsorption process into separate HCl and SO$_2$ theoretically permits the use of calcium-containing reagents for HCl recovery.

The scrubber has been reported to be effective for mercury and mercury salt recovery where the reagent is sodium hypochlorite coupled with a chelating agent. (See discussion of mercury under "Heavy Metals.")

System Performance

Performance data for various types of emission control systems now operational in conjunction with municipal waste incinerators are indicated in Table 11.64. The system performance is grouped by equipment arrangement:

Electrostatic precipitator	ESP
Dry scrubber–fabric filter	DV-FF
Quench reactor–electrostatic precipitator	QR-ESP
Quench reactor–fabric filter	QR-FF

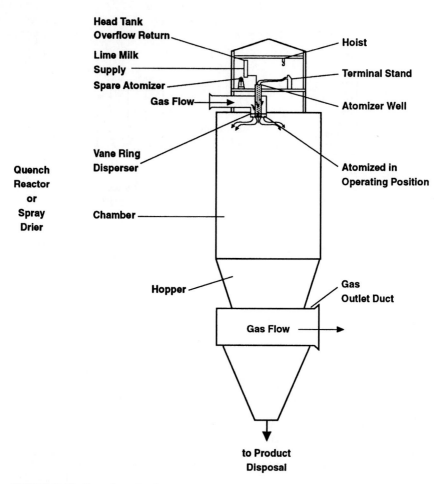

FIGURE 11.58   Spray dryer absorber.

Quench reactor–dry venturi–fabric filter	QR-DV-FF
Electrostatic precipitator–wet scrubber	ESP-WS
Condensation–scrubbing	C-S
Furnace injection–fabric filter	DSI-FF
Furnace injection–electrostatic precipitator	DSI-ESP

It has been established that the performance of an emission control system is a function of:

Type of system

Operating temperature

Humidity

Reaction time, affected by type of collector and residence time for reagents

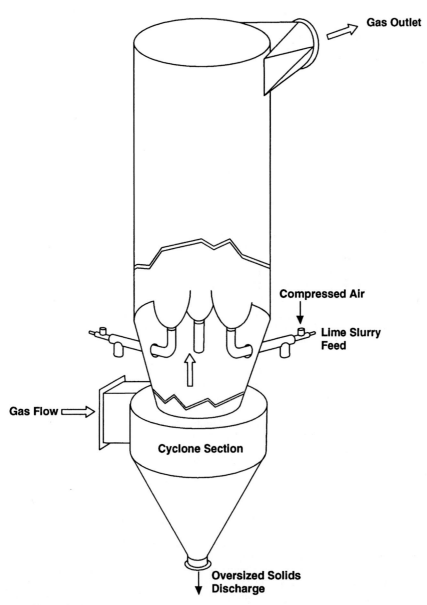

FIGURE 11.59   Upflow quench reactor.

Stoichiometric ratio of acid-neutralizing reagent

Reagents additional to those used for acid gas neutralization

Synergistic effect of hydrogen chloride on both sulfur dioxide and mercury recovery

Not always is this full range of data available. Thus, only generalized or trend comparisons can be made for the various types of systems now in commercial operation.

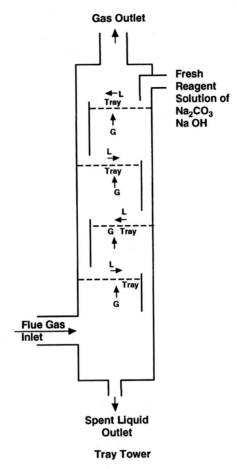

FIGURE 11.60   Tray tower.

Electrostatic precipitators were the primary control units in the early days of incineration emission control. As a result they are the only control component in about one-third of the total installations in the United States. Electrostatic precipitators cannot remove acid gases, exhibit a proclivity to increase dioxin and furan emissions (Tables 11.72 and 11.73), and are not as effective as desired in reducing particulate emission (inclusive of heavy metals) for the new MACT emission requirements (Fig. 11.62) (Table 11.64). Inasmuch as no new installations project the use of electrostatic precipitators as the only control equipment, no discussion of relative performance with systems of more recent technology is included.

Particulate Emissions.   The total particulate emissions comparisons (Fig. 11.63) indicate that the lowest emissions are achieved when the fabric filter is a component of the emission control system. The integration of a quench reactor into a control system enhances the reduction in the total particulate emission, most probably related to the agglomeration and inertial impact in the humid conditions and turbulence existing in the quench reactor. The quench reactor–fabric filter system has demonstrated total

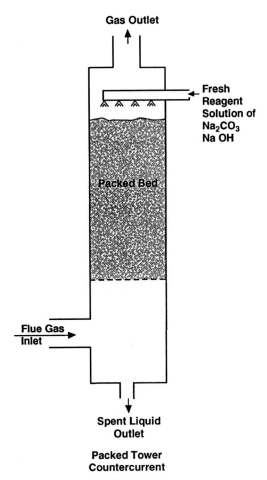

FIGURE 11.61   Packed tower, countercurrent.

particulate reduction to less than 12 mg/Nm^3 (12% CO_2) with a median value of 5 mg/Nm^3 (12 percent CO_2). Critically the incorporation of the dry venturi has demonstrated reduction of the respirable range particulate (less than 2.5 μm) with efficiencies equal to or greater than that of the larger particulates, in all cases exceeding 99.9 percent.

    Acid Gases.   The comparison of system performance for acid gas control (Figs. 11.64 and 11.65) must be tempered by the operating temperature required by a specific equipment arrangement and the humid content of the gas. With these limitations the trends that are indicated are as follows:

1. *Arrangement QR-DV-FF.*   In the operating range

Temperature	107–135°C
Humid content	13–25% v/v (vapor by volume)

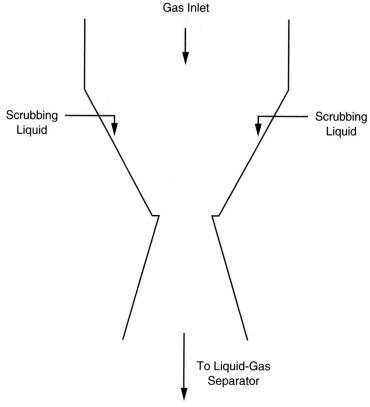

FIGURE 11.62   Venturi wet scrubber.

the QR-DV-FF arrangement provides a reduction in both HCl and SO_2 in the range of 98 to more than 99 percent, almost independent of the stoichiometric ratio of lime to acid gases when the stoichiometric ratio α exceeds 1.3. The high efficiency of acid gas removal does occur at 200°C, where the humid content of the gas is 25 percent, compared with the 15 to 18 percent water content normally encountered in North American municipal waste.

2. *Arrangement QR-FF.*   In the operating range

Temperature	107–135°C
Humid content	17–18% v/v

the QR-FF arrangement provides an HCl reduction in the order of 96 to 97 percent and the SO_2 reduction efficiency is of the order of 90 percent, both for the stoichiometric range 1.5 to 1.7.

3. *Arrangement DR-FF.*   In the operating range

Temperature	140–260°C
Humid content	12–18% v/v

TABLE 11.64 Performance of Commercial Systems

System type	Flow, DSCFM	Temperature, °F Inlet	Temperature, °F Stack	Lime	Moisture stack, % v/v	ΔP, in H_2O gauge	Particulate mg/Nm³ Inlet	Particulate mg/Nm³ Stack	HCl mg/Nm³ Inlet	HCl mg/Nm³ Stack	SO_2 mg/Nm³ Inlet	SO_2 mg/Nm³ Stack
ESP:												
Andover		500										
DS-FF:												
Malmo	36,500			1.2–1.4				36.0	1100	261–390		
				1.7				18	1100	366		
				3.5				4				
Skovde		285	285	4.6	7.0		280	2	861	97		91
Claremont	−14,500	400–525	400–525	3.4				35	(815)	29		21
Dutchess	23,000	550	395	2	12.8			5.9	456	42	61	63
Tsushima	15,000		380	3	20–25	8.5		5	315	23	129	
Iori	8,600	390	380	16	20	>6	2020	<1		9.5	60	<1
QR-ESP:												
Munich	93,166	550	330	2.6	16.9		6485	42	1092	9.5		
Millbury	23,500	436	254	1.4	25	1.7		19	1792	375	814	168
Shizuoka		550	410	2.5	25				1874	280	229	143
Sakaii		450	306	3.3	25–30				601	60	163	40
QR-FF:												
Bridgeport		275		3.4	13		5513	5.6				
Mid-Conn		245						9.1				
Warren	31,000	365	225	1.4–2.1	17.5	10		5.5	740	34	417	1.4
									1422	47	426	40
Skagit	10,000	344	265	1.6	17.3	7.3	2400	12.8	750	23		
Jackson	11,000	370	250	1.3–1.8	17–18	7.5	2600	5.8	670	23	251	31
Maine	41,000	400	280–300		14–17				916	10		
West Phalis	29,400								1100	50	300	10
QR-DV-ESP:												
Fujisawa	16,000	570	360	2.9	26	4			1141	57		

TABLE 11.64 Performance of Commercial Systems (*Continued*)

System type	Flow, DSCFM	Temperature, °F		Lime	Moisture stack, % v/v	ΔP, in H₂O gauge	Particulate mg/Nm³		HCl mg/Nm³		SO₂ mg/Nm³	
		Inlet	Stack				Inlet	Stack	Inlet	Stack	Inlet	Stack
QR-DV-FF:												
Marion	29,000	480	260	1.3	13–17	8.7	2020	5.2	970	26	334	22
		480	330						1077	199	714	187
Commerce	52,000	460	260	1.3			5700	11.2	1877	18.6	1006	4.9
	48,000		247	1.2			4600	9.2	1729	5.2	303	12.3
Tsushima	15,000	570	390	2.7	20–23	8.5		4	815	16.3	186	<14
		570	380	2.4	20–25	9			394	6.5	69	<3
Hennepin	48,000	350	243	1.8		8.5		6.4	(800)	10.1	−430	14.6
								5	(800)	3.7		56.6
ESP-W:												
Kure	17,000		140				4305	98	1092	99	154	20
Condensation:												
Uppsala	9,000								163	81	2571	57

System type	Mercury μg/Nm³		Dioxins, furans		Heavy metals, stack μg/Nm³													Stack ppm	
	Inlet	Stack	Inlet	Stack	As	Be	Cd	Cu	Cr	Mn	Mo	Pb	Ni	Se	Sa	V	Zn	Co	NOₓ
ESP:																			
Andover				6.20	8.5		20.8		−8				38						
DS-FF:																			
Malmo	294	13–65		0.20	<0.18		0.53					11.6							
Skovde	10,170	9																	
Claremont		25		0.45		<0.19													
Dutchess		56										16						161	93
Tsushima																			
Iori				0.50		9													

QR-ESP																	
Munich	217	12.7	914	<4.3	29.1	1.2	<15.4	321	<95	32.3	46.3	36.4	20.7	<3.8	4.4	0.49	
Millbury																	921
Shizuoka																	
Sakaii																	845
QR-FF:																	
Bridgeport											<8.6		<2.6	<3.8	0.75	0.65	289
Mid-Conn							409	<11									49
Warren	188.7	191.4															416
Skagit												3.6	10.5		8.4		
Jackson																	
Maine																	
West Phalis																	1
QR-DV-ESP:																	
Fujisawa																	
QR-DV-FF:																	
Marion	150–360		16				3	18			<0.01	2.5	2.5	<0.002		<0.031	
Commerce	<100	50	38.5			<2.72	6.3	1.97			<0.7	<540	2	<0.19	<0.16	<0.063	406
			<120			<1.4	<2.8	3.1			2.3	<4.1	<4.1	<0.7	<2.6	<0.039	54
																<0.029	
Tsushima			13			<0.1	90	6.1			1.6	5.2	0.3		<0.1		12
Hennepin								3.5							0.16		34
ESP-W:																	
Kure																	
Condensation:																	
Uppsala	60–230																119–20

11.147

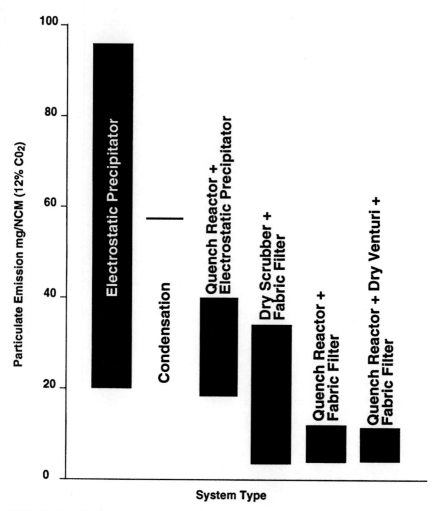

FIGURE 11.63 Total particulate emissions.

the HCl removal efficiency is of the order of 95 percent for $2 \leq \alpha < 3.5$. Where the humidity is lower (7 to 10 percent), the HCl removal efficiency is significantly lower with indications of 70 percent at $\alpha = 1.7$ and 89 percent at $\alpha = 4.6$. The SO_2 removal efficiency is highly dependent on the stoichiometric ratio in DR-FF systems, with efficiencies as low as 51 percent reported.

4. *Arrangement DSI-FF or ESP.* In the operating range

Temperature	150–400°C
Humid content	12–187% v/v
	7–10% v/v

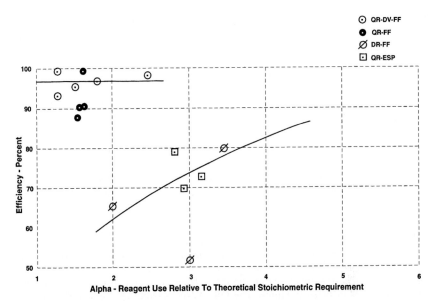

FIGURE 11.64   Sulfur dioxide removal efficiency, dry and semidry systems.

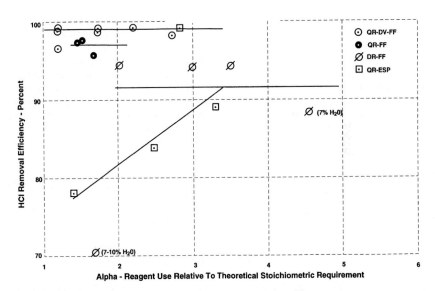

FIGURE 11.65   Hydrochloric acid removal efficiency, dry and semidry systems.

the performance for acid gas control has not been found to be consistent. The efficiencies reported range from 40 to 95 percent HCl and 40 to 70 percent SO_2 at high humidities (about 15 percent) at $\alpha > 2.4$. The behavior of the totally dry system compared to the slurry reaction or gas-quenched system reflects the dependency of SO_2 recovery on the humid content of the reactive lime. Where the slurry is used, the system operates with a drying rather than a dry solid, at a higher humid content in the gas and at a lower temperature and vapor pressure of the water, thus creating a higher water content in the solid. Where quenching is used, the advantage of higher gas humidity and enhanced adsorption of water by the reagent is realized.

5. *Arrangement QR-ESP.* In the operating range

Temperature	120–210°C
Humid content	16.9–35% v/v

The reduction in HCl and SO_2 is highly dependent on the quench reactor efficiency inasmuch as the reaction in the electrostatic precipitator is low. A maximum efficiency achieved on the ESP passage is 35 percent for SO_2 and 50 percent for HCl compared with an excess of 90 percent for both gases in the fabric filter. The efficiency of removal of the acid gases by the QR-ESP system for HCl ranges from about 80 percent at $\alpha = 1.5$ to 90 percent at $\alpha = 3.3$. The SO_2 efficiency is significantly lower in the range of 70 to 80 percent in the range of $\alpha = 3$.

Dioxins, Furans, and Other Organics. The reduction of these organic compounds is more readily achieved by reduction in temperature to reduce the adsorbed phase vapor pressure (nv s) and significant residence time for adsorption by fly ash or activated carbon. The systems utilizing the fabric filter have demonstrated the most efficient reduction in these organics. The QR-DV-FF system performance established capability to meet the 0.1 ng/Nm³ TEF requirement without the necessity of a secondary activated carbon adsorption system.

The ESP systems, often operating above 200°C, have been reported to increase the concentrations of dioxins and furans by denovo synthesis, the formation of dioxins and furans in the boiler system, downstream of the combustion zone.

Heavy Metals. Reductions of heavy metals, exclusive of mercury, exceed 99.9 percent for fabric filter systems. Because the heavy metals are concentrated in the less than 2-μm particulate, ESP systems require a reduction of total particulate to the order of 10 to 15 mg/Nm³ in order to achieve equivalent results. The removal of mercury requires the presence of activated carbon or activated coke either within the existing emission control system or with the addition to the system of an external carbon bed. Because the carbon content in the fly ash varies widely, the inherent adsorption of mercury by fly ash is too variable to provide a basis for reliable design.

Wet Scrubber Performance. Wet scrubbers that can provide high efficiency for acid gas removal—the tray and packed towers—do not have inherent fine-particle removal capabilities. The effective separation capability is limited to particles larger than 3 μm,[55] unless the particles are hydrophilic and the wet bulb temperature of the gas exceeds 65°C.[56]

Although wet scrubber efficiencies for which both inlet and outlet data are available indicate only 85 to 90 percent efficiency, they can readily be increased to levels exceeding 99 percent. In incineration emission control, where the absorption of acid gases is accompanied by neutralization, the efficiency of a wet scrubber can be modified by increasing the length or number of contacts [number of transfer units (NTU)], and/or increasing the quantity of neutralizing reagent dissolved in the scrubbing liquid.

For a constant excess reagent ratio, the relationship for absorption neutralization in the lean system conditions occurring with incineration flue gas, the relationship is $NTU = \ln [y_1 \text{ (inlet)}/y_2 \text{ (outlet)}]$. Therefore, to improve the recovery efficiency from 90 to 99 percent, the required increase in NTU is from 2.2 to 4.6. For the same liquid irrigation rate and excess reagent ratio, the requirement is a 110 percent increase in the packed bed length or in the number of trays in a tray tower.

Particulate.   A major concern in the utilization of wet systems is that the gas discharge is saturated with water, resulting in a visible plume. The presence of particulate in the gas stream results in nucleation by the particulate, water condensation on the particulate, causing an increase in plume density. In most cases reheating of the stack gas is desirable to achieve public acceptance. The wet scrubber capable of recovering fine particulates is the venturi scrubber. The particle size cut achievable as a function of the gas phase pressure drop is as follows:

<table>
<tr><td colspan="2" align="center">Venturi Pressure Drop
Required to Capture Particles
(Liquid to gas weight ratio = 1)</td></tr>
<tr><td>Particle size, μm</td><td>Pressure drop, mm H_2O</td></tr>
<tr><td align="center">0.6</td><td align="center">600</td></tr>
<tr><td align="center">1.0</td><td align="center">200</td></tr>
<tr><td align="center">2.0</td><td align="center">50</td></tr>
</table>

However, the venturi does not have the potential for high efficiency acid recovery of the SO_2 component of the flue gas. Studies for SO_2 recovery indicate a maximum efficiency of about 60 percent.[3] Inasmuch as particulate recovery is achievable at significantly lower energy consumption by an electrostatic precipitator or fabric filter, wet scrubbers are generally preceded by either of the dry particulate separators.

Acid Gases.   The acid gases, HCl, SO_2, and HF, are readily adsorbed by alkaline solutions. Because the calcium-based reagents and the products of calcium ion recovery of SO_2 and HF are solids and relatively insoluble, the potential of plugging of the internals of wet scrubbers is high when using lime reagents. Thus, the more expensive sodium reagents are preferred. Dual scrubbing systems using calcium hydroxide slurries for HCl absorption and sodium reagents for the SO_2 and HF have been attempted to minimize costs.

Mercury.   High efficiencies have been indicated for wet scrubbers using NAClO as reagent.

Organics.   The recovery of organics such as dioxins and furans is difficult in an aqueous solvent system because of low solubilities. Even if nucleation condensation of these compounds occurs, the particulate size formed is generally less than 2 μm, below the recovery capability for low-pressure-drop wet scrubbers.

Dry and Semidry Systems.   In this section, the behavior of dry and semidry systems is treated. The factors influencing the performance of these systems are temperature, humidity, and reaction time.

Temperature Effect.   Temperature affects recovery of acid gases and dioxins-furans, possibly also mercury.

ACID GAS.   The efficiency of the recovery of the acid gases in lime neutralization systems increases with reduction in the temperature of reaction at constant humidity conditions. As a corollary, the efficiency of reagent utilization increases with reduction in temperature.

The data obtained from commercial operation of the quench reactor–dry venturi–fabric filter system (Fig. 11.66), indicates that the efficiency of recovery of SO_2 is more dependent on temperature than that of HCl. This reflects the greater dependency of the kinetics of the SO_2-$Ca(OH)_2$ reaction on the humid content of the solid. The reduction in temperature is related to the decrease in the adsorbed-phase

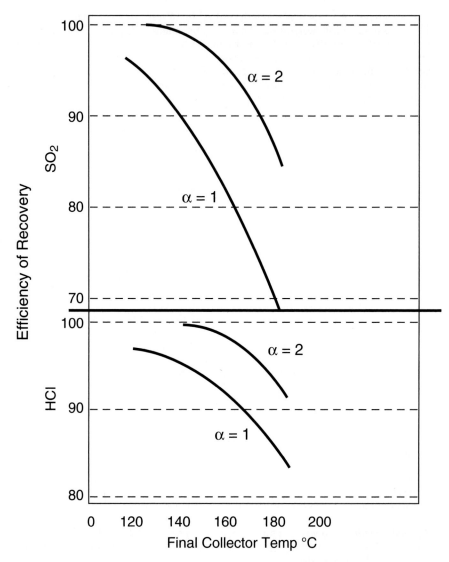

FIGURE 11.66 Acid gas recovery as a function of temperature, system QR-DV-FF.

vapor pressure of the water contained in the solid phase, resulting in higher water concentrations in the solid phase.

With increase in stoichiometric ratio, the temperature effect is reduced (Fig. 11.67). Thus, the reduction in reagent utilization for $\alpha = 1$ decreases from 95.8 to 81.5 percent over the range of 120°C to 180°C. At $\alpha = 2$ the reagent utilization decreases from 49.7 to 46.9 percent for the same temperature range. Table 11.65 compares performance and reagent utilization as a function of temperature and stoichiometric ratio at a humid content of 20 percent v/v.

DIOXINS AND FURANS.   The effect of temperature of adsorption on dioxin-furan reduction is twofold. Below the temperature of 200°C, the adsorption on fly ash increases with decrease in temperature as a function of the adsorbed-phase vapor pressure. In the temperature range 120 to 130°C, the adsorption by fly ash in a thick fabric filter cake results in a reduction of dioxins-furans to less than toxic equivalent TEQ = 0.1 ng/Nm^3.

Above 200°C, synthesis of dioxins and furans occurs as a result of recombination of molecular precursors such as carbonaceous solids, hydrogen chloride, phenols, chlorophenols, and aromatic hydrocarbons adsorbed on the fly ash. This mechanism

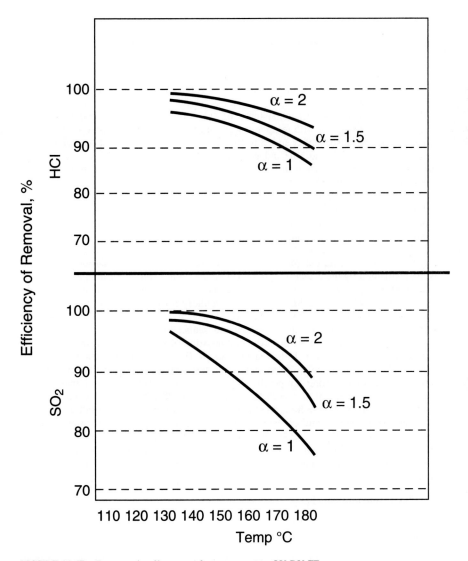

FIGURE 11.67   Temperature effect on performance, system QV-DV-FF.

TABLE 11.65   Effect of Temperature on Acid Gas Removal System—QR-DV-FF

(Water content 20% v/v)

Temperature, °C	Stoichiometric ratio α	Efficiency		Reagent utilization, %
		HCl	SO$_2$	
120	1	96	95	95.8
150	1	92.5	85	90.8
180	1	85	70	81.5
120	2	99.4	99.5	49.7
150	2	98.2	97	49.0
180	2	92.5	85	46.9

generally occurs in the boiler section and is referred to as *denovo synthesis.* See "Dioxins-Furans" below.

MERCURY.   There is little consistency of data indicating a relationship of mercury emissions (Hg + HgCl$_2$) with temperature of operation of the emission control system when no specific adsorbent is added to the system. The reason is that the carbon content of the fly ash is a variable independent of the emission control system temperature. With 20 percent carbon in the fly ash, a 60 percent reduction in mercury emissions has been demonstrated at 150°C.

With activated carbon addition, a 99 percent reduction of mercury and mercuric chloride emissions has been achieved at 140°C (see "Heavy Metals," below), with reductions to the range of 5 to 25 μg/Nm^3 from inlets as great as 10,000 μg/Nm^3 of combined mercury and mercuric chloride.

Effect of Humid Content.   Humid content is the percent water content by volume. Comparison of the acid gas removal efficiency of dry and semidry scrubbing systems is made difficult by the effect of differences in the humid content of the gas in the systems being compared. The kinetics of the adsorption-reaction mechanism for the recovery of hydrochloric acid, and more particularly for the sulfur dioxide, are highly dependent on the humid content of the typical reagent, lime. This humid content of the solid phase is related to the humid content and temperature of the gas stream.

In pilot studies with an all-dry system consisting of injection of hydrated lime and Tesisorb into a dry venturi, followed by a fabric filter, the humid content of the gas was controlled from zero to 35 percent. The tests, conducted with a stoichiometric ratio of 2 and a fabric filter temperature of 125 to 150°C, indicated an increase in HCl removal from 60 percent at a zero humid content to 98 percent at 35 percent humid content (Fig. 11.68), where humid content is water vapor percent by volume.

Below 18 percent humid content, the rise in efficiency E, is almost linear with the humid content H, with $E \propto 1.6H$. The rate of increase diminishes, becoming asymptomatic at the 98 to 99 efficiency level at 35 percent humid content.

The humid content of the flue gases from the incineration process varies as a function of

Water content of the charge

Excess air used in combustion

Mode of cooling of the incineration flue gas

The water content of the charge varies with source—municipal, industrial, or medical—and within these categories, country of origin. For example, the humid content

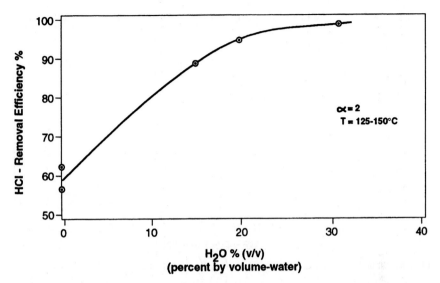

FIGURE 11.68   Hydrochloric acid removal efficiency for dry system (dry venturi–fabric filler)—effect of gas humidity.

of the combustion gas from MSW incineration is of the order of 9 to 15 percent for the United States and 20 to 25 percent for Japan. Thus, for the same system components, reagent use, and mode of operation, the efficiency of acid gas reduction in Japan can be greater than that achieved in the United States.

Reaction Time Effect.   The reaction time for acid gas neutralization occurs in two zones within the dry or semidry system: (1) the reactor and (2) the collector.

The major time of exposure of the reagent to the flue gas occurs in the fabric filter, if that is the collector form. Residence time for multicompartment fabric filters expressed as the cleaning cycle has been reported to range from 3 min to 6 h.

The reduction of the acid gases is a function of the wave front number W:

$$W \propto (FF)(A/C)(T)$$

where   FF = stoichiometric ratio at the inlet to the fabric filter
      A/C = air to cloth ratio
      T = one-half the cleaning cycle

establishes that a 95 percent recovery of both HCl and SO_2 is provided at a wave front number exceeding 10 at 200°C and 22 percent humid content[55] (Figs. 11.69 and 11.70).

The relationship of system efficiency for HCl recovery for an all-dry system[56] (Fig. 11.71) is similar to the baghouse alone. The efficiencies are lower for the same wave front, apparently a result of the higher temperature and lower humid content.

Heavy Metals

Heavy metals emitted in the incineration process have been identified in both the particulate and vapor forms, with the nonvolatile metals predominant. The volatile metals are arsenic, cadmium, mercury, and zinc (Table 11.56). Regulatory requirements have been established in Europe for arsenic, mercury, and cadmium, individually. Both the

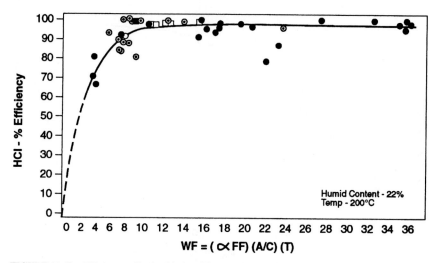

FIGURE 11.69 Efficiency of hydrochloric acid removal in fabric filter.

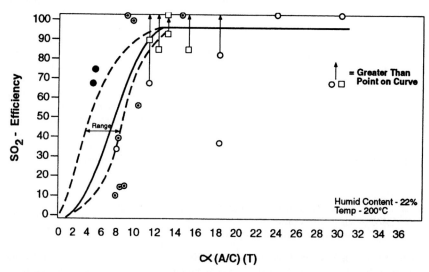

FIGURE 11.70 Efficiency of sulfur dioxide removal in fabric filter.

U.S. EPA and European countries have established emission limits for the total heavy metals (Table 11.58). The heavy metals present in the solid phase, associated with the fly ash suspended in the flue gas, are concentrated in the particulate size of less than 2 μm[63-66] (Table 11.66). Thus they are concentrated in the respirable particulate size range.

The partially volatile heavy metals exhibit vapor-phase concentrations significantly less than would be evident from the pure metal or compound vapor pressures, even though a solid phase exists, primarily because of adsorption on fly ash components. For

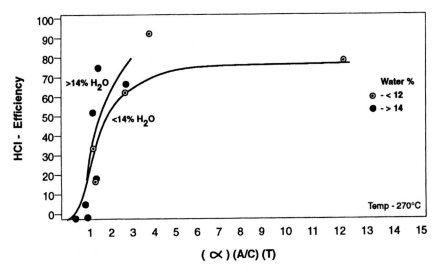

FIGURE 11.71   System efficiency of dry scrubber–fabric filter.

TABLE 11.66   Heavy Metals Distribution on Particulate (Fly Ash)

Percent of total heavy metal appearing in specific size range

	Munich incinerator[66]		Greenburg,[65]
	Inlet, <1.9 μm component	Outlet, <1.3 μm component	inlet, <1.6 μm component
Antimony	93	70	
Arsenic	45	100	
Barium	6	81	
Beryllium	4	100	
Cadmium	77	86	80
Chromium	96	100	
Cobalt	69	100	
Copper	58	78	95
Manganese	20	100	
Nickel	94		
Lead	81	75	95
Tin	78	95	
Vanadium	20	95	
Zinc	67	84	84

example, at 200°C, mercury and its predominant compound, mercuric chloride, exhibit average concentrations of 300 to 500 μg/Nm^3 in untreated flue gas, when a solid adsorbed phase is present. This is approximately 10^{-5} to 10^{-4} of the concentration equivalent to the vapor pressure of the pure compounds. In the case of cadmium, the range of observed concentration is about 10^{-1} that of the equivalent vapor pressure.

For reduction of the nonvolatile heavy metals, the less than 2-μm particulate must be effectively removed from the flue gas. Conventional modes of particulate separation, because of gradation in particulate collection as a function of particle size, require reduction of emissions to the range of 4 mg/Nm^3 in order to achieve a 99.9 percent reduction in the solid phase and the adsorbed-phase heavy metals (Table 11.67).

Where enhanced particulate separation is achieved by a thick-cake baghouse operation, the efficiency of particulate recovery appears to be independent of particle size (Table 11.68). In this case, in excess of 99.9% of the heavy metals are removed at a pressure drop across the fabric filter of less than 5 in. w.c. (Table 11.69). In the case of the volatile heavy metals arsenic, cadmium, mercury, and zinc, only the most volatile, mercury, was reduced by less than 99.9 percent. The others were removed both by capture of particulate less than 2 μm in size and by adsorption on the fly ash in the filter cake. The capture of mercury is extremely variable, depending on the carbon content of the fly ash. For uniform control, an adsorptive additive must be used. This is discussed further below.

Emissions and Regulations for Mercury. Mercury emissions from the incineration process exist in the form of the solid and vapor phases of the metal form (Hg) and mercuric chloride ($HgCl_2$). The predominant species, 65 to 98 percent, is mercuric chloride. The total mercury concentrations in the flue gas with no preseparation have been reported to range from 40 to 7600 μg/Nm^3 (Ref. 75) with a typical value of 300 to 700 μg/Nm^3. About 60 to 70 percent of the mercury that enters the combustion zone is that present in batteries and instruments.[83,84] The remainder is present in paint, plastics, paper, and fluorescent bulbs.

TABLE 11.67 Relationship of Total Particulate Emissions to Heavy Metals Reduction

Total solid particulate emissions, mg/Nm^3	Percent reduction of heavy metals in particulate form			
	Zinc	Lead	Cadmium	Mercury
36	98.6	97.8	97.8	98.5
18	99.4	99.4	99.5	98.8
4	99.9	99.9	~100	~100

Source: Ref. 67.

TABLE 11.68 Summary of Stack Particle Size Results

(Report of Commerce Facility)

	% Less than stated size			
Stated size, μm	2–1	1 stack	2 stack	Average
10.3	5.3	23	17	15
4.2	4.3	20	14	13
2.0	3.2	15	10	9
1.1	2.3	11	6	6

Average baghouse efficiency in stated size range:

    >10.3 μm   99.95%
    <10.3 μm   99.92%
    <2.0 μm   99.90%
    <1.1 μm   99.91%

Source: Ref. 68.

TABLE 11.69　Commerce Resource to Energy Facility (1988), Heavy Metal Recovery

Concentration in $\mu m/Nm^3$ (12% CO_2)

	Boiler emissions	Stack emissions	Efficiency, %
Aluminum	178,000	<16.2	>99.9
Antimony	822	0.29	99.96
Arsenic	78	<0.16	>99.98
Barium	4,695	117	97.5
Beryllium	6.88	<0.19	>99.5
Bismuth	31.4	<0.16	>99.5
Cadmium	1,680	2.0	99.88
Calcium	193,500	56	99.97
Chromium	3,620	2.33	99.94
Cobalt	111	0.34	99.70
Copper	8,818	<54	99.40
Iron	84,167	<54.1	>99.94
Lead	18,133	1.97	99.99
Magnesium	89,933	<270	>99.70
Manganese	3,235	0.96	99.97
Mercury (no additive)	475	41.4	91.30
Molybdenum	522	<12.5	>97.60
Nickel	4,240	6.3	99.85
Selenium	<84	<2.72	96.8
Silicon	1,860	66	96.5
Tin	800	<2	>99.75
Vanadium	257	<0.09	>99.96
Zinc	90,933	38.5	99.96
Total	686,001	<745.2	>99.9
Total without Ca, Al+	402,658	<419	>99.9
Total without Ca, Al, Mg	312,725	149	>99.9

Source:　Ref. 68.

The emission limits in the flue gas established by Germany and the Netherlands is 50 $\mu g/Nm^3$ (11 percent O_2). The World Health Organization has recommended an emission limit of 15 $\mu g/Nm^3$. New Jersey has indicated an emission limit of 27 $\mu g/Nm^3$ (11 percent O_2). The solid leachate limit established by the U.S. EPA is 0.2 mg/L by TCLP.

Reduction of Emissions.　The two approaches utilized for reduction of gasborne emissions of mercury and its compounds are preseparation and capture in the emission control process. Preseparation conducted at the Detroit, Michigan, incinerator provided a maximum reduction of 80 percent.[71] Inasmuch as the reduction in mercury emissions achieved by preseparation results in emissions greater than permissible by the regulatory requirements, postcombustion recovery is required.

Three types of technologies or combinations thereof have been used to reduce mercury emissions. They are

1. Adsorption
2. Chemical reaction
3. Filtration

The adsorptive process is based on both physical adsorption and chemisorption in the presence of HCl on carbon surfaces. The technique is primarily applied to dry or semidry systems.

It was noted by operators of MSW incinerators that significant, though variable, reductions in mercury and mercury-compound emissions occurred with no addition of adsorbents when a fabric filter was used as a final collector. The phenomenon was attributed to the presence of carbon in the fly ash. The effect was quantified by Shager.[72] The mechanism, requiring the presence of HCl on the carbon, was confirmed by Stepanov[73] and Hissayoshi[74] and had been used in Germany in 1925.

In the case of incineration gases, HCl is present, thus providing in situ activation of the carbon. However, the variability of the carbon content in the fly ash precludes reproducible recoveries.

The introduction of carbon into the gas stream[76,79] with final collection in an electrostatic precipitator, but preferably in a fabric filter, or by passage of the cleaned gas, through a carbon bed,[80] has produced effective reduction of mercury and its compounds (Table 11.70). The dose rate of activated carbon is reported to be in the range of 0.44 to 1.8 lb/ton waste. The fabric filter is a more effective contact device for removal of mercury and its compounds because of the longer and more effective contact of the gas with the carbon. The lower the final operating temperature for adsorption, the lower the carbon dose rate required. Fixed bed carbon units such as proposed by de Jong[80] are in use, primarily in Europe, generally downstream of the emission control system and often providing multiple services, including removal of nitrogen oxides (see "Nitrogen Oxides").

Chemical Reaction for Mercury. Chemical conversion of the vapor phase mercury and its compounds to nonvolatile compounds is achieved in semidry systems by the addition of sodium sulfide solutions in the spray drier or quench reactor. The mercury and its compounds are converted in a nonvolatile mercuric sulfide that is effectively recovered as a particulate.[81] A report to the EPA Mercury Conference (1989) indicated the removal capabilities (see Table 11.71).

The mercury group of compounds as nonvolatile suspended particulate, HgS, is then filtered, normally in a fabric filter in conjunction with other particulates. In wet scrubber operation,[82,83] the mercury group is removed by adding NaOCl and a chelating agent to the scrubbing solution, thus oxidizing and solubilizing the mercury components. A reduction of 90 percent is reported for this technique. Removal of the mer-

TABLE 11.70 Mercury Removal with Carbon Injection

Reference	System	Temperature, °C	Concentration, $\mu g/Nm^3$ Inlet	Outlet	Efficiency, %	Dosage rate, lb/ton waste
76	DV-FF	140	294	9	96.9	0.59
		140	10,510	25	99.8	0.59
		140	922	6.5	99.3	0.59
		140	514	9.5	98.2	0.59
77	QR-ESP	110	486	68	86.0	0.8–1.8
		110	411	141	65.7	
		140	395	85	78.5	
		140	537	390	27.4	
77	QR-FF	200	293	52	82.3	0.59
78	ESP	260	185	6	96.8	1.62
78	FF	140	502	33	93.4	0.44

TABLE 11.71 Mercury Removal in Na$_2$S System

Concentration, $\mu g/Nm^3$

Inlet	Outlet	Efficiency, %
400	90	77.5
406	21	94.8
250	13	94.8
280	2	99.3
471	68	85.6

cury by precipitation with trimercapto-s-triacin[83] was found to be effective. An alternative approach[84] requires mixing the scrubbing liquor with fly ash, followed by aeration and filtration.

Dioxins-Furans

Dioxins and furans are created in both the combustion process and by catalytic synthesis (denovo synthesis) when the flue gas is in the temperature range 200 to 450°C, in the presence of fly ash.[110–112] The concentrations generally reported are presented in both the total concentration of the cogeners and as the toxic equivalent (TEQ) based on internationally accepted toxic equivalent factors (TEF). The most recent European and Japanese regulations are based on TEQ (Table 11.72), reflecting the toxicity of the dioxin-furan emissions. The TEF factors are listed in Table 11.72. In general, the TEQ concentrations emitted by incinerators ranges from 1 to 5 percent of the total dioxin-furan cogener concentration.[85,86]

The uncontrolled emissions of dioxins and furans from MSW incineration vary widely within a reported range of 18 to 276 ng/Nm^3 in U.S. incinerators, equivalent to a TEQ range of 0.18 to 13 ng/Nm^3. The establishment of incineration processes without dioxin-furan control has resulted in a 10- to 15-fold increase of dioxins in soil samples in southeast England.[87] In Japan about 80 to 88 percent of the total national emissions of dioxins and furans are initiated by the incineration process[88] compared to 30 to 37 percent of the total emissions in Sweden that emanate from the incineration process.[89]

The dioxin-furan emissions from the furnace can be reduced to nondetectable levels by controlled combustion at 900 to 1000°C with a residence time of 1 to 2 sec.[90,91] However, the dioxin-furan precursors, aromatic structures on particulate carbon, aromatic chlorides, and aromatic hydroxides, are generally more resistant to thermal destruction and can recombine to form dioxins and furans in the presence of fly ash in the temperature range 200 to 450°C (denovo synthesis).[92–96]

The accumulation of fly ash on the boiler tubes after extended operation resulted in one case in an increase in dioxin emissions at the boiler exit from a total cogener concentration of 28.5 ng/Nm^3 in the clean condition of 739 ng/Nm^3 after 18 months of operation, with the CO content remaining the same. The continued synthesis of dioxins-furans downstream of the combustion zone, as a function of temperature and the presence of fly ash, is indicated by studies of dioxin-furan emissions from electrostatic precipitators used only as a particulate collection device (Table 11.73).[87,97–99] The continued synthesis of dioxins and furans in both the boiler and collection device in the 200 to 500°C range based on data obtained on an incinerator is indicated in Table 11.74.

TABLE 11.72   Toxic Equivalent Factors
(Rappe DXN Conf. Kyoto 1990)

Dioxins		Furans	
Cogener	Factor	Cogener	Factor
2,3,7,8 TCDD	1.0	2,3,7,8 TCDF	1.0
1,2,3,7,8 PCDD	0.5	1,2,3,7,8 PCDF	0.05
1,2,3,4,7,8 HCDD	0.1	2,3,4,7,8 PCDF	0.5
1,2,3,6,7,8 HCDD	0.1	1,2,3,4,7,8 HCDF	0.1
1,2,3,7,8,9, HCDD	0.1	1,2,3,6,7,8 HCDF	0.1
1,2,3,4,6,7,8 HpCDD	0.01	1,2,3,7,8,9 HCDF	0.1
OctaCDD	0.001	2,3,4,6,7,8 HCDF	0.1
		1,2,3,4,6,7,8 HpCDF	0.01
		1,2,3,4,7,8,9, HpCDF	0.01
		Octa CDF	0.001

TCD = tetrachlorodibenzo
PCD = pentachlorodibenzo
HCD = hexachlorodibenzo
HpCD = heptochlorodibenzo
OctaCD = octachlorodibenzo
D = dioxin
F = furan

TABLE 11.73   Electrostatic Precipitators, Dioxin-Furan
Emissions, ng/Nm^3

CO 10–100 ppm

Exit temperature, °C	Okajima[97]	Takeda[98]	Ide[99]
150			10
160		100	
210		180	
220	1.0		
240			110
300	100	800	

TABLE 11.74   Dioxin-Furan Concentrations,
Boiler and Precipitator*

*Concentrations in TEQ, ng/Nm^3, stack tempera-
ture 300°C*

Furnace outlet	ESP inlet	Stack
39.8	31.1	41.2
4.3	15.2	28.6
2.2	5.3	3.8

Source: Ref. 88.

As a result of the heterogeneity of the fuel, both in composition and water content, and the heterogeneity of the combustion process (leading to puffing), the destruction of dioxins and furans in the furnace by achieving an average residence time of 1 to 2 s and an average temperature of 900 to 1000°C and CO content below 50 ppm has not reduced the outlet dioxin-furans reproducibly. Emissions ranging from 5 to 700 ng/Nm^3 have been reported for the operating characteristics in the recommended range.[85] Inasmuch as synthesis in the boiler (denovo synthesis) has been established as a producer of dioxins and furans, downstream removal of dioxins and furans is required in order to achieve the emission limitation of 0.1 ng/Nm^3 (11 percent O$_2$).

Dioxin-Furan Control. The reduction of dioxin-furan emissions to levels of less than 0.1 ng/Nm^3 (11 percent O$_2$) has been achieved in operating systems by the chemisorption method within the existing emission control system (Table 11.64). Alternative technologies utilizing external system control beyond the acid gas–particulate–heavy metals emission control system consist of adsorption on activated carbon or coke[113,114] and oxidation simultaneous with NO$_x$ removal by selective catalytic reduction.[115]

Technologies employed for the reduction of dioxins and furans formed in both the combustion zone and by de novo synthesis in the boiler zone are adsorption on fly ash[116] and adsorption on externally introduced activated carbon,[107] both within the existing emission control system.

The destruction using catalytic destruction[100,101,115,116] or carbon or coke adsorption[113,114] require prior removal of the acid gases and particulates inclusive of heavy metals.

Chemisorption. The chemisorptive behavior of the fly ash for the recovery of dioxins and furans results in a reduction of the vapor pressure of the cogeners by a factor of 10^8 (Fig. 11.72).[93,103] The adsorption phase vapor pressure, or pseudo vapor pressure, was estimated on the basis of molecular structure and a limited range of energy of vaporization.[94] From the estimated adsorbed phase vapor pressure, the concentration of the 2,3,7,8 TCDD at 120°C is of the order of less than 0.01 ng/Nm^3 at equilibrium.[93,96]

The adsorption kinetics are satisfied by exposure of the gas in the final collector, the fabric filter, to a cake thickness in the range of 3 to 12 mm with a gas velocity of 2 to 3 ft/min (0.6 to 1 m/min). The kinetic equivalence can be achieved with a thicker cake at higher velocities (Fig. 11.73).[104]

The chemisorptive recovery of dioxins and furans on the extended filter cake resulted in emissions of less than 0.040 ng/Nm^3 (TEQ) (12 percent CO$_2$) with inlets ranging from 8.9 to 162 ng/Nm^3 (12 percent CO$_2$) total or 0.65 to 15.9 ng/Nm^3 (TEQ) (12 percent CO$_2$).[108] The efficiency of recovery, on the basis of TEQ, exceeded 96 percent at low inlet loading and 99.7 percent at the higher inlet loading. This system is integrated with the existing acid gas and particulate (inclusive of heavy metals) equipment and requires no additional capital equipment.

Adsorption. Activated carbon has been used as an adsorbent for dioxins and furans. The commercial installation reporting data uses carbon downstream of an existing wet scrubber, electrostatic precipitator emission control system.[107] Reduction of dioxins and furans in excess of 99% with outlet less than 0.1 ng/Nm^3 (TEQ) was reported. Brown coke has also been used for the removal of dioxins and furans.[113,114]

Catalytic Destruction. Catalytic destruction of dioxins and furans present in incinerator flue gas has been demonstrated. Reductions to less than 0.1 ng/m^3 (TEQ) in pilot operation was achieved at 300 to 350°C with inlets up to 2.5 ng/m^3 (TEQ) using TiO$_2$-based catalysts and selective catalytic reduction NO$_x$ catalysts.[100,101] The

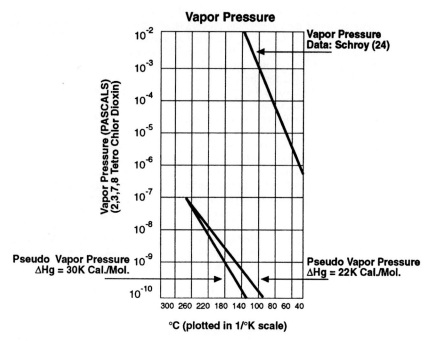

FIGURE 11.72   Effect of the presence of fly ash on the effective vapor pressure of dioxin.

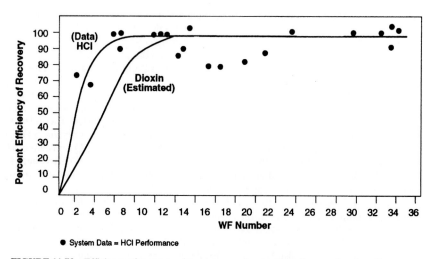

FIGURE 11.73   Efficiency of recovery of moving wave front with baghouse cake adsorption.

flue gas exposure to the catalyst ranged from 0.3 to 0.7 s. Pilot tests using a platinum catalyst honeycomb achieved in excess of 99 percent reduction of PCDD and PCDF at 250°C and a space velocity of less than 2500 h^{-1} with catalysts having a surface area of 860 to 910 m^2/m^3.[102] Simultaneous reduction of NO$_x$ was achieved with ammonia introduction. In commercial operation a 60 percent reduction in dioxins and furans was achieved at 275°C, and 95 percent at 450°C.[115]

Nitrogen Oxides

Nitrogen oxide content in incineration flue gas is subject to regulation in Europe and some states in the United States. Regulations are also proposed by the U.S. EPA. The Netherlands limits emissions to 70 mg/Nm^3 (11 percent O$_2$) and the recommended EPA regulation is 99.5 mg/Nm^3 (11 percent O$_2$).

The nitrogen oxide emissions from the MSW incineration process have been found to range from 100 to 1000 mg/Nm^3 with most data in the 100 to 300 mg/Nm^3 range. It is noted (Ref. 121) that NO$_x$ concentrations increase with reduction in CO content of the flue gas with data from Quebec studies showing a rise of NO$_x$ from 70 to 225 ppm when the CO concentration decreased from 120 to 50 ppm.

The technologies for decreasing NO$_x$ emissions are

1. *Selective noncatalytic reduction (SNCR).* Homogeneous phase reaction with either ammonia or urea in the furnace exhaust in the temperature range 870 to 1060°C.

2. *Selective catalytic reduction (SCR).* Reaction of NO$_x$ with ammonia on a metal–metal oxide catalyst downstream of the particulate and acid gas emission control system. The temperature range is 280 to 400°C.

3. *Coke bed catalysis.* Reaction of NO$_x$ with ammonia over an activated coke catalyst in the temperature range of 100 to 200°C with simultaneous reduction of dioxin-furan and mercury contaminants. This system consists of a packed bed of the activated coke downstream of the particulate–acid gas emission control system.

4. *Flue gas recirculation.* Operation of the combustion zone with flue gas or low excess air.

The use of urea or ammonia, the predominant method for reduction of NO$_x$ emissions utilizing the reaction $4NO + 4NH_3 + O_2 = 4N_2 + 6H_2O$ (Ref. 117) can create secondary problems. Where the ammonia slip (unreacted ammonia) is in excess of 3 ppm and the unrecovered SO$_2$ and HCl have a combined concentration in excess of 15 ppm at the stack, potential exists for combination of ammonia with the acid gases to form submicron ammonium salts as the plume cools. The result is an opaque white plume whose opacity can exceed the regulatory limits. In addition, the method of analysis for particulate emission, EPA 5, achieves recovery of the second half catch gases. HCl and SO$_2$ will react with NH$_3$ to form a solid under this condition and add to the second-half catch, at a magnitude even greater than that resulting from poststack combination.

Selective Noncatalytic Reduction. SNCR requires installation of ammonia or urea injectors at several furnace elevations to achieve the capability of introducing the reagent ammonia in the proper operating temperature range of 880 to 1060°C. Reductions of NO$_x$ emissions have been reported to range from 44 to 69 percent[118,119] as a function of both the stoichiometric ratio of ammonia to NO$_x$ and the inlet concentration of NO$_x$.

The increase in efficiency of NO_x removal is also related to the "slip," the exhaust of unreacted ammonia. As a result, increased efficiency for NO_x removal that results in higher ammonia losses or slip requires increased removal of acid gases to prevent plume formation and excessive second-half catch.

The use of urea rather than ammonia as the de-NO_x reagent does permit a slightly wider temperature zone of operation. This technology, although extensively used in Japan, has only recently been utilized in incinerator application in the United States. An NO_x reduction of 50 to 70 percent has been reported,[120] with low ammonia slip. The temperature range was the same as for ammonia SNCR.

Selective Catalytic Reduction.   The SCR type of control system utilizes a catalyst downstream of an acid gas–particulate emission control system to prevent clogging or poisoning the catalyst. Among the catalysts used are platinum and vanadium compounds supported on ceramics or other oxides (i.e., TiO_2). The temperatures of operation in existing installations range from 250 to 320°C.[121,122]

The catalyst system is located downstream of a dry absorption–fabric filter emission control system to provide catalyst protection from poisoning. The NO_x emissions were reduced to less than 35 ppm and the total acid gas content to less than 15 ppm in order to inhibit plume formation using SCR at 320°C.[122]

The low-temperature catalyst, operational in Japan at 200°C to minimize reheat requirements, has achieved up to 85 percent reduction in NO_x emissions with outlets of the order of 50 ppm. The data[121] regarding ammonia consumption are subject to question inasmuch as the stoichiometric ratio NH_3/NO_x reported was often less than the efficiency of reduction achieved. The data could be acceptable only if a less significant reaction (in the absence of oxygen), $6NO + 4NH_3 = 5N_2 + 6H_2O$, had occurred.[118]

The stoichiometric ratio is generally related to the molar ratio required for the reaction where 1 mol of NO is theoretically consumed per mol of NH_3 utilized (e.g., 1). If the stoichiometric ratio mols NH_3/mol NO is less than the efficiency of NO_x reduction achieved, then it would be required that reaction occur in the absence of oxygen.[118]

Costs

Capital cost estimates (Fig. 11.74) are provided only for those systems that are in operation in an adequate number to provide cost data and that have demonstrated the capability to meet the proposed EPA emission requirements of:

Particulate	0.015 GR/DSCF
Hydrochloric acid	90% reduction
Sulfur oxides	70% reduction

These requirements limit cost considerations to:

Quench reactor–electrostatic precipitator

Quench reactor–fabric filter

Quench reactor–dry venturi–fabric filter

If the European requirements, reducing particulate emissions to less than 0.01 GR/DSCF and restricting acid gas emissions to concentrations alone instead of providing the alternative of percent removal, then the QR-ESP alternative would not be available to the designer-operator, if only because of the rapid increase in capital and

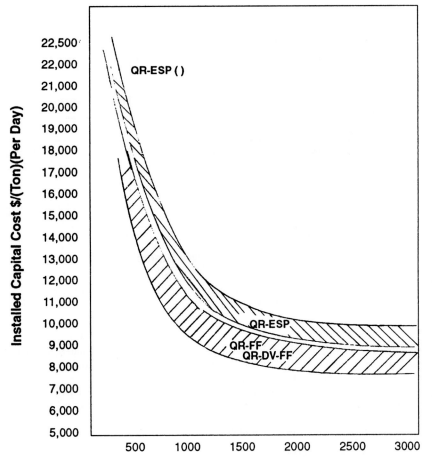

Performance

Type	Particulate GR/DSCF	Efficiency	
		HCl%	SO$_2$%
QR-ESP	0.015	90	70
QR-FF	0.01-0.015	90	70
QR-DV-FF	0.007	95	90

FIGURE 11.74 Capital cost of incinerator emission control systems.

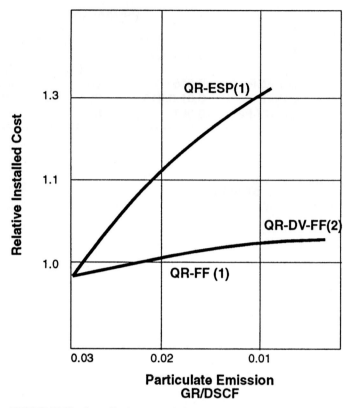

FIGURE 11.75   Cost of incinerator emission control system: effect of particulate emission requirement on capital cost.

annualized costs with reduction in particulate emission (Figs. 11.75 and 11.76). Only one QR-ESP system, modified by the addition of a dry venturi, has complied with the more restrictive requirements (Table 11.64).

The estimated capital costs are presented as a function of the capacity of the incinerator in tons per day of MSW combustion. The costs should be used only as a guideline for capital investment.

As Table 11.75 shows, both capital and operating costs will vary with:

Excess air used in the incinerator

Water content of the municipal waste

Flue gas temperature

Acid gas content range in the flue gas

For example, an increase in excess air from 87 to 125 percent will increase capital cost about 15 percent. The electrical costs for the fan operation will be linear with the total flue gas flow, increasing about 22 percent.

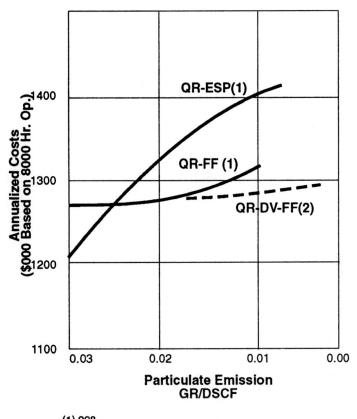

FIGURE 11.76   Cost of incinerator emission control system: estimated annualized cost for 250 tons/day capacity (1990).

Increased water content in the municipal waste has two effects:

- Increases gas flow rate, increasing system size and fan operating cost.
- Decreases the consumption rate of lime required for a specific efficiency of acid gas reduction.

An increase in flue gas temperature increases the quantity of water evaporated, thereby increasing gas flow rate. It also increases the humidity of the gas, thus decreasing lime consumption for a given efficiency. An increase in the acid gas content of the flue gas increases the rate of lime consumption in proportion to the neutralization requirement for each of the acid gases. These factors significantly affect operating costs (Table 11.76), with a range of 28 percent for the same incinerator capacity.

TABLE 11.75 Effect of Excess Air, Temperature, Humidity, Efficiency, and Acid Gas Concentration on Consumption and Costs

Percent excess air	80	80	125	80	80	80
Flow DSCFM	30,000	30,000	36,600	30,000	30,000	30,000
System inlet:						
Temperature, °C	232	232	232	177	177	177
Humid content, %	15	7	15	15	15	7
SO_2 ppm	300	300	246*	400	400	400
HCl ppm	600	600	492	800	800	800
Requirement:						
SO_2 reduction, %	70	70	70	70	90	70
HCl reduction, %	90	90	90	90	95	90
Particulate GR/DSCF	<0.01	<0.01	<0.01	<0.01	<0.01	<0.01
Stack conditions:						
Temperature, °C	127	127	127	127	127	127
Humid content, %	20.4	12.9	20.7	17.7	17.7	9.7
ACFM	51,300	46,900	62,600	49,600	49,600	45,200
Relative system size	100	91	122	97	97	88
Relative lime requirement	100	113	100	133	141	159
Relative fan energy	100	91	122	97	97	88

*Effect of dilution.

The additional costs related to the requirement for mercury and dioxin control vary with the system design; this can range from $0.50/ton MSW to $2.00/ton MSW. NO_x removal is significantly more costly prior to capital investment and operating cost. The capital cost for a 1000 tons/day facility for SCR installation is of the order of $3000 per ton per day MSW and an SNCR (thermal de-NO_x) is of the order of $800 per ton per day MSW. Operating costs are of the order of $1.50/ton MSW.

TABLE 11.76 Operating Cost—Effect of Excess Air, Humidity, Labor, and Utility Costs

	Case 1	Case 2	Case 3	Case 4
Capacity, tons/day	300	300	300	300
Flow DSCFM	30,000	30,000	30,000	36,600
Temperature, °C				
Inlet	232	232	232	232
Stack	127	127	127	127
Humid content, % v/v:				
Inlet	15	7	15	15
Stack	20.4	12.9	20.4	20.4
SO_2, ppm DV:				
Inlet	300	300	400	400
Stack	30	30	40	40
HCl, ppm DV:				
Inlet	600	600	800	800
Stack	30	30	40	40
Stack flow, ACFM	51,300	46,900	51,300	62,600
System DP in w.q.	10	10	10	10
Labor cost $/worker-year	35,000	35,000	50,000	50,000
Power cost $/kWh	0.06	0.06	0.08	0.08
	1.6	1.9	1.6	1.6
Lime CaO, tons/year	1079	1281	1437	1437
Lime CaO, tons/ton MSW	0.0108	0.0128	0.0144	0.0144
Lime, $/ton MSW (at $80/ton)	0.864	1.024	1.152	1.152
Fan horsepower	147	134	147	180
Mechanical horsepower	100	100	100	100
Total horsepower	247	234	247	280
Power, $/ton MSW	1.482	1.404	1.976	2.240
Worker cost at 4 worker-years/ installation, $/ton	1.40	1.40	2.00	2.00
Supervision, maintenance, and insurance, $/ton	3.60	3.60	3.60	4.14
Operating cost exclusive of amortization, $/ton	5.86	6.02	6.75	7.53

PART D

T. Randall Curlee

Energy and Global Change Analysis Section
Energy Division, Oak Ridge National Laboratory
Oak Ridge, Tennessee

11.4 THE SOCIOECONOMICS OF WASTE-TO-ENERGY IN THE UNITED STATES*

Although less than 1 percent of all municipal solid waste (MSW) was burned to retrieve its heat content in 1970, waste-to-energy (WTE) grew to account for 16 percent of MSW in 1990.[126] Many observers of this trend, including this author, forecasted that WTE would continue to grow and would be used to manage as much as one half of all MSW by the turn of the century.[123] Those predictions are now called into question by numerous WTE project cancellations and recent, widely publicized objections to WTE by various groups. Government Advisory Associates[125] reports that 207 WTE projects in the planning phases were canceled between 1986 and 1990 (compared to a total of 140 operational U.S. facilities in 1990).

The future of WTE in the United States depends on a complex set of technical and socioeconomic conditions. Technological issues related to the use of WTE are discussed in other parts of this chapter. This section focuses specifically on the socioeconomics of WTE and reviews the results of recent work at the Oak Ridge National Laboratory (referred to as the Oak Ridge study) to assess the socioeconomic factors that have contributed to local decisions about WTE projects during the 1980s and early 1990s.† More specifically, this section discusses (1) the socioeconomic characteristics that may contribute to local decisions about WTE projects, (2) financial conditions that may have hindered the adoption of WTE, and (3) findings of four in-depth case studies of WTE projects to better understand the decision-making process at the local level. This section also gives a brief discussion of the current status of WTE in the United States and some of the issues that have been debated at the local and national levels.

———
*Prepared by the Oak Ridge National Laboratory Oak Ridge, Tennessee 37831, managed by Martin Marietta Energy Systems, Inc. for the U.S. Department of Energy. The submitted manuscript has been authored by a contractor of the U.S. Government under contract No. DE-AC05-84OR21400. Accordingly, the U.S. Government retains a nonexclusive, royalty-free license to publish or reproduce the published form of this contribution, or allow others to do so, for U.S. Government purposes.
†This section is based on findings presented in a 1994 book from Quorum Books entitled *Waste-To-Energy in the United States: A Social and Economic Assessment,* by T. Randall Curlee, Susan M. Schexnayder, David P. Vogt, Amy K. Wolfe, Michael P. Kelsay, and David L. Feldman.[124]

The Debate over WTE

WTE became the focus of debate at the national, state, and local levels during the 1980s as environmental and other groups opposed the further adoption of WTE. As is reported by Government Advisory Associates (1989–90 and 1991), many projects in the conceptual and advanced planning phases were canceled and supporters of WTE were increasingly placed in a defensive position. The debate has focused on numerous issues including atmospheric emissions, ash management, uncertainties about future environmental and financial regulations, the cost of WTE as compared to other waste-management options, the appropriate role of WTE as a component in an integrated waste management system, the rapid rate of technological advance in WTE designs and performance, markets for energy produced at WTE facilities, obstacles to financing capital-intensive WTE facilities, and uncertainties about the local decision-making process.

At the national level WTE has been discussed in terms of the energy that potentially could be provided. The Oak Ridge study estimated that WTE contributed about 0.3 quadrillion Btu (quads) of energy in 1990 and projected that the energy output from WTE facilities in the United States would increase to about 0.6 quads by the year 2000. The Oak Ridge study found that other projections for the year 2000 range from as little as 0.2 to as much as 1.2 quads. Although WTE is not expected to exceed 1 percent of all energy consumed in the United States by the year 2000, WTE could become a major player in the generation of electricity. Note that all WTE facilities under construction or in the advanced stages of planning will produce electricity in total or in combination with other energy forms.

WTE and Local Socioeconomic Characteristics

Government Advisory Associates (1991)[125] reports that as of 1990 there were 202 WTE facilities that were operational, under construction, in shakedown, or in the advanced stages of planning. Facilities in the advanced planning phase numbered 62 with the remaining facilities categorized as existing sites. The South has the largest number of existing facilities (36.4 percent), with the Northeast, North Central region, and West accounting for 28.6, 23.6, and 11.4 percent, respectively. The distribution of planned facilities is somewhat different, with the Northeast accounting for 53.2 percent, the South having 21.0 percent, the North Central region at 16.1 percent, and the West at 9.4 percent. Existing facilities are located in 38 different states. Minnesota, Florida, and New York have 14, 14, and 12 facilities, respectively. Advance-planned facilities are in 24 different states, with New Jersey, New York, and Pennsylvania accounting for 10, 8, and 8, respectively.

The Oak Ridge study examined the 1982 to 1990 time frame and identified a total of 354 counties that have had WTE initiatives—i.e., facilities being constructed, in operation or shakedown, or in the conceptual or advanced stages of planning. As part of the study, detailed socioeconomic data were obtained for each U.S. county. Statistical techniques were used to assess the socioeconomic differences between counties that have had a WTE initiative as compared to counties that have never formally considered a WTE project. Differences between counties that have completed WTE facilities and counties that have planned but later canceled WTE projects were also assessed.

The study does not claim to have established causal relationships between any socioeconomic characteristics and decisions about WTE projects, but did identify significant differences between counties with initiatives and those counties with no initiatives. Counties with initiatives are generally wealthier, more educated (i.e., larger percentage of population completing high school), less blue collar (i.e., smaller percentage of workers in manufacturing, mining, and construction), less rural (i.e., smaller percentage of population in a rural environment), and have a higher percentage of individuals in what is termed the *family formation* stage (i.e., individuals between 22 and 39 years of age). The study also found that counties with WTE initiatives are more likely to have access to recycling programs. Somewhat surprisingly, there was no statistically significant relationship between having a WTE initiative and population growth. Counties with initiatives tend to have higher costs of municipal waste disposal, have a larger percentage of individuals that are members of environmental groups, and are more likely to be out of attainment with respect to one or more U.S. EPA atmospheric criteria pollutants. (Note that the study investigated the county's attainment or nonattainment at the time the WTE facility was being considered. The study makes no suggestion that WTE facilities are responsible for nonattainment conditions.) Counties within states that have stronger environmental regulations and incentives for recycling were found to be more likely to have had a WTE initiative. The study found that counties that have had initiatives have average populations of about 385,000 as compared to counties with no initiatives at about 41,000.

While the socioeconomic differences between counties with and without WTE initiatives are quite clear, the study identified virtually no differences between counties with completed facilities and counties with facilities that had been planned but later scratched. In other words, the Oak Ridge study did not identify socioeconomic characteristics that are strong statistical predictors of whether an initiative will result in a completed or a scratched WTE facility. If socioeconomic characteristics contributed to the cancellation of WTE projects in recent years, they are factors other than those considered in the Oak Ridge study.

WTE and Changes in Financial Conditions

The Oak Ridge study identified three key financial trends that altered the attractiveness of WTE in the late 1980s and early 1990s. First, the average costs of WTE facilities increased because of the move toward large, mass burn facilities and because of requirements for more sophisticated environmental controls. The study found that, when adjusted for inflation, the mean cost of facilities in the advanced stage of planning increased by 40.6 percent between 1982 and 1990, i.e., from $62.1 million to $87.3 million. The capital costs of existing and advance-planned facilities combined increased from an average of $43.8 million in 1982 to $52.4 million in 1990 in constant 1982 dollars.

Second, the enactment of the United States Tax Reform Act in 1986 (TRA86) placed limits on a community's ability to finance WTE and other waste facilities with federally tax-exempt financing. More specifically, TRA86 placed new restrictions on the types of local debt that qualified for tax-exempt financing and placed volume caps on a state's allotment of tax-exempt, private-activity bonds. To the extent that tax-exempt financing cannot be used, the overall cost of a WTE facility will be increased.

Third, the Oak Ridge study found that increased demands for other environmental infrastructures, limitations on state and local taxes and expenditures, and difficulties in accessing national capital markets hindered some WTE financial packages. Demands for improved wastewater treatment, drinking water, and solid waste projects mean that some local governments must make difficult decisions about which projects are most important. While capital markets can in principle accommodate the increased local demands for capital financing, large capital-intensive environmental projects can crowd out other local investments.

Although financial obstacles have grown in recent years, the Oak Ridge study found that many municipalities are adjusting successfully to the altered conditions. For example, some communities are using a combination of several financial mechanisms in their financial packages to spread the financial risk. Innovative and new methods of finance are being adopted. In addition, private-sector participation in financing is growing, which allows local-sector resources to be allocated for other public-good consumption.

The Oak Ridge study found somewhat mixed results with respect to the impacts of altered financial conditions on WTE project cancellations. The study found that most successful projects used multiple and innovative financing methods, while innovative methods of finance were not present in any of the financial packages of facilities that were canceled. It is unclear if the absence of innovative financing contributed to any project cancellations or if abandoned projects simply did not get far enough along the development path to consider these less obvious financial methods.

The impacts of the volume caps imposed by TRA86 are also less than clear. Some states that had several project cancellations also fully utilized or came close to fully utilizing their available tax-exempt volumes or caps. However, other states with several cancellations did not use a high percentage of their caps.

It is quite clear that the impending restrictions imposed by TRA86 escalated the rate at which WTE projects were introduced in the 1980s. Evidence suggests that some projects that might have moved along at a slower pace were developed more quickly to avoid the impending financial restrictions associated with TRA86. To the extent that projects were hurried along, it can be argued that some projects that were canceled would have not become formal projects in the absence of TRA86. Therefore, TRA86 may have indirectly contributed to some project cancellations.

WTE and the Local Decision-Making Process

The socioeconomic and financial-assessment portions of the Oak Ridge study do not draw strong conclusions about the factors that have contributed to the decisions of communities to cancel WTE projects in recent years. Socioeconomic differences are not significant between counties that cancel and complete WTE projects, and, while financial obstacles may have contributed to some cancellations, financial constraints do not appear to have been the primary force motivating WTE cancellations. The primary obstacle faced by planned WTE facilities may be the decision-making process itself.

To better understand the decision-making process, the Oak Ridge study undertook four in-depth case studies of WTE projects. At two case-study sites—Oakland County, Michigan, and Broward County, Florida—a WTE facility was approved, and in the case of Broward County two facilities are now operational. At two other case-study sites—Monmouth County, New Jersey, and Knox County, Tennessee—facilities in the planning stages were abandoned. The case studies involved extensive background reading related to each site, trips to each community where structured interviews were conducted with numerous proponents and opponents of the facility, and telephone follow-ups with the interviewees to clarify remaining issues. The interviewees were also allowed to comment on the draft findings of the interviews.

Of particular interest in the case studies were questions about the sequence of decision events; the participation of different groups at different steps in the decision process; the degree of agreement at each decision step; the effects of mitigation and compensation; the effectiveness of different siting procedures; implications of alternative ownership methods; public attitudes about WTE technologies, costs, environmental impacts, and the decision-making process itself; and difficulties that may arise when two or more local governments are forced to cooperate or form compacts in order to site a facility.

The findings of the case studies suggest numerous obstacles that can arise in the decision-making process, and, although the study's small sample size does not support strong conclusions, several preliminary findings do stand out. For example, a perception of competition between recycling and WTE for a limited waste stream was a significant obstacle at all but the Broward County site, where WTE facilities were planned about 3 years before the formal planning process began at the other three sites. The mid-to-late 1980s marked the emergence of strong anti-incineration groups that began national campaigns against WTE. Anti-incineration activities at the local and national levels were found to influence the decision-making process at the three sites initiated after 1985.

Additional findings of the case studies include the following:

1. The selection of ownership options was difficult and time-consuming. However, facility ownership was not found to be a strong factor in the community's decision to continue or abandon a planned facility.

2. The size of the proposed facility was an important decision-making factor. Size decisions were found to be important not only because size has an impact on facility cost, but also because oversized facilities offer the potential for waste importation and actual or perceived implications for other waste management methods, especially recycling. The timing of WTE implementation relative to the implementation of recycling and source reduction activities was also found to be important.

3. Somewhat surprisingly, the study did not show any clear link between siting activities or site characteristics and the decision to build or abandon a project.

4. Historic patterns of mistrust among the member municipalities that were jointly planning a facility did have an impact on the decision-making process. Patterns of mistrust that resulted from matters unrelated to WTE appear to have impacted WTE planning and outcome.

5. State-level support for WTE was a positive influence toward acceptance of the facility. State-level support, whether or not the guidelines for support are encoded as law, tend to facilitate WTE adoption.

6. The perception of need for additional waste management capacity and the urgency of that need appear to motivate WTE adoption. Without a perceived or actual urgency, decision makers are more likely to delay action.

7. There was quite strong agreement among the interviewees that public participation in the decision-making process must occur early in the process and be extensive. Dissatisfaction with decision making that excludes or makes public participation difficult exacerbates opposition to the facility.

8. Impending financial constraints associated with TRA86 and selected impending state regulations were found to hasten project activities. However, uncertainty about future environmental legislation and regulations appear to have had little consequence at the study's selected sites.

9. There is evidence to suggest that past experiences with different waste management approaches have an impact on WTE decisions. For example, past experience with incinerators, contaminated landfills, and limited opportunities for waste exportation point toward greater acceptance of WTE technologies.

Once a WTE project enters the planning process, the decision to proceed or abandon that project appears to largely depend on the dynamics of the decision-making process and the interactions of the parties involved in the process. Future work should build on the findings of the Oak Ridge study and other works to identify methods to assist communities in making decisions about a very complex and often divisive issue.

REFERENCES

1. *Incinerator Standards,* Incinerator Institute of America, New York, 1972.

2. Black, R., and A. Klee, "The National Solid Wastes Survey: An Interim Report," presented at the 1968 Annual Meeting of the Institute of Solid Wastes of the American Public Works Association, Miami Beach, October 1968.

3. Brunner, C., and S. Schwarz, *Energy and Resource Recovery from Waste,* Noyes, Park Ridge, N.J., 1983.

4. *Source Category Survey: Industrial Incinerators,* U.S. EPA 450 3-80-013, May 1980.

5. Danielson, J., *Air Pollution Engineering Manual,* County of Los Angeles, Air Pollution Control District, AP-40, May 1973.

6. *Recommended Methods of Reduction, Neutralization, Recovery or Disposal of Hazardous Waste,* vol. 3, Disposal Process Descriptions, Ultimate Disposal, Incineration and Pyrolysis Processes, U.S. EPA 670/2-73-053C, August 1973.

7. Brunner, C. R., *Incineration Systems: Selection and Design,* Incinerator Consultants Incorporated, Reston, Va., 1988.

8. Brunner, C. R., *Handbook of Hazardous Waste Incineration,* McGraw-Hill, New York, 1993.

9. Ontario Ministry of the Environment, "Incinerator Design and Operating Criteria," vol. II, *Biomedical Waste Incinerators,* October 1986.

10. Doyle, B. W., "The Smoldering Question of Hospital Wastes," *Pollution Engineering,* July 1985.

11. Brunner, C. R., "Biomedical Waste Incineration," Monograph, presented at the Air Pollution Control Association Annual Conference, New York, June 1987.

12. Brunner, C. R., "Hospital Waste Disposal by Incineration," *Journal of the Air Pollution Control Association,* vol. 38, no. 10, October 1988, pp. 1297–1309.

13. Keenan, J., and F. Keyes, *Thermodynamic Properties of Steam,* John Wiley & Sons, New York, 1957.

14. Hasselriis, F., *Processing Refuse-derived Fuel,* Butterworth, Stoneham, Mass., 1985.

15. Standards developed by Committee E-38-01 Energy, ASTM, Philadelphia.

16. Studies of waste composition, Gershman, Brickner & Bratton, Inc., Falls Church, Va., 1990.

17. Chesner, W. H., R. J. Collins, and T. Fung, "Assessment of the potential suitability of Southwest Brooklyn incinerator residue in asphaltic concrete mixes," NYSERDA report 90-15, 1988, State of New York, Albany.

18. Thome-Kozmiensky, K., "Measures to reduce incinerator emissions," *Recycling International,* Berlin, 1989, p. 1009.

19. Forrester, K. E., and R. W. Goodwin, "MSW ash field study: achieving optimal disposal characteristics," *Journal of Envir. Engineering,* paper no. 25094, vol. 116, no. 5, September/October 1990.

20. Forrester, K., "State-of-the-Art in Refuse-to-Energy facility ash residue characteristics, handling, reuse, landfill design and management—a summary," *Proc. of 2d Int. Conf. on MSW Combustion Ash Utilization,* November 1989, Resource Recovery Report, Washington, D.C., pp. 167 ff.

21. Goodwin, R. W., "Chemical treatment of utility and industrial waste," *ASCE Nat. Conf. on Environmental Engineering,* July 1982, Minneapolis.

22. Abe, S., "Melting treatment of municipal waste," *Recycling International,* 1984, p. 173 ff.

23. Donnelly, J. E., E. Jons, and P. Mahoney, "By-product disposal from MSW flue gas cleaning systems," *1987 APCA Annual Meeting,* New York, paper 87-94A.3.

24. Shinn, C. E., "Toxicity Characteristic Leaching Procedure (TCLP), Extraction Procedure Toxicity (EPTox) and deionized water leaching characteristics of lead from municipal waste incinerator ash," Department of Environmental Quality, Portland, Ore., 1987.

25. Hjelmar, O., "Leachate from incinerator ash disposal sites," *Int. Workshop on Municipal Waste Incineration,* Montreal, 1982.

26. Lebedur, H. et al., "Emission reduction in the second domestic waste incinerator plant in the RZR Herten," *Recycling International,* 1989, Berlin, p. 1220.

27. Hahn, J., R. G. Rumba, G. T. Hunt, and J. Wadsworth, "Fugitive particulate emissions associated with MSW ash handling—results of a full scale field program," *83d A&WMA Annual Meeting*; Pittsburgh, June 1990.

28. Hasselriis, F., et al., "The removal of metals by washing of incinerator ash," *84th Annual Meeting of the Air and Waste Management Assoc.,* Vancouver, B.C., June 1991.

29. Exner, R., et al., "Thermal effluent treatment for flue gas treatment systems in refuse incineration plants from the point of view of residue minimization by recycling," *Recycling International,* Berlin, 1989, pp. 1372–1392.

30. Holland, P., et al., "Evaluation of leachate properties and assessment of heavy metal immobilization from cement and lime amended incinerator residues," *Proc. Int. Conf. on Municipal Waste Combustion,* vol. 1, April 1989, Hollywood, Fla.

31. Eighmy, T. T., et al., "Theoretical and applied methods of lead and cadmium stabilization in combined ash and scrubber residues," *Proc. 2d Int. Conf. on MSW Combustor Ash Utilization,* November 1989, Resource Recovery Report, Washington, D.C.

32. Wakamura, Y., and K. Nakazato, "Technical approach for flyash stabilization in Japan," *85th A&WMA Annual Meeting,* Kansas City, Mo., June 1992, paper 92-20.08.

33. Fahlenkamp, H., and C. Hemmer, "Effects of the new refuse incineration plant ordinance on the choice of process for noxious gas separation and treatment of residual matter," *Recycling International,* Berlin, 1989, pp. 1244–1262.

34. DeCesare, R., and A. Plumley, "Results from the ASME/US Bureau of Mines investigation—vitrification of residue from municipal waste combustion," *1991 ASME National Waste Processing Conference,* Detroit, May 1992. Also *85th A&WMA Annual Meeting,* Kansas City, Mo., June 1992, paper 92-47.03.

35. Francis, C. W., "Leaching characteristics of resource recovery ash in municipal waste landfills," Oak Ridge National Lab., DOE ERD-83-289, no. 2456, 1984.

36. Goodwin, R. W., and K. E. Forrester, "MSW ash field study: achieving optimal disposal characteristics," *ASCE J. Environmental Engineering Division,* September/October 1990; paper 25094, vol. 116, no. 5, pp. 880–889.

37. EPA, "National Secondary Drinking Water Regulations," 40-CFR 143, September 1988.

38. Roffman, H. K., "Major findings of the U.S.EPA/CORRE MWC-ash study," *Municipal Waste Combustion,* April 1991, A&WMA VIP-19.

39. Clark, R. A., "Summary of the field leachate generated from Northern States Power Company's Red Wing RDF ash disposal facility," *85th Annual A&WMA Meeting,* Kansas City, Mo., June 1992, paper 92-47.02.

40. Cundari, K. L., and J. M. Lauria, "The laboratory evaluation of expected leachate quality from a resource recovery ashfill," *1986 Triangle Conference for Environmental Technology,* Chapel Hill, N.C.

41. Hartlen, J., and P. Elander, "Residues from waste incineration—chemical and physical properties," SGI Varia 172, Swedish Geotechnical Institute, Linkoping, Sweden, 1986.

42. U.S. EPA, "Hazardous waste management system: identification and listing of hazardous waste, toxicity test Extraction Procedure (EP)," *Federal Register,* 45(98), 33063-33285, May 19, 1980.

43. Francis, C. W., and M. P. Maskarinec, "Leaching of metals from alkaline wastes by municipal waste leachate," Oak Ridge National Lab., ORNL/T-10050, publ. no. 2846, March 1987.

44. Fiesinger, T., "Sampling of incinerator ash," *Proc. 2d Int. Conf. on MSW Combustor Ash Utilization,* November 1989, Resource Recovery Report, Washington, D.C., p. 51.

45. Hasselriis, F., "Variability of municipal solid waste and emissions from its combustion," *1984 ASME SWPD Conference,* Orlando, Fla., 1984.

46. Feder, W. A., and J. S. Mika, "Summary update of research projects with incinerator bottom ash residue," Executive Office of Environmental Affairs, Commonwealth of Massachusetts, February 1982.

47. Fiesinger, T., "Incinerator ash management: knowledge and information gaps to 1987," report 92-6, New York State Energy Research and Development Authority, Albany, 1992.

48. Hjelmar, O., "Regulatory and environmental aspects of MSW ash utilization in Denmark," *3d Int. Conf. on Ash Utilization and Stabilization (ASH III),* November 1990, Resource Recovery Report, Washington, D.C.

49. DiPietro, J. V., M. R. Collins, and T. T. Eighmy, "Evaluation of Leachate Properties from Various MSW Incinerator Residues and Assessment of Geochemical Modeling Predictions," *Proc. 2d Int. Conf. on MWC Ash Utilization,* Alexandria, Va., 1989.

50. Chesner, W. A., "Aggregate-related characteristics of MSW combustion residue," *Proc. 2d Int. Conf. on MSW Combustor Ash Utilization,* November 1989, Resource Recovery Report, Washington, D.C.

51. Chesner, W. A., "Working toward beneficial use of waste combustor ash," *Solid Waste & Power,* vol. VII/no. 5, September/October 1993.

52. Roethel, F. J., et al., "The fixation of incinerator residues," Marine Sciences Research Center 87-3, State University of New York at Stony Brook, N.Y.

53. Wexerll, D. R., "Cold crown vitrification of municipal waste combustor flyash," A&WMA 86th Annual Meeting, Denver, June 1993, paper 93-RA-117.02.

54. *Sampling and Testing Methods for Solid Waste,* SW846, U.S. EPA, 1982.

55. Chapman, R., Fifth Ann. Symp on Ceramics, Am. Cer. Soc., Cincinnati, Ohio, April–May 1991; NUS Proj. 9S83, R33-9-11, EPA, February 1990.

56. Council of European Communities Directive, "New Municipal Waste Plants," 89/369/EEC—OJL163, 1989, and "Municipal Waste Plants," 89/429/EEC—OJL203, 1989.

57. *Federal Register* 40 CFR, parts 51, 52, 60, Feb. 11, 1991.

58. *Federal Register,* vol. 51, no. 114, pp. 21648–21693, June 1986.

59. McIlvane, *Fabric Filter News Letter,* no. 182, 1990.

60. Personal communication, E. Wheless, LASAN, 1988.

61. Svensberg, S., "Emission Standards in Selected European Countries and EEC," DXN Conf. Kyoto, 1990.

62. Teller, A. J., *Solid Waste and Power,* vol. 6, no. 12, 1991.

63. Jacko, R., et al., JAPCA, vol. 27, no. 10, 1977, p. 989.

64. Jacko, R., *Env. Sci. & Tech.,* vol. 16, no. 3, 1982, p. 150.

65. Greenburg, R., et al., *Env. Sci. & Tech.,* vol. 12, no. 12, 1978, p. 1329.

66. Hahn, J., Cooper Eng. Report on Munich Incinerator, 1986.

67. Driftstudie Av Sysav:S, Aufallswerki Malmo, June 1983, Svenska Rehallungs Verksforeningen, Pub. 83:7.

68. ESA Report to Los Angeles County Sanitation district, December 1988.

69. Zemba, L., et al., *Solid Waste & Power,* May–June 1992, pp. 38–44.

70. Clarke, M., *Waste Age,* December 1987.

71. Jones, K., EPA Mercury Conf., 1989.

72. Schager, P., Institution for Organisk Kerni, Chalmers Teleniska Höyskola, Goteborg, November 1990.

73. Stepanov, A. S., et al., *Prom. Sanit. Ochistka Gazov,* vol. 5, no. 10, 1979.

74. Hisayoshi, S., et al., *Nippon Kagaka Kaishi,* no. 10, 1976, p. 1596.

75. Johnston, E., Ind. Studies Branch ESD(MDB), May 18, 1990.

76. Teller, A. J., et al., Int. Conf. on Incineration, Miami, 1988.

77. Brown, B., Joy-Niro, presentation to EPA Mercury Conf., 1989.

78. Darco Bulletin AN-74-3, 1990.

79. Moller, J. T., et al., U.S. Patent 4889698, 1989.

80. de Jong, G., U.S. Patent 4196173, 1980.

81. Guest, T. L., et al., *Proc. AWMA,* 91/103.33, 1991.

82. Nakazato, K., *Proc. Nat. Waste Proc. Conf.,* 163-9, 1990.

83. Urabe, T., *Seiso Gibo,* vol. 16, pp. 7–23, 1991.

84. Albert, F. W., *UGB Kraftwerstech,* vol. 71, no. 8, p. 776, 1991.

85. Rappe, C., personal communication.

86. Yoshida, H., DXN, Kyoto, 1991, pp. 31–39.

87. Hiraoka, M., DXN, Kyoto, 1991, p. 125.

88. Hiraoka, M., DXN, Kyoto, 1991, pp. 1–9.

89. Rappe, C., DXN, Kyoto, 1991, pp. 11–29.

90. Bozeka, G. G., *Proc. Nat. Waste Proc. Conf.,* 1976, p. 215.

91. Rordoff, B. F., *Chemosphere,* vol. 14, no. 6-7, 1985, p. 885.

92. Stieglitz, L., *Dioxin,* vol. 90, pp. 169 and 173.

93. Teller, A. J., "Emissions from Combustion Process," *Proc. Int. Conf. on Combustion* (1988), Clement & Kagel, Lewis Press, 1990, pp. 271–291.

94. Teller, A. J., and J. D. Lauber, *Proc. 76th Annual Meeting APCA,* 1984.

95. Boin, J. G. P., et al., *Chemosphere,* vol. 19, 1989, pp. 1629–1633.

96. Teller, A. J., DXN, Kyoto, 1991, pp. 185–191.

97. Okajima, S., et al., DXN, Kyoto, 1991, pp. 95–104.

98. Takeda, M., et al., DXN, Kyoto, 1991, pp. 149–155.

99. Ide, Y., et al., DXN, Kyoto, 1991.

100. Hagenmaier, H., et al., *Dioxin,* vol. 90, pp. 65–68.

101. Hagenmaier, H., et al., *VGB Kraftwerkstechnik,* vol. 70, 1990, pp. 491–493.

102. Hiraoka, M., et al., DXN, Kyoto, 1991, pp. 201–205, 217–224.

103. Schroy, J., et al., "Aquatic Toxicology Hazard Assessment," ASTM spec. tech. pub. 891, 1985, pp. 409–421.

104. Teller, A. J., U.S. Patent 4502346, 1985.

105. Rordoff B. F., *Chemosphere,* vol. 15, no. 9-12, 1986, p. 1325.

106. Eiceman, G. A., et al., *Chemosphere,* vol. 11, no. 9, 1982, p. 832.

107. Vicard, J. F., et al., DXN, Kyoto, 1991, pp. 201–205.

108. Energy Systems Assoc. for County Sanitation Districts of Los Angeles County, ESA 20522-444, 20534-621, 1987, 1988.

109. Naikwadi, K. P., et al., *Chemosphere,* vol. 19, 1989, pp. 299–304.

110. Lasagni, M., et al., *Chemosphere,* vol. 23, nos. 8–10, 1991, p. 1245.

111. Stieglitz, L., et al., *Chemosphere,* vol. 23, nos. 8–10, 1991, p. 1255.

112. Addink, R., et al., *Chemosphere,* vol. 23, nos. 8–10, 1991, p. 1205.

113. Dijkgraff, A., *Polytech. Tijdschr. Procestech.,* vol. 46, no. 9, 1991, pp. 52–56.

114. Richter, E., *Chem. Eng. Tech.,* vol. 64, no. 2, 1992, pp. 125–136.

115. Fahlenkamp, H., et al., *VGB Kraftwerkstechnik,* vol. 71, no. 7, 1991, pp. 671–674.

116. Boss, R., et al., *Chemosphere,* vol. 22, nos. 5–6, 1991, pp. 569–575.

117. Lavalin Inc., Nat Incinerator Testing and Evaluation Program, prepared for Env. Prot. Serv., Env. Canada, September 1987.

118. McDannel, M. D., et al., ESA, Air Emissions.

119. Hahn, S. L., "Ogden Project, Results from Stanislaus CA Resource Recovery Facility," Int. Conf. on Municipal Waste Comb., April 1989.

120. Jones, D. G., et al., *Proc. AWMA* (1990), vol. 2, p. 32.3.

121. U.S. EPA, 450/3-89-27d, Municipal Waste Combusters—Control of NO_x Emissions.

122. Herrlander, B., *Proc. AWMA* (1990), vol. 2, p. 25.4.

123. Curlee, T. R., "The Potential for Energy from the Combustion of Municipal Solid Waste," *Journal of Environmental Systems,* vol. 20, no. 4, 1991, pp. 303–322.

124. Curlee, T. R., S. M. Schexnayder, D. P. Vogt, A. K. Wolfe, M. P. Kelsay, and D. L. Feldman, *Waste-To-Energy in the United States: A Social and Economic Assessment,* Quorum Books, Westport, Conn., 1994.

125. Government Advisory Associates, *Resource Recovery Yearbook,* New York, 1982, 1984, 1986–87, 1988–89, 1991.

126. U.S. Environmental Protection Agency, *Characteristics of Municipal Solid Waste in the United States: 1992 Update,* prepared by Franklin Associates, Ltd., July 1992.

CHAPTER 12
LANDFILLING*

George Tchobanoglous
Professor of Civil and Environmental Engineering
University of California at Davis

Philip R. O'Leary
Co-Director, Solid and Hazardous Waste Education Center
University of Wisconsin–Madison

The safe and reliable disposal of municipal solid waste and solid waste residues is an important component of integrated waste management. Solid waste residues are waste components that are not recycled, that remain after processing at a materials recovery facility, or that remain after the recovery of conversion products and/or energy. Historically, solid waste has been placed on or in the surface soils of the earth or deposited in the oceans. Ocean dumping of municipal solid waste was officially abandoned in the United States in 1933. *Landfill* is the term used to describe the physical facilities used for the disposal of solid wastes and solid waste residuals in the surface soils of the earth. Since the turn of the century, the use of landfills, in one form or another, has been the most economical and environmentally acceptable method for the disposal of solid wastes, both in the United States and throughout the world. Today, landfill management incorporates the planning, design, operation, environmental monitoring, closure, and postclosure control of landfills.

Although many landfills have been constructed in the past with little or no thought for the long-term protection of public health and the environment, the focus of this chapter is modern landfilling practice. In the last 20 years, practices have changed substantially so that recently constructed landfills have overcome the problems formerly associated with "dumps." The major topics covered in this chapter include:

1. A description of the landfill method of solid waste disposal, including environmental concerns, regulatory requirements, and siting considerations
2. Generation, composition, control, and management of landfill gases
3. Formation, composition, and management of leachate
4. Intermediate and final landfill cover

---

*Adapted from G. Tchobanoglous, H. Theisen, and S. A. Vigil, *Integrated Solid Waste Management, Engineering Principles and Management Issues,* McGraw-Hill, Inc., New York, 1993 and P. O'Leary and P. Walsh, *Solid Waste Landfills Correspondence Course,* University of Wisconsin—Madison, 1992.

5. Landfill structural characteristics and settlement
6. Landfill design considerations
7. Development of landfill operation plan
8. Environmental quality monitoring
9. Landfill closure, postclosure care, and remediation

Additional details on the subjects covered in this chapter may be found in Refs. 2, 8, 39, and 52.

12.1 THE LANDFILL METHOD OF SOLID WASTE DISPOSAL

Landfilling is the term used to describe the process by which solid waste and solid waste residuals are placed in a landfill. In the past, the term *sanitary landfill* was used to denote a landfill in which the waste placed in the landfill was covered at the end of each day's operation. Today, *sanitary landfill* refers to an engineered facility for the disposal of MSW designed and operated to minimize public health and environmental impacts. Landfills for individual waste constituents such as combustion ash, asbestos, and other similar wastes are known as *monofills*. Landfills for the disposal of hazardous wastes are called *secure landfills*.

Definition of Terms

The general features of a sanitary landfill are illustrated in Fig. 12.1. Some terms commonly used to describe the elements of a landfill are described below. The term *cell* is used to describe the volume of material placed in a landfill during one operating period, usually one day (see Fig. 12.1b). A cell includes the solid waste deposited and the daily cover material surrounding it. *Daily cover* usually consists of 6 to 12 in of native soil or alternative materials such as compost, foundry sand, or auto shredder fluff that are applied to the working faces of the landfill at the end of each operating period. Historically, daily cover was to prevent rats, flies, and other disease vectors from entering or exiting the landfill. Today, daily landfill cover is used primarily to control the blowing of waste materials and to control the entry of water into the landfill during operation. A *lift* is a complete layer of cells over the active area of the landfill (see Fig. 12.1b). Typically, landfills comprise a series of lifts. A *bench* (or *terrace*) is typically used where the height of the landfill will exceed 50 to 75 ft. Benches are used to maintain the slope stability of the landfill, for the placement of surface water drainage channels, and for the location of landfill gas recovery piping. The *final lift* includes the landfill cover layer.

Landfill liners are materials (both natural and man-made) that are used to line the bottom area and below-grade sides of a landfill (see Fig. 12.1a). Liners usually consist of successive layers of compacted clay and/or geosynthetic material designed to prevent migration of landfill leachate and landfill gas. The final *landfill cover* layer is applied over the entire landfill surface after all landfilling operations are complete (see Fig. 12.1c). Landfill covers consist of successive layers of compacted clay and/or geosynthetic material designed to prevent the migration of landfill gas and to limit the entry of surface water into the landfill.

The liquid that forms at the bottom of a landfill is known as *leachate*. In general, leachate is a result of the percolation of precipitation, uncontrolled runoff, and irriga-

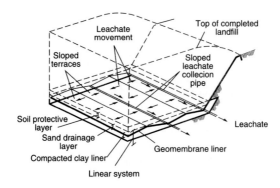

(a)

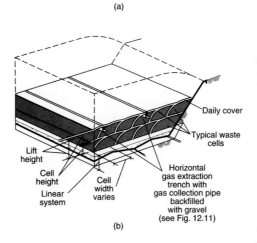

(b)

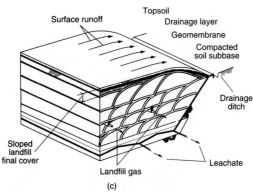

(c)

FIGURE 12.1 Cutaway views of a sanitary landfill: (*a*) after geomembrane liner has been installed over compacted clay layer and before drainage and soil protective layers have been installed, (*b*) after two lifts of solid waste have been completed, and (*c*) completed landfill with final cover installed.

tion water into the landfill. Leachate will also include water initially contained in the waste. Leachate contains a variety of chemical constituents derived from the solubilization of the materials deposited in the landfill and from the products of the chemical and biochemical reactions occurring within the landfill. *Landfill gas* is the term applied to the mixture of gases found within a landfill. The bulk of landfill gas consists of methane (CH_4) and carbon dioxide (CO_2), the principal products of the anaerobic biological decomposition of the biodegradable organic fraction of the MSW in the landfill.

Environmental monitoring involves the activities associated with collection and analysis of water and air samples used to monitor the movement of landfill gases and leachate at the landfill site. *Landfill closure* is the term used to describe the steps that must be taken to close and secure a landfill site once the filling operation has been completed. *Postclosure care* refers to the activities associated with the long-term maintenance of the completed landfill (typically 30 to 50 years). *Remediation* refers to those actions necessary to stop and clean up unplanned contaminant releases to the environment.

Classification of Landfills

Although a number of landfill classification systems have been proposed over the years, the classification system adopted by the state of California in 1984 is perhaps the most widely accepted classification system for landfills. In the California system, as reported below, three classifications are used.

Class	Type of waste
I	Hazardous waste
II	Designated waste
III	Municipal solid waste (MSW)

The majority of the landfills throughout the United States are designed for commingled MSW. In many of these Class III landfills, limited amounts of nonhazardous industrial wastes and sludge from water and wastewater treatment plants are also accepted. In many states, treatment plant sludges are accepted if they are dewatered to a solids content of 51 percent or greater and contain no free-flowing liquids. The acceptance of liquid wastes into MSW landfills is now banned by federal regulations.

An alternative method of landfilling that is being tried in several locations throughout the United States involves shredding of the solid wastes before placement in a landfill. Shredded (or milled) waste can be placed at up to 35 percent greater density than unshredded waste, and possibly receive an exemption from daily cover requirements in some state regulations. Blowing litter, odors, flies, and rats have not been significant problems.

Another approach is to bale the MSW for placement in the landfill. This method has the advantage of easier handling and eliminates the need for compaction equipment. The bales are prepared at a production facility located in either an off-site transfer station or at an unloading station on the landfill property. The bales are moved to the working face on flatbed vehicles and stacked with forklifts or similar equipment. Cover is applied as a lift is completed but daily covering may not always be required.

Designated wastes are nonhazardous wastes that may release constituents in concentrations that are in excess of applicable water quality objectives established by various state and federal agencies. Waste constituents such as combustion ash, asbestos, and other similar wastes, often identified as designated wastes, are typically placed in lined monofills to isolate them from materials placed in municipal landfills.

Landfilling Methods

The principal methods used for the landfilling of MSW may be classified as (1) excavated cell/trench, (2) area, and (3) canyon. The principal features of these types of landfills, illustrated in Fig. 12.2, are described below. Landfill design details are presented later in the chapter.

Excavated Cell/Trench Method.   The cell/trench method of landfilling (see Fig. 12.2*a*) is ideally suited to areas where an adequate depth of cover material is available at the site and where the water table is not near the surface. Typically, solid wastes are placed in cells or trenches excavated in the soil (see Fig. 12.2*a*). The soil excavated from the site is used for daily and final cover. The excavated cells or trenches are lined with synthetic membrane liners or low-permeability clay or the combination of the two to limit the movement of both landfill gases and leachate. Excavated cells are typically square, up to 1000 ft in width and length, with side slopes of 2:1 to 3:1. Trenches vary from 200 to 1000 ft in length, 3 to 10 ft in depth, and 15 to 50 ft in width.

Area Method.   The area method is used when the terrain is unsuitable for the excavation of cells or trenches in which to place the solid wastes (see Fig. 12.2*b*). High groundwater conditions, such as occur in many parts of Florida and elsewhere, necessitate the use of area-type landfills. Site preparation includes the installation of a liner and leachate management system. Cover material must be hauled in by truck or earthmoving equipment from adjacent land or from borrow-pit areas. As noted above, in locations with limited material that can be used as cover, compost produced from yard wastes and MSW, foundry sand, and auto shredder fluff has been used successfully as intermediate cover material. Other techniques include the use of movable temporary cover materials such as soil and geosynthetics. Soil and geosynthetic blankets, placed temporarily over a completed cell, can be removed before the next lift is begun.

Canyon/Depression Method.   Canyons, ravines, dry borrow pits, and quarries have been used for landfills (see Fig. 12.2*c*). The techniques to place and compact solid wastes in canyon/depression landfills vary with the geometry of the site, the characteristics of the available cover material, the hydrology and geology of the site, the type of leachate and gas control facilities to be used, and the access to the site. Control of surface drainage often is a critical factor in the development of canyon/depression sites. Typically, filling starts at the head end of the canyon and ends at the mouth, so as to prevent the accumulation of water behind the landfill. Canyon/depression sites are filled in multiple lifts, and the method of operation is essentially the same as described above. If a canyon floor is reasonably flat, the initial landfilling may be carried out using the excavated cell/trench method discussed previously.

Other Types of Landfills.   In addition to the conventional methods of landfilling already described, other specialized methods of landfilling designed to enhance different goals of landfill management are being developed. Landfills designed to reduce the time required to stabilize the biodegradable organic matter in the landfill and to maximize the recovery of landfill gas are prime examples.

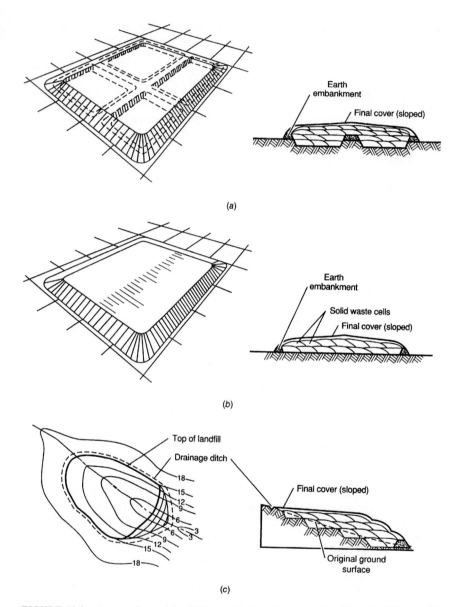

FIGURE 12.2 Commonly used landfilling methods: (*a*) excavated cell/trench; (*b*) area; (*c*) canyon/depression. (*From Ref. 52.*)

Reactions Occurring in Landfills

Solid wastes placed in a sanitary landfill undergo a number of simultaneous and inter-related biological, chemical, and physical changes. The most important biological reactions occurring in landfills are those related to the conversion of the organic material in MSW leading to the evolution of landfill gases and, eventually, leachate. Important chemical reactions that occur within the landfill include: dissolution and suspension of landfill materials and biological conversion products in the liquid percolating through the waste, evaporation and vaporization of chemical compounds and water into the evolving landfill gas, sorption of volatile and semivolatile organic compounds into the landfilled material, dehalogenation and decomposition of organic compounds, and oxidation-reduction reactions affecting metals and the solubility of metal salts. Among the more important physical changes in landfills are the settlement caused by consolidation and decomposition of landfilled material. The reactions occurring in landfills are discussed in greater detail in Secs. 12.2 and 12.3.

Concerns with the Landfilling of Solid Wastes

Concerns with the landfilling of solid waste are related to:

1. The uncontrolled release of landfill gases that might migrate off-site and cause odor and other potentially dangerous conditions
2. The impact on the uncontrolled discharge of landfill gases on the greenhouse effect in the atmosphere
3. The uncontrolled release of leachate that might migrate down to underlying groundwater or to surface streams
4. The breeding and harboring of disease vectors in improperly managed landfills
5. The health and environmental impacts associated with the release of the trace gases found in landfills arising from the hazardous materials that were often placed in landfills in the past

The goal for the design and operation of a modern landfill is to eliminate or minimize the impacts associated with the aforementioned concerns.

Federal and State Regulations for Landfills

In planning for the implementation of a new landfill, special attention must be paid to the many federal and state regulations that have been enacted to improve the performance of sanitary landfills. The principal federal requirements for municipal solid waste landfills are contained in Subtitle D of the Resource Conservation and Recovery Act (RCRA) and in Environmental Protection Agency (EPA) Regulations on Criteria for Classification of Solid Waste Disposal Facilities and Practices (40 CFR Parts 257 and 258).[54] The final version of Part 258—Criteria For Municipal Solid Waste Landfills (MSWLFs) was signed on Sept. 11, 1991. The subparts of Part 258 deal with the following areas:

Subpart A	General
Subpart B	Location Restrictions
Subpart C	Operating Criteria

Subpart D	Design Criteria
Subpart E	Groundwater Monitoring and Corrective Action
Subpart F	Closure and Post-Closure Care
Subpart G	Financial Assurance

Additional details on the implementation requirements for the above subparts are summarized in Table 12.1. Many state environmental protection agencies have parallel regulatory programs that deal specifically with their unique geologic and soil conditions and environmental and public policy issues. Landfill owners, operators, and persons contemplating siting landfills must study their state's regulations carefully and become aware of the public policy issues affecting landfill regulation. It should also be noted that aforementioned landfill regulations necessitate extensive record keeping to document compliance.

The Clean Air Act also contains provisions dealing with the air emissions from landfills. In addition to the federal government, many of the states have also adopted regulations governing the design, operation, closure, and long-term maintenance of landfills. In many cases, the regulations that have been adopted by the individual states have been more restrictive than the federal requirements.

Landfill Siting Considerations

One of the most difficult tasks faced by public agencies and private waste management firms in implementing an integrated waste management program is the siting of new landfills. Factors that must be considered in evaluating potential sites for the long-term disposal of solid waste include:

1. Haul distance
2. Location restrictions
3. Available land area
4. Site access
5. Soil conditions and topography
6. Climatalogical conditions
7. Surface-water hydrology
8. Geologic and hydrogeologic conditions
9. Existing land use patterns
10. Local environmental conditions
11. Potential ultimate uses for the completed site

Final selection of a disposal site usually is based on the results of a detailed site survey, results of engineering design and cost studies, the conduct of one or more environmental impact assessments, and the outcome of public hearings. An overlay procedure for assembling and displaying the relevant site selection information is illustrated in Fig. 12.3. Often an extensive public information and negotiation process must be conducted concurrently with the technical development activities to site a new landfill successfully. It is interesting to note that the up-front development costs for new landfills now varies from $10 to 20 million (1993) before the first load of waste is placed in the landfill.

TABLE 12.1 Summary of U.S. Environmental Protection Agency Regulations for Municipal Solid Waste Landfills

Item	Requirement
Applicability	All active landfills that receive municipal solid waste (MSW) after October 9, 1993. Certain requirements also apply to landfills which received MSW after October 9, 1991, but closed within 2 years. Certain exemptions for very small landfills. Some requirements are waived for existing landfills. New landfills and landfill cells must comply with all requirements.
Location requirements	Airport separation distances of 5000, 10,000, and in some instances greater than 10,000 feet are required. Landfills located on floodplains can operate only if flood flow is not restricted. Construction and filling on wetlands is restricted. Landfills over faults require special analysis and possibly construction practices. Landfills in seismic impact zones require special analysis and possibly construction practices. Landfills on unstable soils require special analysis and possibly construction practices.
Operating criteria	Landfill operators must conduct a random load checking program to ensure exclusion of hazardous waste. Daily cover with 6 in of soil or other suitable materials is required. Disease vector control is required. Permanent monitoring probes are required. Probes must be tested every 3 months. Methane concentrations in occupied structures cannot exceed 1.25 percent. Methane migration offsite must not exceed 5 percent at the property line. Clean Air Act criteria must be satisfied. Access must be limited by fences or other structures. Surface water drainage run-on to the landfill and runoff from the working face must be controlled for 25-year rainfall events. Appropriate permits must be obtained for surface water discharges. Liquid wastes or wastes containing free liquids cannot be landfilled. Extensive landfill operating records must be maintained.
Liner design criteria	Geomembrane and soil liners or equivalent are required under most new landfill cells. Groundwater standards may be allowed as the basis for liner design in some states.
Groundwater monitoring	Groundwater monitoring wells must be installed at many landfills. Groundwater monitoring wells must be sampled at least twice per year. A corrective action program must be initiated where groundwater contamination is detected.

TABLE 12.1   Summary of U.S. Environmental Protection Agency Regulations for Municipal Solid Waste Landfills (*Continued*)

Item	Requirement
Closure and postclosure care	Landfill final cover must be placed within 6 months of closure. The type of cover is soil or geomembrane and must be less permeable than the landfill liner. Postclosure care and monitoring of the landfill must continue for 30 years.
Financial assurance	Sufficient financial reserves must be established during the site operating period to pay for closure and postclosure care amounts.

Source: 40 CFR Parts 257 and 258, 1991.

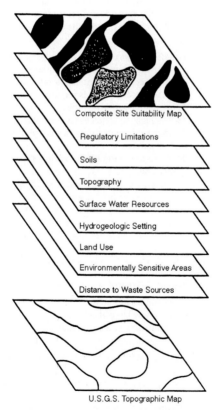

Composite Site Suitability Map

Regulatory Limitations

Soils

Topography

Surface Water Resources

Hydrogeologic Setting

Land Use

Environmentally Sensitive Areas

Distance to Waste Sources

U.S.G.S. Topographic Map

FIGURE 12.3   Overlay maps of various site criteria used in the screening of potential landfill sites. (*From Ref. 36.*)

12.2 GENERATION AND COMPOSITION OF LANDFILL GASES

A solid waste landfill can be conceptualized as a biochemical reactor, with solid waste and water as the major inputs, and with *landfill gas* and *leachate* as the principal outputs. Material stored in the landfill includes partially biodegraded organic material and the other inorganic waste materials originally placed in the landfill. Landfill gas control systems are employed to prevent unwanted movement of landfill gas into the atmosphere. The recovered landfill gas can be used to produce energy or can be flared under controlled conditions to eliminate the discharge of harmful constituents to the atmosphere. These topics are considered in greater detail below.

Generation of Landfill Gases

The generation of the principal landfill gases (CO_2 and CH_4), the variation in their rate of generation with time, and the sources of trace gases in landfills is considered in the following discussion.

Principal Landfill Gases. The generation of principal landfill gases is thought to occur in five more or less sequential phases as illustrated in Fig. 12.4. Each of these phases is described briefly below; additional details may be found in Refs. 6, 11, 12, 37, 40, and 41. Additional details on the anaerobic digestion process may be found in Ref. 20.

Phase I—Initial Adjustment. Phase I is the *initial adjustment phase* in which the organic biodegradable components in municipal solid waste begin to undergo bacterial decomposition soon after they are placed in a landfill. In Phase I, biological decomposition occurs under aerobic conditions because a certain amount of air is trapped within the landfill. The principal source of both the aerobic and the anaerobic organisms responsible for waste decomposition is the soil material that is used as a daily and final cover. Digested wastewater treatment plant sludge, disposed of in many MSW landfills, and recycled leachate are other sources of organisms.

Phase II—Transition Phase. In Phase II, identified as the *transition phase,* oxygen is depleted and anaerobic conditions begin to develop. As the landfill becomes anaerobic, nitrate and sulfate which can serve as electron acceptors in biological conversion reactions are often reduced to nitrogen gas and hydrogen sulfide. The onset of anaerobic conditions can be monitored by measuring the oxidation/reduction potential. Reducing conditions sufficient to bring about the reduction of nitrate and sulfate occur at about -50 to -100 mV. The production of methane occurs when the oxidation/reduction potential values are in the range from -150 to -300 mV. As the oxidation/reduction potential continues to decrease, members of the consortium of microorganisms responsible for the conversion of the organic material in MSW to methane and carbon dioxide begin the three-step process in which the complex organic material is converted to organic acids and other intermediate products as described in Phase III. In Phase II, the pH of the leachate, if formed, starts to drop due to the presence of organic acids and the effect of the elevated concentrations of CO_2 within the landfill (see Fig. 12.3).

Phase III—Acid Phase. In Phase III, known as the *acid phase,* the bacterial activity initiated in Phase II is accelerated with the production of significant amounts of organic acids and lesser amounts of hydrogen gas. The first step in the three-step process involves the enzyme-mediated transformation (hydrolysis) of higher-molecular mass compounds (e.g., lipids, organic polymers, and proteins) into compounds

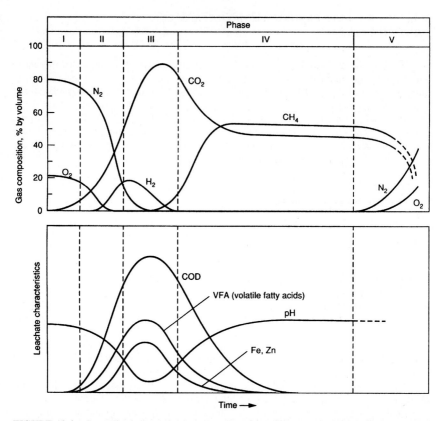

FIGURE 12.4   Generalized phases in the generation of landfill gases (I—Initial Adjustment, II— Transition Phase, III—Acid Phase, IV—Methane Fermentation, and V—Maturation Phase). (*Adapted from Refs. 12, 37, 40, and 41.*)

suitable for use by microorganisms as a source of energy and cell carbon. The second step in the process (acidogenesis) involves the bacterial conversion of the compounds resulting from the first step into lower molecular-weight intermediate compounds as typified by acetic acid (CH_3COOH) and small concentrations of fulvic and other more complex organic acids. CO_2 is the principal gas generated during Phase III. Smaller amounts of hydrogen gas (H_2) will also be produced. The microorganisms involved in this conversion, described collectively as nonmethanogenic, consist of facultative and obligate anaerobic bacteria. These microorganisms are often identified in the literature as *acidogens* or *acid formers.*

Because of the acids produced during Phase III, the pH of the liquids held within the landfill will drop. The pH of the leachate, if formed, will often drop to a value of 5 or lower because of the presence of the organic acids and the effect of the elevated concentrations of CO_2 within the landfill. The biochemical oxygen demand (BOD_5), the chemical oxygen demand (COD), and the conductivity of the leachate will increase significantly during Phase III due to the dissolution of the organic acids in the leachate. Also, because of the low pH values in the leachate, a number of inorganic constituents, principally heavy metals, will be solubilized during Phase III. Many essential nutrients are also removed in the leachate in Phase III. If leachate is not recycled, the essential nutrients will be lost from the system. It is important to note that if leachate is not formed, the conversion products produced during Phase III will remain within the landfill as sorbed constituents and in the water held by the waste as defined by the field capacity (see Sec. 12.4).

Phase IV—Methane Fermentation Phase. In Phase IV, known as the *methane fermentation phase,* a second group of microorganisms which converts the acetic acid and hydrogen gas formed by the acid formers in the acid phase to methane (CH_4) and CO_2 becomes more predominant. In some cases, these organisms will begin to develop toward the end of Phase III. The bacteria responsible for this conversion are strict anaerobes and are called methanogenic. Collectively, they are identified in the literature as *methanogens* or *methane formers.* In Phase IV, both methane and acid fermentation proceed simultaneously, although the rate of acid fermentation is considerably reduced.

Because the acids and the hydrogen gas produced by the acid formers have been converted to CH_4 and CO_2 in Phase IV, the pH within the landfill will rise to more neutral values in the range of 6.8 to 8. In turn, the pH of the leachate, if formed, will rise, and the concentration of BOD_5 and COD and the conductivity value of the leachate will be reduced. With higher pH values, fewer inorganic constituents are solubilized; as a result, the concentration of heavy metals present in the leachate will also be reduced.

Phase V—Maturation Phase. Phase V, known as the *maturation phase,* occurs after the readily available biodegradable organic material has been converted to CH_4 and CO_2 in Phase IV. As moisture continues to migrate through the waste, portions of the biodegradable material that were previously unavailable will be converted. The rate of landfill gas generation diminishes significantly in Phase V, because most of the available nutrients have been removed with the leachate during the previous phases and the substrates that remain in the landfill are slowly biodegradable. The principal landfill gases evolved in Phase V are CH_4 and CO_2. Depending on the landfill closure measures, small amounts of nitrogen and oxygen may also be found in the landfill gas. During maturation phase, the leachate will often contain higher concentrations of humic and fulvic acids, which are difficult to process further biologically.

Duration of Phases. The duration of the individual phases in the production of landfill gas will vary depending on the distribution of the organic components in landfill, the availability of nutrients, the moisture content of waste, moisture routing through the waste material, and the degree of initial compaction. For example, if several loads of brush are compacted together, the carbon/nitrogen ratio and the nutrient balance may not be favorable for the production of landfill gas. The generation of landfill gas will be retarded if sufficient moisture is not available. Increasing the density of the material placed in the landfill will decrease the availability of moisture to some parts of the waste and thus reduce the rate of bioconversion and gas production. Typical data on the percentage distribution of principal gases found in a newly completed landfill as a function of time are reported in Table 12.2.

Volume of Gas Produced. The general anaerobic transformation of the organic portion of the solid waste placed in a landfill can be described by the following equation.

$$\text{Organic matter} + H_2O + \text{nutrients} \rightarrow \text{new cells} + \begin{array}{c} \text{resistant} \\ \text{organic} \\ \text{matter} \end{array} + CO_2$$

$$+ CH_4 + NH_3 + H_2S + \text{heat} \tag{12.1}$$

Assuming methane, carbon dioxide, and ammonia are the principal gases that are produced, Eq. (12.1) can be represented with the following equation:[42]

TABLE 12.2   Typical Percentage Distribution of Landfill Gases during the First 48 Months

Time interval since cell completion, months	Average percent by volume		
	Nitrogen, N_2	Carbon dioxide, CO_2	Methane, CH_4
0–3	5.2	88	5
3–6	3.8	76	21
6–12	0.4	65	29
12–18	1.1	52	40
18–24	0.4	53	47
24–30	0.2	52	48
30–36	1.3	46	51
36–42	0.9	50	47
42–48	0.4	51	48

Source: Ref. 34.

$$C_aH_bO_cN_d \rightarrow nC_wH_xO_yN_z + mCH_4 + sCO_2 + rH_2O + (d - nx)NH_3 \quad (12.2)$$

where $s = a - nw - m$ and $r = c - ny - 2s$. The terms $C_aH_bO_cN_d$ and $C_wH_xO_yN_z$ are used to represent (on a molar basis) the composition of the organic material present at the start and the end of the process, respectively. If it is assumed that the biodegradable portion of the organic waste is stabilized completely, the corresponding expression is

$$C_aH_bO_cN_d + \left(\frac{4a - b - 2c + 3d}{4}\right)H_2O \rightarrow \left(\frac{4a + b - 2c - 3d}{8}\right)CH_4$$

$$+ \left(\frac{4a - b + 2c + 3d}{8}\right)CO_2 + dNH_3 \quad (12.3)$$

An important point to note is that the reaction given by Eq. (12.3) requires the presence of water. Landfills lacking sufficient moisture content have been found in a "mummified" condition, with decades-old newsprint still in readable condition. Hence, although the total amount of gas that will be produced from solid waste derives straightforwardly from the reaction stoichiometry, the rate and the period of time over which that gas production takes place will vary significantly with local hydrologic conditions.

The volume of the gases released during anaerobic decomposition can be estimated in a number of ways. For example, if the individual organic constituents found in MSW (with the exception of plastics) are represented with a generalized formula of the form $C_aH_bO_cN_d$, then the total volume of gas can be estimated by using Eq. (12.3). In general, the organic materials present in solid wastes can be divided into two classifications: (1) those materials that will decompose rapidly (3 months to 5 years) and (2) those materials that will decompose slowly (up to 50 years or more). The rapidly decomposable components of the organic fraction of MSW include food waste, newspaper, cardboard, and a portion of the yard wastes. The slowly decomposable components of the organic fraction of MSW include rubber, leather, the woody portions of yard waste, and wood. The theoretical amount of gas that would be expected under optimum conditions from the conversion of the rapidly and slowly biodegradable organic wastes in a landfill will vary from 12 to 15 and 14 to 16 ft^3/lb of biodegradable organic solids destroyed, respectively. However, because the biodegradable fraction of the organic waste depends to a large extent on the lignin content of the waste, not all of the organic matter will be degraded at the same rate. Widely varying rates have been observed in the field, with the typical values ranging between 1 and 4 ft^3/lb of MSW.

Variation in Gas Production with Time.   The overall rate at which the organic material in a landfill will be decomposed biologically will, as noted previously, depend on the distribution of the organic components in landfill, the availability of nutrients, the moisture content of waste, the routing of moisture through the fill, and the degree of initial compaction. Under normal conditions, the rate of decomposition of mixed organic wastes deposited in a landfill, as measured by gas production, reaches a peak within the first 2 years and then slowly tapers off, continuing in many cases for periods up to 25 years or more. If moisture is not added to the wastes in a well-compacted landfill, it is not uncommon to find materials in their original form years after they were buried.

The variation in the rate of gas produced from the anaerobic decomposition of the rapidly (5 years or less—some highly biodegradable wastes are decomposed within days of being placed in a landfill) and slowly (5 to 50 years) biodegradable organic materials in MSW can be modeled as shown in Fig. 12.5. As shown in Fig. 12.5, the yearly rates of decomposition for rapidly and slowly decomposable material are based

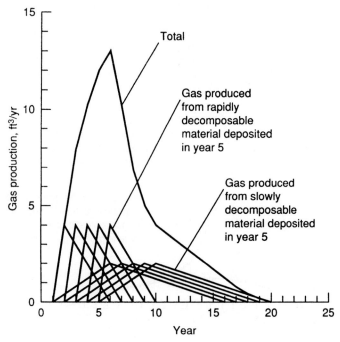

FIGURE 12.5   Graphical representation of gas production over a 5-year period from the rapidly and slowly decomposable organic materials placed in a landfill. (*From Ref. 52.*)

on a triangular gas production model in which the peak rate of gas production occurs 1 and 5 years, respectively, after gas production starts. Gas production is assumed to start at the end of the first full year of landfill operation. Although a triangular gas production model is used in Fig. 12.5, it should be noted that a variety of different models have been used, including a first-order model.

The total rate of gas production from a landfill in which wastes were placed for a period of 5 years is obtained graphically by summing the gas produced from the rapidly and slowly biodegradable portions of the MSW deposited each year (see Fig. 12.5). The total amount of gas produced corresponds to the area under the rate curve. As noted previously, in many landfills the available moisture is insufficient to allow for the complete conversion of the biodegradable organic constituents in the MSW. The optimum moisture content for the conversion of the biodegradable organic matter in MSW is on the order of 45 to 60 percent. Also, in many landfills, the moisture that is present is not distributed uniformly. When the moisture content of the landfill is limited, the gas production curve is more flattened out and is extended over a greater period of time. An example of the effect of reduced moisture content on the production of landfill gas is illustrated in Fig. 12.6. The production of landfill gas over extended periods of time is of great significance with respect to the management strategy to be adopted for postclosure maintenance. The goal of leachate recirculation is to enhance the rate of gas production and thus reduce the time required to stabilize the biodegradable organic matter in the landfill.

Sources of Trace Gases. Trace constituents in landfill gases have two basic sources. They may be brought to the landfill with the incoming waste or they may be produced by biotic and abiotic conversion reactions occurring within the landfill. Trace compounds mixed with the incoming waste are typically in liquid form, but tend to volatilize. As noted previously, the occurrence of significant concentrations of volatile organic compounds (VOCs) in landfill gas is associated with older landfills which accepted industrial and commercial wastes that contained VOCs and other organic compounds from which VOCs can be derived. In newer landfills, where the disposal

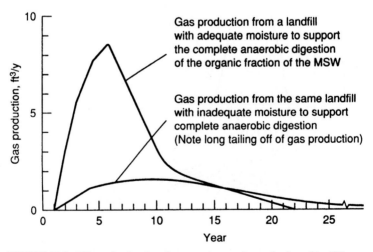

FIGURE 12.6   Effect of reduced moisture content on the production of landfill gas. (*From Ref. 52.*)

of hazardous waste has been banned, the concentrations of VOCs in the landfill gas have been reduced significantly.

Composition of Landfill Gas

Landfill gas comprises a number of gases that are present in large amounts (the principal gases) and in very small amounts (the trace gases). The principal gases are produced from the decomposition of the biodegradable organic fraction of MSW. Trace gases, although present in small percentages, may be toxic and could present risks to public health.

Principal Landfill Gas Constituents. Gases found in landfills include ammonia (NH_3), carbon dioxide (CO_2), carbon monoxide (CO), hydrogen (H_2), hydrogen sulfide (H_2S), methane (CH_4), nitrogen (N_2), and oxygen (O_2). The typical percentage distribution of the gases found in the landfill is reported in Table 12.3. Data on the molecular weight and density are presented in Table 12.4. As shown in Table 12.3, methane and carbon dioxide are the principal gases produced from the anaerobic decomposition of the biodegradable organic waste components in MSW. When methane is present in the air in concentrations between 5 and 15 percent, it is explosive. Because only limited amounts of oxygen are present in a landfill when methane concentrations reach this critical level, there is little danger that the landfill will explode. However, methane mixtures in the explosive range can be formed if landfill gas migrates off-site and is mixed with air. The concentration of these gases that may be expected in the leachate will depend on their concentration in the gas phase in contact with the leachate.

TABLE 12.3 Typical Constituents Found in and Characteristics of Landfill Gas

Component	Percent (dry volume basis)
Methane	45–60
Carbon dioxide	40–60
Nitrogen	2–5
Oxygen	0.1–1.0
Ammonia	0.1–1.0
Sulfides, disulfides, mercaptans, etc.	0–1.0
Hydrogen	0–0.2
Carbon monoxide	0–0.2
Trace constituents	0.01–0.6

Characteristic	Value
Moisture content	Saturated
Specific gravity	1.02–1.06
Temperature, °F	100–160
High heating value, Btu/std ft^3	475–550

Source: Adapted in part from Refs. 15, 26, and 37.

TABLE 12.4   Molecular Weight and Density of Gases Found in
Sanitary Landfill at Standard Conditions (0°C, 1 atm)

Gas	Formula	Molecular weight	Density g/L	Density lb/ft³
Air		28.97	1.2928	0.0808
Ammonia	NH_3	17.03	0.7708	0.0482
Carbon dioxide	CO_2	44.00	1.9768	0.1235
Carbon monoxide	CO	28.00	1.2501	0.0781
Hydrogen	H_2	2.016	0.0898	0.0056
Hydrogen sulfide	H_2S	34.08	1.5392	0.0961
Methane	CH_4	16.03	0.7167	0.0448
Nitrogen	N_2	28.02	1.2507	0.0782
Oxygen	O_2	32.00	1.4289	0.0892

Note: For ideal gas behavior, the density is equal to mp/RT where m is the
molecular weight of the gas, p is the pressure, R is the universal gas constant, and
T is the temperature.
Source: Adapted from Ref. 38.

Trace Landfill Gas Constituents.   Summary data on the concentration of trace compounds found in landfill gas samples from 66 landfills are reported in Table 12.5. In
another study conducted in England, gas samples were collected from three different
landfills and analyzed for 154 compounds. A total of 116 organic compounds was
found in landfill gas.[58] Many of the compounds found would be classified as VOCs.
The data presented in Table 12.5 are representative of the trace compounds found at
most MSW landfills. The presence of these gases in the leachate that is removed from
the landfill will depend on their concentration in the landfill gas in contact with the
leachate. It should be noted that the occurrence of significant concentrations of volatile
organic compounds in landfill gas is associated with older landfills which accepted
industrial and commercial wastes that contained VOCs. In newer landfills in which the
disposal of hazardous waste has been banned, the concentrations of VOCs in the landfill gas have been extremely low. Even at these low concentrations, some landfill operators must install emission control facilities for VOCs to achieve compliance with air-
quality protection standards imposed by health-based risk assessment.

Movement of Landfill Gases

Under normal conditions, gases produced in soils are released to the atmosphere by
means of molecular diffusion. In the case of an active landfill, the internal pressure is
usually greater than atmospheric pressure and landfill gas will be released by both
convective (pressure-driven) flow and diffusion. Other factors influencing the movement of landfill gases include the sorption of the gases into liquid or solid components
and the generation or consumption of a gas component through chemical reactions or
biological activity.

Movement of Principal Gases.   Although most of the methane escapes to the atmosphere, both methane and carbon dioxide have been found at concentrations up to 40
percent each, at lateral distances of up to 400 ft from the edges of unlined landfills.
Methane concentrations over 5 percent have been measured at a distance of 1000 feet.
If methane is allowed to migrate underground in an uncontrolled manner, it can accu-

TABLE 12.5 Typical Concentrations of Trace Compounds Found in Landfill Gas at 66 California MSW Landfills

Compound	Concentration, ppb by volume		
	Median	Mean	Maximum
Acetone	0	6,838	240,000
Benzene	932	2,057	39,000
Carbon dioxide	330,000,000	10,000,000	534,000,000
Chlorobenzene	0	82	1,640
Chloroform	0	245	12,000
1,1-dichloroethane	0	2,801	36,000
Dichloromethane	1,150	25,694	620,000
1,1-dichloroethene	0	130	4,000
Diethylene chloride	0	2,835	20,000
1,2-trans-dichloroethane	0	36	850
2,3-dichloropropane	0	0	0
1,2-dichloropropane	0	0	0
Ethylene bromide	0	0	0
Ethylene dichloride	0	59	2,100
Ethylene oxide	0	0	0
Ethyl benzene	0	7,334	87,500
Hydrogen sulfide	0	0	0
Hydrogen	0	0	4
Methane	440,000,000	70,000,000	740,000,000
Nitrogen	12	26	98
Oxygen	1	2	17
1,1,2-trichloroethane	0	0	0
1,1,1-trichloroethane	0	615	14,500
Trichloroethylene	0	2,079	32,000
Toluene	8,125	34,907	280,000
1,1,2,2-tetrachloroethane	0	246	16,000
Tetrachloroethylene	260	5,244	180,000
Vinyl chloride	1,150	3,508	32,000
Methyl ethyl ketone	0	3,092	130,000
Styrenes	0	1,517	87,000
Vinyl acetate	0	5,663	240,000
Xylenes	0	2,651	38,000

Source: Adapted from Ref. 5.

mulate (because its specific gravity is less than that of air) below buildings or in other enclosed spaces at, or close to, a sanitary landfill. With proper venting, methane, with the exception of the fact that it is a greenhouse gas, should not pose a problem by itself. Odorous compounds and VOCs mixed with the methane may lead to odor complaints and the need for emission controls. Carbon dioxide, on the other hand, which is about 1.5 times as dense as air and 2.8 times as dense as methane (see Table 12.3), tends to move toward the bottom of the landfill. As a result, the concentration of carbon dioxide in the lower portions of a landfill may be high for years.

Upward Migration of Landfill Gas. The principal gases, methane and carbon dioxide, can be released through the landfill cover into the atmosphere by convection and diffusion. The diffusive flow through the cover can be estimated by using Eq. (12.4):

$$N_A = -D\alpha^{4/3}\, \frac{(C_{A_{atm}} - C_{A_{fill}})}{L} \tag{12.4}$$

where N_A = gas flux of compound A, g/cm² • s (lb • mol/ft² • d)
D = effective diffusion coefficient, cm²/s (ft²/d)
α = total porosity, cm³/cm³ (ft³/ft³)
$C_{A_{atm}}$ = concentration of compound A at the surface of the landfill cover, g/cm³ (lb • mol/ft³)
$C_{A_{fill}}$ = concentration of compound A at bottom of the landfill cover, g/cm³ (lb • mol/ft³)
L = depth of the landfill cover, cm (ft)

Typical values for the coefficient of diffusion for methane and carbon dioxide are 0.20 cm²/s (18.6 ft²/d) and 0.13 cm²/s (12.1 ft²/d), respectively.[28] It is also common to assume dry soil conditions, thus $\alpha_{gas} = \alpha$. Assuming dry soil conditions introduces a safety factor in that any infiltration of water into the landfill cover will reduce the gas-filled porosity and thus reduce the vapor flux from the landfill. Typically porosity values for different types of clay vary from 0.010 to 0.30.

Downward Migration of Landfill Gas. Ultimately, carbon dioxide, because of its density, can accumulate in the bottom of a landfill. If a clay or soil liner is used, the carbon dioxide can move from there downward primarily by diffusive transport through the liner and the underlying formation until it reaches the groundwater (note the movement of carbon dioxide can be limited with the use of a geomembrane liner). Because carbon dioxide is readily soluble in water, it usually lowers the pH, which in turn can increase the hardness and mineral content of the groundwater through solubilization.

Movement of Trace Gases. In a manner similar to that outlined above for the principal gases, the movement of trace gases due to diffusion can be estimated using the following equation:

$$N_i = \frac{D\alpha^{4/3}\,(C_{i(s)}W_i)}{L} \tag{12.5}$$

Estimated values of the diffusion coefficient D, for 12 trace compounds are reported in Table 12.6 for temperatures varying from 0 to 50°C. Porosity values typically vary from 0.010 to 0.30 for different types of clay. The term $C_{i(s)}W_i$ corresponds to the concentration of the compound in question just below the cover at the top of the landfill. If the value of the term $C_{i(s)}W_i$ is to be estimated in the field, measurements should be taken by inserting a gas probe through the landfill cover, to a point just beyond the bottom of the cover, and recording both the concentration of the compound and the temperature at this point in the landfill. By obtaining actual field measurements, an estimate of the average emission rate can be obtained very quickly. If field measurements are not available, the value of the term $C_{i(s)}W_i$ can be estimated using the data given in Table 12.6 for $C_{i(s)}$ and a value of 0.001 as an estimate for W_i.

Active and Passive Control of Landfill Gases

The release of landfill gases is controlled to reduce atmospheric emissions; to minimize the release of odorous emissions; to minimize subsurface gas migration, in unlined landfills; and to allow for the recovery of energy from methane. Control sys-

TABLE 12.6 Selected Physical Properties for Twelve Trace Compounds Found in Landfills

Compound	0°C D*	vp†	C_s‡	10°C D	vp	C_s	20°C D	vp	C_s	30°C D	vp	C_s	40°C D	vp	C_s	50°C D	vp	C_s
Ethyl benzene	0.052	2.0	12.48	0.055	3.9	23.47	0.059	7.3	42.44	0.062	13	73.08	0.066	22	119.7	0.069	36	189.9
Toluene	0.056	6.7	36.26	0.060	12	62.65	0.064	22	110.9	0.068	37	180.4	0.073	59	278.5	0.077	92	420.9
Tetrachloroethene	0.053	4.1	39.95	0.057	7.9	74.27	0.061	15.6	127.1	0.065	24	210.7	0.069	40	340.0	0.073	63	581.9
Benzene	0.066	27	123.9	0.070	47	208.1	0.075	76	325.0	0.081	122	504.6	0.086	185	740.7	0.091	274	1063
1,2-Dichloroethane	0.063	24	139.6	0.068	41	230.0	0.072	62	363.0	0.077	107	560.7	0.082	164	831.9	0.088	243	1194
Trichloroethene	0.059	20	154.5	0.063	36	268.4	0.067	60	424.8	0.072	94	654.5	0.077	146	984.1	0.082	217	1417
1,1,1-Trichloroethane	0.058	36	282.2	0.062	61	461.3	0.067	100	715.9	0.071	153	1061	0.076	231	1580	0.081	338	2240
Carbon tetrachloride	0.058	32	289.3	0.062	54	470.9	0.066	90	741.2	0.071	138	1124	0.075	209	1648	0.080	308	2353
Chloroform	0.065	61	427.9	0.070	100	676.7	0.075	160	1026	0.080	240	1517	0.085	354	2166	0.090	508	3012
1,2-Dichloroethene	0.077	110	626.7	0.082	175	961.8	0.087	269	1428	0.092	399	2048	0.097	576	2862	0.102	810	3901
Dichloromethane	0.074	155	773.6	0.080	242	1165	0.085	349	1702	0.091	536	2410	0.097	763	3322	0.103	1060	4472
Vinyl chloride	0.080	1280	4701	0.085	1810	6413	0.091	2548	8521	0.098	3350	11090	0.104	4410	14130	0.110	5690	17660

*Diffusion coefficient, cm²/s.

†Vapor pressure, mm Hg.

‡Saturation vapor concentration, g/m³.

Source: From Ref. 18.

12.21

tems can be classified as active or passive. In active gas control systems, energy in the form of an induced vacuum is used to control the flow of gas. In passive gas control systems, the pressure of the gas which is generated within the landfill serves as the driving force for the movement of the gas. For both the principal and trace gases, passive control during times when the principal gases are being produced at a high rate can be achieved by providing paths of lower permeability to guide the gas flow in the desired direction. A gravel-packed trench, for example, can serve to channel the gas to a flared vent system. When the production of the principal gases is limited, passive controls are not very effective because molecular diffusion will be the predominant transport mechanism. However, at this stage in the life of the landfill it may not be as important to control the residual emission of the methane in the landfill gas. Control of VOC emissions may, however, necessitate the use of both active and passive gas control facilities.

Active Control of Landfill Gas.   Both vertical and horizontal gas wells have been used for the extraction of landfill gas from within landfills. In some installations both types of wells have been used. The management of the condensate that forms when landfill gas is extracted is also an important element in the design of gas recovery systems.

   Vertical Gas Extraction Wells.   A typical gas recovery system using vertical gas extraction wells is illustrated in Fig. 12.7. The wells are spaced so that their radii of influence overlap. For completed landfills, the radius of influence for gas wells is sometimes determined by conducting gas drawdown tests in the field. Typically, an extraction well is installed along with gas probes at regular distances from the well, and the vacuum within the landfill is measured as a vacuum is applied to the extraction well. Both short-term and long-term extraction tests can be conducted. Because the volume of gas produced will diminish with time, some designers prefer to use a uniform well spacing and to control the radius of influence of the well by adjusting the vacuum at the well head. For deep landfills, with a composite cover containing a geomembrane, a 150 to 200 ft spacing is common for landfill gas extraction wells. In landfills with clay and/or soil covers, a closer spacing (e.g., 100 ft) may be required to

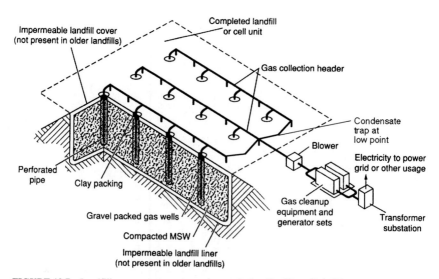

FIGURE 12.7   Landfill gas recovery system using vertical wells. (*From Ref. 52.*)

avoid pulling atmospheric gases into the gas recovery system. The entry of air introduces oxygen into the landfill, which may affect the methane-producing bacteria, and can, by spontaneous combustion, result in the development of an internal landfill fire.

Vertical gas extraction wells are usually installed after the landfill or portions of the landfill have been completed. In older landfills, vertical wells are installed both to recover energy and to control the movement of gases to adjacent properties. The typical extraction well design consists of 4- to 6-in pipe casing [usually polyvinylchloride (PVC) or polyethylene (PE)] set in an 18- to 36-in borehole (see Fig. 12.8). The bot-

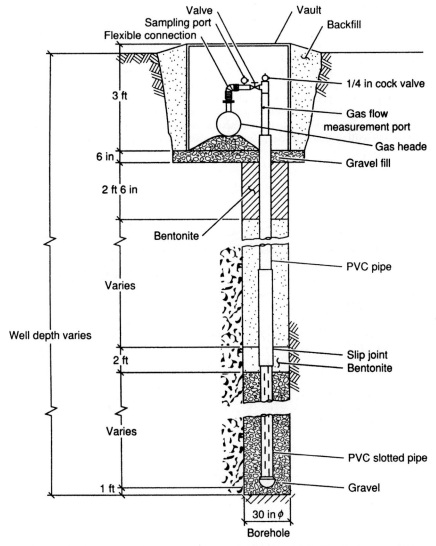

FIGURE 12.8 Typical landfill gas extraction well. (*Courtesy of California Integrated Waste Management Board.*)

tom third to half of the casing is perforated and set in a gravel backfill. The remaining length of the casing is not perforated and is backfilled with soil and sealed with a clay.[47] Landfill gas recovery wells are typically designed to penetrate 80 percent of the depth of the waste in the landfill, because their radius of influence will extend to the bottom of the landfill. However, to allay the public's fear concerning the escape of landfill gas, some designers now place gas recovery wells all the way to the bottom of the landfill. In instances where effective wells cannot be developed inside the landfill due to well clogging or small radii of influence, wells may be placed in the ground immediately adjacent to the landfill. The available vacuum in the collection manifold at the well head is typically 10 in of water.

   Horizontal Gas Extraction Wells.   An alternative to vertical gas recovery wells is horizontal wells. The use of horizontal wells was pioneered and developed by the County Sanitation Districts of Los Angeles County (see Figs. 12.9 and 12.10). The use of vertical perimeter wells in conjunction with horizontal gas extraction wells is also illustrated in Figs. 12.9 and 12.10. Horizontal wells are installed after two or more lifts have been completed. The horizontal gas extraction trench is excavated in the solid waste by a backhoe. The trench is then backfilled halfway with gravel, and a perforated pipe with open joints is installed (see Fig. 12.11). The trench is then filled with gravel and capped with solid waste. By using a gravel-filled trench and a perforated pipe with open joints, the gas extraction trench remains functional even with the differential settling that will occur in the landfill with the passage of time (see Fig.

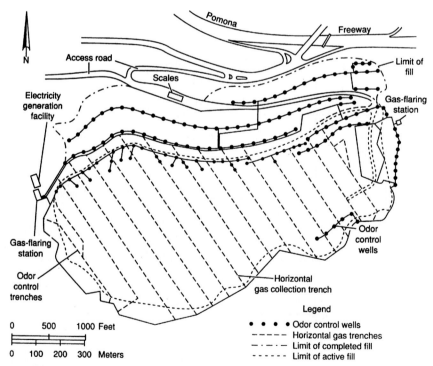

FIGURE 12.9   Plan view of gas collection facilities at Puente Hills landfill. (*Courtesy County Sanitation Districts of Los Angeles County.*)

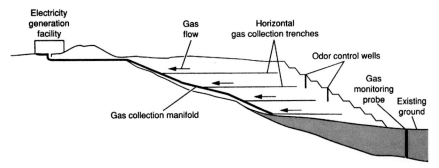

FIGURE 12.10 Sectional view through Puente Hills landfill showing horizontal gas collection trenches. (*Courtesy County Sanitation Districts of Los Angeles County.*)

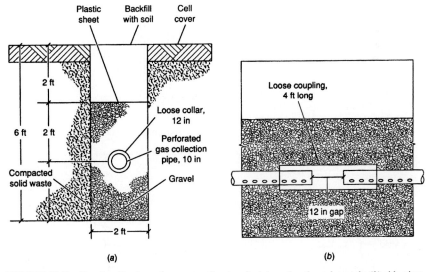

FIGURE 12.11 Details of horizontal gas extraction trench: (*a*) section through trench; (*b*) side view. (*Courtesy County Sanitation Districts of Los Angeles County.*)

12.11*b*). The horizontal trenches are installed at approximately 80-ft vertical intervals and on 200-ft horizontal intervals.[48]

Condensate Management. Condensate forms when the warm landfill gas is cooled as it is transported in the header leading to the blower. Gas collection headers are usually installed with a minimum slope of 3 percent to allow for differential settlement. Because headers are constructed in sections that slope up and down throughout the extent of the landfill, condensate traps are installed at the low spots in the line (see Fig. 12.7). A typical condensate trap in which the condensate is collected in a holding tank is shown in Fig. 12.12. Condensate from the holding tanks is pumped out periodically and recirculated with leachate to the landfill, transported to an authorized disposal facility, treated on-site prior to disposal, or discharged to a local sewer. In some states, the direct return of condensate to the landfill is allowed.

Passive Control of Landfill Gas.   One of the most common passive methods for the control of landfill gases is based on the fact that the lateral migration of landfill gas can be reduced by relieving gas pressure within the landfill interior. For this purpose, vents are installed through the final landfill cover extending down into the solid waste mass (see Fig. 12.13). Gas moves through the vent system to the landfill exterior. Due to relatively low gas pressures, many landfills, equipped with passive vents and capped with

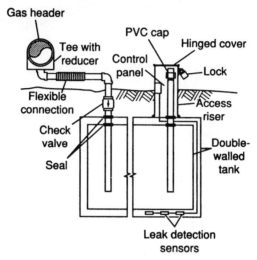

FIGURE 12.12   Typical condensate trap with holding tank. (*From Ref. 52.*)

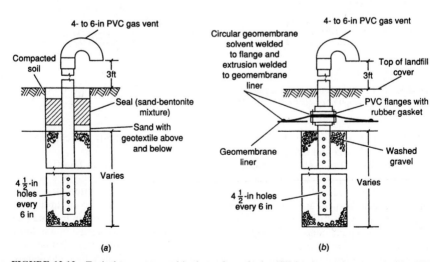

FIGURE 12.13   Typical gas vents used in the surface of a landfill for the passive control of landfill gas: (*a*) gas vent for landfill with a cover that does not contain a geomembrane liner; (*b*) gas vent for a landfill with a cover that contains a synthetic membrane liner. (*From Ref. 52.*)

soil covers, have experienced vegetative stress on the landfill cover or underground gas migration outside the landfill, indicating that only a portion of the gas is flowing through the passive vents. Where landfills are located near occupied buildings active control systems are usually necessary to achieve adequate migration control.

If the methane in the venting gas is of sufficient concentration, several vents can be connected together and equipped with a gas burner (see Fig. 12.14). Where waste gas burners are used it is recommended that the well penetrate into the upper waste cells. The height of the waste burner can vary from 10 to 20 ft above the completed fill. The burner can be ignited either by hand or by a continuous pilot flame. To derive maximum benefit from the installation of a waste gas burner, a pilot flame is recommended. It should be noted, however, that passive vents with burners may not achieve the VOC and odor destruction efficiencies that are required by many urban air quality control agencies, and, thus, their use is not considered good practice. Gas burners are considered later in this section.

Management of Landfill Gas

Typically, landfill gases that have been recovered from an active landfill are either flared or used for the recovery of energy in the form of electricity, or both. More recently, the separation of the carbon dioxide from the methane in landfill gas has been suggested as an alternative to the production of heat and electricity.

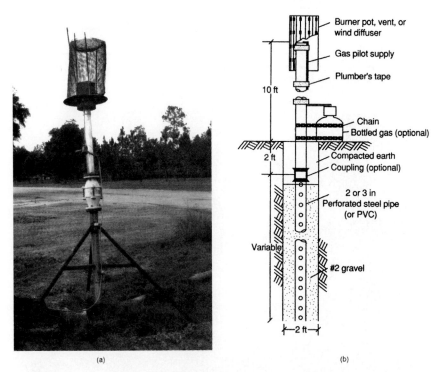

(a)   (b)

FIGURE 12.14 Typical candlestick-type waste gas burner used to flare landfill gas from a well vent or several interconnected well vents. (*a*) Without pilot flame; (*b*) with pilot flame.

Flaring of Landfill Gases.   A common method of treatment for landfill gases is thermal destruction in which the methane and any other trace gases (including VOCs) are combusted in the presence of oxygen to CO_2, sulfur dioxide (SO_2), oxides of nitrogen, and other related gases. The thermal destruction of landfill gases is usually accomplished in a specially designed flaring facility (see Figs. 12.15 and 12.16). Because of concerns over air pollution, modern flaring facilities are designed to meet rigorous

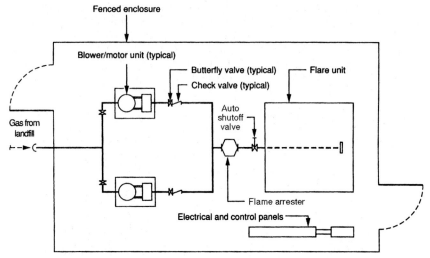

FIGURE 12.15   Schematic layout of blower/flare station for the flaring of landfill gas. (*Courtesy of California Integrated Waste Management Board.*)

FIGURE 12.16   Large array of ground effects flares used to flare landfill gas.

operating specifications to ensure effective destruction of VOCs and other similar compounds that may be present in the landfill gas. For example, a typical requirement might be a minimum combustion temperature of 1500°F and a residence time of 0.3 to 0.5 s, along with a variety of controls and instrumentation in the flaring station. Typical requirements for a modern flaring facility are summarized in Table 12.7. Where the landfill gas contains less than 15 percent methane, supplemental natural gas or propane may need to be supplied to the flare to sustain combustion. Installation of a carbon filter is an alternative approach to flaring for control of VOCs.

Landfill Gas Energy Recovery Systems.   Landfill gas is usually converted to electricity (see Fig. 12.17). In smaller installations, it is common to use dual-fuel internal combustion piston engines (see Fig. 12.17*a*). In larger installations, the use of turbines is common (see Fig. 12.17*b*). Where piston-type engines are used, the landfill gas must be processed to remove as much moisture as possible to limit damage to the cylinder heads. If the gas contains hydrogen sulfide (H_2S), the combustion temperature must be controlled carefully to avoid corrosion problems. Alternatively, the landfill gas can be passed through a scrubber containing iron shavings, or other proprietary scrubbing devices, to remove the H_2S before the gas is combusted.

Combustion temperatures will also be critical where the landfill gas contains VOCs released from wastes placed in the landfill before the disposal of hazardous waste was banned in municipal landfills. The typical service cycle for dual-fuel engines running on landfill gas varies from 3000 to 10,000 h before the engine must be overhauled. In most installations, low-Btu landfill gas is compressed under high pressure so that it can be used more effectively in the gas turbine. The typical service cycle for gas turbines running on landfill gas is approximately 10,000 h.

TABLE 12.7   Important Design Elements for Enclosed Ground Level Landfill Gas Flares

Item	Comments
Automatically controlled combustion air louvers	Used to control the amount of combustion air and the temperature of the flame
Automatic pilot restart system	To ensure continuous operation
Failure alarm with an automatic isolation system	The alarm and isolation system is used to isolate the flare from the landfill gas supply line, shut off the blower, and notify a responsible party of the shutdown.
Heat shield	A heat shield should be provided around the top of the flare shroud for use during source testing.
Source test ports with adequate and safe access provided	Test ports used for sampling
Temperature indicator and recorder	Used to measure and record gas temperature in the flare stack. Whenever the flare is in operation, a temperature of 1500°F or greater must be maintained in the stack as measured by the temperature indicator 0.3 s after passing through the burner.
View ports	A sufficient number of view ports must be available to allow visual inspection of the temperature sensor location within the flare.

Source: Adapted from Ref. 47.

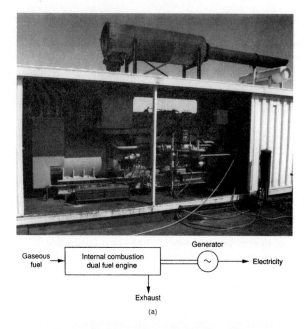

(a)

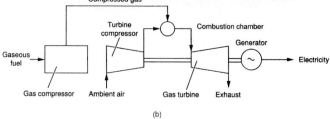

(b)

FIGURE 12.17 Schematic flow diagrams for the recovery of energy from gaseous fuels. (*a*) Using internal combustion engine; (*b*) using a gas turbine.

Other energy recovery methods are also available or under development. Landfill gas can be used to fuel utility boilers at institutional or industrial facilities located near the landfill. After scrubbing, the landfill gas is piped directly from the landfill to the boiler. Another option implemented by some municipalities is to operate vehicles in their service fleet with compressed landfill gas. Fuel cell technology is being developed in an effort to achieve higher conversion efficiencies and lower emissions.

Gas Purification and Recovery.   Where there is a potential use for the CO_2 contained in the landfill gas, the CH_4 and CO_2 in landfill gas can be separated. The separation of the CO_2 from the CH_4 can be accomplished by physical adsorption, chemical adsorption, or membrane separation. In physical and chemical adsorption, one component is adsorbed preferentially by a suitable solvent. Membrane separation involves the use of a semipermeable membrane to remove the CO_2 from the methane. Semipermeable membranes have been developed that allow CO_2, H_2S, and H_2O to pass while the CH_4 molecule is retained. Membranes are available as flat sheets or as hollow fibers.

12.3   FORMATION, COMPOSITION, AND MANAGEMENT OF LEACHATE

Leachate may be defined as liquid that has percolated through solid waste and has extracted dissolved or suspended materials. In most landfills, leachate is composed of the liquid that has entered the landfill from external sources, such as surface drainage and rainfall and the liquid produced from the decomposition of the wastes, if any.

Formation of Leachate in Landfills

The potential for the formation of leachate can be assessed by preparing a water balance on the landfill.[13] The water balance involves summing the amounts of water entering the landfill and subtracting the amount of water consumed in chemical reactions and the quantity leaving as water vapor. The potential leachate quantity is the quantity of water in excess of the moisture-holding capacity of the landfill material.

Preparation of Landfill Water Balance.   The components that make up the water balance for a landfill cell are illustrated in Fig. 12.18. As shown in the figure, the principal components involved in the water balance are (1) the water entering the landfill cell from above, the moisture in the solid waste, the moisture in the cover material, and the moisture in the sludge, if the disposal of sludge is allowed and (2) the water leaving the landfill as part of the landfill gas, as saturated water vapor in the landfill gas, and as leachate.

The terms that comprise the water balance can be put into equation form:

$$\Delta S_{SW} = W_{SW} + W_{TS} + W_{CM} + W_{A(R)} - W_{LG} - W_{WV} - W_E + W_{B(L)} \qquad (12.6)$$

where ΔS_{SW} = change in the amount of water stored in solid waste in landfill, lb/yd^3
$\quad\quad\quad W_{SW}$ = water (moisture) in incoming solid waste, lb/yd^3
$\quad\quad\quad W_{TS}$ = water (moisture) in incoming treatment plant sludge, lb/yd^3
$\quad\quad\quad W_{CM}$ = water (moisture) in cover material, lb/yd^3

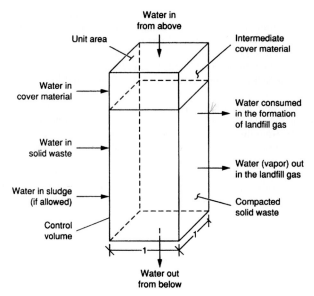

FIGURE 12.18 Definition sketch for water balance used to assess leachate formation in a landfill (*From Ref. 52.*)

$W_{A(R)}$ = water from above (for upper landfill layer water from above corresponds to rainfall or water from snowfall), lb/yd^2

W_{LG} = water lost in the formation of landfill gas, lb/yd^3

W_{WV} = water lost as saturated water vapor with landfill gas, lb/yd^3

W_E = water lost due to surface evaporation, lb/yd^2

$W_{B(L)}$ = water leaving from bottom of element (for the cell placed directly above a leachate collection system, water from bottom corresponds to leachate), lb/yd^2

Water in Solid Waste. Water entering the landfill with the waste materials is the moisture that is inherent in the waste material and moisture that has been absorbed from the atmosphere or from rainfall where the storage containers are not sealed properly. In dry climates, some of the inherent moisture contained in the waste can be lost, depending on the conditions of the storage. The moisture content of residential and commercial MSW varies from about 15 to 35 percent, depending on the season.

Water in Cover Material. The amount of water entering with the cover material will depend on the type and source of the cover material and the season of the year. The maximum amount of moisture that can be contained in the cover material is defined by the field capacity (FC) of the material. The field capacity is defined as the liquid which remains in the pore space subject to the pull of gravity. Typical values for soils range from 6 to 12 percent for sand to 23 to 31 percent for clay loams (see Table 12.11).

Water from Above. For the upper layer of the landfill, the water from above corresponds to the precipitation that has percolated through the cover material. For the layers below the upper layer, water from above corresponds to the water that has percolated through the solid waste above the layer in question. In landfills with leachate recirculation the water from above will also include the recirculated leachate. One of the most critical aspects in the preparation of a water balance for a landfill is to determine the amount of the rainfall that actually percolates through the landfill cover layer. Where a geomembrane is not used, the amount of rainfall that percolates through the landfill cover can be determined using the latest version of the Hydrologic Evaluation of Landfill Performance (HELP) model.[44,45] A simplified method for estimating the amount of percolation that can be expected is presented in Sec. 12.5.

Water Lost in the Formation of Landfill Gas. Water is consumed during the anaerobic decomposition of the organic constituents in MSW. The amount of water consumed by the decomposition reaction can be estimated by using Eq. (12.3). The amount of water consumed per cubic foot of gas produced is typically in the range from 0.012 to 0.015 lb H_2O/ft^3 of gas

Water Lost as Water Vapor. Landfill gas usually is saturated in water vapor. The quantity of water vapor escaping the landfill is determined by assuming the landfill gas is saturated with water vapor. The numerical value for the mass of water vapor contained per cubic foot of landfill gas at 90°F is about 0.0022 lb H_2O/ft^3 landfill gas.

Water Lost Due to Evaporation. There will be some loss of moisture to evaporation as the waste is being landfilled. The amounts are not large, and are often ignored. The decision to include these variables in the water balance analysis will depend on local conditions.

Water Leaving from Below. Water leaving from the bottom of the *first* cell of the landfill is termed *leachate*. As noted previously, water leaving the bottom of the second and subsequent cells corresponds to the water entering from above for the cell below the cell in question.

Field Capacity of Solid Waste. Water entering the landfill that is not consumed and does not exit as water vapor may be held within the landfill or may appear as leachate. Both the waste material and the cover material are capable of holding water against the pull of gravity. The quantity of water that can be held against the pull of gravity is referred to as field capacity (FC). The potential quantity of leachate is the amount of moisture within the landfill in excess of the landfill field capacity. The field capacity, which varies with the overburden weight, can be estimated using the following equation:[22,23]

$$FC = 0.6 - 0.55 \left[\frac{W}{10,000 + W} \right] \qquad (12.7)]$$

where FC = field capacity (i.e., the fraction of water in the waste based on the dry
          weight of the waste)
       W = overburden mass calculated at the mid height of the waste in the lift in
          question, lb

The landfill water balance is prepared by adding the mass of water entering a unit area of a particular layer of the landfill during a given time increment to the moisture content of that layer at the end of the previous time increment, and subtracting the mass of water lost from the layer during the current time increment. The result is referred to as the available water in the current time increment for the particular layer of the landfill. To determine whether any leachate will form, the field capacity of landfill is compared to the amount of water that is present. If the field capacity is less than the amount of water present, then leachate will be formed.

In general, it has been found that the quantity of leachate is a direct function of the amount of external water entering the landfill. In fact, if a landfill is constructed properly, the production of leachate can be reduced substantially. When wastewater treatment plant sludge is added to the solid wastes to increase the amount of methane produced, leachate control facilities must be provided. In those landfills where leachate treatment may be required, the most common approach is to transport the leachate to a municipal wastewater treatment plant.

Composition of Leachate

When water percolates through solid wastes that are undergoing decomposition, both biological materials and chemical constituents are leached into solution. Typical data on the characteristics of leachate are reported in Table 12.8 for both new and mature landfills. Because the range of the observed concentration values for the various constituents reported in Table 12.8 is rather large, especially for new landfills, great care should be exercised in using the typical values that are given.

Variations in Leachate Composition. It should be noted that the chemical composition of leachate will vary greatly depending on the age of landfill and the history of events proceeding the time of sampling. For example, if a leachate sample is collected during the acid phase of decomposition (see Fig. 12.4), the pH value will be low and

TABLE 12.8 Typical Data on the Composition of Leachate from New and Mature Landfills

	Value, mg/L*		
	New landfill (less than 2 years)		Mature landfill (greater than 10 years)
Constituent	Range†	Typical‡	
BOD$_5$ (5-day biochemical oxygen demand)	2,000–30,000	10,000	100–200
TOC (total organic carbon)	1,500–20,000	6,000	80–160
COD (chemical oxygen demand)	3,000–60,000	18,000	100–500
Total suspended solids	200–2,000	500	100–400
Organic nitrogen	10–800	200	80–120
Ammonia nitrogen	10–800	200	20–40
Nitrate	5–40	25	5–10
Total phosphorus	5–100	30	5–10
Ortho phosphorus	4–80	20	4–8
Alkalinity as CaCO$_3$	1,000–10,000	3,000	200–1000
pH	4.5–7.5	6	6.6–7.5
Total hardness as CaCO$_3$	300–10,000	3,500	200–500
Calcium	200–3,000	1,000	100–400
Magnesium	50–1,500	250	50–200
Potassium	200–1,000	300	50–400
Sodium	200–2,500	500	100–200
Chloride	200–3,000	500	100–400
Sulfate	50–1,000	300	20–50
Total iron	50–1200	60	20–200

*Except pH, which is unitless.

†Representative range of values. Higher maximum values have been reported in the literature for some of the constituents.

‡Typical values for new landfills will vary with the metabolic state of the landfill.

Source: Developed from Refs. 2, 7, 8, 10, 49, and 50.

the concentrations of BOD_5, TOC, COD, nutrients, and heavy metals will be high. If, on the other hand, a leachate sample is collected during the methane fermentation phase (see Fig. 12.4), the pH will be in the range from 6.5 to 7.5, and the BOD_5, TOC, COD, and nutrient concentration values will be significantly lower. Similarly the concentrations of heavy metals will be lower because most metals are less soluble at neutral pH values. The pH of the leachate will not only depend on the concentration of the acids that are present, it will also depend on the partial pressure of the CO_2 in the landfill gas which is in contact with the leachate.

The biodegradability of the leachate will also vary with time. Changes in the biodegradability of the leachate can be monitored by checking the BOD_5/COD. Initially, the BOD_5/COD ratios will be in the range of 0.5. Ratios in the range from 0.4 to 0.6 are taken as an indication that the organic matter in the leachate is readily biodegradable. In mature landfills, the BOD_5/COD ratio is often in the range of 0.05 to 0.2. The reason that the BOD_5/COD ratio drops is that the leachate from mature landfills typically contains humic and fulvic acids which are not readily biodegradable.

Because of the variability in the characteristics of leachate, the design of leachate treatment systems is complicated. For example, the type of treatment plant designed to treat a leachate with the characteristics reported for a new landfill would be quite different from one designed to treat the leachate from a mature landfill. The problem of analysis is complicated further by the fact that the leachate that is being generated at any point in time is a mixture of leachate derived from solid waste of different ages.

Trace Compounds. The presence of trace compounds (some of which may pose health risks), will depend on the concentration of these compounds in the gas phase within the landfill. The expected concentrations can be estimated using Henry's law. It is interesting to note that, as more communities and operators of landfills institute programs to limit the disposal of hazardous wastes with MSW, the quality of the leachate from new landfills is improving with respect to the presence of trace constituents.

Movement of Leachate in Unlined Landfills

Under normal conditions, leachate is found in the bottom of landfills. From there its movement in unlined landfills is through the underlying strata, although some lateral movement may also occur, depending on the characteristics of the surrounding material. As leachate percolates through the underlying strata, many of the chemical and biological constituents originally contained in it will be removed by the filtering and adsorptive action of the material composing the strata. In general, the extent of this action depends on the characteristics of the soil, especially the clay content. Because of the potential risk involved in allowing leachate to percolate to the groundwater, best practice calls for its elimination or containment.

Control of Leachate in Landfills

Landfill liners are now commonly used to limit or eliminate the movement of leachate and landfill gases from the landfill site. To date (1993), the use of clay as a liner material has been the favored method of reducing or eliminating the seepage (percolation) of leachate from landfills. Clay is favored for its ability to adsorb and retain many of the chemical constituents found in leachate and for its resistance to the flow of leachate. However, the use of combination composite geosynthetic and clay liners is gaining in popularity, especially because of the resistance afforded by geomembranes to the movement of both leachate and landfill gases. Typical specifications for geomembrane liners are given in Table 12.9.

Liner Systems for MSW. The objective in the design of landfill liners is to minimize the infiltration of leachate into the subsurface soils below the landfill to substantially reduce the potential for the groundwater contamination. A number of liner designs have been developed to minimize the movement of leachate into the subsurface below the landfill. Some of the many types of liner designs that have been proposed are illustrated in Fig. 12.19. In the multilayer landfill liner designs illustrated in Fig. 12.19, each of the various layers has a specific function. For example, in Fig. 12.19a the clay layer and the geomembrane serve as a composite barrier to the movement of leachate and landfill gas. The sand layer serves as a collection and drainage layer for any leachate that may be generated within the landfill. The geotextile layer is used to minimize the intermixing of the soil and sand layers. The final soil layer is used to protect the drainage and barrier layers. A modification of the liner design shown in Fig. 12.19a involves the installation of leachate collection pipes in the leachate collection layer. Composite liner designs employing a geomembrane and clay layer provide more protection and are hydraulically more effective than either type of liner alone.

TABLE 12.9 Performance Tests Used to Measure Properties of Synthetic Liners and Typical Values

Test	Test method	Typical values
Chemical resistance:		
Resistance to chemical waste mixtures	EPA method 9090	10% tensile strength change over 120 days
Resistance to pure chemical reagents	ASTM D543	10% tensile strength change over 7 days
Durability:		
Carbon black percent	ASTM D1603	2%
Carbon black dispersion	ASTM D3015	A-1
Accelerated heat aging	ASTM D573, D1349	Negligible strength change after 1 month at 110°C
Strength category:		
Tensile properties	ASTM D638, Type IV;	2400 lb/in^2
Tensile strength at yield	dumbbell 2 in/min	4000 lb/in^2
Tensile strength at break		15%
Elongation at yield		700%
Elongation at yield		
Stress cracking resistance:		
Environmental stress crack resistance	ASTM D1693, condition C	1500 h
Toughness:		
Tear resistance initiation	ASTM D1004 die C	45 lb
Puncture resistance	FTMS 101B, method 2031	230 lb
Low-temperature brittleness	ASTM D746, procedure B	$-94°F$

Source: Adapted from Refs. 2 and 56.

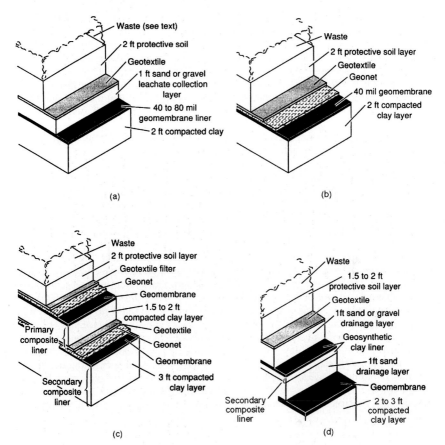

FIGURE 12.19 Typical landfill liners. (*a, b*) Single composite barrier types; (*c, d*) double composite barrier types. Note in the double liner systems the first composite liner is often identified as the primary liner or as the leachate collection system while the second composite liner is identified as the leachate detection layer. Leachate detection probes are normally placed between the first and second liners. (*From Ref. 52.*)

In Fig. 12.19*b*, a specifically designed open weave plastic mesh (geonet) and geo-textile filter cloth are placed over the geomembrane, which, in turn, is placed over a compacted clay layer. A protective soil layer is placed above the geotextile. The geonet and the geotextile function together as the drainage layer to convey leachate to the leachate collection system. The permeability of the liner system composed of a drainage layer and filter layer is equivalent to that of coarse sand or gravel (see Table 12.11). When preparing a liner system design, the long-term reliability of manufactured materials for drainage media must be compared to the characteristics of soils with regard to biofouling and clogging.

In the liner system shown in Fig. 12.19*c*, two composite liners, commonly identified as the primary and secondary composite liners, are used. The primary composite liner is used for the collection of leachate while the secondary composite liner serves as a leak detection system and a backup for the primary composite liner. A modification of the liner system, shown in Fig. 12.19c, involves replacing the sand drainage layer with a geonet drainage system as shown in Fig. 12.19b. The two-layer composite design shown in Fig. 12.19d is the same as the liner shown in Fig. 12.19c, with the exception that the clay layer below the first geomembrane liner is replaced with a geosynthetic clay liner (GCL). A manufactured product, the GCL is made from a high-quality bentonite clay (from Wyoming) and an appropriate binding material. The bentonite clay is essentially a sodium montmorillonite mineral that has the capacity to absorb as much as 10 times its weight in water. As the clay absorbs water, it becomes putty-like and very resistant to the movement of water. Permeabilities as low as 10^{-10} cm/s have been observed. Available in large sheets (12 to 14 by 100 ft), GCLs are overlapped in the construction of a liner system.

Liner Systems for Monofills. Liner systems for monofills are usually composed of two geomembranes, each provided with a drainage layer and a leachate collection system (see Fig. 12.19c and d). A leachate detection system is placed between the first and second liner as well as below the lower liner. In many installations, a thick (3- to 5-ft) clay layer is used below the two geomembranes for added protection.

Construction of Clay Liners. In all the liner designs illustrated in Fig. 12.19, great care must be exercised in the construction of the clay layer. Perhaps the most serious problem with the use of clay is its tendency to form cracks due to desiccation. It is critical that the clay not be allowed to dry out as it is being placed. To ensure that the clay liner performs as designed, the clay liner should be laid in 4- to 6-in layers with adequate compaction between the placement of succeeding layers (see Fig. 12.20b). Laying the clay in thin layers limits the possibility of leaks due the alignment of clods that could occur if the clay layer is applied in a single pass. Another problem that has been encountered when clays of different types have been used is cracking due to differential swelling. To avoid differential swelling, the same type of clay must be used in the construction of the liner. Detailed specifications usually are prepared to describe the construction and testing procedures necessary to achieve a high-quality clay liner.

Leachate Collection Systems

A leachate collection system comprises the landfill liner, the leachate collection system, the leachate removal facilities, and the leachate holding facilities.

Landfill Liner. The type of landfill liner selected will depend to a large extent on the local geology and environmental requirements of the landfill site. For example, in locations where there is no groundwater, a single compacted clay liner has been sufficient. In locations where both leachate and gas migration must be controlled, the use of a composite liner composed of a clay liner and a geosynthetic liner with an appropriate drainage and soil protection layer will be necessary. Federal regulations for MSW landfills now mandate either the construction of some type of liner that is equivalent to a geomembrane and clay composite liner, or placing the landfill over a soil formation that will severely restrict leachate movement to protect groundwater quality. New cells added to existing landfills must also comply with this standard. Special wastes are regulated by state standards.

Leachate Collection Facilities. Collection of the leachate that accumulates in the bottom of a landfill is usually accomplished by using a series of sloped terraces and a

FIGURE 12.20   Preparation of compacted clay layer before geomembrane liner is placed.

system of collection pipes. As shown in Fig. 12.21a, the terraces are sloped so that the leachate that accumulates on the surface of the terraces will drain to leachate collection channels. Perforated pipe, placed in each leachate collection channel (see Fig. 12.21b), is used to convey the collected leachate to a central location, from which it is removed for treatment or reapplication to the surface of the landfill.

The cross slope of the terraces is usually 1 to 5 percent, and the slope of the drainage channels is 0.5 to 1.0 percent. The configuration and slope of the drainage system can be analyzed using the equations developed by Wang.[24] The cross slope and flow length of the terraces determine the depth of leachate above the liner. Flatter and

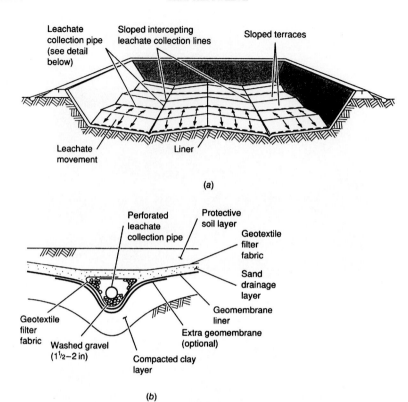

FIGURE 12.21   Leachate collection system with graded terraces. (*a*) Pictorial view; (*b*) detail of typical leachate collection pipe. (*From Ref. 52.*)

longer slopes result in higher head buildup. The design objective is not to allow the leachate to pond in the bottom of the landfill so as to create a significant hydraulic head on the landfill liner (less than 1 ft at the highest point as specified in the new federal Subtitle D landfill regulations). The depth of flow in the perforated drainage pipe increases continually from the upper reaches of the drainage channel to the lower reaches. In very large landfills, the drainage channels will be connected to a larger cross-collection system.

Leachate Removal and Holding Facilities.   Two methods have been used for the removal of leachate which accumulates within a landfill. In Fig. 12.22*a*, the leachate collection pipe is passed though the side of the landfill. Where this method is used, great care must be taken to ensure that the seal where the pipe penetrates the landfill liner is sound. An alternative method used for the removal of leachate from landfills involves the use of an inclined collection pipe located within the landfill (see Fig. 12.22*b*). Leachate collection facilities are used where the leachate is to be recycled from or treated at a central location. A typical leachate collection access vault is shown in Fig. 12.23*a*. In some locations, the leachate removed from the landfill is collected in a holding tank such as shown in Fig. 12.23*b*. The capacity of the holding tank will depend on the type of treatment facilities that are available and the maximum allowable discharge rate to the treatment facility. Typically, leachate holding tanks are

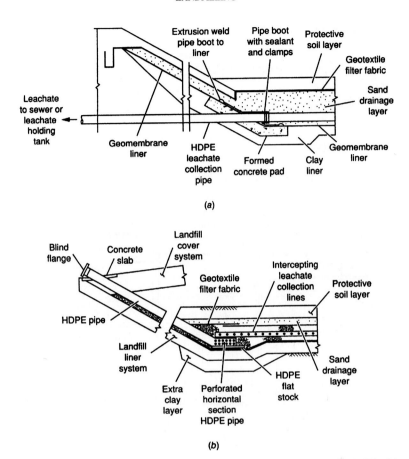

FIGURE 12.22 Typical systems used to collect and remove leachate form landfills. (*a*) Leachate collection pipe passed through side of landfill; (*b*) inclined leachate collection pipe and pump located within landfill. (*From Ref. 52.*)

designed to hold from 1 to 3 days of leachate production during the peak leachate production period. Double-walled tanks are preferred because of the added safety they afford compared to a single-walled tank. Tank materials must be carefully selected to resist corrosion.

Leachate Management

The management of leachate, when and if it forms, is key to the elimination of the potential for a landfill to pollute underground aquifers. A number of alternatives have been used to manage the leachate collected from landfills including: (1) leachate recycling, (2) leachate evaporation, (3) treatment followed by spray disposal, and (4) discharge to municipal wastewater collection systems. These options are discussed briefly below.

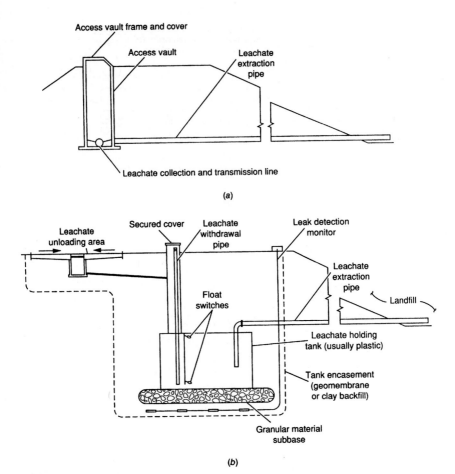

FIGURE 12.23 Examples of leachate collection facilities. (a) Leachate collection and transmission access vault; (b) leachate holding tank. (*From Ref. 52.*)

Leachate Recycling. An effective method for the treatment of leachate is to recirculate the leachate through the landfill (see Fig. 12.24). During the early stages of landfill operation the leachate will contain significant amounts of total dissolved solids (TDS), BOD_5, COD, nutrients, and heavy metals (see Sec. 12.3). When the leachate is recirculated, the constituents are attenuated by the biological activity and by other chemical and physical reactions occurring within the landfill. For example, the simple organic acids present in the leachate will be converted to CH_4 and CO_2. Because of the rise in pH within the landfill when CH_4 is produced, metals will be precipitated and retained within the landfill. An additional benefit of leachate recycling is the recovery of landfill gas that contains CH_4.

Typically, the rate of gas production is greater in leachate recirculation systems. To avoid the uncontrolled release of landfill gases when leachate is recycled for treatment, the landfill should be equipped with a gas recovery system. Ultimately, it will be necessary to collect, treat, and dispose of the residual leachate. In large landfills it may be necessary to provide leachate storage facilities.

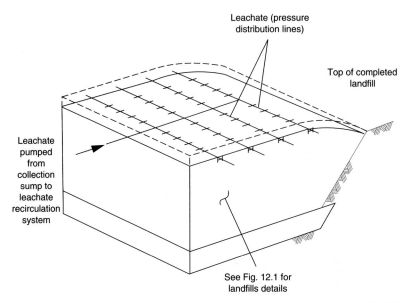

Leachate (pressure distribution lines)

Top of completed landfill

Leachate pumped from collection sump to leachate recirculation system

See Fig. 12.1 for landfills details

FIGURE 12.24 Schematic of leachate recirculation system used to apply leachate to landfill for treatment. In addition to the leachate distribution system placed on top of the final lift of the landfill as shown, leachate distributions systems are often placed on intermediate lifts.

Leachate Evaporation. One of the simplest leachate management systems involves the use of lined leachate evaporation ponds. Leachate that is not evaporated is sprayed on the completed portions of the landfill. In locations with high rainfall, the lined leachate storage facility is covered with a geomembrane during the winter season to exclude rainfall. The accumulated leachate is disposed of by evaporation during the warm summer months by uncovering the storage facility and by spraying the leachate on the surface of the operating and completed landfill. Odorous gases that may accumulate under the surface cover are vented to a compost or soil filter.[3,51] Soil beds are typically 2 to 3 ft deep, with organic loading rates of about 0.1 to 0.25 lb/ft^3 of soil. During the summer when the pond is uncovered, surface aeration may be required to control odors. If the storage pond is not large it can be left covered year round. Another example involves treatment of the leachate (usually biologically) with winter storage and spray disposal of the treated effluent on nearby lands during the summer. If enough land is available, spraying of effluent can be carried out on a continuous basis, even when it is raining.

Leachate Treatment. Where leachate recycling and evaporation is not used, and the direct disposal of leachate to a treatment facility is not possible, some form of pretreatment or complete treatment will be required. Because the characteristics of the collected leachate can vary so widely, a number of options have been used for the treatment of leachate. The principal biological and physical/chemical treatment operations and processes used for the treatment of leachate are summarized in Table 12.10. The treatment process or processes selected will depend to a large extent on the contaminant(s) to be removed. Typical examples of the types of the biological and physical/chemical processes that have been used for the treatment of leachate are shown in Fig. 12.25. Design details on the treatment options reported in Table 12.10 may be found in Ref. 43.

TABLE 12.10   Commonly Used Leachate Treatment Processes

Treatment process	Application	Comments
Biological processes		
Activated sludge	Removal of organics from leachate	Defoaming additives may be necessary; separate clarifier needed
Sequencing batch reactors	Removal of organics	Similar to activated sludge, but no separate clarifier needed; only applicable to relatively low flow rates
Aerated stabilization basins	Removal of organics	Requires large land area
Fixed film processes (trickling filters, rotating biological contactors)	Removal of organics	Commonly used on industrial effluents similar to leachates, but untested on actual landfill leachates
Anaerobic lagoons and contactors	Removal of organics	Lower power requirements and sludge production than aerobic systems; requires heating; greater potential for process instability; slower than aerobic systems
Nitrification/ denitrification	Removal of nitrogen	Nitrification/denitrification can be accomplished simultaneously with the removal of organics
Physical/chemical		
Sedimentation/flotation	Removal of suspended matter	Of limited applicability alone; may be used in conjunction with other treatment processes
Filtration	Removal of suspended matter	Useful only as a polishing step
Air stripping	Removal of ammonia or volatile organics	May require air pollution control equipment
Steam stripping	Removal of volatile organics	High energy costs; condensate steam requires further treatment
Adsorption	Removal of organics	Proven technology; variable costs depending on leachate
Ion exchange	Removal of dissolved inorganics	Useful only as a polishing step
Ultrafiltration	Removal of bacteria and high-molecular-weight organics	Subject to fouling; of limited applicability to leachate
Reverse osmosis	Dilute solutions of inorganics	Costly; extensive pretreatment necessary
Neutralization	pH control	Of limited applicability to most leachates
Precipitation	Removal of metals and some anions	Produces a sludge, possibly requiring disposal as a hazardous waste
Oxidation	Removal of organics; detoxification of some inorganic species	Works best on dilute waste streams; use of chlorine can result in formation of chlorinated hydrocarbons
Evaporation	Where leachate discharge is not permissible	Resulting sludge may be hazardous; can be costly except in arid regions
Wet air oxidation	Removal of organics	Costly; works well on refractory organics

Source: Adapted from Refs. 46 and 51.

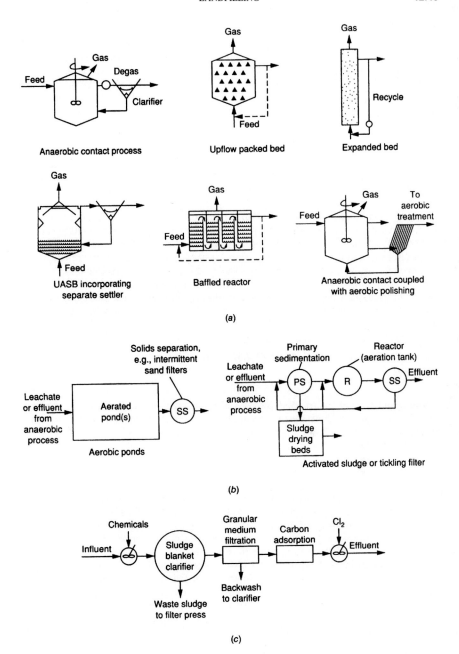

FIGURE 12.25 Typical processes used for the treatment of leachate. (*a*) Anaerobic processes; (*b*) aerobic processes; (*c*) chemical treatment process for the removal of heavy metals and selected organics. (*From Ref. 52.*)

The type of treatment facilities used will depend primarily on the characteristics of the leachate and secondarily on the geographic and physical location of the landfill. Leachate characteristics of concern include TDS, COD, SO_4^{-2}, and heavy metals, as well as nonspecific toxic constituents. Leachate containing extremely high TDS concentrations (e.g., >50,000 mg/L) may be difficult to treat biologically. High COD values favor anaerobic treatment processes, as aerobic treatment is expensive. High sulfate concentrations may limit the use of anaerobic treatment processes due the production of odors from the biological reduction of sulfate to sulfide. Heavy metal toxicity is also a problem with many biological treatment processes. Another important question is how large should the treatment facilities be. The capacity of the treatment facilities will depend on the size of the landfill and the expected useful life. The presence of nonspecific toxic constituents is often a problem with older landfills that received a variety of wastes, before environmental regulations governing the operation of landfills were enacted.

Discharge to Wastewater Treatment Plant.   In those locations where a landfill is located near a wastewater collection system or where a pressure sewer can be used to connect the landfill leachate collection system to a wastewater collection system, leachate is often discharged to the wastewater collection system. In many cases pretreatment, using one or more of the methods reported in Table 12.10, or leachate recirculation may be required to reduce the organic content before the leachate can be discharged to the sewer. In locations where sewers are not available, and evaporation and spray disposal are not feasible, complete treatment followed by surface discharge may be required.

12.4   INTERMEDIATE AND FINAL LANDFILL COVER

The use and type of intermediate cover and the design and performance of the final landfill cover are critical issues in the implementation of the landfill method of waste disposal.

Intermediate Cover

Intermediate cover layers are used to cover the wastes placed each day to enhance the aesthetic appearance of the landfill site, to limit the amount of surface infiltration, and to eliminate the harboring of disease vectors. The greatest amount of water that enters a landfill and ultimately becomes leachate enters during the period when the landfill is being filled. Some of the water, in the form of rain and snow, enters while the wastes are being placed in the landfill. Water also enters the landfill by first infiltrating and subsequently percolating through the intermediate landfill cover. Thus, the materials and method of placement of the intermediate cover can limit the amount of surface water that enters the landfill. The question of whether an intermediate cover layer is even needed, or should be required, is currently the subject of renewed debate, especially in landfills designed to recirculate leachate.[57]

Materials Used for Intermediate Cover Layers.   The types of materials that have been used as intermediate landfill cover include a variety of native soils, composted MSW, composted yard waste, yard waste mulch, agricultural residues, old carpets, synthetic foam, geomembranes, and construction and demolition waste. Of the materials listed, only the native soils and geomembranes are the most effective in limiting the entry of surface water into the landfill. In general, synthetic foam works well, except when it rains. To be effective, the intermediate cover, using the materials cited above, must be sloped properly to enhance surface water runoff. In some landfill operations, a very thick layer of soil (3 to 6 ft) is placed temporarily over the completed cell. Any rainfall that infiltrates the intermediate cover layer is retained by virtue of its field capacity. When a second lift is to be placed over the first lift, the soil is removed and stockpiled before filling begins. By using the operating technique of temporarily storing additional cover material over a completed cell, the amount of water entering the landfill can be limited significantly.

Intermediate Cover Layers Using Compost. Where the amount of native soil available for use as intermediate cover material is limited, alternative waste materials have been used for the purpose. Suitable materials that can be used as a substitute for native soil include compost and mulch produced from yard wastes and compost produced from MSW (see Fig. 12.26). An important advantage of using compost and mulch produced from MSW is that the landfill volume that would have been occupied by the soil used for intermediate cover is now available for the disposal of waste materials. In locations where the amount of cover material is limited, the use of composted MSW can increase the capacity of the landfill significantly. Excess compost produced at the landfill site can be stored until it is needed. Cured compost placed on the MSW deposited in the landfill also serves as an odor filter. The use of composted MSW for intermediate cover is expected to increase significantly in the coming years, as the conservation of landfill capacity becomes a more important issue.[35]

Final Landfill Cover

The primary purposes of the final landfill cover are (1) to minimize the infiltration of water from rainfall and snowfall after the landfill has been completed, (2) to limit the uncontrolled release of landfill gases, (3) to suppress the proliferation of vectors, (4) to limit the potential for fires, (5) to provide a suitable surface for the revegetation of the site, and (6) to serve as the central element in the reclamation of the site. To meet these purposes the landfill cover must:[17,25]

1. Be able to withstand climatic extremes (e.g., hot/cold, wet/dry, and freeze/thaw cycles)
2. Be able to resist water and wind erosion
3. Have stability against slumping, cracking and slope failure, and downslope slippage or creep
4. Resist the effects of differential landfill settlement caused by the release of landfill gas and the compression of the waste and the foundation soil
5. Resist failure due to surcharge loads resulting from the stockpiling of cover material and the travel of collection vehicles across completed portions of the landfill
6. Resist deformations caused by earthquakes
7. Withstand alterations to cover materials caused by constituents in the landfill gas
8. Resist the disruptions caused by plants, burrowing animals, worms, and insects

The cover must be configured in such a manner that it can be maintained efficiently and be amenable to relatively easy repair. It is important to note that under current legislation all of these purposes and attributes must continue to be satisfied far into the

FIGURE 12.26   Yard waste used as intermediate landfill cover. (*a*) Size of yard waste is reduced using a tub grinder; (*b*) ground-up waste is applied to face of landfill.

future. Federal regulations also establish minimum standards for MSW landfill covers, specifying permeability and construction materials. The general features of a landfill cover, some typical types of landfill cover designs, and the long-term performance requirements for landfill covers are considered below.

General Features of Landfill Covers.   A modern landfill cover, as shown in Fig. 12.27, is made up of a series of layers, each of which has a special function. The subbase soil layer is used to contour the surface of the landfill and to serve as a subbase for the barrier layer. In some cover designs, a gas collection layer is placed below the soil layer to transport landfill gas to gas management facilities. The barrier layer is used to restrict the movement of liquids into the landfill and the release of landfill gas through the cover. The drainage layer is used to transport rainwater and snowmelt that percolates through the cover material away from the barrier layer and to reduce the water pressure on the barrier layer. The protective layer is used to protect the drainage and barrier layers. The surface layer is used to contour the surface of the landfill and to support the plants that will be used in the long-term closure design of the landfill.

It should be noted that not all of the layers will be required in each location. Sometimes the subbase layer can also be used as the gas collection layer. Of the layers identified in Fig. 12.27, the barrier layer is the most critical for the reasons cited above.[17,25] Although clay has been used in many existing landfills as the barrier layer, a number of problems are inherent with its use. For example, clay is difficult to compact on a soft foundation, compacted clay can develop cracks due to desiccation, clay can be damaged by freezing, clay will crack due to differential settling, the clay layer in a landfill cover is difficult to repair once damaged, and finally, the clay layer does not restrict the movement of landfill gas to any significant extent. As a consequence, use of a geomembrane in combination with clay or two or more geomembranes is recommended over the use of clay alone as a barrier layer in landfill covers.

Typical Landfill Cover Designs.   Some of the many types of cover designs that have been proposed and used are illustrated in Fig. 12.28. In Fig. 12.28a, the geotextile filter cloth is used to limit the intermixing of the soil with the sand layer. If the available topsoil at the landfill site is not suitable for plant growth, a suitable topsoil must be brought to the site or the available topsoil should be amended to improve its characteristics for plant growth. The use of a composite barrier design comprising a geomembrane and clay layer is illustrated in Fig. 12.28b. In the cover design illustrated in Fig.

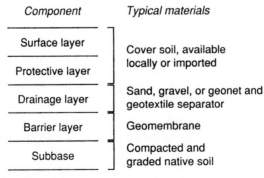

Component	Typical materials
Surface layer	Cover soil, available locally or imported
Protective layer	
Drainage layer	Sand, gravel, or geonet and geotextile separator
Barrier layer	Geomembrane
Subbase	Compacted and graded native soil

FIGURE 12.27   Typical components that comprise a landfill cover. (*From Ref. 52.*)

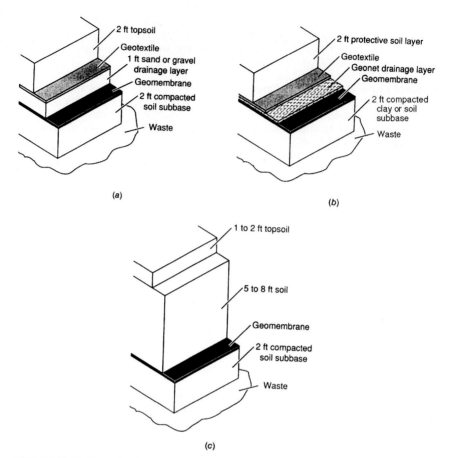

FIGURE 12.28   Examples of landfill final cover configurations. (*From Ref. 52.*)

12.28*c*, a 6- to 10-ft-thick layer of soil is used as the cover layer. Functionally, the soil layer is sloped adequately to maximize surface runoff. The depth of soil is used to retain rainfall that does not run off and infiltrates into the soil cover. The flexible membrane liner is used to limit the release of landfill gases. In an another design, the waste is first covered with a base layer of old carpets or similar materials. A flexible membrane liner is placed over the base layer. A layer of astroturf is placed over the flexible membrane liner. Use of the astroturf is advantageous because the amount of maintenance required is minimized.

Long-Term Performance and Maintenance of Landfill Covers.   Regardless of the design of the final landfill cover, the following question must be considered. How will the integrity and performance of the landfill cover be maintained as the landfill settles, as a result of the loss of weight due to the production of landfill gas and by long-term consolidation? For example, how will a composite liner be repaired to maintain adequate drainage? Typically, if settlement occurs, the landfill cover material is stripped back, soil or composted waste is added to adjust the grade, and the various layers replaced. Where a thick soil cover is used, proper surface drainage may be restored by regrading the cover layer. In drier climates, where vegetation is planted on the soil cover layer, a sprinkler system may be required to sustain the vegetation during the summer. In landfills where astroturf is used, when the turf starts to fall apart, the landfill cover is opened, the used turf is placed in the landfill, the flexible membrane repaired, and a new astroturf layer is added to the top.

Percolation through Intermediate and Final Cover Layers

If it is assumed (1) that the cover material is saturated, (2) that a thin layer of water is maintained on the surface, and (3) that there is no resistance to flow below the cover layer, then the theoretical amount of water expressed in gallons that could enter the landfill per unit area in a 24-h period for various cover materials is given in Table 12.11 in column 3. Clearly, these data are only theoretical values, but they can be used in assessing the worst possible situation. In practice, the amount of water entering the landfill will depend on local hydrological conditions, the design of the landfill cover, the final slope of the cover, and whether vegetation has been planted. In general, landfill cover designs employing a flexible membrane liner are designed to eliminate the percolation of rainwater or snowmelt into the waste below the landfill cover.

Estimation of the percolation of rainwater or snowmelt through the soil layer above the drainage layer (see Fig. 12.29a) or through a cover layer composed of soil only (see Fig. 12.29b) is usually accomplished by using one of the many available hydrologic simulation programs. Perhaps the best known is the Hydrologic Evaluation of Landfill Performance (HELP) model which is being revised continuously.[44,45] Percolation through the landfill cover layer can also be estimated using a standard hydrological water balance. Referring to Fig. 12.29, the water balance for a soil landfill cover is given by the following expression:

$$\Delta S_{LC} = P - R - ET - PER_{SW} \tag{12.8}$$

TABLE 12.11 Typical Permeability Coefficients for Various Soils (Laminar Flow)

	Coefficient of permeability, K	
Material	ft/day	gal/ft^2 • day
Uniform coarse sand	1333	9970
Uniform medium sand	333	2490
Clean, well graded sand and gravel	333	2490
Uniform fine sand	13.3	100
Well-graded silty sand and gravel	1.3	9.7
Silty sand	0.3	2.2
Uniform silt	0.16	1.2
Sandy clay	0.016	0.12
Silty clay	0.003	0.022
Clay (30 to 50% clay sizes)	0.0003	0.0022
Colloidal clay	0.000003	0.000022

Note: ft/day $\times$ 0.3048 = m/day
gal/ft^2 • day $\times$ 0.0408 = m^3/m^2 • day
Source: Adapted from Refs. 9 and 43.

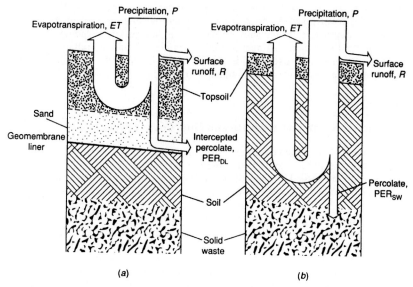

FIGURE 12.29 Definition sketch for water balance for landfill. (*a*) For landfill cover containing a drainage layer and geomembrane liner; (*b*) for landfill with no drainage layer or geomembrane liner. (*From Ref. 52.*)

where ΔS_{LC} = change in the amount of water held in storage in a unit volume of landfill cover, in

P = amount of precipitation per unit area, in

R = amount of runoff per unit area, in

ET = amount of water lost through evapotranspiration per unit area, in

PER_{SW} = amount of water percolating through unit area of landfill cover into compacted solid waste, in

The total amount of water that can be stored in a unit volume of soil will depend on the field capacity (FC) and the permanent wilting percentage (PWP). The FC is defined as the amount of water that is retained in a soil against the pull of gravity. Soil moisture tension at FC is typically between $\frac{1}{10}$ and $\frac{1}{3}$ atm (Ref. 16). The PWP is defined as the amount of water left in a soil when plants are no longer able to extract any more. Soil moisture tension at PWP is approximately 15 atm (Ref. 16). The difference between the FC and the PWP represents the amount of water that can be stored in a soil. Typical FC and the PWP values for representative soils are given in Table 12.12. If a layered landfill cover is used, the field capacity of each layer must be considered in the analysis. Typical runoff coefficients for completed landfill covers are given in Table 12.13. Monthly precipitation and evapotranspiration data are site specific, but local weather bureau data are usually acceptable.

12.5 STRUCTURAL AND SETTLEMENT CHARACTERISTICS OF LANDFILLS

The structural characteristics and settlement of the landfill must be considered in the design of gas collection and surface water drainage facilities, during filling opera-

TABLE 12.12 Typical Field Capacity (FC) and Permanent Wilting Point (PWP) Values for Various Soil Classifications

Soil classification	Value, %			
	Field capacity		Permanent wilting point	
	Range	Typical	Range	Typical
Sand	6–12	6	2–4	4
Fine sand	8–16	8	3–6	5
Sandy loam	10–18	14	4–8	6
Fine sandy loam	0	0	0	0
Loam	18–26	22	8–12	10
Silty loam	0	0	0	0
Light clay loam	0	0	0	0
Clay loam	23–31	27	11–15	12
Silty clay	27–35	31	12–17	15
Heavy clay loam	0	0	0	0
Clay	31–39	35	15–19	17

Source: Adapted from Refs. 16, 29, and 53.

TABLE 12.13 Typical Runoff Coefficients for Storms of 5- to 10-Year Frequency

Type of cover	Slope, %	Runoff coefficient			
		With grass		Without grass	
		Range	Typical	Range	Typical
Sandy loam	2	0.05–0.10	0.06	0.06–0.14	0.10
	3–6	0.10–0.15	0.12	0.14–0.24	0.18
	7	0.15–0.20	0.17	0.20–0.30	0.24
Silt loam	2	0.12–0.17	0.14	0.25–0.35	0.30
	3–6	0.17–0.25	0.22	0.35–0.45	0.40
	7	0.25–0.36	0.30	0.45–0.55	0.50
Tight clay	2	0.22–0.33	0.25	0.45–0.55	0.50
	3–6	0.30–0.40	0.35	0.55–0.65	0.60
	7	0.40–0.50	0.45	0.65–0.75	0.70

Source: Developed in part from Refs. 14, 29, and 55.

tions, and before a decision is reached on the final use to be made of a completed landfill.

Structural Characteristics

When solid waste is initially placed in a landfill, it behaves in a manner that is quite similar to other fill material. The nominal angle of repose for waste material placed in a landfill is approximately 1.5 to 1. Because solid waste has a tendency to slip when the slope angle is too steep, the slopes used for the completed portions of a landfill will vary from 2.5:1 to 4:1, with 3:1 being the most common. Because of the problems encountered with slippage due to settlement, many landfills are benched (see Fig. 12.1) where the height of the landfill will exceed 50 ft. In addition to helping to maintain slope stability, benches are also used for the placement of surface water drainage channels and for the location of landfill gas recovery piping.

In general, the construction of permanent facilities on completed landfills is not recommended because of the uneven settlement characteristics, varying bearing capacity of the upper layers of the landfill, and the potential problems that can result from gas migration, even with the use of gas collection facilities. When the final use of the landfill is known before waste placement begins, it is possible to control the deposition of certain materials during the operation of the landfill. For example, relatively inert materials such as construction and demolition wastes can be placed in those locations where buildings and/or other physical facilities are to be placed in the future. Recent regulatory trends have further limited the placement of structures on completed landfills.

Settlement of Landfills

As the organic material in landfill is decomposed and weight is lost as landfill gas and leachate components, the landfill settles. Settlement also occurs as a result of increasing overburden mass as landfill lifts are added and as water percolates into and out of the landfill. Landfill settlement results in ruptures of the landfill surface and cover, breaks and misalignments of gas recovery facilities, cracking of manholes, and interference with subsequent use of the landfill after closure.

Effect of Waste Decomposition. Once placed in a landfill, the organic components of the waste will decompose, resulting in loss of as much as 30 to 40 percent of the original mass. The rate of decomposition is directly related to the moisture content of the waste, with wet waste decomposing the fastest. The loss of mass results in a loss of volume, which becomes available for refilling with new waste. The volume that is lost is usually filled in when higher lifts are subsequently placed over the initial lifts. Weight and volume will also be lost after a landfill is closed.

Effect of Overburden Pressure (Height). The density of the material placed in the landfill will increase with the weight of the material placed above it, so that the average specific weight of waste in a lift depends on the depth of the lift. The maximum specific weight of solid waste residue in a landfill under overburden pressure will vary from 1750 to 2150 lb/yd^3 (Refs. 22 and 23). The following relationship can be used to estimate the increase in the specific weight of the waste as a function of the overburden pressure:

$$D_{Wp} = D_{Wi} + \frac{p}{a + bp} \qquad (12.9)$$

where D_{Wp} = specific weight of the landfill material at pressure p, lb/yd^3
   D_{Wi} = initial compacted specific weight of the waste, lb/yd^3
   p = overburden pressure, lb/ft^2
   a = empirical constant, yd^3/ft^2
   b = empirical constant, yd^3/lb

Typical specific weights versus applied pressure curves for compacted solid waste for several initial specific weights are shown in Fig. 12.30. The increase in the specific weight of the waste material in the landfill is important in (1) determining the actual

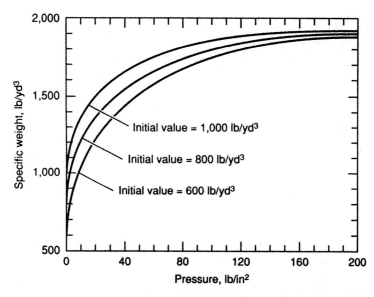

FIGURE 12.30 Specific weight of solid waste placed in landfill as function of the initial compacted specific weight of the waste and the overburden pressure.

amount of waste that can be placed in a landfill up to a given grade limitation and (2) in determining the degree of settlement that can be expected in a completed landfill after closure.

Extent of Settlement. The extent of settlement depends on the initial compaction, the characteristics of wastes, the degree of decomposition, the effects of consolidation when water and air are forced out of the compacted solid waste, and the height of the completed fill. Representative data on the degree of settlement to be expected in a landfill as a function of the initial compaction are shown in Fig. 12.31. It has been found in various studies that about 90 percent of the ultimate settlement occurs within the first 5 years.

12.6 LANDFILL DESIGN CONSIDERATIONS

Among the important topics that must be considered in the design of a landfill, though not necessarily in the order given, are the following:

1. Layout of landfill site
2. Types of wastes that must be handled
3. The need for a convenience transfer station
4. Estimation of landfill capacity
5. Evaluation of the local geology and hydrogeology of the site
6. Selection of leachate management facilities

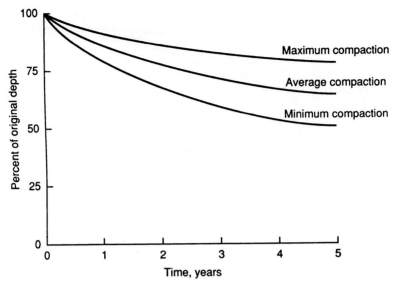

FIGURE 12.31 Surface settlement of compacted landfills. (*From Ref. 52.*)

7. Selection of landfill cover
8. Selection of landfill gas control facilities
9. Surface water management
10. Aesthetic design considerations
11. Development of landfill operation plan
12. Determination of equipment requirements
13. Environmental monitoring
14. Public participation
15. Closure and postclosure care

The development of an operational plan for a landfill and the determination of equipment requirements is considered in following sections. Environmental monitoring is considered in Sec. 12.8. Closure and postclosure care is considered in Sec. 12.9. Important factors that must be considered in the design of landfills are reported in Table 12.14. Throughout the development of the engineering design report, careful consideration must be given to the final use or uses to be made of the completed site. Land reserved for administrative offices, buildings, and parking lots should be filled with dirt only. Buildings should be protected from the underground gas migration and should be sealed. Protection can be accomplished with membrane seals or soil gas extraction systems.

Layout of Landfill

In planning the layout of a landfill site, the location of the following must be determined:

TABLE 12.14 Important Factors That Must Be Considered in the Design of Landfills

Factors	Remarks
Access	Paved all-weather access roads to landfill site; temporary roads to unloading areas.
Cell design and construction	Each day's wastes should form one cell; cover at end of day with 6 in of earth or other suitable material; typical cell width, 10 to 30 ft; typical lift height including intermediate cover, 10 to 14 ft; slope of working faces, 3 to 1.
Completed landfill characteristics	Finished slopes of landfill, 3 to 1; height to bench, if used, 50 to 75 ft; slope of final landfill cover, 3 to 6%.
Environmental requirements	Install vadose zone gas- and liquid-monitoring facilities; install up- and downgradient groundwater monitoring facilities; locate ambient air monitoring stations.
Equipment requirements	Number and type of equipment will vary with the type of landfill and the capacity of the landfill.
Final cover	Use multilayer design; slope of final landfill cover, 3 to 6%.
Fire prevention	Water on site; if nonpotable, outlets must be marked clearly; proper cell separation prevents continuous burn-through if combustion occurs.
Groundwater protection	Divert any underground springs; if required, install perimeter drains, well point system, or other control measures.
Intermediate cover material	Maximize use of on-site soil materials; other materials such as compost produced from yard waste and MSW can also be used to maximize the landfill capacity; typical waste-to-cover ratios vary from 5 to 1 to 10 to 1.
Land area	Area should be large enough to hold all community wastes for a minimum of 5 years, but preferably 10 to 25 years; area for buffer strips or zones must also be included.
Landfill gas management	Develop landfill gas management plan including extraction wells, manifold collection system, condensate collection facilities, the vacuum blower facilities, and flaring facilities and or energy production facilities. Operating vacuum at well head, 10 in of water.
Landfill liner	Single clay layer (2 to 4 ft) or multilayer design incorporating the use of a geomembrane. Cross slope for terrace type leachate collection systems, 1 to 2%; maximum flow distance over terrace, 100 ft; slope of drainage channels, 0.5 to 0.75%. Slope for piped type leachate collection system, 1 to 2%; size of perforated pipe, 4 in.; pipe spacing, 20 ft.
Landfilling method	Landfilling method will vary with terrain and available cover; most common methods are excavated cell/trench, area, canyon.
Leachate collection	Determine maximum leachate flow rates and size leachate collection pipe and/or trenches; size leachate pumping facilities; select collection pipe materials to withstand static pressures corresponding to the maximum height of the landfill.
Leachate treatment	On basis of expected quantities of leachate and local environmental conditions, select appropriate treatment process.
Surface drainage	Install drainage ditches to divert surface water runoff; maintain 3 to 6% grade on finished landfill cover to prevent ponding; develop plan to divert stormwater from lined, but unused portions of landfill.

Source: From Ref. 52.

1. Access roads
2. Equipment shelters
3. Scales, if used
4. Office space
5. Location of convenience transfer station, if used
6. Storage and/or disposal sites for special wastes
7. Identification of areas to be used for waste processing (e.g., composting)
8. Definition of the landfill areas and areas for stockpiling cover material
9. Drainage facilities
10. Location of landfill gas management facilities
11. Location of leachate treatment facilities, if required
12. Location of monitoring wells
13. Placement of barrier berms or structures to limit sight lines into the landfill
14. Plantings

A typical layout for a landfill disposal site is shown in Fig. 12.32. Because site layout is specific for each case, Fig. 12.32 is meant to serve only as a guide. However, the items identified on Fig. 12.32 can be used as a checklist of the areas that must be addressed in the preliminary layout of a landfill.

Types of Wastes

Knowledge of the types of wastes to be handled is important in the design and layout of a landfill, especially if special wastes are involved. It is usually best to develop separate disposal sites or monofills for designated and special wastes such as asbestos or incinerator ash because, under most conditions, special treatment of the site will be necessary before these wastes can be landfilled. The associated disposal costs are often significant, and it is wasteful to use this landfill capacity for wastes that do not require special precautions. If significant quantities of demolition wastes are to be handled, it may be possible to use them for embankment stabilization. In some cases, it may not be necessary to cover demolition wastes on a daily basis.

Need for a Convenience Transfer Station

Because of safety concerns and the many new restrictions governing the operation of landfills, many operators of landfills have constructed convenience transfer stations at the landfill site for the unloading of wastes brought to the site by individuals and small-quantity haulers (see Fig. 12.33). By diverting private individuals and small-quantity haulers to a separate transfer facility, the potential for accidents at the working face of the landfill is reduced significantly. Operating efficiency at the working face is also enhanced by excluding small vehicles. In some locations, transfer facilities are also used for the recovery of recyclable materials. Waste materials are usually emptied into large transfer trailers each of which is hauled to the disposal site, emptied, and returned to the transfer station. The need for a convenience transfer station will depend on the physical characteristics and the operation of the landfill and

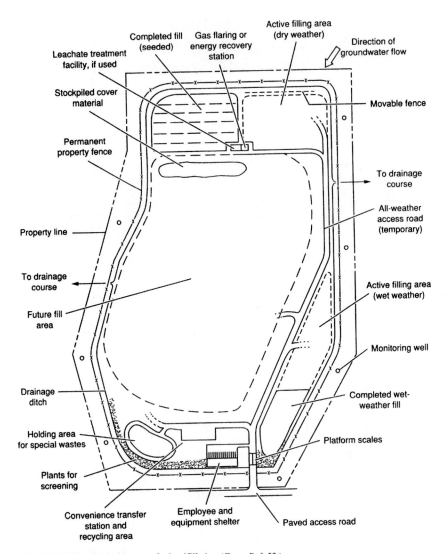

FIGURE 12.32 Typical layout of a landfill site. (*From Ref. 52.*)

whether there is a separate location where the public can be allowed to dispose of waste safely.

Estimation of Landfill Capacity

The nominal volumetric capacity of a proposed landfill site is determined by first laying out several different landfill configurations, taking into account appropriate design

FIGURE 12.33   Small direct-load convenience transfer station located at landfill.

criteria including the planned thickness of the liner and final cover systems (see Figs. 12.32 and 12.47). The next step is to determine the surface area for each lift. The nominal volume of the landfill is determined by multiplying the average area between two adjacent contours by the height of the lift and summing the volume of successive lifts. If the cover material will be excavated from the site, then the computed volume corresponds to the volume of solid waste that can be placed in the site. If the cover material has to be imported, then the computed capacity must be reduced by a factor to account for the volume occupied by the cover material. For example, if a cover-to-waste ratio of 1:5 is adopted, then the capacity reported must be multiplied by a factor of 0.833 ($\frac{5}{6}$).

The nominal volumetric capacity of the landfill is used as a preliminary estimate of landfill capacity for estimating purposes. The actual total capacity of the landfill to accept waste on a weight basis will depend on the initial density at which the residual solid waste is placed in the landfill, on the subsequent compaction of the waste material due to overburden pressure, and loss of mass as a result of biological decomposition. The impacts of these factors on the capacity of the landfill are considered below.

Impact of Compactibility of Solid Waste Components.   The initial density of solid wastes placed in a landfill varies with the mode of operation of the landfill, the compactibility of the individual solid waste components, and the percentage distribution of the components. If the waste placed in the landfill is spread out in thin layers and compacted against an inclined surface, a high degree of compaction can be achieved. With minimal compaction, the initial density will be somewhat less than the compacted density in a collection vehicle. In general, the initial density of solid waste placed in a landfill will vary from 550 to 1200 lb/yd^3, depending on the degree of initial compaction given to the waste. Typical compactibility data for the components found in MSW are reported in Table 12.15. Volume-reduction factors are given for both normally compacted and well-compacted landfills.

Impact of Cover Material.   Federal regulations mandate that some type of permanent or temporary daily cover always be placed over MSW. The cover material, typically soil, is incorporated into a landfill at each stage of its construction. Alternative materi-

TABLE 12.15 Typical Compaction Factors for Various Solid Waste Components Placed in Landfills

| Component | Range | Compaction factors for components in landfills* | |
		Normal compaction	Well compacted
Food wastes	0.2–0.5	0.35	0.33
Paper	0.1–0.4	0.2	0.15
Cardboard	0.1–0.4	0.25	0.18
Plastics	0.1–0.2	0.15	0.10
Textiles	0.1–0.4	0.18	0.15
Rubber	0.2–0.4	0.3	0.3
Leather	0.2–0.4	0.3	0.3
Garden trimmings	0.1–0.5	0.25	0.2
Wood	0.2–0.4	0.3	0.3
Glass	0.3–0.9	0.6	0.4
Tin cans	0.1–0.3	0.18	0.15
Nonferrous metals	0.1–0.3	0.18	0.15
Ferrous metals	0.2–0.6	0.35	0.3
Dirt, ashes, brick, etc.	0.6–1.0	0.85	0.75

*Compaction factor $= V_f/V_i$ where V_f = final volume of solid waste after compaction and V_i = initial volume of solid waste before compaction.

Source: From Ref. 52.

als, such as foams or blankets, may also be used. Daily cover, consisting of 6 in to 1 ft of soil, is applied to the working faces of the landfill at the close of operation each day to stop material from blowing from the working face and to control disease vectors such as insects and rats. Certain clay and silt soils, when used as daily cover, function satisfactorily as cover. However, later when new MSW is landfilled on top of the cover, downward movement of leachate is impeded. If the waste cell is located near the outside edge of the landfill, hydrostatic pressure may result in the leachate leaking through the side of the landfill. To overcome this problem some operators remove and stockpile the daily cover before placing new MSW. This saves landfill space, conserves cover soil, and prevents leachate seeps through the side of the landfill.

Interim cover is a thicker layer of daily cover material applied to areas of the landfill that will not be worked for some time. Final covers usually are 3 to 6 ft thick and include several layers, as discussed previously, to enhance drainage and support surface vegetation. The quantity of cover material necessary for operation of the landfill is an important factor in determining the capacity of a landfill site. Usually, daily and interim cover needs are expressed as a waste/soil ratio, defined as the volume of waste deposited per unit volume of cover provided. Typically, waste/soil ratios range from 4:1 to 10:1.

Impact of Waste Decomposition and Overburden Height. The loss of mass through biological decomposition results in a loss of volume, which becomes available for refilling with new waste. In the preliminary assessment of site capacity, only compaction due to overburden is considered. At later stages of landfill design, the loss of landfill material to decomposition should be considered. Timing is also important. An area that is filled rapidly to the topmost height allowed by a regulatory permit will not have sufficient time to achieve maximum settlement and forestall the opportunity to refill before closure.

Evaluation of Local Geology and Geohydrology

To evaluate the geologic and hydrogeological characteristics of a site that is being considered for a landfill, core samples must be obtained . Sufficient borings should be made so that the geologic formations under the proposed site can be established from the surface to (and including) the upper portions of the bedrock or other confining layers. At the same time, the depth to the surface water table should be determined along with the piezometric water levels in any bedrock or confined aquifers that may be found. The resulting information is then used to determine:

1. The general direction of groundwater movement under the site
2. Whether any unconsolidated or bedrock aquifers are in direct hydraulic connection with the proposed landfill site
3. The type of liner system that will be required
4. The suitability of soils available at the site for use as liner and cover materials

A portion of the borings are usually converted to permanent monitoring wells from which samples are collected periodically and tested to determine variations in background groundwater quality. One year's worth of data is frequently the minimum required. Geophysical data collection techniques can be used to characterize large areas quickly as a preliminary reconnaissance step before commencing borings. Electromagnetic, resistivity, and seismic techniques can provide the designer with preliminary information that can be used to reduce the number of sites under consideration before engaging in subsurface borings.

The presence of wetlands should also be noted during site evaluation in preparation for application to the U.S Army Corps of Engineers for a wetland exemption. Permits for constructing landfills in wetland areas can be controversial and mitigation may be required. The operators of landfills located over faults, in seismic impact zones, or over unstable soils are required, by federal regulation, to demonstrate that the landfill can operate continuously in an environmentally safe manner. Maps from state highway departments and the U.S. Geological Survey identify faults and seismic areas. Unstable soils must be identified through borings.

Selection of Leachate Management Facilities

The principal leachate management facilities required in the design of a landfill include the landfill liner, the leachate collection system, and the leachate treatment facilities. To provide assurances to the public that leachate will not contaminate underground waters, most states under federal mandates now require some type of liner for all new landfill cells. The current trend is toward the use of composite liners including a geomembrane and clay layer. In extremely arid areas where no possibility exists of contaminating the groundwater, it may be possible to develop a landfill without a liner. Nevertheless, the use of a liner system is a critical factor in siting new landfills. Further, the relative cost of a liner system is not great considering the potential environmental benefits. Multimillion dollar costs can result where groundwater pollution occurs. To determine the size of the leachate collection and treatment facilities, the quantity of leachate must be estimated using the methods outlined in Sec. 12.3. As noted previously, the most common alternatives that have been used to manage the leachate collected from landfills include: (1) leachate recycling, (2) treatment followed by disposal, and (3) discharge to municipal wastewater collection systems. The particular option used will depend on local conditions.

Selection of Landfill Cover

As discussed previously, a landfill cover usually comprises several layers, each with a specific function (see Fig. 12.27). The use of a geomembrane liner as a barrier layer is becoming more common to limit the entry of surface water and to control the release of landfill gases. The specific cover configuration selected will depend on the location of the landfill and the local climatalogical conditions. For example, to allow for regrading, the use of a deep layer of soil is favored by some designers. To ensure the rapid removal of rainfall from the completed landfill and to avoid the formation of puddles, the final cover should have a minimum slope of about 3 to 5 percent. Cover slope stability must be considered during design when the slope is greater than 25 percent.

Selection of Gas Control Facilities

Because the uncontrolled release of landfill gas, especially methane, contributes to the greenhouse effect, and because landfill gas can migrate laterally underground to potentially cause explosions or kill vegetation and trees, most new landfills are equipped with gas collection and treatment facilities. To determine the size of the gas collection and processing facilities, the quantity of landfill gas must first be estimated using the methods outlined in Sec. 12.3. Because the rate of gas production varies, depending on the operating procedures (e.g., without or with leachate recycle), several rates should be analyzed. The next step is to determine the rate of gas production with time. The decision to use horizontal or vertical gas recovery wells depends on the design and capacity of the landfill. The decision to flare or to recover energy from the landfill gas depends on the capacity of the landfill site and the opportunity to sell power produced from the conversion of landfill gas to electrical energy or the availability of utility boilers for energy recovery. In many small landfills located in remote areas, gas collection equipment is not used routinely.

Surface Water Management

Elimination or reduction of the amount of surface water that enters the landfill is of fundamental importance in the design of a sanitary landfill because surface water is the major contributor to the total volume of leachate. Storm water runoff from the surrounding area must not be allowed to enter the landfill, and surface water runoff (from rainfall) must not be allowed to accumulate on the surface of the landfill. The proper management of storm water runoff that flows away from the landfill is also important. The total and peak runoff flow rates will be increased significantly when a relatively impermeable and sloping cover is placed over a landfill that is located on previously level land. Increased sediment loading to nearby surface water bodies may also occur.

The location and operation of landfills on floodplains and floodways is restricted by federal and many state regulations. Provisions must be made to minimize obstruction of floodwater flow and reduction in floodplain storage capacity by landfilling activities. Landfills not capable of excluding floodwaters from entry may be required to close under certain circumstances.

Surface Water Drainage Facilities. An important step in the design of a landfill is to develop an overall drainage plan for the area that shows the location of storm

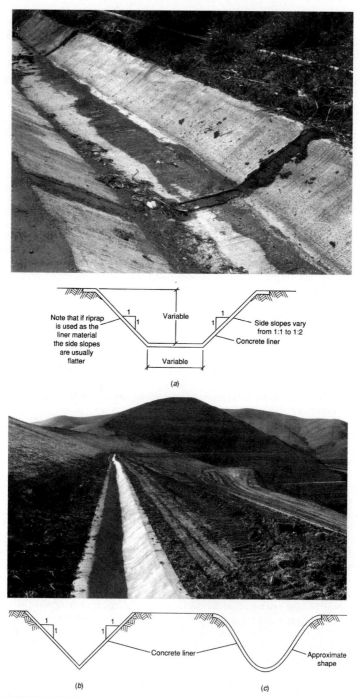

FIGURE 12.34   Typical drainage facilities used at landfills. (*a*) Trapezoidal lined ditch; (*b*) vee lined ditch; (*c*) shaped-vee lined ditch. Note the trapezoidal ditch cross section is expandable to accommodate a wide range of flows. (*From Ref. 52.*)

drains, culverts, ditches, and subsurface drains as the filling operation proceeds. In those locations where storm water runoff from the surrounding areas can enter the landfill (e.g., landfills located in canyons), the site must be graded appropriately and properly designed drainage facilities must be installed (see Fig. 12.34). The drainage facilities may be designed to remove the runoff from the surrounding area only, or from the surrounding area as well as the surface of the landfill. In current federal regulations, a 25-year storm event must be used as the basis for design. In locations where the entire leachate liner system is installed at one time, the design of the liner must allow for the diversion of storm water not falling on the wastes being landfilled. In locations where only the surface water from the top of the landfill must be removed, the drainage facilities should be designed to limit the travel distance of the surface water. In many designs, a series of interceptor ditches are used. Flow from the interceptor ditches is routed to a larger main ditch for removal from the site.

Storm Water Storage Basins. Depending on the location and configuration of the landfill and the capacity of the natural drainage courses, it may be necessary to install a storm water storage or retention basin. In many cases, it may be necessary to construct storm water storage basins to contain the diverted storm water flows so as to minimize downstream flooding. Typically storm water must be collected from the completed portions of the landfill as well as from areas yet to be filled. An example of a large storm water retention/storage basin is illustrated in Fig. 12.35. Standard hydrological procedures are followed in sizing the storm water basins.[19,29,30] Discharges from storm water facilities must have a federal or state discharge permit.

FIGURE 12.35 View of large storm water retention/storage basin at a large landfill. The size of the basin can be estimated from the size of the vehicles parked in the bottom of the basin.

Environmental Monitoring Facilities

Monitoring facilities are required at new landfills and at selected existing landfills for (1) gases and liquids in the vadose zone, (2) for groundwater quality both upstream and downstream of the landfill site, and (3) for air quality at the boundary of the landfill and from any processing facilities (e.g., flares). The specific number of monitoring stations will depend on the configuration and size of the landfill and the requirements of the local air and water pollution control agencies. Environmental monitoring is considered in greater detail in Sec. 12.8.

Aesthetic Design Considerations

Aesthetic design considerations relate to minimizing the impact of the landfilling operation on nearby residents as well as the public that may be passing by the landfill.

Screening of Landfilling Areas. Screening of the daily landfilling operations from nearby roads and residents with berms, plantings, and other landscaping measures is one of the most important examples of an aesthetic design consideration (see Fig. 12.36*a*). Screening of the active areas in the landfill must be taken into account in the preliminary design and layout of the landfill.

The Control of Birds. Birds at the landfill site are not only a nuisance; they can cause serious problems if the landfill site is located near an airport. Federal regulations limit landfill development within 10,000 ft of an airport, and in some instances

(a)   (b)

(c)   (d)

FIGURE 12.36 Aesthetic considerations in landfill design. (*a*) View of landscaped landfill in which filling operations are not visible from nearby freeway; (*b*) overhead wire system used to control seagulls at landfills; (*c*) wire screen used to control blowing papers and plastic; (*d*) daily cover used to control vectors at landfills.

greater separation distances are required. Techniques to control birds at landfill sites include the use of noisemakers, recordings of the sounds made by birds of prey, and overhead wires. The control of birds at reservoirs and fish ponds with overhead wires dates back to the early 1930s.[1,33] The use of overhead wires to control seagulls at landfills was pioneered by the County Sanitation Districts of Los Angeles County in the early 1970s (see Fig. 12.36b). Because seagulls descend in a circular pattern when landing, it appears that the wires may interfere with the birds' guidance system. The poles are typically spaced 50 to 75 ft apart, with line spans from 500 to 1200 ft.[32] Crisscrossing improves the effectiveness of the wire system. Typically, 100-lb-test monofilament fish line is used, although stainless steel wire has also been used.

The Control of Blowing Materials.   Depending on the location, windblown paper, plastics, and other debris can be a problem at some landfills. The most common solution is to use portable screens near the operating face of the landfill (see Fig. 12.36c). To avoid problems with vectors, the material accumulated on the screens must be removed daily. Prompt pickup of paper that is not retained by portable screens is important for maintaining an image of good landfill operation.

Open topped vehicles hauling waste to the landfill should be covered with a tarpaulin to prevent paper and dust from falling onto the highway. Waste that does fall out of a truck should be picked up promptly by either the vehicle operator or the landfill personnel. Periodic collection of all litter along the roads leading to the landfill is also a good approach to help the landfill operator enhance community relations.

The Control of Pests and Vectors.   The principal vectors of concern in the design and operation of landfills are pests, including mosquitos and flies, and rodents, such as rats and other burrowing animals. Flies and mosquitos are controlled by the placement of daily cover and by the elimination of standing water. Standing water can be a problem in areas where white goods and used tires are stored for recycling. The use of covered facilities for the storage of these materials will eliminate most problems. Rats and other burrowing animals are controlled by the use of daily cover (see Fig. 12.36d).

Public Participation

Landfill development is often controversial. For privately owned sites there is no mandate to discuss development with representatives of the public until permit applications are submitted and hearings are conducted. Often this approach leads to very time-consuming protracted hearings after which a ruling may or may not allow landfill construction. Usually legal appeals are filed which result in extended delays.

The development of publicly owned sites is often discussed by boards and councils who must decide how to proceed. This approach while more open to public scrutiny can also become embroiled in public controversy. A better approach is to establish a protocol that involves the various public interests in an educated decision-making process. The protocol shown in Fig. 12.37, from Ref. 21, defines an issue evolution and education intervention model that seeks to enhance understanding and decision making.

12.7 LANDFILL OPERATION

The development of a workable operating schedule, a filling plan for the placement of solid wastes, an estimate of the equipment requirements, development of landfill operating records and billing information, a load inspection for hazardous waste, traffic

1 CONCERN
Help audiences understand
existing conditions.
Show how different groups
are affected. Help people look
beyond symptoms.
Help separate facts and myths
and clarify values.

8 EVALUATION
Help monitor and evaluate
policies. Inform people
about formal evaluations and
their results. Help stakeholders participate in formal
evaluations.

2 INVOLVEMENT
Identify decision makers
and others affected. Stimulate
involvement. Encourage
communication among
decision makers, supporters,
and opponents.

7 IMPLEMENTATION
Inform people about new
policies and how they and
others are affected.
Explain how and why they
were enacted. Help people
understand how to ensure
proper implementation.

3 ISSUE
Help clarify goals or interests.
Help understand goals or
interests of others and points
of disagreement. Help get the
issue on the agenda.

6 CHOICE
Explain where and when
decisions will be made
and who will make them.
Explain how decisions are
made and influenced.
Enable audiences to design
realistic strategies.

4 ALTERNATIVES
Identify alternatives,
reflecting all sides of the issue
and including "doing
nothing." Help locate or invent
additional alternatives.

5 CONSEQUENCES
Help predict and analyze
consequences, including
impacts on values as well as
objective conditions.
Show how consequences vary
for different groups. Facilitate
comparison of alternatives.

FIGURE 12.37   Issue evolution/education intervention model. (*From Ref. 21.*)

control on highways leading to the landfill, and a site safety and security program are important elements of a landfill operation plan. An ongoing community relations program is also a part of managing a landfill. Other factors that must be considered in the operation of a landfill are reported in Table 12.16.

Landfill Operating Schedule

Factors that must be considered in developing operating schedules include:

1. Arrival sequences for collection vehicles
2. Traffic patterns at the site
3. The time sequence to be followed in the filling operations
4. Effects of wind and other climatic conditions
5. Commercial and public access

For example, because of heavy truck traffic early in the morning, it may be necessary to restrict public access to the site until later in the morning. Also, because of adverse winter conditions, the filling sequence should be established so that the landfill opera-

TABLE 12.16 Important Factors That Must Be Considered in the Operation of Landfills

Factors	Remarks
Communications	Telephone for emergencies.
Days and hours of operation	Usual practice is 5 to 6 days/week and 8 to 10 h/day
Employee facilities	Rest rooms and drinking water should be provided.
Equipment maintenance	A covered shed should be provided for field maintenance of equipment.
Litter control	Use movable fences at unloading areas; crews should pick up litter at least once per month or as required.
Operation plan	With or without the codisposal of treatment plant sludges and the recovery of gas.
Operational records	Tonnage, transactions, and billing if a disposal fee is charged.
Salvage	No scavenging; salvage should occur away from the unloading area; no salvage storage on site.
Scales	Essential for record keeping if collection trucks deliver wastes; capacity to 100,000 lb.
Security	Provide locked gates and fencing, lighting of sensitive areas.
Spreading and compaction	Spread and compact waste in layers less than 2 ft thick.
Unloading area	Keep small, generally under 100 ft on a side; operate separate unloading areas for automobiles and commercial trucks.

Source: From Ref. 52.

tions are not impeded by unusual weather conditions. If it is not possible to control blowing paper during high wind conditions, closing the landfill when winds exceed, for example, 35 mi/h may be necessary.

Solid Waste Filling Plan

Once the general layout of the landfill site has been established, it will be necessary to select the placement method to be used and to lay out and design the individual solid waste cells. The specific method of filling will depend on the characteristics of the site, such as the amount of available cover material, the topography, and the local hydrology and geology. Details on the various filling methods were presented earlier in this chapter in Sec. 12.1. To assess future development plans, it will be necessary to prepare a detailed plan for the layout of the individual solid waste cells. A typical example of such a plan is shown in Fig. 12.38.

On the basis of the characteristics of the site or the method of operation (e.g., gas recovery), it may be necessary to incorporate special features for the control of the movement of gases and leachate from the landfill. These might include the use of horizontal and vertical gas extraction wells, composite liners, and special extraction facilities.

Equipment Requirements

The type, size, and amount of equipment required will depend on the size of the landfill and the method of operation. The types of equipment that have been used at sanitary landfills include crawler tractors, scrapers, compactors, draglines, and motor-

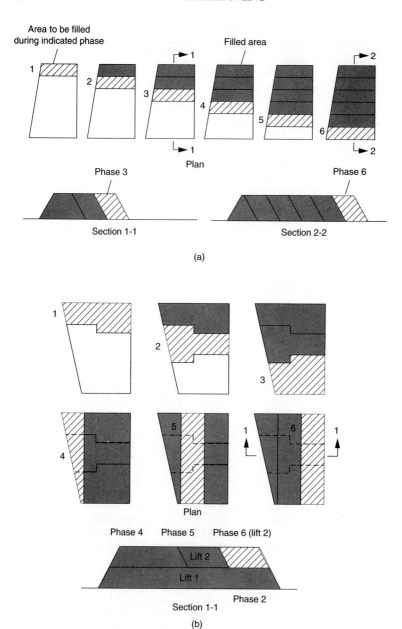

FIGURE 12.38 Typical examples of solid waste filling plans. (*a*) filling plan for single-lift landfill; (*b*) filling plan for a multilift landfill.

FIGURE 12.39 Views of equipment used at landfills. (*a*) Crawler tractor with dozer blade; (*b*) high track crawler tractor with trash blade; (*c*) steel wheel compactor with trash blade; (*d*) self-loading scraper; (*e*) water wagon; (*f*) drag line.

graders (see Fig. 12.39). Of these, crawler tractors are most commonly used. Properly equipped tractors can be used to perform all the necessary operations at a sanitary landfill, including spreading, compacting, covering, trenching, and even hauling cover materials.[4] The size and amount of equipment will depend primarily on the size of the landfill operation. Local site conditions will also influence the size of the equipment. Average equipment requirements that may be used as a guide for landfill operations are reported in Table 12.17.

TABLE 12.17 Typical Equipment Requirements for Sanitary Landfills

Approximate population	Daily wastes, tons	Equipment			Accessory*
		Number	Type	Size, lb	
0–20,000	0–50	1	Tractor, crawler	10,000–30,000	Dozer blade Front-end loader (1 to 2 yd^3) Trash blade
20,000–50,000	50–150	1	Tractor, crawler	30,000–60,000	Dozer blade Front-end loader (2 to 4 yd^3) Bullclam Trash blade
		1	Scraper or dragline		
		1	Water truck		
50,000–100,000	150–300	1–2	Tractor, crawler	30,000+	Dozer blade Front-end loader (2 to 5 yd^3) Bullclam Trash blade
		1	Scraper or dragline†		
		1	Water truck		
100,000	300‡	1–2	Tractor, crawler	45,000+	Dozer blade Front-end loader Bullclam Trash blade
		1	Steel wheel compactor		
		1	Scraper or dragline†		
		1	Water truck		
		*	Road grader		

*Optional, depends on individual needs.
†The choice between a scraper or dragline will depend on local conditions.
‡For each 500-ton increase add one each of each piece of equipment.
Source: From Ref. 52.

Landfill Operating Records

To determine the quantities of waste that are disposed, an entrance scale and gate-house will be required. The gatehouse would be used by personnel who are responsible for weighing the incoming and outgoing trucks. The sophistication of the weighing facilities will depend on the number of vehicles that must be processed per hour and the size of the landfill operation. In some larger landfills, weigh stations are equipped with radiation detectors, to detect the presence of radioactive substances in the incoming wastes. Many weigh stations are monitored with continuously recording video systems. Some examples of weighing facilities are shown in Fig. 12.40. If the weight of the solid wastes delivered is known, then the in-place specific weight of the wastes can be determined and the performance of the operation can be monitored. The weight records would also be used as a basis for charging participating agencies and private haulers for their contributions.

Load Inspection for Hazardous Waste

Load inspection is the term used to describe the process of unloading the contents of a collection vehicle near the working face or in some designated area, spreading the wastes out in a thin layer, and manually inspecting the wastes to determine whether any hazardous wastes are present (see Fig. 12.41). Federal standards mandate randomly selecting MSW loads for inspection. The presence of radioactive wastes can be detected with a hand-held radiation measuring device or at the weigh station, as described above. If hazardous wastes are found, the waste collection company is responsible for removing the hazardous materials or is billed for its removal. At some landfills, if a company is caught bringing in hazardous wastes a second time, a high fine is levied. If caught a third time, the company is banned from discharging wastes at the landfill.

Public Health and Safety

Public health and safety issues are related to worker health and safety and to the health and safety of the public.

Health and Safety of Workers.   The health and safety of the workers at landfills is critical in the operation of a landfill. The types of accidents that have occurred at landfills include puncture wounds from sharp objects, equipment rollovers, laborers being run over, personnel falling into holes, and asphyxiation in confined spaces. The federal government through the Occupational Safety and Health Administration (OSHA) and state-instituted OSHA-type programs have established requirements for a comprehensive health and safety program for the workers at landfill sites. Because the requirements for these programs change continually, the most recent regulations should be consulted in the development of worker health and safety programs. Depending on the activities at the landfill, careful attention must be given to the types of protective clothing and boots, air filtering head gear, and punctureproof gloves supplied to the workers.

Safety of the Public.   As noted previously, safety concerns and the many new restrictions governing the operation of landfills have forced landfill operators to reexamine past operational practices with respect to public safety and site security. As a result,

(a)

(b)

FIGURE 12.40 Typical truck weighing facilities. (*a*) At large landfill; (*b*) at small landfill.

(a)

(b)

FIGURE 12.41 Inspection of solid waste unloaded at landfill for the presence of hazardous wastes at the Frank R. Bowerman landfill in Orange County, Calif. (*a*) Residential load; (b) commercial load.

the use of a convenience transfer station at the landfill site to minimize the public contact with the working operations of the landfill is gaining in popularity.

Site Safety and Security

The increasing number of lawsuits over accidents at landfill sites has caused landfill operators to improve security at landfill sites significantly. Most sites now have restricted access and are fenced and posted with *no trespassing* and other warning signs. In some locations, television cameras are used to monitor landfill operations and landfill access.

Community Relations

An important and often overlooked aspect of landfill operation is sustaining good community relations. The landfill manager must maintain a dialog with neighbors, municipal leaders, community activists, and state governmental representatives in an effort to build trust through honest communications. While community relations activities do not guarantee continued support for the landfilling operation, poor relations almost certainly will result in complaints and problems.

12.8 ENVIRONMENTAL QUALITY MONITORING AT LANDFILLS

Environmental monitoring is conducted at sanitary landfills to ensure that no contaminants that may affect public health and the surrounding environment are released from the landfill. The monitoring required may be divided into three general categories: (1) vadose zone monitoring for gases and liquids, (2) groundwater monitoring, and (3) air quality monitoring. Environmental monitoring involves the use of both sampling and nonsampling methods. Sampling methods involve the collection of a sample for analysis, usually at an off-site laboratory. Nonsampling methods are used to detect chemical and physical changes in the environment as a function of an indirect measurement such as a change in electrical current. Representative devices that have been used to monitor landfill sites are listed in Table 12.18. The typical instrumentation of a landfill for environmental monitoring is illustrated in Fig. 12.42.

Vadose Zone Monitoring

The vadose zone is defined as that zone from the ground surface to where the permanent groundwater is found. An important characteristic of the vadose zone is that the pore spaces are not filled with water, and that the small amounts of water that are present coexist with air. Vadose zone monitoring at landfills involves both liquids and gases.

Liquid Monitoring in the Vadose Zone. Monitoring for liquids in the vadose zone is necessary to detect any leakage of leachate from the bottom of a landfill. In the vadose zone, moisture held in the interstices of the soil particles or within porous rock is

TABLE 12.18 Representative Devices Used to Monitor Landfill Gases and Leachate at Landfills

Type	Application/description
Sampling methods†	
Air quality:	
Evacuated flask	Collection of air grab samples for analysis.
Gas syringe	Collection of air grab samples for analysis.
Air collection bag	Collection of air grab samples for analysis.
Active air sampler	Continuous collection and analysis of gas samples.
Groundwater:	
Monitoring wells; single- and multiple-depth	Used to collect groundwater samples. Multiple extraction wells are used to collect samples from different depths.
Piezometers	Used to collect groundwater samples.
In landfills:	
Piezometers\	Used to collect leachate samples. Piezometers can be installed before filling of the landfill is initiated or after the landfill has been completed.
Vadose zone:	
Collection lysimeter	Used to collect liquid samples below landfill liners.
Soil gas probes; single- and multiple-depth	Used to monitor landfill gases and volatile organic compounds in the soil. The gas may be analyzed in situ using a portable gas chromatograph or tested in a laboratory after absorption in charcoal.
Suction cup lysimeter	Used to obtain liquid samples from the vadose zone.
Nonsampling methods‡	
Groundwater:	
Conductivity cells	Used to monitor changes in groundwater conductivity. Conductivity cells are often located in or near monitoring wells.
In landfills:	
Piezometer	Used to measure the depth of leachate in landfills.
Temperature blocks	Used to measure temperature.
Temperature probes	Used to measure temperature.
Vadose zone:	
Electrical probes	Used to determine the salinity of the vadose zone. A four-probe array is installed so that conductivity of the soil can be measured.
Electrical resistance blocks	Used to measure changes in water content of the vadose zone. Electrode blocks embedded in porous material are installed in the soil. Electrical properties of the blocks change with the changing water content of the vadose zone.
Gamma ray attenuation probes	Used for detecting changes in moisture content of the vadose zone. Based on gamma ray transmission and scattering. In the transmission method, two wells are installed at a known distance apart. A single well is used in the scattering method. Usually limited to shallow depth because of difficulties in installing parallel wells.
Heat dissipation sensors	Used to monitor water content of the vadose zone by measuring the rate of heat dissipation from the block to the surrounding soil.
Neutron moisture meter	Used to obtain a profile of the moisture content of the soil below the landfill. Meter can be installed below a landfill or moved through a borehole next to the landfill.

TABLE 12.18   Representative Devices Used to Monitor Landfill Gases and Leachate at Landfills (*Continued*)

Type	Application/description
	Nonsampling methods‡
Salinity sensors	Used to monitor soil salinity. Electrodes attached to a porous ceramic cup are installed in the soil.
Thermocouple psychrometers	Used to detect changes in moisture content. Operation is based on cooling of a thermocouple junction by the Peltier effect. Wet bulb and dew point. The dew point method is used more commonly in landfill monitoring.
Tensiometers	Used to measure the matric potential of soil. Tensiometers measure the negative pressure (capillary pressure) that exists in unsaturated soil.
Time-domain reflectrometry (TDR)	Based on the difference in dielectric properties of water and soil. Bandwidth and short-pulse length, which are sensitive to the high-frequency electrical properties of the material, are measured.
Wave-sensing devices	Use of seismic or acoustic wave propagation properties for leak detection. In the seismic wave technique, the difference in travel time of Rayleigh waves between the source and geophones is used to detect leaks. In the acoustic emission monitoring (AEM) technique, sound waves generated by flowing water from a leak are utilized in leak detection.

†Methods involving the collection of samples for subsequent laboratory analysis.
‡Methods involving physical and electrical measurements.
Source: From Ref. 52.

always held at pressures below atmospheric pressure. To remove the moisture it is necessary to develop a negative pressure or vacuum to pull the moisture away from the soil particles. Because suction must be applied to draw moisture out of the soil in the vadose zone, conventional wells or other open cavities cannot be used to collect samples in this zone. The sampling devices used for sample extraction in the unsaturated zone are called *suction lysimeters.* Three commonly used classes of lysimeters are (1) the ceramic cup, (2) the hollow fiber, and (3) the membrane filter.[2, 46]

The most commonly used device for obtaining samples of moisture in the vadose zone is the ceramic cup sampler (see Fig. 12.43), which consists of a porous cup or ring made of ceramic material which is attached to a short section of nonporous tubing (e.g., PVC). When the cup is placed in the soil, the pores become an extension of the pore space of the soil. Soil moisture is drawn in through the porous ceramic element by the application of a vacuum. When a sufficient amount of water has collected in the sampler, the collected sample is pulled to the surface through a narrow tube by the application of a vacuum or is pushed up by air pressure.

Gas Monitoring in the Vadose Zone.   Monitoring for gases in the vadose zone is necessary to detect the lateral movement of any landfill gases. A typical example of a vadose zone gas monitoring probe is illustrated in Fig. 12.44. In many monitoring systems, gas samples are collected from multiple depths in the vadose zone. Where landfills are located near occupied buildings, testing as frequently as twice per week has been necessary to monitor landfill gas migration adequately.

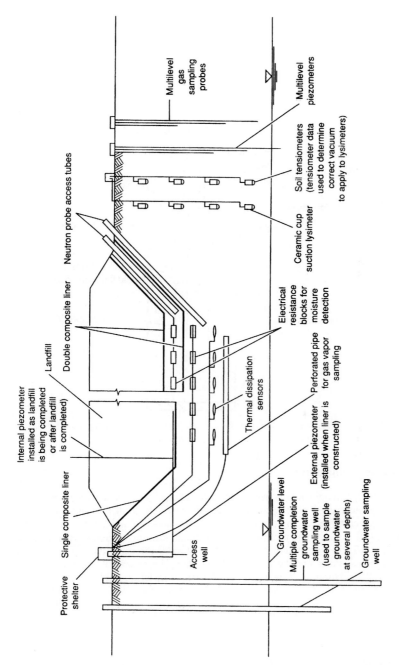

FIGURE 12.42 Instrumentation of a landfill for the collection of environmental monitoring data. Not all of the devices and instrumentation shown would be used at an individual landfill. (*From Ref. 52.*)

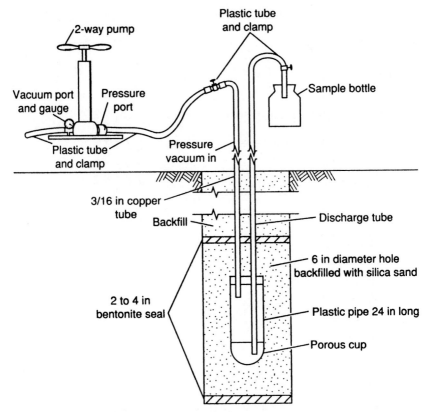

FIGURE 12.43 Porous cup suction lysimeter for the collection of liquid samples from the vadose zone. (*Courtesy of California Integrated Waste Management Board.*)

Groundwater Monitoring

Monitoring of the groundwater is necessary to detect changes in water quality that may be caused by the escape of leachate and landfill gases. Both down- and upgradient wells are required to detect any contamination of the underground aquifer by leachate from the landfill. An example of a well used for the monitoring of groundwater is illustrated in Fig. 12.45. To obtain a representative sample, the same type of equipment should be used each time and the well must be purged prior to sample collection.

By federal regulation, all new MSW landfills must install groundwater monitoring facilities. Existing sites have a number of years to implement the 1993 requirements for monitoring. There are also extensive regulations for sample collection, testing, and data analysis.

Landfill Air Quality Monitoring

Air quality monitoring at landfills involves (1) the monitoring of ambient air quality at and around the landfill site, (2) the monitoring of landfill gases extracted from the

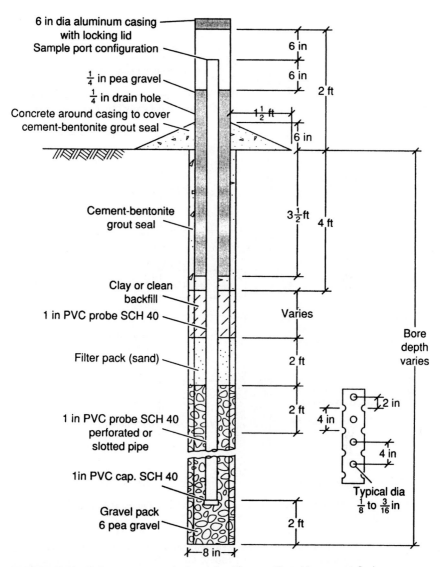

FIGURE 12.44 Vadose zone gas monitoring probe. (*Courtesy Waste Management, Inc.*)

landfill, and (3) the monitoring of the offgases from any gas processing or treatment facilities.

Monitoring Ambient Air Quality. Ambient air quality is monitored at landfill sites to detect the possible movement of gaseous contaminants from the boundaries of the landfill site. Gas sampling devices can be divided into three categories: (1) passive, (2) grab, and (3) active. Passive sampling involves the collection of a gas sample by passing a stream of gas through a collection device in which the contaminants con-

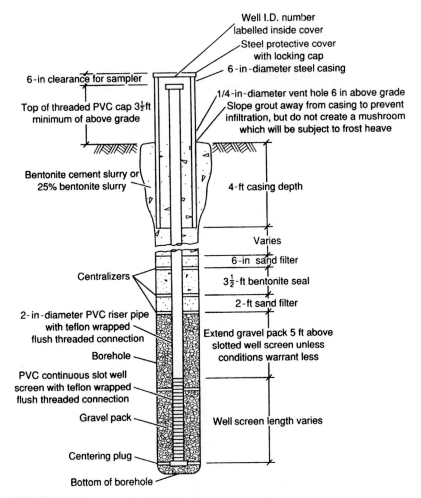

FIGURE 12.45 Typical groundwater monitoring well. (*Courtesy Waste Management, Inc.*)

tained in the gas stream are removed for subsequent analysis. Commonly used in the past, passive sampling is seldom used today. Grab samples are collected using an evacuated flask, gas syringe, or an air collection bag made of a synthetic material (see Fig. 12.46). An active sampler involves the collection and analysis of a continuous stream of gas.

Monitoring Extracted Landfill Gas. Landfill gas is monitored to assess the composition of the gas and to determine the presence of trace constituents that may pose a health or environmental risk.

Monitoring Offgases. Monitoring offgases from treatment and energy recovery facilities is done to determine compliance with local air pollution control requirements. Both grab and continuous sampling has been used for this purpose.

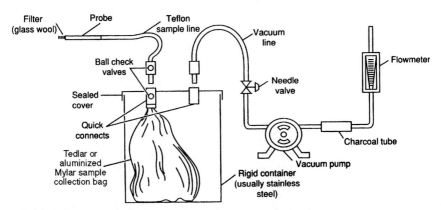

FIGURE 12.46 Sampling apparatus for the collection of air grab samples at landfills. (*From Ref. 52.*)

12.9 LANDFILL CLOSURE, POSTCLOSURE CARE, AND REMEDIATION

Landfill closure and postclosure care are the terms used to describe what is to happen to completed landfill in the future. To ensure that completed landfills will be maintained 30 to 50 years into the future, many states and the federal government have passed legislation that requires the operator of a landfill to put aside enough money so that when the landfill is completed, the facility can be closed, maintained, and monitored properly for 30 to 40 years. Greater long-term care periods are being considered in some European countries.

Development of Long-Term Closure Plan

Perhaps the most important element in the long-term maintenance of a completed landfill is the availability of a closure plan in which the requirements for closure are delineated clearly. A closure plan must include a design for the landfill cover and the landscaping of the completed site. Closure must also include long-term plans for runoff control, erosion control, gas and leachate collection and treatment, and environmental monitoring. The closure plan for the landfill layout given in Fig. 12.32 is presented in Fig. 12.47.

Cover and Landscape Design. The landfill cover must be designed to divert surface runoff and snowmelt from the landfill site and to support the landscaping design selected for the landfill. Increasingly, the final landscaping design is based on local plant and grass species as opposed to nonnative plant and grass species. In many water-short locations in the southwest, a desert type of landscaping is favored.

Control of Landfill Gases. The control of landfill gases is a major concern in the long-term maintenance of landfills. Because of the concern over the uncontrolled release of landfill gases, most modern landfills have some sort of gas control system installed before the landfill is completed. Older completed landfills without gas collection systems are being retrofitted with gas collection systems.

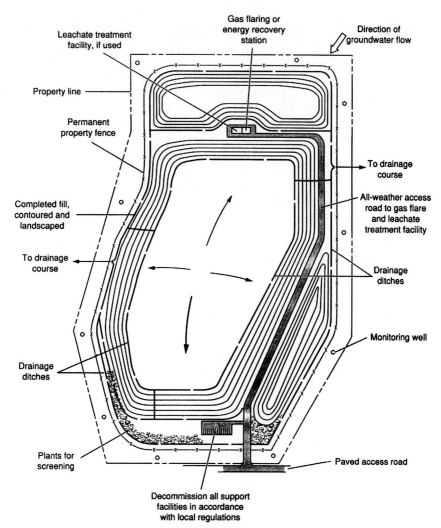

FIGURE 12.47 Plan view of completed landfill showing all of the elements involved in closure and postclosure care. (*From Ref. 52.*)

Collection and Treatment of Leachate. As with the control of landfill gas, the control of leachate discharges is another major concern in the long-term maintenance of landfills. Again, most modern landfills have some sort of leachate control system as discussed above. Older completed landfills without leachate collection systems, are being retrofitted with leachate collection systems. These retrofitted collection systems are similar in construction to vertical gas wells in which leachate pumps are installed. The leachate head wells, as they are commonly referred to, can be difficult to install due to obstructions encountered during drilling. Poor hydraulic flow conditions through the waste may also limit their effectiveness.

Environmental Monitoring Systems.   To be able to conduct long-term environmental monitoring after a landfill has been completed it will be necessary to install monitoring facilities. The monitoring required at completed landfills usually involves: (1) vadose zone monitoring for gases and liquids, (2) groundwater monitoring, and (3) air quality monitoring. The required facilities have been described previously.

Postclosure Care

Postclosure care involves the routine inspection of the completed landfill site, maintenance of the infrastructure, and environmental monitoring. These subjects are considered briefly below.

Routine Inspections.   A routine inspection program must be established to monitor continually the condition of the completed landfill. Criteria must be established to determine when corrective action must be taken. For example, how much settlement will be allowed before regrading must be undertaken?

Infrastructure Maintenance.   Infrastructure maintenance typically involves the continued maintenance of surface water diversion facilities; landfill surface grades; the condition of liners in covers, where used; revegetation; and maintenance of landfill gas and leachate collection equipment. The amount of regrading that will be required will depend on the amount of settlement (see Fig. 12.48). In turn, the rate of settlement will depend on the rate of gas formation and the degree of initial compaction achieved in the placement of the waste materials in the landfill. The amount of equipment that must be available at the site will depend on the extent of the landfill and the nature of the facilities that must be maintained.

Environmental Monitoring Systems.   Long-term environmental monitoring is conducted at completed landfills to ensure that there is no release of contaminants from the landfill that may impact health or the surrounding environment. The monitoring required at completed landfills usually involves: (1) vadose zone monitoring for gases and liquids, (2) groundwater monitoring, and (3) air quality monitoring. The number of samples collected for analysis and the frequency of collection will usually depend on the regulations of the local air pollution and water pollution control agencies. EPA has developed a baseline procedure for sampling of groundwater that should be reviewed (40 CFR 258).

Remediation

Remedial actions may be necessary if unacceptable levels of environmental emissions are detected in the postclosure monitoring program. Remedial actions may be the result of landfill gas migration, toxic air emissions, leachate polluting the groundwater, or some other unforeseen event. The severity of the problem will determine the intensity of the remedial action and the long-term cost.

Migration Control.   Federal regulations specify that methane concentrations cannot exceed 5 percent methane at the property boundary of the MSW landfill. Some states require even lower concentrations. Landfill gas migration may unexpectedly extend

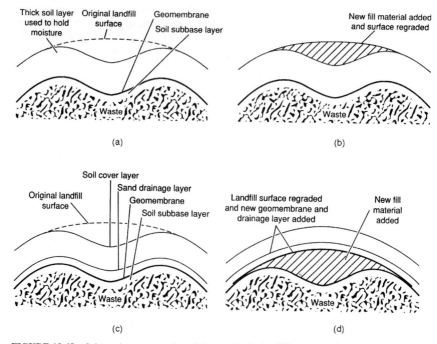

FIGURE 12.48 Schematic representation of the repair of a landfill cover employing a geomembrane to restore drainage. (*a*) Landfill after closure and settlement; (*b*) landfill repair procedure; (*c*) landfill employing a drainage layer and a geomembrane after closure and settlement; (*d*) landfill after repair to restore surface drainage. (*From Ref. 52.*)

into areas on which there are occupied buildings. Emergency measures to secure the area and evacuate buildings are the first steps that must be initiated without delay. Local fire departments usually have the appropriate equipment to measure for the presence of methane in buildings. Wells are then usually installed not only on or adjacent to the landfill to stop gas movement away from the site, but also in the vicinity of the buildings to remove the gas from the ground. The wells in or adjacent to the landfill will likely be operated for years until the concentrations of methane being generated is determined to not be a threat. The wells located near the occupied buildings will temporarily operate, usually on the order of months, until the methane in the vadose zone is reduced to safe levels.

Toxic Air Emissions. A number of landfill operators have unexpectedly found it necessary to install landfill gas control and recovery systems to limit the release of toxic compounds into the atmosphere. The technology and system configurations are described earlier in this chapter. The necessary duration for operating these systems is unknown.

Groundwater Remediation. Unlined landfills and landfills without leachate collection systems are the most likely to have a deleterious effect on groundwater quality. A remedial action program is instituted under federal or state regulations where contamination is detected in groundwater monitoring wells. As shown in Fig. 12.49, the first remediation step is placing a new, highly impermeable cap over the landfill to reduce water draining through the waste. Subsequent measures are designed to limit or cut off

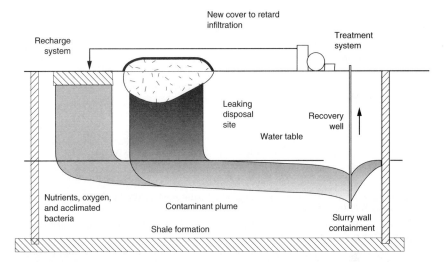

FIGURE 12.49 Typical example of a groundwater remediation system involving the use of slurry wall containment, recover wells, treatment of contaminated groundwater, and groundwater recharge with nutrient addition to achieve in situ remediation.

the movement of contaminated groundwater away from the landfill by the installation of bentonite slurry walls or the operation of recovery wells which control subsurface hydraulics. Contaminated groundwater in the aquifer surrounding the site is treated in above ground facilities and either reinjected, sprayed onto nearby land, or discharged to a surface water. In situ bioremediation techniques may also be utilized to remove contaminants from the groundwater. Complete restoration will likely take years or decades to complete.

REFERENCES

1. Amling, W., "Exclusion of Gulls from Reservoirs in Orange County, California," *Progressive Fish Culturist,* vol. 43, no. 3, 1981.

2. Bagchi, A., *Design, Construction, and Monitoring of Sanitary Landfill,* John Wiley & Sons, New York, 1990.

3. Bohn, H., and R. Bohn, "Soil Beds Weed Out Air Pollutants," *Chemical Engineering,* vol. 95, no. 6, 1988.

4. Brunner, D. R., and D. J. Keller, "Sanitary Landfill Design and Operation," U.S. Environmental Protection Agency, Publication SW-65ts, Washington, D.C., 1972.

5. California Waste Management Board, "Landfill Gas Characterization," Sacramento, 1988.

6. Christensen, T. H., and P. Kjeldsen, "2.1. Basic Biochemical Processes in Landfills," in T. H. Christensen, R. Cossu, and P. Stegmann (eds.): *Sanitary Landfilling: Process, Technology and Environmental Impact,* Academic Press, Harcourt Brace, Jovanavich, Publishers, London, 1989.

7. County of Los Angeles, Department of County Engineer, Los Angeles, and Engineering-Science, Inc., *Development of Construction and Use Criteria for Sanitary Landfills, An Interim Report,* U.S. Department of Health, Education, and Welfare, Public Health Service, Bureau of Solid Waste Management, Cincinnati, 1969.

8. Crawford, J. F., and P. G. Smith, *Landfill Technology,* Butterworths, London, 1985.

9. Davis, S. N., and R. J. M. DeWiest, *Hydrogeology,* John Wiley & Sons, New York, 1966.

10. Ehrig, H. J., "Leachate Quality," in T. H. Christensen, R. Cossu, and P. Stegmann (eds.): *Sanitary Landfilling: Process, Technology and Environmental Impact,* Academic Press, Harcourt Brace, Jovanavich, Publishers, London, 1989.

11. Emcon Associates, *Methane Generation and Recovery from Landfills,* Ann Arbor Science, Ann Arbor, Mich., 1980.

12. Farquhar, C. J., and F. A. Rovers, "Gas Production During Refuse Decomposition," *Water, Air and Soil Pollution,* vol. 2, pp. 483–495, 1973.

13. Fenn, D. G., K. J. Hanley, and T. V. DeGeare, "Use of Water Balance Method for Predicting Leachate Generation from Solid Waste Disposal Sites," EPA/530/SW-168, 1975.

14. Frevert, R. K., G. O. Schwab, T. W. Edminster, and K. K. Barnes, *Soil and Water Conservation Engineering,* John Wiley & Sons, Inc., New York, 1963.

15. Ham, R. K., et al., "Recovery, Processing and Utilization of Gas from Sanitary Landfills," EPA-600/2-79-001, 1979.

16. Hansen, V. E., O. W. Israelsen, and G. E. Stringham, *Irrigation Principles and Practices,* John Wiley & Sons, New York, 1979.

17. Hatheway, A. W., P. Geol, and C. C. McAneny, "An In-Depth Look at Landfill Covers," *Waste Age,* vol. 17 , no. 8, 1987.

18. Herrera, T. A., R. Lang, and G. Tchobanoglous, "A Study of the Emissions of Volatile Organic Compounds Found in Landfills," *Proc. 43d Annual Purdue Industrial Waste Conference,* Lewis Publishers, Inc., Chelsea, Mich., pp. 229–238, 1989.

19. Hjelmfelt, A. T., Jr., and J. J. Cassidy, *Hydrology for Engineers and Planners,* Iowa State University Press, Ames, 1975.

20. Holland, K. T., J. S. Knapp, and J. G. Shoesmith, *Anaerobic Bacteria,* Chapman and Hall, New York, 1987.

21. House, V. W., and A. A. Young, "Working with Our Publics, Module 6, Education for Public Decisions," Agricultural Extension Service and Department of Adult and Community College Education, North Carolina State University, Raleigh, 1989.

22. Huitric, R. L., "In-Place Capacity of Refuse to Absorb Liquid Wastes," 2d Nat. Conf. on Hazardous Material Management, San Diego, Calif., 1979.

23. Huitric, R. L., S. Raksit, and R. T. Haug, *Moisture Retention of Landfilled Solid Waste,* County Sanitation Districts of Los Angeles County, Los Angeles, 1980.

24. Kmet, P., K. J. Quinn, and C. Slavik, "Analysis of Design Parameters Affecting the Collection Efficiency of Clay Lined Landfills," 4th Annual Madison Conf. of Applied Research and Practice on Municipal and Industrial Waste, University of Wisconsin-Extension, Madison, Wisc., Sept. 28–30, 1981.

25. Koerner, R. M., and D. E. Daniel, "Better Cover-Ups," *Civil Engineering,* vol. 62, no. 5, 1992.

26. Lang, R. J., et al., *Trace Organic Constituents in Landfill Gas,* prepared for the California Waste Management Board, Department of Civil Engineering, University of California, Davis, 1987.

27. Lang, R. J., et al., *Summary Report: Movement of Gases in Municipal Solid Waste Landfills,* prepared for the California Waste Management Board, Department of Civil Engineering, University of California, Davis, 1989.

28. Lang, R. J. and G. Tchobanoglous, *Movement of Gases in Municipal Solid Waste Landfills: Appendix A Modelling the Movement of Gases in Municipal Solid Waste Landfills,* prepared for the California Waste Management Board, Department of Civil Engineering, University of California, Davis, 1989.

29. Linsley, R. K., M. A. Kohler, and J. H. Paulhus, *Hydrology for Engineers,* McGraw-Hill, New York, 1958.

30. Linsley, R. K., J. B. Franzini, D. Fryberg, and G. Tchobanoglous, *Water Resources Engineering,* 4th ed., McGraw-Hill, New York, 1991.

31. Mackay, K. M., P. V. Roberts, and J. A. Cherry, "Transport of Organic Contaminants in Groundwater," *Environmental Science and Technology,* vol. 19, no. 5, pp. 384–392, 1985.

32. Mathias, S. L., "Discouraging Seagulls: The Los Angeles Approach," *Waste Age,* vol. 15, no. 11, 1984.

33. McAtee, W. L., "Excluding Birds from Reservoirs and Fishponds," leaflet 120, U.S. Department of Agriculture, Washington, D.C., 1936.

34. Merz, R. C., and R. Stone, *Special Studies of a Sanitary Landfill,* U.S. Department of Health, Education, and Welfare, Washington, D.C., 1970.

35. Moshiri, G. A., and C. C. Miller, "An Integrated Solid Waste Facility Design Involving Recycling, Volume Reduction, and Wetlands Leachate Treatment," in G. A. Moshiri (ed.), *Constructed Wetlands for Water Quality Improvement,* Lewis Publishers, Boca Raton, Fla., 1993.

36. O'Leary, P., and P. Walsh, *Solid Waste Landfills Correspondence Course,* University of Wisconsin, Madison, 1992.

37. Parker, A., "Behaviour of Wastes in Landfill-Leachate" and "Behaviour of Wastes in Landfill-Methane Generation," Chaps. 7 and 8 in J. R. Holmes (ed.), *Practical Waste Management,* John Wiley & Sons, Chichester, England, 1983.

38. Perry, R. H., D. W. Green, and J. O. Maloney, *Perry's Chemical Engineers' Handbook,* 6th ed., McGraw-Hill, New York, 1984.

39. Pfeffer, J. T., *Solid Waste Management Engineering,* Prentice-Hall, Englewood Cliffs, N.J., 1992.

40. Pohland, F. G., *Critical Review and Summary of Leachate and Gas Production from Landfills,* EPA/600/S2-86/073, U.S. EPA, Hazardous Waste Engineering Research Laboratory, Cincinnati, 1987.

41. Pohland, F. G., "Fundamental Principles and Management Strategies for Landfill Codisposal Practices," *Proc. Sardinia 91, 3d Int. Landfill Symp.,* vol. II, pp. 1445–1460, Grafiche Galeati, Imola, Italy, 1991.

42. Rich, L. G., *Unit Processes of Environmental Engineering,* Wiley & Sons, New York, 1963.

43. Salvato, J. A., W. G. Wilkie, and B. E. Mead, "Sanitary Landfill-Leaching Prevention and Control," *Journal WPCF,* vol. 43, no. 10, pp. 2084–2100, 1971.

44. Schroeder, P. R., et al., "The Hydrologic Evaluation of Landfill Performance (HELP) Model," *User's Guide for Version I,* EPA/530/SW-84-009, 1, U.S. EPA Office of Solid Waste and Emergency Response, Washington, D.C., 1984.

45. Schroeder, P. R., et al., "The Hydrologic Evaluation of Landfill Performance (HELP) Model," *Documentation for Version I,* EPA/530/SW-84-010, 2, U.S. EPA Office of Solid Waste and Emergency Response, Washington, D.C., 1984.

46. SCS Engineers, Inc., *Procedural Guidance Manual for Sanitary Landfills,* vol. I: *Landfill Leachate Monitoring and Control Systems,* California Waste Management Board, Sacramento, 1989.

47. SCS Engineers, Inc., *Procedural Guidance Manual for Sanitary Landfills,* vol. II: *Landfill Gas Monitoring and Control Systems,* California Waste Management Board, Sacramento, 1989.

48. Stahl, J. F., M. Moshiri, and R. Huitric, *Sanitary Landfill Gas Collection and Energy Recovery,* County Sanitation Districts of Los Angeles County, Los Angeles, 1982.

49. State Water Pollution Control Board, *Report on the Investigation of Leaching of a Sanitary Landfill,* California State Water Pollution Control Board, publication 10, Sacramento, 1954.

50. State Water Resources Control Board, *In-Sites Investigation of Movement of Gases Produced from Decomposing Refuse,* Final Report, The Resources Agency, publication 35, State of California, Sacramento, 1967.

51. Tchobanoglous, G., and F. L. Burton, *Wastewater Engineering: Treatment, Disposal, Reuse,* 3d ed., McGraw-Hill, New York, pp. 1–1350, 1991.

52. Tchobanoglous, G., H. Theisen, and S. A. Vigil, *Integrated Solid Waste Management, Engineering Principles and Management Issues,* McGraw-Hill, New York, 1993.

53. U.S. Army Corp of Engineers, *Snow Hydrology,* North Pacific Division, Portland, Ore., 1956.

54. U.S. Environmental Protection Agency, 40 CFR Parts 257 and 258, "Solid Waste Disposal Facility Criteria, Final Rule," *Federal Register,* October 9, 1991.

55. Water Pollution Control Federation and American Society of Civil Engineers (Joint Committee), *Design and Construction of Sanitary and Storm Sewers,* WPCF Manual of Practice No. 9, Washington, D.C., 1969.

56. "World Wastes, Equipment Catalog," Communication Channels, Inc., Atlanta, 1986.

57. Wright, T. D., "To Cover or Not to Cover?" *Waste Age,* vol. 17, no. 3, 1986.

58. Young, P. J., and L. A. Heasman, "An Assessment of the Odor and Toxicity of the Trace Components of Landfill Gas," *Proc. GRCDA 8th Int. Landfill Gas Symp.,* 1985.

CHAPTER 13

SITING SOLID WASTE FACILITIES IN THE UNITED STATES

Lawrence Susskind

*Professor of Urban and Environmental Planning, MIT, and
Director MIT–Harvard Public Disputes Program
Cambridge, Massachusetts*

David Laws

*Doctoral Candidate, Department of Urban Studies and
Planning, MIT
Cambridge, Massachusetts*

13.1 INTRODUCTION

Solid waste management is likely to utilize recycling, landfilling, and incineration for the foreseeable future. New facilities will have to be built. When the siting of such facilities is not viewed as part of a process of building consensus on solid waste management goals, public opposition can lead to long delays and often to the rejection of proposed facilities. Continued failure to site needed solid waste management facilities will compromise our national ability to meet basic environmental protection and resource management needs.

The difficulties that plague siting efforts usually stem from the failure to take adequate account of the concerns that affected groups have whenever new solid waste management facilities are proposed. People bordering on the site are likely to resist siting efforts because they are afraid that their property values or quality of life could be adversely affected. Local business leaders are likely to resist new facilities because they may increase taxes or operating costs. Environmental groups will push for source reduction and extensive recycling before they support additional landfills, and they are likely to oppose siting incinerators on the grounds that they pose potential risks to human health and safety. There may well be other groups with additional concerns, depending on the type, scale, location, and cost of proposed facilities.

This chapter reviews the political, technical, economic, and ethical aspects of siting that are prominent sources of public concern, discusses how they arise and are usually handled in the series of choices involved in a typical siting process, and, finally, presents a siting credo that responds effectively to public concerns and embodies the best practical advice available for anyone who will manage or participate in an effort to site a solid waste management facility.[1]

13.2 UNDERSTANDING THE SOURCES OF PUBLIC CONCERN

Someone charged with selecting a site for a solid waste management facility ought to expect opposition. While the sources of resistance are often characterized as NIMBYism, a not-in-my-backyard attitude, this characterization assumes that opposition to new facilities is based solely on selfish desires to shunt the burden of public responsibility elsewhere, but this is often not the case. While there may be residents who would prefer no change of any kind, the most vigorous opposition usually comes from those who have legitimate concerns or favor alternative methods of solid waste management, other "more appropriate" sites, or different ways of making decisions.

Anyone reading in the newspaper one morning that an empty field at the end of the street was being considered for an incinerator or a landfill would have concerns. What risks would such a facility pose to family and neighbors? How should such risks be calculated? Will the operators of the proposed facility meet their obligations? Will the technology they have chosen work as expected? What will happen if something goes wrong? What kinds of impacts will the proposed facility have on the immediate area? Will neighbors move away? What about those who have no choice about moving? Could the facility affect groundwater quality? Who will pay if it does? What kind of traffic and noise will the facility generate? What might it do to property values? Will it affect schools? Why should this particular neighborhood be asked to bear the burden so that others can dispose of their trash? Will wastes from other communities be trucked in? Is someone going to make a profit at the expense of others? How was this site selected? These concerns and others might readily lead potential neighbors to the conclusion that their best strategy is to be first and loudest in raising objections in order to increase the chances that another site will be selected instead.

Solid waste management facilities represent long-term commitments of public resources that can dramatically alter the quality of life in a community. Thus, neighbors of proposed facilities and members of the public have legitimate grounds for concern. They also have reason to expect that every effort will be made to ensure that wise decisions are made.

Political Aspects of the Siting Process

Siting decisions hinge on the trust that citizens have in government, technology, and business. To the extent that corruption, closed decision-making processes, and highly publicized accidents involving advanced technologies (like Three Mile Island, the Challenger, or the Exxon Valdez) have undermined public confidence, siting decisions have become that much more difficult. For the public official, corporate representative, or technical consultant involved in the siting process, this general lack of trust often translates into skepticism, an unwillingness to accept assertions at face value, requirements that extra margins of safety be met, and demands for risk reduction or compensation.

Advocacy groups and citizen leaders have developed considerable sophistication in putting forward such demands. There are national networks they can tap for advice and legal assistance. While they rarely have the power to veto a siting decision, they can often mount legal and political challenges that will tie up a project indefinitely.

Healthy skepticism is not necessarily inappropriate and should perhaps even be encouraged when decisions of great significance must be made, but when profound distrust coupled with unreasonable demands produce political gridlock, everyone is hurt. Without some willingness to "suspend disbelief" and engage in a community-wide dialogue, unwise decisions are likely to result.

Technical Aspects of the Siting Process

There are several types of technical analysis that are important to making wise siting decisions, although they are rarely precise enough to produce definitive answers. Good technical analysis can clarify operating assumptions, specify areas of uncertainty, highlight data deficiencies, and spell out the sensitivity of key findings to slight variations in underlying assumptions. The key technical analyses required to evaluate solid waste management options involve forecasts of demand, site suitability analysis, impact and mitigation analysis, risk assessment, and specification of monitoring and management standards. Because nonobjective judgments play an important role in all of these analyses, it is imperative that groups concerned about the impacts of proposed solid waste management facilities become involved in the decision-making process.

Economic Aspects of the Siting Process

From an economic perspective, a solid waste management facility represents a stream of benefits and costs. The benefits flow from the ability to dispose of waste, although for most people such a service is often taken for granted. Benefits of this sort are usually distributed fairly evenly across all users of a facility. Costs include disposal fees and taxes, indirect costs such as traffic, noise, odor, changes in property values, and elevated risks to human health and safety. Except for certain out-of-pocket costs, these are likely to be distributed *unevenly*. They tend to fall hardest on those closest to a facility, and drop off across a gradient that correlates roughly with distance from it.

The obvious economic imperative is to find a technological and locational option that is efficient: one that provides the greatest level of *net* benefit. As with technical analysis, however, we suggest that the definition of costs and benefits hinges on a great many subjective judgments. Again, unless concerned groups are involved in preparing such analyses, they will be likely to question the legitimacy of the results.

Moreover, even if a facility is efficient, and the overall benefits to the "gainers" greatly outweigh the costs, resistance can be anticipated. The potential losses faced by the relatively small group of "losers," while modest in aggregate terms, are likely to be significant to individuals and provide them with a substantial (even pressing) incentive to act. And these individuals will face only minimal organizational costs if they wish to act collectively, since the losers are likely to be concentrated geographically and may even know each other. The beneficiaries, on the other hand, cannot be expected to support a proposal since, though numerous, they receive only modest individual benefits and face large organizational costs if they wish to act as a group.[2]

Ethical Aspects of the Siting Process

The siting of solid waste management facilities raises ethical questions of various kinds. The distribution of costs and benefits may or may not be fair. When solid waste facilities serve regional needs or where facilities accept waste hauled in from long distances for high fees, questions of fairness to the host community are almost always raised. If disproportionate burdens fall on those who appear to be targeted because they are poor, racially or ethnically distinct, or disadvantaged in some way, issues of fairness are often reframed as questions of justice or abridgment of rights. When historical patterns of siting appear to repeatedly burden a particular community, legal and political charges of discrimination will probably be raised. These have been some of the motivating concerns behind calls for "environmental justice" which have become prominent in siting and other environmental disputes. A recent issue of *EPA Journal* was devoted to exploring concerns about environmental justice,[3] and authors such as Bullard[4] have presented the case that "environmental racism" exists. Godsil's review of recent legal history suggests that it may be possible to challenge siting decisions on the grounds of discrimination.[5] While the evidence currently available is not conclusive, more than enough examples are available to fuel the argument on several fronts.

All of this suggests that fairness should (and will) be a primary concern in facility siting. While it is difficult to define fairness in a way that will satisfy all stakeholding groups in every situation, steps should at least be taken to ensure that a siting process is viewed as fair by as many groups as possible. This, of course, raises additional questions about what constitutes *procedural fairness*. At minimum, a siting process needs to permit all stakeholders to participate in open deliberations in which they have an opportunity to understand the issues under discussion, ask questions, voice concerns, and propose ways of resolving those concerns in order to fulfill the basic requirements of fairness.[6] In effect, a siting process needs to be *transparent*. Siting decisions should be taken only after there has been sufficient opportunity to explore alternatives that might meet the interests of all stakeholding parties.

13.3 A TYPICAL SITING CHRONOLOGY

A typical siting process begins with a determination that a facility of some sort is needed. Then, technological options are either explicitly or implicitly considered. These two steps can occur either before alternative sites are considered or at the same time that specific sites are reviewed. In either case, the next step is to forecast and assess ways of mitigating potential impacts. This is generally undertaken as part of an environmental impact process. Finally, operational guidelines for the management of the facility are prepared, usually as the product of all the prior steps.

Determining Need

Need is the pivotal question in any siting process. If a community accepts the need for a facility, a siting process has a much better chance of succeeding. If doubts linger about the way need was determined, those opposed—for whatever reasons—are likely to find allies to join in blocking actions. While it is clear that judgments about need should be based on an analysis of past practice and future requirements, it is also clear that determination of need rests heavily on a great many nonobjective judgments, such as estimates of the size and composition of future waste streams, which are always debatable.

An assessment of need involves forecasting population growth (including both rates and absolute levels), consumption levels, and the prospect of compliance with new regulations. Forecasts hinge on assumptions about how people will live in the future and how much waste they will generate. They also require judgments about the changing composition of the waste stream. Economists, demographers, civil engineers, sociologists, and others can offer advice on how to make such estimates, but it is clear that judgments, more than facts, dominate.

A needs assessment must also take account of what disposal costs are likely to be, and how the behavior of individuals and firms will be influenced by price. The not-in-anybody's-backyard movement is a challenge to the way in which the need for new waste disposal facilities has traditionally been determined. This movement focuses on blocking all new facilities, not to push construction to other locations, but rather to increase the pressure for source reduction and recycling. Proponents argue that source reduction and recycling goals will be undermined if new facilities are built. In short, they question the need for any new solid waste management facilities.

Choosing a Technology

Suppose that the parties involved in a siting decision are able to agree that, indeed, a new facility of some sort is required. This may include a shared understanding that some reduction and recycling goals must also be met. Given such an agreement—and assuming that the volume and composition of the waste stream have been specified—it may be tempting to think that the choice of technology will be obvious. Unfortunately, this is not the case. The choice of a "best" technology is as open to challenge as the determination of need. The magnitude of the impacts (in terms of costs and risks) that alternative technologies might have, as well as probable impacts of alternative technologies on ecology, human health, safety, and welfare are likely to all be quite sensitive to assumptions that are open to challenge.

Some technological impacts are relatively easy to forecast, such as the increase in traffic that will occur during the construction and operation of a new facility. The level of noise associated with the given volume of traffic can also be extrapolated. The tolerance of individuals to different traffic and noise levels is, however, much harder to forecast with certainty. Assumptions must be made about whether traffic and noise will be constant or intermittent, what the hours of operation will be, and how sensitive different households will be. While ballpark estimates may be feasible, the individuals involved will demand the final word.

It may be possible to forecast the likely impacts the various technologies will have on human health. Estimates of the level of toxic metals in stack emissions from the incineration of municipal solid waste may be feasible, for example.[7] In many cases, however, the absence of historical, or baseline, data makes this very difficult. Moreover, the relevance of historical or comparable situations will be questioned by those who see things differently.[8]

Forecasting the mobility of chemicals in the ash from a municipal solid waste incinerator requires the use of a leaching test. Several such tests are available. The forecast of how hazardous the ash from a particular facility may be depends on which test is used, as well as the standard for extrapolation. It also depends on assumptions about the content of the waste stream. Estimates of the costs (and health risks) associated with disposing of the ash depend on how federal regulations governing household waste are presumed to apply. Moreover, assessments of the health effects of alternative disposal technologies require a detailed study of possible exposure pathways. Some of these are not yet well understood. Estimating the magnitude of a human dose requires assumptions about body weight, rates of respiration, consumption of food and water, incidental soil ingestion, and the extent of dermal absorption.[9] Variations in these assumptions substantially influence estimates of health effects. The use of average or standard figures, often substituted for empirical data, can obscure differences important to certain segments of the local population.

Even when the risks associated with alternative technologies can be estimated with reasonable certainty, the acceptability of such risks usually varies among different parts of the population. This may be because of the character of the risks (e.g., unfamiliar, involuntary, undetectable risks are usually less acceptable than familiar, voluntary, and detectable risks), or the way they are distributed across the population (e.g., risks to children may be treated more seriously than risk to adults).[10] Any of these items is sufficient to alter the assessment of a particular technology.

Site Selection

Many aspects of site suitability can only be assessed with a particular technology in mind. The converse is also true. The question that is most often asked in evaluating sites is, "How suitable is this particular site for the activity we have in mind?" Those responsible for selecting a site may deem some locations inappropriate on the basis of exclusionary criteria (related to health and safety standards) that do not require contingent analysis. Asking the question this way, however, leaves a variety of sites that may be acceptable, although each will present a different combination and distribution of costs and risks. Thus, there is a strong temptation (as with the choice of technology) to ask which is the best or optimal site. Unfortunately, answers to this question will probably reveal more about the perspective that the analyst adopts than about the objective characteristics of the situation.

Any number of factors can be taken into account in assessing the suitability of a site. Transportation access, soil capability, adjacent land uses, and land ownership patterns are usually given consideration in assessing sites. If a landfill is the technology of choice, the community will want to consider hydrology and geology. If an incinerator is planned, air circulation patterns around the site will be important. Impacts on flora and fauna will need to be considered. Historical patterns of siting may be brought up, as well as the location of other facilities that pose possible health threats. Thus, there is no all-purpose list of factors that can substitute for good judgment exercised with respect to knowledge of local conditions.

As more factors are added, it becomes increasingly difficult to amalgamate all the concerns identified. As the analysis becomes more comprehensive, complexities and tradeoffs make it more difficult to justify a final decision. Unfortunately, there is no generally accepted method for drawing together all these considerations into a single metric that allows for sites to be compared or ranked.

It is also unlikely that there will be agreement on how the many characteristics of sites ought to be weighted in making a final decision. Different groups are likely to have their own opinions about the relative importance of each consideration. There is no objective system for weighting the relative importance of various features of alternative sites. This explains why the task of identifying a best or optimal site is so difficult. The only way that this challenge can be met is to build a consensus among all the stakeholding parties that synthesizes all siting considerations.

Site selection is also haunted by the political polarization that accompanies the actual designation of candidate sites. Once sites have been announced, the majority of residents in a region breathe a sigh of relief. Those who have been designated to host a facility, however, often feel stigmatized. They may feel enormous pressure to mount whatever political resistance they can muster in the hope of coming off the list quickly. A willingness to participate in reasoned deliberations may be interpreted as a sign of weakness. While a careful analysis of alternative sites would surely benefit from the inclusion of numerous options for purposes of comparison, the political pressure to eliminate sites quickly will be enormous. If sites under consideration are not announced—in an effort to avoid premature political battles—neighbors of those sites that remain after site comparisons are complete will challenge the legitimacy of the analysis on the grounds that they had no chance to contribute crucial information that only those in the area could possibly have.

Assessing and Mitigating Impacts

The assessment of impacts is an integral part of site selection. A thorough understanding of prospective sites can only be gained by studying the likely impacts of the proposed activity and the prospects of mitigating them. It is rare, however, for impacts to be studied before a favored site has been selected. Impact assessments are more commonly undertaken to comply with federal or state regulations *after* a site has been selected. In this context, such assessments tend to be narrowly focused in ways that cast the favored site and technology in a positive light.[11]

The models available for forecasting environmental impacts are usually derived and calibrated using data from other times and places. It is sometimes unclear whether these are appropriate to new conditions and how they should be recalibrated. The costs of building new models or recalibrating old ones usually exceed the resources available for most impact assessments. Longitudinal data that would help in determining prospective sites are usually unavailable.

Environmental impact assessments are expected to enrich our understanding of the tradeoffs associated with each solid waste management alternative. Yet, like the analyses involved in the earlier steps of the siting process, they are unlikely to be conclusive.[12] The sources of uncertainty are too numerous, the data too sparse or coarse (if available at all), and the analytic methods themselves dependent on choices that cannot be made solely on objective grounds.

Environmental impact assessment becomes even more complex when an attempt is made to assess possible mitigation measures. Suppose a landfill is considered for a site adjacent to a wetland. The functions of that wetland are likely to be complex (and not completely understood). The importance of the wetland for the survival of neighboring plant and animal populations may not be well understood. Finally, it is difficult to judge (much less allocate responsibility for) the cumulative impacts on a wetland that the proposed facility will have, given other changes that might also occur in the area.[13] There is a possibility that beyond a certain threshold a process of catastrophic, irreversible change will occur, but, prior to that point, impacts will be minimal. If the threshold is breached, however, mitigation will not be possible. The boundary that separates environmentally divergent outcomes may be difficult to predict or, in the worst case, recognizable only after it has been passed.

If a portion of a wetland is adversely affected by a new facility, is rehabilitating an adjacent wetland or creating a new one some distance away a suitable response? Without a clear sense of the wetland's functions, it is difficult to answer such a question. What if there is also a chance that groundwater will be contaminated by a proposed landfill? Is securing an alternative supply of water a satisfactory mitigation measure? What does securing mean? Is an insurance policy satisfactory, or should an actual supply of water be set aside? And for how long should a supply be secured? Groundwater contamination, even if stopped, can persist for an extended period.

Even when mitigation measures are straightforward and relatively easy to quantify (such as providing guarantees that property values will not decline), the acceptability of such measures is not necessarily clear. Some residents may be able to move. But dislocating families from their homes may produce stresses that are difficult to predict, much less to value. Are cash payments that permit residents to move an appropriate mitigation measure? Such payments, like artificial wetlands, offset some of the effects of change, but they cannot guarantee that the future will be like the past, and they may have very different meaning to some of the affected parties.

Managing the Facility

Assumptions about operating standards must be made before a forecast of the impacts of a facility can be completed. Thus, clarifying how a facility will be managed is an integral part of the siting process. Yet it is difficult to develop reliable or precise estimates of future organizational performance because many of the factors most likely to affect it—operational procedures, management structures, and individual behavior—are difficult to quantify and forecast.[14]

Perceptions of the risk associated with each type of waste management option will eventually form the basis for voluntary consent or imposition of a new facility. In the former case, technology, site, and impact assessments will permit reasonable people to feel secure. In the latter (which we do not favor) assessments will provide a justification that elected officials or the court can use to impose a choice. The expectations that people have about how a facility will be operated, and about the effectiveness of monitoring and response systems, will also play an influential role in the judgments they make about their security and the acceptability of a facility. In an atmosphere of minimal trust (or even healthy skepticism) neighbors and other affected parties may demand guarantees that provide for independent monitoring and give the host community a substantial role in overseeing management performance. Communities may even wish to bind a facility operator through covenants that require incremental improvements in performance as new knowledge becomes available over time.

It is important that management arrangements satisfy the concerns of affected citizens and build trust in the operator and the siting process. For a facility's neighbors such arrangements are the only guarantee against threats to their health, safety, and welfare. If an operator believes in its ability to manage a facility within specified standards, then the risks associated with providing contractual guarantees on performance ought to be minimal.

13.4 A SITING CREDO

In practice, solid waste management facilities are usually proposed by government agencies, although private corporations are often the prime movers behind such announcements. The staffs of these organizations must respond to internal (i.e., political or institutional) pressures that translate into time and cost limitations. If agency and corporate staff are adept, they will realize that they are operating in a highly political context.

Unfortunately, the most common response to such constraints is to adopt a "decide-announce-defend" strategy.[15] The agency draws on a small set of professionals with pertinent expertise. This group draws heavily on current ideas about best practice, responds to current legal requirements (like health and safety regulations), and tries to take account of political realities. They provide few opportunities for public comment *prior* to deciding what they think ought to be done. Opening up the decision process *before* the experts have announced their proposed strategy is tantamount, in the eyes of many elected and appointed officials, to admitting that political consid-erations are more important than technical factors. This is viewed as an unwise admis-sion (even if it is true!). Once the agency and its experts (in consultation with indus-try) reach a decision about a site and a technology, they may well allow hearings or other occasions for unhappy groups to "sound off."

Proposals are typically presented as the technically best way of meeting a need or responding to a crisis (i.e., an incinerator for Bloomsbury or a landfill for Brixton.) Once the announcement has been made, agency personnel shift into a defensive mode. Their justifications are (1) that they used the best available evidence and analyses, (2) that they followed all required procedures, and (3) that the outcome is fair, since it was based on an objective process and criteria. Interactions with angry or dissatisfied groups are likely to be polarized, with communication limited to legalistic exchanges as everyone prepares for lawsuits.

Proponents of facilities may feel that they are acting responsibly and providing leadership by getting good advice, making the tough choice, and standing by their decisions. Indeed, this view periodically lead to calls for preemptive legislation as a way to avoid siting problems. While such proposals may be well-intentioned, avoiding procedural safeguards and opportunities for public scrutiny and deliberation seems inconsistent with the complexity of siting decisions and the influence they can have on people's lives. And efforts to preempt public involvement may only prompt affected parties to recast their challenges or move them to other forums. Short-circuiting estab-lished review processes or providing government agencies with the ability to site by fiat is likely to only erode trust further and, in the end, make siting more difficult.[16]

Moreover, most siting decisions cannot be justified on technical grounds alone. At each stage, from the determination of need, through the selection of a technology and a site, to the specification of impact mitigation strategies and management practices, the one thing that is clear is that nonobjective judgments can *and should* come into play. Questions of fairness may prove to be pivotal. The key is to acknowledge these difficulties and involve representatives of all stakeholding groups in open discussions about the tradeoffs and choices involved.

There is an alternative to the decide-announce-defend approach. The Facility Siting Credo outlined below offers what we think is the best advice currently available on siting facilities. It was distilled from the experience of dozens of siting experts and practitioners who participated in the National Workshop on Facility Siting sponsored by MIT and the University of Pennsylvania's Wharton School of Business in 1989–1990. The credo is an integrated set of propositions and needs to be adopted as a package. Its 13 points suggest that those managing siting processes should take the following actions.

Seek Consensus

Much of our discussion to this point has focused on characteristics of the siting process that provide legitimate sources of disagreement and suggest how people concerned about siting decisions can be seen to be acting in reasonable ways. Yet we cannot allow these characteristics to produce inertia when there are legitimate and pressing needs to move forward with solid waste management programs. Responsible leadership will openly acknowledge sources of concern, but seek to move forward by building consen-sus on the choices that must be made. The push for consensus grows out of a hope that agreement can be reached through reasoned dialogue that takes account of the com-plexity of the problem. The practical justification for this approach rests in the ability of many affected parties to block facilitating siting efforts with which they disagree.

The ethical justification lies in the right of affected parties to participate in decisions that affect them.

Thus, efforts to seek consensus should begin with active solicitation of all parties affected by a siting decision. Organized groups representing certain stakeholding interests can be readily identified. Other interests may need help organizing themselves. While an open invitation may seem untenable, the commitments involved are likely to limit participation to those with a sincere interest in extended deliberations. In fact, care must be taken to ensure that no affected party is barred from participation because of a lack of resources. This inevitably leads to charges of discrimination that can undermine a siting effort.

The technical aspects of siting should be treated as an opportunity for joint inquiry. The best available advice should be sought throughout the process, but experts providing advice should be acceptable to all the stakeholding parties, including public officials. Moreover, the knowledge and experience of stakeholders should be seen as still another source of information, rather than as a challenge to the authority of experts and officials. If the invitation to participate is genuine, and the solicitation is made before proceedings bog down in confrontation, adversarial "science" can be avoided without sacrificing diversity of opinion. Vigorous debate can be seen as an appropriate response to uncertainty.

Consensus should reflect the substantive concerns of all affected parties. It should also reflect the concerns of officials who have the responsibility for making decisions. These are not irreconcilable objectives. Elected leaders can attempt to accommodate the concerns of affected parties while still reserving the right to veto any proposal that does not meet their own sense of what the public interest requires. If, however, a proposal can be identified that meets all statutory requirements *and* the interests of affected parties, it should clearly be preferred.

Managing a consensus-building procedure is a complex undertaking, particularly if the numbers of participants is large. Specialized facilitation skills are required. If officials lack such experience, which is usually the case, the burdens of building consensus may best be assigned to a professional mediator of facilitator. Such a neutral party, acceptable to all participants, can help to ensure that the process is perceived as fair. They should also be able to handle any disruptions that arise.

Work to Develop Trust

Siting efforts usually take place in a climate of distrust—the result of past decisions that have placed disproportionate burdens on disadvantaged communities, or closed debates that have fueled charges of "backdoor politics" or unsavory deals of one kind or another. The lack of trust poses what is probably the greatest challenge to any consensus-building effort.

It is harder to say what contributes to rebuilding trust. Typically it must begin with a suspension of disbelief for some parties to agree to try again. Beyond this, trust can probably only be rebuilt slowly, with each participant demonstrating trustworthy behavior. Honesty and directness are also important, even where this requires acknowledging doubts and uncertainties. Acknowledging past mistakes or inequities may also help, since they are probably apparent to interested parties anyway.

Set Realistic Timetables

Consensus building involves a substantial commitment of time. Only a full airing of views, substantial debate, and a careful exploration of the sources of disagreement will help. Cutting corners is unlikely to work and may actually undermine any effort already made.

Consensus-building procedures usually require realistic deadlines to keep them on track. Parties who wish to participate may have only limited resources; explicit timetables allow them to gauge the resources they will need to commit. If such resources are being supplied or underwritten by a public agency, that entity will also need a timetable to make cost estimates. Without agreed-on timetables, participants may be held hostage when a single group prolongs discussion as a blocking tactic. Deadlines should allow sufficient time for careful deliberation, but ensure that incremental milestones are reached at specified intervals. An experienced mediator or facilitator can be helpful in constructing such a timetable and ensuring that it is met.

Get Agreement That the Status Quo is Unacceptable

A new facility should be proposed only when there is agreement that solid waste management needs require action. This will undoubtedly require public exploration of the risks and costs associated with doing nothing as well as with alternative methods of addressing waste management needs. Until there is agreement that a new facility is needed, it does not make sense to move ahead.

Choose the Design That Best Addresses the Problem

The difficulties of selecting a technology and a site should be addressed through a review of alternatives. The long- and short-term implications of each alternative, including doing nothing, must be analyzed carefully. Since it is unlikely that an indisputably "best" option will emerge, no matter how much technical analysis is done, "best" should be judged relative to the extent to which the interests of all stakeholders are likely to be met.

Guarantee That Stringent Safety Standards Will Be Met

Health and safety are bound to be among the primary concerns of residents likely to be affected by the construction of a new solid waste management facility. They may not wish to discuss other issues until they are satisfied that their health and safety concerns will be met. Residents should not be asked to compromise legitimate health and safety concerns so that a facility can be built.

At minimum, this means that facilities must meet all applicable federal and state standards and conform to current ideas about best practice. It means that health and safety concerns must be given substantial weight in assessing alternative sites and technologies. In making such choices, no one should be permitted to compromise health and safety in exchange for financial or other compensation.

Deliberations about health and safety should be informed by joint investigations of the risks associated with alternatives, including the risks posed by taking no action. Consideration should also be given to possible mitigation measures. Risks that can be mitigated (such as arrangements for alternative supplies of water in case of contamination) may be differentiated from other, more problematic risks. Management provisions that provide community oversight and control (e.g., through enforceable shutdown provisions) may assuage fears tied to doubts about an operator's ability to meet required operating standards.

Once sites have been identified, host communities may wish to raise still other considerations that have to do with the unique characteristics of the site or the potential interaction of the proposed facility with other activities in the community. These should be attended to in the same manner as generic health and safety concerns. An effective response to such concerns may require more detailed knowledge of local needs and conditions. The outcome should be an agreement that permits the host community to feel that its health and safety will not be significantly or unfairly compromised.

Fully Compensate for the Negative Effects of a Facility

Even after all health and safety concerns have been attended to and all impacts have been mitigated to the fullest extent possible, there may be impacts or costs that cannot be handled to the satisfaction of all affected parties. In such instances appropriate compensation may be necessary. This may include insurance policies to guarantee property values, or investments in services threatened by the new facility. It may also involve the rehabilitation or creation of equivalent habitats to offset the adverse effects of a new facility. The use of contingent agreements may be important in developing acceptable compensation arrangements.

Use Contingent Agreements

Some concerns about the management of facilities can be resolved by contingent agreements spelling out what will be done in case of accidents, interruptions of service, changes in standards, or the emergence of new scientific information about risks or impacts. Such agreements should specify the conditions under which the facility will be shut down temporarily or permanently. They should also describe the triggers for closure, who has responsibility for taking action, and the means of guaranteeing that contingent promises will be met at no additional cost to those adversely affected. This can be accomplished through the use of nearly self-enforcing contracts, similar to liquidating bonds, or permits that must to be renewed periodically in light of performance.

Keep Multiple Options on the Table at All Times

It is easy for debates over the siting of waste management facilities to become polarized. This is particularly true if a community or neighborhood is singled out as the only possible site. A community may feel so threatened that it will doubt even well-intentioned and responsible analysis. It may escalate its demands unreasonably or cast about frantically for grounds on which to justify a legal challenge. When this happens, a sponsor may feel that it is being held hostage and become suspicious of even reasonable requests from a host community or affected parties. Such polarization colors all communication and undercuts the possibility of constructive dialogue.

These problems can be minimized by keeping multiple options under consideration for as long as possible, ideally until the very end of a siting process. When parties on both sides feel that they are being treated fairly and openly, communication may improve, producing a more creative and acceptable outcome. If the costs of keeping multiple options open seem onerous, they should be compared with the costs of picking a site quickly and then having to justify the choice in protracted legal and political battles.

Make the Host Community Better off

The value of the benefits associated with new solid waste management facilities is usually greatly underestimated. This may be because beneficiaries take waste disposal service for granted or because previous alternatives were either poorly built or badly managed. When a facility responds to real social needs, the level of benefits should be more than large enough for the beneficiaries to actually make a host community better off (in its own terms) than it would be without the facility.

Compensation is the means by which the benefits of a new facility can be used to offset the costs. A comprehensive package of benefits could include anything from reducing risks to health and safety in a region to reducing waste disposal taxes or fees, to providing social amenities such as parks or public buildings, or even direct cash payments to selected residents. The net effect of a package of benefits should be that the host community knows it is better off accepting the facility than rejecting it.

Seek Acceptable Sites through a Volunteer Process

We have pointed to the importance of fairness in siting decisions and the difficulties involved in using any kind of technical analysis to justify the selection of a "best" site. In the face of such difficulties, consent gained through a process of soliciting volunteer sites may offer the best means of ensuring fairness.

For volunteers to step forward, the need for a facility and the costs associated with hosting it must be clear. All possible efforts to avoid or mitigate adverse impacts as well as the compensation for impacts that cannot be avoided should be spelled out. If these conditions are met, and the benefits from a facility are going to be shared, communities will volunteer. A lack of success may mean that the level of benefits being offered to host communities is too low.

For communities to even consider exploring the option of hosting a facility, it is critical that an expression of interest not be interpreted as a tacit admission that the risks associated with a facility are acceptable. Expressions of interest should be viewed as nothing more than an invitation to negotiate. They should also not be viewed as a promise to meet all demands. Any process of soliciting volunteer sites should be preceded by a screening process that eliminates all sites that are technically *un*acceptable.

Consider a Competitive Siting Process

If a process treats participants fairly and responsibly, and the level of benefits is significant, it may be possible that more than one jurisdiction will consider volunteering. Communities currently compete for industrial development that creates adverse impacts but brings in desirable tax revenues. If the level of benefits is high enough, it may be possible to do the same for solid waste management facilities. A competitive process would help to prevent the prospect of offering an unreasonably high benefits package by causing potential host sites to bid against each other. Care should be exercised, though, to ensure that the process of competitive bidding does not reduce benefits below the level needed to fairly compensate nonmitigatable impacts.

Work for Geographic Fairness

Compensation and incentive payments are unlikely to address all concerns about fairness, even if a community volunteers to host a facility. There may also be concerns that no single community should bear the risks associated with solid waste disposal, regardless of the level of compensation they are offered. Some jurisdictions may wish to address such concerns by developing fair-share formulas that limit the burdens that can be imposed on a single community or neighborhood. New York City uses a point system to allocate burdens for a variety of social needs.[17] Since it is unlikely that exact equivalencies among different types of facilities will be attainable, working for geographic fairness will not do away with the need to provide compensation and benefit sharing. While it may not emerge as an issue in every siting effort, at some point a threshold will be crossed and geographic fairness will have to be addressed as an issue in siting solid waste management facilities.

13.5 CONCLUSIONS

We have presented a view of facility siting that takes account of the complex substantive and procedural concerns involved. Good technical advice is critical to making wise choices about the need for and location of new solid waste management facilities, but it is equally important to recognize the importance of the political and ethical concerns that drive many parties' involvement in solid waste management decisions. The credo provides a practical famework for responding to such concerns.

The credo was tested by Kunreuther, Aarts, and Fitzgerald[18] in a survey of participants in 29 siting processes across the United States, including efforts to site new landfills, expand existing landfills, and site new municipal solid waste incinerators. They found clear support for the principles embodied in the credo from all groups— government officials, business leaders, and environmental and citizens groups. The results of the survey suggest that voluntary siting processes aimed at building consensus are more likely to produce results that are viewed as fair and wise.

Acknowledgments

We wish to thank our colleagues Howard Kunreuther, Professor of Decision Sciences at the University of Pennsylvania's Wharton School of Business and Director of the Wharton Risk and Decision Processes Center, and Tom Aarts and Kevin Fitzgerald, doctoral candidates at the Wharton School, for their continued collaboration in developing and promulgating the Facility Siting Credo. Portions of this chapter appeared in an earlier form in Laws and Susskind.[19]

REFERENCES

1. National Workshop on Facility Siting, MIT, Cambridge, Mass., November 1989, and the Wharton School, University of Pennsylvania, Philadelphia February 1990.

2. O'Hare, M., L. S. Bacow, and D. Sanderson, *Facility Siting and Public Opposition,* Van Nostrand Reinhold, New York, 1983.

3. *EPA Journal,* vol. 18, no. 1, March/April, 1992.

4. Bullard, R. D., *Dumping in Dixie: Race, Class, and Environmental Quality,* Westview, Boulder, Colo., 1990.

5. Godsil, R., "Remedying Environmental Racism," *Michigan Law Review,* November 1991.

6. Cohen, J., "Deliberation and Democratic Legitimacy," in A. Hamlin and P. Pettit, eds., *The Good Polity,* Basil Blackwell, Oxford, 1989. Susskind, L. and J. Cruikshank, *Breaking the Impasse: Consensual Approaches to Resolving Public Disputes,* Basic Books, New York, 1987.

7. Washburn, S. T., J. Brainard, and R. H. Harris., "Human Health Risks of Municipal Solid Waste Incineration," *Environmental Impact Assessment Review,* vol.9 no. 3, 1989.

8. Ozawa, C., *Recasting Science: Consensual Procedures in Public Policy Making,* Westview, Boulder, Colo., 1991.

9. Washburn, S. T., J. Brainard, and R. H. Harris, "Human Health Risks of Municipal Solid Waste Incineration," *Environmental Impact Assessment Review,* vol.9, no. 3, 1989.

10. Sandman, P. M., "Getting to Maybe: Some Communications Aspects of Siting Hazardous Waste Facilities," in R. W. Lake, ed., *Resolving Locational Conflict,* Rutgers Center for Urban Policy Research, New Brunswick, N.J., 1987.

11. Barzok, L., "The Role of Impact Assessment in Environmental Decision-Making in New England: A Ten Year Retrospective," *Environmental Impact Assessment Review,* vol. 6, no. 2, 1986. Susskind, L. E., "It's Time to Shift Our Attention from Impact Assessment to Strategies for Resolving Environmental Disputes." *Environmental Impact Assessment Review 2,* October 1978.

12. Elliott, M. L., "Pulling the Pieces Together: Amalgamation in Environmental Impact Assessment," *Environmental Impact Assessment Review,* vol. 2, no. 1, 1981.

13. Contant, C. K., and L. L. Wiggins, "Defining and Analyzing Cumulative Environmental Impacts," *Environmental Impact Assessment Review,* vol. 11, no. 4, 1991. Dickert, T. G., and A. E. Tuttle, "Cumulative Impact Assessment in Environmental Planning: A Coastal Watershed Example," *Environmental Impact Assessment Review,* vol. 5, no. 1, 1985.

14. Elliott, M. L., "Improving Community Acceptance of Hazardous Facilities Through Alternative Systems for Mitigating and Managing Risk," *Hazardous Waste,* vol. 1, no. 3, 1984.

15. Duscik, D., *Electricity Planning and the Environment: Toward a New Role for Government in the Decision Process,* Ph.D. dissertation, Department of Civil Engineering, MIT, Cambridge, Mass., January 1978. O'Hare, M., L. S. Bacow, and D. Sanderson, *Facility Siting and Public Opposition,* Van Nostrand Reinhold, New York, 1983.

16. Susskind, L. E., and S. R. Cassella, "The Dangers of Preemptive Legislation," *Environmental Impact Assessment Review,* vol. 1, 1980.

17. Rose, J. B., "A Critical Assessment of of New York City's Fair Share Criteria," *Journal of the American Planning Association,* vol. 59, no. 1, 1993. Weisberg, B., "One City's Approach to NIMBY: How New York City Developed a Fair Share Siting Process," *Journal of the American Planning Association,* vol. 59, no. 1, 1993.

18. Kunreuther, H., T. D. Aarts, and K. Fitzgerald, "Siting Noxious Facilities: A Test of the Facility Siting Credo," Wharton Risk and Decision Processes Center, U. of Pennsylvania, Report 92-03-01, 1992.

19. Laws, D. and L. Susskind, "Changing Perspectives on the Siting Process," *Maine Policy Review,* vol. 1, no. 1, 1991.

CHAPTER 14
FINANCING AND LIFE-CYCLE COSTING OF SOLID WASTE MANAGEMENT SYSTEMS

Nicholas S. Artz

Senior Environmental Engineer and Principal, Franklin Associates, Ltd.
Prairie Village, Kansas

Jacob E. Beachey

Senior Environmental Scientist and Principal, Franklin Associates, Ltd.
Prairie Village, Kansas

The concept of integrated solid waste management is being applied in many communities in the United States. There are numerous waste-to-energy facilities, material recovery facilities (MRFs), and other waste management facilities in operation and the number is growing rapidly. Most states have some form of regulations or incentives to encourage recycling; many states promote composting and waste minimization as well.

Changes in solid waste management have had a significant effect on public works operations and will continue to have an impact for years to come. The waste management technologies become more complex as we move away from the traditional method of simply collecting the waste in packer trucks and disposing of it in the municipal landfill. With the increasing complexity in waste management technology comes an increased complexity in the requirements for financing the new programs. Not only is there a need for greater capital expenditures, which usually means financing through borrowed funds, but also there is a need to finance multiple facilities in addition to the traditional MSW landfill. For example, communities choosing to implement an integrated solid waste management system may include recyclables processing, composting, incineration, and landfilling in their system. The need for multiple facilities in such an integrated system often leads to system financing rather than individual facility financing.

The increased complexity of integrated solid waste management has also resulted in a movement toward privatization of services. Municipalities do not wish to become involved in operations where they lack experience and often contract with private firms that specialize in such services. This desire for limited public involvement is also a factor in the financing of solid waste management facilities and systems.

The purpose of this chapter is to help public works officials deal with some of the changes in solid waste management as they relate to financing alternatives. The following sections summarize the options available in the 1990s for financing integrated solid waste management systems, review some of the issues involved in selecting the best financing mechanism for the local situation, and present a list of steps typically needed for securing system financing. In addition, the last section describes, and presents examples of, life-cycle cost analysis.

14.1 FINANCING OPTIONS

A number of options are available for financing solid waste management facilities. Choosing from among these options will involve consideration of several issues discussed later in this chapter. The "best" financing option is obviously not the same for all communities. The discussions below describe the more prominent financing options from which choices may be made and provide a basic understanding of their structure and applicability. Clearly, not all of the options described are available for every financing need. Also, combinations of these options are often used to finance solid waste management projects.

Private Equity

Privately owned facilities may be financed in total or in part by the use of equity—i.e., the owner's cash. The owner may be the vendor who builds and operates the facility or a third party who contributes equity in anticipation of a sound investment return.

In general, privately owned solid waste management facilities have been financed with a combination of equity and tax-exempt project revenue bonds (discussed later).[1] The equity is often for that portion of the facility that does not qualify for tax-exempt debt, which may be 10 to 20 percent of the facility cost. For example, that portion of a waste-to-energy facility used to produce an energy product cannot normally be financed with tax-exempt revenue bonds.

Owners are often hesitant to provide more than a minimal amount of equity because some of the investment returns are fixed (i.e., they do not increase as the equity increases above the minimum).[2] The owner is allowed the tax benefit of an accelerated depreciation schedule on the full value of the facility even though the amount of equity may be only 10 to 20 percent of the facility's cost. Also, the owner retains the residual value of the facility after the debt is retired.

Third-party investors have a favorable effect on a project in some instances. Such investors sometimes have greater potential for maximizing use of the tax benefits generated by the project and/or may require a lower rate of return on their investment than the project vendor. The net effect can be a less expensive project.

Some solid waste management facilities are financed entirely by owner equity. This avoids the time and expense of obtaining debt financing and is the easiest means of introducing a new technology. It is often the choice for financing less capital-intensive operations including small recyclables-processing facilities.

In other instances, the amount of private owner equity in a solid waste project may be set in the service contract. For example, municipalities sometimes require an owner-operator to post considerable equity in a facility or system to assure the continued interest of the owner in meeting the terms of a service agreement.

Traditional Loans

Solid waste management facilities may be financed through traditional loans between the borrower and lending institutions. Different lenders will market loans for different periods of time or for different phases of a project.[1,3]

Commercial banks, finance companies, and thrift institutions generally provide construction loans on a project. These lenders are interested in short-term loans of 1 to 3 years and do not usually participate in the permanent financing of a project.

Permanent lenders provide financing after the project is operational. These long-term lenders include insurance companies, pension funds, and other financial institutions with long-term sources of cash. Permanent lenders may provide project financing for 20 years or longer, depending on the expected life of the facility (or facilities) included in the financing.

Construction loans and permanent loans can be structured and committed back-to-back if project sponsors, investors, or construction lenders are unwilling to risk refinancing at the end of construction. In the solid waste industry, construction and permanent financing commitments are usually required at the beginning of the project. This avoids the risk of the permanent financing being too costly or unavailable when needed. Typically, under this arrangement, the permanent lender repays the construction loan when the project is at a preagreed acceptance stage.[1]

Traditional loans may be used to finance solid waste projects where tax-exempt financing is not readily available. Such loans must generally be accompanied by owner equity as part of the loan collateral. Traditional loan financing is more common with private ownership than public ownership.

Tax-exempt Bonds

Tax-exempt bonds issued by a governmental agency are an alternative to taxable debt on some solid waste management projects. Because the interest paid on funds raised from these bonds is exempt from federal taxation, the interest rate may be 2 or 3 percentage points lower than that on taxable bonds.[4]

Two basic types of tax-exempt bonds may be issued by a state or local government to finance solid waste projects: general obligation bonds and project revenue bonds. Each of these bonds and their various forms is discussed.

General Obligation Bonds. General obligation (GO) bonds are tax-exempt certificates of indebtedness that may be used by local governments to finance their capital expenditures. The local government pledges its full faith and credit and taxing power as the security behind the debt service on the bonds. GO bonds are generally considered the most secure form of debt which, coupled with their tax-exempt status, results in the lowest interest rate on a project.

The use of GO bonds requires voter approval and is limited by the general obligation debt capacity of the municipality or other governmental unit choosing to use them. They are not typically used to finance large solid waste management projects because of the need to preserve a community's GO debt capacity and the availability of other financing mechanisms. Also, public ownership of the project is required when GO bonds are used.

Project Revenue Bonds.   Revenue bonds are more commonly used than GO bonds to finance solid waste management projects. These bonds are also tax-exempt, but are not as secure as GO bonds, and, therefore, generally have higher interest rates. As the name implies, revenue bonds are largely secured by the revenues from the project being financed. A project mortgage and other guarantees may be pledged as well, but the credit and taxing power of a local government is not included. Two types of tax-exempt project revenue bonds exist as a result of the Tax Reform Act of 1986: government-purpose bonds (GPBs) and private activity bonds (PABs).

  Government-Purpose Bonds.   While the defining characteristics of GPBs are somewhat complex, the basic criteria for their use in solid waste management projects are as follows:[1,4,5]

- The project must be publicly owned
- Limitations on the sale of project outputs to private business must be met
- Private operations of any part of the project must not exceed 5 years and may be canceled after 3 years

GPBs may be beneficial in financing publicly owned and operated solid waste management projects. They generally carry a lower interest rate than PABs, since PAB interest is included in the calculations of the alternative minimum tax for individuals and corporations.[5] However, the restrictions on the use of GPBs, particularly with respect to private sector involvement, results in more solid waste projects financed with PABs.

  Private Activity Bonds.   PABs are also subject to certain restrictions, but allow private ownership and/or long-term private operation of a solid waste project. Privately owned projects desiring to use PAB financing must obtain a portion of the state's annual allotment of PABs. The annual state ceiling is equal to $50 multiplied by the state's population or $150 million—whichever is greater.

  Competition with other projects for a PAB allocation may lead to public ownership, which is exempted from the state's allocation cap. However, PABs are the only means of obtaining tax-exempt financing for privately owned projects. PABs for private use must be issued through a public agency and the funds from the bonds passed on to the private owner through a loan or other ancillary agreement.[6]

  Whether PABs are used for publicly or privately owned solid waste projects, they allow much greater private involvement than GPBs. The ability to enter into long-term service agreements with a vendor allows a local government to share project risks and responsibilities in a manner not available with GPB financing.

  PABs may not be used for certain expenditures in solid waste projects such as the energy-generating equipment in a waste-to-energy facility. This factor plus demand for equity as additional debt service security normally results in PABs being used in conjunction with other funds to finance solid waste projects.

Taxable Bonds

Taxable bonds—in particular, taxable municipal bonds—may be used for all or partial financing of solid waste projects. Taxable municipal bonds (TMBs) are commonly used to finance costs that do not qualify for PAB financing in a publicly owned project.[1] They may be used for that purpose in privately owned projects, as well, when the non-qualifying costs are not all covered by equity. In some instances, TMBs may be substituted for PABs when the tax-exempt bond allocation for private use is not available.

Although TMBs require higher interest rates than tax-exempt bonds, they afford a private owner more favorable depreciation periods on solid waste equipment. This benefit has the effect of at least partially offsetting the higher interest costs.

Federal/State Grants and Loans

State and federal sources of financial assistance to solid waste projects are limited and vary over time. However, money in the form of grants or loans has periodically become available for projects that can show a demonstration or research function.

In cases where state or federal funding may be available, local funding may also be required at some level.

Public Funds

A local government (i.e., county or municipality) may sometimes use general or special reserve funds it possesses to pay for a publicly owned project. This form of equity financing may reduce some or all of the numerous steps necessary to obtain debt financing.

Public funds are typically used to finance projects that are less capital-intensive or portions of projects not qualifying for PABs.[1,6] Material recovery facilities used to process recyclables and composting operations are examples of solid waste facilities that might be financed in total with public funds. These facilities are generally lower in capital cost than waste-to-energy facilities, for example, and are more difficult to finance with debt because of the uncertainty in prices for their products.

The availability of public funds for solid waste project financing may depend on whether a means of collecting money specifically for solid waste management services is available. In many areas, various forms of surcharges for services are being assessed to provide public funds for expanded or new solid waste projects.

14.2 ISSUES IN FINANCING CHOICES

Choosing between financing options may involve a variety of project issues. These issues are addressed below along with their potential effects on financing solid waste management projects.

Facility/System Financing

As previously indicated, solid waste management systems are becoming more complex; they often include several types of facilities to accomplish the necessary or desired waste processing and disposal. Historically, waste-to-energy facilities and landfills have been financed individually with debt payable from revenues derived through tipping fees and energy sales. The recycling and composting facilities included in many solid waste management systems today, however, are not as amenable to individual facility financing. The uncertainties of markets for recovered materials and compost coupled with difficulties in predicting waste composition make the economic feasibility of these facilities difficult to demonstrate.

The movement toward integrated solid waste management and the importance of demonstrating economic viability to attract capital for solid waste facilities have resulted in more system financings. System financings rely on the strength and diversity of all facilities in the system to secure the repayment of debt or equity. If one facility in the system does not meet expectations, another may take up the slack. For example, if a material recovery facility is not paying for itself, revenues from another facility in the system (e.g., the landfill) will, it is hoped, cover the deficit.

In general, long-term debt financing of recycling/composting facilities will require a system financing structure if revenues are the principal means of securing the debt.[6] It will be necessary to provide assurances that a shortage in revenues from the recycling/composting facility can be covered by revenues from another element in the system, such as a landfill or waste-to-energy facility. Without a system financing structure, an MRF or composting facility will probably be excluded from project revenue bond financing.

Ownership

Ownership of solid waste management facilities may be either public or private. Public ownership is usually through a municipal government unit, authority, or agency. Private ownership can be through a private corporation, partnership, or sole proprietorship.

The choice between public and private ownership affects not only financing choices, but also project implementation including options for procurement and operation.[7] Features of solid waste management projects under public versus private ownership are shown in Table 14.1.

In the past, private ownership of capital-intensive waste management facilities was sometimes chosen to avoid public agency involvement and risk in an unfamiliar area. Also, the private ownership tax benefits were substantial prior to the Tax Reform Act of 1986 and were often judged to result in a lower-cost project.

Currently, public ownership of highly capitalized waste management facilities is frequently recommended as the most cost-effective and practical approach. Publicly owned projects are reported to require less time to finance and implement and may involve little, if any, additional public risk. Recent comparisons suggest that risk allocation between the public and private sectors in a solid waste project is virtually irrespective of ownership.

Tax-exempt debt financing is easier to obtain with public ownership and is one of the reasons public ownership is more often used than in the past. Whereas PABs issued for private use are limited, no such limit is set for public use, and PABs are frequently used to finance publicly owned solid waste projects. For public projects with limited private sector involvement operationally or otherwise, GPBs provide even lower-cost financing than PABs. GO bonds provide the lowest-cost form of public debt financing but, as noted previously, limitations on their use has resulted in infrequent use of GO bonds to finance solid waste projects.

Other financing options with public ownership include the use of public funds, federal/state grants and loans (as available), taxable municipal bonds, and traditional loans from lending institutions. These options may be used in combination with each other or with the tax-exempt bonds described above.

TABLE 14.1   Features of Public versus Private Ownership of Solid Waste Management Facilities

	Public ownership	Private ownership
Procurement options	A/E Turnkey Full service	Full-service
Financing options	General obligation bonds (GO) Government-purpose bonds (GPB) Private activity bonds (PAB) Taxable municipal bonds Traditional loans Federal/state grants Public funds	Private activity bonds (PAB) Taxable bonds Private equity Traditional loans
Operation	Public (typically) with A/E Public/private with turnkey Private with full service	Private
Public risk	Similar*	Similar*
Implementation time	Less than with private ownership	Greater than with public ownership

*Applies primarily to facilities/systems financed with large bond issues.
Source: Franklin Associates, Ltd.

Private ownership financing options for solid waste projects include private equity, traditional loans, taxable bonds, and PABs. Some combination of these options is typically used. In some instances, taxable municipal bonds may be issued to assist in private project financing. Private financing without the use of tax-exempt bonds (i.e., PABs) allows the owner more favorable equipment depreciation periods for tax purposes. This added tax benefit without tax-exempt debt must be considered in comparing financing options under private ownership.

Procurement and Operation

Three basic forms of procurement are used for solid waste management projects:

- Architectural/engineering (A/E)
- Turnkey
- Full service

The A/E procurement method is the standard approach that governmental bodies use to build most public facilities. A consulting engineer is retained by the governmental entity to prepare the facility design and a contractor is hired through a bidding process to build the facility. The facility is publicly owned and publicly operated in most cases.

With a turnkey arrangement, a single contractor will have responsibility for both designing and building the facility. Turnkey procurements usually involve public ownership. The completed facility may be operated either publicly or privately. The turnkey contractor, who is intimately familiar with the facility design and construction, is often hired to operate the facility.

A full-service procurement involves one private entity accepting project responsibility for design, construction, and operation. This form of procurement is normally considered mandatory for private ownership of a capital-intensive waste management facility, but may be used with public ownership as well.

Most waste management facility or system procurements follow one of the three basic options described above or close variations thereof. Any of these options may be used with public ownership, while full service would usually be the only acceptable procurement for private ownership.

Procurement and operation of waste management facilities are related to financing insofar as they affect the choice of public or private ownership. For example, public operation is incompatible with private ownership, although private operation and public ownership are compatible with a turnkey or full-service procurement. An A/E procurement will require public ownership and, in general, public operation.

Risk Allocation

Financing solid waste projects generally requires that the risks be allocated between the public and private participants.[1] Lenders, including bond investors, are interested in obtaining maximum security on their investment. The credit rating of a project will determine the availability of lenders and the interest rate.

Most solid waste project financings require similar allocations of risks regardless of whether the project is publicly or privately owned. The vendor accepts the completion and technical risks of construction and the responsibility of operating the facility properly to meet certain performance standards. The local government/public agency guarantees the waste supply including payment for any shortfalls. Generally, the public entity also assumes the risk of force majeure events and the risk of changes in laws that affect operation.

In addition to project revenues backed by waste stream guarantees, lender/bond holder security may include a project mortgage, letter of credit, bond insurance, and a company guaranty, if the company is sufficiently strong.[5] A financially strong vendor may be willing to assume some risks normally borne by the governmental unit, but a substantial price will usually be charged.

A governmental unit assumes the highest level of risk when it issues GO bonds for a project. This gives bondholders the highest degree of security because the full taxing power of the local government is pledged to the repayment of the bonds.

Implementation Time

In general, the time required for financing a project is least when no debt is required and is greatest when debt is used.

Solid waste projects may be expected to require more time to implement with private ownership than public ownership. Long negotiation periods are often associated with arranging private ownership. In addition, tax-exempt PAB financing is more time-consuming when issued for private use. PABs for private use require obtaining a governmental issuer and obtaining a portion of the state's annual allotment of PABs. If the needed allotment is not available in the year requested, the request would not be allowed until, at least, the following year.

Issuing tax-exempt bonds for public use is often less complex and may, therefore, be less time-consuming. However, GO bonds cannot be issued without a public vote, which adds to the time requirement. Further, if the GO bond issue fails, other financing must be arranged.

Cost of Financing

Since the loss of most of the tax benefits of private ownership following the Tax Reform Act of 1986, public ownership financing is often recommended for solid waste projects. However, where tax-exempt PABs are available for private use, there may be no cost advantage with public ownership.

Clearly, the least expensive financing usually involves the use of tax-exempt bonds, which often carry interest rates of 2 or 3 percentage points below that of taxable debt. GO bonds are the least expensive because of their comparatively low risk. Of the project revenue bonds, GPBs carry lower interest rates than PABs. Interest on PABs is subject to the alternative minimum tax calculations for individuals and corporations and must, therefore, offer somewhat higher rates than GPBs.

For solid waste projects financed over 20 years, debt service payments may be 10 to 20 percent lower with tax-exempt bonds than with taxable debt. This advantage must be compared with the tax benefits of private ownership where tax-exempt debt is *not* used. The shorter equipment depreciation periods for tax purposes (5 to 7 years versus 10 years with tax-exempt financing) are of some added benefit in lowering project costs. In most cases, however, they will probably not be worth giving up the cost savings from tax-exempt bonds.

14.3 STEPS TO SECURE SYSTEM FINANCING

Once a decision to proceed with implementation of a facility for solid waste management has been made, financing must be considered. The steps necessary to secure financing will vary, depending on the financing options chosen and the party obtaining the financing. In general, PAB financing on behalf of a private company involves the greatest effort, and the steps necessary to secure this form of financing are emphasized in this section. The assistance of technical consultants, bond counsel, and an investment banking firm usually will be needed in the financing process.

The steps generally required in the more complex financing processes are summarized below. They are presented in an order in which they might normally occur, but specific financings may dictate variations in this order. References 3 and 5 were used in developing the process descriptions.

Decisions on Issues and Options in Financing

The first step in financing involves choosing between financing options. This will be done in view of the issues attendant to financing choices, as described previously. Decisions on ownership, procurement, operation, cost, etc., may enter into the final determination of what financing option or combination of options will be chosen. Quite often, pending the availability of grants, equity, or public funds, tax-exempt bonds are chosen for the permanent financing of most of a proposed system/facility. If GO bonds are chosen, in conjunction with public ownership, their use will require voter approval. However, project revenue bonds—PABs, usually—are chosen far more frequently to finance solid waste projects.

Feasibility Study and Plan

A study and plan providing details on project feasibility and the role of the proposed facility will also be needed early in the financing process. Bond counsel will use the description of the facility to be financed to make an initial determination that the project will qualify under the Internal Revenue Code for the tax-exempt financing desired. The feasibility of the project will also be closely studied by the investment banking firm and others prior to preparing the financing documents.

Determine Issuer of Bonds

If the project is determined to qualify for tax-exempt PAB funding, bond counsel must then determine what state agency or local government can act as issuer of the bonds and which state statutes apply to the financing. A private company seeking financing should contact the chosen governmental issuer to officially apply for assistance with the financing. The company will need to convince the prospective issuer to perform this service.

Prepare and Adopt Bond Resolution

Once an issuer for the bonds has been identified, bond counsel will draft a bond resolution to be adopted by the governmental issuer. A final resolution signifying the governmental issuer's intention to issue bonds for the project will need to be adopted. Major project expenditures that are incurred prior to the resolution's adoption may not be paid for by bond proceeds. Thus, it is important to obtain adoption of this resolution as early as possible.

Structure Bond Security

The investment banking firm selected to underwrite the bonds will suggest sources of repayment or security for the bondholders. In addition to pledging the project revenues from tipping fees and the sale of recovered materials/energy products, other potential sources of collateral to back the bonds include:

- Project mortgage
- Company guarantee, if the company is financially strong
- Flow control guarantees through contracts or ordinances that assure waste delivery and tipping fees to the project
- Letter of credit, surety bond, or bond insurance from an institution with a high credit rating

The greater the security of the bonds, the easier they will be to market. At minimum, guarantees of waste delivery will generally need to accompany project revenue pledges when revenues are the principal source of bond security. Without waste delivery guarantees, the revenues from a solid waste project can be very uncertain and revenue bonds may not be marketable.

Prepare Financing Agreements

After further review of project feasibility, bond counsel may begin drafting the agreements needed to issue the bonds; for PABs issued on behalf of a private company these will include:

- The agreement between the governmental issuer and the company
- The indenture stating the terms under which the bonds will be issued
- The bond purchase agreement providing for the sale of the bonds

Prepare Project Report

Factors relevant to obtaining financing for the project should be addressed in this report. The purpose, costs, and function of each portion of the project should be described. Bond counsel should determine which elements of the project qualify for bond financing and the amount of financing needed. The status of permits needed to construct and operate the project and other factors necessary to the project should be reviewed, as well.

Obtain PAB Volume Cap Allocation

Private company ownership of the project will require obtaining an allocation from the state's annual allotment of PABs if tax-exempt bonds are to be used. Bond counsel may need to assist the company and the governmental issuer of the bonds in the timing and method for obtaining the allocation.

Provide Official Statement

The investment banking firm (underwriter) will use an official statement or other disclosure documents to offer the bonds to the public or to private investors. The official statement summarizes the project, the financing arrangements, the forms of security offered to bondholders, etc. Substantial detail on the bonds being offered is provided in the statement. A bond issue rating may be included if obtained from a national rating agency.

Final Execution

Before PABs are offered for sale on behalf of a private company, a public hearing must be conducted and final documents and closing papers must be executed by all parties.

14.4   LIFE-CYCLE COSTING

This section provides an overview of the basic concepts of life-cycle costing (LCC) analysis. LCC is a method of comparing new projects by taking into account relevant costs over time, including the project's initial investment, future replacement costs, operation and maintenance costs, project revenues, and salvage or resale values. All the costs and revenues over the life of the project are adjusted to a consistent time basis and combined to account for the time value of money. This analysis method provides a single cost-effectiveness measure that makes it easy to compare projects directly.

Time Value of Money

The value of money changes, depending on when it is spent or received. There are two reasons for this: inflation and the "opportunity cost" of money. Inflation erodes the buying power of money over time, and results in dollars spent today buying fewer goods and services than they did a few years ago. The opportunity cost reflects the fact that money invested has the opportunity to yield a return over time, even in the absence of inflation.

Since the value of money changes with time, cash flows from one year cannot be combined directly with flows from another in a meaningful way, but must first be "discounted" to a common year, usually the first year of the project. These discounted values can then be summed to obtain the total life-cycle cost, which can be compared with the total life-cycle cost of an alternative project that may have different proportions of initial costs and net annual operating costs.

Discount Factors

The formula for discounting a future value F to a present value P is

$$P = F \times \frac{1}{[(1 + d/100)]^n} = F \times \text{PWF } (d, n)$$

where d is the discount rate expressed in percent and n is the number of years in the future. The effect of discounting is to reduce the costs of the future to today's values. The present worth factors PWF (d, n), that convert future year values into present values for various discount rates and years have been calculated and are shown in Table 14.2.

If all investments yielded the same rate of return, then all future cash flows would be discounted at that rate. However, since different investments yield different rates, the choice of rate to use is sometimes difficult to determine. The discount rate commonly used is the cost of capital, which is the weighted average rate at which the borrowing agency is financed.

Capital Recovery Factors

The cost of a waste management system is generally made up of two parts: the capital required to purchase land, buildings, and equipment and the annual costs to operate the system. Capital investments are costs incurred at the beginning of the project. These costs are frequently financed with borrowed funds. The borrowed money and accrued interest are repaid with income received later in the project from the sale of energy or materials, from tipping fees, or from taxes. The constant annual payment required to repay the financed amount is determined by multiplying the borrowed amount by a capital recovery factor CRF (d, n) which is calculated by

$$\text{CRF } (d, n), = \frac{d}{1 - (1 + d)^{-n}}$$

where d is the interest rate expressed as a decimal and n is the number of interest periods. Table 14.3 lists capital recovery factors per thousand dollars as a function of interest rate and length of financing term.

LCC Case Studies

In this section, two example analyses are shown to illustrate the LCC method. While every project has unique features, these generalized examples illustrate the methodology. Two types of systems are examined: a privately owned waste-to-energy (WTE) system for MSW and a privately owned material recovery facility for recyclables separately collected from households. Both systems are assumed to be financed with private activity bonds.

The WTE facility for this analysis has a capacity of 1000 tons per day, and generates revenue from the sale of electricity. Revenue from the 200-ton/day MRF is derived from the sale of processed recyclables. In both cases, the revenues are supple-

TABLE 14.2 Single Present-Worth Factors PWF (d, n)*

Year n	Discount rate d, %							
	3	4	5	6	7	8	9	10
1	0.9709	0.9615	0.9524	0.9434	0.9346	0.9259	0.9174	0.9091
2	0.9426	0.9246	0.9070	0.8900	0.8734	0.8573	0.8417	0.8264
3	0.9151	0.8890	0.8638	0.8396	0.8163	0.7938	0.7722	0.7513
4	0.8885	0.8548	0.8227	0.7921	0.7629	0.7350	0.7084	0.6830
5	0.8626	0.8219	0.7835	0.7473	0.7130	0.6806	0.6499	0.6209
6	0.8375	0.7903	0.7462	0.7050	0.6663	0.6302	0.5963	0.5645
7	0.8131	0.7599	0.7107	0.6651	0.6227	0.5835	0.5470	0.5132
8	0.7894	0.7307	0.6768	0.6274	0.5820	0.5403	0.5019	0.4665
9	0.7664	0.7026	0.6446	0.5919	0.5439	0.5002	0.4604	0.4241
10	0.7441	0.6756	0.6139	0.5584	0.5083	0.4632	0.4224	0.3855
11	0.7224	0.6496	0.5847	0.5268	0.4751	0.4289	0.3875	0.3505
12	0.7014	0.6246	0.5568	0.4970	0.4440	0.3971	0.3555	0.3186
13	0.6810	0.6006	0.5303	0.4688	0.4150	0.3677	0.3262	0.2897
14	0.6611	0.5775	0.5051	0.4423	0.3878	0.3405	0.2992	0.2633
15	0.6419	0.5553	0.4810	0.4173	0.3624	0.3152	0.2745	0.2394
16	0.6232	0.5339	0.4581	0.3936	0.3387	0.2919	0.2519	0.2176
17	0.6050	0.5134	0.4363	0.3714	0.3166	0.2703	0.2311	0.1978
18	0.5874	0.4936	0.4155	0.3503	0.2959	0.2502	0.2120	0.1799
19	0.5703	0.4746	0.3957	0.3305	0.2765	0.2317	0.1945	0.1635
20	0.5537	0.4564	0.3769	0.3118	0.2584	0.2145	0.1784	0.1486

*The factor for finding the present value P worth of a future amount F, is $[1 + (d/100)]^{-n}$

Source: Franklin Associates, Ltd.

TABLE 14.3 Capital Recovery Factors CRF (*d, n*)*

Interest	Years										
	3	4	5	6	7	8	9	10	15	20	30
5.0	367.21	282.01	230.97	197.02	172.82	154.72	140.69	129.50	96.34	80.24	65.05
5.5	370.65	285.29	234.18	200.18	175.96	157.86	143.84	132.67	99.63	83.68	68.81
6.0	374.11	288.59	237.40	203.36	179.14	161.04	147.02	135.87	102.96	87.18	72.65
6.5	377.58	291.90	240.63	206.57	182.33	164.24	150.24	139.10	106.35	90.76	76.58
7.0	381.05	295.23	243.89	209.80	185.55	167.47	153.49	142.38	109.79	94.39	80.59
7.5	384.54	298.57	247.16	213.04	188.80	170.73	156.77	145.69	113.29	98.09	84.67
8.0	388.03	301.92	250.46	216.32	192.07	174.01	160.08	149.03	116.83	101.85	88.83
8.5	391.54	305.29	253.77	219.61	195.37	177.33	163.42	152.41	120.42	105.67	93.05
9.0	395.05	308.67	257.09	222.92	198.69	180.67	166.80	155.82	124.06	109.55	97.34
9.5	398.58	312.06	260.44	226.25	202.04	184.05	170.20	159.27	127.74	113.48	101.68
10.0	402.11	315.47	263.80	229.61	205.41	187.44	173.64	162.75	131.47	117.46	106.08
10.5	405.66	318.89	267.18	232.98	208.80	190.87	177.11	166.26	135.25	121.49	110.53
11.0	409.21	322.33	270.57	236.38	212.22	194.32	180.60	169.80	139.07	125.58	115.02
11.5	412.78	325.77	273.98	239.79	215.66	197.80	184.13	173.38	142.92	129.70	119.56
12.0	416.35	329.23	277.41	243.23	219.12	201.30	187.68	176.98	146.82	133.88	124.14

*The constant annual payment, in dollars, required to repay a present amount of $1000, as a function of the compound interest rate and number of years shown.

Source: Franklin Associates, Ltd.

mented by tipping fees or taxes to pay for the facilities. The estimated costs used for these examples are thought to be typical for the central United States, but may not apply to any specific community.

The two analyses differ in the way collection of incoming material is handled. Since a WTE facility does not require a separate collection system (i.e., the vehicles that collect the waste for disposal simply deliver the waste to a new site), collection costs are not included in that analysis. However, the haul distance will be affected if the distances to the new and old facilities are different. Adding a MRF to an existing system, on the other hand, generally requires additional equipment and staff for collecting recyclables and delivering them to the MRF. These additional costs may be partially offset by avoided MSW collection and disposal costs; however, the avoided costs are usually not proportional to the reduction in quantities disposed and are often quite small.

The life-cycle cost of a project is determined by annualizing the capital costs and then summing all discounted annual capital and operating costs for the life of the system. This life-cycle costing approach is a particularly useful tool for comparing total costs of alternative waste management scenarios over a 20-year period, where one scenario has higher capital requirements than the other.

A listing of typical capital cost elements for a financed waste-to-energy facility, MRF, composting facility, or landfill is shown in Fig. 14.1. The costs over and above the direct construction costs may increase the total bond issue requirement for a large WTE facility by 50 percent or more. An explanation of these additional costs is given below:

- *Start-up costs* are funds used to operate the facility during the testing and shakedown period, before revenues are routinely generated. The start-up time depends on the type and complexity of the system. For a large WTE facility, the start-up time is typically 6 months to a year. The time for getting a MRF into commercial operation can vary from one month or less for a manual sorting station to more than 6 months for a large mechanized processing facility.

- Direct construction costs
  - Land
  - Site development
  - Buildings, with utilities
  - Process equipment
  - Mobile equipment
  - Design and engineering
  - Delivery and installation
  - Construction supervision
  - Contingencies/profit
- Interest during construction
- Start-up costs
- Legal and financial fees
- Debt service reserve fund

FIGURE 14.1 Typical capital cost components for solid waste management facilities (when financed with borrowed funds).

- *Interest during construction and start-up* is money included in the bond issue to pay the interest costs during construction and start-up of the facility, when there may be reduced or no revenues. The construction time depends on the complexity of the facility, ranging from 1 or 2 months for a manually operated MRF with minimal equipment to 2 years or more for a large WTE facility.
- *Legal and financing fees* are for legal counsel and financial advice. These costs are typically in the neighborhood of 4 percent of the total bond issue.
- *The debt service reserve fund* is money set aside to pay for unanticipated problems. It is more likely to be required for the more complex or unproven technologies, and may amount to a year's debt service payment.

The 20-year life-cycle cost analyses for the two case studies are shown in Tables 14.4 and 14.5. The assumptions used for the analysis are listed as footnotes. The MRF costs were derived from a recent survey of 10 operating MRFs.[8] The present values of the net costs are developed for each year of the 20 years assumed in the analysis. An 8 percent annual cost of capital is used for discounting. This is the same rate assumed for bond interest.

As shown in Table 14.4, the first-year cost of the WTE facility is $66.56 per ton. This cost is higher than landfilling in most communities, and analyzed on a first-year basis, one may conclude that WTE is a much more costly option. However, since a rather large component of the WTE cost is capital investment debt service, which remains fixed for the 20 years, the average discounted life-cycle cost is much lower ($38.29 per ton). This value is a better number to compare with other systems over a 20-year life cycle. Usually, a life-cycle cost analysis of the continuation of the existing system will also be conducted, including capital and operation and maintenance (O&M) costs. Then the life-cycle costs can be compared directly.

The MRF costs shown in Table 14.5 show that the first-year costs (including collection of recyclables) are about $143 per ton, or about $2.33 per household per month. The annual cost in the twentieth year, discounted to present-value dollars, becomes $59 per ton. These are net costs after subtracting the revenues from the sale of recyclables (based on 1993 market prices). The MRF costs would be expected to be

TABLE 14.4 Twenty-Year Life-Cycle Cost Analysis of Waste-to-Energy Facility

(Production of electricity for sale)
1000 ton/day capacity

	Costs, $1000				Revenues	Net cost (tipping fee required)		Present value of net cost	
Year	Capital (debt service)	Operation and maintenance	Residue disposal	Total cost	Energy sales, $1000	$1000 per year	$ per ton	$1000 per year	$ per ton
1	14,259	13,000	2250	29,509	9,540	19,969	66.56	19,969	66.56
2	14,259	13,455	2329	30,043	9,874	20,169	67.23	18,675	62.25
3	14,259	13,926	2410	30,595	10,219	20,376	67.92	17,469	58.23
4	14,259	14,413	2495	31,167	10,577	20,590	68.63	16,345	54.48
5	14,259	14,918	2582	31,759	10,947	20,812	69.37	15,297	50.99
6	14,259	15,440	2672	32,372	11,331	21,041	70.14	14,320	47.73
7	14,259	15,980	2766	33,005	11,727	21,278	70.93	13,409	44.70
8	14,259	16,540	2863	33,662	12,138	21,524	71.75	12,559	41.86
9	14,259	17,119	2963	34,341	12,562	21,778	72.59	11,766	39.22
10	14,259	17,718	3067	35,043	13,002	22,041	73.47	11,026	36.75

11	14,259	18,338	3174	35,771	13,457	22,314	74.38	10,336	34.45
12	14,259	18,980	3285	36,524	13,928	22,596	75.32	9,691	32.30
13	14,259	19,644	3400	37,303	14,416	22,888	76.29	9,089	30.30
14	14,259	20,331	3519	38,110	14,920	23,189	77.30	8,527	28.42
15	14,259	21,043	3642	38,944	15,442	23,502	78.34	8,002	26.67
16	14,259	21,780	3770	39,808	15,983	23,826	79.42	7,511	25.04
17	14,259	22,542	3901	40,703	16,542	24,160	80.53	7,052	23.51
18	14,259	23,331	4038	41,628	17,121	24,507	81.69	6,623	22.08
19	14,259	24,147	4179	42,586	17,720	24,866	82.89	6,223	20.74
20	14,259	24,993	4326	43,577	18,341	25,237	84.12	5,848	19.49

Total life-cycle cost in discounted dollars 229,737

Average life-cycle cost in discounted dollars 38.29

Assumptions:

Total capital required	$140,000,000	O&M cost (year 1)	$13,000,000
PAB interest rate	8%	Residue quantity	75,000 tons/year
Inflation rate	3.5%	Residue disposal cost (year 1)	30 dollars/ton
Discount rate	8%	Salable electricity	530 kWh/ton
Facility financing period	20 years	Electricity revenue (year 1)	6 cents/kWh
MSW throughput	300,000 tons/year	O&M cost (year 1)	43 dollars/ton
		Financing cost (year 1)	48 dollars/ton

Source: Franklin Associates, Ltd.

14.17

TABLE 14.5 Twenty-Year Life-Cycle Cost Analysis of Material Recovery Facility
(Including collection of cummingled recyclables) 200 tons/day capacity

	Collection costs			MRF costs			
Year	Capital[2] debt service	Operation and maintenance	Total collection	Capital[3] debt service	Operation and maintenance	Residue disposal	Total MRF
1	921,800	3,992,400	4,914,200[5]	875,000	1,368,400	56,900	2,300,300[6]
2	921,800	4,132,100	5,053,900	875,000	1,416,300	58,900	2,350,200
3	921,800	4,276,700	5,198,500	875,000	1,465,900	61,000	2,401,900
4	921,800	4,426,400	5,348,200	875,000	1,517,200	63,100	2,455,300
5	921,800	4,581,300	5,503,100	875,000	1,570,300	65,300	2,510,600
6	921,800	4,741,600	5,663,400	875,000	1,625,300	67,600	2,567,900
7	921,800	4,907,600	5,829,400	875,000	1,682,200	70,000	2,627,200
8	1,124,400	5,079,400	6,203,800	982,200	1,741,100	72,500	2,795,800
9	1,124,400	5,257,200	6,381,600	982,200	1,802,000	75,000	2,859,200
10	1,124,400	5,441,200	6,565,600	982,200	1,865,100	77,600	2,924,900
11	1,197,300	5,631,600	6,828,900	982,200	1,930,400	80,300	2,992,900
12	1,197,300	5,828,700	7,026,000	982,200	1,998,000	83,100	3,063,300
13	1,197,300	6,032,700	7,230,000	982,200	2,067,900	86,000	3,136,100
14	1,197,300	6,243,800	7,441,100	982,200	2,140,300	89,000	3,211,500
15	1,455,100	6,462,300	7,917,400	1,118,600	2,215,200	92,100	3,425,900
16	1,455,100	6,688,500	8,143,600	1,118,600	2,292,700	95,300	3,506,600
17	1,455,100	6,922,600	8,377,700	1,118,600	2,372,900	98,600	3,590,100
18	1,455,100	7,164,900	8,620,000	1,118,600	2,456,000	102,100	3,676,700
19	1,455,100	7,415,700	8,870,800	1,118,600	2,542,000	105,700	3,766,300
20	1,455,100	7,675,200	9,130,300	1,118,600	2,631,000	109,400	3,859,000

Assumptions:

Material throughput	162 tons/day
	42,100 tons/year
MRF operation	260 days/year
Total households in collection area	215,000
Participation rate	75%
Setout rate	50%
MRF building cost (year 1)	$4,725,000
MRF equipment cost (year 1)	$2,050,000
Collection trucks and recycling bins (year 1)	$5,306,200
PAB interest rate	8%
Inflation rate	3.5%
Discount rate	8%
Residue quantity	1895 tons/year
Residue disposal cost (year 1)	30 dollars/ton
Material sales revenue (year 1)	30 dollars/ton

Total costs	Revenues	Net cost[1]			Net present value	
Collection and MRF	Material sales	Dollars/ year	Dollars/ ton	Dollars/ hh/month[4]	Dollars/ year	Dollars/ ton
7,214,500	1,206,200	6,008,300	142. 71	2.33	6,008,300	142.71
7,404,100	1,248,400	6,155,700	146.22	2.39	5,699,700	135.39
7,600,400	1,292,100	6,308,300	149.84	2.45	5,408,400	128.46
7,803,500	1,337,300	6,466,200	153.59	2.51	5,133,100	121.93
8,013,700	1,384,100	6,629,600	157.47	2.57	4,873,000	115.75
8,231,300	1,432,500	6,798,800	161.49	2.64	4,627,100	109.91
8,456,600	1,482,700	6,973,900	165.65	2.70	4,394,700	104.39
8,999,600	1,534,600	7,465,000	177.32	2.89	4,355,800	103.46
9,240,800	1,588,300	7,652,500	181.77	2.97	4,134,400	98.20
9,490,500	1,643,900	7,846,600	186.38	3.04	3,925,300	93.24
9,821,800	1,701,400	8,120,400	192.88	3.15	3,761,300	89.34
10,089,300	1,761,000	8,328,300	197.82	3.23	3,571,900	84.84
10,366,100	1,822,600	8,543,500	202.93	3.31	3,392,700	80.59
10,652,600	1,886,400	8,766,200	208.22	3.40	3,223,300	76.56
11,343,300	1,952,400	9,390,900	223.06	3.64	3,197,200	75.94
11,650,200	2,020,700	9,629,500	228.73	3.73	3,035,600	72.10
11,967,800	2,091,500	9,876,300	234.59	3.83	2,882,800	68.48
12,296,700	2,164,700	10,132,000	240.67	3.93	2,738,400	65.04
12,637,100	2,240,400	10,396,700	246.95	4.03	2,601,800	61.80
12,989,300	2,318,900	10,670,400	253.45	4.14	2,472,500	58.73
		Total life-cycle cost in discounted dollars			79,437,300	
		Average life-cycle cost in discounted dollars				94.34

[1]The net cost may be partially offset by avoided MSW collection and disposal costs.

[2]Financing period for trucks is 7 years and recycling bins, 10 years.

[3]Financing period for building is 20 years and MRF equipment, 7 years.

[4]Net cost distributed to all households (hh) in the collection area (not just the participants).

[5]First year collection costs are $117 per ton.

[6]First year MRF costs are $55 per ton processed.

at least partially offset by the savings experienced in the existing system collection and disposal costs. The average life-cycle cost for the recycling/MRF operation in discounted dollars is $94 per ton.

14.5 SUMMARY

The requirements for financing solid waste management projects can be substantial when tax-exempt PABs are used. Other forms of financing—particularly those where little, if any, debt is included—can be easier to arrange. The steps described in this chapter provide a general description of the process required when PABs are issued for private use. They are generic in nature and the specifics of a given project may result in more or less effort than indicated. Professional assistance will be needed with most solid waste management project financings. With the advent of system versus facility financings, the complexities of financing solid waste management projects are even greater than before.

The last section of this chapter describes the process of life-cycle cost analysis. The tables provide hypothetical examples of life-cycle costs over 20 years for waste-to-energy and recovery of materials for recycling. The importance of life-cycle costs in comparing solid waste management alternatives is demonstrated.

REFERENCES

1. Chen, P. M., G. D. France, and S. A. Sharpe, "Financing Solid Waste Disposal Projects in the 1990s," presented by S. E. Howard at the National Conference of State Legislatures, Kansas City, Mo., May 1992.
2. Turbeville, W. C., "Cutting Facility Finance Costs." *Waste Age,* May 1990.
3. Lee, W. B., and E. T. Ashdown, "Financing Waste Facilities During the Credit Crunch," *World Wastes,* March 1992.
4. MacCarthy, R. N., "Financing Recycling Facilities," *Waste Age,* March 1991.
5. Ollis, R. W., "Financing Recycling Programs," *Waste Age,* March 1992.
6. Horning, C., "Laws Give New Shape to Solid Waste Contracts and Finance," *Solid Waste & Power,* October 1991.
7. Artz, N. S., "Integrated Solid Waste Planning for a Regional Area," Franklin Associates, Ltd., presented at the First U.S. Conference on Municipal Solid Waste Management, Washington D.C., June 1990.
8. "The Cost to Recycle at a Materials Recovery Facility," National Solid Wastes Management Association, 1992.

CHAPTER 15

THE U.S. ENVIRONMENTAL PROTECTION AGENCY'S ROLE IN MUNICIPAL SOLID WASTE MANAGEMENT

Bruce Weddle

Director, Municipal and Industrial Solid Waste Division
Office of Solid Waste, U.S. Environmental Protection Agency
Washington, D.C.

15.1 INTRODUCTION

Since its inception in 1970, the U.S. Environmental Protection Agency (EPA) has been involved in municipal solid waste (MSW) management. In the 1970s, EPA played a pivotal role in advancing land disposal, waste-to-energy, and recycling technology, resulting in safer waste management. In the early 1980s, EPA turned its attention to the nation's hazardous waste problem. By the late 1980s, however, EPA had renewed its efforts to improve the management of MSW. Since that time, the Agency's role has primarily been one of setting the course for environmentally sound waste management and fostering cooperation among all the different sectors involved—individuals, businesses, industries, environmental groups, and government at all levels.

The Solid Waste Dilemma: An Agenda for Action is EPA's waste management blueprint for the nation. Published in 1989, this document was developed in conjunction with trade associations, environmental groups, government organizations, and citizens. It identifies the roles that every individual and institution plays in generating and managing MSW. It also sets forth goals and objectives for both EPA and the nation as a whole to address the waste management problems of today and tomorrow. This chapter briefly describes EPA's relationship with other levels of government in managing waste, and examines the role of the federal government for each objective established in the *Agenda for Action*.

15.2 *LOCAL GOVERNMENT EFFORTS*

The challenge of MSW management is far from new. Archaeological and historical evidence suggests that as far back as 6000 years ago in ancient Middle Eastern urban developments, people began to recognize the problems that could result from simply discarding waste out the back door. To control vermin and reduce the stench, residents began removing their waste to prescribed sites at the edges of their cities; these sites became the world's first garbage dumps. Leaders of these early communities evidently realized that a certain level of waste management was necessary to support their development. As civilization evolved and population centers grew, communities continued to seek more effective ways to manage their waste.

Today, local governments remain the institutions responsible for ensuring that the trash generated by their residents and commercial establishments is collected and properly managed. Local governments are well suited for this role. Because the types and amounts of solid waste generated vary considerably from community to community, local governments are in the best position to determine the most effective means for managing their waste. Day-to-day waste management decisions depend heavily on other local factors as well, such as available landfill and combustor capacity, public attitudes and behaviors, applicable regulations, and financial constraints.

In recent years, as waste management has become more complex in many areas of the country, the roles and responsibilities of local solid waste officials are changing to keep pace. Nationwide, waste generation is on the rise,[2] while in some areas disposal capacity is becoming more limited.[3] This has prompted many communities to look for more creative solutions to manage MSW. Some communities are taking steps to prevent the generation of waste in the first place. Others are building on their waste collection and transportation infrastructures to implement recycling and composting programs that reduce the amount of waste requiring disposal. In many areas, local officials also are sponsoring household hazardous waste collections to divert potentially toxic materials from the MSW stream and to help improve the safety of combustion and landfilling.[4]

As the technology of waste management is changing in this country, the related costs are escalating. Ensuring the economic viability of waste management has become a major issue in communities across the nation. Increasingly, communities attempting to implement comprehensive solid waste management programs must balance the cost of such programs against available resources. In response, many local governments are developing regional strategies that allow them to distribute costs and share resources with other municipalities, a concept known as *regionalization.*

Typical regionalization strategies include siting a single, large regional landfill, building a materials recovery facility that accepts waste from a number of nearby communities, or operating a waste-to-energy facility that serves several cities. In addition, some communities are banding together to market their recyclables or use their collective purchasing power to buy goods with recycled content. Regionalization is helping many communities develop a comprehensive joint approach to waste management. In the Portland, Oregon, area, several local governments forged METRO, an organization that functions as a regional government, to help them tackle a variety of issues, including MSW management.[5] METRO currently is evaluating the waste management difficulties faced by member communities and considering how to use recycling, household hazardous waste collections, public education campaigns, and other programs to address these problems. In another example, 20 mayors from communities in six states (Arizona, Colorado, Nevada, New Mexico, Texas, and Utah) joined forces in 1991 to form the Southwest Public Recycling Association.[6] Together, association members work to expand these markets for both collected recyclable materials and manufactured end products. The association also encourages and supports local recycling programs.

An increasing number of communities also are taking advantage of their role as waste management providers to create economic incentives to reduce waste. Traditionally, since consumers have been charged a fixed fee no matter how much waste they generate, rather than a fee based on the amount they produce, they have no incentive to limit their waste generation. To address this, communities like Seattle, Washington, have adopted unit pricing systems whereby households pay for waste management based directly on the amount of trash set out for disposal.[7] Under Seattle's program, residents may use either a regular garbage can or a smaller can one-third the size. The city then bills the residents according to the size and number of cans they put out for disposal. Items set out for recycling are collected without charge. This provides an incentive for individuals to change their buying habits by encouraging them to purchase items that are recyclable and to find ways (for example, through reuse) to produce less waste. As a result, this program has reduced the average number of trash cans left at the curb from three and one-half to just over one can per week.

15.3 STATE GOVERNMENT EFFORTS

While local governments bear the primary responsibility for managing MSW, state leadership and involvement also is critical. In recent years, state governments have assumed a steadily more active role in this area as they search for waste management options that are both economically and environmentally sound. The late 1980s saw a dramatic rise in state legislation related to solid waste issues. In 1989 alone, states passed 125 solid waste laws.[8] Some of these laws aimed to reduce packaging; mandate statewide recycling programs; encourage the procurement of goods with recycled content; or ban the landfilling or combustion of certain recyclable, compostable, or hard-to-manage wastes (such as yard trimmings). The Council of Northeastern Governors (CONEG) launched a ground-breaking initiative by drafting model legislation for states to use when writing their own laws on waste reduction.[8] The model legislation bans certain heavy metals from packaging and requires the use of less toxic substitutes. Nine Northeastern states already have enacted legislation based on CONEG's model.

Although targeted toward specific goals, these state laws often have a broader objective in mind—to encourage decision-makers to plan for waste management on the local, regional, tribal, or state level. Such planning is critical to successful implementation of waste reduction, recycling, and disposal activities. For example, Connecticut, Michigan, Oregon, Vermont, and Washington have developed legislation requiring the development of state solid waste management plans. In addition, many states provide direct technical assistance to local and regional governments during the development of their plans. Draft plans are carefully reviewed to ensure consistency with established state waste reduction goals. Table 5.5 lists the recycling goals established by 36 states.

States also have important roles in market development, procurement, research, and information dissemination. A major goal in most states is to encourage the development of markets for recycled materials. To stabilize and expand recycling markets, states are using a variety of incentives, including tax credits, low-interest loans, and grants for recycling firms and industries using recycled materials in their products.[9] Some states have banded together to form coalitions that share information and work to stimulate and stabilize markets. Many state agencies, recognizing their influence in the marketplace, also are using their buying power to purchase goods with recycled content. In addition, states are taking responsibility for filling gaps in solid waste research and facilitating information transfer among the different sectors involved in solid waste management. Finally, many states are revising their facility standards and permit requirements to reduce the environmental and health risks from landfilling and combustion. A major undertaking for states will be to implement the new federal standards for municipal solid waste landfills.

15.4   FEDERAL ROLE IN MUNICIPAL SOLID WASTE MANAGEMENT

Federal activity concerning solid waste management began with legislation in the 1960s. The Solid Waste Disposal Act (SWDA) was passed in 1965.[10] It was the first federal law to focus on improving household, municipal, commercial, and industrial solid waste handling and disposal practices in this country. The act was amended in 1970 with the Resource Recovery Act[11]) and again in 1976 with the passage of the Resource Conservation and Recovery Act (RCRA).[12] While recognizing that the primary responsibility for MSW management belonged with state and local governments, RCRA gave EPA significant hazardous waste regulatory responsibilities.

In 1984, Congress amended RCRA with the Hazardous and Solid Waste Amendments, expanding the scope and the requirements of the law considerably.[13] Among other requirements, the law directed EPA to revise existing regulations for MSW landfills. In 1991, EPA promulgated new municipal solid waste landfill standards governing location restrictions, facility design and operation, ground-water monitoring and corrective action, closure and postclosure care, and financial assurance.[14]

In addition to writing and implementing regulations, the federal government has a responsibility to provide leadership for the nation through its actions. By establishing programs to purchase items with recycled content in its own agencies, the federal government can help to develop markets for products with recycled content while also serving as a role model for the nation. EPA procurement guidelines require government agencies to buy products made with recovered materials. These guidelines (listed in Fig. 15.1) apply to all federal, state, and local agencies (and their contractors) using appropriated federal funds that purchase more than $10,000 worth of a guideline item in one year.

To unify and coordinate the growing number of federal recycling programs, the General Services Administration (GSA) established the National Federal Recycling Program in 1989. Under this program, all GSA owned and operated facilities with 100 or more employees must collect office paper for recycling. Through the program, a national recycling team has been established and a recycling coordinator designated to help ensure rapid implementation. In addition, interagency partnerships have formed to share expertise. For example, the Department of Defense is working with EPA to help design effective recycling programs for several military and research bases. Under the auspices of EPA and the GSA, annual Federal Agency Recycling Conferences also are being held to increase federal employees' awareness of recycling and procurement of recycled goods issues.

The U.S. Postal Service is another good example of how federal agencies are developing creative and effective ways to reduce, reuse, and recycle. In 1990, the Postal Service kicked off an ambitious recycling program at its Washington, D.C.,

---

Procurement Guideline for Paper and Paper Products, 40 CFR 250; 53 FR 23546 (June 22, 1988)

Procurement Guideline for Lubricating Oils Containing Re-Refined Oil, 40 CFR 252; 53 FR 24699 (June 30, 1988)

Procurement Guideline for Retread Tires, 40 CFR 253; 53 FR 46588 (Nov. 17, 1988)

Procurement Guideline for Building Insulation Products Containing Recovered Materials, 40 CFR 248; 54 FR 7328 (Feb. 17, 1989)

Procurement Guideline for Cement and Concrete Containing Fly Ash, 40 CFR 249; 48 FR 4230 (Jan. 28, 1983)

The procurement guidelines apply to all federal, state, and local agencies (and their contractors) using federal funds that purchase more than $10,000 worth of a guideline item in one year. Affected agencies must develop affirmative procurement programs, including appropriate specification revisions. The guidelines do not have to be followed if doing so would result in unreasonable cost, inadequate competition, unreasonable delays, or inability to meet reasonable performance standards. On April 20, 1994 EPA proposed comprehensive changes to the procurement guideline development process. Also proposed for addition were 21 construction, landscaping, and office products.

FIGURE 15.1 EPA procurement guidelines.

headquarters and at more than 40,000 post offices across the nation. The program, called *Saving of America's Resources* (SOAR), calls for the recycling of paper, newspaper, cardboard, and aluminum. Through SOAR, the Postal Service procures recycled paper for its newsletters, notices to households, and other publications; searches for alternatives to plastic windows in envelopes to make envelopes more recyclable; develops stamps with glues that are compatible with recycling processes; and minimizes waste in its 160,000-vehicle maintenance operations.

Federal efforts were expanded again on October 31, 1991, when President Bush signed Executive Order 12780, entitled "Federal Agency Recycling and the Council on Federal Recycling and Procurement Policy."[15] The order requires all federal agencies to develop a program to reduce the amount of waste produced, recycle as much material as practical, and support recycling markets by focusing their purchasing power on products containing recycled materials. The order also established the Council on Federal Recycling and Procurement. Consisting of representatives of key federal agencies, the Council disseminates information to agencies on waste reduction techniques and helps them develop and implement their plans. In addition, at the direction of the Joint Committee on Printing, the Government Printing Office (GPO) is favoring the purchase of paper made with recovered rather than virgin materials. This effectively makes GPO one of the largest purchasers of recycled photocopying, printing, and writing paper in the world.

The federal government also is conducting research to help increase our understanding of how the nation's waste can best be managed. Prior to 1989, this research focused on collecting data necessary to support the development of regulations for MSW landfills. Since then, however, the emphasis has shifted to programs that help advance the utilization of source reduction, recycling, and resource recovery. One major component of EPA's research program has been the Municipal Innovative Technology Evaluation (MITE) program. Under this program, EPA conducts impartial technical, economic, and environmental performance evaluations of emerging technologies and waste management concepts.

15.5   EPA'S ROLE IN MSW MANAGEMENT

EPA's present efforts to improve MSW management grew out of the growing environmental awareness in this country, and concern in particular about the problems associated with solid waste management. *The Solid Waste Dilemma: An Agenda for Action,*[1] documents EPA's strategy for addressing the nation's solid waste management problems. A fundamental precept of this strategy is that MSW is primarily a local responsibility. The report outlines the steps necessary not just for the federal government, but also for state and local governments, businesses, private organizations, and individual citizens, to manage safely and effectively the nation's municipal solid waste stream. Since all individuals and sectors of society generate solid waste, the report emphasizes that everyone has a role in finding and supporting solutions. Underscoring this emphasis are the three broad, national goals outlined in the *Agenda for Action:* (1) finding a safe and permanent way to eliminate the gap between waste generation and available capacity in landfills, combustors, and secondary materials markets; (2) reducing and recycling 25 percent of the nation's solid waste stream by the end of 1992; and (3) increasing the safety of the remaining disposal options, combustion and landfilling.

The *Agenda for Action* also urges the adoption of integrated waste management, the complementary use of different waste management practices designed to handle MSW safely and effectively. It presents EPA's recommended hierarchy for the most efficient use of integrated waste management (Fig. 15.2). The first and most favored option in the hierarchy is source reduction, which can be defined as the design, manufacture, purchase, or reuse of materials or products to reduce the amount or toxicity of waste before it enters the MSW stream. Source reduction is essentially a preventive approach. Recycling, including composting of food scraps and yard trimmings, is the next preferred alternative. Recycling saves energy and natural resources, provides useful products from our discards and preserves landfill space. While listed below source reduction and recycling on the integrated waste management hierarchy, safe combustion and/or landfilling nonetheless will remain vital components of waste management planning in many communities.

Responsibility for municipal solid waste management is shared primarily by three major components within the agency: the Municipal and Industrial Solid Waste Division in the Office of Solid Waste, the agency's 10 regional field offices, and the Office of Research and Development. Table 15.1 contains recent budget distributions among these three major components of the agency.

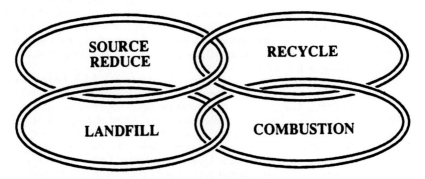

FIGURE 15.2   The EPA hierarchy for MSW management.

TABLE 15.1 EPA's Municipal Solid Waste Budget

	Fiscal year '88		Fiscal year '89		Fiscal year '90		Fiscal year '91		Fiscal year '92	
	FTE*	$	FTE	$	FTE	$	FTE	$	FTE	$
Headquarters	19	5,455,000	24	5,323,800	25	5,000,000	22	5,400,000	26	5,100,000
Regions			20	1,800,000	25	3,800,000	50	4,300,000	50	4,300,000
Office of Research and Development	8	2,475,000	8	2,475,000	12	4,700,000	20	7,244,000	24	5,800,000
Total	27	7,930,000	52	9,598,000	62	13,500,000	92	16,944,000	100	15,200,000

*Full-time equivalent employees.

An organizational chart for the Municipal and Industrial Solid Waste Division is presented in Fig. 15.3. As the "headquarters" component of EPA, its responsibility is to develop national regulations, policies, strategies, information tools, and programs.

Municipal solid waste research activities are conducted, for the most part, in the Risk Reduction Engineering Laboratory in Cincinnati, Ohio. Other laboratories with responsibility for municipal solid waste include the Air and Energy Engineering Research Laboratory in Research Triangle Park, N.C. and the Environmental Criteria and Assessment Office in Cincinnati, Ohio. Table 15.2 presents a list of the major EPA labs with municipal solid waste responsibilities.

Figure 15.4 shows the ten EPA regional offices and the states that each serves. The role of each regional office is to facilitate effective solid waste management, including implementation of the MSW landfill criteria through technical assistance, training, and outreach and education efforts. The regional offices are the hands-on contacts for states, tribes, and local communities.

15.6 EPA ACTIVITIES IN MUNICIPAL SOLID WASTE MANAGEMENT

The six objectives identified in the *Agenda for Action* established the framework for EPA's current efforts on behalf of improved solid waste management. A brief description of the federal program addressing each of these objectives is described below.

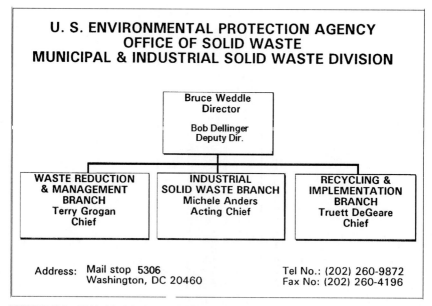

FIGURE 15.3 EPA's Municipal and Solid Waste Division.

TABLE 15.2   EPA Research Laboratories with MSW Responsibility

Name and address	Areas of responsibility
Risk Reduction Engineering Laboratory U.S. EPA, Office of Research and Development 5995 Center Hill Rd. Cincinnati, OH 45224 Tel: (513) 569-7871	Land disposal, recycling, source reduction, evaluations of innovative technologies
Center for Environmental Research Information U.S. EPA, Office of Research and Development 26 West Martin Luther King Dr. Cincinnati, OH 45268 Tel: (513) 569-7391	Development and dissemination of informational materials, training
Environmental Criteria and Assessment Office U.S. EPA, Office of Research and Development 26 West Martin Luther King Dr. Cincinnati, OH 45268 Tel: (513) 569-7531	Risk assessments
Air and Energy Engineering Research Laboratory U.S. EPA, Office of Research and Development Mail Stop MD-60 Research Triangle Park, NC 27711 Tel: (919) 541-2821	Combustion research
Environmental Monitoring Systems Laboratory U.S. EPA, Office of Research and Development Mail Stop MD-75 Research Triangle Park, NC 27711 Tel: (919) 541-2106	Air monitoring
Environmental Monitoring Systems Laboratory U.S. EPA, Office of Research and Development P.O. Box 93478 Las Vegas, NV 89193-3478 Tel: (702) 798-2100	Groundwater monitoring

Increase Research and Information Dissemination

Day-to-day management of MSW is a local responsibility. However, a key role for the federal government is to help local managers make the best decisions for their communities. Conducting basic research, collecting data, evaluating technology, and disseminating that information are important federal activities. Technical assistance (in the form of hands-on assistance, manuals, conferences, and other tools) ensures that all types of waste handlers have the information they need to manage solid waste safely and effectively. Among the technical assistance tools employed by EPA are peer matches whereby local government officials grappling with waste management issues in their communities can gain access to other local decision makers from other communities who have successfully dealt with these issues. Assistance can be provided in a variety of forms, including site visits, written correspondence, telephone conversations, referrals, library resources and the personal experiences of the peer. Educational materials increase the awareness of suitable waste management practices and encourage an environmental ethic. Data collection and research and development expand the boundaries of our knowledge, providing new information, technologies, and solutions.

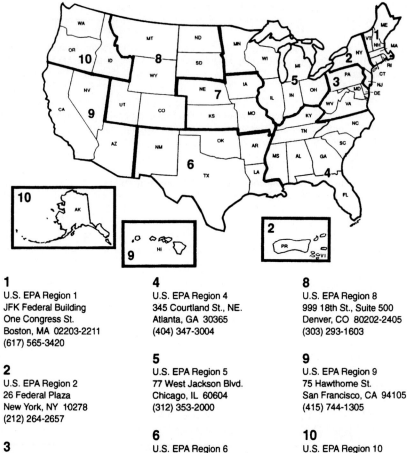

1
U.S. EPA Region 1
JFK Federal Building
One Congress St.
Boston, MA 02203-2211
(617) 565-3420

2
U.S. EPA Region 2
26 Federal Plaza
New York, NY 10278
(212) 264-2657

3
U.S. EPA Region 3
841 Chestnut St.
Philadelphia, PA 19107
(215) 597-9800

4
U.S. EPA Region 4
345 Courtland St., NE.
Atlanta, GA 30365
(404) 347-3004

5
U.S. EPA Region 5
77 West Jackson Blvd.
Chicago, IL 60604
(312) 353-2000

6
U.S. EPA Region 6
1445 Ross Ave.
Dallas, TX 75202-2733
(214) 655-6444

7
U.S. EPA Region 7
726 Minnesota Ave.
Kansas City, KS 66101
(913) 551-7000

8
U.S. EPA Region 8
999 18th St., Suite 500
Denver, CO 80202-2405
(303) 293-1603

9
U.S. EPA Region 9
75 Hawthorne St.
San Francisco, CA 94105
(415) 744-1305

10
U.S. EPA Region 10
1200 Sixth Ave.
Seattle, WA 98101
(206) 553-4973

FIGURE 15.4   EPA's regional field offices.

   Since the publication of the *Agenda for Action,* EPA has printed dozens of technical and educational publications for the different sectors involved in waste management. These documents range from a pamphlet for car owners on safely changing their used oil to a handbook for businesses on implementing an office paper recycling program to a guidance manual for state and local governments on used household battery collection programs. In 1989, EPA published a comprehensive educational package for schoolteachers in kindergarten through grade 12. It provides course and resource materials on the importance of reducing and recycling waste. Also in 1989, EPA launched a quarterly newsletter, *Reusable News,* to help raise awareness and foster improvement of solid waste management. Another ongoing newsletter, *Native American News,* focuses on solid waste issues of concern to Native American tribes. In addition to its own documents, EPA has provided technical assistance or funding to numerous organizations for the development of additional "how-to" documents and research studies. A number of these documents are referenced throughout this chapter, and the bibliography at the end of this chapter presents a list of publications developed by EPA on solid waste management that are either available at no cost from EPA's hotline (800-424-9346) or at a nominal cost from the National Technical Information Service (800-336-4700). EPA has also established or supports several information networks to provide data and assistance to anyone interested in MSW issues. These networks are listed in Fig. 15.5.

EPA also organizes and conducts solid waste management conferences to encourage professionals to share ideas and learn from expert speakers. EPA has already sponsored two international conferences on MSW management since 1990; another conference is planned for 1994. Since 1985, EPA also has been sponsoring annual Household Hazardous Waste Management Conferences to encourage solid waste professionals to share ideas on managing these types of waste.

Municipal Solid Waste Planning

Planning is essential to effective MSW management. States, tribes, and municipalities hold the primary responsibility for planning in their jurisdictions. Through careful planning, state, tribal, and local officials can fully understand their short- and long-

---

Resource Conservation and Recovery Act (RCRA) Hotline

EPA's information clearinghouse for regulations and publications on both solid and hazardous waste. Call (800) 424-9346 weekdays from 8:30 a.m. to 7:30 p.m., Eastern Standard Time.

Solid Waste Assistance Program (SWAP)

Operated with EPA funding by the Solid Waste Association of North America, SWAP is designed to help government agencies, industries, citizen groups, and all other interested parties gather important solid waste management information. To order material or to request a list of SWAP's available publications, call (800) 677-9424 weekdays from 8:30 a.m. to 5 p.m., Eastern Standard Time.

Solid Waste Information Clearinghouse (SWICH)

An on-line electronic library system and bulletin board that can be accessed through a computer and modem for an annual user fee. Created through an EPA grant to the Solid Waste Association of North America, SWICH lists over 6,500 volumes—journals, reports, studies, newsletters, proceedings, films, and videos—that can be ordered for a minimal charge. Users can do a computer "search" for literature on a huge range of solid waste topics. SWICH's electronic bulletin board posts meeting and conference announcements, employment leads, and the quarterly SWICH newsletter. In addition, users can network with other solid waste professionals to share information and get answers to specific technical questions. Call (800) 677-9424 weekdays from 8:00 a.m. to 5:30 p.m., Eastern Standard Time.

FIGURE 15.5   Solid waste information networks.

term solid waste capacity needs and devise a comprehensive approach for ensuring that all waste, including hard-to-manage wastes (such as used oil, tires, and batteries) are effectively handled.

The federal government has assumed an important assistance role to help state and local government planning efforts by providing information and guidance. EPA's *Decision-Makers Guide to Solid Waste Management*[16] is an example of the type of information EPA has provided. The first volume of this guide, published in 1990, helps local officials understand the nature of the solid waste issues they are confronting and identify and evaluate the different management options. Roles that may be new to some decision makers also are presented, such as educating the community, learning about recyclable materials markets, and encouraging waste generators to reduce and recycle their waste. EPA currently is working on the second volume of this guide, which will contain more technical information for managers responsible for implementing and integrating their area's chosen waste management approaches.

An essential element of planning is ensuring adequate disposal capacity for the solid waste generated within a state or community. The siting of solid waste management facilities is an issue that has received much attention in recent years. Recognizing that siting has the potential to become a contentious issue in a community, EPA published a guidebook in 1990 to assist solid waste decision makers in the siting process. The document, entitled *Sites for Our Solid Waste: A Guidebook for Effective Public Involvement,*[17] is designed to help communities develop a local consensus on a siting project by eliciting full and effective community participation. In addition, EPA worked with the International City/County Management Association to produce a report detailing examples of successful siting projects.[18]

Source Reduction

Source reduction is defined as reducing the amount and toxicity of waste generated. It is important to note that all other options (recycling, combustion, and landfilling) manage waste after it is generated, while source reduction actually prevents the generation of waste in the first place. While the manufacturing and design industry plays perhaps the most pivotal role in launching source reduction initiatives, EPA also has a responsibility to provide leadership, guidance, and consistency as waste generators embark on source reduction projects.

Source reduction is not a widely understood management technique. The federal government can play an important role in educating citizens, businesses and local governments on how they can reduce waste volumes and toxicity. EPA actively supported the research and development of the 1991 World Wildlife Fund and Conservation Foundation report entitled *Getting at the Source: Strategies for Reducing Municipal Solid Waste.*[19] Prepared under an EPA grant, the report analyzes the role of source reduction in MSW management and presents recommendations for increasing source reduction activities in the nation. EPA is beginning to implement several of the recommendations called for in *Getting at the Source,* including supporting the development of mechanisms to measure source reduction efforts.

EPA is conducting research to identify toxic substances entering the waste stream and to explore potential substitutes for these constituents. One study documents the levels of lead and cadmium in the U.S. solid waste stream, identifying lead-acid batteries as the primary sources of lead and rechargeable batteries and the stabilizers used in plastics as the leading contributors of cadmium.[20] A separate study of products containing mercury documents the declining use of mercury in the most commonly used battery types, and examined the effect of such trends on MSW management.[21]

EPA also is supporting research into the development of life-cycle analysis, a tool that offers to significantly improve the effectiveness of source reduction programs. Used to measure the total environmental impact of a given product from raw material extraction from the earth to the product's ultimate disposal, life-cycle analyses allow solid waste decision makers to focus their efforts on the specific products and waste-generating processes that are truly wasteful. While life-cycle studies tracing a product's manufacture, use, and disposal have been conducted for years, environmental impacts have not been quantified. These studies have therefore been limited to subjective comparisons between different options for product use and disposal. EPA is planning a series of documents addressing the gap between these early studies and scientifically quantifiable analysis.

To help consumers understand their vital role in source reduction, EPA published *The Consumer's Handbook for Reducing Solid Waste*[22] in 1992. The handbook describes simple steps that citizens can take to reduce the amount and toxicity of the waste they generate. It encourages individuals to consider source reduction in their everyday lives—while working, shopping, or simply relaxing at home.

Building consumer confidence is another area especially suited for the federal government. In 1992, the Federal Trade Commission, with support from EPA, issued guidelines[23] concerning the use of environmental claims in advertising and labeling. The guidelines are intended to reduce consumer confusion and to help manufacturers responsibly advertise their products' beneficial environmental attributes. Under the guidelines, marketers are advised to make only those claims they can substantiate. According to the new labeling guidelines, any claims about the product's environmental impact:

- Must be clear and prominent enough to prevent deception
- Should clearly state whether they apply to the product, its packaging, or both
- Should not overstate environmental benefits

The guidelines also specify appropriate, nondeceptive uses for claims such as "environmentally friendly," "degradable," "biodegradable," "recyclable," and "ozone-safe." In addition, EPA, the Federal Trade Commission, and the U.S. Office of Consumer Affairs jointly published a pamphlet to help consumers evaluate the accuracy of such advertising claims.[24] The pamphlet warns consumers against broad or overly vague claims, such as "environmentally safe" and "eco-friendly." For example, a consumer may be attracted to a specific product because the label "recycled content" appears on the package; however, this claim does not clearly state whether it applies to the product, its packaging, or both.

Recycling

As stated earlier, one function that is particularly well-suited to the federal government is the development and dissemination of information. This has been one of EPA's primary activities in the recycling area. In particular, EPA has developed information for the general public, technical information for state and local governments, and technical training and guidance.

In 1988, EPA, in an effort to educate the general public, provided funding to help the Environmental Defense Fund launch a nationwide advertising campaign to promote recycling. This continuing campaign was created in conjunction with a private advertising company and the Ad Council. The ads received free air time worth millions of dollars and generated hundreds of thousands of requests from individuals wanting more information on recycling.

In 1989, EPA published a booklet called *Recycling Works!*[25] that describes a variety of successful recycling programs at the state and local level. This publication provides solid waste officials with real-life information that can prove useful as they consider options in their areas. Building on that premise, EPA is currently conducting a study of 30 municipal recycling programs to help officials embark on or expand existing programs. The report will function as a database of information on starting up and operating collection programs, containing details on collection methods, publicity, and costs. It also will examine the strengths and weaknesses of each of the case studies included and identify factors common to nearly all successful programs. Again, this information will help to improve decision making by local governments.

Another area that holds tremendous recovery potential is the composting of yard trimmings and other organic materials in MSW. More and more communities are turning to composting to help them meet state or local recovery goals, to divert materials from the MSW stream, and to comply with regulations that ban certain materials (particularly yard trimmings) from landfills. Many municipalities already have programs in place to compost yard debris and lawn trimmings. With regard to mixed municipal solid waste composting, approximately 20 facilities currently operate, but over 140 new facilities are in the planning stages. To help communities utilize composting as part of their integrated waste management strategy, EPA is developing a comprehensive guide to composting. This document will describe the composting process (for both mixed waste and yard waste) and offer guidance to state and local officials on planning, siting, and operating a compost facility in their community.

Even as more and more communities begin recycling programs in their jurisdictions, the fluctuation of market availability for many secondary materials remains an impediment to expanded recycling. As with any commodity, the market for secondary materials is driven by supply and demand. Since the release of the *Agenda for Action,* EPA recycling efforts have increasingly turned to the need for market development. Because understanding the current and future markets for recycled materials is critical to the success of recycling, EPA is developing a series of market studies that will provide decision makers with data to plan for effective recycling. Reports have been published on scrap tires,[26] glass,[27] and aluminum.[28] Additional studies will focus on the current and future markets for paper and compost.

Procurement of recycled goods is an important way to stimulate and strengthen the marketplace for goods made of recycled content. The federal government, as a major purchaser of goods and services, can have a direct impact on markets for goods made with recycled materials. To date, EPA has established procurement guidelines for the purchase of paper and paper products, lubricating oils, retread tires, building insulation products, and cement and concrete containing fly ash (see Fig. 15.1). These guidelines are directed to federal agencies and others using federally appropriated funds. EPA plans to revise the recycled content standards of its paper procurement guideline to reflect increases in the post-consumer content of many recycled paper products. EPA also intends to develop additional guidelines for building and construction materials, compost from yard waste, and rubberized asphalt paving. In addition, feasibility studies will be conducted to evaluate other commodities and products that could be targeted for procurement.

EPA has sponsored several other procurement workshops, bringing together federal agency procurement officials and hundreds of representatives from business and industry. EPA also is working to increase recycled goods procurement by state and local governments, businesses, and individuals. Many businesses, in particular, represent potentially larger purchases of recycled-content goods. One example of businesses taking a leading role in purchasing goods made with recycled materials is the Buy Recycled Campaign, launched last year by the National Recycling Coalition with financial support from EPA.[29] Through the campaign, many U.S. corporations are committing themselves to buy recycled goods.

Recognizing the importance of business involvement in procurement of recycled goods, EPA is providing funding to the Institute for Local Self-Reliance to develop case studies that illustrate successful programs that aid the siting of new recycling businesses in our large urban centers. These types of programs also can create jobs within the nation's cities, thereby helping our unemployment problem while creating a recycling infrastructure.

Combustion

Combustion with energy recovery and land disposal are the cornerstones for managing the waste that cannot be prevented or economically recycled. An important role for the federal government is to assure that these practices are conducted in a way that protects both human health and the environment. EPA is therefore taking steps to improve the effectiveness of these practices. In the area of combustion, EPA promulgated a set of comprehensive air emissions standards for new municipal waste combustors in 1991.[30] These regulations, which are summarized in Fig. 15.6, mandate proper operation of municipal waste combustors, including emissions limits for toxic metals and organics, acid gases, and nitrogen oxides; operating standards to ensure optimum combustion; and an operator certification requirement. In accordance with the Clean Air Act Amendments of 1990, EPA is broadening these regulations to include limits on lead, mercury, and cadmium, and is considering extending the limits on nitrogen oxides established under the standards to existing facilities as well. EPA is also broadening these regulations to include all but the very smallest facilities.

EPA also has conducted studies of solid waste combustors in this country to better understand current practices and identify areas of potential concern. In 1987 EPA surveyed the MSW combustion industry nationwide, generating data on facilities, waste types received, and residue handling and disposal practices.[31] Another study[32] evaluated current ash management practices and assessed the risk posed by any resulting fugitive dust emissions. As a result of these and other studies, EPA has determined that disposal of ash generated from the combustion of nonhazardous municipal solid waste under the Subtitle D land disposal criteria is protective of human health and the environment. EPA also believes that Section 3001 (i) of RCRA (12) exempts MWC from regulation under RCRA Subtitle C.[33] However, because of conflicting judgements by the Second Circuit Court of Appeals of New York, and the Seventh Circuit Court of Appeals of Chicago, the U.S. Supreme Court has been asked to rule on which Section (C or D) of RCRA is applicable.

Landfilling

To reduce potential environmental and health risks associated with land disposal, EPA issued new comprehensive landfill design, location and operating standards in 1991.[14] States are required by law to adopt their own rules and regulations to implement these requirements. To facilitate efficient implementation of the criteria, EPA is providing technical training sessions on landfill design and permitting for states, tribes and local governments. The sessions explain the criteria's minimum standards, discuss ways the flexibility provisions may be used, show landfill design advances such as computer modeling software, and provide additional sources of information. A handbook has

SPECIFIC REQUIREMENTS OF THE NEW SOURCE PERFORMANCE STANDARDS
AND EMISSIONS GUIDELINES FOR UNITS LARGER THAN 250 TPD
(All emission levels are at 7% oxygen, dry basis)

1. Good combustion practices.

   Must not exceed 110% of maximum load level demonstrated during dioxin/furan performance tests.

   Temperature at inlet to APC cannot exceed maximum demonstrated temperature by more than 30°F.

   CO must not exceed 50 to 150 ppmv depending on type of unit for new units, and 50 to 250 ppmv for existing units.

   Certification of plant managers and supervisors, and training for plant operators.

   Continuous emissions monitoring of sulfur dioxide, NOx, opacity, CO, load, and temperature.

   Annual stack testing for particulates, dioxins/furans, and HCL.

2. Particulate emission limits

   34 mg/dscm (0.015 gr/dscf). [For existing facilities smaller than 1100 TPD the standards is 69 mg/dscm (0.03 gr/dscf)]

3. Organic emission limits

   Dioxins and furans are measured as total tetra-through-octa-chlorinated dibenzo-p-dioxins and dibenzofurans, and not as toxic equivalents.

New sources	30 ng/dscm
Existing facilities >1100 TPD	60 ng/dscm
Existing facilities <1100 TPD	125 ng/dscm

4. Acid gas controls

   Sulfur dioxide:

	The higher of:
New sources	80% reduction or 30 ppmv
Existing facilities >1100 TPD	70% reduction or 30 ppmv
Existing facilities <1100 TPD	50% reduction or 30 ppmv

   Hydrochloric acid:

	The higher of:
New sources	95% reduction or 25 ppmv
Existing facilities >1100 TPD	90% reduction or 25 ppmv
Existing facilities <1100 TPD	50% reduction or 25 ppmv

5. Nitrous oxide (NOx) emission limits:

   New sources only: 180 ppmv

FIGURE 15.6   New source performance standards and emission guidelines for MWCs.

been prepared to help owners and operators of public and private landfills understand the new federal standards.[34] It offers a summary of the regulations and how they should improve landfill design and operation. The handbook also discusses how landfills will be permitted by states and tribes. EPA has also published a similar booklet directed toward the general public describing how the standards can improve landfill safety in their community.[35]

REFERENCES*

1. *The Solid Waste Dilemma: An Agenda for Action—Final Report of the Municipal Waste Task Force,* EPA/530-SW-89-019, U.S. EPA, Washington, D.C., 1989.

2. *Characterization of Municipal Solid Waste in the United States: 1992 Update,* EPA/530-R-92-019, NTIS: PB92-207 166, U.S. EPA, Washington, D.C., 1992.

3. Repa, E. W., and S. K. Sheets, "Landfill Capacity in North America," *Waste Age,* vol. 23, no. 5, May, 1992, NSWMA, Washington, D.C., pp. 18–28.

4. *Proceedings of the Seventh National USEPA Conference on Household Hazardous Waste Management,* Dec. 8–12, 1992, The Waste Watch Center, Andover, Mass., pp. 650–653.

5. Abbott, Carl, and M. P. Abbott, "Historical Development of the Metropolitan Service District," Metro Home Rule Charter Committee, Portland, Ore., 1991.

6. Olson, Gary, "The State of Markets in the Southwest," *Proceedings of the Third International Recycling Symposium,* Solid Waste Association of North America, Silver Spring, Md., 1992.

7. Skumatz, L., and C. Breckinridge, *Variable Rates in Solid Waste: Handbook for Solid Waste Officials; Volume I—Executive Summary,* EPA/530-SW-90-084A; *Volume II—Detailed Manual,* EPA/530-SW-90-084B. U.S. EPA, Washington, D.C., 1990.

8. *Toxics in Packaging Legislation: A Comparative Analysis,* CONEG Policy Research Center, Inc., Washington, D.C., July, 1992.

9. Miller, Chaz, "Recycling in the States: 1992 Update," *Waste Age,* vol. 24, no. 3, NSWMA, Washington, D.C., March, 1993, pp. 26–32.

10. Pub. L. 89-272, title II, Oct. 20, 1965.

11. Pub. L. 91-512, Oct. 26, 1970.

12. Pub. L. 94-580, Oct. 21, 1976.

13. Pub. L. 98-616, Nov. 8, 1984.

14. Solid Waste Disposal Facility Criteria, Final Rule, 40 CFR Parts 257 and 258, *Federal Register,* vol. 56, no. 196, pp. 50978–51119, Washington, D.C., October 9, 1991.

15. Bush, George, Executive Order 12780 of October 31, 1991, Federal Agency Recycling and the Council on Federal Recycling and Procurement Policy, *Federal Register,* vol. 56, no. 213, pp. 56289–56292, Washington, D.C., November 4, 1991.

16. *Decision Maker's Guide to Solid Waste Management,* EPA/530-SW-89-072, U.S. EPA, Washington, D.C., 1989.

17. *Sites for Our Solid Waste: A Guidebook for Effective Public Involvement,* EPA/530-SW-90-019, U.S. EPA, Washington, D.C., 1990.

18. "Siting Solid Waste Facilities: Seven Case Studies," *MIS Report,* vol. 24, no. 10, International City Management Association, Washington, D.C., Oct., 1992.

19. *Getting at the Source: Strategies for Reducing Municipal Solid Waste,* World Wildlife Fund and the Conservation Foundation, Washington, D.C., 1991.

20. *Characterization of Products Containing Lead and Cadmium in Municipal Solid Waste in the United States from 1970–2000—Study,* NTIS: PB89-151 039, U.S. EPA, Washington, D.C., 1988.

21. *Characterization of Products Containing Mercury in Municipal Solid Waste in the United States from 1970–2000—Study,* EPA/530-R-92-013, NTIS: PB92-162 569. U.S. EPA, Washington, D.C., 1992.

---

*References cited with an EPA publication number may be available for free from the RCRA Hotline (800-424-9346). Those with an NTIS number may be purchased from NTIS by calling 800-336-4700.

22. *The Consumer's Handbook for Reducing Solid Waste,* EPA/530-K-92-003. U.S. EPA, Washington, D.C., 1992.

23. Guides for the Use of Environmental Marketing Claims, Final Guides, 16 CFR Part 260, *Federal Register,* vol. 57, no. 157, pp. 36363–36369, Washington, D.C., August 13, 1992.

24. *Green Advertising Claims,* EPA/530-F-92-024, U.S. EPA, Washington, D.C., October, 1992.

25. *Recycling Works! State and Local Solutions to Solid Waste Management Problems,* EPA/530-SW-89-014, U.S. EPA, Washington, D.C., 1989.

26. *Markets for Scrap Tires,* NTIS Publ. No. PB92-115 252. U.S. EPA, Washington, D.C., 1991.

27. *Markets for Recovered Glass,* NTIS Publ. No. PB93-169-845. U.S. EPA, Washington, D.C., 1992.

28. *Markets for Recovered Aluminum,* NTIS Publ. No. PB93-170-132. U.S. EPA, Washington, D.C., 1992.

29. Meade, Kathleen. "Notes from NRC: National Recycling Coalition's Buy Recycled Campaign Means Business," *Resource Recycling Magazine,* vol. XII, no. 3, Portland, Ore., March, 1993.

30. Standards of Performance for New Stationary Sources and Final Emission Guidelines, Final Rule, 40 CFR Parts 51, 52, and 60, *Federal Register,* vol. 56, no. 28, pp. 5488–5527, Washington, D.C., February 11, 1991.

31. *Municipal Solid Waste Combustion Study,* 9 vols., NTIS Publ. No. PB87-206 066 (volumes may also be ordered individually), U.S. EPA, Washington, D.C., 1987.

32. *Characterization of MWC Ashes and Leachates from MSW Landfills, Monofills, and Co-Disposal Sites,* 7 vols., NTIS Publ. No. PB88-127 931 (volumes may also be ordered individually), U.S. EPA, Washington, D.C., 1987.

33. Reilly, William K., Unpublished memorandum to all (EPA) Regional Administrators, U.S. EPA, Washington, D.C., September 18, 1992.

34. *Criteria for Solid Waste Disposal Facilities, A Guide for Owners/Operators,* EPA/530-SW-91-089, U.S. EPA, Washington, D.C., March, 1993.

35. *Safer Disposal for Solid Waste, The Federal Regulations for Landfills,* EPA/530-SW-91-092, U.S. EPA, Washington, D.C., March, 1993.

SUGGESTIONS FOR FURTHER READING

Many of the following documents are available at no cost from EPA's Superfund/RCRA Hotline at (800) 424-9346.

About the Municipal Solid Waste Stream—Environmental Fact Sheet, EPA/530-SW-90-042B, Washington, D.C., 1990.

America's War on Waste—Environmental Fact Sheet, EPA/530-SW-90-002, Washington, D.C., 1990.

Be an Environmentally Alert Consumer, EPA/530-SW-90-034A, Washington, D.C., 1990.

Bibliography of Municipal Solid Waste Management Alternatives, EPA/530-SW-89-055, Washington, D.C., 1989.

A Business Guide for Reducing Solid Waste, EPA/530-K-92-004, Washington, D.C., 1992.

Characterization of Municipal Solid Waste in the United States: 1992 Update, Executive Summary—EPA/530-S-92-019, Fact Sheet—EPA/530-F-92-019, Washington, D.C., 1992.

Characterization of Products Containing Mercury in Municipal Solid Waste in the United States from 1970–2000, Executive Summary—EPA/530-F-92-013, Fact Sheet—EPA/530-F-92-017, Washington, D.C., 1992.

Decision-Makers Guide to Solid Waste Management—Promotional Brochure, EPA/530-SW-89-073, Washington, D.C., 1990.

The Facts on Recycling Plastics, EPA/530-SW-90-017E, Washington, D.C., 1990.

Final Cover Requirements for Municipal Solid Waste Landfills—Environmental Fact Sheet, EPA/530-SW-91-084, Washington, D.C., 1992.

Household Hazardous Waste: Bibliography of Useful References and List of State Experts, EPA/530-SW-88-014, Washington, D.C., 1988.

How Businesses Are Saving Money by Reducing Waste, EPA/530-K-92-005, Washington, D.C., 1992.

How to Set Up a Local Program to Recycle Used Oil, EPA/530-SW-89-039A, Washington, D.C., 1989.

Let's Reduce and Recycle: A Curriculum for Solid Waste Awareness, EPA/530-SW-90-005, Washington, D.C., 1990.

Methods to Manage and Control Plastic Wastes—Executive Summary, EPA/530-SW-89-051A, Washington, D.C., 1989.

Native American Network: A RCRA Information Exchange, EPA/530-SW-90-079, Washington, D.C., 1990. Spring 1991—EPA/530-SW-91-001. Summer/Fall 1991—EPA/530-SW-91-002.

Plastics: The Facts on Source Reduction, EPA/530-SW-90-017C, Washington, D.C., 1990.

Recycle, EPA/530-F-92-003, Washington, D.C., 1992.

Recycling in Federal Agencies, EPA/530-SW-90-082, Washington, D.C., 1990.

Recycling Used Oil: 10 Steps to Change Your Oil, EPA/530-SW-89-039C. Washington, D.C., 1989.

Recycle Today: Educational Materials for Grades K–12, EPA/530-SW-90-025, Washington, D.C., 1990.

Recycling Used Oil: What Can You Do? EPA/530-SW-89-093B, Washington, D.C., 1989.

Reusable News. Winter 1990—EPA/530-SW-90-018, Spring 1990—EPA/530-SW-90-039, Summer 1990—EPA/530-SW-90-055, Fall 1990—EPA/530-SW-90-056, Winter 1991—EPA/530-SW-91-020, Spring/Summer 1991—EPA/530-SW-91-021, Fall 1991—EPA/530-SW-91-022, Winter 1992—EPA/530-SW-91-085, Washington, D.C.

School Recycling Programs: A Handbook for Educators, EPA/530-SW-90-023, Washington, D.C., 1990.

Siting Our Solid Waste: Making Public Involvement Work, EPA/530-SW-90-020, Washington, D.C., 1990.

Small Communities and the Municipal Landfill Regulations—Environmental Fact Sheet, EPA/530-SW-91-067, Washington, D.C., 1991.

SWICH: EPA's National Solid Waste Information Clearinghouse—Environmental Fact Sheet, EPA/530-SW-91-025, Washington, D.C., 1991.

Unit Pricing: Providing an Incentive to Reduce Municipal Solid Waste, EPA/530-SW-91-005, Washington, D.C., 1991.

The EPA documents listed below are available from NTIS for a fee by calling (703) 487-4650.

Addendum for the Regulatory Impact Analysis for the Final Criteria for Municipal Solid Waste Landfills, EPA/530-SW-91-073B, NTIS: PB92-100 858, Washington, D.C., 1991.

Charging Households for Waste Collection and Disposal: The Effects of Weight- or Volume-Based Pricing on Solid Waste Management, EPA/530-SW-90-047, NTIS: PB91-111 484, Washington, D.C., 1990.

Criteria for Municipal Solid Waste Landfills: Updated Review of Selected Provisions of Solid Waste Regulations, EPA/530-SW-88-039, NTIS: PB88-242 458, Washington, D.C., 1988.

Guidance Manual for the Classification of Solid Waste Disposal Facilities, EPA/SW-199C, NTIS: PB81-218 505, Washington, D.C., 1981.

List of Municipal Waste Landfills, NTIS: PB87-178 331, Washington, D.C., 1986.

Markets for Scrap Tires, EPA/530-SW-90-074A, NTIS: PB92-115 252, Washington, D.C., 1991.

Municipal Waste Combustion Study, Complete Set, EPA/530-SW-87-021, NTIS: PB87-206 066, Washington, D.C., 1987.

National Survey of Solid Waste (Municipal) Landfill Facilities, EPA/530-SW-88-034, NTIS: PB89-118 525, Washington, D.C., 1988.

Promoting Source Reduction and Recyclability in the Marketplace, EPA/530-SW-89-066, NTIS: PB90-163 122, Washington, D.C., 1989.

Regulatory Impact Analysis for the Final Criteria for Municipal Solid Waste Landfills, EPA/530-SW-91-073A, NTIS: PB92-100 841, Washington, D.C., 1992.

Subtitle D Municipal Landfill Survey Report, EPA/530-SW-91-070, NTIS: PB91-242 396, Washington, D.C., 1991.

Variable Rates in Solid Waste: Handbook for Solid Waste Officials, vol. II—Detailed Manual, EPA/530-SW-90-084B, NTIS: PB90-272 063, Washington, D.C., 1990.

Additional Publications

Yard Waste Management-A Planning Guide for New York State, New York State Department of Environmental Conservation, Division of Solid Waste—Bureau of Resource Recovery, 1991. This publication can be obtained by calling the Bureau of Resource Recovery at (518) 457-7336.

APPENDIX A
GLOSSARY

absorption   The penetration of one substance into or through another.

acid gas scrubber   A device that removes particulate and gaseous impurities from a gas stream. This generally involves the spraying of an alkaline solid or liquid, and sometimes the use of condensation or absorbent particles.

activated carbon   A highly absorbent form of carbon used to remove odors and toxic substances from gaseous emissions or to remove dissolved organic material from wastewater.

adhesion   Molecular attraction which holds the surfaces of two substances in contact, such as water and rock particles.

adsorption   The attachment of the molecules of a liquid or gaseous substance to the surface of a solid.

aeration   The process of exposing bulk material, such as compost, to air. *Forced aeration* refers to the use of blowers in compost piles.

aerobic   A biochemical process or environmental condition occurring in the presence of oxygen.

aerobic digestion   The utilization of organic waste as a substrate for the growth of bacteria which function in the presence of oxygen to stabilize the waste and reduce its volume. The products of this decomposition are carbon dioxide, water, and a remainder consisting of inorganic compounds, undigested organic material, and water.

aerosol   A particle of solid or liquid matter that can remain suspended in the air because of its small size.

air-cooled wall   A refractory wall with a lane directly behind it through which cool air flows.

air emissions   Solid particulates (such as unburned carbon) and gaseous pollutants (such as oxides of nitrogen or sulfur) or odors. These can result from a broad variety of activities including exhaust from vehicles, combustion devices, landfills, compost piles, street sweepings, excavation, demolition, etc.

air pollutant   Dust, fumes, smoke and other particulate matter, vapor, gas, odorous substances, or any combination thereof. Also, any air pollution agent or combination of such agents, including any physical, chemical, biological, radioactive substances, or matter which is emitted into or otherwise enters the ambient air.

air pollution   The presence of unwanted material in the air in excess of standards. The term "unwanted material" here refers to material in sufficient concentrations, present for a sufficient time to interfere significantly with health, comfort, or welfare of persons, or with the full use and enjoyment of property.

ambient air   The portion of the atmosphere external to buildings to which the general public has access.

anaerobic   A biochemical process occurring in the absence of oxygen.

anaerobic digestion   The utilization of organic waste as a substrate for the growth of bacteria which function in the absence of oxygen to reduce the volume of waste. The bacteria consume the carbon in the waste as their energy source and convert it to gaseous products. Properly controlled, anaerobic digestion will produce a mixture of methane and carbon dioxide, with a sludge remainder consisting of inorganic compounds, undigested organic material, and water.

ash   The residue that remains after a fuel or solid waste has been burned. (See also *bottom ash* and *fly ash*.)

avoided costs   Cost savings resulting from a recycling, incineration, or energy conservation program. A cost saving can be avoided disposal fees.

backyard composting   The controlled biodegradation of leaves, grass clippings, and/or other yard wastes on the site where they were generated.

bacteria   Single-cell, microscopic organisms with rigid cell walls. They may be aerobic, anaerobic, or facultative anaerobic; some can cause disease; and some are important in the stabilization and conversion of solid wastes.

baffles   Deflector vanes, guides, grids, grating, or other similar devices constructed or placed in air or gas flow systems, flowing water, or slurry systems to effect a more uniform distribution of velocities; absorb energy, divert, guide, or agitate fluids; and check eddies.

bagasse   An agricultural waste material consisting of the dry pulp residue that remains after juice is extracted from sugar cane or sugar beets.

baghouse   An air pollution abatement device used to trap particulates by filtering gas streams through large fabric bags usually made of cloth or glass fibers.

baler   A machine used to compress recyclables into bundles to reduce volume. Balers are often used on newspaper, plastics, and corrugated cardboard.

biodegradable   A substance or material which can be broken down into simpler compounds by microorganisms or other decomposers such as fungi.

biological waste   Waste derived from living organisms.

biomass   The amount of living matter in the environment.

blowdown   The minimum discharge of recirculating water for the purpose of discharging materials contained in the process, the further buildup of which would cause concentrations or amounts exceeding limits established by best engineering practice.

bottle bill   Legislation requiring deposits on beverage containers; appropriately called beverage container deposit law (BCDL).

bottom ash   The nonairborne combustion residue from burning fuel or waste in a boiler. The material falls to the bottom of the boiler and is removed mechanically. Bottom ash constitutes the major portion (about 90 percent) of the total ash created by the combustion of solid waste.

Btu (British thermal unit)   Unit of measure for the amount of energy a given material contains (e.g., energy released as heat during the combustion is measured in Btu). Technically, 1 Btu is the quantity of heat required to raise the temperature of one pound of water one degree Fahrenheit.

burning rate   The volume of solid waste incinerated or the amount of heat released during incineration. The burning rate is usually expressed in pounds of solid waste per square foot of burning area per area or in British thermal units per cubic foot of furnace volume per hour.

buy-back recycling center   A facility which pays a fee for the delivery and transfer of ownership to the facility of source-separated materials for the purpose of recycling or composting.

capital costs   Those direct costs incurred in order to acquire real property assets such as land, buildings, and machinery, and equipment.

carbon dioxide (CO_2)   A colorless, odorless, nonpoisonous gas that forms carbonic acid when dissolved in water. It is produced during the thermal degradation and microbial decomposition of solid wastes and contributes to global warming.

carbon monoxide (CO)   A colorless, poisonous gas that has an exceedingly faint metallic odor and taste. It is produced during the thermal degradation and microbial decomposition of solid wastes when the oxygen supply is limited.

carbonaceous matter   Pure carbon or carbon compounds present in solid wastes.

carcinogenic   Capable of causing the cells of an organism to react in such a way as to produce cancer.

centrifugal collector   A mechanical system using centrifugal force to remove aerosols from a gas stream.

chain grate stoker   A stoker with a moving chain as a grate surface. The grate consists of links mounted on rods to form a continuous surface that is generally driven by a shaft with sprockets.

charcoal   A dark or black porous carbon prepared from vegetable or animal substances (as from wood by charring in a kiln from which air is excluded).

charge   The amount of solid waste introduced into a furnace at one time.

classification   The separation and rearrangement of waste materials according to composition (e.g., organic or inorganic), size, weight, color, shape, and the like, using specialized equipment.

Clean Air Act   Act passed by Congress to have the air "safe enough to protect the public's health" by May 31, 1975. Required the setting of National Ambient Air Quality Standards (NAAQS) for major primary air pollutants.

Clean Water Act   Act passed by Congress to protect the nation's water resources. Requires EPA to establish a system of national effluent standards for major water pollutants, requires all municipalities to use secondary sewage treatment by 1988, sets interim goals of making all U.S. waters safe for fishing and swimming, allows point-source discharges of pollutants into waterways only with a permit from EPA, requires all industries to use the best practicable technology (BPT) for control of conventional and nonconventional pollutants and to use the best available technology (BAT) that is reasonable and affordable.

coal refuse   Waste products of coal mining, cleaning, and coal preparation operations and containing coal, matrix material, clay, and other organic and inorganic material.

cocollection   The collection of ordinary household garbage in combination with special bags of source-separated recyclables.

coding   In the context of solid waste, coding refers to a system to identify recyclable materials. The coding system for plastic packaging utilizes a three-sided arrow with a number in the center and letters underneath. The number and letters indicate the resin from which each container is made: 1 = PETE (polyethylene terephthalate), 2 = HDPE (high-density polyethylene), 3 = V (vinyl), 4 = LDPE (low-density polyethylene), 5 = PP (polypropylene), 6 = PS (polystyrene), and 7 = other/mixed plastics. Noncoded containers are recycled through mixed plastics processes. To help recycling sorters, the code is molded into the bottom of bottles with a capacity of 16 oz or more and other containers with a capacity of 8 oz or more.

codisposal   Burning of municipal solid waste with other material, particularly sewage sludge; the technique in which sludge is combined with other combustible materials (e.g., refuse, refuse-derived fuel, coal) to form a furnace feed with a higher heating value than the original sludge.

cofiring or coburning   Combustion of municipal solid waste along with other fuel, especially coal.

cogeneration   Production of electricity as well as heat from one fuel source.

collection   The act of picking up and moving solid waste from its location of generation to a disposal area, such as a transfer station, resource recovery facility, or landfill.

collection routes   The established routes followed in the collection of commingled and source-separated wastes from homes, businesses, commercial and industrial plants, and other locations.

collection systems   Collectors and equipment used for the collection of commingled and source-separated waste. Waste collection systems may be classified from several points of view, such as the mode of operation, the equipment used, and the types of wastes collected. In this text, collection systems have been classified according to their mode of operation in two categories: hauled container systems and stationary container systems.

combustible Various materials in the waste stream which are burnable, such as paper, plastic, lawn clippings, leaves, and other organic materials; materials that can be ignited at a specific temperature in the presence of air to release heat energy.

combustion The chemical combining of oxygen with a substance, which results in the production of heat.

combustion air The air used for burning a fuel.

combustion gases The mixture of gases and vapors produced by burning.

commercial sector One of the four sectors of the community that generates garbage. Designed for profit.

commercial waste All types of solid wastes generated by stores, offices, restaurants, warehouses, and other nonmanufacturing activities, excluding residential and industrial wastes.

commingled recyclables A mixture of several recyclable materials in one container.

commingled waste Mixture of all waste components in one container.

compaction The unit operation used to increase the specific weight (density in metric units) of waste materials so that they can be handled, stored, and transported more efficiently.

compactor Any power-driven mechanical equipment designed to compress and thereby reduce the volume of wastes.

compactor collection vehicle A large vehicle with an enclosed body having special power-driven equipment for loading, compressing, and distributing wastes within the body.

composite liner A liner composed of both a plastic and soil component for a landfill.

composition A set of identified solid waste materials, categorized into waste categories and waste types.

compost A relatively stable mixture of organic wastes partially decomposed by an aerobic and/or anaerobic process. Compost can be used as a soil conditioner.

composting The controlled biological decomposition of organic solid waste materials under aerobic or anaerobic conditions. Composting can be accomplished in windrows, static piles, and enclosed vessels (known as in-vessel composting).

concentration The amount of one substance contained in a unit of another substance.

conservation The planned management of a natural resource to prevent exploitation, destruction, or neglect.

construction and demolition waste The waste building materials, packaging, and rubble resulting from construction, remodeling, and demolition operations on pavements, houses, commercial buildings, and other structures. The materials usually include used lumber, miscellaneous metal parts, packaging materials, cans, boxes, wire, excess sheet metal, and other materials.

consumer waste Materials used and discarded by the buyer, or consumer, as opposed to wastes created and discarded in-plant during the manufacturing process.

consumption The amount of any resource (material or energy) used.

controlled air incinerator An incinerator with excess or starved air having two or more combustion chambers in which the amounts and distribution of air are controlled. EPA prefers to use the term "combustor" instead of "incinerator".

conversion The transformation of wastes into other forms; for example, transformation by burning or pyrolysis into steam, gas, or oil.

conversion products Products derived from the first-step conversion of solid wastes, such as heat from combustion and gas from biological conversion.

corrosive Defined for regulatory purposes as a substance having a pH level below 2 or above 12.5, or a substance capable of dissolving or breaking down other substances, particularly metals, or causing skin burns.

corrugated container According to SIC Code 2653, a paperboard container fabricated from two layers of kraft linerboard sandwiched around a corrugating medium. Kraft linerboard means paperboard made from wood pulp produced by a modified sulfate pulping process, with basis weight ranging from 18 to 200 lb, manufactured for use as facing material for corrugated or solid fiber containers. Linerboard also may mean that material which is made from reclaimed paper stock.

cost-effective A measure of cost compared with an unvalued output (e.g., the cost per ton of solid waste collected) such that the lower the cost, the more cost-effective the action.

cover material Soil or other material used to cover compacted solid wastes in a sanitary landfill.

crusher A mechanical device used to break secondary materials such as glass bottles into smaller pieces.

cullet Clean, generally color-sorted, crushed glass used in the manufacture of new glass products.

curbside collection Collection of recyclable materials at the curb, often from special containers, to be brought to various processing facilities. Collection may be both separated and/or mixed wastes.

curbside-separation To separate commingled recyclables prior to placement in individual compartments in truck providing curbside collection service; this task is performed by the collector.

cyclone separator A separator that uses a swirling airflow to sort mixed materials according to the size, weight, and density of the pieces.

decomposition The breakdown of organic wastes by bacterial, chemical, or thermal means. Complete chemical oxidation leaves only carbon dioxide, water, and inorganic solids.

decontamination or detoxification Processes which will convert pesticides into nontoxic compounds or the selective removal of radioactive material from a surface or from within another material.

degradable plastics Plastics specifically developed for special products that are formulated to break down after exposure to sunlight or microbes. By law, six-pack rings are degradable; however, they degrade only gradually, causing litter and posing a hazard to birds and marine animals.

degradation (Also biodegradation) A natural process that involves assimilation or consumption of a material by living organisms.

deinking The removal of ink, filler, and other nonfibrous material from printed waste paper.

demolition wastes Wastes produced from the demolition of buildings, roads, sidewalks, and other structures. These wastes usually include large, broken pieces of concrete, pipe, radiators, ductwork, electrical wire, broken-up plaster walls, lighting fixtures, bricks, and glass.

densified refuse-derived fuel (d-RDF) Refuse-derived fuel which has been compressed or compacted through such processes as pelletizing, briquetting, or extruding, causing improvements in certain handling or burning characteristics.

deposit Matter deposited by a natural process; a natural accumulation of iron ore, coal; money paid as security.

dioxin The generic name for a group of organic chemical compounds formally known as polychlorinated dibenzo-p-dioxins. Heterocyclic hydrocarbons that occur as toxic impurities, especially in herbicides.

discards Include the municipal solid waste remaining after recovery for recycling and composting. These discards are usually combusted or disposed of in landfills, although some MSW is littered, stored, or disposed of on site, particularly in rural areas.

dispersion technique The use of dilution to attain ambient air quality levels including any intermittent or supplemental control of air pollutants varying with atmospheric conditions.

disposable Something that is designed to be used once and then thrown away.

disposal The activities associated with the long-term handling of (1) solid wastes that are collected and of no further use and (2) the residual matter after solid wastes have been processed and the recovery of conversion products or energy has been accomplished. Normally, disposal is accomplished by means of sanitary landfilling.

disposal facility A collection of equipment and associated land area which serves to receive waste and dispose of it. The facility may incorporate one or more disposal methods.

diversion rate A measure of the amount of material now being diverted from landfilling for reuse and recycling compared with the total amount of waste that was thrown away previously.

DOT Department of Transportation.

draft The pressure difference between an incinerator (or combustor) and the atmosphere.

drag conveyer A conveyer that uses vertical steel plates fastened between two continuous chains to drag material across a smooth surface.

drop-off center A location where residents or businesses bring source-separate recyclable materials. Drop-off centers range from single-material collection points (e.g., easy access "igloo" containers) to staffed, multimaterial collection centers.

dump A site where mixed wastes are indiscriminately deposited without controls or regard to the protection of the environment. Dumps are now illegal.

ecosystem A system made up of a community of living things and the physical and chemical environment with which they interact.

eddy-current separation An electromagnetic technique for separating aluminum from a mixture of materials.

effluent Waste materials, usually waterborne, discharged into the environment, treated or untreated; the liquid leaving wastewater treatment systems.

electrostatic precipitator (ESP) A gas-cleaning device that collects entrained particulates by placing an electrical charge on them and attracting them onto oppositely charged collecting electrodes. They are installed in the back end of the incineration process to reduce air pollution.

embedded energy The sum of all the energy involved in product development, transportation, use, and disposal.

emission rate The amount of pollutant emitted into atmospheric circulation per unit of time.

encapsulation The complete enclosure of a waste in another material in such a way as to isolate it from external effects such as those of water or of air.

energy Ability to do work by moving matter or by causing a transfer of heat between two objects at different temperatures.

energy recovery The conversion of solid waste into energy or a marketable fuel. A form of resource recovery in which the organic fraction of waste is converted to some form of usable energy, such as burning processed or raw refuse, to produce steam.

environment Water, air, land, and all plants and human and other animals living therein, and the interrelationships which exist among them.

environmental impact statement (EIS) A document prepared by EPA or under EPA guidance (generally a consultant hired by the applicant and supervised by EPA) which identifies and analyzes in detail the environmental impacts of a proposed action. Individual states also may prepare and issue an EIS as regulated by state law. Such state documents may be called environmental impact reports (EIR).

environmental quality   The overall health of an environment determined by comparison with a set of standards.

EPA   U.S. Environmental Protection Agency; a federal agency created in 1970 and charged with the enforcement of all federal regulations having to do with air and water pollution, radiation and pesticide hazard, ecological research, and solid waste disposal.

evaporation   The physical transformation of a liquid to a gas.

exhaust system   The system comprised of a combination of components which provides for enclosed flow of exhaust gas from the furnace exhaust port to the atmosphere.

external costs   A cost relating to, or connected with outside expenses.

facility operator   Full-service contractors or other operators of a part of a resource recovery system.

fee   Dollar amount charged by a community to pay for services; e.g., tipping fee at a landfill.

ferrous metals   Metals composed predominantly of iron. In the waste materials, these metals usually include tin cans, automobiles, refrigerators, stoves, and other appliances. In resource recovery, often used to refer to materials that can be removed from the waste stream by magnetic separation.

filter   A membrane or porous device through which a gas or liquid is passed to remove suspended particles or dust.

firebrick   Refractory brick made from fireclay.

fireclay   A sedimentary clay containing only small amounts of fluxing impurities, high in hydrous aluminum and capable of withstanding high temperatures.

fixed grate   A grate without moving parts, also called a *stationary grate.*

flammable waste   A waste capable of igniting easily and burning rapidly.

flash point   The minimum temperature at which a liquid or solid gives off sufficient vapor to form an ignitable vapor-air mixture near the surface of the liquid or solid.

flow control   A legal or economic means by which waste is directed to particular destinations. For example, an ordinance requiring that certain wastes be sent to a combustion facility is waste flow control.

flow diagram of a process   A diagram which shows the assemblage of unit operations, facilities, and manual operations used to achieve a specified waste separation goal.

flue   Any passage designed to carry combustion gases and entrained particulates.

flue gas   The products of combustion, including pollutants, emitted to the air after a production process or combustion takes place.

fluidized bed combustion   Oxidation of combustible material within a bed of solid, inert (noncombustible) particles which under the action of vertical hot airflow will act as a fluid.

fly ash   All solids, including ash, charred papers, cinders, dusty soot, or other matter that rise with the hot gases from combustion rather than falling with the bottom ash. Fly ash is a minor portion (about 10 percent) of the total ash produced from combustion of solid waste, is suspended in the flue gas after combustion, and can be removed by pollution control equipment.

food wastes   Animal and vegetable wastes resulting from the handling, storage, sales, preparation, cooking, and serving of foods; commonly called *garbage.*

forced draft   The positive pressure created by the action of a fan or blower which supplies the primary or secondary combustion air in an incinerator.

fossil fuel   Natural gas, petroleum, coal, and any form of solid, liquid, or gaseous fuel derived from such materials for the purpose of creating useful heat.

front-end loader (1) A solid waste collection truck which has a power-driven loading mechanism at the front; (2) a vehicle with a power-driven scoop or bucket at the front, used to load secondary materials into processing equipment or shipping containers.

front-end recovery The salvage of reusable materials, most often the inorganic fraction of solid waste, prior to the processing or combusting of the organic fraction. Some processes for front-end recovery are grinding, shredding, magnetic separation, screening, and hand sorting.

fuel Any material which is capable of releasing energy or power by combustion or other chemical or physical means.

full material recovery facility (MRF) A process for removing recyclables and creating a compost-like product from the total of full mixed municipal solid waste (MSW) stream. Differs from a "clean" MRF which processes only commingled recyclables (see *waste recovery facility*).

furnace A combustion chamber; an enclosed structure in which heat is produced.

garbage Solid waste consisting of putrescible animal and vegetable waste materials resulting from the handling, preparation, cooking, and consumption of food, including waste materials from markets, storage facilities, handling and sale of produce, and other food products. Generally defined as wet food waste, but not synonymous with "trash," "refuse," "rubbish," or solid waste.

gas control system A system at a landfill designed to prevent explosion and fires due to the accumulation of methane concentrations and damage to vegetation on final cover of closed portions of a landfill or vegetation beyond the perimeter of the property on which the landfill is located and to prevent objectionable odors off site.

gas scrubber A device where a caustic solution is contacted with exhaust gases to neutralize certain combustion products, primarily sulfur oxides (SO_x) and secondary chlorine (Cl).

gaseous emissions Waste gases released into the atmosphere as a by-product of combustion.

generation Refers to the amount (weight, volume, or percentage of the overall waste stream) of materials and products as they enter the waste stream and before material recovery, composting, or combustion takes place.

generation rate Total tons diverted, recovered, and disposed per unit of time divided by the population. The annual per capita generation rate is the total tons generated in 1 year divided by the population of residents.

generator Any person, by site or location, whose act or process produces a solid waste; the initial discarded of a material.

grain loading The rate at which particles are emitted from a pollution source, in grains per cubic foot of gas emitted, 7000 gr = 1 lb.

grate A device used to support the solid fuel or solid waste in a furnace during drying, ignition, or combustion. Openings are provided for passage of combustion air.

gravity separation The separation of mixed materials based on the differences of material size and specific gravity.

gross national product (GNP) The total market value of all the goods and services produced by a nation during a specified time period.

groundwater Water beneath the surface of the earth and located between saturated soil and rock. It is the water that supplies wells and springs.

growth rate Estimation of progressive development; the rate at which a population or anything else grows.

Hammermill A type of crusher used to break up waste materials into smaller pieces or particles, which operates by using rotating and flailing heavy hammers.

haul distance The distance a collection vehicle travels (1) after picking up a loaded container (hauled container system) or from its last pickup stop on collection route (stationary container system) to a materials recovery facility, transfer station, or sanitary landfill, and (2) the distance the collection vehicle travels after unloading to the location where the empty container is to be deposited or to the beginning of a new collection route.

haul time   The elapsed or cumulative time spent transporting solid wastes between two specific locations.

haulers   Those persons, firms, or corporations or governmental agencies responsible (under either oral or written contract, or otherwise) for the collection of solid waste within the geographic boundaries of the contract community(ies) or the unincorporated county and the transportation and delivery of such solid waste to the resource recovery system as directed in the plan of operations.

hazard   Having one or more of the characteristics that cause a substance or combination of substances to qualify as a hazardous material.

hazardous waste   A waste, or combination of wastes, that may cause or significantly contribute to an increase in mortality or an increase in serious irreversible or incapacitating illness or that pose a substantial present or potential hazard to human health or the environment when improperly treated, stored, transported, disposed of, or otherwise managed. Hazardous wastes include radioactive substances, toxic chemicals, biological wastes, flammable wastes, and explosives.

HDPE (high-density polyethylene)   A recyclable plastic, used for items such as milk containers, detergent containers, and base cups of plastic soft drink bottles.

heat balance   An accounting of the distribution of the heat input and output of an incinerator or boiler, usually on an hourly basis.

heavy metals   Hazardous elements including cadmium, mercury, and lead which may be found in the waste stream as part of discarded items such as batteries, lighting fixtures, colorants, and inks.

high-grade paper   Relatively valuable types of paper such as computer printout, white ledger, and tab cards. Also used to refer to industrial trimmings at paper mills that are recycled.

household hazardous waste   Those wastes resulting from products purchased by the general public for household use which, because of their quantity, concentration, or physical, chemical, or infectious characteristics, may pose a substantial known or potential hazard to human health or the environment when improperly treated, disposed, or otherwise managed.

household hazardous waste collection   A program activity in which household hazardous wastes are brought to a designated collection point where the household hazardous wastes are separated for temporary storage and ultimate recycling, treatment, or disposal.

hydrocarbon   Any of a vast family of compounds containing carbon and hydrogen in various combinations, found especially in fossil fuels.

ignition temperature   The lowest temperature of a fuel at which combustion becomes self-sustaining.

impermeable   Restricts the movement of products through the surface.

incineration   An engineered process involving burning or combustion to thermally degrade waste materials. Solid wastes are reduced by oxidation and will normally sustain combustion without the use of additional fuel. Incineration is occasionally referred to as *combustion* in this text.

industrial unit   A site zoned for an industrial business and which generates industrial solid wastes.

industrial waste   Materials discarded from industrial operations or derived from industrial operations or manufacturing processes, all nonhazardous solid wastes other than residential, commercial, and institutional. Industrial waste includes all wastes generated by activities such as demolition and construction, manufacturing, agricultural operations, wholesale trade, and mining. A distinction should be made between scrap (those materials that can be recycled at a profit) and solid wastes (those that are beyond the reach of economical reclamation).

infectious waste Waste containing pathogens or biologically active material which because of its type, concentration, or quantity is capable of transmitting disease to persons exposed to the waste.

infrastructure A substructure or underlying foundation; those facilities upon which a system or society depends; for example, roads, schools, power plants, communication networks, and transportation systems.

inorganic Not composed of once-living material (e.g., minerals); generally, composed of chemical compounds not principally based on the element carbon.

integrated solid waste management The management of solid waste based on a combination of source reduction, recycling, waste combustion and disposal. The purposeful, systematic control of the functional elements of generation; waste handling, separation, and processing at the source; collection; separation and processing and transformation of solid waste; transfer and transport; and disposal associated with the management of solid wastes from the point of generation to final disposal.

intermediate processing center (IPC) Usually refers to a facility that processes residentially collected mixed recyclables into new products for market; often used interchangeably with *materials recovery facility (MRF)*. A facility where recyclables which have been separated from the rest of the waste are brought to be separated and prepared for market (crushed, baled, etc.). An IPC can be designed to handle commingled or separated recyclables or both.

internal costs Expenses of, relating to, or occurring within the confines of an organized structure.

investment tax credit A reduction in taxes permitted for the purchase and installation of specific types of equipment and other investments.

jurisdiction The city or county responsible for preparing any one or all of the following: the countywide integrated waste management plan or the countywide siting element.

kraft paper A comparatively coarse paper noted for its strength and used primarily as a wrapper or packaging material.

landfill, sanitary An engineered method of disposing of solid wastes on land in a manner that protects human health and the environment. Waste is spread in thin layers, compacted to the smallest practical volume, and covered with soil or other suitable material at the end of each working day, or more frequently, as necessary.

large-quantity generator Sources, such as industries and agriculture, that generate more than 1000 kg of hazardous waste per month.

leachate Liquid that has percolated through solid waste or another medium and has extracted, dissolved, or suspended materials from it, which may include potentially harmful materials. Leachate collection and treatment is of primary concern at municipal waste landfills.

liner Impermeable layers of heavy plastic, clay, and gravel that protect against groundwater contamination through downward or lateral escape of leachate. Most sanitary landfills have at least two plastic liners or layers of plastic and clay. Also refers to the material used on the inside of a furnace wall to ensure that a chamber is impervious to escaping gases.

litter That highly visible portion of solid wastes that is generated by the consumer and carelessly discarded outside the regular disposal system. Litter accounts for only about 2 percent of the total solid waste volume.

long-term impact Future effect of an action, such as an oil spill.

low-grade paper Less valuable types of paper such as mixed office paper, corrugated paperboard, and newspaper.

LULU Locally unwanted land use, e.g., landfills.

magnetic separator Equipment usually consisting of a belt, drum, or pulley with a permanent or temporary electromagnet and used to attract and remove magnetic materials from other materials.

mandatory recycling   Programs which, by law, require consumers to separate trash so that some or all recyclable materials are not burned or dumped in landfills.

manual separation   The separation of wastes by hand. Sometimes called "hand picking" or "hand sorting," manual separation is done in the home or office by keeping food wastes separate from newspaper, or in a materials recovery facility by picking out large cardboard and other recoverable materials.

market development   A method of increasing the demand for recovered materials so that end markets for the materials are established, improved, or stabilized and thereby become more reliable.

mass-burn facility   A type of incinerator (or combustor) that burns solid waste without any attempt to separate recyclables or process waste before burning.

mass combustion   The burning of as-received, unprocessed, commingled refuse in furnaces designed exclusively for solid waste disposal and energy recovery. Sometimes referred to as *mass burn.*

material recovery   Extraction of materials from the waste stream for reuse or recycling. Examples include source separation, front-end recovery, in-plant recycling, postcombustion recovery, leaf composting, etc.

materials balance   An accounting of the weights of materials entering and leaving a processing unit, such as an incinerator, usually on an hourly basis.

materials recovery/transfer facilities (MR/TFs)   Multipurpose facilities which may include the functions of a drop-off center for separated wastes, a materials recovery facility, a facility for the composting and bioconversion of wastes, a facility for the production of refuse-derived fuel, and a transfer and transport facility.

mechanical separation   The separation of waste into various components using mechanical means, such as cyclones, trommels, and screens.

metal   A mineral source that is a good conductor of electricity and heat, and yields basic oxides and hydroxides.

methane (CH_4)   An odorless, colorless, flammable, and asphyxiating gas that can explode under certain circumstances and that can be produced by solid wastes undergoing anaerobic decomposition. Methane emitted from municipal solid waste landfills can be used as fuel.

microorganisms   Microscopically small living organisms, including bacteria, yeasts, simple fungi, actinomycetes, some algae, slime molds, and protozoans, that digest decomposable materials through metabolic activity. Microorganisms are active in the composting process.

mixed paper   A waste type which is a mixture, unsegregated by color or quality, of at least two of the following paper wastes: newspaper, corrugated cardboard, office paper, computer paper, white paper, coated paper stock, or other paper wastes.

mixed refuse   Garbage or solid waste that is in a fully commingled state at the point of generation.

mixed-waste processing facility   A facility which processes mixed solid waste to remove recyclables and, sometimes, refuse-derived fuel and/or a compost substrate.

mixing chamber   A chamber usually placed between the primary and secondary combustion chamber and in which the products of combustion are thoroughly mixed by turbulence that is created by increased velocities of gases, checkerwork, or turns in the direction of the gas flow.

mixture   Any combination of two or more chemical substances if the combination does not occur in nature and is not, in whole or in part, the result of a chemical reaction.

modular incinerator   Smaller-scale waste combustion units prefabricated at a manufacturing facility and transported to the municipal waste combustor facility site.

moisture content   The weight loss (expressed in percent) when a sample of solid wastes is dried to a constant weight at a temperature of 100 to 105°C.

mulch   Any material, organic or inorganic, applied as a top-dressing layer to the soil surface. Mulch is also placed around plants to limit evaporation of moisture and freezing of roots and to nourish the soil.

municipal incinerator (or combustor)   A privately or publicly owned incinerator (or combustor) primarily designed and used to burn residential and commercial solid wastes within a community.

municipal solid waste (MSW)   Includes all the wastes generated from residential households and apartment buildings, commercial and business establishments, institutional facilities, construction and demolition activities, municipal services, and treatment plant sites.

municipal solid waste composting   The controlled degradation of municipal solid waste including after some form of preprocessing to remove noncompostable inorganic materials.

National Ambient Air Quality Standards   Federal standards which limit the concentration of particulates, sulfur dioxide, nitrogen dioxide, ozone, carbon monoxide, and lead in the atmosphere.

natural resource   Material or energy obtained from the environment that is used to meet human needs; material or energy resources not made by humans.

NIMBY (not in my back yard)   Refers to the fact that people want the convenience of products and proper disposal of the waste generated by their use of products, provided the disposal area is not located near them.

nitrogen   A tasteless, odorless gas that constitutes 78 percent of the atmosphere by volume. One of the essential ingredients of composting.

nonferrous metals   Any metal scraps that have value and that are derived from metals other than iron and its alloys in steel, such as aluminum, copper, brass, bronze, lead, zinc, and other metals, and to which a magnet will not adhere.

nonpoint source   Undefined wastewater discharges such as runoff from urban, agricultural, or strip-mined areas which do not originate from a specific point.

nonrecyclable   Not capable of being recycled or used again.

nonrenewable (resource)   Not capable of being naturally restored or replenished; resources available in a fixed amount (stock) in the earth's crust; they can be exhausted either because they are not replaced by natural processes (copper) or because they are replaced more slowly than they are used (oil and coal).

odor threshold   The lowest concentration of an airborne odor that a human being can detect.

office wastes   Discarded materials that consist primarily of paper waste including envelopes, ledgers, and brochures.

oil   Oil of any kind or in any form including, but not limited to, petroleum, fuel oil, sludge, oil refuse, and oil mixed with wastes other than dredged spoil.

old newspaper (ONP)   Any newsprint which is separated from other types of solid waste or collected separately from other types of solid waste and made available for reuse and which may be used as a raw material in the manufacture of a new paper product.

opacity   Degree of obscuration of light; e.g., a window has zero opacity, while a wall has 100 percent opacity.

operational costs   Those direct costs incurred in maintaining the ongoing operation of a program or facility. Operational costs do not include capital costs.

organic materials   Chemical compounds containing carbon, excluding carbon dioxide, combined with other chemical elements. Organic materials can be of natural or anthropogenic origin. Most organic compounds are a source of food for bacteria and are usually combustible.

oscillating-grate stoker   A stoker whose entire grate surface oscillates to move the solid waste and residue over the grate surface.

packaging Any of a variety of plastics, papers, cardboard, metals, ceramics, glass, wood, and paperboard used to make containers for food, household, and industrial products.

packed tower A pollution control device that forces dirty gas through a tower packed with crushed rock, wood chips, or other packing while liquid is sprayed over the packing material. Pollutants in the gas stream either dissolve in or chemically react with the liquid.

paper The term for all kinds of matted or felted sheets of fiber. Made from the pulp of trees, paper is digested in a sulfurous solution, bleached and rolled into long sheets. Acid rain and dioxin are standard by-products in this manufacturing process. Specifically, as one of the two subdivisions of the general term, paper refers to materials that are lighter in basic weight, thinner, and more flexible than paperboard, the other subdivision.

paperboard A type of matted or sheeted fibrous product. In common terms, paperboard is distinguished from paper by being heavier, thicker, and more rigid. (See also *special wastes.*)

partially allocated costs The costs of adding a recycling program to an existing operation such as a waste hauling company or public works department. Also known as *incremental costs.*

participant Any household that contributes any materials at least once during a specified tracking period.

participation rate A measure of the number of people participating in a recycling program or other similar program, compared with the total number of people that could be participating.

particulate matter (PM) Tiny pieces of partially incinerated matter, resulting from the combustion process, that can have harmful health effects on those who breathe them. Pollution control at municipal waste combustor facilities is designed to limit particulate emissions.

pathogen An organism capable of causing disease. The four major classifications of pathogen found in solid waste are bacteria, viruses, protozoans, and helminths.

permeable Having pores or openings that permit liquids or gases to pass through.

permits The official approval and permission to proceed with an activity controlled by the permitting authority. Several permits from different authorities may be required for a single operation.

PET (polyethylene terephthalate) A plastic resin used to make packaging, particularly soft drink bottles.

petroleum A mineral resource that is a complex mixture of hydrocarbons, an oily, flammable bituminous liquid, occurring in many places in the upper strata of the earth.

photodegradable Refers to plastics which will decompose if left exposed to sunlight.

pickup time For a hauled container system, it represents the time spent driving to a loaded container after an empty container has been deposited, plus the time spent picking up the loaded container and the time required to redeposit the container after its contents have been emptied. For a stationary container system, it refers to the time spent loading the collection vehicle, beginning with the stopping of the vehicle prior to loading the contents of the first container and ending when the contents of the last container to be emptied have been loaded.

plastics Synthetic materials consisting of large molecules called polymers derived from petrochemicals (compared with natural polymers such as cellulose, starch, and natural rubbers).

point of generation The physical location where the generator discards material (mixed refuse and/or separated recyclables).

point source Specific, identifiable end-of-pipe discharges of wastes into receiving bodies of water; for example, municipal sewage treatment plants, industrial wastewater treatment systems, and animal feedlots.

pollutant Dredged spoil, solid waste, incinerator residue, sewage, garbage, sewage sludge, munitions, chemical wastes, biological materials, radioactive materials, heat, wrecked or discarded equipment, rock, sand, cellar dirt, and industrial, municipal, and agricultural waste discharged into the environment. Any solid, liquid, or gaseous matter which is in excess of natural levels or established standards.

pollution The presence of matter or energy whose nature, location, or quantity produces undesired environmental effects. Also, the artificial or human-introduced alteration of the chemical, physical, biological, and radiological integrity of water.

polyethylenes A group of resins created by polymerizing ethylene gas. The two major categories are high-density polyethylene and low-density polyethylene.

polyvinyl chloride (PVC) A plastic made by polymerization of vinyl chloride with peroxide catalysts. A typically insoluble plastic used in packaging, pipes, detergent bottles, wraps, etc.

porosity The ratio of the volume of pores of a material to the volume of its mass.

postconsumer recycling The reuse of materials generated from residential and commercial waste, excluding recycling of material from industrial processes that has not reached the consumer, such as glass broken in the manufacturing process.

precycling Activities such as source and size reduction, material selection when shopping, and reducing toxicity of products in manufacturing prior to recycling, which helps reduce the amounts of municipal solid wastes generated.

primary materials Virgin or new materials used for manufacturing basic products. Examples include wood pulp, iron ore, and silica sand.

primary standard A natural air emissions standard intended to establish a level of air quality that, with an adequate margin of error, will protect public health.

privatization The assumption of responsibility for a public service by the private sector, under contract to local government or directly to the receivers of the service.

process waste Any designated toxic pollutant which is inherent to or unavoidable resulting from any manufacturing process, including that which comes into direct contact with or results from the production or use of any raw material, intermediate product, finished product, by-product, or waste product.

processing Any method, system, or other means designated to change the physical form or chemical content of solid wastes.

program The full range of source reduction, recycling, composting, special waste, or household hazardous waste activities undertaken by or in the jurisdiction or relating to management of the jurisdiction's waste stream to achieve the objectives identified in the source reduction, recycling, composting, special waste, and household hazardous waste components, respectively.

pulp A moist mixture of fibers from which paper is made.

PURPA The Public Utilities Regulatory Policies Act of 1978. A federal law whose key provision mandates private utilities to buy power commissions equal to the "avoid cost" power production to the utility. The act is intended to guarantee a market for small producers of electricity at rates equal or close to the utilities' marginal production costs.

pyrolysis The chemical decomposition of organic matter through the application of heat in an oxygen-deficient atmosphere.

rack collection The collection of old newspapers at the same time as residential waste collection. The waste paper is placed in a side or front rack attached to the waste collection truck.

radioactive A substance capable of giving off high-energy particles or rays as a result of spontaneous disintegration of atomic nuclei.

rate structure That set of prices established by a jurisdiction, special district (as defined in Government Code section 56036), or other rate-setting authority to compensate the jurisdiction, special district, or rate-setting authority for the partial or full costs of the collection, processing, recycling, composting, and/or transformation or landfill disposal of solid wastes.

rated incinerator capacity The number of tons of solid waste that can be processed in an incinerator per 24-h period when specified criteria prevail.

raw materials   Substances still in their natural or organic state, before processing or manufacturing; or the starting materials for a manufacturing process.

RCRA   Resource Conservation and Recovery Act of 1976; requires states to develop solid waste management plans and prohibits open dumps; identifies lists of hazardous wastes and sets the standards for their disposal. This law amends the Solid Waste Disposal Act of 1965 and expands on the Resource Recovery Act of 1970 to provide a program to regulate hazardous waste.

reactive   For regulatory purposes, defined as a substance which tends to react spontaneously with air or water, to explode when dropped, or to give off toxic gases.

reclamation   The restoration to a better or more useful state, such as land reclamation by sanitary landfilling, or the extraction of useful materials from solid wastes.

recoverable resources   Materials that still have useful physical or chemical properties after serving a specific purpose and can therefore be reused or recycled for the same or other purposes.

recovery   Refers to materials removed from the waste stream for the purpose of recycling and/or composting. Recovery does not automatically equal recycling and composting, however. For example, if markets for recovered materials are not available, the materials that were separated from the waste stream for recycling may simply be stored or, in some cases, sent to a landfill or combustor. The extraction of useful materials or energy from waste.

recycle   To separate a given material from waste and process it so that it can be used again in a form similar to its original use; for example, newspapers recycled into newspapers or cardboard.

recycled material   A material that is used in place of a primary, raw, or virgin material in manufacturing a product and consists of material derived from postconsumer waste, industrial scrap, material derived from agricultural wastes, and other items, all of which can be used in the manufacture of new products. Also referred to as *recyclables.*

recycling   Separating a given waste material (e.g., glass) from the waste stream and processing it so that it may be used again as a useful material for products which may or may not be similar to the original.

recycling program   Should include the following: types of collection equipment used, collection schedule, route configuration, frequency of collection per household, whether curbside set-out containers are provided by the program, publicity and educational activities, and budget, financial evaluation (costs, revenues, and savings), processing and handling procedures, market prices, ordinances, and enforcement activities.

refuse   All solid materials which are discarded as useless. A term often used interchangeably with the term *solid waste.*

refuse-derived fuel (RDF)   The combustible, or organic, portion of municipal waste that has been separated out and processed for use as fuel.

renewable resources   A naturally occurring raw material or form of energy, such as the sun, wind, falling water, biofuels, fish, and trees, derived from an endless or cyclical source, where, through management of natural means, replacement roughly equals consumption ("sustained yield").

request for bid (RFB)   A mechanism for seeking bidders to supply recycling goods and services or to purchase secondary materials.

request for proposal (RFP)   A mechanism for seeking qualified firm or individuals to supply recycling goods or services.

request for qualifications (RFQ)   A mechanism for determining the experience, skills, financial resources, or expertise of a potential bidder or proposer.

residential wastes   Wastes generated in houses and apartments, including paper, cardboard, beverage and food cans, plastics, food wastes, glass containers, and garden wastes.

residual oil   A general term used to indicate a heavy viscous fuel oil.

residual wastes   Those solid, liquid, or sludge substances from human activities in the urban, agricultural, mining, and industrial environments remaining after collection and necessary treatment.

residue   The solid or semisolid materials remaining after processing, incineration, composting, or recycling have been completed. Residues are usually disposed of in landfills.

resource conservation   Reduction of the amounts of wastes generated, reduction of overall consumption, and utilization of recovered resources.

reusability   The ability of a product or package to be used more than once in its same form.

reverse vending machine   A machine which accepts empty beverage containers (or other items) and rewards the donor with a cash refund.

rotary kiln stoker   A cylindrical, inclined device, utilized for the combustion of materials at high temperatures, that rotates, thus causing the solid waste to move in a slow cascading and forward motion.

scrap   Products that have completed their useful life, such as appliances, cars, construction materials, ships, and postconsumer steel cans; also includes new scrap materials that result as byproducts when metals are processed and products are manufactures. Steel scrap is recycled in steel mills to make new steel products.

screening   A unit operation that is used to separate mixtures of materials of different sizes into two or more size fractions by means of one or more screening surfaces.

scrubber   A device for removing unwanted dust particles, liquids, or gaseous substances from an airstream by spraying the airstream with a liquid (usually water or a caustic solution) or forcing the air through a series of baths; common antipollution device that uses a liquid or slurry spray to remove acid gases and particulates from municipal waste combustion facilities flue gases.

secondary burner   A burner installed in the secondary combustion chamber of an incinerator to maintain a minimum temperature and to complete the combustion of incompletely burned gas.

secondary combustion air   The air introduced above or below the fuel (waste) bed by a natural, induced, or forced draft.

secondary material   A material that is used in place of a primary or raw material in manufacturing a product.

secure landfill   A landfill designed to prevent the entry of water and the escape of leachate by the use of impermeable liners.

separation   To divide wastes into groups of similar material, such as paper products, glass, food wastes, and metals. Also used to describe the further sorting of materials into more specific categories, such as clear glass and dark glass. Separation may be done manually or mechanically with specialized equipment.

set-out   A quantity of material placed for collection. Usually a set-out denotes one household's entire collection of recyclable materials, but in urban areas, where housing density makes it difficult to identify ownership of materials, each separate container or bundle is counted as a set-out. A single household, for example, may have three set-outs: commingled glass, metals, and newspapers.

sewage sludge   A semiliquid substance consisting of settled sewage solids combined with varying amounts of water and dissolved materials.

shredder   A machine used to break up waste materials into smaller pieces by cutting, tearing, shearing, and impact action.

shrinkage   The difference in the purchase weight of a secondary material and the actual weight of the material when consumed.

SIC code   The standards published in the U.S. Standards Industrial Classification Manual (1987).

silo   A storage vessel, generally tall relative to its cross section, for dry solids; materials are fed into the top and withdrawn from the bottom through a control mechanism.

size reduction, mechanical   The mechanical conversion of solid wastes into small pieces. In practice, the terms *shredding, grinding,* and *milling* are used interchangeably to describe mechanical size reduction operations.

sludge   Any solid, semisolid, or liquid waste generated from a municipal, commercial, or industrial wastewater treatment plant, water supply treatment plant, or air pollution control facility, or any other such waste having similar characteristics and effects. Must be processed by bacterial digestion or other methods, or pumped out for land disposal, incineration, or composting.

slurry   A pumpable mixture of solids and fluid.

small quantity generator   Sources such as small businesses and institutions that generate less than 1000 kg of hazardous waste per month.

smoke   Particles suspended in air after incomplete combustion of materials containing carbon.

soil liner   Landfill liner composed of compacted soil used for the containment of leachate.

solid waste disposal facility   Any solid waste management facility which is the final resting place for solid waste, including landfills and incineration facilities that produce ash from the process of incinerating municipal solid waste.

solid waste management   See *integrated solid waste management.*

solid wastes   Any of a wide variety of solid materials, as well as some liquids in containers, which are discarded or rejected as being spent, useless, worthless, or in excess, including contained gaseous material resulting from industrial, commercial, mining, and agricultural operations, and from community activities. (See also *commercial, construction and demolition, hazardous, industrial, municipal,* and *residential wastes.*)

source reduction   Reduction of the amount of materials entering the waste stream by voluntary or mandatory programs to eliminate the generation of waste. The design, manufacture, acquisition, and reuse of materials so as to minimize the toxicity of the waste generated.

source separation   The separation of waste materials from other commingled wastes at the point of generation.

special wastes   Special wastes include bulky items, consumer electronics, white goods, yard wastes that are collected separately, hazardous wastes, concrete, batteries, used oil, asphalt, and tires. Special wastes are usually handled separately from other residential and commercial wastes.

spray chamber   A chamber equipped with water sprays that cool and clean the combustion products passing through it.

stack   Any chimney, flue, vent, roof monitor, conduit, or duct arranged to discharge emissions to the ambient air.

stack emissions   Air emissions from combustion facility stacks.

stationary container systems   Collection systems in which the containers used for the storage of wastes remain at the point of waste generation, except for occasional short trips to the collection vehicle.

statistically representative   Those representative and random samples of units that are taken from a population sample. For the purpose of this definition, population sample includes, but is not limited to, a sample from a population of solid waste generation sites, solid waste facilities and recycling facilities, or a population of items of materials and solid wastes in a refuse load of solid waste.

stoichiometric air   The amount of air theoretically required to provide the exact amount of oxygen for total combustion of a fuel. Municipal solid waste incineration technologies make use of both substoichiometric and excess air processes.

styrofoam Also known as polystyrene, a synthetic material consisting of large molecules called polymers derived from petrochemicals. Experts agree that styrofoam will never decompose.

Subtitle C The hazardous waste section of the Resource Conservation and Recovery Act (RCRA).

Subtitle D The solid, nonhazardous waste section of the Resource Conservation and Recovery Act (RCRA).

Subtitle F Section of the Resource Conservation and Recovery Act (RCRA) requiring the federal government to actively participate in procurement programs fostering the recovery and use of recycled materials and energy.

Superfund Common name for the Comprehensive Environmental Response, Compensation and Liability Act (CERCLA) to clean up abandoned or inactive hazardous waste dump sites.

tare The weight of extraneous material, such as pallets, strapping, bulkhead, and sideboards, that is deducted from the gross weight of a secondary material shipment to obtain net weight.

thermal efficiency The ratio of heat used to total useful energy generated.

threshold dose The minimum application of a given substance to produce a measurable effect.

tipping fee A fee, usually dollars per ton, for the unloading or dumping of waste at a landfill, transfer station, recycling center, or waste-to-energy facility. Also called a disposal or service fee.

tipping floor Unloading area for wastes delivered to an MRF, transfer station, or waste combustor.

tire-derived fuel (tdf) A form of fuel consisting of scrap tires shredded into chips.

ton A unit of weight in the U.S. Customary System of Measurement, an avoirdupois unit equal to 2000 lb. Also called a short ton or net ton; equals 0.907 metric tonnes.

toxic Defined for regulatory purposes as a substance containing poison and posing a substantial threat to human health and/or the environment.

transfer station A place or facility where wastes are transferred from smaller collection vehicles (e.g., compactor trucks) into larger transport vehicles (e.g., over-the-road and off-road tractor trailers, railroad gondola cars, or barges) for movement to disposal areas, usually landfills. In some transfer operations, compaction or separation may be done at the station.

transport The transport of solid wastes transferred from collection vehicles to a facility or disposal site for further processing or action.

trash Wastes that usually do not include food wastes but may include other organic materials, such as plant trimmings. Generally defined as dry waste material, but in common usage, it is a synonym for rubbish or refuse.

trommel A perforated, rotating, horizontal cylinder that may be used in resource recovery facilities to break open trash bags, to remove glass and such small items as stone and dirt, and to remove cans from incinerator residue.

turbidity Cloudiness of a liquid.

unacceptable waste Motor vehicles, trailers, comparable bulky items of machinery or equipment, highly inflammable substances, hazardous waste, sludges, pathological and biological wastes, liquid wastes, sewage, manure, explosives and ordinance materials, and radioactive materials. Also includes any other material not permitted by law or regulation to be disposed of at a landfill, unless such landfill is specifically designed, constructed, and licensed or permitted to receive such material. None of such material constitutes either processable waste or unprocessable waste.

unprocessable waste   That portion of the solid waste stream that is predominantly noncombustible and therefore should not be processed in a mass burn resource recovery system; includes, but is not limited to, metal furniture and appliances, concrete rubble; mixed roofing materials; noncombustible building debris; rock, gravel, and other earthen materials; equipment; wire and cable; and any item of solid waste exceeding 6 ft in any one of its dimensions or being in whole or in part of a solid mass, the solid mass portion of which has dimensions such that a sphere with a diameter of 8 in could be contained within such solid mass portion, and processable waste (to the extent that it is contained in the normal unprocessable waste stream); excludes unacceptable waste.

vapor   The gaseous phase of substances that are liquid or solid at atmospheric temperature and pressure, e.g., steam.

vibrating screen   A mechanical device which sorts material according to size.

virgin material   Any basic material for industrial processes that has not previously been used, for example, wood-pulp trees, iron ore, silica sand, crude oil, and bauxite. (See also *primary materials, secondary material.*)

vitrification   A process whereby high temperatures effect permanent chemical and physical change in a ceramic body.

volatile solid (VS)   The portion of the organic material that can be released as a gas when organic material is burned in a muffle furnace at 550°C (1022°F).

volume   A three-dimensional measurement of the capacity of a region of space or a container. Volume is commonly expressed in terms of cubic yards or cubic meters. Volume is not expressed in terms of mass or weight.

volume-based rates   A system of charging for garbage pickup that charges the waste generator rates based on the volume of waste collected, so that the greater the volume of waste collected, the higher the charge. "Pay-by-the-bag" systems and variable can rates are types of volume-based rates.

volume reduction   The processing of wastes so as to decrease the amount of space they occupy. Reduction is presently accomplished by three major processes: (1) mechanical, which used compaction techniques (baling, sanitary landfills, etc.) and shredding; (2) thermal, which is achieved by heat or combustion (incineration) and can reduce volume by 80 to 90 percent; and (3) biological, in which the organic waste fraction is degraded by bacterial action (composting, etc.).

voluntary separation   The willing participation in waste recycling as opposed to mandatory recycling.

waste   Unwanted materials left over from manufacturing processes, or refuse from places of human or natural habitation.

waste categories   The grouping of solid wastes with similar properties into major solid waste classes, such as grouping together office, corrugated, and newspaper as a paper waste category, as identified by a solid waste classification system, except where a component-specific requirement provides alternative means of classification.

waste composition   The relative amount of various types of materials in a specific waste stream.

waste diversion   To divert solid waste, in accordance with all applicable federal, state, and local requirements, from disposal at solid waste landfills or transformation facilities through source reduction, recycling, or composting.

waste generator   Any person whose act or process produces solid waste, or whose act first causes solid waste to become subject to regulation.

waste management   See *integrated solid waste management.*

waste minimization   An action leading to the reduction of waste generation, particularly by industrial firms.

waste recovery facility (WRF)   A facility for separating recyclables and creating a compost-like material from the total of full mixed municipal solid waste stream.

waste reduction   The prevention or restriction of waste generation at its source by redesigning products or the patterns of production and consumption.

waste sources   Agricultural, residential, commercial, and industrial activities, open areas, and treatment plants where solid wastes are generated.

waste stream   A term describing the total flow of solid waste from homes, businesses, institutions, and manufacturing plants that must be recycled, burned, or disposed of in landfills; or any segment thereof, such as the "residential waste stream" or the "recyclable waste stream." The total waste produced by a community or society, as it moves from origin to disposal.

waste transformation   The transformation of waste materials involving a phase change (e.g., solid to gas). The most commonly used chemical and biological transformation processes are combustion and aerobic composting.

wastewater   Water carrying dissolved or suspended solids from homes, farms, businesses, institutions, and industries.

water table   Level below the earth's surface at which the ground becomes saturated with water. Landfills and composting facilities are designed with respect to the water table in order to minimize potential contamination.

waterwall furnace   Furnace constructed with walls of welded steel tubes through which water is circulated to absorb the heat of combustion. These furnaces can be used as incinerators. The stream of hot water thus generated may be put to a useful purpose or simply used to carry the heat away to the outside environment.

waterwall incinerator   An incinerator whose furnace walls consist of vertically arranged metal tubes through which water passes and absorbs the radiant energy from burning solid waste.

weight-based rates   A system of charging for garbage pickup that charges based on weight of garbage collected, so that the greater the weight collected, the higher the charge. The logistics of implementing this system are currently being experimented with.

wetland   Area that is regularly wet or flooded and has a water table that stands at or above the land surface for at least part of the year. Coastal wetlands extend back from estuaries and include salt marshes, tidal basins, marshes, and mangrove swamps. Inland freshwater wetlands consist of swamps, marshes, and bogs. Federal regulations apply to landfills sited at or near wetlands.

wet scrubber   Antipollution device in which a lime slurry (dry lime mixed with water) is injected into the flue gas stream to remove acid gases and particulates.

white goods   Large worn-out or broken household, commercial, and industrial appliances, such as stoves, refrigerators, dishwashers, and clothes washers and dryers.

windrow   A large, elongated pile of composting material.

yard waste   Leaves, grass clippings, prunings, and other natural organic matter discarded from yards and gardens. Yard wastes may also include stumps and brush, but these material are not normally handled at composting facilities.

REFERENCES

This glossary was compiled and edited from the following sources:
California Integrated Waste Management Board, Title 14, Chapter 9, Article 3, Definitions.

National Recycling Coalition, *Measurement Standards and Reporting Guidelines*, Washington, D.C.

National Resource Recovery Association, The United States Conference of Mayors, *A Solid Waste Management Glossary*, Washington, D.C.

Resource Recycling, Inc., *Glossary of Recycling Terms and Acronyms.*

Tchobanoglous, G., et. al., *Integrated Solid Waste Management,* McGraw-Hill, New York, 1993.

APPENDIX B
ABBREVIATIONS

ADF Advanced disposal fee

ANSI American National Standards Institute

BAT Best available technology

BPT Best practical technology

Btu British thermal unit

C&D Construction and demolition materials

CERCLA Comprehensive Environmental Response, Compensation and Liability Act

CPO Computer printout

CSWS Council for Solid Waste Solution

DEP Department of Environmental Protection

DER Department of Environmental Regulation

DOT Department of Transportation

EIS Environmental impact statement

EPA Environmental Protection Agency

GNP Gross national product

HDPE High-density polyethylene

HHW Household hazardous waste

HSWA Hazardous and Solid Waste Act of 1984

IPC Intermediate processing center

kW Kilowatt

LDPE Low-density polyethylene

MRF Materials recovery facility

MSW Municipal solid waste

MSW-RDF Municipal solid waste and refuse-derived fuel processing facility

MW Megawatt

MWC Municipal waste combustor

NAAQS National Ambient Air Quality Standards

NAPCOR National Association of Plastic Container Recovery

NSPS New Source Performance Standards

NSWMA   National Solid Waste Management Association

OCC   Old corrugated containers

ONP   Old newspapers

OSHA   Occupational Safety and Health Act

PET   Polyethylene terephthalate

PP   Polypropylene

PPB   Parts per billion

PPM   Parts per million

PVC   Polyvinyl chloride

RAN   Revenue anticipation note

RCRA   Resource Conservation and Recovery Act

RDF   Refuse-derived fuel

RFP   Request for proposals

RFQ   Request for qualifications

SARA   Superfund Amendment and Recovery Act

SWANA   Solid Waste Association of North America

SWDA   Solid Waste Disposal Act

tdf   Tire derived fuel

TPD   Tons per day

U.S. EPA   United States Environmental Protection Agency

WPF   Waste processing facility

WRF   Waste recovery facility

APPENDIX C
SOLID WASTE MANAGEMENT ORGANIZATIONS

Air and Waste Management Association
P.O. Box 2861
Pittsburgh, PA 15320
(412) 232-3444

Aluminum Recycling Association
1000 - 16th St., N.W., Suite 400
Washington, DC 20036
(202) 785-0951

American Council for an Energy Efficient
   Economy
1001 Connecticut Ave., N.W., Suite 801
Washington, DC 20036
(202) 429-8873

American Forest & Paper Association
1250 Connecticut Ave., N.W., Suite 210
Washington, DC 20036
(202) 463-2420

American Institute of Chemical Engineers
(Center for Waste)
345 E. 47th St.
New York, NY 10017
(212) 705-7338

American Iron and Steel Institute
1101 17th St., N.W., Suite 1300
Washington, DC 20036
(202) 452-7100

American Petroleum Institute
1220 L Street, N.W., Suite 900
Washington, DC 20005
(202) 682-8000

American Plastics Council
1275 K St., N.W., Suite 500
Washington, DC 20005
(202) 371-5319

American Public Works Association
Institute for Solid Wastes
106 W. 11th St., Suite 1800
Kansas City, MO 64105-1806
(816) 472-6100

American Society of Civil Engineers
345 E. 47th St.
New York, NY 10017
(212) 705-7496

American Society of Mechanical Engineers
345 E. 47th St.
New York, NY 10017
(212) 705-7722

Asphalt Recycling and Reclaiming
   Association
No. 3 Church Circle, Suite 250
Annapolis, MD 21401
(410) 267-0023

Association of Petroleum Re-Refiners
P.O. Box 605
Buffalo, NY 14205
(716) 855-2757

Association of State and Territorial
   Solid Waste Management Officials
444 N. Capitol St., N.W., Suite 388
Washington, DC 20001
(202) 624-5828

Automotive Dismantlers and Recyclers
   Association
3975 Fair Ridge Dr., Suite 20, Terrace Level
   North
Fairfax, VA 22033
(703) 385-1001

Battery Council International
401 N. Michigan Ave.
Chicago, IL 60611-4267
(312) 644-6610

Biomass Energy Research Association
1825 K St., N.W., Suite 503
Washington, DC 20006
(202) 785-2856

Can Manufacturers Institute, Inc.
1625 Mass. Ave., N.W., Suite 500
Washington, DC 20036
(202) 232-4677

Cement Kiln Recycling Association
1101 - 30th St., N.W., Suite 500
Washington, DC 20007
(202) 625-3440

Citizens for the Environment
1250 H St., N.W., Suite 700
Washington, DC 20005
(202) 783-3870

Coalition for Responsible Waste Incineration
1133 Connecticut Ave., N.W., Suite 1200
Washington, DC 20036
(202) 775-9869

Coalition of Northeastern Governors'
  Source Reduction Task Force
400 N. Capitol St., Suite 382
Washington, DC 20001
(202) 624-8450

Committee for Environmentally Effective
  Packaging
601 13th St., N.W., Suite 510 South
Washington, DC 20005
(202) 783-5588

Council of State Governments
P.O. Box 11910
Lexington, KY 40578
(606) 231-1939

Council on Plastics & Packaging in the
  Environment
1001 Connecticut Ave., N.W., Suite 401
Washington, DC 20036
(202) 331-0099

Electric Power Research Institute's
  Hydroelectric Generation and Renewable
  Fuels Program
P.O. Box 10412
Palo Alto, CA 94303
(415) 855-2179

Environmental Action
6930 Carroll Ave., 6th Floor
Takoma Park, MD 20912
(202) 745-4870

Environmental Defense Fund
257 Park Ave.
New York, NY 10010
(212) 387-3500

Environmental Law Institute
1616 P St., N.W., Suite 200
Washington, DC 20036
(202) 328-5150

Flexible Packaging Association
1090 Vermont Ave., N.W., Suite 500
Washington, DC 20005
(202) 842-3880

Foodservice and Packaging Institute
1025 Connecticut Ave., N.W., Suite 513
Washington, DC 20036
(202) 822-6420

Friends of the Earth
218 D St., S.E.
Washington, DC 20003
(202) 544-2600

Gas Research Institute
8600 W. Bryn Mawr Ave.
Chicago, IL 60631
(312) 399-8100

General Federation of Women's Clubs
1734 N St., N.W.
Washington, DC 20036
(202) 347-3168

Glass Packaging Institute
1801 K St., N.W., Suite 1105-L
Washington, DC 20006
(202) 887-4850

Greenpeace U.S.A.
1436 U St., N.W.
Washington, DC 20009
(202) 462-1177

Institute of Scrap Recycling Industries
1325 G St., N.W., Suite 1000
Washington, DC 20005
(202) 466-4050

International City Management Association
777 N. Capitol St., N.E., Suite 500
Washington, DC 20002
(202) 289-4262

Keep America Beautiful, Inc.
9 W. Broad St.
Stamford, CT 06902
(203) 323-8987

League of Women Voters of the U.S.
1730 M St., N.W.
Washington, DC 20036
(202) 429-1965

Municipal Waste Management Association
U.S. Conference of Mayors
1620 Eye St., N.W.
Washington, DC 20006

National Association for Environmental
  Management
1440 New York Ave., N.W., Suite 300
Washington, DC 20005
(202) 966-0019

National Association for Plastic Container
  Recovery
100 N. Tryon St., Suite 3770
Charlotte, NC 28202
(704) 358-8882

National Association of Counties
440 1st St., N.W., 8th Floor
Washington, DC 20001
(202) 393-6226

National Association of State Energy Officials
505 - 11th St., S.E.
Washington, DC 20003
(202) 546-2200

National Audubon Society
666 Penn. Ave., S.E.
Washington, DC 20003
(202) 547-9009

National Conference of State Legislatures
1560 Broadway, Suite 700
Denver, CO 80202
(303) 830-2200

National Corn Growers Association
1000 Executive Pkwy., Suite 105
St. Louis, MO 63141-6397
(314) 275-9915

National Governors Association
444 N. Capitol St., N.W., Suite 250
Washington, DC 20001
(202) 624-5300

National League of Cities
1301 Penn. Ave., N.W., 6th Floor
Washington, DC 20004
(202) 626-3000

National Paperbox and Packing Association
1201 E. Abingdon Dr., Suite 203
Alexandria, VA 22314
(703) 684-2212

National Recycling Coalition
1101 30th St., N.W., Suite 305
Washington, DC 20007
(202) 625-6406

National Soft Drink Association
State & Environmental Affairs
1101 16th St., N.W.
Washington, DC 20036
(202) 463-6700

National Solid Wastes Management
  Association
1730 Rhode Island Ave., N.W., Suite 1000
Washington, DC 20036
(202) 659-4613

National Tire Dealers and Retreaders Assoc.
1250 I St., N.W., Suite 400
Washington, DC 20005
(202) 789-2300

National Wood Energy Association
777 N. Capitol St., Suite 805
Washington, DC 20002
(202) 408-0664

Natural Resources Defense Council
40 West 20th Street
New York, NY 10011
(212) 727-2700

Paperboard Packaging Council
1101 Vermont Ave., N.W., Suite 411
Washington, DC 20005
(202) 289-4100

Plastics Institute of America, Inc.
277 Fairfield Rd., Suite 100
Fairfield, NJ 07004
(201) 808-5950

Polystyrene Packaging Council
1025 Connecticut Ave., N.W., Suite 515
Washington, DC 20036
(202) 822-6424

Recycled Paperboard Technical Association
350 S. Kalamazoo Mall, Suite 207
Kalamazoo, MI 49007
(616) 344-0394

Rubber Manufacturers' Association
1400 K St., N.W.
Washington, DC 20005
(202) 682-1338

Sierra Club
408 C St., N.E.
Washington, DC 20002
(202) 547-1144

Society of the Plastics Industry
1275 K St., N.W., Suite 400
Washington, DC 20005
(202) 371-5200

Solid Waste Association of North America
8750 Georgia Ave., Suite 140
Silver Spring, MD 20910
(301) 585-2898

Solid Waste Composting Council
601 Penn. Ave., N.W., Suite 900
Washington, DC 20004
(202) 638-0182

Steel Can Recycling Institute
680 Andersen Drive
Pittsburgh, PA 15220
(800) 876-7274

The Alliance to Save Energy
1725 K Street, N.W., Suite 914
Washington, DC 20006-1401
(202) 331-9588

The National Resource Center for State Laws
and Regulations
3600 Glenwood Avenue, Suite 100
Raleigh, NC 27612

U.S. Chamber of Commerce
1615 H St., N.W.
Washington, DC 20062
(202) 463-5531

U.S. Environmental Protection Agency Office
of Solid Waste
401 M St., S.W., Mail Stop OS-301
Washington, DC 20460
(202) 260-6261

Union of Concerned Scientists
26 Church St.
Cambridge, MA 02238
(617) 547-5552

APPENDIX D

SCHOOLS AND ORGANIZA-
TIONS OFFERING COURSES
AND INFORMATION ON SOLID
WASTE MANAGEMENT*

Carleton University
Civil & Environmental Engineering
1125 Colonel By Drive
Ottowa, Ontario, Canada K1S 5B6
Phone: (613) 788-3663

Cornell University
Cornell Waste Management Institute
Cornell Cooperative Extension
Hollister Hall
Ithaca, NY 14853
Phone: (601) 855-5940

Drexel University
School of Engineering
Environmental Department
31st & Chestnut Streets
Abbots Bldg. West, 3rd Floor
Philadelphia, PA 19104
(215) 895-2000

Duke University
School of Engineering
Civil and Environmental Engineering
Durham, NC 27706
Phone: (919) 660-5204

Humboldt State University
Environmental Resources Engineering Dept.
Arcata, CA 95521
Phone: (707) 826-3619

Montana College of Mineral Science &
Technology
Environmental Engineering
Admissions Office
Butte, MT 59701
Phone: (406) 496-4178

North Carolina State University
Civil Engineering
Box 7908
Raleigh, NC 27695-7908
Phone: (919) 515-7676

Rochester Institute of Technology
Environmental Management
Eastman Bldg., P.O. Box 9887
Rochester, NY 14623-0887
Phone: (716) 475-7318

Rutgers University Cook College
Dept. of Environmental Sciences
New Brunswick, NJ 08903-0231
Phone: (908) 932-9735

San Diego State University
College of Extended Studies
S. W. Mgmt for Business and Industry
San Diego, CA 92182-0722
Phone: (619) 594-5669

*Abstracted from *MSW Management, The Journal for Municipal Solid Waste Professionals*, vol. 3, July–August 1993.

San Francisco State University
Extended Education
Integrated Waste Management Program
1600 Holloway Ave.
San Francisco, CA 94132
Phone: (415) 904-7738

San Jose State University
Office of Continuing Education
Environmental Studies
One Washington Square
San Jose, CA 95192-0135
Phone: (408) 924-2600

State University of New York, Stony Brook
The Waste Management Institute
Marine Sciences Research Center
Stony Brook, NY 11794-5000
Phone: (516) 632-8704

State University of New York, Syracuse
College of Enviro Science & Forestry
Environmental & Resource Engineering
312 Bray Hall
Syracuse, NY 13210

Texas A&M University
Texas Agricultural Extension Service
Consumer and Family Sciences
237 Special Services [2251]
College Station, TX 77843
Phone: (409) 845-1332

Tufts University
Civil Engineering
Anderson Hall
Medford, MA 02155
Phone: (617) 627-3640

University of California, Berkeley
Environmental Resources Engineering
Civil Engineering
631 Davis Hall
Berkeley, CA 94720
Phone: (415) 642-6464

University of California, Berkeley
UC Berkeley Extension
Environmental Hazard Management
2223 Fulton St.
Berkeley, CA 94720

University of California, Davis
University Extension
Hazardous Management Program
Davis, CA 95616-8727
Phone: (916) 757-8878

University of California, Davis
Civil & Environmental Engineering
Davis, CA 95616
Phone: (916) 752-5671

University of California, Irvine
University Extension
Environmental Auditing
P.O. Box 6050
Irvine, CA 92716-6050
Phone: (714) 856-5414

University of California, Los Angeles
UCLA Extension
Humanities, Sciences & Social Sciences
P.O. Box 24901
Los Angeles, CA 90024

University of California, Santa Cruz
UC Extension
Environmental Science
3120 De la Cruz Blvd.
Santa Clara, CA 95054
Phone: (408) 748-7380

University of Central Florida, Orlando
Environmental Engineering
4000 Central Florida Blvd.
Orlando, FL
(407) 823-2841

University of Florida
Institute of Food and Ag. Sciences
3002 McCarty Hall
Gainesville, FL 32611-0130
Phone: (904) 392-1945

University of Florida
Center for Biomass Energy Systems
Gainesville, FL 32611-0130
Phone: (904) 392-1511

University of Florida
Solid Waste Program
Environmental Engineering Sciences
3002 McCarty Hall
Gainesville, FL 32611-0130

University of Florida
Division of Continuing Education
Center for Trng. Research and Ed. for
  Environmental Occupations (TREEO)
3900 SW 63rd Blvd.
Gainesville, FL 32608-3848

University of Florida
Dept. of Engineering
3002 McCarty Hall
Gainesville, FL 32611-0130
Phone: (904) 392-0945

University of Illinois, Champaign
Dept. of Civil Engineering
205 N. Mathews Ave.
Urbana, IL 61801
Phone: (217) 333-3812 Fax: (217) 333-9464

University of Louisville
Chemical Engineering
Louisville, KY 40292
Phone: (502) 588-6347

University of Maryland
Dept. of Horticulture
College Park, MD 20742-5611
Phone: (301) 314-9308

University of Michigan
College of Engineering
2340 G. G. Brown Bldg.
Ann Arbor, MI 48109-2125

University of New Orleans
Urban Waste Management and Research Center
New Orleans, LA 70148
Phone: (504) 286-6000

University of Oklahoma
School of Civil Engineering & Environmental
  Science
202 West Boyd Street, Room 334
Norman, OK 73019-0631
Phone: (405) 325-5911

University of Pennsylvania
Systems Department
220 S. 33rd St., Room 113
Philadelphia, PA 19104-6315
Phone: (215) 898-5000

University of Pittsburgh
School of Engineering
Civil Engineering
949 Benedum Hall
Pittsburgh, PA 15261-2294

University of Tennessee, Knoxville
Waste Management Research and Ed. Institute
327 South Stadium Hall
Knoxville, TN 37996-0710
Phone: (615) 974-4251

University of Tennessee, Knoxville
Master of Public Works Program
316 Morgan Hall
Knoxville, TN 37996
Phone: (615) 974-2503

University of Tennessee, Knoxville
Municipal Technology
Advisory Service
316 Morgan Hall
Knoxville, TN 37996
Phone: (615) 974-4251

University of Tennessee, Knoxville
Agricultural Economics Program
316 Morgan Hall
Knoxville, TN 37996
Phone: (615) 974-7231

University of Texas, Austin
Civil Engineering, Bldg. ECJ
Austin, TX 78712
Phone: (512) 471-4921

University of Wisconsin, Madison
University Extension
Solid & Hazardous Waste Ed. Center
610 Langdon St., Room 529
Madison, WI 53703

University of Wisconsin, Madison
College of Engineering
Engineering Professional Development
432 N. Lake St.
Madison, WI 53706

University of Wisconsin, Milwaukee
Center for Continuing Engineering Ed.
College of Engineering & Applied Sci.
929 North Sixth Street
Milwaukee, WI 53203

Utah State University
College of Engineering
UMC 6000
Logan, UT 84322-6000
Phone: (801) 750-2775

Virginia Polytechnic Institute and State
  University
College of Agriculture and Life Sciences
Crop and Soil Environmental Sciences
330 Smyth Hall
Blacksburg, VA 24061-0404
Phone: (703) 231-3431

Widener University
Civil Engineering
One University Place
Chester, PA 19013-5692
Phone: (215) 499-4000

Yale University
School of Forestry & Enviro Studies
Program on Solid Waste Policy
205 Prospect St.
Sage Hall
New Haven, CT 96511

Organizations offering information regarding solid waste management:

Academy of Environmental Engrs.
130 Holiday Court, Suite 100
Annapolis, MD 21401
Phone: (410) 266-3311

California Resource Recovery Assn.
CRRA Gambier Ct.
Sunnyvale, CA 96511-2189

Environmental Education Assoc., Inc.
National Workshop Series
2000 P Street, NW, Suite 515
Washington, DC 20036
Phone: (202) 296-4572

Government Institutes Inc.
4 Research Place, Suite 200
Rockville, MD 20850
Phone: (301) 921-2345

Hazardous Materials Control Resources
  Institute
7237 Hanover Parkway
Greenbelt, MD 20770-3602
Phone: (301) 982-9500

Olds College Composting
Olds, Alberta, Canada T0M 1P0
Phone: (403) 556-4650

Omni Environmental Corporation
Groundwater Pollution and Hydrology
The Princeton Corporate Center
Three Independence Way
Princeton, NJ 08540
Phone: (609) 243-9399

SWANA
P.O. Box 6126
Silver Spring, MD 20916
Phone: (301) 585-2898

APPENDIX E

UNIT CONVERSIONS

Factors for the Conversion of U.S. Engineering Units to the International System (SI) of Units

Multiply the U.S. engineering unit		By	To obtain the corresponding SI unit	
Name	Abbreviation		Name	Symbol
acre	acre	4047	square meter	m^2
acre	acre	0.4047	hectare	ha*
British thermal unit	Btu	1.055	kilojoule	kJ
British thermal units per cubic foot	Btu/ft^3	37.259	kilojoules per cubic meter	kJ/m^3
British thermal units per hour per square foot	$Btu/h \cdot ft^2$	23.158	joules per second per square meter	$J/s \cdot m^2$
British thermal units per kilowatthour	Btu/kWh	1.055	kilojoules per kilowatt hour	kJ/kWh
British thermal units per pound	Btu/lb	2.326	kilojoules per kilogram	kJ/kg
British thermal units per ton	Btu/ton	0.00116	kilojoules per kilogram	kJ/kg
degree Celsius	°C	plus 273	kelvin	K
cubic foot	ft^3	0.0283	cubic meter	m^3
cubic foot	ft^3	28.32	liter	L*
cubic feet per minute	ft^3/min	0.0004719	cubic meters per second	m^3/s
cubic feet per minute	ft^3/min	0.4719	liters per second	L*/s
cubic feet per second	ft^3/s	0.0283	cubic meters per second	m^3/s
cubic yard	yd^3	0.7646	cubic meter	m^3
day	d	86.4	kilosecond	ks
degree Fahrenheit	°F	0.555 (°F − 32)	degree Celsius	°C
foot	ft	0.3048	meter	m
feet per minute	ft/min	0.00508	meters per second	m/s
feet per second	ft/s	0.3048	meters per second	m/s
gallon	gal	0.003785	cubic meter	m^3
gallon	gal	3.785	liter	L*
gallons per minute	gal/min	0.0631	liters per second	L*/s
grain	gr	0.0648	gram	g
horsepower	hp	0.746	kilowatt	kW
horsepower-hour	hp-h	2.684	megajoule	MJ
inch	in	2.54	centimeter	cm
inch	in	0.0254	meter	m
kilowatthour	kWh	3.600	megajoule	MJ

Factors for the Conversion of U.S. Engineering Units to the International System (SI) of Units
(*Continued*)

Multiply the U.S. engineering unit		By	To obtain the corresponding SI unit	
Name	Abbreviation		Name	Symbol
million gallons per day	Mgal/d	0.04381	cubic meters per second	m^3/s
miles	mi	1.609	kilometer	km
miles per hour	mi/h	1.609	kilometers per hour	km/h
miles per hour	mi/h	0.447	meters per second	m/s
miles per gallon	mi/gal	0.425	kilometers per liter	km/L*
ounce	oz	28.35	gram	g
pound (force)	lb_f	4.448	newton	N
pound (mass)	lb_m	0.4536	kilogram	kg
pounds per acre	lb/acre	0.1122	grams per square meter	g/m^3
pounds per acre	lb/acre	1.122	kilograms per hectare	kg/ha
pounds per capita per day	lb/capita • d	0.4536	kilograms per capita per day	kg/capita • d
pounds per cubic foot	lb/ft^3	16.019	kilograms per cubic meter	kg/m^3
pounds per cubic yard	lb/yd^3	0.5933	kilograms per cubic meter	kg/m^3
pounds per square foot	lb/ft^2	47.88	newtons per square meter	N/m^2
pounds per square inch (psi)	lb/in^2	6.895	kilonewtons per square meter	kN/m^3
square foot	ft^2	0.0929	square meter	m^2
square mile	mi^2	2.590	square kilometer	km^2
square yard	yd^2	0.8361	square meter	m^2
ton (2000 pounds mass)	ton (2000 lb_m)	907.2	kilogram	kg
watthour	Wh	3.60	kilojoule	kJ
yard	yd	0.9144	meter	m

*Not an SI unit, but a commonly used term.

INDEX

ABOUT THE EDITOR IN CHIEF

Dr. Frank Kreith, P.E., is president of Kreith Engineering, Inc., a consulting firm specializing in waste management, solar energy, and conservation. He is Professor Emeritus of Mechanical and Chemical Engineering at the University of Colorado and currently serves as the ASME Legislative Fellow for Energy, Environment, and Waste Management at the National Conference of State Legislatures where he provides technical assistance to state governments. He is the author or editor of 15 books and more than 150 papers on heat transfer and a variety of energy and waste management topics.